Applied Calculus
for Business, Life, and Social Sciences

Denny Burzynski

with the assistance of
Guy Sanders & Kate Duffy Pawlik

SECOND PRINTING — July 2014

- Added corrections from Ross Rueger and Alicia Li

*xyz*textbooks

Applied Calculus
for Business, Life, and Social Sciences

Denny Burzynski, Guy Sanders, Kate Duffy Pawlik

Publisher: XYZ Textbooks

Editor: Anne Scanlan-Rohrer
Two Ravens Editorial Services

Project Manager: Matthew Hoy

Editorial Assistants: Kendra Paulding, Joshua Wilbur, Judy Barclay, Molly Stites, Lauren Reeves, Octabio Garcia

Composition: Aaron Kroeger, Rachel Hintz, Jennifer Thomas

Sales: Amy Jacobs, Richard Jones, Rachael Hillman, Bruce Spears, Katherine Hofstra

Cover Design: Kyle Schoenberger, Rachel Hintz

Cover Image: © Matthew Bowden

Printed in the United States of America

ISBN-13: 978-1-936368-33-4 / ISBN-10: 1-936368-33-1

For product information and technology assistance, contact us at
XYZ Textbooks, 1-877-745-3499

For permission to use material from this text or product,
e-mail: **info@mathtv.com**

XYZ Textbooks
1339 Marsh Street
San Luis Obispo, CA 93401
USA

For your course and learning solutions, visit **www.xyztextbooks.com**

Preface to the Instructor

We have designed this book to help solve challenges that you may encounter in the classroom. Applied Calculus is the perfect book for students who need to satisfy a one-semester calculus requirement for their degree in business or one of the social sciences. The book emphasizes modeling and uses a variety of technologies to enhance the modeling process. Students will leave class with confidence in their skills and a greater understanding of how calculus is used to model real-world situations.

Solutions to Your Challenges and Tickets to Success

Challenge **Some students may ask, "What are we going to use this for?"**
Solution Chapter and section openings feature real-world examples, which show students how the material they are learning appears in the world around them.

Challenge **Getting students to read the book.**
Solution At the end of each section, under the heading *Getting Ready for Class*, are questions for students to answer from the reading. Even a minimal attempt to answer these questions enhances the students' in-class experience.

Challenge **Getting students to connect the topics.**
Solution At the conclusion of the problem set for each section are a series of problems under the heading *Getting Ready for the Next Section*. These problems are designed to bridge the gap between topics covered previously, and topics introduced in the next section. Students intuitively see how topics lead into, and out of, each other.

Challenge **Improving students' study skills.**
Solution Study skills and success skills appear throughout the book, as well as online at MathTV.com. Students learn the skills they need to become successful in this class, and in their other courses as well.

Challenge **Getting students to maintain skills.**
Solution We have designed this textbook so that no topic is covered and then discarded. Throughout the book, features such as *Getting Ready for the Next Section*, *Maintaining Your Skills*, the *Chapter Summary*, and the *Chapter Test* continually reinforce the skills students need to master. If students need still more practice, there are a variety of worksheets online at MathTV.com.

Challenge **Connecting MathTV to active learning.**
Solution The Matched Problems worksheets (available online at MathTV.com) contain problems similar to the video examples. Assigning the Matched Problems worksheets ensures that students will be actively involved with the videos.

Preface to the Student

We often find our students asking themselves the question "Why can't I understand this stuff the first time?" The answer is "You're not expected to." Learning a topic in mathematics isn't always accomplished the first time around. There are many instances when you will find yourself reading over new material a number of times before you can begin to work problems. That's the way things are in mathematics. And this is calculus. This is sophisticated mathematics involving big ideas. Understanding mathematics can take time. The process of understanding requires reading the book, studying the examples, working problems, and getting your questions answered.

How to Be Successful in Mathematics

1. **If you are in a lecture class, be sure to attend all class sessions on time.** You cannot know exactly what goes on in class unless you are there. Missing class and then expecting to find out what went on from someone else is not the same as being there yourself. In class, your instructor can pass on valuable insights into the mathematics that may not appear in the text.
2. **Read the book.** It is best to read the section that will be covered in class beforehand. Reading in advance, even if you do not understand everything you read, is still better than going to class with no idea of what will be discussed.
3. **Work problems every day and check your answers.** The key to success in calculus is working problems. The more problems you work, the better you will become at working them. The answers to the odd-numbered problems are given in the back of the book. When you have finished an assignment, be sure to compare your answers with those in the book. If you have made a mistake, find out what it is, and correct it.
4. **Do it on your own.** Don't be misled into thinking someone else's work is your own. Having someone else show you how to work a problem is not the same as working the same problem yourself. It is okay to get help when you are stuck. As a matter of fact, it is a good idea. Just be sure you do the work yourself. You won't learn how to fly an airplane by just watching someone else do it in a video, you have to try it and practice it yourself.
5. **Review every day.** After you have finished the problems your instructor has assigned, take another 15 minutes and review a section you have already completed. The more you review, the longer you will retain the material you have learned.
6. **Don't expect to understand every new topic the first time you see it.** Sometimes you will understand everything you are doing, and sometimes you won't. That's just the way things are in mathematics. Expecting to understand each new topic the first time you see it can lead to disappointment and frustration. The process of understanding takes time. It requires that you read the book, work problems, and get your questions answered.
7. **Spend as much time as it takes for you to master the material.** No set formula exists for the exact amount of time you need to spend on mathematics to master it. You will find out as you go along what is or isn't enough time for you. If you end up spending 2 or more hours on each section in order to master the material there, then that's how much time it takes; trying to get by with less will not work.
8. **Enjoy.** Calculus is a major achievement of human beings. Morris Kline, one of twentieth century's great mathematicians, notes that mathematics has given direction to our philosophies, our religions, our economics and politics as well as our art, architecture, music, and literature. Now is your chance to experience this significant achievement. Enjoy it.

Supplements and Resources

Emphasis on Modeling Modeling Examples and Problems are labeled with a puzzle icon and are found throughout the text.

Staying Current with Technology Graphing Calculators, Wolfram|Alpha, QR codes, are all integrated throughout the text.

Extensive Video Library Every example in the book is done in videos available online at MathTV.com. Students can choose from a variety of instructors and see and hear the examples done in both English and Spanish. MathTV has over 8,000 videos for students to watch, including the instructional videos, videos on study skills, interviews with student instructors, enrichment problems, and more.

Online Homework XYZ Homework provides powerful online instructional tools for faculty and students. Randomized questions provide unlimited practice and instant feedback with all the benefits of automatic grading. Tools for instructors can be found at www.xyzhomework.com.

Online Textbook Access to our eBooks come with the purchase of a new print book. For students wanting an eBook only, a 1-year subscription is $30.

QR Codes Throughout This book and your students' smart devices work together to make studying from our books a visual, animated experience.

Complete Course Template Whether you are teaching in-class or online, our Complete Course Template creates a foundation for your course. The lessons in the Complete Course Template match the sections in the textbook. Most lessons include a welcome video, objectives, a reading and written assignment, an electronic assignment, and a Take Five video covering an enrichment topic, a study skill, or a success skill. Many instructors use our Complete Course Template as is the first time they teach their the course.

Student Solutions Manual The Student Solutions Manual contains complete solutions to selected problems from each problem set.

Using Technology In each section of the book there are optional exercises that demonstrate how students can use graphing calculators and online resources such as Wolfram|Alpha to enhance their understanding of the topics being covered.

Wolfram|Alpha

We are pleased to offer Wolfram|Alpha problems in this textbook. Wolfram|Alpha is a trademark of Wolfram Alpha LLC. All logos, illustrations and search results are used with permission and are the copyrighted material of Wolfram Alpha LLC. The problems presented in this text using Wolfram|Alpha were not suggested by Wolfram|Alpha, nor does Wolfram|Alpha endorse these problems.

We have designated some of the more challenging problems with a chili pepper icon

Acknowledgements

Coauthors

We are extremely pleased to have the assistance of coauthors Guy Sanders and Kate Duffy Pawlik on this project. Guy's assistance and hard work on the original manuscript have been invaluable. His experience in the classroom and his ability to translate that into written words helped get the book off to a good start. Kate started as the developmental editor on the book. Her work was so impressive and extended beyond the normal duties of development that she elevated herself to coauthor status. Our sincere thank you to both these people. This book would not have been possible without them.

XYZ Textbooks Crew

Production Team: Matthew Hoy, Rachel Hintz, Aaron Kroeger, Jennifer Thomas, Ross Rueger   Our fantastic production team shepherded this book from handwritten manuscript to the final form you see today. Their attention to detail and ideas for making this book user-friendly for students is greatly appreciated.

Editing and Proofreading: Kate Pawlik, Octabio Garcia, Judy Barclay, Robert Schwennicke, Molly Stites   Their eye for detail and ferreting out even the most seemingly trivial error never ceased to amaze us. This book is far better off both mathematically and grammatically due to their invaluable assistance.

Office Staff and Customer Support: Rachael Hillman, Katherine Hofstra, Anne Gentilucci   Our office staff are reliable, pleasant, and efficient. Plus they are lots of fun to work with.

Sales Department: Amy Jacobs, Rich Jones, Bruce Spears, Rachael Hillman   Our award-winning, responsive sales staff is always conscientious and hard-working.

Technology Department: Stephen Aiena, Lauren Barker   Stephen and Lauren are the brains behind the XYZ Textbooks' family of websites, including MathTV.com and XYZHomework.com.

XYZ Homework: Patrick McKeague, Matthew Hoy, Stephen Aiena, Mike Landrum   From the big concepts to the little details, our XYZ Homework team has provided a solid, dependable online homework system that works.

MathTV Student Peer Instructors: Gordon Kirby, Cynthia Ruiz, Edwin Martinez, Lauren Reeves, Joshua Wilbur, Octabio Garcia, Molly Stites, Betsy Andrews   These students and their genuine love for math have brightened my days. The videos they've made have helped countless students improve their math skills.

Acknowledgements

Focus Group Participants

Many thanks to the following instructors who participated in one of our online focus groups. Your suggestions were invaluable to us!

Kari Arnoldsen, *Snow College*
Elsie M. Campbell, *Angelo State University*
Linda Chandler, *Sandhills Community College*
Julane Crabtree, *Johnson County Community College*
Marsha Driskill, *Aims Community College*
Rob Eby, *Blinn College*
Elaine Fitt, *Bucks County Community College*
Dianne Hendrickson, *Becker College*
Steven Howard, *Rose State College*
Lynette Meslinsky, *Erie Community College*
Ashod Minasian, *El Camino College*
Rhoda Oden, *Gadsden State Community College*
Dennis Reisig, *Suffolk County Community College*
William H. Remele, *Brunswick Community College*
Stephen Rodi, *Austin Community College*
Christian J. Roldán, *Santos Black Hawk College*
Melinda Rudibaugh, *Scottsdale Community College*
Sal Sciandra, *Niagara County Community College*
Ben Smith, *Los Angeles Pierce College*
Cal Smith, *Gadsden State Community College*
Linda Tansil, *Southeast Missouri State University*
Richard Uchida, *Sinclair Community College*
Joyce Wagner, *Santiago Canyon College*
Willem Wallinga, *Fisher College*
Betty Weiss, *West Valley College*
Jim Wolper, *Idaho State University*
David Zeigler, *California State University, Sacramento*

Brief Contents

Contents

7 Calculus of Functions of Several Variables 577

Appendix 687

AN Answers AN-1

I Index I-1

Functions, Limits, and Rates of Change

1

Chapter Outline

A lot of competitors were ugly to me.
What they didn't realize was they were
making it worse for themselves.
The angrier they made me, ...
the better I was in the car.

Shirley Muldowney
Hot Rod Magazine, April 2009

Shirley Muldowney made a name for herself in the sport of drag racing. Her first race was in 1976. In 1977 she won the Top Fuel World Championship, which she won three more times in her career. A movie of her life in racing, *Heart Like a Wheel*, starring Bonnie Bedelia (above), was released in 1983. The function and graph shown below are models of her speed in one of the races from that movie.

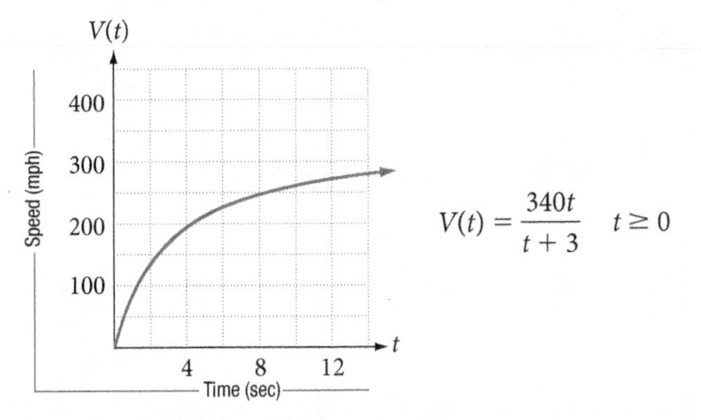

$$V(t) = \frac{340t}{t + 3} \qquad t \geq 0$$

Note When you see this icon next to an example or problem in this chapter, you will know that we are using the topics in this chapter to model situations in the world around us.

As you know from your previous math classes, we use functions, graphs, tables, and words to model situations that we find in the world around us. The type of graph shown here is a good model for the speed of a dragster during a race. The graph shows that the speed increases quickly at the beginning of the race, then more slowly towards the end of the race. These changes in the speed are quantities that we can model with calculus. Here are two questions that show the difference between what you have done previously in algebra, and what you will be doing in calculus.

Algebra What is her speed four seconds after the race begins?

Calculus How fast is her speed changing four seconds after the race begins?

We begin this chapter with a review of functions and graphing, which allow us to answer the algebra question above. Then we introduce our first calculus topic, limits. Once we have some experience with limits, we will go on to look at instantaneous rates of change, which is the tool we need in order to answer the calculus question above.

Study Skills

Some of the students enrolled in our applied calculus classes develop difficulties early in the course. Their difficulties are not associated with their ability to learn mathematics; they all have the potential to pass the course. Students who get off to a poor start do so because they have not developed the study skills necessary to be successful in mathematics. Here is a list of things you can do to begin to develop effective study skills.

1. **Put Yourself on a Schedule** The general rule is that you spend 2 hours on homework for every hour you are in class. Make a schedule for yourself in which you set aside 2 hours each day to work on calculus. Once you make the schedule, stick to it. Don't just complete your assignments and stop. Use all the time you have set aside. If you complete an assignment and have time left over, read the next section in the book, and then work more problems.

2. **Find Your Mistakes and Correct Them** There is more to studying calculus than just working problems. You must always check your answers with the answers in the back of the book. When you have made a mistake, find out what it is and correct it. Making mistakes is part of the process of learning mathematics. In the prologue to *The Book of Squares*, Leonardo Fibonacci (ca. 1170–ca. 1250) had this to say about the content of his book:

> I have come to request indulgence if in any place it contains something more or less than right or necessary; for to remember everything and be mistaken in nothing is divine rather than human ...

   Fibonacci knew, as you know, that human beings make mistakes. You cannot learn calculus without making mistakes.

3. **Gather Information on Available Resources** You need to anticipate that you will need extra help sometime during the course. One resource is your instructor; you need to know your instructor's office hours and where the office is located. Another resource is the math lab or study center, if they are available at your school. It also helps to have the phone numbers of other students in the class, in case you miss class. You want to anticipate that you will need these resources, so now is the time to gather them together.

Introduction to Functions and Relations 1.1

The diagram below is called the *spiral of roots*. The spiral of roots mimics the shell of the chambered nautilus, an animal that has survived largely unchanged for millions of years. The mathematical diagram is constructed using the Pythagorean theorem. The table gives the lengths of the diagonals in the spiral of roots, accurate to the nearest hundredth.

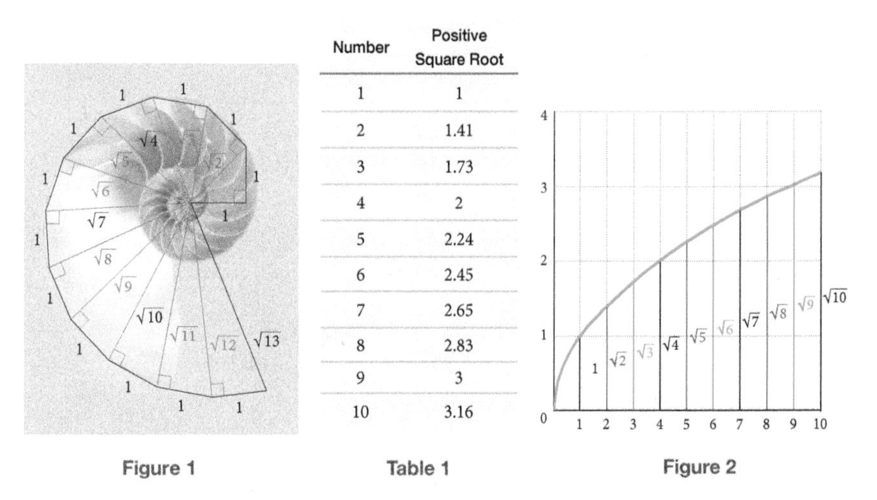

Number	Positive Square Root
1	1
2	1.41
3	1.73
4	2
5	2.24
6	2.45
7	2.65
8	2.83
9	3
10	3.16

Figure 1 Table 1 Figure 2

If we take each of the diagonals in the spiral of roots and place it above the corresponding whole number on the x-axis and then connect the tops of all these segments with a smooth curve, we have the graph shown in Figure 2. This curve is also the graph of the equation

$$y = \sqrt{x}$$

In this chapter we begin our work with functions. The diagrams, table, and text above all refer to the **square root function**. The figures give us a **visual** representation of the square root function; the table is a **numeric** representation, and the equation $y = \sqrt{x}$ is a **symbolic** representation. If you were to describe this relationship to a friend by saying "If we input a whole number, or any positive number, this function gives us (outputs) the square root of that number," you would be giving a **verbal** description as well. Giving different representations for the same function is something we will do throughout this chapter.

An Informal Look at Functions

HELP WANTED

YARD PERSON

Full time 40 hrs. with weekend work required. Cleaning & loading trucks. $8.50/hr.

Valid CDL with clean record & drug screen required.

Submit current MVR to KCI, 225 Suburban Rd., SLO 93405 (805) 555-3304.

BABY SITTER

The ad shown here appeared in the Help Wanted section of the local newspaper the day we were writing this section of the book. If you held the job described in the ad, you would earn \$8.50 for every hour you worked. The amount of money you make in one week depends on the number of hours you work that week. In mathematics, we say that your weekly earnings are a **function** of the number of hours you work.

Suppose you have a job that pays \$8.50 per hour and that you work anywhere from 0 to 40 hours per week. If we let the variable x represent hours and the variable y represent the money you make, then the relationship between x and y can be written as

$$y = 8.5x \quad \text{for} \quad 0 \le x \le 40$$

VIDEO EXAMPLES

SECTION 1.1

Example 1 Construct a table and graph for the function

$$y = 8.5x \qquad \text{for} \qquad 0 \le x \le 40$$

Solution Table 2 gives some of the paired data that satisfy the equation $y = 8.5x$. Figure 3 is the graph of the equation with the restriction $0 \le x \le 40$.

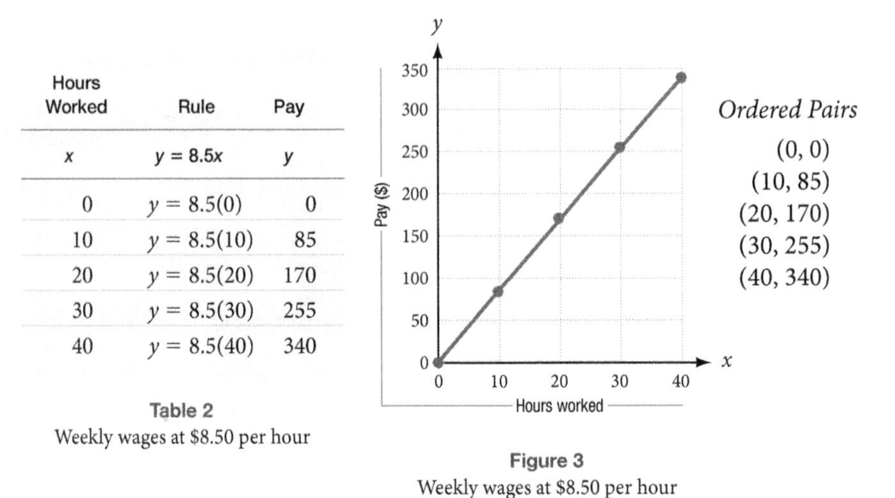

Hours Worked	Rule	Pay
x	$y = 8.5x$	y
0	$y = 8.5(0)$	0
10	$y = 8.5(10)$	85
20	$y = 8.5(20)$	170
30	$y = 8.5(30)$	255
40	$y = 8.5(40)$	340

Table 2
Weekly wages at \$8.50 per hour

Figure 3
Weekly wages at \$8.50 per hour

The equation $y = 8.5x$ with the restriction $0 \le x \le 40$, Table 2, and Figure 3 are three ways to describe the same relationship between the number of hours you work in one week and your gross pay for that week. In all three, we **input** values of x, and then use the function rule to **output** values of y. ■

Domain and Range of a Function

Note The vertical line used in domain and range notation is read "such that."

We began this discussion by saying that the number of hours worked during the week was from 0 to 40, so these are the values that x can assume. From the line graph in Figure 3, we see that the values of y range from 0 to 340. We call the complete set of values that x can assume the **domain** of the function. The values that are assigned to y are called the **range** of the function.

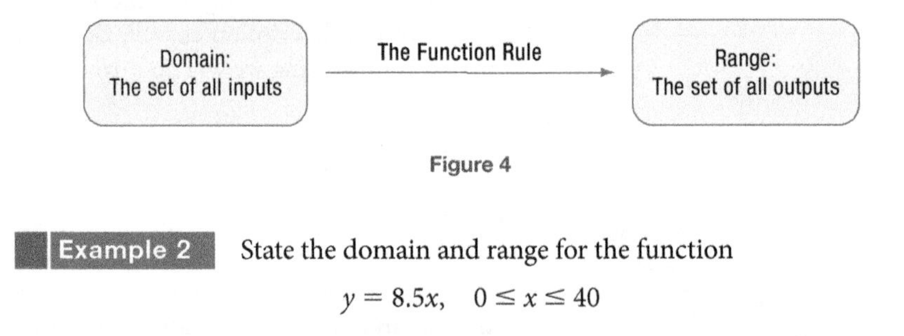

Figure 4

Example 2 State the domain and range for the function

$$y = 8.5x, \quad 0 \le x \le 40$$

Solution From the previous discussion, we have

$$\text{Domain} = \{x \mid 0 \le x \le 40\}$$
$$\text{Range} = \{y \mid 0 \le y \le 340\}$$

■

Function Maps

Another way to visualize the relationship between x and y is with the diagram in Figure 5, which we call a **function map**.

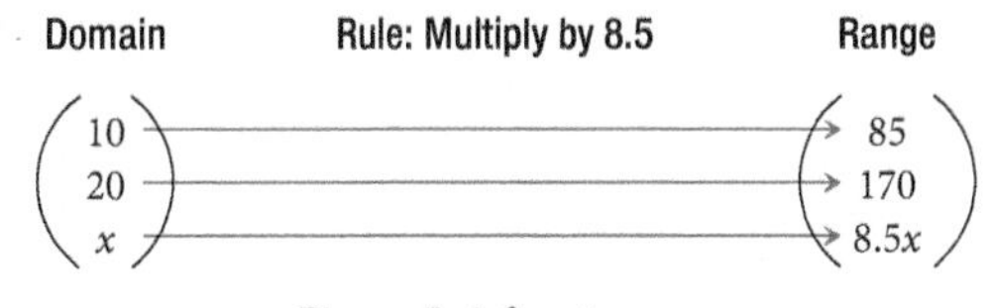

Figure 5 A function map

Although the diagram in Figure 5 does not show all the values that x and y can assume, it does give us a visual description of how x and y are related. It shows that values of y in the range come from values of x in the domain according to a specific rule (multiply by 8.5 each time).

A Formal Look at Functions

We are now ready for the formal definition of a function.

> **Function**
> A **function** is a rule that pairs each element in one set, called the **domain,** with exactly one element from a second set, called the **range.**

In other words, a function is a rule for which each input is paired with exactly one output.

Modeling: Softball Toss

Example 3 Kendra tosses a softball into the air with an underhand motion. The distance of the ball above her hand is given by the function in symbolic form as

$$h = 32t - 16t^2 \qquad \text{for} \qquad 0 \leq t \leq 2$$

where h is the height of the ball in feet and t is the time in seconds. Construct a table that gives the height of the ball at quarter-second intervals, starting with $t = 0$ and ending with $t = 2$, then graph the function.

Solution We construct Table 3 using the following values of t: $0, \frac{1}{4}, \frac{1}{2}, \frac{3}{4}, 1, \frac{5}{4}, \frac{3}{2}, \frac{7}{4}, 2$. Then we construct the graph in Figure 6 from the table. The graph appears only in the first quadrant because neither t nor h can be negative.

Input		Output
Time (sec) t	Function Rule $h = 32t - 16t^2$	Distance (ft) h
0	$h = 32(0) - 16(0)^2 = 0 - 0 = 0$	0
$\frac{1}{4}$	$h = 32\left(\frac{1}{4}\right) - 16\left(\frac{1}{4}\right)^2 = 8 - 1 = 7$	7
$\frac{1}{2}$	$h = 32\left(\frac{1}{2}\right) - 16\left(\frac{1}{2}\right)^2 = 16 - 4 = 12$	12
$\frac{3}{4}$	$h = 32\left(\frac{3}{4}\right) - 16\left(\frac{3}{4}\right)^2 = 24 - 9 = 15$	15
1	$h = 32(1) - 16(1)^2 = 32 - 16 = 16$	16
$\frac{5}{4}$	$h = 32\left(\frac{5}{4}\right) - 16\left(\frac{5}{4}\right)^2 = 40 - 25 = 15$	15
$\frac{3}{2}$	$h = 32\left(\frac{3}{2}\right) - 16\left(\frac{3}{2}\right)^2 = 48 - 36 = 12$	12
$\frac{7}{4}$	$h = 32\left(\frac{7}{4}\right) - 16\left(\frac{7}{4}\right)^2 = 56 - 49 = 7$	7
2	$h = 32(2) - 16(2)^2 = 64 - 64 = 0$	0

TABLE 3 Tossing a Softball into the Air

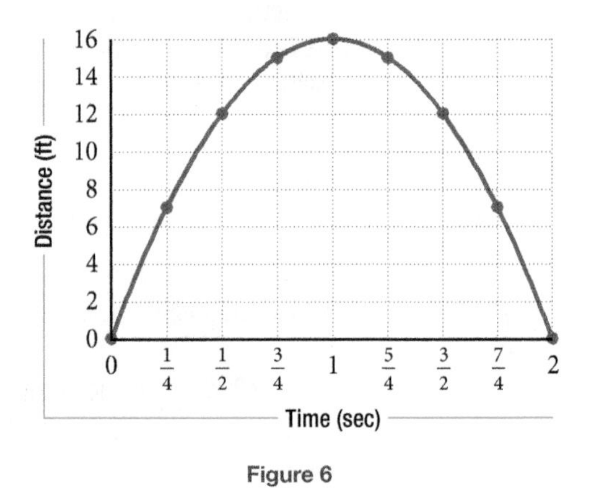

Figure 6

Here is a summary of what we know about functions as it applies to Example 3: We input values of t and output values of h according to the function rule

$$h = 32t - 16t^2 \qquad \text{for} \qquad 0 \le t \le 2$$

The domain is given by the inequality that follows the equation; it is

$$\text{Domain} = \{t \mid 0 \le t \le 2\}$$

The range is the set of all outputs that are possible by substituting the values of t from the domain into the equation. From our table and graph, it seems that the range is

$$\textit{Range} = \{h \mid 0 \le h \le 16\}$$

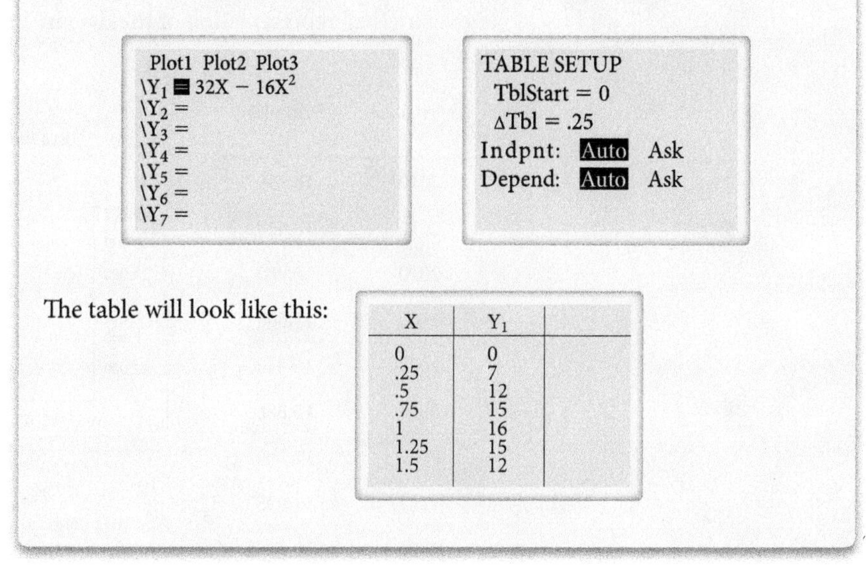

TECHNOLOGY NOTE *More About Example 3*

Most graphing calculators can easily produce the information in Table 2. Simply set Y1 equal to $32X - 16X^2$. Then set up the table so it starts at 0 and increases by an increment of 0.25 each time. (On a TI-82/83, press [2nd] [WINDOW] to set up the table.)

The table will look like this:

X	Y_1
0	0
.25	7
.5	12
.75	15
1	16
1.25	15
1.5	12

Functions as Ordered Pairs

As you can see from the examples we have done to this point, the function rule produces ordered pairs of numbers. We use this result to write an alternate definition for a function.

Function (Alternate Definition)

A **function** is a set of ordered pairs in which no two different ordered pairs have the same first coordinate. The set of all first coordinates is called the **domain** of the function. The set of all second coordinates is called the **range** of the function.

The restriction on first coordinates in the alternate definition keeps us from assigning a number in the domain to more than one number in the range.

A Relationship That Is Not a Function

You may be wondering if any sets of paired data fail to qualify as functions. The answer is yes, as the next example reveals.

Example 4 Table 4 shows the prices of used Ford Mustangs that were listed in the local newspaper. The diagram in Figure 7 is called a **scatter diagram.** It gives a visual representation of the data in Table 4. Why is this data not a function?

Year x	Price ($) y
2010	18,999
2010	18,420
2010	16,980
2009	17,600
2009	16,840
2008	15,888
2007	12,900
2007	11,995
2006	10,985

Table 4
Used Mustang Prices

Used Mustang Prices

Price ($): 0, 3,000, 6,000, 9,000, 12,000, 15,000, 18,000, 21,000

Year: 2006, 2007, 2008, 2009, 2010

Figure 7
Scatter diagram of data in Table 4

Ordered Pairs

(2010, 18,999)
(2010, 18,420)
(2010, 16,980)
(2009, 17,600)
(2009, 16,840)
(2008, 15,888)
(2007, 12,900)
(2007, 11,995)
(2006, 10,985)

Solution In Table 4, the year 2010 is paired with three different prices: $18,999, $18,420, and $16,980. That is enough to disqualify the data from belonging to a function. For a set of paired data to be considered a function, each number in the domain must be paired with exactly one number in the range. ■

Still, there is a relationship between the first coordinates and second coordinates in the used car data. It is not a function relationship, but it is a relationship. To classify all relationships specified by ordered pairs, whether they are functions or not, we include the following two definitions.

Relation
A **relation** is a rule that pairs each element in one set, called the **domain**, with one or more elements from a second set, called the **range.**

Relation (Alternate Definition)
A **relation** is a set of ordered pairs. The set of all first coordinates is the **domain** of the relation. The set of all second coordinates is the **range** of the relation.

Here are some facts that will help clarify the distinction between relations and functions:

1. Any rule that assigns numbers from one set to numbers in another set is a relation. If that rule makes the assignment so no input has more than one output, then it is also a function.
2. Any set of ordered pairs is a relation. If none of the first coordinates of those ordered pairs is repeated, the set of ordered pairs is also a function.
3. Every function is a relation.
4. Not every relation is a function.

Example 5 Sketch the graph of $x = y^2$.

Solution Without going into much detail, we graph the equation $x = y^2$ by finding a number of ordered pairs that satisfy the equation, plotting these points, then drawing a smooth curve that connects them. A table of values for x and y that satisfy the equation follows, along with the graph of $x = y^2$ shown in Figure 8.

x	y
0	0
1	1
1	−1
4	2
4	−2
9	3
9	−3

Table 5

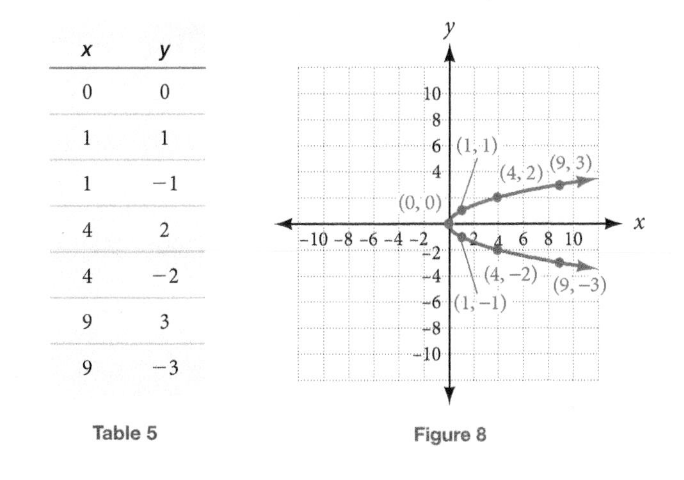

Figure 8

As you can see from looking at the table and the graph in Figure 8, several ordered pairs whose graphs lie on the curve have repeated first coordinates, for instance $(1, 1)$ and $(1, -1)$, $(4, 2)$ and $(4, -2)$, as well as $(9, 3)$ and $(9, -3)$. The graph is therefore not the graph of a function. ■

Notice that the points $(4, 2)$ and $(4, -2)$ on the graph in Figure 8 lie on the same vertical line. Since this is true for any points with the same first coordinate, it allows us to write the following test that uses the graph to determine whether a relation is also a function.

Vertical Line Test

If a vertical line crosses the graph of a relation in more than one place, the relation cannot be a function. If no vertical line can be found that crosses a graph in more than one place, then the graph represents a function.

If we look back to the graph of $h = 32t - 16t^2$ as shown in Figure 6, we see that no vertical line can be found that crosses this graph in more than one place. The graph shown in Figure 6 is therefore the graph of a function. If we look at Figure 7 or Figure 8, however, we can easily see that the data do not constitute a function.

Example 6 Graph the equation $y = \frac{1}{x}$.

Solution Notice that since y is equal to 1 divided by x, y will be positive when x is positive. (The quotient of two positive numbers is a positive number.) Likewise, when x is negative, y will be negative. In other words, x and y always will have the same sign.

Next, notice that the expression $\frac{1}{x}$ will be undefined when x is 0, meaning that there is no value of y corresponding to $x = 0$. This tells us that the domain will include all real numbers except $x = 0$. Because of this, the graph will not cross the y-axis. Further, the graph will not cross the x-axis either. If we try to find the x-intercept by letting $y = 0$, we have

$$0 = \frac{1}{x}$$

But there is no value of x to divide into 1 to obtain 0. Therefore, since there is no solution to this equation, our graph will not cross the x-axis.

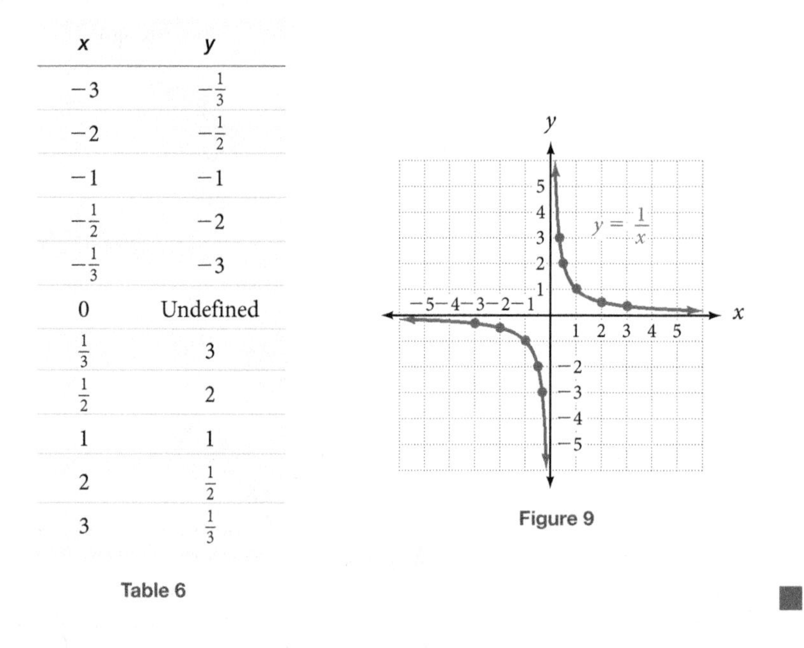

x	y
-3	$-\frac{1}{3}$
-2	$-\frac{1}{2}$
-1	-1
$-\frac{1}{2}$	-2
$-\frac{1}{3}$	-3
0	Undefined
$\frac{1}{3}$	3
$\frac{1}{2}$	2
1	1
2	$\frac{1}{2}$
3	$\frac{1}{3}$

Table 6

Figure 9

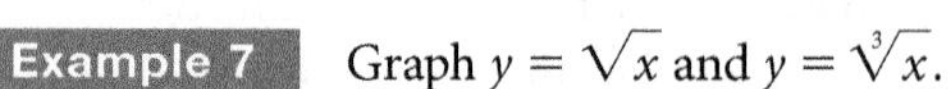

Example 7 Graph $y = \sqrt{x}$ and $y = \sqrt[3]{x}$.

Solution The graphs are shown in Figures 10 and 11. Notice that the graph of $y = \sqrt{x}$ appears in the first quadrant only, because in the equation $y = \sqrt{x}$, x and y cannot be negative.

The graph of $y = \sqrt[3]{x}$ appears in Quadrants I and III because the cube root of a positive number is also a positive number, and the cube root of a negative number is a negative number. That is, when x is positive, y will be positive, and when x is negative, y will be negative.

The graphs of both equations will contain the origin, because $y = 0$ when $x = 0$ in both equations.

Note Because of these restrictions on $y = \sqrt{x}$, the domain can be stated as $\{x \mid x \geq 0\}$. Similarly, we would express the range as $\{y \mid y \geq 0\}$.

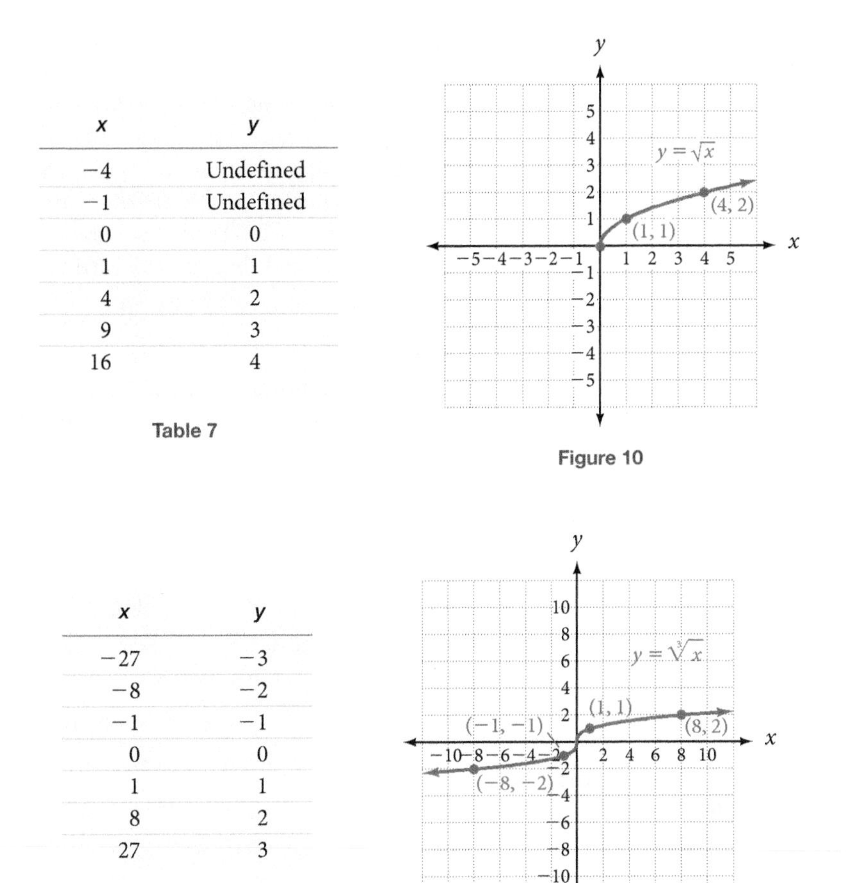

x	y
−4	Undefined
−1	Undefined
0	0
1	1
4	2
9	3
16	4

Table 7

Figure 10

x	y
−27	−3
−8	−2
−1	−1
0	0
1	1
8	2
27	3

Table 8

Figure 11

Function Notation

Let's return to the discussion that introduced us to functions. If a job pays \$8.50 per hour for working from 0 to 40 hours a week, then the amount of money, y, earned in one week is a function of the number of hours worked, x. The exact relationship between x and y is written

$$y = 8.5x \quad \text{for} \quad 0 \le x \le 40$$

Because the amount of money earned, y, depends on the number of hours worked, x, we call y the **dependent variable** and x the **independent variable**. Furthermore, if we let f represent all the ordered pairs produced by the equation, then we can write

$$f = \{(x, y) \mid y = 8.5x \quad \text{and} \quad 0 \le x \le 40\}$$

Once we have named a function with a letter, we can use an alternative notation to represent the dependent variable y. The alternative notation for y is $f(x)$. It is read "f of x" and can be used instead of the variable y when working with functions. The notation y and the notation $f(x)$ are equivalent. That is,

$$y = 8.5x \Leftrightarrow f(x) = 8.5x$$

When we use the notation $f(x)$ we are using **function notation**. The benefit of using function notation is that we can write more information with fewer symbols than we can by using just the variable y. For example, asking how much money a person will make for working 20 hours is simply a matter of asking for $f(20)$. Without function notation, we would have to say, "Find the value of y that corresponds to a value of $x = 20$." To illustrate further, using the variable y, we can say "y is 170 when x is 20." Using the notation $f(x)$, we simply say "$f(20) = 170$." Each expression indicates that you will earn \$170 for working 20 hours.

Example 8 If $f(x) = 8.5x$, find $f(0)$, $f(10)$, and $f(20)$.

Solution To find $f(0)$, we substitute 0 for x in the expression $8.5x$ and simplify. We find $f(10)$ and $f(20)$ in a similar manner—by substitution.

$$\text{If} \qquad f(x) = 8.5x$$

$$\text{then} \qquad f(0) = 8.5(0) = 0$$

$$f(10) = 8.5(10) = 85$$

$$f(20) = 8.5(20) = 170$$

If we changed the example in the discussion that opened this section so the hourly wage was \$9.50 per hour, we would have a new equation to work with, namely,

$$y = 9.5x \qquad \text{for} \qquad 0 \le x \le 40$$

Suppose we name this new function with the letter g. Then

$$g = \{(x, y) \mid y = 9.5x \quad \text{and} \quad 0 \le x \le 40\}$$

and

$$g(x) = 9.5x$$

Note Some students like to think of functions as machines. Values of x are put into the machine, which transforms them into values of $f(x)$, which are then output by the machine.

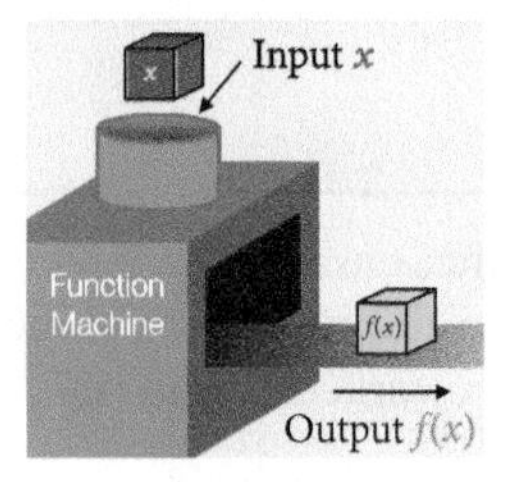

If we want to talk about both functions in the same discussion, having two different letters, f and g, makes it easy to distinguish between them. For example, since $f(x) = 8.5x$ and $g(x) = 9.5x$, asking how much money a person makes for working 20 hours is simply a matter of asking for $f(20)$ or $g(20)$, avoiding any confusion over which hourly wage we are talking about.

The diagrams shown in Figure 12 further illustrate the similarities and differences between the two functions we have been discussing.

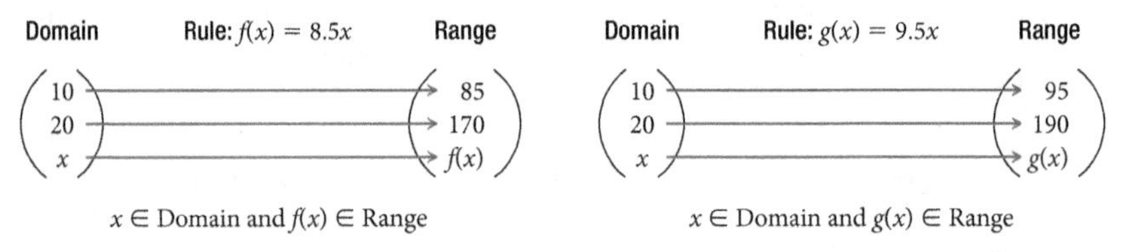

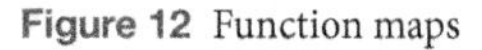
Figure 12 Function maps

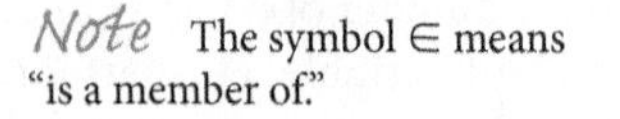
Note The symbol $\in$ means "is a member of."

Example 9 If $f(x) = 4x - 1$ and $g(x) = x^2 + 2$, then

$$f(5) = 4(5) - 1 = 19 \quad \text{and} \quad g(5) = 5^2 + 2 = 27$$

$$f(-2) = 4(-2) - 1 = -9 \quad \text{and} \quad g(-2) = (-2)^2 + 2 = 6$$

$$f(0) = 4(0) - 1 = -1 \quad \text{and} \quad g(0) = 0^2 + 2 = 2$$

$$f(z) = 4z - 1 \quad \text{and} \quad g(z) = z^2 + 2$$

$$f(a) = 4a - 1 \quad \text{and} \quad g(a) = a^2 + 2$$

$$f(a + 3) = 4(a + 3) - 1 \quad \text{and} \quad g(a + 3) = (a + 3)^2 + 2$$

$$= 4a + 12 - 1 \quad \text{and} \quad = (a^2 + 6a + 9) + 2$$

$$= 4a + 11 \quad \text{and} \quad = a^2 + 6a + 11$$

TECHNOLOGY NOTE *More About Example 9*

Most graphing calculators can use tables to evaluate functions. To work Example 9 using a graphing calculator table, set Y1 equal to 4X − 1 and Y2 equal to $X^2 + 2$. Then set the independent variable in the table to Ask instead of Auto. Go to your table and input 5, −2, and 0. Under Y1 in the table, you will find $f(5)$, $f(-2)$, and $f(0)$. Under Y2, you will find $g(5)$, $g(-2)$, and $g(0)$.

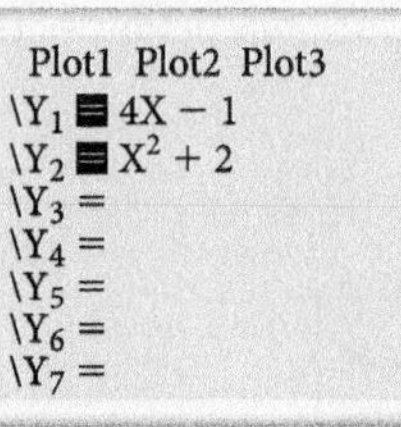

TABLE SETUP
TblStart = 0
ΔTbl = 1
Indpnt: Auto **Ask**
Depend: **Auto** Ask

The table will look like this:

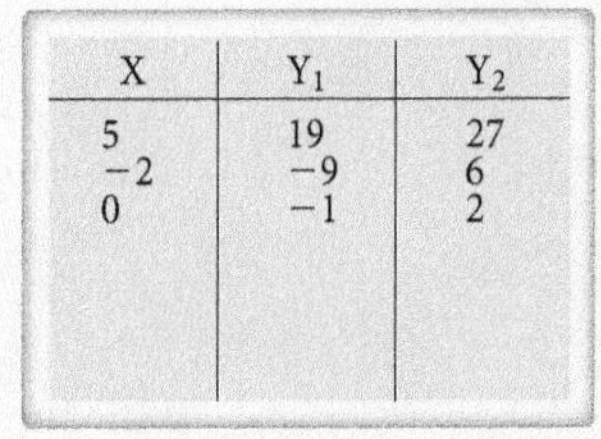

X	Y_1	Y_2
5	19	27
−2	−9	6
0	−1	2

Although the calculator asks us for a starting value and a table increment, these values don't matter because we are inputting the X values ourselves.

Function Notation and Graphs

We can visualize the relationship between x and $f(x)$ on the graph of the function. Figure 13 shows the graph of $f(x) = 8.5x$ along with two additional line segments that form a triangle. The length of the horizontal line segment is 20, and the length of the vertical line segment is $f(20)$. (Note that the domain is restricted to $0 \le x \le 40$.)

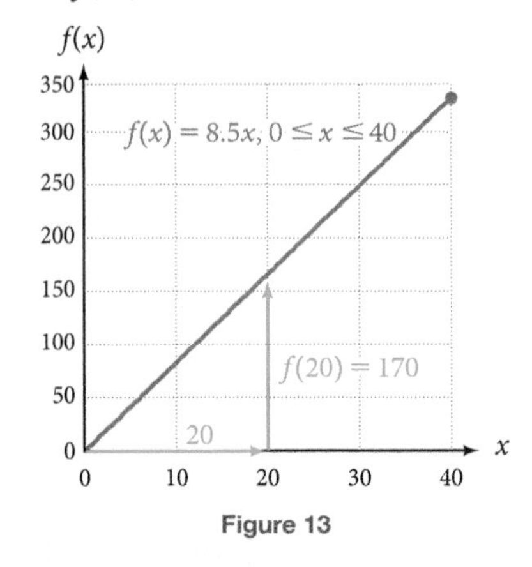

Figure 13

We can also use functions and function notation to talk about numbers in many of the charts and graphs we see in the media. For example, consider the chart on gasoline prices shown below. Let's let x represent one of the years in the chart.

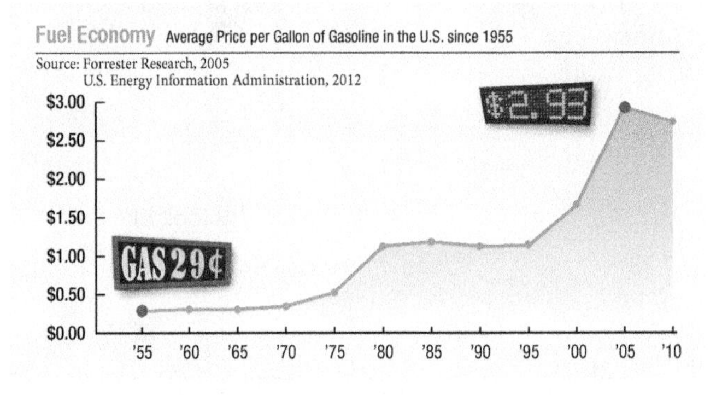

If the function f pairs each year in the chart with the average price of regular gasoline for that year, then each statement below is true:

$$f(1955) = \$0.29$$

The domain of $f =$

$$\{1955, 1960, 1965, 1970, 1975, 1980, 1985, 1990, 1995, 2000, 2005, 2010\}$$

In general, when we refer to the function f we are referring to the domain, the range, and the rule that takes elements in the domain and outputs elements in the range. When we talk about $f(x)$ we are talking about the rule itself, or an element in the range, or the variable y.

The function f

Domain of f	$y = f(x)$	Range of f
Inputs	*Rule*	*Outputs*

Modeling: Using Function Notation

The remaining examples in this section show a variety of ways to use and interpret function notation.

Example 10 If it takes Lorena t minutes to run a mile, then her average speed s, in miles per hour, is given by the formula

$$s(t) = \frac{60}{t} \quad \text{for} \quad t > 0$$

Find $s(10)$ and $s(8)$, and then explain what they mean.

© Andrew Rich/ iStockPhoto

Solution To find $s(10)$, we substitute 10 for t in the equation and simplify:

$$s(10) = \frac{60}{10} = 6$$

In words When Lorena runs a mile in 10 minutes, her average speed is 6 miles per hour.

We calculate $s(8)$ by substituting 8 for t in the equation. Doing so gives us

$$s(8) = \frac{60}{8} = 7.5$$

In words Running a mile in 8 minutes is running at a rate of 7.5 miles per hour. ■

Example 11 A painting is purchased as an investment for \$125. If its value increases continuously so that it doubles every 5 years, then its value is given by the function

© XYZ Textbooks

$$V(t) = 125 \cdot 2^{t/5} \quad \text{for} \quad t \geq 0$$

where t is the number of years since the painting was purchased, and V is its value (in dollars) at time t. Find $V(5)$ and $V(10)$, and explain what they mean.

Solution The expression $V(5)$ is the value of the painting when $t = 5$ (5 years after it is purchased). We calculate $V(5)$ by substituting 5 for t in the equation $V(t) = 125 \cdot 2^{t/5}$. Here is our work:

$$V(5) = 125 \cdot 2^{5/5} = 125 \cdot 2^1 = 125 \cdot 2 = 250$$

In words After 5 years, the painting is worth \$250.

The expression $V(10)$ is the value of the painting after 10 years. To find this number, we substitute 10 for t in the equation:

$$V(10) = 125 \cdot 2^{10/5} = 125 \cdot 2^2 = 125 \cdot 4 = 500$$

In words The value of the painting 10 years after it is purchased is \$500. ■

Example 12 A balloon has the shape of a sphere with a radius of 3 inches. Use the following formulas to find the volume and surface area of the balloon.

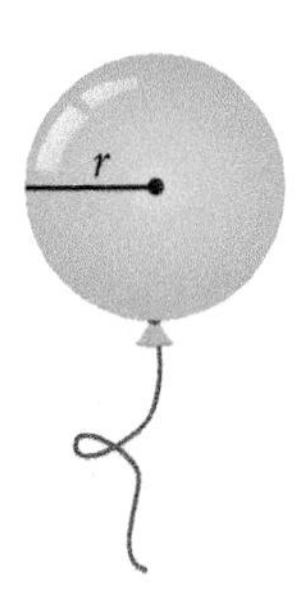

$$V(r) = \frac{4}{3}\pi r^3 \qquad S(r) = 4\pi r^2$$

Note Notice how important the units are in this problem. The numerical part of each answer is the same. It is the units that distinguish them.

Solution As you can see, we have used function notation to write the formulas for volume and surface area, because each quantity is a function of the radius. To find these quantities when the radius is 3 inches, we evaluate $V(3)$ and $S(3)$:

$$V(3) = \frac{4}{3}\pi \cdot 3^3 = \frac{4}{3}\pi \cdot 27$$

$$= 36\pi \text{ cubic inches, or } 113 \text{ cubic inches (to the nearest whole number)}$$

$$S(3) = 4\pi \cdot 3^2$$

$$= 36\pi \text{ square inches, or } 113 \text{ square inches (to the nearest whole number)}$$

The fact that $V(3) = 36\pi$ means that the ordered pair $(3, 36\pi)$ belongs to the function V. Likewise, the fact that $S(3) = 36\pi$ tells us that the ordered pair $(3, 36\pi)$ is a member of function S. ■

TECHNOLOGY NOTE *More About Example 12*

If we look at Example 12, we see that when the radius of a sphere is 3, the numerical values of the volume and surface area are equal. How unusual is this? Are there other values of r for which $V(r)$ and $S(r)$ are equal? We can answer this question by looking at the graphs of both V and S.

To graph the function $V(r) = \frac{4}{3}\pi r^3$, set Y1 = $4\pi X^3/3$. To graph $S(r) = 4\pi r^2$, set Y2 = $4\pi X^2$. Graph the two functions in each of the following windows:

Window 1: X from −4 to 4, Y from −2 to 10

Window 2: X from 0 to 4, Y from 0 to 50

Window 3: X from 0 to 4, Y from 0 to 150

Then use the TRACE and ZOOM features of your calculator to locate the point in the first quadrant where the two graphs intersect. How do the coordinates of this point compare with the results in Example 12?

Using Technology 1.1

Many times, graphs of functions can be constructed more accurately and efficiently using technology. Here are some examples using Wolfram|Alpha.

Example 13 Use Wolfram|Alpha to construct the graphs of the functions $f(x) = x^2$ and $f(x) = x^2 + 2$, on the same coordinate system.

Solution Online, go to www.wolframalpha.com. Here you will see:

In the entry field, enter

graph f(x)=x^2, f(x)=x^2+2

Click the = sign at the far right of the entry field or press the return or enter key on your computer keyboard. You will see the graph of both functions on the same coordinate system. This is what you will see:

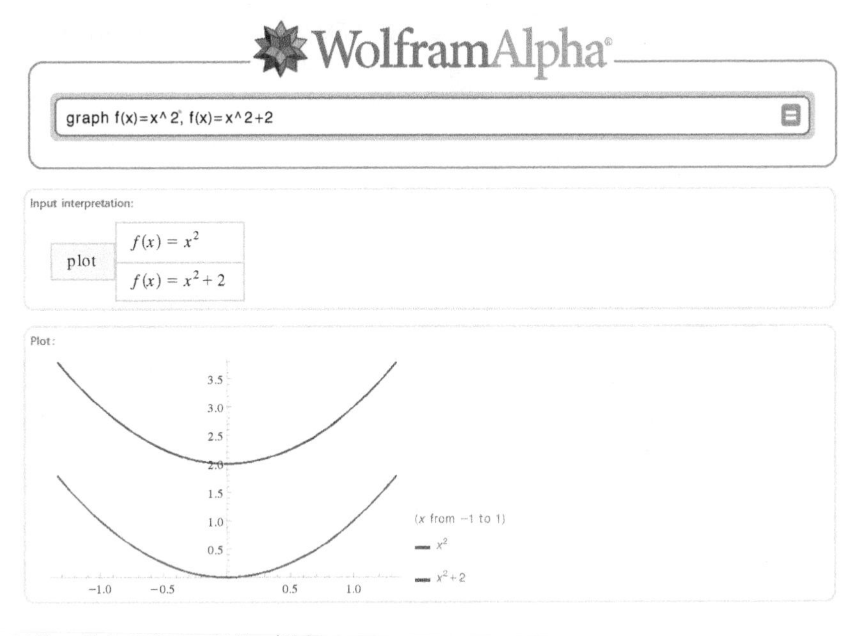

Wolfram Alpha LLC. 2012. Wolfram|Alpha
http://www.wolframalpha.com/
(access June 18, 2012)

If you want to specify the domain, or see the graph for certain values of x, you can specify that in the command line.

Example 14 We know the domain for the function $f(x) = \frac{x+3}{x-2}$ is all real numbers except 2, since 2 would make the denominator zero and the expression would be undefined. Use Wolfram|Alpha to find the domain.

Solution Enter the following into the command line at Wolfram|Alpha

find the domain for f(x)=(x+3)/(x-2)

Notice how Wolfram|Alpha specifies the domain.

Result:

$\{x \in \mathbb{R} : x \neq 2\}$

(assuming a function from reals to reals)

Wolfram Alpha LLC. 2012. Wolfram|Alpha
http://www.wolframalpha.com/
(access June 18, 2012)

It is telling us that the domain is all x such that x is a real number and x is not equal to 2. Notice that Wolfram|Alpha also returns the range for the function, though we are not going to show it here.

Example 15 If $f(x) = x^2 - 4$, use Wolfram|Alpha to find x so that $f(x) = 0$.

Solution We could do some of the work ourselves and simply say solve $x^2 - 4 = 0$, but with Wolfram|Alpha, that is unnecessary. Enter the following into the Wolfram|Alpha command line:

if f(x)=x^2-4, solve f(x)=0

Notice that Wolfram|Alpha solves this problem by finding the intersection of the graphs of $y = x^2$ and $y = 4$. We know this because part of the information that is returned to us is this graph:

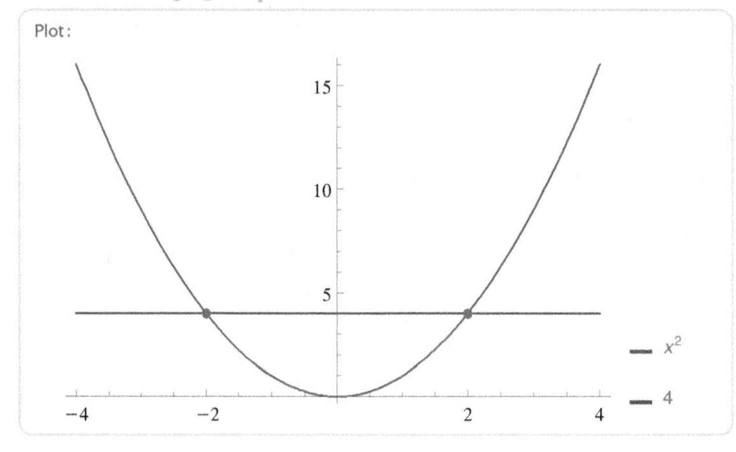

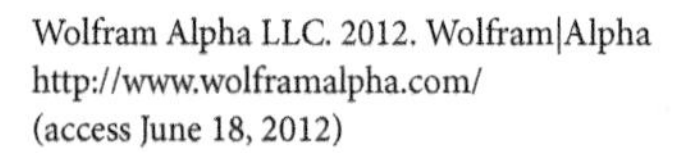
Wolfram Alpha LLC. 2012. Wolfram|Alpha
http://www.wolframalpha.com/
(access June 18, 2012)

Getting Ready for Class

After reading through the preceding section, respond in your own words and in complete sentences.

A. What is a function?

B. The range of a function is usually associated with what variable, input or output?

C. Which variable is usually associated with the domain of a function?

D. If $f(6) = 0$ for a particular function f, then you can immediately graph one of the intercepts. Explain.

Problem Set 1.1

Skills Practice

For each of the following relations, give the domain and range, and indicate which are also functions.

1. $(1, 2), (3, 4), (5, 6), (7, 8)$

2. $(2, 1), (4, 3), (6, 5), (8, 7)$

3. $(2, 5), (3, 4), (1, 4), (0, 6)$

4. $(0, 4), (1, 6), (2, 4), (1, 5)$

5. $(a, 3), (b, 4), (c, 3), (d, 5)$

6. $(a, 5), (b, 5), (c, 4), (d, 5)$

7. $(a, 1), (a, 2), (a, 3), (a, 4)$

8. $(a, 1), (b, 1), (c, 1), (d, 1)$

Check Your Understanding

State whether each of the following graphs represents a function.

9. **10.** **11.** **12.**

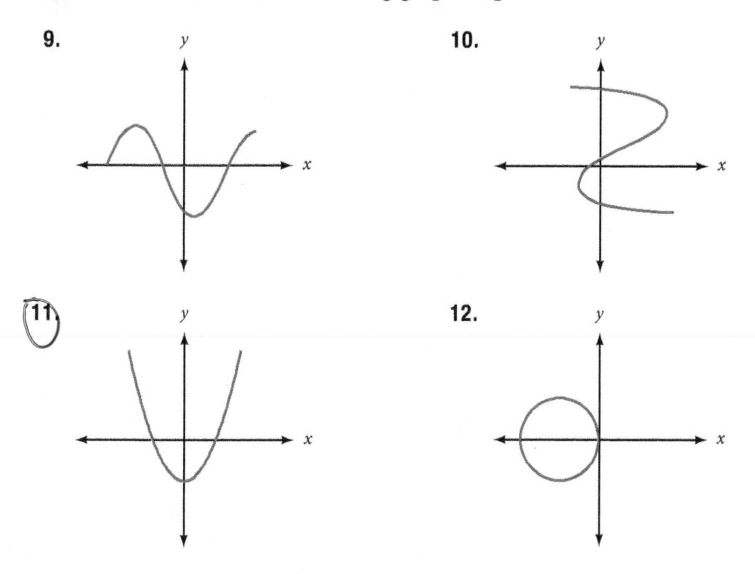

Determine the domain and range of the following functions. Assume the *entire* function is shown.

13. **14.**

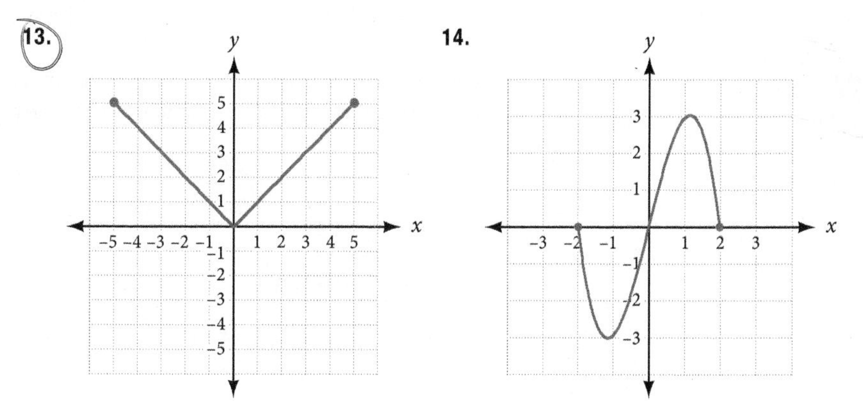

15. **16.**

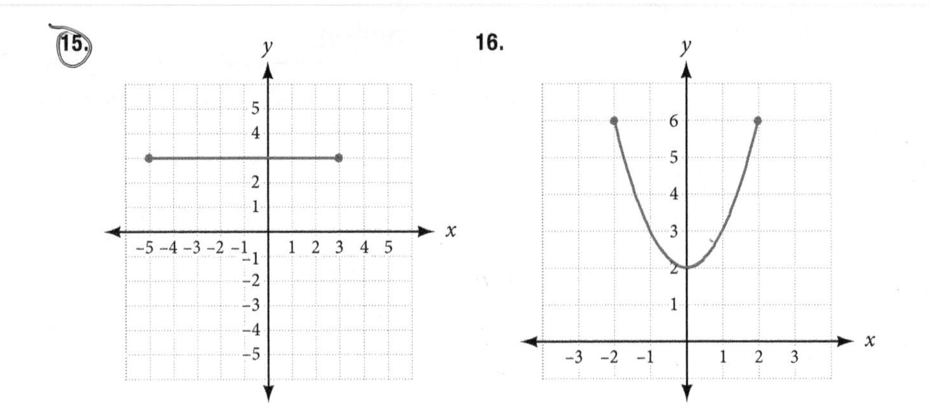

Let $f(x) = 2x - 5$ and $g(x) = x^2 + 3x + 4$. Evaluate the following.

17. $f(2)$ **18.** $f(3)$ **19.** $f(-3)$ **20.** $g(-2)$

21. $g(-1)$ **22.** $f(-4)$ **23.** $g(-3)$ **24.** $g(2)$

25. $g(a)$ **26.** $f(a)$ **27.** $f(a + 6)$ **28.** $g(a + 6)$

Let $f(x) = \dfrac{1}{x+3}$ and $g(x) = \dfrac{1}{x} + 1$. Evaluate the following.

29. $f\left(\frac{1}{3}\right)$ **30.** $g\left(\frac{1}{3}\right)$ **31.** $f\left(-\frac{1}{2}\right)$

32. $g\left(-\frac{1}{2}\right)$ **33.** $f(-3)$ **34.** $g(0)$

35. For the function $f(x) = x^2 - 4$, evaluate each of the following expressions.

a. $f(a) - 3$ **b.** $f(a - 3)$ **c.** $f(x) + 2$

d. $f(x + 2)$ **e.** $f(a + b)$ **f.** $f(x + h)$

36. For the function $f(x) = 3x^2$, evaluate each of the following expressions.

a. $f(a) - 2$ **b.** $f(a - 2)$ **c.** $f(x) + 5$

d. $f(x + 5)$ **e.** $f(a + b)$ **f.** $f(x + h)$

Graph each of the following relations. In each case, use the graph to find the domain and range, and indicate whether the graph is the graph of a function.

37. $y = x^2 - 1$ **38.** $y = x^2 + 4$ **39.** $x = y^2 - 1$ **40.** $y = (x + 2)^2$

41. $x = (y + 1)^2$ **42.** $x = 3 - y^2$

Graph each function. In each case, note any restrictions on the domain.

43. $y = \dfrac{-4}{x}$ **44.** $y = \dfrac{4}{x}$ **45.** $y = \dfrac{8}{x}$ **46.** $y = \dfrac{-8}{x}$

Graph each function. In each case, note any restrictions on the domain.

47. $y = \sqrt{x} - 2$ **48.** $y = \sqrt{x - 2}$ **49.** $y = \sqrt[3]{x} + 3$ **50.** $y = \sqrt[3]{x + 3}$

51. Graph the function $f(x) = \frac{1}{2}x + 2$. Then draw and label the horizontal and vertical line segments that correspond to $x = 4$ and $f(4)$. (See Figure 13.)

52. Graph the function $f(x) = -\frac{1}{2}x + 6$. Then draw and label the horizontal and vertical line segments that correspond to $x = 4$ and $f(4)$. (See Figure 13.)

53. Graph the function $f(x) = x^2$. Then draw and label the horizontal and vertical line segments that correspond to $x = 1$ and $f(1)$, $x = 2$ and $f(2)$ and, finally, $x = 3$ and $f(3)$. (See Figure 13.)

54. Graph the function $f(x) = x^2 - 2$. Then draw and label the horizontal and vertical line segments that correspond to $x = 2$ and $f(2)$ and the line segments corresponding to $x = 3$ and $f(3)$. (See Figure 13.)

Modeling Practice

55. Business: Hourly Pay Suppose you have a job that pays $9.50 per hour and you work anywhere from 10 to 40 hours per week.

a. Write an equation, with a restriction on the variable x, that gives the amount of money, y, you will earn for working x hours in one week.

b. Use the function rule you have written in part a to complete the table.

© Richard Bowden/ iStockPhoto

Weekly Wages

Hours Worked	Function Rule	Gross Pay ($)
x		y
10		
20		
30		
40		

c. Construct a line graph from the information in the table.

d. State the domain and range of this function.

e. What is the minimum amount you can earn in a week with this job? What is the maximum amount?

56. Marketing: Camera Phone Sales The chart shows the estimated number of camera phones and non-camera phones sold from 2004 to 2010. Using the chart, list all the values in the domain and range for the total phones sales.

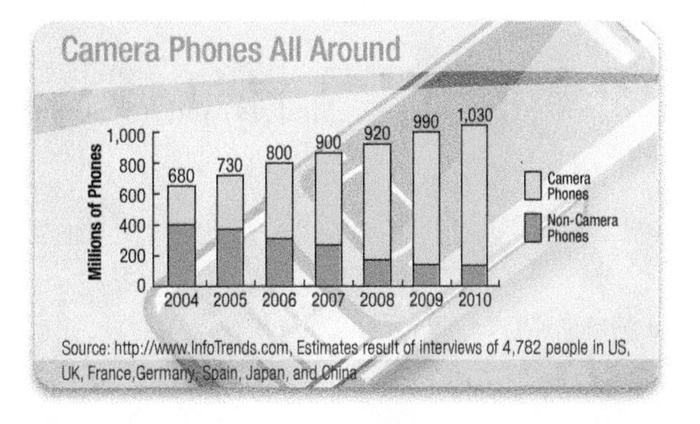

57. Consumer Awareness: Light Bulb Efficiency The chart shows a comparison of power usage between incandescent and energy efficient light bulbs. Use the chart to state the domain and range of the function for an energy efficient bulb.

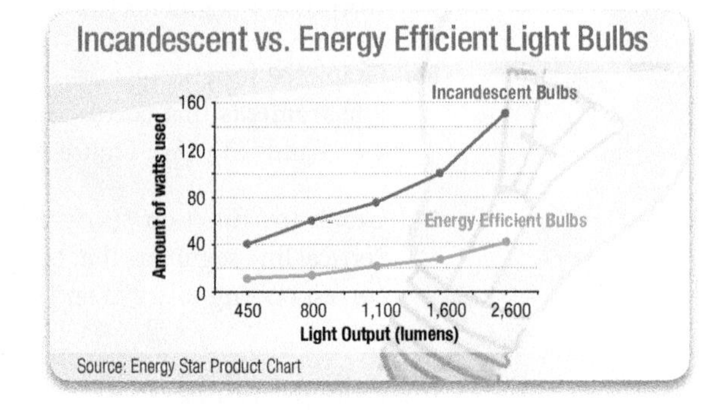

58. Economics: Investing in Art A painting is purchased as an investment for \$150. If its value increases continuously so that it doubles every 3 years, then its value is given by the function

$$V(t) = 150 \cdot 2^{t/3} \quad \text{for} \quad t \geq 0$$

where t is the number of years since the painting was purchased, and $V(t)$ is its value (in dollars) at time t. Find $V(3)$ and $V(6)$, and then explain what they mean.

59. Business: Online Advertising Revenue Social networking and video-sharing sites, such as Facebook, dating sites, and YouTube, often include rich media and display advertising placed around user-generated content. Suppose that, for a particular social networking company, the annual revenue from rich media advertisements, in millions of dollars, for the years 2007 through 2012 can be approximated with the model $R(x) = -x^4 + 11x^3 - 39x^2 + 45x$, where x is the number of years from the beginning of 2007.

Source: eMarketer April 2008

a. Find the company's revenue from rich media advertisements at the beginning of the years 2007, 2008, 2009, 2010, 2011, and 2012. Summarize the results in a table.

b. If the company adheres to its present business model, can it expect its revenues to rise after 2012? (Hint: Evaluate $R(x)$ for the years 2013 and 2023; that is, find $R(6)$ and $R(16)$.)

60. Physical Science: Average Speed If it takes Minke t minutes to run a mile, then her average speed $s(t)$, in miles per hour, is given by the formula

$$s(t) = \frac{60}{t} \quad \text{for} \quad t > 0$$

Find $s(4)$ and $s(5)$, and then explain what they mean.

61. Health Science: Antidepressant Sales Suppose x represents one of the years in the chart. Suppose further that we have three functions f, g, and h that do the following:

f pairs each year with the total sales of Zoloft in billions of dollars for that year.
g pairs each year with the total sales of Effexor in billions of dollars for that year.
h pairs each year with the total sales of Wellbutrin in billions of dollars for that year.

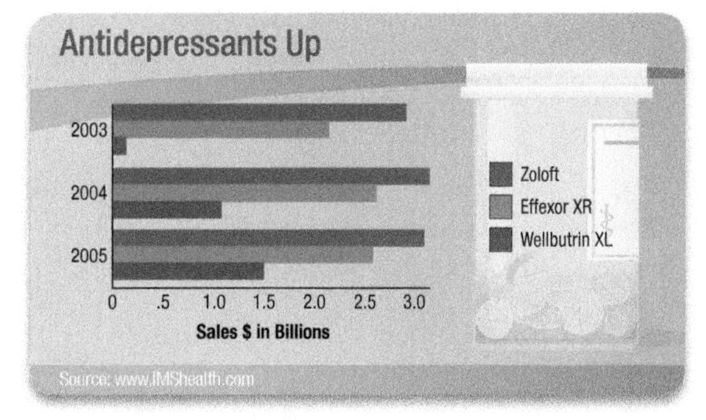

For each statement below, indicate whether the statement is true or false.

a. The domain of g is {2003, 2004, 2005}

b. The domain of g is $\{x \mid 2003 \leq x \leq 2005\}$

c. $f(2004) > g(2004)$

d. $h(2005) > 1.5$

e. $h(2005) > h(2004) > h(2003)$

Straight-Line Depreciation Straight-line depreciation is an accounting method used to help spread the cost of new equipment over a number of years. It takes into account both the cost when new and the salvage value, which is the value of the equipment at the time it gets replaced.

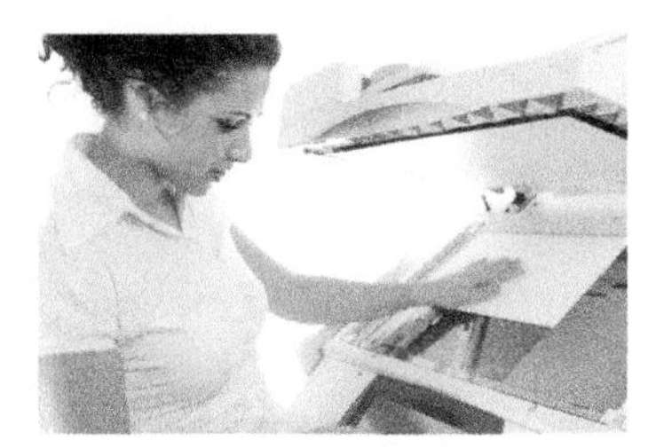

© Nicholas Wave/iStockPhoto

62. Business: Value of a Copy Machine The function

$$V(t) = -3{,}300t + 18{,}000$$

where V is value and t is time in years, can be used to find the value of a large copy machine during the first 5 years of use.

a. What is the value of the copier after 3 years and 9 months?

b. What is the salvage value of this copier if it is replaced after 5 years?

c. State the domain of this function.

d. Sketch the graph of this function.

e. What is the range of this function?

f. After how many years will the copier be worth only $10,000?

© esolla/iStockPhoto

63. Health Science: Dieting The function below can be used to model the progress of a person on a diet. The quantity $W(x)$ is the weight (in pounds) of the person after x weeks of dieting. Use the function to fill in the table, then place your results on the graph. Round to one decimal place.

$$W(x) = \frac{80(2x + 15)}{x + 6}$$

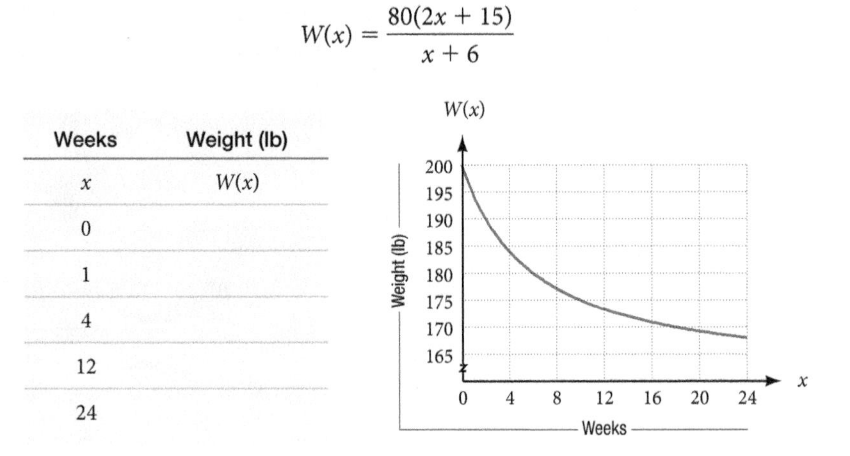

Weeks	Weight (lb)
x	$W(x)$
0	
1	
4	
12	
24	

© Shaun Lowe Photographic/ iStockPhoto

64. Physical Science: Drag Racing The following rational function gives the speed $V(t)$, in miles per hour, of a dragster at each second t during a quarter-mile race. Use the function to fill in the table, then place your results on the graph. Round to one decimal place.

$$V(t) = \frac{340t}{t + 3}$$

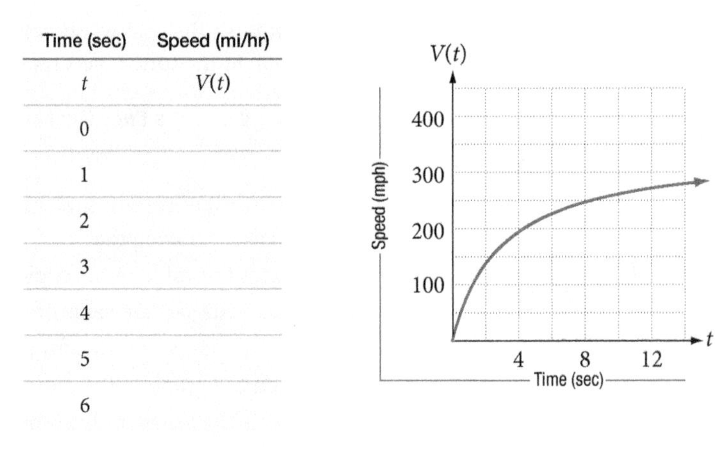

Time (sec)	Speed (mi/hr)
t	$V(t)$
0	
1	
2	
3	
4	
5	
6	

© Warchi/iStockPhoto

65. Business: Average Cost A producer and distributor of gift cards produces x boxes of cards per day at an average cost $\overline{C}(x)$ (in dollars) of

$$\overline{C}(x) = \frac{6x + 8{,}000}{2x + 300}$$

a. What is the domain for this function?

b. Find $\overline{C}(280)$.

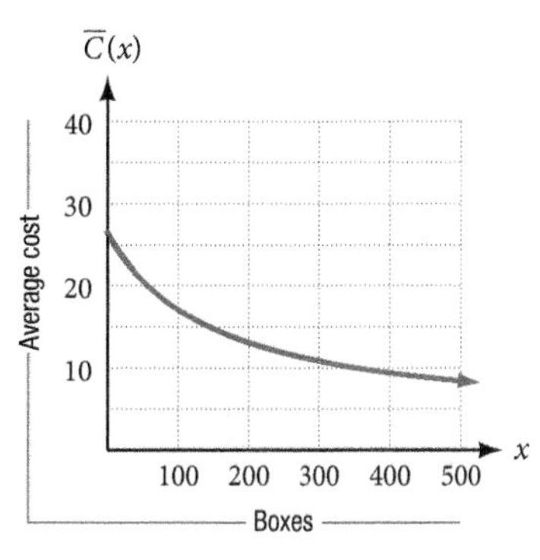

Using Technology Exercises

Use Wolfram|Alpha for the following problems.

66. If $f(x) = x^3 - 4x$, find x so that $f(x) = 0$.

67. If $f(x) = x^4 - 4x^2$, find x so that $f(x) = 0$.

68. If $f(x) = x^2 - 2$, find x so that $f(x) = 0$.

69. If $f(x) = x^2 - 2$, find x so that $f(x) = x$.

Use Wolfram|Alpha for the following problems.

70. On the same coordinate system, construct the graphs of

$$f(x) = 2x, f(x) = 2^x, \text{ and } f(x) = x^2$$

71. On the same coordinate system, construct the graphs of

$$y = x, \quad y = \sqrt{x}, \quad \text{and } y = x^2$$

72. Construct the graphs of

$$y = x, \quad y = |x|, \quad \text{and } y = \sqrt{x^2}$$

73. Find the domain for each function.

a. $y = \sqrt{x}$ **b.** $y = \sqrt{x - 2}$ **c.** $y = \sqrt{x + 2}$

On your graphing calculator, graph each equation and build a table as indicated.

74. $y = 64t - 16t^2$ TblStart $= 0$ ΔTbl $= 1$

75. $y = \frac{1}{2}x - 4$ TblStart $= -5$ ΔTbl $= 1$

76. $y = \frac{12}{x}$ TblStart $= 0.5$ ΔTbl $= 0.5$

77. $y = 2^x$ TblStart $= -2$ ΔTbl $= 1$

Getting Ready for the Next Section

Problems under this heading, "Getting Ready for the Next Section", are problems that you must be able to work in order to understand the material in the next section. They are exactly the type of problems you will see in the explanations and examples in the next section.

Multiply.

78. $x(35 - 0.1x)$

79. $0.6(M - 70)$

80. $(4x - 3)(x - 1)$

81. $(4x - 3)(4x^2 - 7x + 3)$

Simplify.

82. $(35x - 0.1x^2) - (8x + 500)$

83. $(4x - 3) + (4x^2 - 7x + 3)$

84. $(4x^2 + 3x + 2) - (2x^2 - 5x - 6)$

85. $(4x^2 + 3x + 2) + (2x^2 - 5x - 6)$

86. $4(2)^2 - 3(2)$

87. $4(-1)^2 - 7(-1)$

Algebra and Composition with Functions

© Dean Bertoncelj/iStockphoto

People tend to drink more water on hot days than they do on cold days. Because most drinking water is sold in plastic bottles, as the temperature goes up, there is an increase in the number of plastic bottles recycled. We can model this situation with the following diagram.

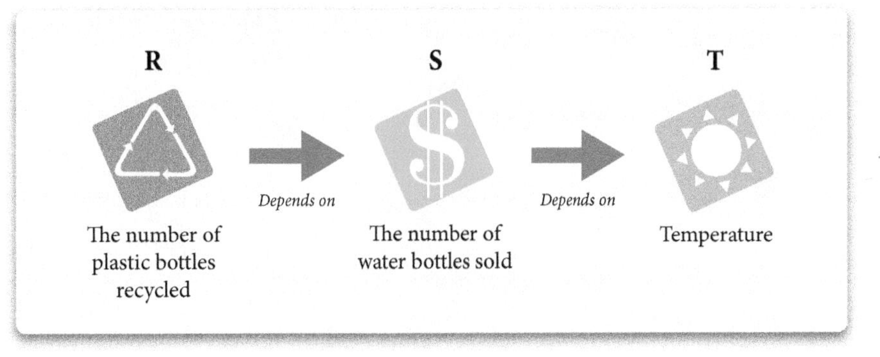

This situation involves a **composite function** because the number of plastic bottles recycled depends on the number of plastic bottles sold, which in turn, depends on the temperature. Composition of functions is one of the topics we will study in this section.

Algebra with Functions

If we are given two functions f and g with a common domain, we can define four other functions as follows.

Note Take caution: this notation, $(f + g)(x)$, does not mean multiplication.

$(f + g)(x) = f(x) + g(x)$	The function $f + g$ is the sum of functions f and g.
$(f - g)(x) = f(x) - g(x)$	The function $f - g$ is the difference of functions f and g.
$(fg)(x) = f(x)g(x)$	The function fg is the product of functions f and g.
$\left(\frac{f}{g}\right)(x) = \frac{f(x)}{g(x)}$	The function $\frac{f}{g}$ is the quotient of functions f and g, where $g(x) \neq 0$.

Let's look at some examples of algebra with functions. In each example that follows, notice that combining functions with addition, subtraction, multiplication, and division produces another function.

VIDEO EXAMPLES

SECTION 1.2

Example 1 Let $f(x) = 4x - 3$, $g(x) = 4x^2 - 7x + 3$, and $h(x) = x - 1$. Find $f + g$, fh, fg, and $\frac{g}{f}$.

Solution The function $f + g$, the sum of functions f and g, is defined by

$$\begin{aligned}(f + g)(x) &= f(x) + g(x) \\ &= (4x - 3) + (4x^2 - 7x + 3) \\ &= 4x^2 - 3x\end{aligned}$$

The function fh, the product of functions f and h, is defined by

$$\begin{aligned}(fh)(x) &= f(x)h(x) \\ &= (4x - 3)(x - 1) \\ &= 4x^2 - 7x + 3 \\ &= g(x)\end{aligned}$$

The function fg, the product of the functions f and g, is defined by

$$\begin{aligned}(fg)(x) &= f(x)g(x) \\ &= (4x - 3)(4x^2 - 7x + 3) \\ &= 16x^3 - 28x^2 + 12x - 12x^2 + 21x - 9 \\ &= 16x^3 - 40x^2 + 33x - 9\end{aligned}$$

The function $\frac{g}{f}$, the quotient of the functions g and f, is defined by

$$\begin{aligned}\left(\frac{g}{f}\right)(x) &= \frac{g(x)}{f(x)} \\ &= \frac{4x^2 - 7x + 3}{4x - 3}\end{aligned}$$

Factoring the numerator, we can reduce to lowest terms:

$$\begin{aligned}&= \frac{(4x - 3)(x - 1)}{4x - 3} \\ &= x - 1 \\ &= h(x)\end{aligned}$$

■

Example 2 If f, g, and h are the same functions defined in Example 1, evaluate $(f + g)(2)$, $(fh)(-1)$, $(fg)(0)$, and $\left(\frac{g}{f}\right)(5)$.

Solution We use the formulas for $f + g$, fh, fg, and $\frac{g}{f}$ found in Example 1:

$$\begin{aligned}(f + g)(2) &= 4(2)^2 - 3(2) \\ &= 16 - 6 \\ &= 10\end{aligned}$$

$$\begin{aligned}(fh)(-1) &= 4(-1)^2 - 7(-1) + 3 \\ &= 4 + 7 + 3 \\ &= 14\end{aligned}$$

$$\begin{aligned}(fg)(0) &= 16(0)^3 - 40(0)^2 + 33(0) - 9 \\ &= 0 - 0 + 0 - 9 \\ &= -9\end{aligned}$$

$$\begin{aligned}\left(\frac{g}{f}\right)(5) &= 5 - 1 \\ &= 4\end{aligned}$$

■

Modeling: Profit, Revenue, and Cost

The most common formula in business is the formula that gives profit as the difference between revenue and cost. It is a simple formula that gives one number to tell how well a company is doing. Here is the formula and a graph that shows one way the three functions could be related graphically:

$$\textbf{Profit} = \textbf{Revenue} - \textbf{Cost}$$

$$P(x) = R(x) - C(x)$$

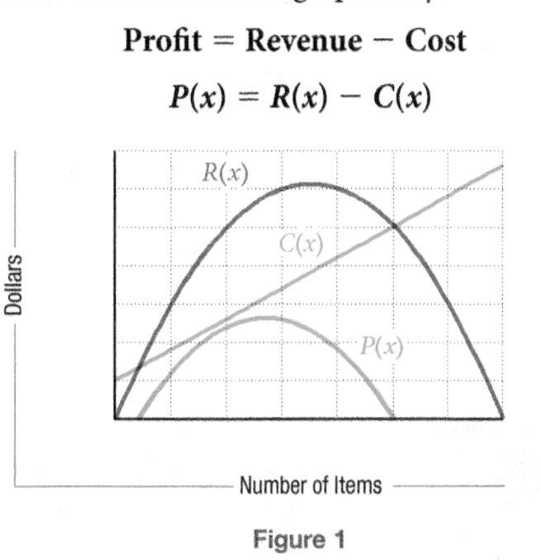

Figure 1

Revenue is the total amount of money a company brings in by selling their product, and cost is the total cost to produce the product they sell. Revenue itself can be broken down further by another formula common in the business world. The revenue obtained from selling x items is the product of the number of items sold and the price per item. That is,

$$\text{Revenue} = (\text{Number of items sold})(\text{Price of each item})$$

For example, if 100 items are sold for \$9 each, the revenue is $100(9) = \$900$. Likewise, if 500 items are sold for \$11 each, then the revenue is $500(11) = \$5{,}500$. In general, if x is the number of items sold and p is the selling price of each item, then we can write

$$R = xp$$

© Simone Becchetti/iStockPhoto

Example 3 A manufacturer of hard shell cases for the iPhone knows that the number of cases she can sell each week is related to the price of the cases by the equation $x = 1{,}300 - 100p$, where x is the number of cases and p is the price per case. What price should she charge for each case if she wants the weekly revenue to be \$4,000?

Solution The formula for total revenue is $R = xp$. Because we want R in terms of p, we substitute $1{,}300 - 100p$ for x in the equation $R = xp$:

$$\begin{array}{ll} \text{If} & R = xp \\ \text{and} & x = 1{,}300 - 100p \\ \text{then} & R = (1{,}300 - 100p)p \end{array}$$

We want to find p when R is 4,000. Substituting for R in the formula gives us

$$4{,}000 = (1{,}300 - 100p)p$$

$$4{,}000 = 1{,}300p - 100p^2$$

This is a quadratic equation. To write it in standard form, we add $100p^2$ and $-1{,}300p$ to each side, giving us

$$100p^2 - 1{,}300p + 4{,}000 = 0$$

$$p^2 - 13p + 40 = 0 \quad \text{Divide each side by 100}$$

$$(p - 5)(p - 8) = 0$$

$$p - 5 = 0 \quad \text{or} \quad p - 8 = 0$$

$$p = 5 \quad \text{or} \quad p = 8$$

If she sells the cases for \$5 each or for \$8 each, she will have a weekly revenue of \$4,000. ■

Composition of Functions and Training Heart Rate

© Rob Cruse/iStockPhoto

In addition to the four operations used to combine functions shown so far in this section, there is a fifth way to combine two functions to obtain a new function. It is called **composition of functions.** To illustrate the concept, here is the definition of training heart rate: training heart rate, in beats per minute, is resting heart rate plus 60% of the difference between maximum heart rate and resting heart rate. If your resting heart rate is 70 beats per minute, then your training heart rate is a function of your maximum heart rate M.

$$T(M) = 70 + 0.6(M - 70) = 70 + 0.6M - 42 = 28 + 0.6M$$

But your maximum heart rate is found by subtracting your age in years from 220. So, if x represents your age in years, then your maximum heart rate is

$$M(x) = 220 - x$$

Therefore, if your resting heart rate is 70 beats per minute and your age in years is x, then your training heart rate can be written as a function of x.

$$T(x) = 28 + 0.6(220 - x)$$

This last line is the composition of functions T and M. We input x into function M, which outputs $M(x)$. Then, we input $M(x)$ into function T, which outputs $T(M(x))$, which is the training heart rate as a function of age x. Here is a diagram, called a function map, of the situation:

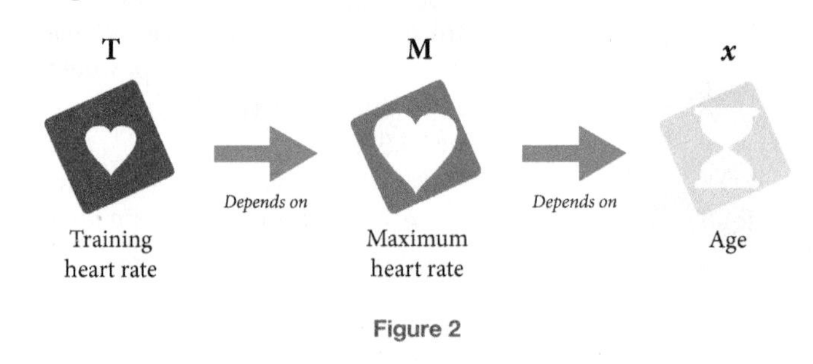

Figure 2

Now let's generalize the preceding ideas into a formal development of composition of functions. To find the composition of two functions f and g, we first require that the range of g have numbers in common with the domain of f. Then the composition of f with g, is defined this way:

$$(f \circ g)(x) = f(g(x))$$

To understand this new function, we begin with a number x, and we operate on it with g, giving us $g(x)$. Then we take $g(x)$ and operate on it with f, giving us $f(g(x))$. The only numbers we can use for the domain of the composition of f with g are numbers x in the domain of g, for which $g(x)$ is in the domain of f. The diagrams in Figure 3 illustrate the composition of f with g.

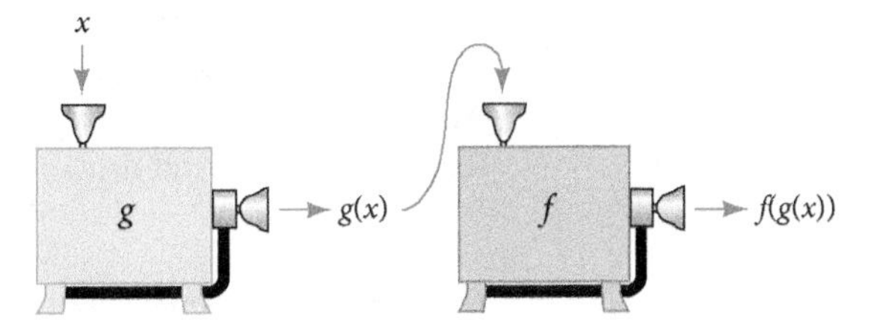

Figure 3 Function machines

Composition of functions is not commutative. The composition of f with g, $f \circ g$, may therefore be different from the composition of g with f, $g \circ f$.

$$(g \circ f)(x) = g(f(x))$$

Again, the only numbers we can use for the domain of the composition of g with f are numbers in the domain of f, for which $f(x)$ is in the domain of g. The diagrams in Figure 4 illustrate the composition of g with f.

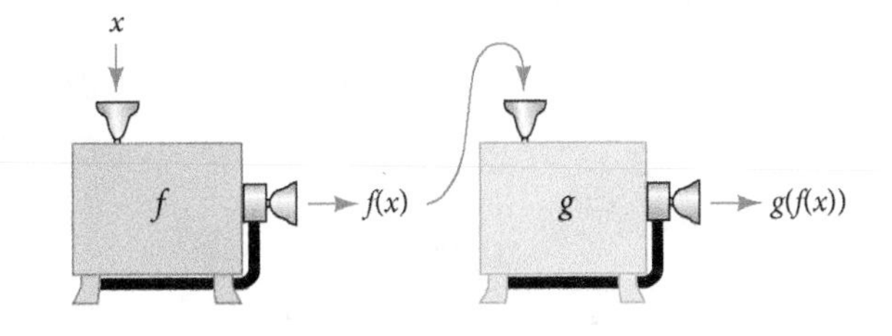

Figure 4 Function machines

Example 4 If $f(x) = x + 5$ and $g(x) = x^2 - 2x$, find $(f \circ g)(x)$ and $(g \circ f)(x)$.

Solution The composition of f with g is

$$\begin{aligned}(f \circ g)(x) &= f(g(x)) \\ &= f(x^2 - 2x) \\ &= (x^2 - 2x) + 5 \\ &= x^2 - 2x + 5\end{aligned}$$

The composition of g with f is

$$\begin{aligned}(g \circ f)(x) &= g(f(x)) \\ &= g(x + 5) \\ &= (x + 5)^2 - 2(x + 5) \\ &= (x^2 + 10x + 25) - 2x - 10 \\ &= x^2 + 8x + 15\end{aligned}$$

Modeling: Temperature and Recycling

Example 5 Suppose that, for temperatures greater than 10 °C, the average number of soft drinks sold each day, s (in hundreds), depends on the temperature, t, as given by

$$s(t) = 1{,}200 + 23t + 600(t - 10)^{1/3}$$

And, the number of soft drink aluminum cans recycled, r (in hundreds), depends on the number of soft drinks sold, s, as

$$r(s) = s^{3/4}$$

a. Give some reasonable numbers for the domain of the function s.

b. How many cans are sold when the temperature is 27 °C?

c. How many cans are recycled when the temperature is 27 °C?

d. Suppose we want to use the Fahrenheit temperature scale, instead of the Celsius scale. How would the problem change?

Solution Here is a diagram of the situation.

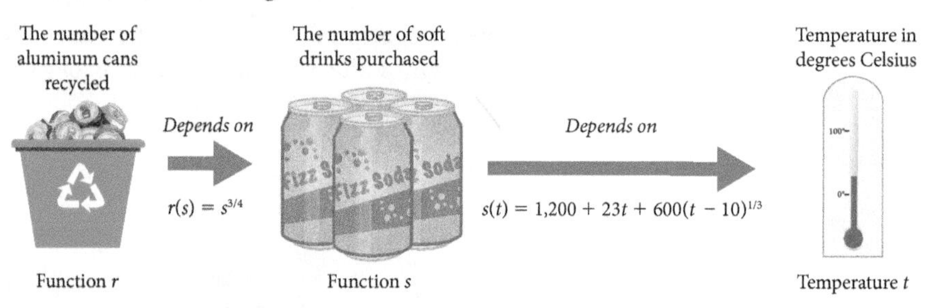

a. People tend to drink more soft drinks on hot days than on cold days. If we restrict the domain of this function so that t ranges from 20 °C (which is 68 °F) to 40 °C (104 °F), we should have a reasonable range of values to input for temperature.

b. To find the number of cans sold when the temperature is 27 °C, we simply substitute 27 for t and use our formulas.

$$\text{When } t = 27, \text{ then } s(27) = 1{,}200 + 23(27) + 600(27 - 10)^{1/3}$$
$$= 1{,}200 + 621 + 600(17)^{1/3}$$
$$\approx 1{,}200 + 621 + 600(2.57)$$
$$\approx 3{,}363$$

Since s represents hundreds of cans, the number of cans sold is about 336,300.

c. To find the number of cans recycled, we substitute $s = 3{,}363$ into the formula for the number of cans recycled, giving us

$$r(3{,}363) = 3{,}363^{3/4} \approx 442$$

Since r is in hundreds of cans, the number of cans recycled when the temperature is 27 °C is about 44,200.

d. Let's use the variable T to represent the temperature in degrees Fahrenheit. If we want to input our temperatures in degrees Fahrenheit, we must add a third function on to the end of the chain that will convert our degrees Fahrenheit (T) to degrees Celsius (t), in the beginning of the problem. Here is a diagram of this situation.

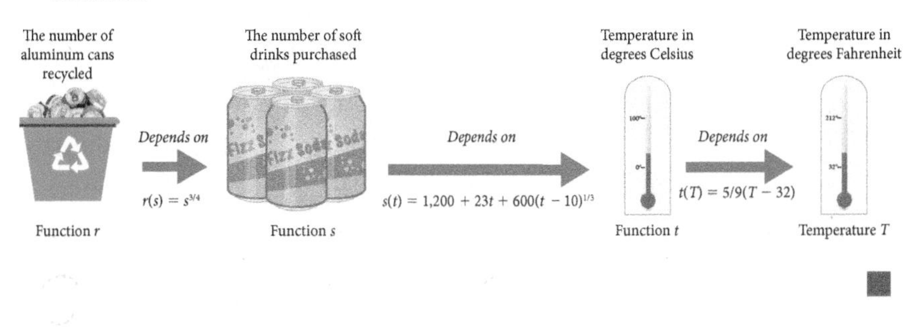

Using Technology 1.2

You can use your graphing calculator to evaluate a composite function.

Example 6 Evaluate $(f \circ g)(2)$ for $f(x) = \sqrt{x-3}$ and $g(x) = 2x + 3$.

Solution Set up your Y-variables this way:

Y1 = √(X-3)

Y2 = 2X+3

In the computation window, enter

Y1(Y2(2))

The calculator responds with the value 2.

When $(g \circ f)(2)$ is evaluated, the calculator responds with an error. Can you get your calculator to respond with an error? Why does it do so?

Getting Ready for Class

After reading through the preceding section, respond in your own words and in complete sentences.

A. How are profit, revenue, and cost related?
B. What is the difference between $(fg)(x)$ and $(f \circ g)(x)$?
C. How do you find maximum heart rate?
D. For functions f and g, how do you find the composition of g with f?

Problem Set 1.2

Skills Practice

Let $f(x) = 4x - 3$ and $g(x) = 2x + 5$. Write a formula for each of the following functions.

1. $f + g$ **2.** $f - g$ **3.** $g - f$ **4.** $g + f$

5. fg **6.** $\dfrac{f}{g}$ **7.** $\dfrac{g}{f}$ **8.** ff

If the functions f, g, and h are defined by $f(x) = 3x - 5$, $g(x) = x - 2$, and $h(x) = 3x^2 - 11x + 10$, write a formula for each of the following functions.

9. $g + f$ **10.** $f + h$ **11.** $g + h$ **12.** $f - g$

13. $g - f$ **14.** $h - g$ **15.** fg **16.** gf

17. fh **18.** gh **19.** $\dfrac{h}{f}$ **20.** $\dfrac{h}{g}$

21. $\dfrac{f}{h}$ **22.** $\dfrac{g}{h}$ **23.** $f + g + h$ **24.** $h - g + f$

25. $h + fg$ **26.** $h - fg$

Let $f(x) = 2x + 1$, $g(x) = 4x + 2$, and $h(x) = 4x^2 + 4x + 1$, and find the following.

27. $(f + g)(2)$ **28.** $(f - g)(-1)$ **29.** $(fg)(3)$ **30.** $\left(\dfrac{f}{g}\right)(-3)$

31. $\left(\dfrac{h}{g}\right)(1)$ **32.** $(hg)(1)$ **33.** $(fh)(0)$ **34.** $(h - g)(-4)$

35. $(f + g + h)(2)$ **36.** $(h - f + g)(0)$ **37.** $(h + fg)(3)$ **38.** $(h - fg)(5)$

39. Let $f(x) = x^2$ and $g(x) = x + 4$, and find

a. $(f \circ g)(5)$ **b.** $(g \circ f)(5)$ **c.** $(f \circ g)(x)$ **d.** $(g \circ f)(x)$

40. Let $f(x) = 3 - x$ and $g(x) = x^3 - 1$, and find

a. $(f \circ g)(0)$ **b.** $(g \circ f)(0)$ **c.** $(f \circ g)(x)$ **d.** $(g \circ f)(x)$

41. Let $f(x) = x^2 + 3x$ and $g(x) = 4x - 1$, and find

a. $(f \circ g)(0)$ **b.** $(g \circ f)(0)$ **c.** $(f \circ g)(x)$ **d.** $(g \circ f)(x)$

42. Let $f(x) = (x - 2)^2$ and $g(x) = x + 1$, and find

a. $(f \circ g)(-1)$ **b.** $(g \circ f)(-1)$ **c.** $(f \circ g)(x)$ **d.** $(g \circ f)(x)$

For each of the following pairs of functions f and g, show that $(f \circ g)(x) = (g \circ f)(x) = x$.

43. $f(x) = 5x - 4$ and $g(x) = \dfrac{x + 4}{5}$

44. $f(x) = \dfrac{x}{6} - 2$ and $g(x) = 6x + 12$

Modeling Practice

© Rob Cruse/iStockPhoto

45. Health Science: Training Heart Rate Find the training heart rate function, $T(M)$, for a person with a resting heart rate of 62 beats per minute, then find the following. Round to the nearest whole number.

a. Find the maximum heart rate function, $M(x)$, for a person x years of age.

b. What is the maximum heart rate for a 24-year-old person?

c. What is the training heart rate for a 24-year-old person with a resting heart rate of 62 beats per minute?

d. What is the training heart rate for a 36-year-old person with a resting heart rate of 62 beats per minute?

e. What is the training heart rate for a 48-year-old person with a resting heart rate of 62 beats per minute?

46. Health Science: Training Heart Rate Find the training heart rate function, $T(M)$, for a person with a resting heart rate of 72 beats per minute, then find the following to the nearest whole number.

a. Find the maximum heart rate function, $M(x)$, for a person x years of age.

b. What is the maximum heart rate for a 20-year-old person?

c. What is the training heart rate for a 20-year-old person with a resting heart rate of 72 beats per minute?

d. What is the training heart rate for a 30-year-old person with a resting heart rate of 72 beats per minute?

e. What is the training heart rate for a 40-year-old person with a resting heart rate of 72 beats per minute?

47. Business: Revenue A company manufactures and sells DVDs. The revenue obtained by selling x DVDs is given by the formula

$$R = 11.5x - 0.05x^2$$

Solve the equation below to find the number of DVDs they must sell to receive \$650 in revenue.

$$650 = 11.5x - 0.05x^2$$

48. Business: Price and Revenue The relationship between the number of pencil sharpeners, x, a company can sell each week and the price of each sharpener, p, is given by the equation $x = 1{,}800 - 100p$. At what price should the sharpeners be sold if the weekly revenue is to be \$7,200?

Using Technology Exercises

Use your graphing calculator to evaluate each composite function. Approximate your answers to one decimal place when necessary.

49. If $f(x) = 5x^2 - x$ and $g(x) = \sqrt{x}$, evaluate both $f(g(16))$ and $g(f(16))$.

50. If $f(x) = x^{2/3}$ and $g(x) = 27x^6$, evaluate both $f(g(1))$ and $g(f(1))$.

51. If $f(x) = \dfrac{x-4}{x+1}$ and $g(x) = x^2$, evaluate both $f(g(2))$ and $g(f(2))$.

52. If $f(x) = 7x - 8$ and $g(x) = \sqrt[5]{x^3}$, evaluate both $f(g(2))$ and $g(f(2))$.

53. **Business: Revenue, Cost, and Profit** The cost, in dollars, to a company to produce x units of a product is modeled by the function

$$C(x) = 0.02x^3 - 0.5x^2 + 10x + 50$$

The revenue, in dollars, the company expects from the sale of x units of the product is modeled by

$$R(x) = -1.1x^2 + 41.5x$$

The profit, in dollars, realized on the sale of x units of the product is just the revenue minus the cost, that is,

$$P(x) = R(x) - C(x)$$

a. Construct a model for the profit function.
b. Find the profit when the number of units sold is 19.
c. Find the profit when the number of units sold is 20.
d. Find the profit when the number of units sold is 21.
e. What is the trend in the profit as the number of units produced and sold increases from 19 through 21?

54. **Business: Inventory Costs** Retailers are concerned about inventory costs. The total inventory cost, $C(x)$, for a lot size of x units, is the sum of the carrying costs, $H(x)$, and the order/reorder costs, $R(x)$. That is, $C(x) = H(x) + R(x)$. All costs are in dollars. Carrying costs are

$$(\text{carrying cost per unit}) \cdot \frac{x}{2}$$

Order/reorder costs are

$$(\text{cost per order}) \cdot \frac{(\text{number of units sold during time period})}{x}$$

The lot size x is the number of units the company expects to sell in some time period. Approximate to 3 decimal places when necessary.

a. Construct a model for the inventory cost function when the carrying cost per unit is \$4, the order/reorder cost is \$15, and the company expects to sell 1,600 units over the next year.
b. Find the inventory cost when the lot size is 109 units.
c. Find the inventory cost when the lot size is 110 units.
d. Find the inventory cost when the lot size is 111 units.
e. Which of these three lot sizes produces the minimum cost?

55. **Business: Retail Sales** A retail store sells very expensive products. On the first day of every month, it sells any item for \$50 off its list price. The store also offers a discount of 20% to any customers who can prove that he or she contributed to a charity the previous month. If the list price of an item is x dollars,
   a. create a function, call it $F(x)$, that models the price of an item on the first day of the month.
   b. create a function, call it $D(x)$, that models the price of an item to a customer who donates to a charity.
   c. create a composite function, call it $D[F(x)]$, that models the price offered to a first-day-of-the-month customer who donates to a charity.
   d. Find the price to a first-day-of-the-month customer who donates to charity on an item with a list price of \$500.

© Nigel Miller/iStockPhoto

56. **Business: Cost of Ocean Oil Leak Clean-up** Oil is leaking from an oil tanker at the rate of 2,500 gallons per day. At this rate the radius of the circular slick of oil is increasing at a rate of 4 miles per day.
   a. Create a function, call it $r(t)$, that models the radius of the circular slick of oil at a time, t, days after the beginning of the leak. That is, find an expression for $r(t)$.
   b. The area of a circle is given by the formula $A = \pi r^2$. Create a function, call it $A(t)$, that models the area of the oil slick at time t. That is, find an expression for $A[r(t)]$.
   c. To the nearest whole number, what will be the area of the oil slick 2 days after the beginning of the leak?
   d. If it costs \$5,000,000 to clean the oil from each square mile of the ocean, what is the cost of this spill 2 days after the beginning of the leak?

57. **Health Science: Carbon Monoxide Levels** A study of a northwestern community indicates that the average daily level C of carbon monoxide in the air, in parts per million (ppm), is approximated by the function

$$C(p) = \sqrt{0.18p^2 + 9.6} \quad 0 < p \leq 20$$

where p represents the population of the community in thousands of people. (Enter this function using Y1.) The population of the community depends on the time t in years from now and is approximated by the function

$$p(t) = 1.8 + 0.04t^2 \quad 0 < t \leq 20$$

(Enter this function using Y2.) Compute the level of carbon monoxide in this community's air 15 years from now in two ways. First, compute $p(15)$ and then substitute this result into $C(p)$. Second, compute $C[p(15)]$ directly using the method shown in Example 6. Compare the two results. If they are equal, write your conclusion in sentence form. If they are unequal, try again.

58. **Health Science: Carbon Monoxide Levels** For the previous exercise, make a single change in the defining expression of Y1 so that when you turn off the graphing capability of Y2, the graph of the composite function $C[p(t)]$ appears in the window $0 < t \leq 20$ and $0 \leq C \leq 20$. (Hint: Replace the X in Y1's equation with Y2.)

Getting Ready for the Next Section

For Problems 59 & 60, write as a fraction with denominator 100.

59. -0.06

60. -0.07

61. If $y = 2x - 3$, find y when $x = 2$.

62. If $y = 2x - 3$, find y when $x = 5$.

Simplify.

63. $\dfrac{1 - (-3)}{-5 - (-2)}$

64. $\dfrac{-3 - 1}{-2 - (-5)}$

65. $\dfrac{-1 - 4}{3 - 3}$

66. $\dfrac{-3 - (-3)}{2 - (-1)}$

Solve for y.

67. $\dfrac{y - b}{x - 0} = m$

68. $2x + 3y = 6$

69. $y - 3 = -2(x + 4)$

70. $y + 1 = -\dfrac{2}{3}(x - 3)$

71. If $y = -\dfrac{4}{3}x + 5$, find y when x is 0.

72. If $y = -\dfrac{4}{3}x + 5$, find y when x is 3.

Slope, Rates of Change, and Linear Functions

1.3

For an intuitive introduction to the slope of a line, imagine that a highway sign tells us we are approaching a 6% downgrade. As we drive down this hill, each 100 feet we travel horizontally is accompanied by a 6-foot drop in elevation.

In mathematics we say the slope of the highway is $-0.06 = -\frac{6}{100} = -\frac{3}{50}$. The slope is the ratio of the vertical change to the accompanying horizontal change.

In defining the slope of a straight line, we want to associate a number with the line. First, we want the slope of a line to measure the "steepness" of the line. That is, in comparing two lines, the slope of the steeper line should have the larger numerical value. Second, we want a line that rises going from left to right to have a **positive** slope. We want a line that falls going from left to right to have a **negative** slope. (A line that neither rises nor falls going from left to right must, therefore, have 0 slope.) Geometrically, we can define the **slope** of a line as the ratio of the vertical change to the horizontal change encountered when moving from one point to another on the line. The vertical change is sometimes called the **rise**. The horizontal change is called the **run**.

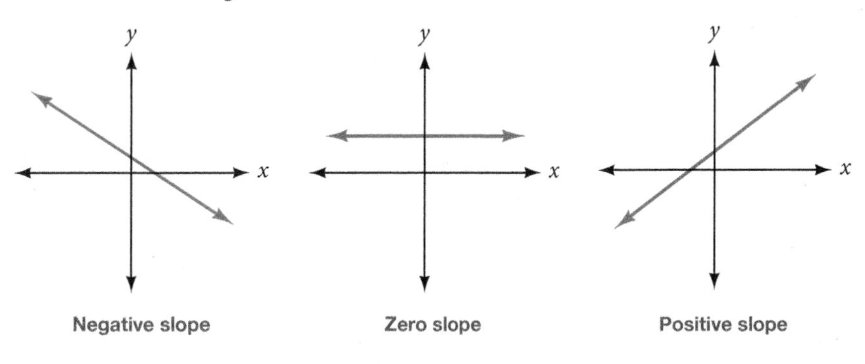

Negative slope   Zero slope   Positive slope

VIDEO EXAMPLES

SECTION 1.3

Example 1 Find the slope of the line $y = 2x - 3$.

Solution To use our geometric definition, we first graph $y = 2x - 3$ (see Figure 1 on the following page). We then pick any two convenient points and find the ratio of rise to run. By convenient points we mean points with integer coordinates. If we let $x = 2$ in the equation, then $y = 1$. Likewise, if we let $x = 4$, then y is 5.

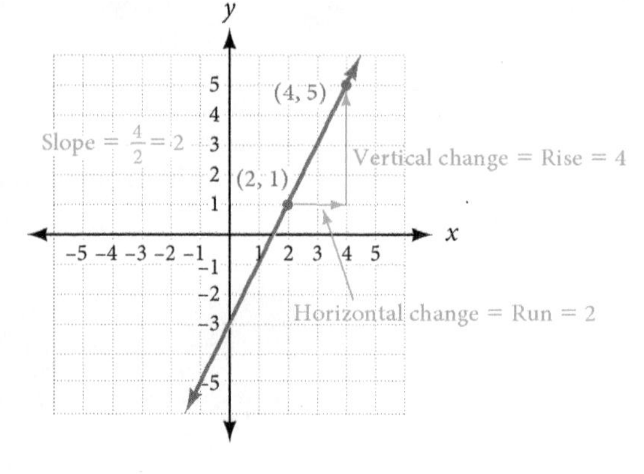

Figure 1

The ratio of vertical change to horizontal change is 4 to 2, giving us a slope of $\frac{4}{2} = 2$. Our line has a slope of 2. ■

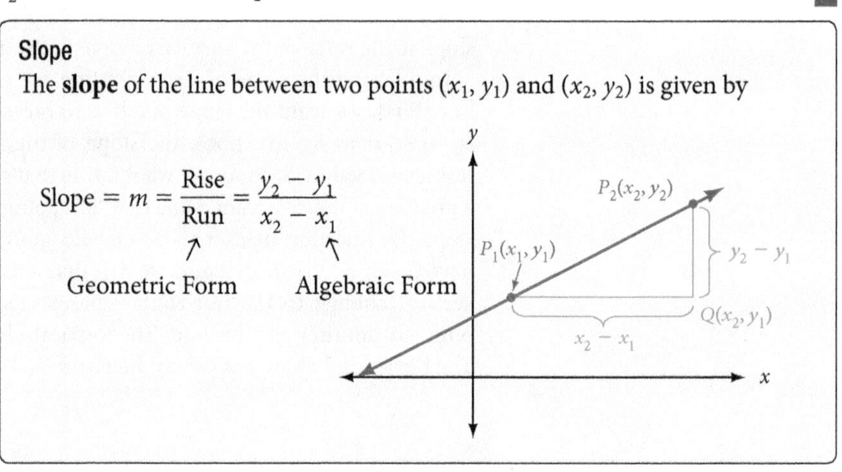

Slope

The **slope** of the line between two points (x_1, y_1) and (x_2, y_2) is given by

$$\text{Slope} = m = \frac{\text{Rise}}{\text{Run}} = \frac{y_2 - y_1}{x_2 - x_1}$$

Example 2 Find the slope of the line through $(-2, -3)$ and $(-5, 1)$.

Solution Let's use the slope formula.

$$m = \frac{y_2 - y_1}{x_2 - x_1} = \frac{1 - (-3)}{-5 - (-2)} = \frac{4}{-3} = -\frac{4}{3}$$

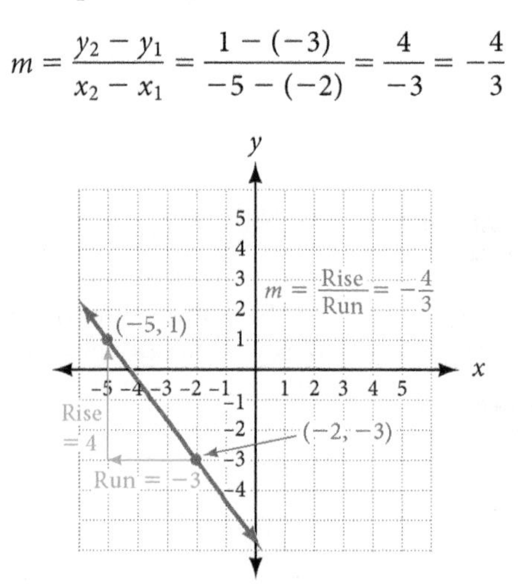

Figure 2

Looking at the graph of the line between the two points (Figure 2), we can see our geometric approach does not conflict with our algebraic approach.

We should note here that it does not matter which ordered pair we call (x_1, y_1) and which we call (x_2, y_2). If we were to reverse the order of subtraction of both the x- and y-coordinates in the preceding example, we would have

$$m = \frac{-3 - 1}{-2 - (-5)} = \frac{-4}{3} = -\frac{4}{3}$$

which is the same as our previous result.

Example 3 Find the slope of the line containing $(3, -1)$ and $(3, 4)$.

Solution Using the definition for slope, we have

$$m = \frac{-1 - 4}{3 - 3} = \frac{-5}{0}$$

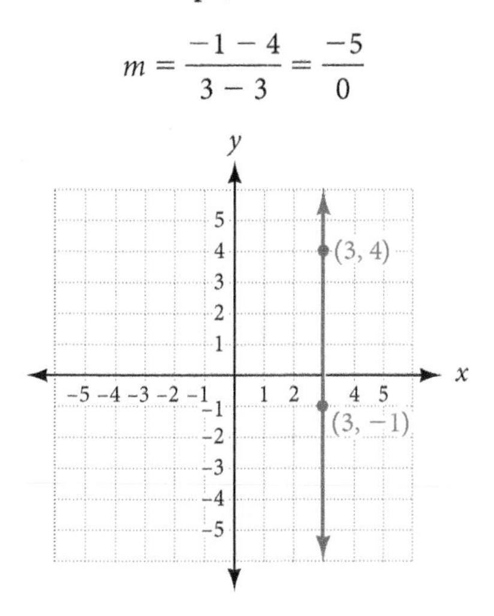

Figure 3

The expression $\frac{-5}{0}$ is undefined. That is, there is no real number to associate with it. In this case, we say the slope is *undefined*, or, we can say the line *has no slope.* Either way, this situation will only occur when our line is vertical.

The graph of our line is shown in Figure 3. Our line with no slope is a vertical line. All vertical lines have no slope. (All horizontal lines, as we mentioned earlier, have 0 slope.)

Modeling: Slope and Rate of Change

So far, the slopes we have worked with represent the ratio of the change in y to the corresponding change in x, or, on the graph of the line, the slope is the ratio of vertical change to horizontal change in moving from one point on the line to another. However, when our variables represent quantities from the world around us, slope can have additional interpretations.

Example 4 On the chart below, find the slope of the line connecting the first point (1955, 0.29) with the highest point (2005, 2.93). Explain the significance of the result.

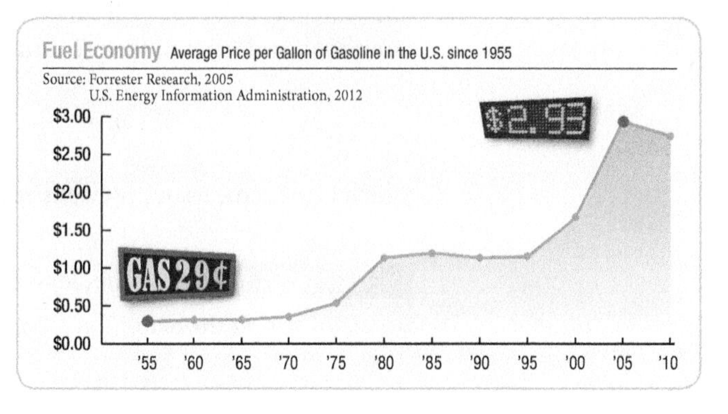

Solution The slope of the line connecting the first point (1955, 0.29) with the highest point (2005, 2.93), is

$$m = \frac{2.93 - 0.29}{2005 - 1955} = \frac{2.64}{50} = 0.0528$$

The units are dollars/year. If we write this in terms of cents we have

$$m = 5.28 \text{ cents/year}$$

which is the average rate of change in the price of a gallon of gasoline over a 50-year period of time.

Likewise, if we connect the points (2005, 2.93) and (2010, 2.74), the line that results has a slope of

$$m = \frac{2.74 - 2.93}{2010 - 2005} = \frac{-0.19}{5} = -0.038 \text{ dollars/year} = -3.8 \text{ cents/year}$$

which is the average rate of change in the price of a gallon of gasoline over the most recent 5-year period. As you can imagine by looking at the chart, the line connecting the first point and highest point is very different from the line connecting the points from 2005 and 2010, and this is what we are seeing numerically with our slope calculations. If we were summarizing this information for an article in the newspaper, we could say, "Although the price of a gallon of gasoline increased 5.28 cents per year from 1955 to its peak in 2005, in the last 5 years the average annual rate has actually decreased at a rate of 3.8 cents per year." When describing rates of change, positive slopes translate into increases, while negative slopes are decreases. ■

Example 5 Kendra tosses a softball into the air with an underhand motion. The distance of the ball above her hand is given by the function in symbolic form as

$$h(t) = 32t - 16t^2 \quad \text{for} \quad 0 \le t \le 2$$

Find the average rate of change of the function from $\frac{1}{4}$ second to 1 second.

Solution To find the average rate of change of the function from $t = \frac{1}{4}$ to $t = 1$, we find the slope of the line that passes through the points $\left(\frac{1}{4}, 7\right)$ and (1, 16).

$$\begin{aligned}\text{Average rate of change} &= \frac{h(1) - h\left(\frac{1}{4}\right)}{1 - \frac{1}{4}} \\ &= \frac{16 - 7}{1 - \frac{1}{4}} \\ &= \frac{9}{\frac{3}{4}} \\ &= 12 \text{ ft/sec}\end{aligned}$$

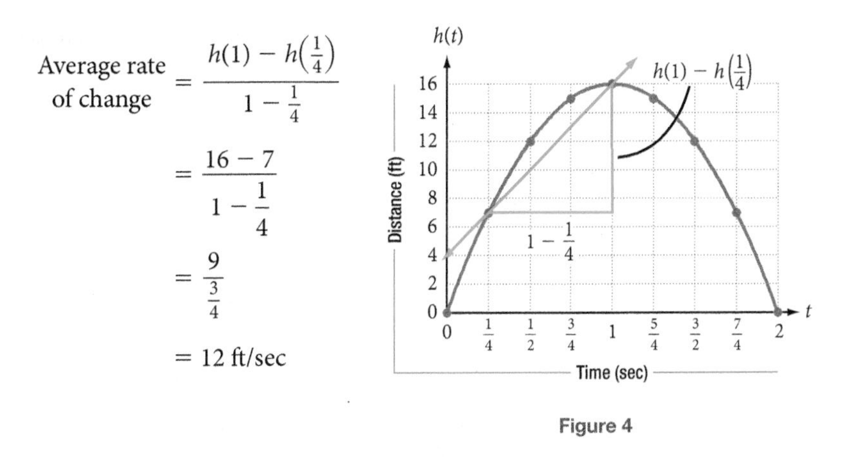

Figure 4

As the ball rises from $t = \frac{1}{4}$ to $t = 1$ the height changes at an average rate of 12 ft/sec.

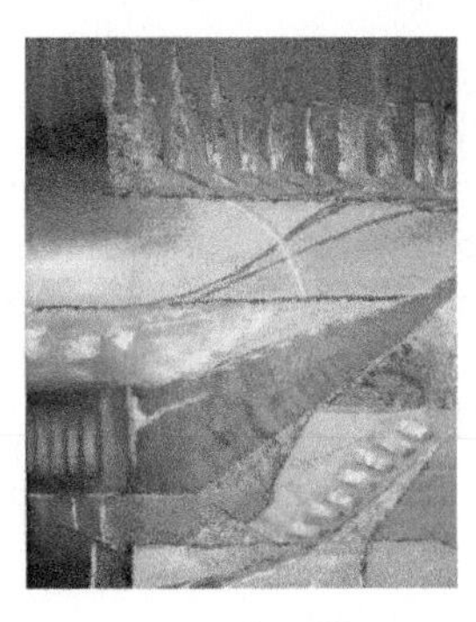

Example 6 A painting is purchased as an investment for $125. If its value increases continuously so that it doubles every 5 years, then its value is given by the function

$$V(t) = 125 \cdot 2^{t/5} \quad \text{for} \quad t \geq 0$$

where t is the number of years since the painting was purchased, and V is its value (in dollars) at time t.

Find the average rate of change of the value of the painting from $t = 0$ to $t = 10$.

Solution To find the average rate of change of the function from $t = 0$ to $t = 10$, we find the slope of the line that passes through the points $(0, V(0))$ and $(10, V(10))$.

$$\begin{aligned}\text{Average rate of change} &= \frac{V(10) - V(0)}{10 - 0} \\ &= \frac{125 \cdot 2^{10/5} - 125 \cdot 2^{0/5}}{10 - 0} \\ &= \frac{375}{10} \\ &= 37.5 \text{ dollars/year}\end{aligned}$$

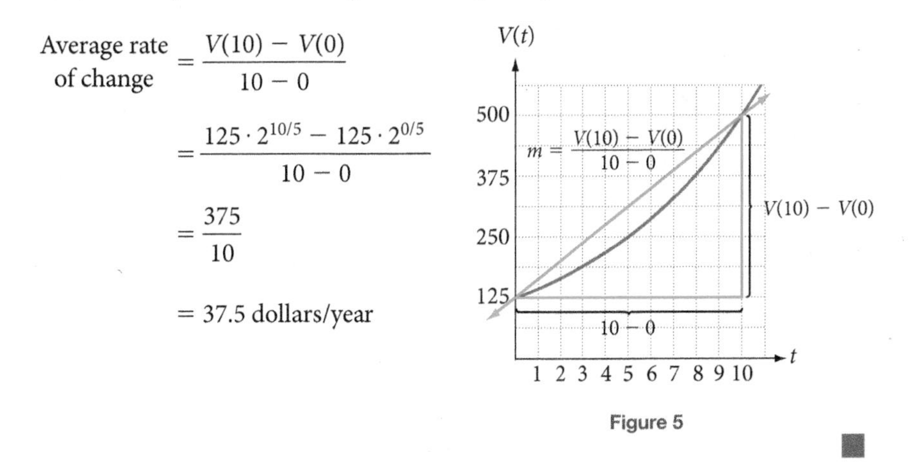

Figure 5

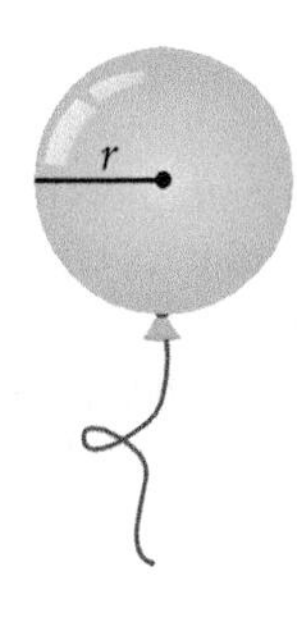

Example 7 A balloon in the shape of a sphere is being inflated. The formula for the surface area of the balloon is

$$S(r) = 4\pi r^2 \quad r \geq 0$$

Find the average rate of change in the surface area of the balloon as the radius increases from 1 inch to 3 inches.

Solution To find the average rate of change in the surface area from $r = 1$ to $r = 3$, we find the slope of the line that passes through the points $(1, S(1))$ and $(3, S(3))$.

$$\begin{aligned}\text{Average rate of change} &= \frac{S(3) - S(1)}{3 - 1} \\ &= \frac{4\pi \cdot 3^2 - 4\pi \cdot 1^2}{3 - 1} \\ &= \frac{36\pi - 4\pi}{2} \\ &= 16\pi \text{ in}^2/\text{in}\end{aligned}$$

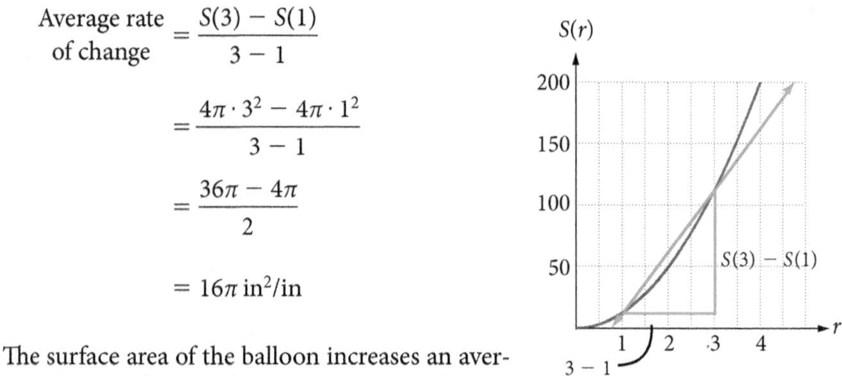

Figure 6

The surface area of the balloon increases an average of 16π in^2/in as the radius increases from 1 inch to 3 inches.

Note that this is one situation in which reducing the units to in^2/in = in makes the result less understandable. ■

Difference Quotients and Average Rate of Change

We can generalize the average rate of change of any function, and, at the same time, begin our work with a formula that is important in calculus. In the diagram below, the slope of the blue line (called a **secant line**) is the average rate of change of the red curve from point P to point Q. The slope of the line passing through the points P and Q is given by the following formula:

$$\text{Slope of line through } P \text{ and } Q = m = \frac{f(b) - f(a)}{b - a}$$

Note A line that passes through two points on a graph is called a secant line. We will see much more of these lines in Section 1.6.

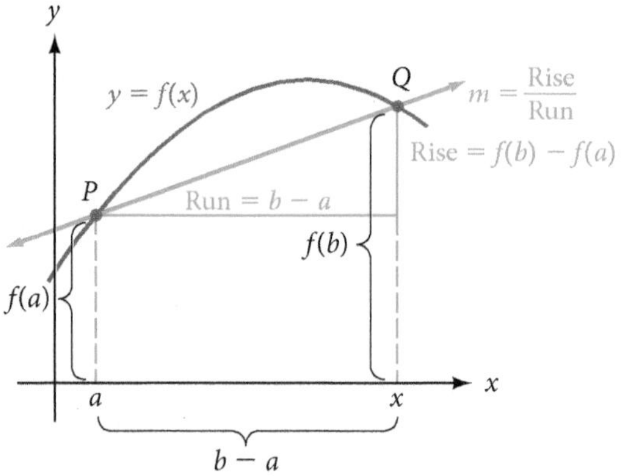

Figure 7

The expression $\frac{f(b) - f(a)}{b - a}$ is called a **difference quotient**. It represents the average rate of change of the function f from the point $(a, f(a))$ to the point $(b, f(b))$. Throughout your study of calculus, you will see the difference quotient in a variety of forms. For instance, Example 8 uses the points (x_1, y_1) and (x_2, y_2) rather than $(a, f(a))$ and $(b, f(b))$. Either way, the difference quotient represents the slope of the secant line and the average rate of change between those points.

Example 8 If $f(x) = 3x - 5$, find $\dfrac{f(x_2) - f(x_1)}{x_2 - x_1}$.

Solution

$$\begin{aligned}\frac{f(x_2) - f(x_1)}{x_2 - x_1} &= \frac{(3x_2 - 5) - (3x_1 - 5)}{x_2 - x_1} \\ &= \frac{3x_2 - 3x_1}{x_2 - x_1} \\ &= \frac{3(x_2 - x_1)}{x_2 - x_1} \\ &= 3\end{aligned}$$

■

Example 9 If $f(x) = x^2 - 4$, find $\dfrac{f(x_2) - f(x_1)}{x_2 - x_1}$ and simplify.

Solution Because $f(x) = x^2 - 4$ and $f(x_1) = x_1^2 - 4$, we have

$$\begin{aligned}\frac{f(x_2) - f(x_1)}{x_2 - x_1} &= \frac{(x_2^2 - 4) - (x_1^2 - 4)}{x_2 - x_1} \\ &= \frac{x_2^2 - 4 - x_1^2 + 4}{x_2 - x_1} \\ &= \frac{x_2^2 - x_1^2}{x_2 - x_1} \\ &= \frac{(x_2 + x_1)(x_2 - x_1)}{x_2 - x_1} && \text{Factor and divide out common factor} \\ &= x_2 + x_1\end{aligned}$$

■

Linear Functions and Equations of Lines

Suppose line l has slope m and y-intercept b. What is the equation of l? Because the y-intercept is b, we know the point $(0, b)$ is on the line. If (x, y) is any other point on l, then using the definition for slope, we have

$$\begin{aligned}\frac{y - b}{x - 0} &= m && \text{Definition of Slope} \\ y - b &= mx && \text{Multiply both sides by } x \\ y &= mx + b && \text{Add } b \text{ to both sides}\end{aligned}$$

This last equation is known as the **slope-intercept form** of the equation of a straight line.

Slope-Intercept Form of the Equation of a Line

The equation of any line with slope m and y-intercept b is given by

$$y = mx + b$$

↗ Slope ↑ y-intercept

When the equation is in this form, the **slope** of the line is always the **coefficient** of x and the y-intercept is always the **constant term**.

Our slope-intercept form of the equation can also be used to classify all **linear functions** together.

> **Linear Function**
> A **linear function** is any function that can be put in the form
> $$y = f(x) = mx + b$$
> where m and b are real numbers.

Example 10 Find the equation of the line with slope $-\frac{4}{3}$ and y-intercept 5.

Solution Substituting $m = -\frac{4}{3}$ and $b = 5$ into the equation $y = mx + b$, we have

$$y = -\frac{4}{3}x + 5$$

Finding the equation from the slope and y-intercept is just that easy. If the slope is m and the y-intercept is b, then the equation is always $y = mx + b$.

Example 11 Give the slope and y-intercept for the line $2x - 3y = 5$.

Solution To use the slope-intercept form, we must solve the equation for y in terms of x:

$$2x - 3y = 5$$
$$-3y = -2x + 5 \qquad \text{Add } -2x \text{ to both sides}$$
$$y = \frac{2}{3}x - \frac{5}{3} \qquad \text{Divide by } -3$$

The last equation has the form $y = mx + b$. The slope must be $m = \frac{2}{3}$ and the y-intercept is $b = -\frac{5}{3}$.

Example 12 Graph the linear function $f(x) = -\dfrac{2}{3}x + 2$ using the slope and y-intercept.

Solution The slope is $m = -\frac{2}{3}$ and the y-intercept is $b = 2$. Therefore, the point (0, 2) is on the graph, and the ratio of rise to run going from (0, 2) to any other point on the line is $-\frac{2}{3}$. If we start at (0, 2) and move 2 units up (that's a rise of 2) and 3 units to the left (a run of -3), we will be at another point on the graph. (We could also go down 2 units and right 3 units and still be assured of ending up at another point on the line because $\frac{2}{-3}$ is the same as $\frac{-2}{3}$.)

> *Note* As we mentioned earlier, the rectangular coordinate system is the tool we use to connect algebra and geometry. Example 12 illustrates this connection, as do many other examples in this chapter. In Example 12, Descartes's rectangular coordinate system allows us to associate the equation $y = -\frac{2}{3}x + 2$ (an algebraic concept) with the straight line (a geometric concept) shown in Figure 8.

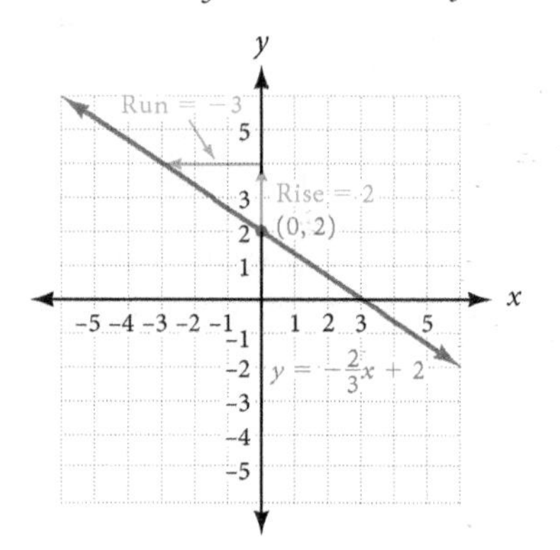

Figure 8

A second useful form of the equation of a line is the point-slope form.

Let line l contain the point (x_1, y_1) and have slope m. If (x, y) is any other point on l, then by the definition of slope we have

$$\frac{y - y_1}{x - x_1} = m$$

Multiplying both sides by $(x - x_1)$ gives us

$$(x - x_1) \cdot \frac{y - y_1}{x - x_1} = m(x - x_1)$$

$$y - y_1 = m(x - x_1)$$

This last equation is known as the **point-slope form** of the equation of a line.

Point-Slope Form of the Equation of a Line

The equation of the line through (x_1, y_1) with slope m is given by

$$y - y_1 = m(x - x_1)$$

This form of the equation of a line is used to find the equation of a line, either given one point on the line and the slope, or given two points on the line.

Example 13 Find the equation of the line with slope -2 that contains the point $(-4, 3)$. Write the answer in slope-intercept form.

Solution

Using	$(x_1, y_1) = (-4, 3)$ and $m = -2$	
in	$y - y_1 = m(x - x_1)$	Point-slope form
gives us	$y - 3 = -2(x + 4)$	Note: $x - (-4) = x + 4$
	$y - 3 = -2x - 8$	Multiply out right side
	$y = -2x - 5$	Add 3 to each side

Figure 9 is the graph of the line that contains $(-4, 3)$ and has a slope of -2. Notice that the y-intercept on the graph matches that of the equation we found.

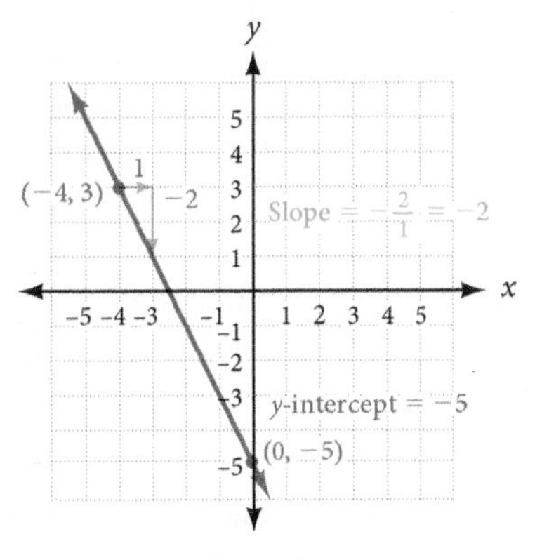

Figure 9

Example 14 Find the equation of the line that passes through the points $(-3, 3)$ and $(3, -1)$.

Solution We begin by finding the slope of the line:

$$m = \frac{3 - (-1)}{-3 - 3} = \frac{4}{-6} = -\frac{2}{3}$$

Using $(x_1, y_1) = (3, -1)$ and $m = -\frac{2}{3}$ in $y - y_1 = m(x - x_1)$ yields

$$y + 1 = -\frac{2}{3}(x - 3)$$

$$y + 1 = -\frac{2}{3}x + 2 \qquad \text{Multiply out right side}$$

$$y = -\frac{2}{3}x + 1 \qquad \text{Add } -1 \text{ to each side}$$

Note We could have used the point $(-3, 3)$ instead of $(3, -1)$ and obtained the same equation. That is, using $(x_1, y_1) = (-3, 3)$ and $m = -\frac{2}{3}$ in $y - y_1 = m(x - x_1)$ gives us

$$y - 3 = -\frac{2}{3}(x + 3)$$

$$y - 3 = -\frac{2}{3}x - 2$$

$$y = -\frac{2}{3}x + 1$$

which is the same result we obtained using $(3, -1)$.

Figure 10 shows the graph of the line that passes through the points $(-3, 3)$ and $(3, -1)$. As you can see, the slope and y-intercept are $-\frac{2}{3}$ and 1, respectively.

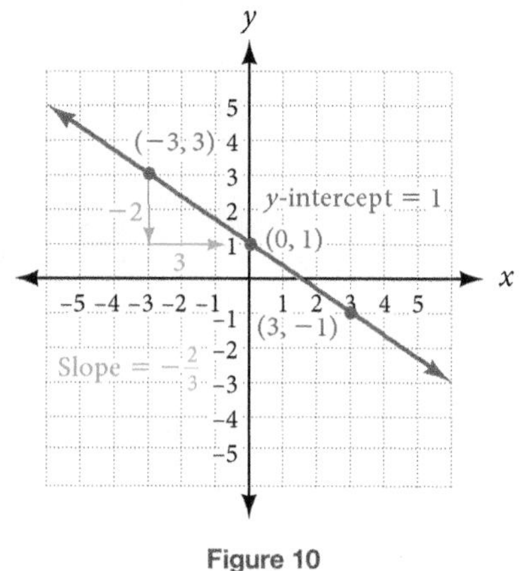

Figure 10

TECHNOLOGY NOTE *Graphing Calculators*

One advantage of using a graphing calculator to graph lines is that a calculator does not care whether the equation has been simplified or not. To illustrate, in Example 14 we found that the equation of the line with slope $-\frac{2}{3}$ that passes through the point $(3, -1)$ is

$$y + 1 = -\frac{2}{3}(x - 3)$$

Normally, to graph this equation we would simplify it first. With a graphing calculator, we add -1 to each side and enter the equation this way:

$$\text{Y1} = -(2/3)(\text{X} - 3) - 1$$

No simplification is necessary. We can graph the equation in this form, and the graph will be the same as that of the simplified form of the equation, which is $y = -\frac{2}{3}x + 1$. To convince yourself that this is true, graph both the simplified form for the equation and the unsimplified form in the same window. As you will see, the two graphs coincide.

The following summary reminds us that all horizontal lines have equations of the form $y = b$, and slopes of 0. Since they cross the y-axis at b, the y-intercept is b; there is no x-intercept. Vertical lines have an undefined, or no slope, and equations of the form $x = a$. Each will have an x-intercept at a, and no y-intercept. Finally, equations of the form $y = mx$ have graphs that pass through the origin. The slope is always m and both the x-intercept and the y-intercept are 0.

FACTS FROM GEOMETRY

Special Equations: Their Graphs, Slopes, and Intercepts

For the equations below, m, a, and b are real numbers.

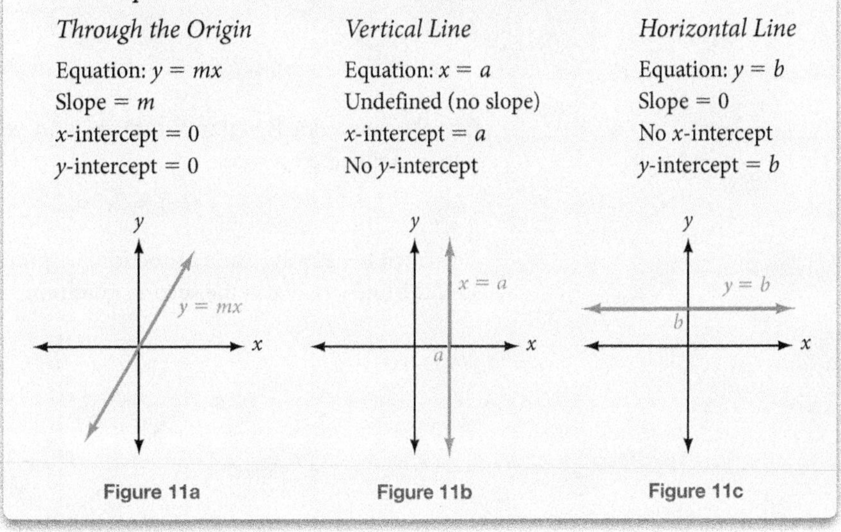

Figure 11a Figure 11b Figure 11c

Piecewise-Defined Functions

A function may be defined by different formulas on different portions of the x-axis. Such a function is said to be defined **piecewise.**

Example 15 Graph the function defined by

$$f(x) = \begin{cases} x + 1 & \text{if } x \leq 1 \\ 3 & \text{if } x > 1 \end{cases}$$

Solution Think of the coordinate system as divided into two regions by the vertical line $x = 1$, as shown in Figure 12. In the left-hand region $(x \leq 1)$, we graph the line $y = x + 1$. Notice that the point $(1, 2)$ is included in the graph. We indicate this with a solid dot at the point $(1, 2)$. In the right-hand region $(x > 1)$, we graph the horizontal line $y = 3$. The point $(1, 3)$ is *not* part of the graph. We indicate this with an open circle at the point $(1, 3)$.

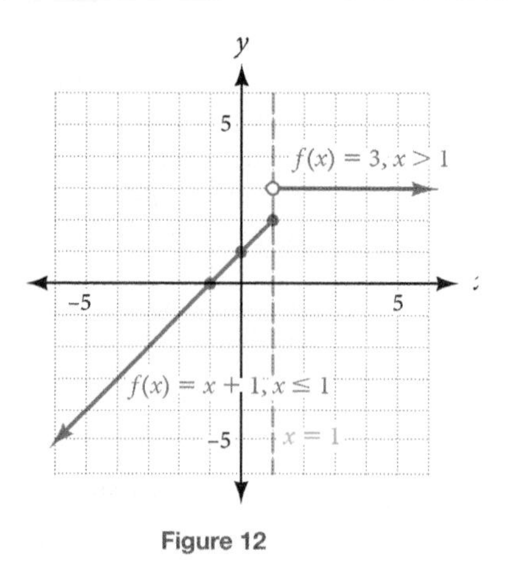

Figure 12

The absolute value function $f(x) = |x|$ is an example of a well-known function that can be defined piecewise.

$$f(x) = |x| = \begin{cases} x & \text{if } x \geq 0 \\ -x & \text{if } x < 0 \end{cases}$$

To sketch the absolute value function, we graph the line $y = x$ in the first quadrant and the line $y = -x$ in the second quadrant.

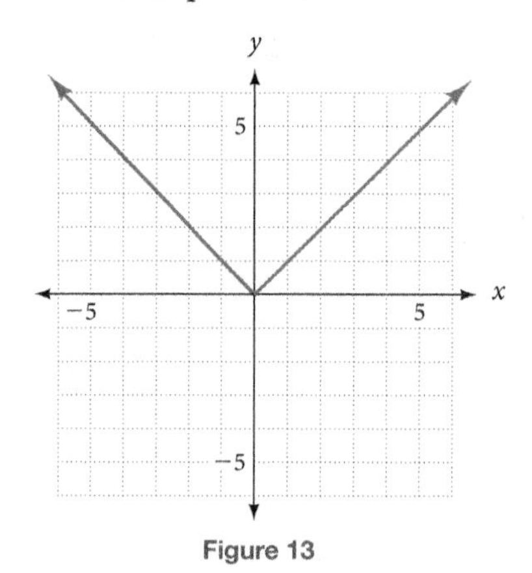

Figure 13

Modeling: Profit Function

Example 16 A college sorority is organizing a fundraising dance. The only expenses will be renting the facility and paying the security guard. All the other work will be done by the sorority members, and all the refreshments have been donated. Tickets will be sold ahead of time for \$20 each. The facility rental is \$100 for the evening, and the security guard is only necessary if more than 150 people attend the dance. The security guard is \$150 for the night. The hall can hold a maximum of 350 people. Here is the function that gives the profit the sorority can expect based on the number of tickets sold, along with the graph of the function.

$$P(x) = \begin{cases} 20x - 100 & 0 \leq x \leq 150 \\ 20x - 250 & 150 < x \leq 350 \end{cases}$$

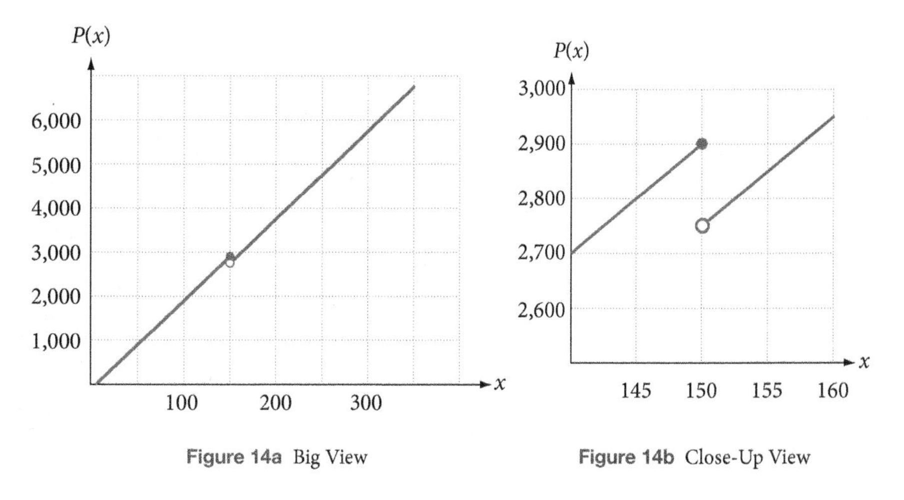

Figure 14a Big View

Figure 14b Close-Up View

Use the function and graphs to find

a. the domain **b.** $P(145)$ **c.** $P(150)$ **d.** $P(155)$

Solution

a. The domain is $0 \leq x \leq 350$, where x is an integer. Notice that we graph each part of the piecewise function as a continuous line, even though the domain consists of integers only. (Of course we can't sell a fraction of a ticket, but analyzing the function is easier this way.)

b. $P(145) = 20(145) - 100 = \$2,800$

c. $P(150) = 20(150) - 100 = \$2,900$

d. $P(155) = 20(155) - 250 = \$2,850$ ■

Using Technology 1.3

You can use your graphing calculator to find the average rate of change of a function. The following entries illustrate one way of computing the average rate of change of the function $f(x) = (5x - 3)^{2/3}$ as x increases from 16 to 20.

Example 17 If $f(x) = (5x - 3)^{2/3}$, find $\dfrac{f(20) - f(16)}{20 - 16}$.

Solution Begin by putting the function in the Y-variables list.

Y1=(5X-3)^(2/3)

Now, in the computation window, enter

(Y1(20)-Y1(16))/(20-16)

The result is 0.7530075664. ■

You can also use your calculator to construct graphs of functions that are defined piecewise. The following entry shows how to construct the graph of a piecewise function that contains both a linear and nonlinear function.

Example 18 Graph the following function.

$$f(x) = \begin{cases} x^2 - 4 & \text{if } x \leq 2 \\ -x + 2 & \text{if } x > 2 \end{cases}$$

Solution Here is how we enter the function into our Y-variables list.

Y1=(X^2-4)(X≤2)+(-X+2)(X>2)

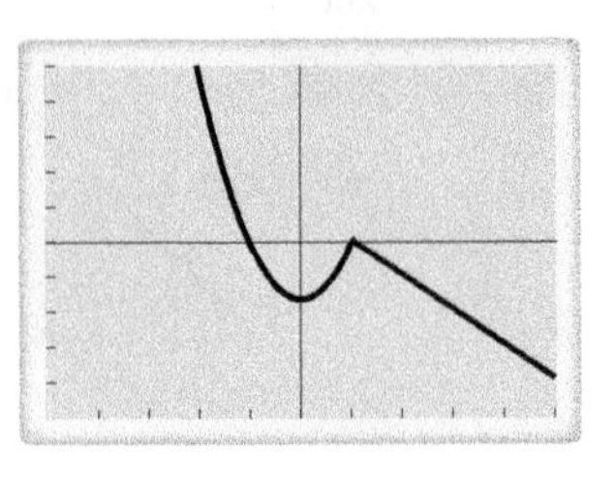

GETTING READY FOR CLASS

After reading through the preceding section, respond in your own words and in complete sentences.

A. Describe the behavior of a line with a negative slope.

B. Would you rather climb a hill with a slope of $\frac{1}{2}$ or a slope of 3? Explain why.

C. What is a difference quotient?

D. Describe how you would find the equation of a line if you knew the slope and the y-intercept of the line.

Problem Set 1.3

Skills Practice

Find the slope of the line through each of the following pairs of points. Then, plot each pair of points, draw a line through them, and indicate the rise and run in the graph in the manner shown in Example 2.

1. $(2, 1), (4, 4)$

2. $(3, 1), (5, 4)$

3. $(1, 4), (5, 2)$

4. $(1, 3), (5, 2)$

5. $(1, -3), (4, 2)$

6. $(2, -3), (5, 2)$

7. $(2, -4)$ and $(5, -9)$

8. $(-3, 2)$ and $(-1, 6)$

9. $(-3, 5)$ and $(1, -1)$

10. $(-2, -1)$ and $(3, -5)$

11. $(-4, 6)$ and $(2, 6)$

12. $(2, -3)$ and $(2, 7)$

13. $(a, -3)$ and $(a, 5)$

14. $(x, 2y)$ and $(4x, 8y)$

Solve for the indicated variable if the line through the two given points has the given slope.

15. $(a, 3)$ and $(2, 6)$, $m = -1$

16. $(a, -2)$ and $(4, -6)$, $m = -3$

17. $(2, b)$ and $(-1, 4b)$, $m = -2$

18. $(-4, y)$ and $(-1, 6y)$, $m = 2$

19. $(2, 4)$ and (x, x^2), $m = 5$

20. $(3, 9)$ and (x, x^2), $m = -2$

21. $(1, 3)$ and $(x, 2x^2 + 1)$, $m = -6$

22. $(3, 7)$ and $(x, x^2 - 2)$, $m = -4$

For each of the equations in Problems 23–26, complete the table, and then use the results to find the slope of the graph of the equation.

23. $2x + 3y = 6$

x	y
0	
	0

24. $3x - 2y = 6$

x	y
0	
	0

25. $y = \frac{2}{3}x - 5$

x	y
0	
3	

26. $y = -\frac{3}{4}x + 2$

x	y
0	
4	

Note This symbol indicates a challenging problem.

For Problems 27-36, evaluate $\dfrac{f(b) - f(a)}{b - a}$.

27. $f(x) = 4x$

28. $f(x) = -3x$

29. $f(x) = 5x + 3$

30. $f(x) = 6x - 5$

31. $f(x) = x^2$

32. $f(x) = 3x^2$

33. $f(x) = x^2 + 1$

34. $f(x) = x^2 - 3$

35. $f(x) = x^2 - 3x + 4$

36. $f(x) = x^2 + 4x - 7$

Give the equation of the line with the following slope and y-intercept.

37. $m = -4, b = -3$

38. $m = -6, b = \frac{4}{3}$

39. $m = -\frac{2}{3}, b = 0$

40. $m = 0, b = \frac{3}{4}$

41. $m = -\frac{2}{3}, b = \frac{1}{4}$

42. $m = \frac{5}{12}, b = -\frac{3}{2}$

Give the slope and y-intercept for each of the following equations. Sketch the graph using the slope and y-intercept.

43. $y = 3x - 2$ **44.** $y = 2x + 3$ **45.** $2x - 3y = 12$

46. $3x - 2y = 12$ **47.** $4x + 5y = 20$ **48.** $5x - 4y = 20$

For each of the following problems, the slope and one point on the line are given. In each case, find the equation of that line. (Write the equation for each line in slope-intercept form.)

49. $(-2, -5); m = 2$ **50.** $(-1, -5); m = 2$ **51.** $(-4, 1); m = -\frac{1}{2}$

52. $(-2, 1); m = -\frac{1}{2}$ **53.** $\left(-\frac{1}{3}, 2\right); m = -3$ **54.** $\left(-\frac{2}{3}, 5\right); m = -3$

55. $(-4, 2); m = \frac{2}{3}$ **56.** $(3, -4); m = -\frac{1}{3}$ **57.** $(-5, -2); m = -\frac{1}{4}$

58. $(-4, -3); m = \frac{1}{6}$

Find the equation of the line that passes through each pair of points. Write the equation in slope-intercept form.

59. $(3, -2), (-2, 1)$ **60.** $(-4, 1), (-2, -5)$ **61.** $\left(-2, \frac{1}{2}\right), \left(-4, \frac{1}{3}\right)$

62. $(-6, -2), (-3, -6)$ **63.** $\left(\frac{1}{3}, -\frac{1}{5}\right), \left(-\frac{1}{3}, -1\right)$ **64.** $\left(-\frac{1}{2}, -\frac{1}{2}\right), \left(\frac{1}{2}, \frac{1}{10}\right)$

65. The equation $3x - 2y = 10$ is a linear equation in standard form. From this equation, answer the following:
- **a.** Find the x and y intercepts.
- **b.** Find a solution to this equation other than the intercepts in Part a.
- **c.** Write this equation in slope-intercept form.
- **d.** Is the point $(2, 2)$ a solution to the equation?

66. The equation $4x + 3y = 8$ is a linear equation in standard form. From this equation, answer the following:
- **a.** Find the x and y intercepts.
- **b.** Find a solution to this equation other than the intercepts in Part a.
- **c.** Write this equation in slope-intercept form.
- **d.** Is the point $(-3, 2)$ a solution to the equation?

Paying Attention to Instructions The next two problems are intended to give you practice reading, and paying attention to, the instructions that accompany the problems you are working.

67. Work each problem according to the instructions given:

- **a.** Solve: $-2x + 1 = -3$
- **b.** Simplify: $-2x + 1 - 3$
- **c.** Find x when y is 0: $-2x + y = -3$
- **d.** Find y when x is 0: $-2x + y = -3$
- **e.** Graph: $-2x + y = -3$
- **f.** Solve for y: $-2x + y = -3$

68. Work each problem according to the instructions given:

a. Solve: $\frac{x}{3} + \frac{1}{4} = 1$

b. Write as a single fraction: $\frac{x}{3} + \frac{y}{4} - 1$

c. Find x when y is 0: $\frac{x}{3} + \frac{y}{4} = 1$

d. Find y when x is 0: $\frac{x}{3} + \frac{y}{4} = 1$

e. Graph: $\frac{x}{3} + \frac{y}{4} = 1$

f. Solve for y: $\frac{x}{3} + \frac{y}{4} = 1$

69. Graph each of the following lines. In each case, name the slope, the x-intercept, and the y-intercept.

a. $y = \frac{1}{2}x$ **b.** $x = 3$ **c.** $y = -2$

70. Graph each of the following lines. In each case, name the slope, the x-intercept, and the y-intercept.

a. $y = -2x$ **b.** $x = 2$ **c.** $y = -4$

71. Find the equation of the line with x-intercept 3 and y-intercept 2.

72. Find the equation of the line with x-intercept 2 and y-intercept 3.

For Problems 73-84, graph the following piecewise-defined functions. Indicate whether the endpoints of each piece are included on the graph.

73. $f(x) = \begin{cases} -2 & \text{if } x \le 1 \\ x - 3 & \text{if } x > 1 \end{cases}$

74. $h(x) = \begin{cases} -x + 2 & \text{if } x \le -1 \\ 3 & \text{if } x > -1 \end{cases}$

75. $G(t) = \begin{cases} 3t + 9 & \text{if } t < -2 \\ -3 - \frac{1}{2}t & \text{if } t \ge -2 \end{cases}$

76. $F(s) = \begin{cases} \frac{1}{3}s + 3 & \text{if } s < 3 \\ 2s - 3 & \text{if } s \ge 3 \end{cases}$

77. $H(t) = \begin{cases} t^2 & \text{if } t \le 1 \\ \frac{1}{2}t + \frac{1}{2} & \text{if } t > 1 \end{cases}$

78. $g(t) = \begin{cases} \frac{3}{2}t + 7 & \text{if } t \le -2 \\ t^2 & \text{if } t > -2 \end{cases}$

79. $k(x) = \begin{cases} |x| & \text{if } x \le 2 \\ \sqrt{x} & \text{if } x > 2 \end{cases}$

80. $S(x) = \begin{cases} \frac{1}{x} & \text{if } x < 1 \\ |x| & \text{if } x \ge 1 \end{cases}$

81. $D(x) = \begin{cases} |x| & \text{if } x < -1 \\ x^3 & \text{if } x \ge -1 \end{cases}$

82. $m(x) = \begin{cases} x^2 & \text{if } x \le \frac{1}{2} \\ |x| & \text{if } x > \frac{1}{2} \end{cases}$

83. $P(t) = \begin{cases} t^3 + 1 & \text{if } t \le 1 \\ t^2 & \text{if } t > 1 \end{cases}$

84. $Q(t) = \begin{cases} t^2 & \text{if } t \le -1 \\ \sqrt[3]{t} & \text{if } t > -1 \end{cases}$

Write a piecewise definition for the function and sketch its graph.

85. $f(x) = |2x - 8|$ **86.** $g(x) = |3x + 6|$ **87.** $g(t) = \left|1 + \frac{t}{3}\right|$

88. $f(t) = \left|\frac{1}{2}t - 3\right|$ **89.** $F(x) = |x^3|$ **90.** $G(x) = \left|1 + \frac{1}{x}\right|$

Check Your Understanding

91. **Finding Slope from Intercepts** Graph the line that has an x-intercept of 3 and a y-intercept of -2. What is the slope of this line?

92. The following lines have slope 2, $\frac{1}{2}$, 0, and -1. Match each line to its slope value.

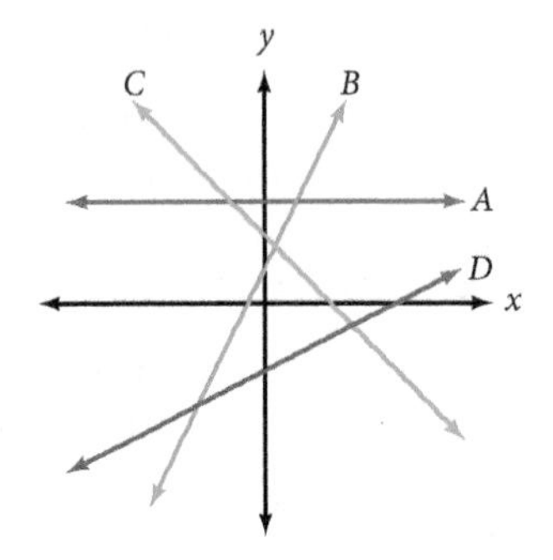

Find the slope of each of the following lines from the given graph.

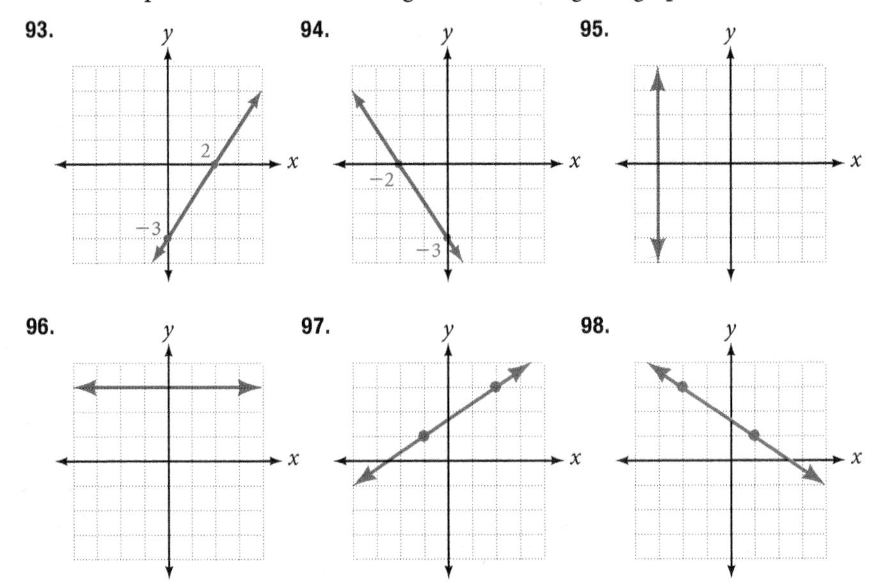

For each of the following lines, name the slope and y-intercept. Then write the equation of the line in slope-intercept form.

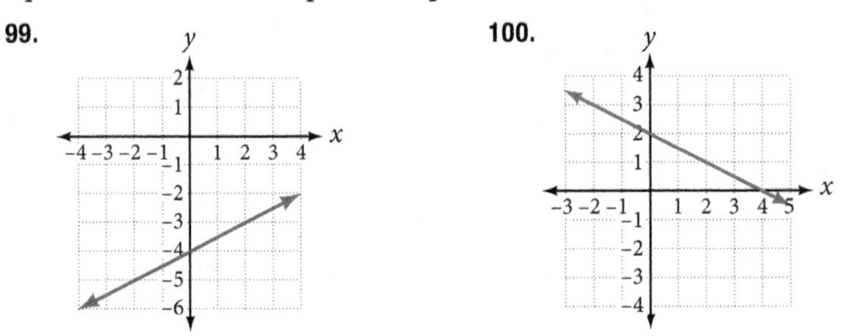

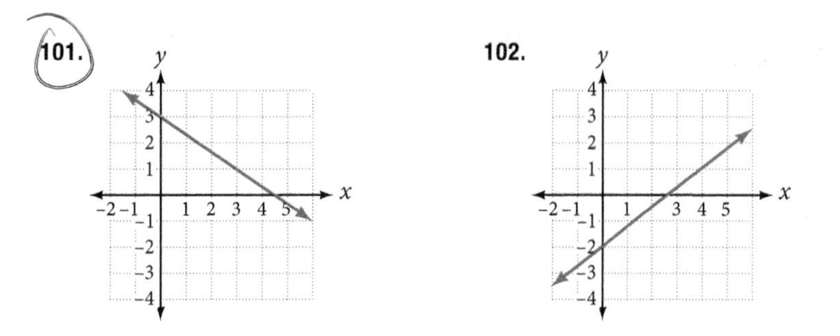

For each of the following lines, name the coordinates of any two points on the line. Then use those two points to find the equation of the line.

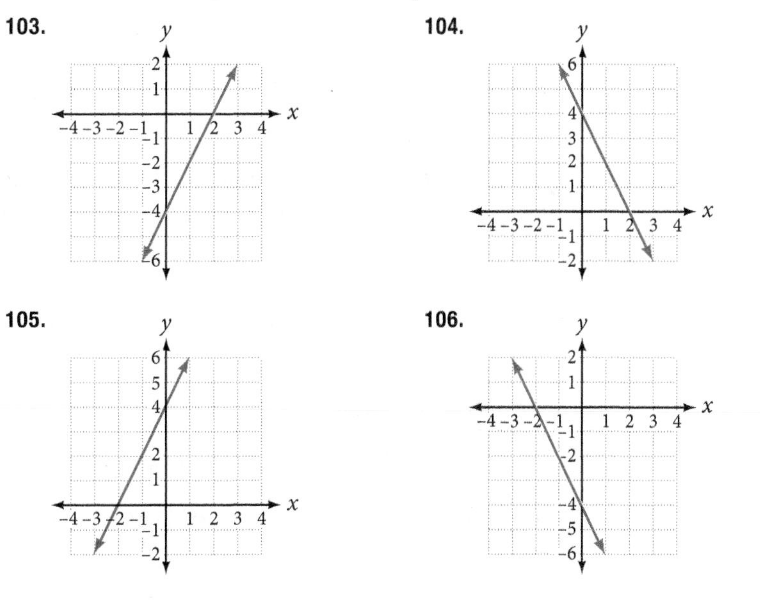

Modeling Practice

Physical Science: Speed An object is traveling at a constant speed. The distance and time data are shown on the given graph. Use the graph to find the speed of the object.

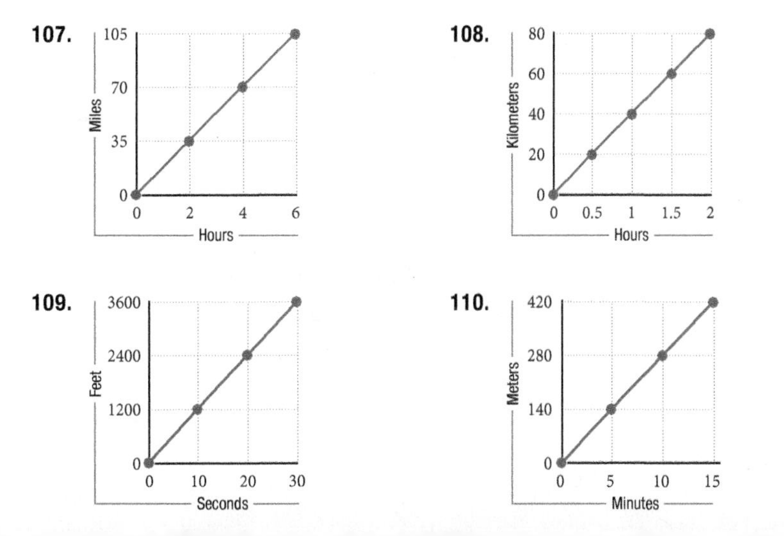

111. Chemistry: Applying Heat to Ice A block of ice with an initial temperature of -20 °C is heated at a steady rate. The graph shows how the temperature changes as the ice melts to become water and the water boils to become steam and water.

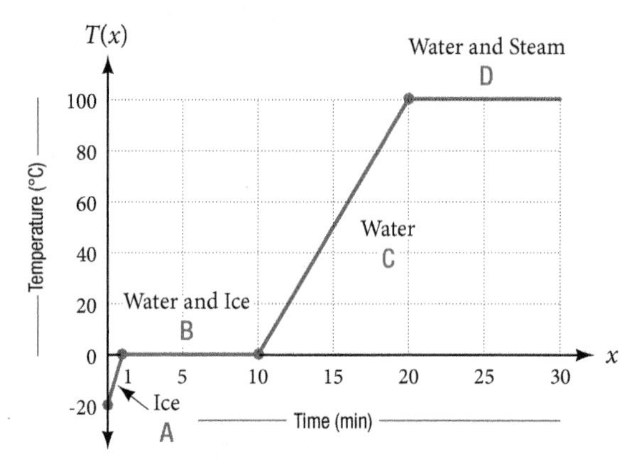

a. How long does it take all the ice to melt?

b. From the time the heat is applied to the block of ice, how long is it before the water boils?

c. Write a piecewise-defined function for this graph.

d. Find the slope of the line segment labeled A. What units would you attach to this number?

e. Find the slope of the line segment labeled C. Be sure to attach units to your answer.

f. Is the temperature changing faster during the 1st minute or the 16th minute?

112. Engineering: Slope of a Highway A sign at the top of the Cuesta Grade, outside of San Luis Obispo, reads "7% downgrade next 3 miles." The following diagram is a model of the Cuesta Grade that takes into account the information on that sign.

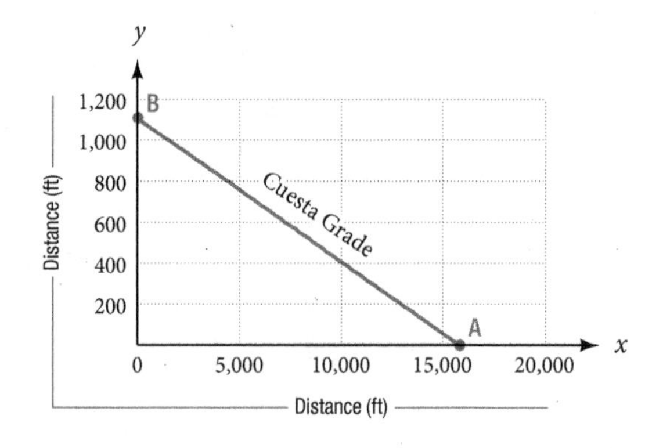

a. At point B, the graph crosses the y-axis at 1,106 feet. How far is it from the origin to point A?

b. What is the slope of the Cuesta Grade?

113. Business: Online Advertising Revenue Suppose that, for a particular social networking company, the annual revenue from rich media advertisements, in millions of dollars, for the years 2007 through 2012 can be approximated with the model $R(x) = -x^4 + 11x^3 - 39x^2 + 45x$, where x is the number of years from the beginning of 2007.

Source: eMarketer April 2008

a. Find the average rate of change in revenue from the beginning of 2007 to the beginning of 2008.

b. Find the average rate of change in revenue from the beginning of 2007 to the beginning of 2009.

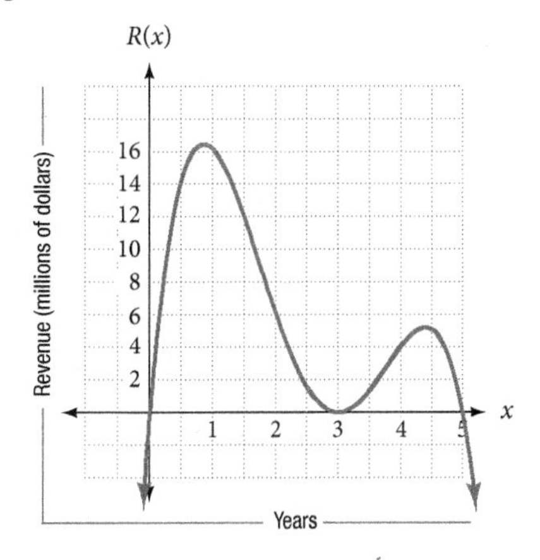

114. Physical Science: Solar Thermal Collectors The graph below shows the annual shipments of solar thermal collectors in the United States. Using the graph below, find the slope of the line connecting the first (1997, 8,000) and last (2006, 20,000) endpoints and then explain in words what the slope represents.

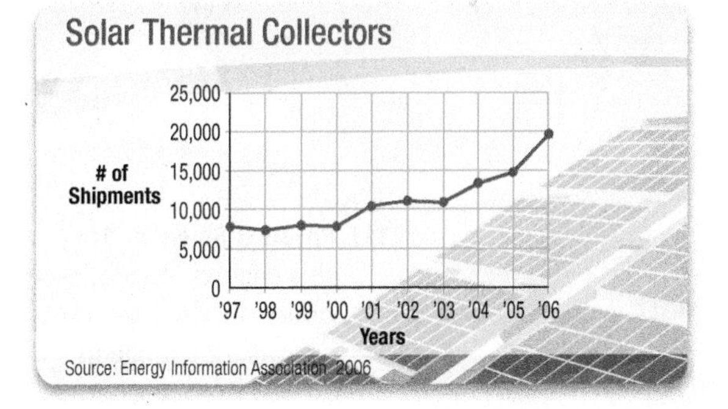

115. Environmental Studies: Energy Efficient Light Bulbs The chart shows a comparison of power usage between incandescent and energy efficient light bulbs. Use the chart to work the following problems involving slope.

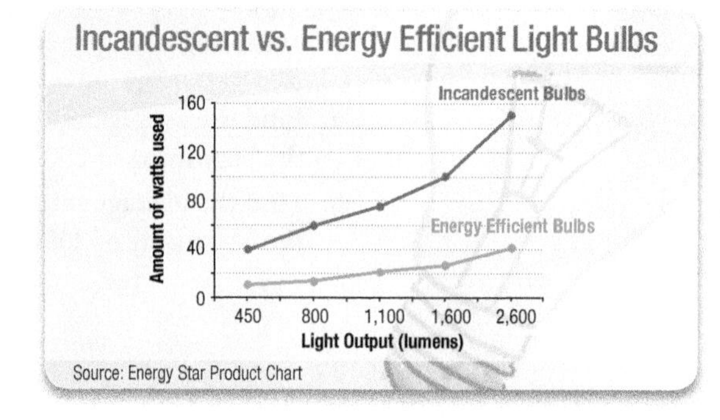

a. Find the slope of the line for the incandescent bulb from the two endpoints and then explain in words what the slope represents.

b. Find the slope of the line for the energy efficient bulb from the two endpoints and then explain in words what the slope represents.

c. Which light bulb is better? Why?

116. Health Science: Dieting The function below can be used to model the progress of a person on a diet. The quantity $W(x)$ is the weight (in pounds) of the person after x weeks of dieting. Find the average rate of change in weight from week 4 to week 14.

$$W(x) = \frac{80(2x + 15)}{x + 6}$$

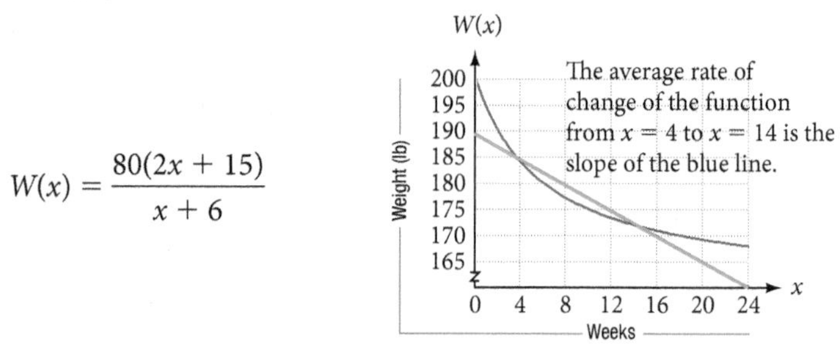

117. Physical Science: Drag Racing Recall the race example from the beginning of this chapter. The following rational function gives the speed $V(t)$, in miles per hour, of a dragster at each second t during a quarter-mile race. Find the average rate of change in speed from the first second to the eighth second.

$$V(t) = \frac{340t}{t + 3}$$

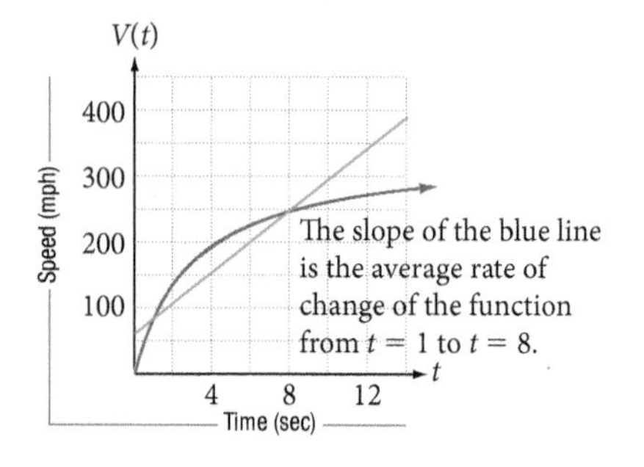

© Hector Mandel/iStockPhoto

118. Physical Science: Sky Diving A skydiver jumps from an airplane. For the first few seconds, the distance she falls in feet is given by the function $f(t) = 16t^2$. Find the average rate of change in the distance the skydiver falls from second 2 to second 5.

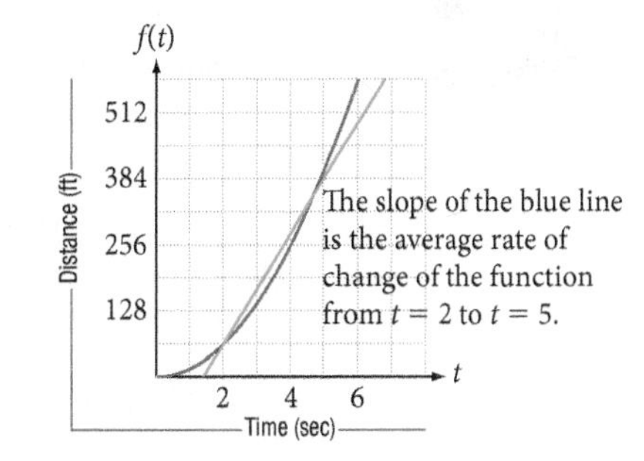

119. Business: Revenue Growth XYZ Textbooks/MathTV is a small publisher of college textbooks (publishing quality material at reasonable prices). Their first year of operations was 2010. At the beginning of that year they had sold 0 books. By the end of the year they had sold 5,000 books. After that, books doubled in sales each year. The function that describes the growth of XYZ Textbooks/MathTV during its first few years of operation is

$$S(t) = \begin{cases} 5{,}000t & 0 \le t < 1 \\ 5{,}000 \cdot 2^{t-1} & t \ge 1 \end{cases}$$

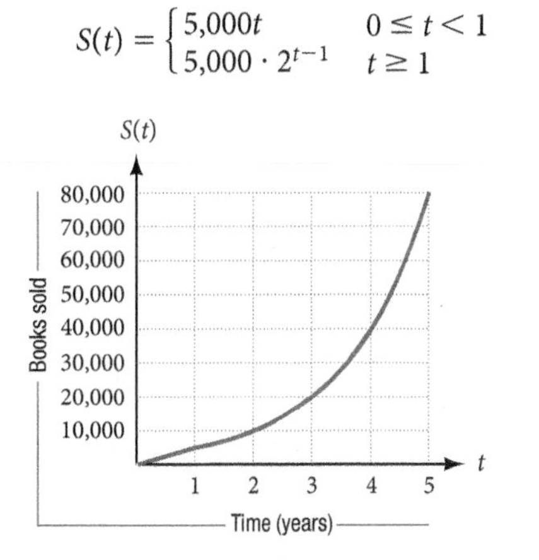

a. Find $S(0)$, $S(1)$, and $S(2)$, and explain the meaning of each one in terms of books sold.

b. Find the average rate of change in the number of books sold from time $t = 0$ to time $t = 2$. How does this compare with $S(2)$ that you found in Part a?

c. Find the average rate of change in the number of books sold from time $t = 1$ to time $t = 3$.

© Dario Egidi/iStockPhoto

120. Business: Textbook Cost To produce a textbook, suppose the publisher spent \$125,000 for typesetting and \$6.50 per book for printing and binding. The total cost to produce and print n books can be written as

$$C = 125{,}000 + 6.5n$$

a. Suppose the number of books printed in the first printing is 10,000. What is the total cost?

b. If the average cost is the total cost divided by the number of books printed, find the average cost of producing 10,000 textbooks.

c. Find the cost to produce one more textbook when you have already produced 10,000 textbooks.

121. Health Science: Training Heart Rate In an aerobics class, the instructor indicates that her students' training heart rate is 60% of their maximum heart rate, where maximum heart rate is 220 minus their age.

a. Determine the equation that gives training heart rate E in terms of age A.

b. Use the equation to find the training heart rate of a 22-year-old student.

c. Sketch the graph of the equation for students from 18 to 80 years of age.

Using Technology Exercises

Use your calculator to solve the following problems.

122. If $f(x) = 4x - 5$, find $\frac{f(8) - f(3)}{8 - 3}$.

123. If $f(x) = 0.85x^2 - 1.82x - 5$, find $\frac{f(6) - f(0)}{6 - 0}$.

124. A company that manufactures a particular product has determined that the cost of producing x units of the product is described by the function $C(x) = 0.035x^2 + 1{,}855$. Find the average rate of change of the cost as the number of units produced increases from 200 to 500.

Use your graphing calculator to construct the graphs of each piecewise-defined function.

125. $f(x) = \begin{cases} 3 - x^2 & \text{if } x \leq 1 \\ x^3 - 4x & \text{if } x > 1 \end{cases}$

126. $f(x) = \begin{cases} x^2 - 0.9x + 3.5 & \text{if } x \leq 3 \\ 2.8x - 4.5 & \text{if } x > 3 \end{cases}$

127. $f(x) = \begin{cases} 2.6 - 1.1x^2 & \text{if } |x| < 6 \\ 2.6 & \text{if } |x| > 6 \end{cases}$

128. $f(x) = \begin{cases} x^3 - 8x & \text{if } -3 \leq x < 3 \\ 10 & \text{if } 3 \leq x \leq 7 \\ \dfrac{10}{x - 7} & \text{if } x > 7 \end{cases}$

Getting Ready for the Next Section

129. For the function $f(x) = \frac{3x^2 - 2x - 8}{x - 2}$, find

a. The domain

b. $f(1), f(1.5)$, and $f(1.99)$

c. The y-intercept

d. The x-intercept

130. For the function $P(x) = \begin{cases} 20x - 100 & \text{if } 0 \leq x \leq 150 \\ 20x - 250 & \text{if } 150 < x \leq 350 \end{cases}$, find

a. The domain

b. $P(100), P(150), P(200)$

c. The y-intercept

d. The x-intercept

131. If $f(x) = \frac{1}{(x - 3)^2}$, find

a. The domain

b. $f(2)$ and $f(5)$

c. The x-intercept

d. The graph

132. How are the graphs of the two functions below different?

$$f(x) = \frac{x^2 - x - 2}{x - 2} \qquad g(x) = x + 1$$

133. Rationalize the denominator.

a. $\frac{1}{\sqrt{2}}$ **b.** $\frac{1}{\sqrt{2} + 1}$ **c.** $\frac{\sqrt{2} + 1}{\sqrt{2} - 1}$

134. Rationalize the numerator.

a. $\frac{\sqrt{x} - 2}{x - 4}$ **b.** $\frac{\sqrt{x} - 3}{x - 9}$ **c.** $\frac{\sqrt{2} + 1}{\sqrt{2} - 1}$

Spotlight on Success

Student Instructor Lauren

There are a lot of word problems in algebra and many of them involve topics that I don't know much about. I am better off solving these problems if I know something about the subject. So, I try to find something I can relate to. For instance, an example may involve the amount of fuel used by a pilot in a jet airplane engine. In my mind, I'd change the subject to something more familiar, like the mileage I'd be getting in my car and the amount spent on fuel, driving from my hometown to my college. Changing these problems to more familiar topics makes math much more interesting and gives me a better chance of getting the problem right. It also helps me to understand how greatly math affects and influences me in my everyday life. We really do use math more than we would like to admit—budgeting our income, purchasing gasoline, planning a day of shopping with friends—almost everything we do is related to math. So the best advice I can give with word problems is to learn how to associate the problem with something familiar to you.

You should know that I have always enjoyed math. I like working out problems and love the challenges of solving equations like individual puzzles. Although there are more interesting subjects to me, and I don't plan on pursuing a career in math or teaching, I do think it's an important subject that will help you in any profession.

Introduction to Limits

1.4

Suppose a bike manufacturer wants to know what the average cost of its product per unit will be as production increases. The manufacturer can observe price trends by evaluating the limit of the function that represents average cost of producing the bicycles. In this section, we will introduce you to limits and their applications, such as this one. Let's begin with an intuitive discussion.

An Intuitive Discussion of Limits

Consider the function represented graphically in Figure 1. We first notice that the x-value c is not in the domain of the function since there is no point on the graph at $x = c$. Although there is a y-value, namely L, on the vertical axis at the same level as the hole in the curve, there is no x-value that will produce it. The x-value c cannot be used and, hence, the y-value L is never realized. That is, $f(c)$ is not defined, and therefore cannot be equal to L.

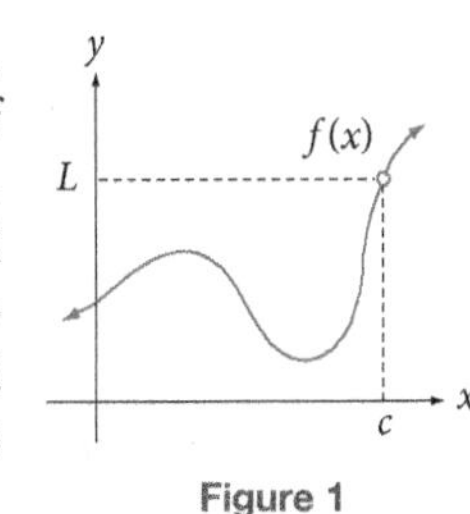

Figure 1

We can investigate the behavior of the function around $x = c$ by observing its graph. We observe the following behavior:

a. As we *approach* $x = c$ from the left side, the function values, as measured by their heights, approach $y = L$ from below. We also observe that as we *approach* $x = c$ from the left side only, the function values, although climbing toward $y = L$, cannot go beyond it; L acts as a limit to the increasing y-values.

b. As we *approach* $x = c$ from the right side, the function values, as measured by their heights, approach $y = L$ from above. We also observe that as we *approach* $x = c$ from the right side only, the function values, although descending toward $y = L$, cannot go beyond it; L acts as a limit to the decreasing y-values.

> *Note* Shorthand notation: The notation $\lim_{x \to c^-} f(x)$ instructs us to find the limit as x approaches c from the left ($x < c$). We use $\lim_{x \to c^+} f(x)$ to indicate the limit as x approaches c from the right ($x > c$).

Thus, regardless from which side we approach $x = c$, the function values approach $y = L$. The values of $f(x)$ could be made very close to L by choosing x-values that are very close to c. The closer x gets to c, the closer y gets to L, and if x approaches c without going beyond it, y will approach L without going beyond it. Here is the notation we use to represent this situation.

> *Note* We point out that we are observing the behavior of the function *near* the x-value c, and not at c itself. Remember, the function is not defined at $x = c$ and, therefore, has no behavior there at all.

The Limit Notation

The limit of $f(x)$, as x approaches c, is L

$$\lim_{x \to c} f(x) = L$$

if we can get as close as we like to L with $f(x)$, by taking x sufficiently close to c.

When you see the symbol "lim," you should think that the word reminds us of the word *approach*. Since we are interested in the behavior of the function only at values near $x = c$, and not at $x = c$ itself, we are interested in the behavior of the function only as we *approach* $x = c$.

In the next two examples we will illustrate the limit concept using actual numerical values and specific functions. The method used in these two examples is computational and is quite useful for complicated functions. The computational method, however, only helps us to *suggest* a limit.

VIDEO EXAMPLES

SECTION 1.4

Note Remember, we are not trying to determine the behavior at 2, but rather at points close to 2. Since $x = 2$ produces an indeterminate form, there is no behavior at $x = 2$.

Example 1 Find the limit as x approaches 2, of the function

$$f(x) = \frac{3x^2 - 2x - 8}{x - 2}$$

Solution We can see that $f(x)$ is defined for all numbers x except 2. At $x = 2$, the denominator is 0. In fact, $f(2) = \frac{0}{0}$ which is a form we call *indeterminate* and which we will discuss later in this section. We will investigate the behavior of $f(x)$ as x approaches 2 from both the left side and the right side by constructing a table of values consisting of several x-values close to 2 and their associated function values, as shown in Tables 1 and 2.

$f(x) = \frac{3x^2 - 2x - 8}{x - 2}$

Where the x values are near 2, the y values are near 10

Figure 2

Approaching 2 from the left

$x \to 2^-$

x	$\frac{3x^2 - 2x - 8}{x - 2}$
1	7
1.5	8.5
1.7	9.1
1.9	9.7
1.99	9.97
1.999	9.997
1.9999	9.9997
1.99999	9.99997

Table 1

Approaching 2 from the right

$x \to 2^+$

x	$\frac{3x^2 - 2x - 8}{x - 2}$
3	13
2.5	11.5
2.3	10.9
2.1	10.3
2.01	10.03
2.001	10.003
2.0001	10.0003
2.00001	10.00003

Table 2

According to our tables, it appears that as the x-values get close to 2 from the left, the $f(x)$ values get close to 10. It appears that as the x-values get close to 2 from the right, the $f(x)$ values get close to 10. Therefore, we write

$$\lim_{x \to 2} f(x) = 10$$

Our next example involves the piecewise function we used to model the profit for the fundraising dance put on by the college sorority. The interesting point on the graph is at $x = 150$. As you will see, we are unable to find a limit of $P(x)$, as x approaches 150, because as we move up to 150 from the left, we approach one number, and as we move down to 150 from above we get a different number.

Example 2 Find each limit for the function below:

$$P(x) = \begin{cases} 20x - 100 & 0 \le x \le 150 \\ 20x - 250 & 150 < x \le 350 \end{cases}$$

a. $\lim_{x \to 145} P(x)$ **b.** $\lim_{x \to 150} P(x)$ **c.** $\lim_{x \to 155} P(x)$

Solution Let's look at the close-up graph of our function that we used in Section 1.3. As you can see from the graph, two of the limits can be found from the graph, and one of the limits does not appear to have a single value.

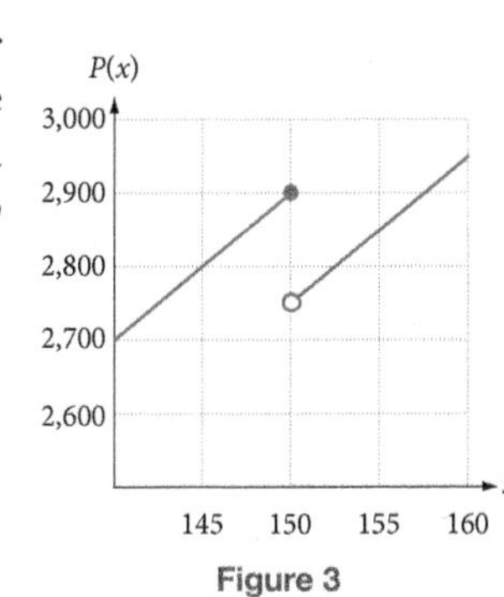

Figure 3

a. $\lim_{x \to 145} P(x) = 2{,}800$

b. $\lim_{x \to 150} P(x)$ does not exist

c. $\lim_{x \to 155} P(x) = 2{,}850$

To further justify our answer to part b, we make two tables. As you can see, as we move up to 150 from the left, we approach one number, and as we move down to 150 from above we get a different number.

x	P(x)
149	2,880
149.5	2,890
149.9	2,898
149.99	2,899.8
149.999	2,899.98
↓	↓
150	2,900

Table 3

x	P(x)
151	2,770
150.5	2,760
150.1	2,752
150.01	2,750.2
150.001	2,750.02
↓	↓
150	2,750

Table 4

In Example 2, we saw that $\lim_{x \to 150^-} = 2{,}900$, while $\lim_{x \to 150^+} = 2{,}750$. Since the limits from the left and right are different, there is no unique value for the limit, and we say $\lim_{x \to 150}$ *does not exist*. This leads us to the following theorem.

Theorem

$\lim_{x \to c} f(x) = L$ if and only if $\lim_{x \to c^-} f(x) = \lim_{x \to c^+} f(x) = L$. That is, $\lim_{x \to c} f(x) = L$ if and only if the limits from the left and right both exist and are equal.

Determining Limits Using the Properties of Limits

Limits of functions can often be determined directly and precisely using the properties of limits. We will state, interpret, and illustrate nine of the properties of limits. We will not prove any of the properties since the proofs rely on the theoretical definition of the limit and we are concerned with only the applications to calculus.

Limit of a Constant

If k is a constant, then

$$\lim_{x \to c} k = k$$

Figure 4 shows the graph of the constant function $f(x) = k$. The graph is the horizontal line through k on the y-axis. As the values of x get near c, the y-values are always equal to k.

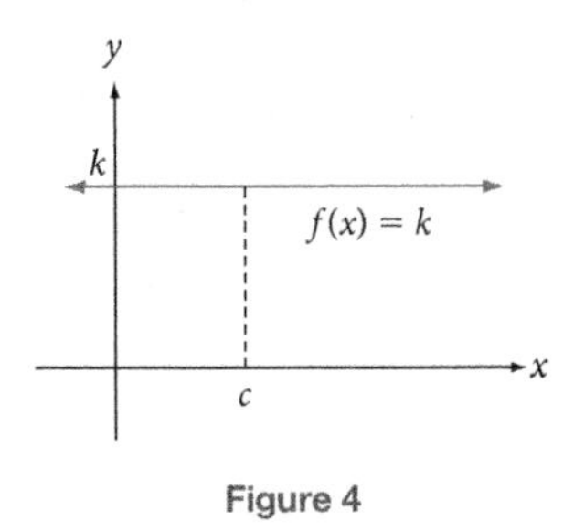

Figure 4

Example 3 If $f(x) = 4$, find $\lim_{x \to 7} f(x)$.

Solution This function always outputs 4 no matter what value of x is input. It is a constant function. Therefore:

$$\lim_{x \to 7} f(x) = \lim_{x \to 7} 4 = 4$$

Limit of a Polynomial Function

If $P(x) = a_n x^n + a_{n-1} x^{n-1} + \cdots + a_1 x + a_0$ is a polynomial function, then

$$\begin{aligned}\lim_{x \to c} P(x) &= P(c) \\ &= a_n c^n + a_{n-1} c^{n-1} + \cdots + a_1 c + a_0\end{aligned}$$

The graphs of polynomial functions are always smooth, unbroken curves. Figure 5 shows the graph of a polynomial function. As the values of x get close to c, the values of y get close to $f(c)$.

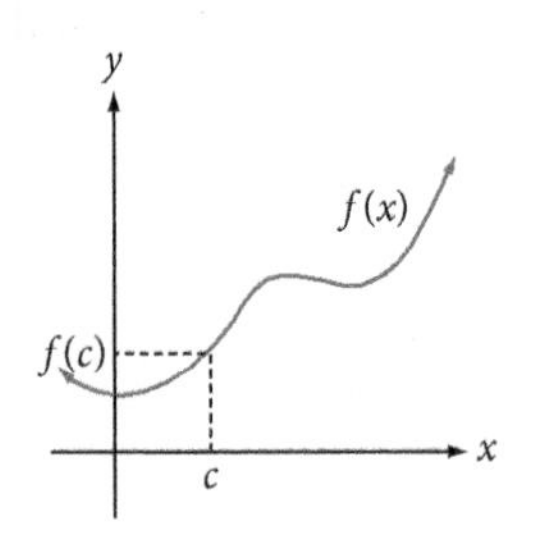

Figure 5

Example 4 If $f(x) = 3x^2 - 5x + 1$, find $\lim_{x \to 2} f(x)$.

Solution Our limit property for polynomial functions tells us we can always find limits with polynomial functions by direct substitution.

$$\begin{aligned}\lim_{x \to 2} f(x) &= \lim_{x \to 2} (3x^2 - 5x + 1)\\ &= 3(2)^2 - 5(2) + 1\\ &= 3 \cdot 4 - 10 + 1\\ &= 3\end{aligned}$$

We can construct a table of values near $x = 2$ to help you intuitively see that the limit is 3. Such a table is illustrated in Table 5.

x	$f(x)$
1.9	2.33
1.99	2.9303
1.999	2.993003
2.001	3.007003
2.01	3.0703
2.1	3.73

Table 5

The limit property of rational functions states the limit of a rational function can be determined by direct substitution only if that substitution does not produce 0 in the denominator.

Limit of a Rational Function

If $R(x) = \dfrac{P(x)}{Q(x)}$, where $P(x)$ and $Q(x)$ are polynomial functions, then

$$\begin{aligned}\lim_{x \to c} R(x) &= \lim_{x \to c} \frac{P(x)}{Q(x)}\\ &= \frac{P(c)}{Q(c)}\\ &= R(c), \quad \text{if } Q(c) \neq 0\end{aligned}$$

Example 5 If $f(x) = \dfrac{x^2 - 3}{x + 5}$, find $\lim_{x \to -2} f(x)$.

Solution Our limit property for rational functions tells us we can find this limit by direct substitution because this function is defined everywhere except at $x = -5$, and we are looking for the limit at $x = -2$.

$$\begin{aligned}\lim_{x \to -2} f(x) &= \lim_{x \to -2} \frac{x^2 - 3}{x + 5}\\ &= \frac{(-2)^2 - 3}{-2 + 5}\\ &= \frac{4 - 3}{3}\\ &= \frac{1}{3}, \quad \text{since } x + 5 \neq 0 \text{ at } x = -2\end{aligned}$$

Limit of a Sum or Difference

If $f(x)$ and $g(x)$ are functions and $\lim_{x \to c} f(x)$ and $\lim_{x \to c} g(x)$ both exist, then

$$\lim_{x \to c} \left[f(x) \pm g(x)\right] = \lim_{x \to c} f(x) \pm \lim_{x \to c} g(x)$$

The limit property of sums or differences states that the limit of a sum or a difference of two functions can be obtained from direct substitution provided the limits of each of the individual functions exist. The limit is obtained by determining the limit of each individual function, then adding or subtracting those limit values.

Example 6 Find $\lim\limits_{x \to -1} \left(4x^2 - 2x + \frac{5}{x}\right)$.

Solution The limit of a sum is the sum of the limits (provided those individual limits exist).

$$\begin{aligned} \lim_{x \to -1} \left(4x^2 - 2x + \frac{5}{x}\right) &= \lim_{x \to -1} 4x^2 - \lim_{x \to -1} 2x + \lim_{x \to -1} \frac{5}{x} \\ &= 4(-1)^2 - 2(-1) + \frac{5}{-1} \\ &= 4 + 2 - 5 \\ &= 1 \end{aligned}$$

Limit of a Constant Times a Function

If $f(x)$ is a function and $\lim\limits_{x \to c} f(x)$ exists, then $\lim\limits_{x \to c} kf(x) = k \lim\limits_{x \to c} f(x)$ for any constant k.

Limit of a Product of Functions

If $f(x)$ and $g(x)$ are functions and $\lim\limits_{x \to c} f(x)$ and $\lim\limits_{x \to c} g(x)$ both exist, then

$$\lim_{x \to c} \left[f(x) \cdot g(x)\right] = \left[\lim_{x \to c} f(x)\right] \cdot \left[\lim_{x \to c} g(x)\right]$$

The limit of a constant times a function is simply the constant times the limit of the function, provided that limit exists. Similarly, the limit property of products states that the limit of a product of two functions can be determined by direct substitution provided the limits of the two individual functions exist. It is obtained by determining the limit of each individual function, then multiplying those limit values together.

Example 7 Find **a.** $\lim\limits_{x \to 1} 5x^2$ and **b.** $\lim\limits_{x \to 1} \left[(2x - 3)(x^2 + 1)\right]$.

Solution

a. $\lim\limits_{x \to 1} 5x^2 = 5 \lim\limits_{x \to 1} x^2 = 5(1)^2 = 5$

b. The limit of a product is the product of the limits (provided those individual limits exist).

$$\begin{aligned} \lim_{x \to 1} \left[(2x - 3)(x^2 + 1)\right] &= \left[\lim_{x \to 1} (2x - 3)\right] \cdot \left[\lim_{x \to 1} (x^2 + 1)\right] \\ &= [2(1) - 3][1^2 + 1] \\ &= [2 - 3][1 + 1] \\ &= [-1][2] \\ &= -2 \end{aligned}$$

Limit of a Quotient of Functions

If $f(x)$ and $g(x)$ are functions and $\lim_{x \to c} f(x)$ and $\lim_{x \to c} g(x)$ both exist and $\lim_{x \to c} g(x) \neq 0$, then

$$\lim_{x \to c} \frac{f(x)}{g(x)} = \frac{\lim_{x \to c} f(x)}{\lim_{x \to c} g(x)}$$

Note Later we will see that substitution may produce 0 in the denominator and the limit may still exist, but other techniques must be applied to produce the limit.

The limit property of quotients states that the limit of a quotient of two functions can be obtained by direct substitution provided the limits of the two individual functions exist and the limit of the denominator is not 0. It is obtained by determining the limit of each individual function, then finding the indicated quotient.

Example 8 Find $\lim_{x \to -2} \frac{3x - 2}{x + 1}$.

Solution Our quotient is defined when $x = -2$, so we can use either our limit property for rational functions, or our limit property for quotients.

$$\begin{aligned} \lim_{x \to -2} \frac{3x - 2}{x + 1} &= \frac{\lim_{x \to -2} (3x - 2)}{\lim_{x \to -2} (x + 1)} \\ &= \frac{3(-2) - 2}{-2 + 1} \\ &= \frac{-6 - 2}{-1} \\ &= \frac{-8}{-1} \\ &= 8 \end{aligned}$$

Limit of a Power of a Function

If $f(x)$ is a function and $\lim_{x \to c} f(x)$ exists, then for any integer n,

$$\lim_{x \to c} [f(x)]^n = \left[\lim_{x \to c} f(x)\right]^n$$

The limit property of a power states that the limit of an integer power of a function can be determined by direct substitution provided that the limit of the function exists, and it is obtained by determining the limit of the function, and raising that value to the nth power.

Example 9 Find $\lim_{x \to 1/5} (5x - 3)^5$.

Solution We have the limit of a power function and our property ensures that we can find this limit by substitution.

$$\begin{aligned} \lim_{x \to 1/5} (5x - 3)^5 &= \left[\lim_{x \to 1/5} (5x - 3)\right]^5 \\ &= \left(5 \cdot \frac{1}{5} - 3\right)^5 \\ &= (-2)^5 = -32 \end{aligned}$$

Limit of a Radical

If $f(x)$ is a function and $\lim_{x \to c} f(x)$ exists, n is a positive integer, and the nth root of this limit is defined, then

$$\lim_{x \to c} \sqrt[n]{f(x)} = \sqrt[n]{\lim_{x \to c} f(x)}$$

The limit property of a radical states that the limit of the nth root of a function under a radical can be determined by direct substitution provided that the limit of the function exists. The limit is obtained by determining the limit of the function, and taking the nth root of that value, if it is defined.

Example 10 Find $\lim_{x \to 8} \sqrt[3]{x^2}$.

Solution Using substitution, we have

$$\begin{aligned}\lim_{x \to 8} \sqrt[3]{x^2} &= \sqrt[3]{\lim_{x \to 8} x^2} \\ &= \sqrt[3]{8^2} \\ &= \sqrt[3]{64} \\ &= 4\end{aligned}$$

■

Determining Limits When the Properties Do Not Directly Apply

All the limits we have examined so far were determined by tables of values or by application of the limit properties and direct substitution. Quite often, however, direct substitution fails. We will examine several such possibilities now.

Undefined Form Limits of rational functions, $\lim_{x \to c} \frac{P(x)}{Q(x)}$, that produce upon direct substitution the form $\frac{k}{0}$ where k is a nonzero constant, will never approach a specific value. Instead, as the values of x approach c, the values of the numerator approach k, and the values of the denominator approach 0. This means the numerator of the fraction is becoming k, and the denominator is getting smaller and smaller. This, in turn, means that the fraction itself is getting bigger and bigger. It gets bigger without bound and, therefore, approaches no specific value. When a rational expression produces the form $\frac{k}{0}$ (an undefined form), we say that the *limit does not exist.* Part b of our next example illustrates this.

Example 11 Find each limit for the function below:

$$f(x) = \frac{1}{(x-3)^2}$$

a. $\lim_{x \to 2} f(x)$ **b.** $\lim_{x \to 3} f(x)$ **c.** $\lim_{x \to 5} f(x)$

Solution We use direct substitution and the graph shown in Figure 6 to find the limits that exist.

a. $\lim_{x\to 2} f(x) = 1$

b. $\lim_{x\to 3} f(x)$ does not exist. The closer we get to 3 with x, the larger $f(x)$ becomes. It never reaches a limit, it just keeps getting larger and larger.

c. $\lim_{x\to 5} f(x) = \dfrac{1}{4}$

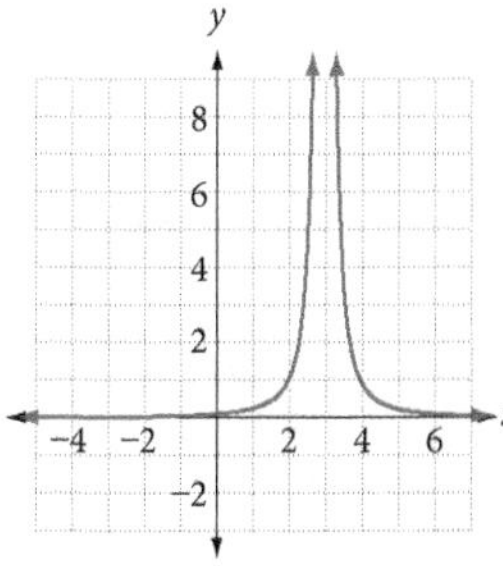

Figure 6

Example 12 Find $\lim_{x\to -1} \dfrac{x^2 - x - 3}{x + 1}$.

Solution Trying direct substitution leads to an undefined expression.

$$\lim_{x\to -1} \frac{x^2 - x - 3}{x + 1} = \frac{(-1)^2 - (-1) - 3}{-1 + 1}$$

$$= \frac{1 + 1 - 3}{0}$$

$$= \frac{-1}{0}, \quad \text{which is undefined.}$$

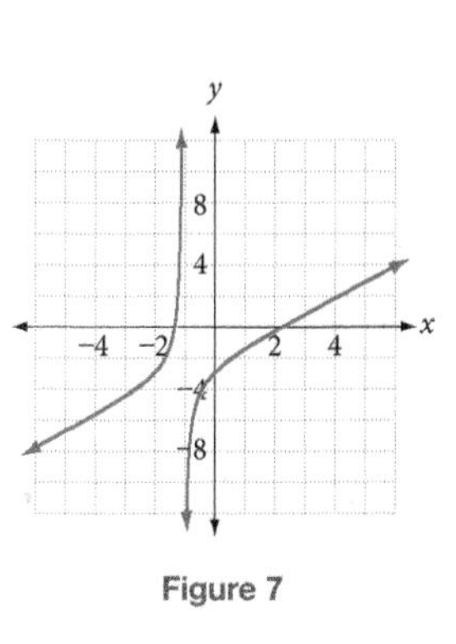

Figure 7

Thus, $\lim_{x\to -1} \dfrac{x^2 - x - 3}{x + 1}$ does not exist.

Notice the difference in the graphs in Figures 6 and 7. When we take the limit as x approaches 3 in Example 11b, we find that $f(x)$ grows in the positive direction without bound (whether we approach from the left or right). In Example 12, however, $f(x)$ gets larger and larger in the positive direction as we approach -1 from the left, but it grows without bound in the *negative* direction as we approach -1 from the right. In both cases, we say the limit does not exist. However, you will often see the first limit written in this way:

$$\lim_{x\to 3} \frac{1}{(x-3)^2} = +\infty$$

We are *not* saying that the actual limit is positive infinity, as infinity is not a real number; we are simply describing the behavior of the graph as x approaches 3.

Indeterminate Form Limits of rational expressions that produce upon direct substitution the *indeterminate* form $\frac{0}{0}$ when computing the limit need further work. Both the numerator and denominator of the fraction are, upon direct substitution, producing 0. This will occur when the numerator and denominator have a common factor that is approaching 0 while x is approaching some value c. The problem can often be eliminated by factoring and dividing out the problem-causing common factor. It is important to note that the factors that are approaching zero can be divided out because they are only close to zero in value and not *equal* to zero. Remember, x is approaching c and is not actually equal to c.

Note The indeterminate form $\frac{0}{0}$ is unlike the form $\frac{k}{0}$, where $k \neq 0$. A function producing the form $\frac{k}{0}$ upon direct substitution of c for x has a graph like that pictured in Figure 6 or 7. It is broken at $x = c$, and the limit does not exist.

Example 13 Find $\lim_{x\to 2} \dfrac{x^2 - x - 2}{x - 2}$.

Solution Direct substitution produces $\frac{2^2 - 2 - 2}{2 - 2} = \frac{0}{0}$, which is indeterminate. The indeterminate form $\frac{0}{0}$, along with a rational function or quotient of functions, alerts us to the fact that there must be a factor common to the numerator and denominator that is approaching 0 while x is approaching 2. We'll try to factor and divide out the problem-causing factors.

Note When the indeterminate form $\frac{0}{0}$ occurs from the direct substitution of c for x, be very careful not to mentally predetermine the limit. Analyze the behavior of the function by factoring and looking for the factor that is common to the numerator and denominator. Divide it out and then try direct substitution again.

$$
\begin{aligned}
\lim_{x\to 2} \frac{x^2 - x - 2}{x - 2} &= \lim_{x\to 2} \frac{(x-2)(x+1)}{x-2} && \text{Divide out the common factor } x - 2 \\
&= \lim_{x\to 2} (x+1) \\
&= 2 + 1 \\
&= 3
\end{aligned}
$$

So, the limit actually does exist! This is a case where the function has a hole at $x = 2$. Figure 8 visually verifies this limit and Table 6 intuitively suggests it.

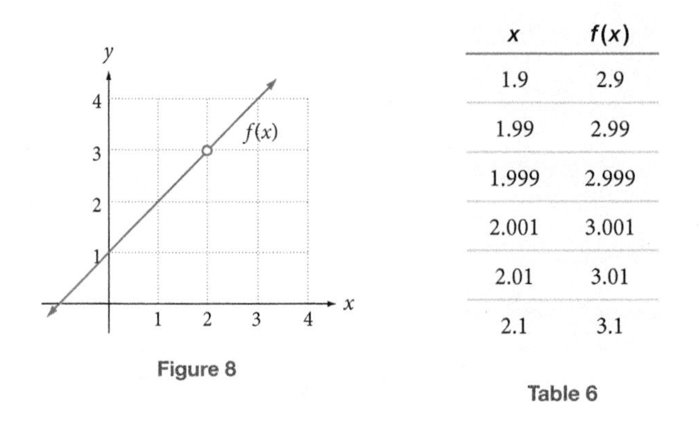

Figure 8

x	$f(x)$
1.9	2.9
1.99	2.99
1.999	2.999
2.001	3.001
2.01	3.01
2.1	3.1

Table 6

As we saw in Example 13, the appearance of the indeterminate form $\frac{0}{0}$ along with a rational expression or quotient alerts us to the technique of factoring and dividing out the problem-causing factors. Similarly, the appearance of the indeterminate form $\frac{0}{0}$ along with a radical fractional expression alerts us to the technique of rationalizing the numerator or denominator and then dividing out the problem-causing factors.

Example 14 Find $\lim_{x\to 4} \dfrac{\sqrt{x} - 2}{x - 4}$.

Solution Direct substitution produces

$$\frac{\sqrt{4} - 2}{4 - 4} = \frac{2 - 2}{4 - 4} = \frac{0}{0}$$

which is indeterminate. We will try to rationalize the numerator and then factor and divide out the problem-causing factors.

$$
\begin{aligned}
\lim_{x\to 4} \frac{\sqrt{x} - 2}{x - 4} &= \lim_{x\to 4} \frac{\sqrt{x} - 2}{x - 4} \cdot \frac{\sqrt{x} + 2}{\sqrt{x} + 2} \\
&= \lim_{x\to 4} \frac{x - 4}{(x-4)(\sqrt{x} + 2)} && \text{Simplify the numerator only} \\
&= \lim_{x\to 4} \frac{x - 4}{(x-4)(\sqrt{x} + 2)} && \text{Divide out the common factor } x - 4
\end{aligned}
$$

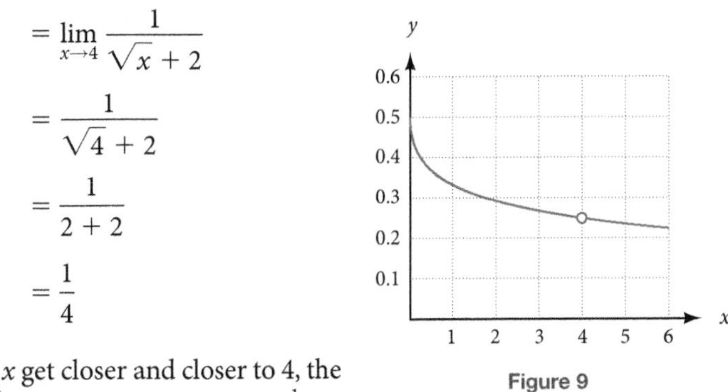

$$= \lim_{x \to 4} \frac{1}{\sqrt{x} + 2}$$

$$= \frac{1}{\sqrt{4} + 2}$$

$$= \frac{1}{2 + 2}$$

$$= \frac{1}{4}$$

Figure 9

As the values of x get closer and closer to 4, the values of $\frac{\sqrt{x} - 2}{x - 4}$ get closer and closer to $\frac{1}{4}$, as Figure 9 illustrates. Notice that there is a hole where the function is undefined. ■

Limits Involving Infinity

© Nicole Waring/iStockPhoto

Recall our discussion of manufacturing bicycles at the beginning of this section. Let's suppose that $\overline{C}(x) = 200 + \frac{100{,}000}{x}$ represents the average cost $\overline{C}(x)$ (in dollars) of producing x bicycles. If we are interested in what will happen to the average cost per bicycle as the number of bicycles produced increases, we can see any trends by considering the limit of $\overline{C}(x)$ as x approaches infinity. To do so, we first note that as the values of x get larger and larger, the values of $\frac{1}{x}$ get smaller and smaller and approach 0.

$$\lim_{x \to \infty} \overline{C}(x) = \lim_{x \to \infty} \left(200 + \frac{100{,}000}{x}\right)$$

$$= \lim_{x \to \infty} (200) + \lim_{x \to \infty} \frac{100{,}000}{x}$$

$$= 200 + 100{,}000 \lim_{x \to \infty} \frac{1}{x}$$

$$= 200 + 100{,}000 \cdot 0$$

$$= 200 + 0 = 200$$

Figure 10

Consequently, we observe that the average cost of each bicycle tends toward \$200 as more and more bicycles are built. Note that the average cost of producing one bicycle, $\overline{C}(1)$, is \$100,200! The cost is so high because of the initial production set-up costs. As the production equipment is used, the cost per bicycle decreases. These facts are illustrated in Figure 10. Although this function is defined only for discrete values of x, that is, $x = 1, 2, 3, \ldots$, we draw the graph as a smooth curve for the simplicity of the sketch and the ease of viewing. Notice that the average cost of each bicycle drops dramatically as the number of bicycles produced increases from 1 to 2,000, but drops gradually after that.

To evaluate limits of this type quickly and accurately, let's consider the following two limit properties. In doing so, we will take the general concept of infinity (∞) and break it into two specific types, positive infinity ($+\infty$), in which x grows without bound in the positive direction, and negative infinity ($-\infty$), in which x grows without bound in the negative direction. The symbol ∞ without an affixed $+$ or $-$ sign represents the general concept of infinity, that is, x growing without bound in either direction.

Property 1

$$\lim_{x\to\infty}(x^n) = \infty \quad \text{(or } \textit{does not exist}\text{)}, \quad \text{where } n > 0$$

Property 2

$$\lim_{x\to\infty}\left(\frac{1}{x^n}\right) = 0 \quad \text{where } n > 0$$

Property 1 tells us that as x grows larger and larger, x^n also grows larger and larger. Property 2 tells us that as x grows larger and larger, the reciprocal of x^n, $\frac{1}{x^n}$, shrinks toward 0, as illustrated in our bicycle production example.

Often we are confronted with limits of rational functions such as

$$\lim_{x\to\infty}\frac{5x^2+7x-2}{8x^2-3x+1}$$

Notice that

$$\lim_{x\to\infty}\frac{5x^2+7x-2}{8x^2-3x+1} = \frac{\lim\limits_{x\to\infty} 5x^2+7x-2}{\lim\limits_{x\to\infty} 8x^2-3x+1}$$

$$= \frac{\infty}{\infty}$$

Note Remember that we are not actually saying that the limits of both the numerator and denominator are equal to infinity. The limits do not exist, but we *can* use the infinity symbol to describe the behavior of each function as x grows larger and larger.

Like $\frac{0}{0}$, $\frac{\infty}{\infty}$ is an indeterminate form. Thus, when we encounter it, we must look further to find the limit, if it exists. A helpful technique is to divide the numerator and denominator by the variable of highest degree. We are, in essence, multiplying by 1.

Example 15 Divide the numerator and denominator of the rational expression by x^2 to help evaluate this limit.

$$\lim_{x\to\infty}\frac{5x^2+7x-2}{8x^2-3x+1}$$

Solution Dividing numerator and denominator of our function by x^2, we have

$$\lim_{x\to\infty}\frac{5x^2+7x-2}{8x^2-3x+1} = \lim_{x\to\infty}\frac{5x^2+7x-2}{8x^2-3x+1}\cdot 1$$

$$= \lim_{x\to\infty}\frac{(5x^2+7x-2)\cdot\frac{1}{x^2}}{(8x^2-3x+1)\cdot\frac{1}{x^2}}$$

$$= \lim_{x\to\infty}\frac{5+\dfrac{7}{x}-\dfrac{2}{x^2}}{8-\dfrac{3}{x}+\dfrac{1}{x^2}}$$

$$= \frac{\lim_{x\to\infty}\left(5 + \frac{7}{x} - \frac{2}{x^2}\right)}{\lim_{x\to\infty}\left(8 - \frac{3}{x} + \frac{1}{x^2}\right)}$$

$$= \frac{5 + 0 - 0}{8 - 0 + 0}$$

$$= \frac{5}{8}$$

Notice that the 5 comes from the term $5x^2$, the dominant (leading) term in the numerator, and that the 8 comes from the term $8x^2$, the dominant term in the denominator. Thus, the limit can also be found by considering only the dominant terms.

$$\lim_{x\to\infty} \frac{5x^2 + 7x - 2}{8x^2 - 3x + 1} = \lim_{x\to\infty} \frac{5x^2}{8x^2} = \lim_{x\to\infty} \frac{5}{8} = \frac{5}{8}$$ ■

Dominant Term Property

If $P(x)$ and $Q(x)$ are polynomials, then

$$\lim_{x\to\infty} \frac{P(x)}{Q(x)} = \lim_{x\to\infty} \frac{\text{Dominant term in } P(x)}{\text{Dominant term in } Q(x)}$$

Example 16 Find $\lim_{x\to\infty} \frac{8x^3 + 7x - 2}{5x^2 + 3x + 11}$.

Solution Initially (from direct substitution), $\lim_{x\to\infty} \frac{8x^3 + 7x - 2}{5x^2 + 3x + 11} = \frac{\infty}{\infty}$. We will look further for the limit by considering the dominant terms.

$$\lim_{x\to\infty} \frac{8x^3 + 7x - 2}{5x^2 + 3x + 11} = \lim_{x\to\infty} \frac{8x^3}{5x^2} = \frac{8x}{5} = \infty$$

and we conclude that the limit does not exist. ■

Notice that in this case the degree of the dominant term in the numerator is greater than the degree of the dominant term in the denominator.

Example 17 Find $\lim_{x\to\infty} \frac{7x - 3}{2x^2 + 5}$.

Solution Initially (from direct substitution), $\lim_{x\to\infty} \frac{7x - 3}{2x^2 + 5} = \frac{\infty}{\infty}$. We'll look further for the limit by considering the dominant terms.

$$\lim_{x\to\infty} \frac{7x - 3}{2x^2 + 5} = \lim_{x\to\infty} \frac{7x}{2x^2} = \lim_{x\to\infty} \frac{7}{2x} = \frac{7}{2} \lim_{x\to\infty} \frac{1}{x} = \frac{7}{2} \cdot 0 = 0$$ ■

Notice that the degree of the dominant term in the numerator is less than the degree of the dominant term in the denominator.

Here is a summary of what we have done so far.

In Case 1, the denominator grows significantly faster than the numerator, and the ratio tends toward 0.

In Case 2, both the numerator and the denominator grow at about the same rate. Thus, the numbers multiplying the larger powers become important.

In Case 3, the numerator grows significantly faster than the denominator, and the ratio tends toward ∞.

If $N(x)$ and $D(x)$ are functions for which $\lim_{x\to\infty} \frac{N(x)}{D(x)} = \frac{\infty}{\infty}$, then one of the cases below will apply:

Case 1: Degree of $N(x)$ < Degree of $D(x)$ As shown in Example 17,

$$\lim_{x\to\infty} \frac{N(x)}{D(x)} = 0$$

Case 2: Degree of $N(x)$ = Degree of $D(x)$ As shown in Example 15,

$$\lim_{x\to\infty} \frac{N(x)}{D(x)} = \text{the ratio of the coefficients of the dominant terms.}$$

Case 3: Degree of $N(x)$ > Degree of $D(x)$ As shown in Example 16,

$$\lim_{x\to\infty} \frac{N(x)}{D(x)} = \infty \text{ or } \textit{does not exist.}$$

Each of these results can be quickly determined by paying attention to only the *dominant terms*. But keep in mind that this process works only when the indeterminate form $\frac{\infty}{\infty}$ occurs upon initial substitution.

Example 18 Each of the following limits results in the indeterminate form $\frac{\infty}{\infty}$ upon direct substitution. Use the Dominant Term Property to evaluate each one.

a. $\lim_{x\to\infty} \frac{3x}{x^2+1}$ **b.** $\lim_{x\to\infty} \frac{2x+1}{x-3}$ **c.** $\lim_{x\to\infty} \frac{x^2}{2x-4}$

Solution

a. $\lim_{x\to\infty} \frac{3x}{x^2+1} = \lim_{x\to\infty} \frac{3x}{x^2} = \lim_{x\to\infty} \frac{3}{x} = 0$

This is an example of Case 1. The degree of the numerator is less than the degree of the denominator. The graph of $f(x) = \frac{3x}{x^2+1}$ is displayed in Figure 11. Notice that as x grows large (or small), $f(x)$ approaches 0.

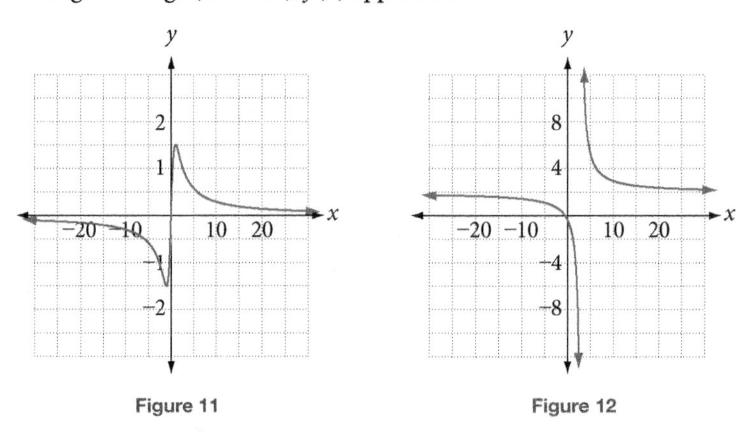

Figure 11 Figure 12

> *Note* What would happen in Example 18c if we replaced ∞ with $-\infty$ in the limit? The final step would be
>
> $$\lim_{x\to-\infty}\frac{x}{2}=\frac{-\infty}{2}=-\infty$$
>
> Of course, we know the limit does not actually exist, but it is interesting to examine this result in terms of the graph of the function. Looking at the graph in Figure 13, we see that, as we might suspect,
>
> $$f(x)=\frac{x^2}{2x-4}$$
>
> grows in the negative direction without bound as x grows in the negative direction without bound.

b. $\displaystyle\lim_{x\to\infty}\frac{2x+1}{x-3}=\lim_{x\to\infty}\frac{2x}{x}=\lim_{x\to\infty}2=2$

This is an example of Case 2. The degree of the numerator is equal to the degree of the denominator. The graph of $f(x)=\frac{2x+1}{x-3}$ is displayed in Figure 12.

c. $\displaystyle\lim_{x\to\infty}\frac{x^2}{2x-4}=\lim_{x\to\infty}\frac{x^2}{2x}=\lim_{x\to\infty}\frac{x}{2}=\infty$

This is an example of Case 3. The degree of the numerator is greater than the degree of the denominator. The graph of $f(x)=\frac{x^2}{2x-4}$ is displayed in Figure 13.

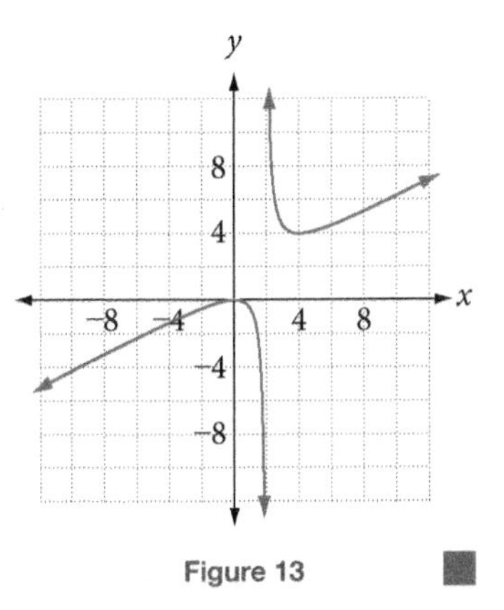

Figure 13

Asymptotes

In your previous math classes, you worked with asymptotes when graphing rational functions, exponential functions, and other functions as well. Now that you have worked with limits, we can give more formal definitions to horizontal and vertical asymptotes.

> **Horizontal Asymptote**
> For the graph of the function $y=f(x)$, a horizontal line that is approached by the graph as the independent variable tends toward infinity is called a horizontal asymptote. That is, if $\lim_{x\to-\infty}f(x)=\lim_{x\to\infty}f(x)=b$, then the horizontal line $y=b$ is a horizontal asymptote for the graph of $y=f(x)$.
>
> **Vertical Asymptote**
> If $\lim_{x\to c}f(x)=\infty$ or $\lim_{x\to c}f(x)=-\infty$ as $x\to c^-$ or $x\to c^+$, then $x=c$ is a vertical asymptote for the graph of $y=f(x)$.

Example 19 The function below was used previously to model the progress of a person on a diet with a starting weight of 200 pounds. The quantity $W(x)$ is the weight (in pounds) of the person after x weeks of dieting.

© esolla/iStockPhoto

$$W(x)=\frac{80(2x+15)}{x+6}\quad\text{if } x\ge 0$$

The function has a natural restriction on the domain because x is the number of weeks on the diet and so x must be 0 or greater, and that is why we have the graph shown in Figure 14a. Figure 14b shows the graph without the restriction on the domain. Find the value of both the vertical asymptote and the horizontal asymptote.

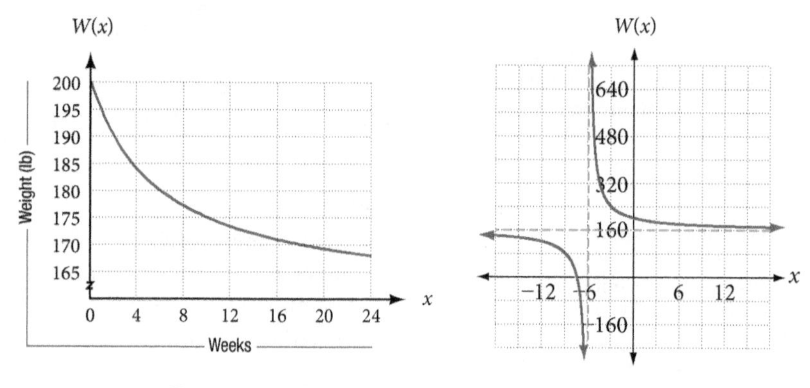

Figure 14a

Figure 14b

Solution We have a vertical asymptote at $x = -6$ because

$$\lim_{x \to -6} \frac{80(2x + 15)}{x + 6}$$

does not exist. Specifically, $\lim_{x \to -6^-} W(x) = -\infty$ and $\lim_{x \to -6^+} W(x) = +\infty$.

To find the horizontal asymptote we take the limit as x goes to infinity:

$$\lim_{x \to \infty} \frac{80(2x + 15)}{x + 6} = \lim_{x \to \infty} \frac{160x + 1{,}200}{x + 6} = \underbrace{\lim_{x \to \infty} \frac{160x}{x}}_{\text{Dominant term property}} = 160$$

We will get the same result if we find the limit as x goes to $-\infty$. Therefore, we have a horizontal asymptote at $y = 160$. Comparing our results with the graph in Figure 14b, we see that they match.

Interpretation

According to this weight-loss model, a person with a starting weight of 200 pounds will lose weight quickly at first, then gradually reach a lower limit of 160 pounds.

Getting Ready for Class

After reading through the preceding section, respond in your own words and in complete sentences.

A. Explain why it is possible for $\lim_{x \to 3} f(x) = 7$, but for $f(3)$ to not exist.

B. $\lim_{x \to 4} \frac{3x + 5}{x - 4}$ does not exist. Explain why this is so in terms of the behavior of the function near $x = 4$.

C. Upon the direct substitution of 6 for x in $\lim_{x \to 6} \frac{x^2 - 3x - 18}{x^2 - 4x - 12}$, the expression $\frac{0}{0}$ results. Explain what this result indicates and what should be done at this point to find the limit, if it exists.

Problem Set 1.4

Skills Practice

Use the graph to find each limit.

1. $\lim_{x \to a} f(x)$

2. $\lim_{x \to a} f(x)$

3. a. $\lim_{x \to 10} f(x)$ b. $\lim_{x \to 3} f(x)$

4. a. $\lim_{x \to 3} f(x)$ b. $\lim_{x \to 6} f(x)$

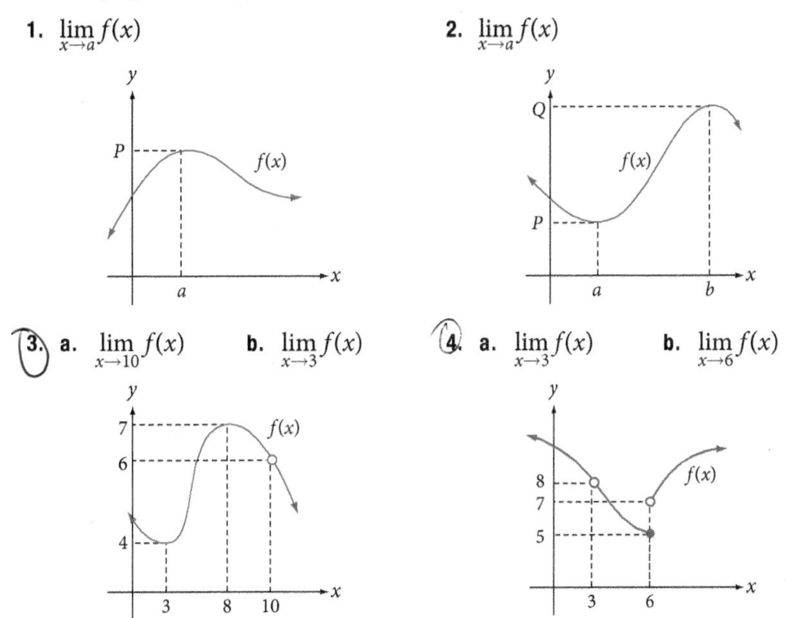

Use each set of tables to find each limit, if it exists.

5. $\lim_{x \to 8} f(x)$

x (from the left)	f(x)
7	12.1
7.5	12.3
7.7	12.95
7.9	12.995
7.99	12.9995
7.999	12.99995

x (from the right)	f(x)
9	13.2
8.5	13.1
8.3	13.01
8.1	13.001
8.01	13.0001
8.001	13.00001

6. $\lim_{x \to -8} f(x)$

x (from the left)	f(x)
−8.5	6.8
−8.3	6.2
−8.1	6.1
−8.01	6.01
−8.001	6.001
−8.0001	6.0001

x (from the right)	f(x)
−7.5	4.2
−7.7	4.7
−7.9	4.9
−7.99	4.99
−7.999	4.999
−7.9999	4.9999

Use the graph to find each limit if it exists.

7. $\lim_{x \to 6} f(x)$ **8.** $\lim_{x \to 3} f(x)$

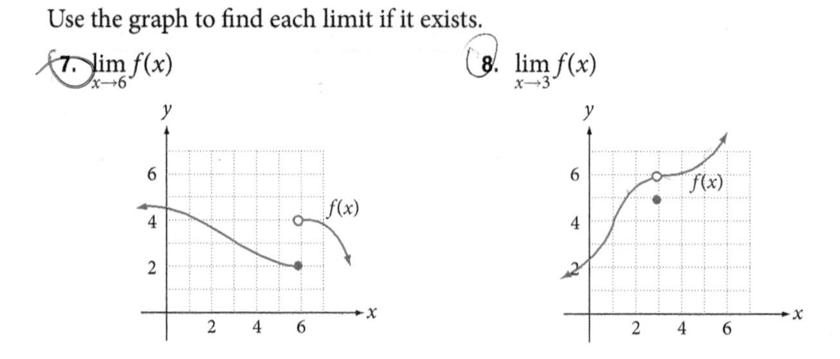

Determine the limit, if it exists, at the specified value of x by completing the tables.

9. $\lim_{x \to 1} (x^2 + 4x - 8)$

x (from the left)	f(x)
0.5	
0.7	
0.9	
0.99	
0.999	

x (from the right)	f(x)
1.5	
1.3	
1.1	
1.01	
1.001	

10. $\lim_{x \to -3} (-5x^2 + 2x + 7)$

x (from the left)	f(x)
−3.5	
−3.3	
−3.1	
−3.01	
−3.001	

x (from the right)	f(x)
−2.5	
−2.7	
−2.9	
−2.99	
−2.999	

Use the graphs and your intuitive knowledge of limits to find each limit if it exists.

11. $\lim_{x \to 3} 5$ **12.** $\lim_{x \to 2} -4$ **13.** $\lim_{x \to 2} x$

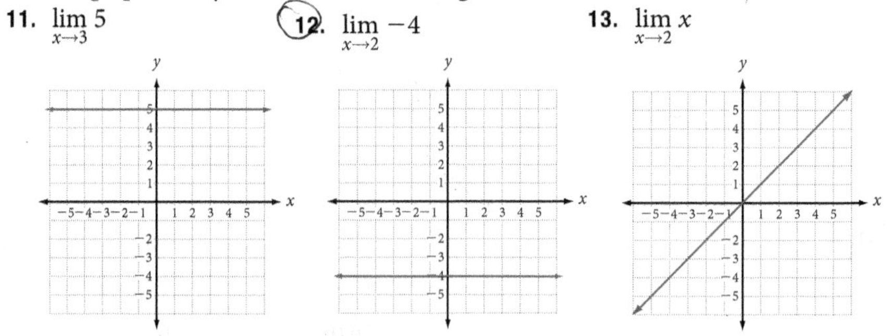

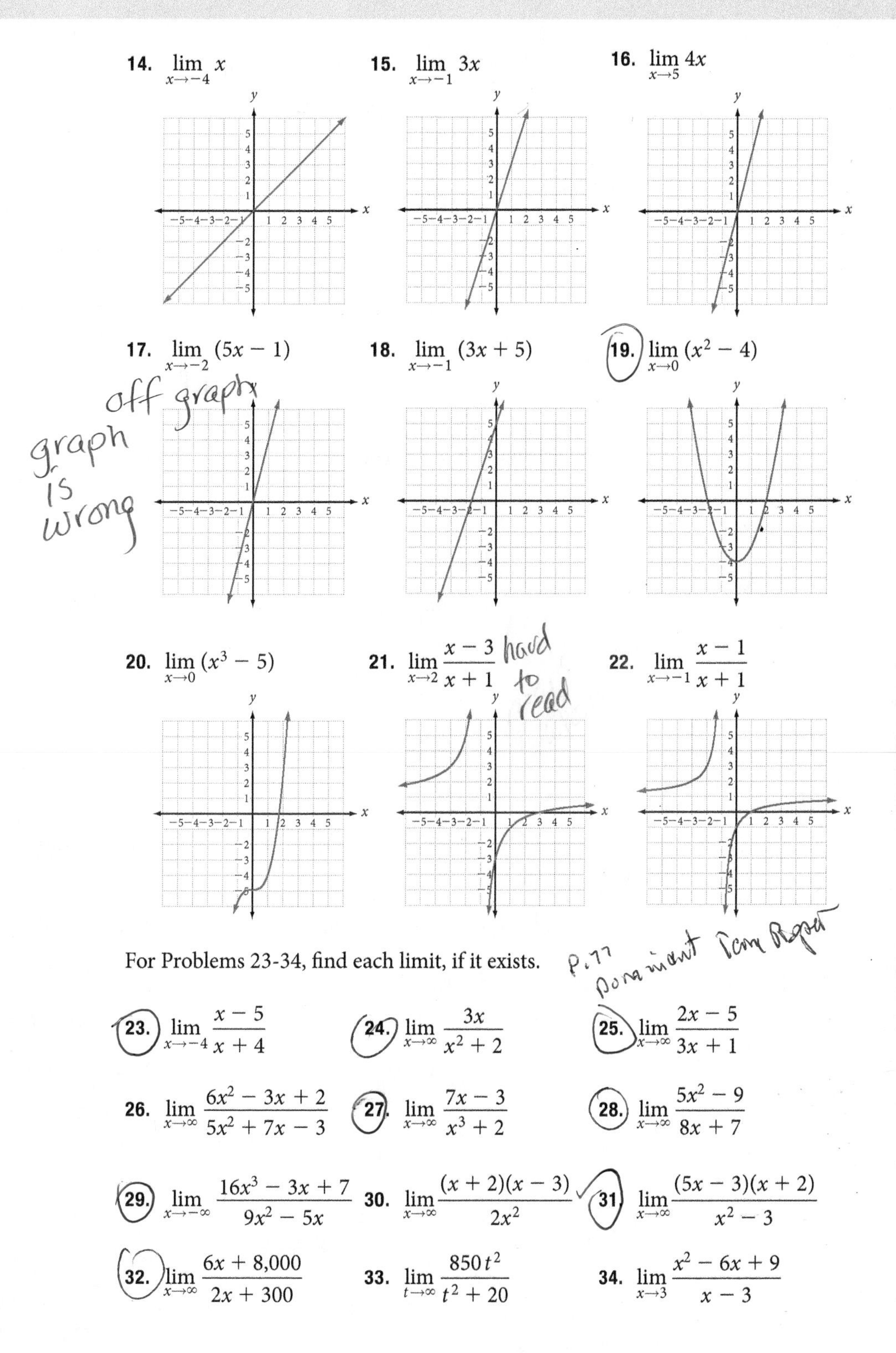

14. $\lim_{x \to -4} x$

15. $\lim_{x \to -1} 3x$

16. $\lim_{x \to 5} 4x$

17. $\lim_{x \to -2} (5x - 1)$

18. $\lim_{x \to -1} (3x + 5)$

19. $\lim_{x \to 0} (x^2 - 4)$

20. $\lim_{x \to 0} (x^3 - 5)$

21. $\lim_{x \to 2} \frac{x - 3}{x + 1}$

22. $\lim_{x \to -1} \frac{x - 1}{x + 1}$

For Problems 23-34, find each limit, if it exists.

23. $\lim_{x \to -4} \frac{x - 5}{x + 4}$

24. $\lim_{x \to \infty} \frac{3x}{x^2 + 2}$

25. $\lim_{x \to \infty} \frac{2x - 5}{3x + 1}$

26. $\lim_{x \to \infty} \frac{6x^2 - 3x + 2}{5x^2 + 7x - 3}$

27. $\lim_{x \to \infty} \frac{7x - 3}{x^3 + 2}$

28. $\lim_{x \to \infty} \frac{5x^2 - 9}{8x + 7}$

29. $\lim_{x \to -\infty} \frac{16x^3 - 3x + 7}{9x^2 - 5x}$

30. $\lim_{x \to \infty} \frac{(x + 2)(x - 3)}{2x^2}$

31. $\lim_{x \to \infty} \frac{(5x - 3)(x + 2)}{x^2 - 3}$

32. $\lim_{x \to \infty} \frac{6x + 8{,}000}{2x + 300}$

33. $\lim_{t \to \infty} \frac{850t^2}{t^2 + 20}$

34. $\lim_{x \to 3} \frac{x^2 - 6x + 9}{x - 3}$

Check Your Understanding

35. In the following figure, find $\lim_{x\to\infty} f(x)$.

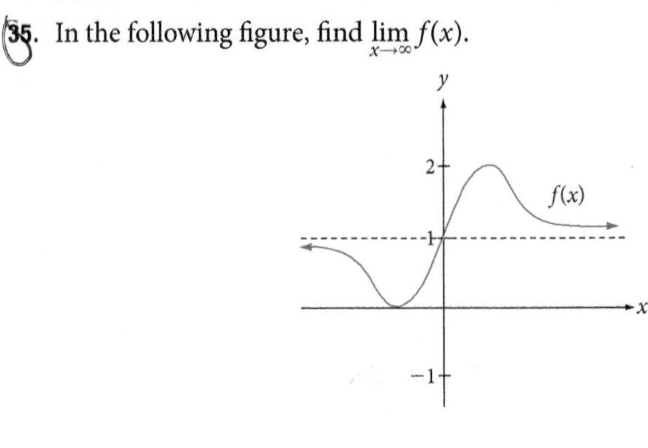

36. In the following figure, find

a. $\lim_{x\to\infty} f(x)$

b. $\lim_{x\to-\infty} f(x)$

c. $\lim_{x\to-1} f(x)$

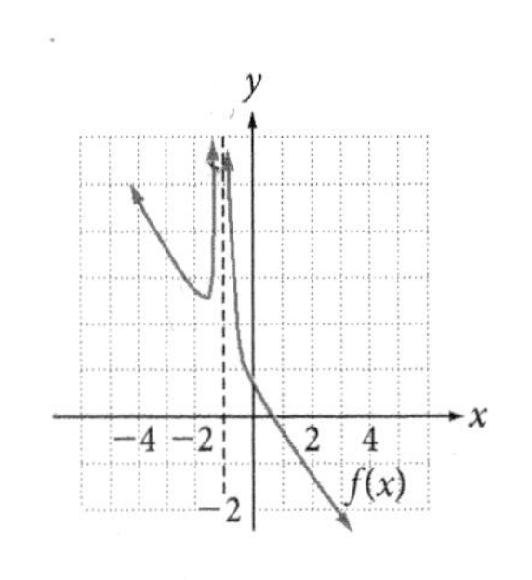

37. In the following figure, find

a. $\lim_{x\to\infty} f(x)$

b. $\lim_{x\to-\infty} f(x)$

c. $\lim_{x\to2} f(x)$

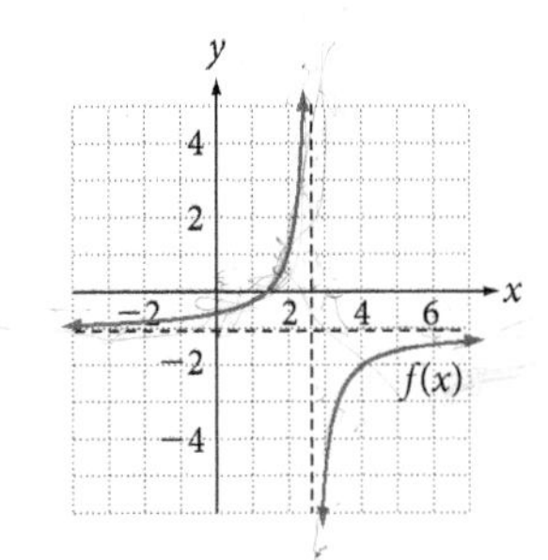

38. In the following figure, find

a. $\lim_{x\to\infty} f(x)$

b. $\lim_{x\to2} f(x)$

c. $\lim_{x\to-2} f(x)$

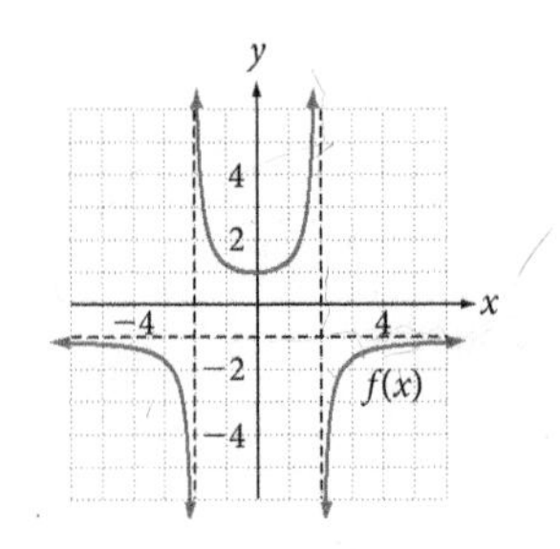

39. In the following figure, find

a. $\lim_{x\to\infty} f(x)$

b. $\lim_{x\to-\infty} f(x)$

c. $\lim_{x\to2} f(x)$

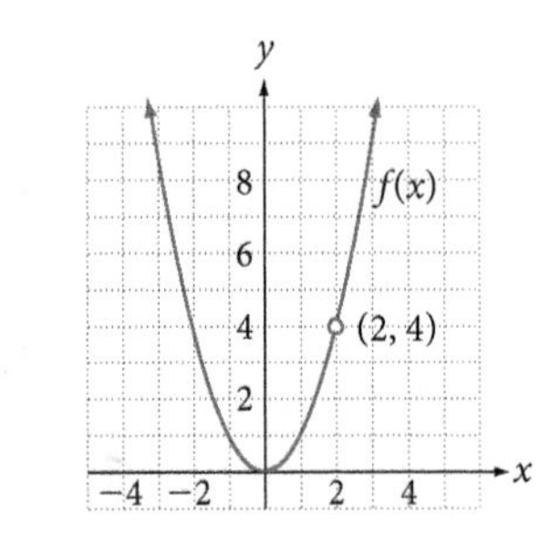

40. The variables x and y are related according to the rule $y = f(x) = x$, the graph of which is shown here.

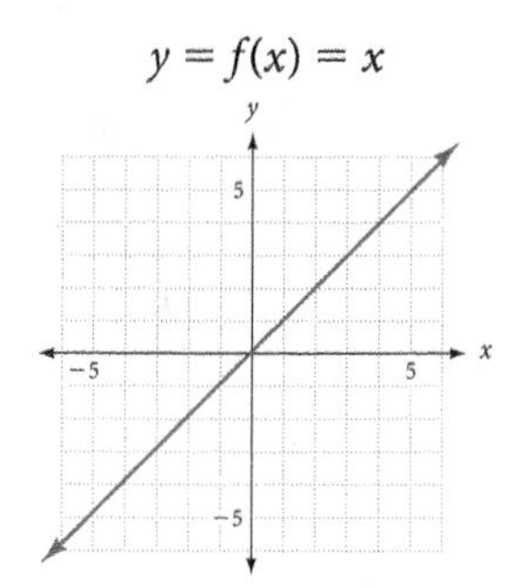

a. As x approaches 3, which number does y approach?

b. Which number is $\lim_{x \to 1} f(x)$?

c. True or false: $\lim_{x \to 0} f(x) = 0$.

41. The variables x and y are related according to the rule $y = f(x) = |x|$ the graph of which is shown here.

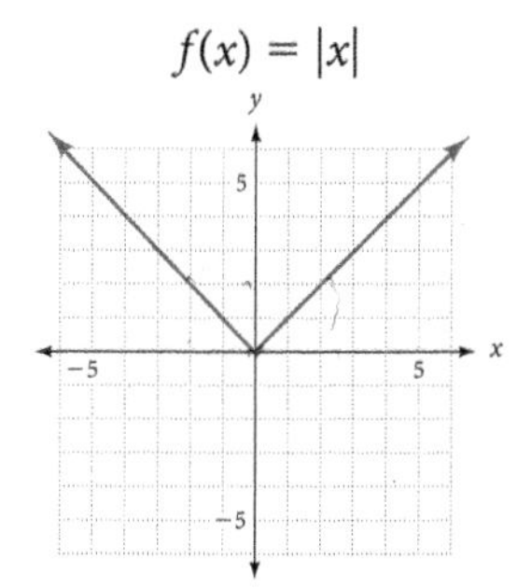

a. As x approaches -2, which number does y approach?

b. Find $\lim_{x \to 0} f(x)$.

c. True or false: $\lim_{x \to 2} f(x) = -2$.

42. The variable y depends on the variable x according to the rule $y = f(x) = x^2$.

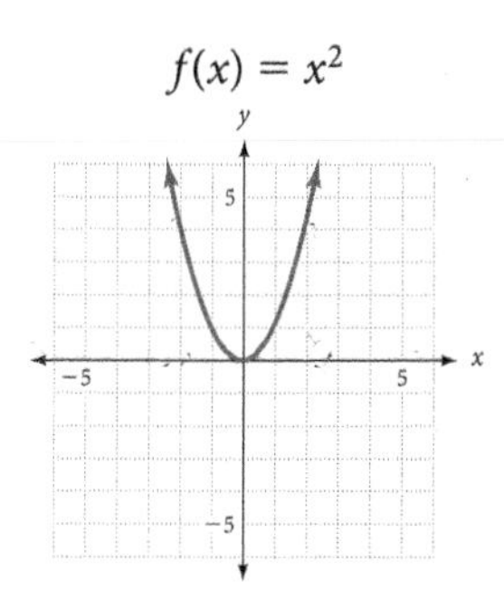

a. Which number does $f(x)$ approach as x approaches 2?

b. Find a if $\lim_{x \to a} f(x) = 4$.

c. True or false: $\lim_{x \to 5} f(x) = 25$.

43. The variable y depends on the variable x according to the rule $y = f(x) = \frac{1}{x}$.

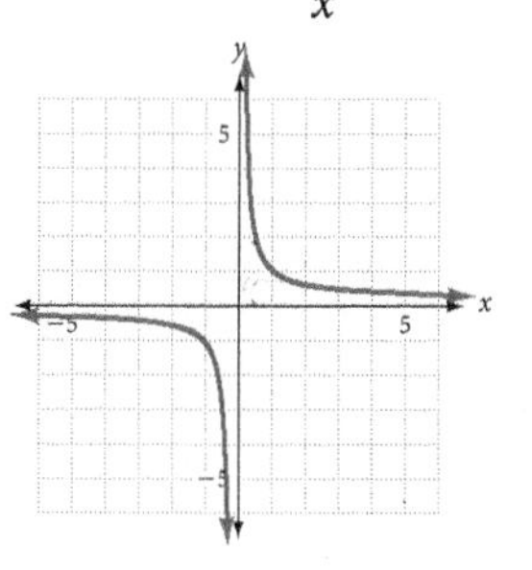

a. Which number does $f(x)$ approach as x approaches $\frac{1}{2}$?

b. Find a if $\lim_{x \to a} f(x) = -1$.

c. True or false: $\lim_{x \to 0} f(x)$ does not exist.

Modeling Practice

44. Economics: City Water The monthly charge, in dollars, for water in a particular city is given by the function

$$C(x) = \begin{cases} 0.3x & \text{if } 0 \le x \le 15 \\ 0.9x - 9 & \text{if } x > 15 \end{cases}$$

where x is the number of cubic feet of water used. Find $\lim_{x \to 15} C(x)$, if it exists, and interpret the result.

45. Economics: Residential Electricity The monthly charge, in dollars, for x kilowatt hours of electricity used by a residential customer in a particular city is given by the function

$$C(x) = \begin{cases} 0.05x & \text{if } 0 \le x < 210 \\ 0.08x & \text{if } 210 \le x < 340 \\ 0.1x - 6.8 & \text{if } 340 \le x < 450 \\ 0.12x - 15.8 & \text{if } x \ge 450 \end{cases}$$

Find the following limits, if they exist, and interpret the results.

a. $\lim_{x \to 210} C(x)$ **b.** $\lim_{x \to 340} C(x)$ **c.** $\lim_{x \to 450} C(x)$

46. Business: Online Advertising Revenue Suppose that, for a particular social networking company, the annual revenue from rich media advertisements, in millions of dollars, for the years 2007 through 2012 can be approximated with the model $R(x) = -x^4 + 11x^3 - 39x^2 + 45x$, where x is the number of years from the beginning of 2007. Find each limit.

Source: eMarketer April 2008

a. $\lim_{x \to 2} P(x)$ **b.** $\lim_{x \to 0} P(x)$ **c.** $\lim_{x \to 3} P(x)$

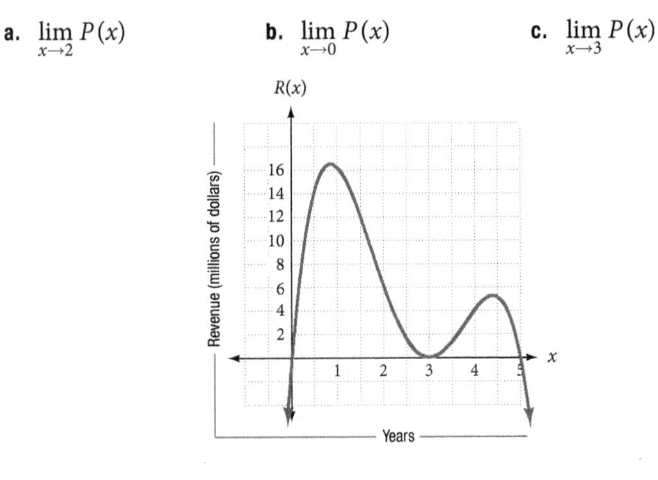

47. Business: Average Cost The average cost $\overline{C}$ (in dollars) per tube of toothpaste incurred by a company in producing x tubes is given by the average cost function

$$\overline{C}(x) = 0.89 + \frac{1{,}500}{x}$$

Find and interpret $\lim_{x \to \infty} \overline{C}(x)$.

48. Economics: Real Estate The Population P (in thousands) of a small valley t years from the beginning of 1990 is given by the function

$$P(t) = \frac{35t^2 + 105t + 140}{t^2 + 7t + 70}$$

a. What is the population of the valley in 1990?

b. What is the population of the valley at the beginning of the year 2000?

c. What is the expected population of the valley in the long term? (Hint: Find and interpret $\lim_{t \to \infty} P(t)$.)

49. Life Science: Fish Population The number N of fish in a pond is related to p, the level of PCB (in parts per million), in the pond by the function

$$N(p) = \frac{500}{1 + p}$$

If the level of PCB were allowed to increase without control, what would happen to the number of fish in the pond in the long run?

50. Life Science: Anatomy The focal length f (in millimeters) of the human eye is related to the distance d (in millimeters) from the lens of the eye to the object by the function

$$f(d) = \frac{25d}{25 + d}$$

What is the focal length of the eye for an object very far away (such as 2 miles)?

51. Business: Average Cost A producer and distributor of gift cards produces x boxes of cards per day at an average cost $\overline{C}(x)$ (in dollars) of

$$\overline{C}(x) = \frac{6x + 8{,}000}{2x + 300}$$

a. What is the average cost to the distributor if she produces 100 boxes?

b. To what number does the average cost per box approach as the number of boxes increases?

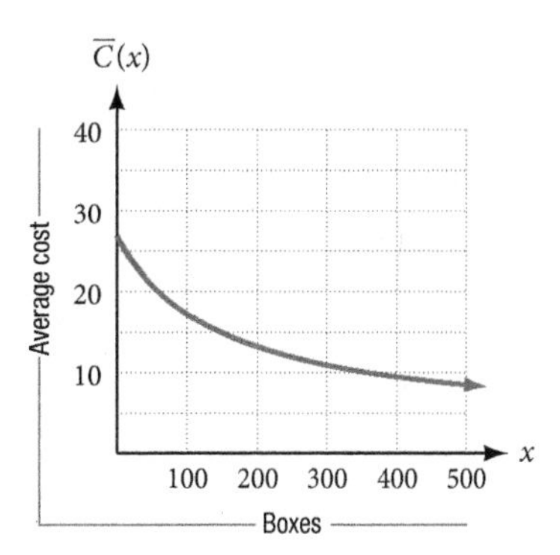

52. Manufacturing: Production Costs The average weekly cost $\overline{C}(n)$ (in dollars) of a computer company to produce n computers is given by the function

$$\overline{C}(n) = \frac{15{,}000 + 240n + 60n^{0.5}}{n}$$

What happens to the average cost in the long run, that is, as the number of computers increases without bound?

53. Medicine: Drug Concentration The concentration C (in milligrams per liter) of a drug in the human body is related to the time t in hours that the drug has been in the body by the function

$$C(t) = \frac{36t}{t^2 + 12}$$

What will be the concentration of the drug in the body in the long run?

54. Business: Pollution The amount $A(t)$ of pollution in a lake (in Formazin Turbidity Units) is related to time t in years by the function

$$A(t) = \frac{4.25t^{1/4} + 5}{5t^{1/4}}$$

What is the expected amount of pollution in the lake in the long run?

Source: U.S. Geological Survey: Science for a Changing World

© Shaun Lowe Photographic/ iStockPhoto

55. Physical Science: Drag Racing The following rational function gives the speed $V(t)$, in miles per hour, of a dragster at each second t during a quarter-mile race.

$$V(t) = \frac{340t}{t + 3} \qquad t \geq 0$$

The function has a natural restriction on the domain because t is the number of seconds since the beginning of the race, so t must be 0 or greater, and that is why we have the graph shown in the figure shown below on the left. The figure on the right shows the graph without the restriction on the domain. Find the value of both the vertical asymptote and the horizontal asymptote.

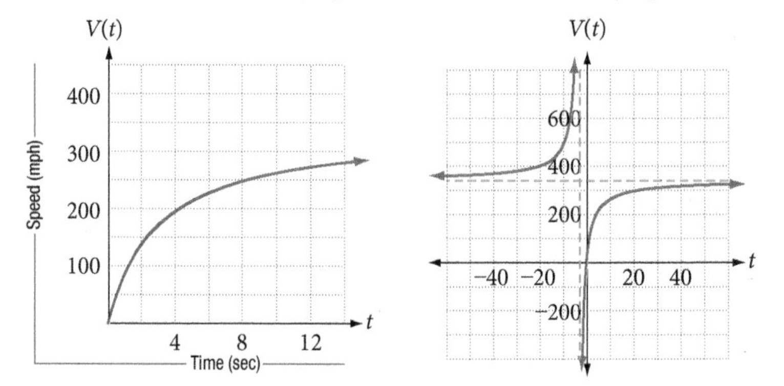

Getting Ready for the Next Section

56. For the function $f(x) = \frac{x^2 - x - 2}{x - 2}$, find the following, if they exist.

a. $f(2)$ **b.** $\lim_{x \to 2^-} f(x)$ **c.** $\lim_{x \to 2^+} f(x)$ **d.** $\lim_{x \to 2} f(x)$

57. For the function $g(x) = \begin{cases} \frac{x^2 - x - 2}{x - 2} & \text{if } x \neq 2 \\ 3 & \text{if } x = 2 \end{cases}$, find the following, if they exist.

a. $g(2)$ **b.** $\lim_{x \to 2^-} g(x)$ **c.** $\lim_{x \to 2^+} g(x)$ **d.** $\lim_{x \to 2} g(x)$

58. For the function $P(x) = \begin{cases} 20x - 100 & 0 \leq x \leq 150 \\ 20x - 250 & 150 < x \leq 350 \end{cases}$, find the following, if they exist.

a. $P(150)$ **b.** $\lim_{x \to 150^-} P(x)$ **c.** $\lim_{x \to 150^+} P(x)$ **d.** $\lim_{x \to 150} P(x)$

Spotlight on **Success**

Student Instructor Cynthia

Each time we face our fear, we gain strength, courage, and confidence in the doing.
—Unknown

I must admit, when it comes to math, it takes me longer to learn the material compared to other students. Because of that, I was afraid to ask questions, especially when it seemed like everyone else understood what was going on. Because I wasn't getting my questions answered, my quiz and exam scores were only getting worse. I realized that I was already paying a lot to go to college and that I couldn't afford to keep doing poorly on my exams. I learned how to overcome my fear of asking questions by studying the material before class, and working on extra problem sets until I was confident enough that at least I understood the main concepts. By preparing myself beforehand, I would often end up answering the question myself. Even when that wasn't the case, the professor knew that I tried to answer the question on my own. If you want to be successful, but you are afraid to ask a question, try putting in a little extra time working on problems before you ask your instructor for help. I think you will find, like I did, that it's not as bad as you imagined it, and you will have overcome an obstacle that was in the way of your success.

Functions and Continuity

1.5

© smithcjb/iStockPhoto

Recall the problem we have worked with previously concerning the fundraising dance put on by the sorority. The only expenses will be renting the facility and paying the security guard. All the other work will be done by the sorority members, and all the refreshments have been donated. Tickets will be sold ahead of time for \$20 each. The facility rental is \$100 for the evening, and the security guard is necessary only if more than 150 people attend the dance. The security guard is \$150 for the night. The hall can hold a maximum of 350 people. Here is the function that gives the profit the sorority can expect based on the number of tickets sold:

$$P(x) = \begin{cases} 20x - 100 & 0 \le x \le 150 \\ 20x - 250 & 150 < x \le 350 \end{cases}$$

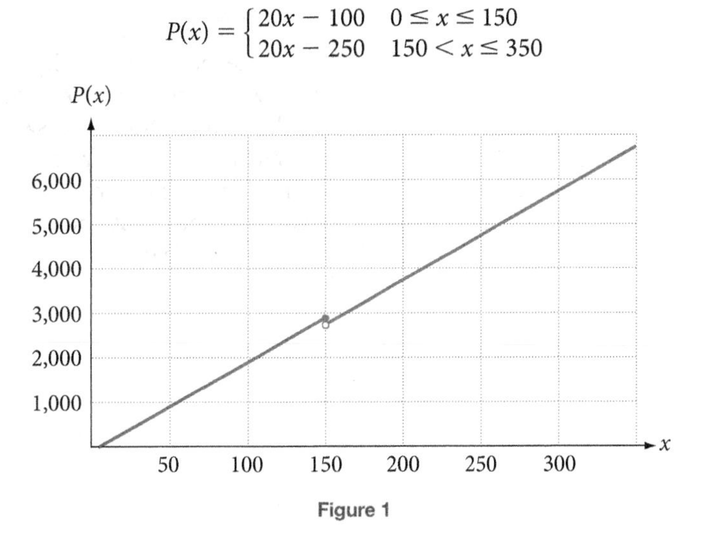

Figure 1

As you can see, there is a break in the graph when 150 people attend the dance. This corresponds to the point at which they need to hire a security guard. In mathematics, we say the function is discontinuous at this point. In this section we examine these types of situations in order to determine when a function is always continuous, and when it is not.

Continuous and Discontinuous Functions

Continuous functions have an essential role in calculus. Nearly all of the rules for the calculus operations assume that the function to be operated on is continuous. In this section we examine the conditions necessary for a function to be continuous at a point and then how to construct an argument that proves or disproves continuity.

Intuitively, a continuous function is a function whose graph runs continuously from one point to another point without any breaks or jumps. That is, the curve is completely unbroken and can be drawn, from start to finish, without lifting the pencil from the paper. Some functions, however, are not continuous. If a hole or a jump appears at some point $x = a$, the function is discontinuous at that point. Figure 2a shows a continuous function, and Figure 2b shows a discontinuous function (with two points of discontinuity).

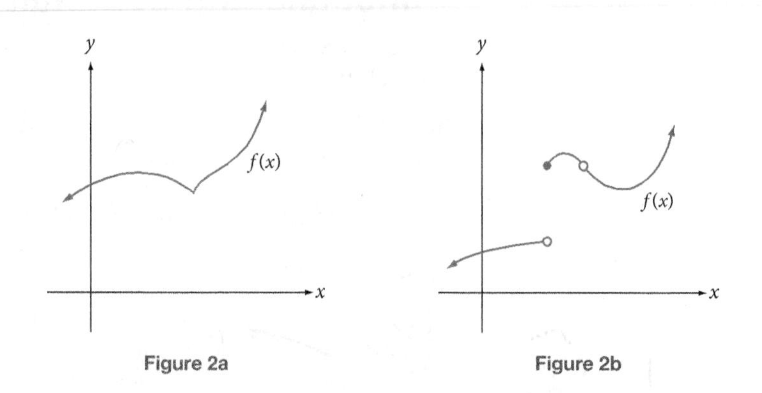

Figure 2a

Figure 2b

By observing the behavior of a function that is continuous at a point $x = a$, we may describe the conditions that are needed for a function to be continuous at a point $x = a$. Consider the continuous function pictured in Figure 3. This function is continuous at the point $x = a$, and it meets the following conditions:

1. A point exists at $x = a$. That is, $f(a)$ is defined. In fact, $f(a) = L$.
2. The branches of the curve come nearer to each other as the x-values get nearer to $x = a$. That is, $\lim_{x \to a^-} f(x) = \lim_{x \to a^+} f(x)$ implies that $\lim_{x \to a} f(x)$ exists. In fact, $\lim_{x \to a} f(x) = L$.
3. Not only is there a point at $x = a$, and not only do the branches of the curve come near each other near $x = a$, but the branches of the curve join together at $x = a$. That is, $\lim_{x \to a} f(x) = f(a)$, $(L = L)$.

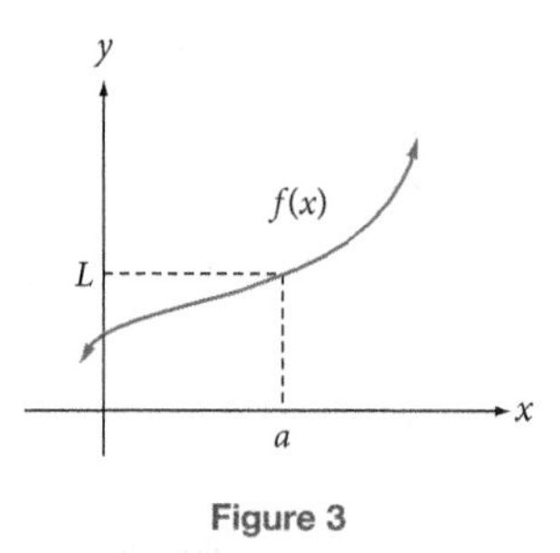

Figure 3

All of these conditions must be satisfied for a function to be continuous at a point $x = a$. We now state them symbolically.

Continuity Conditions

If $f(x)$ is a function, then $f(x)$ is continuous at point $x = a$ if all three of the following conditions are satisfied.

1. $f(a)$ is defined.
2. $\lim_{x \to a} f(x)$ exists.
3. $\lim_{x \to a} f(x) = f(a)$.

Condition 1 means that a point exists at $x = a$.
Condition 2 means that the branches of the curve come near each other at $x = a$.
Condition 3 means that not only is there a point at $x = a$, and not only do the branches of the curve come near each other near $x = a$, but that the branches of the curve actually join together at $x = a$.

To prove that a function is continuous at a point $x = a$, it is necessary to show that all three continuity conditions are satisfied. If one of the conditions fails at $x = a$, the function is discontinuous at that point, and no further investigation is necessary. The following examples illustrate the procedure for determining whether a function is continuous at a point $x = a$. Each example uses the phrase "discuss the continuity of the function at $x = a$." This phrase directs us to specify whether or not $f(x)$ is continuous at $x = a$, and to substantiate our conclusion using the Continuity Conditions.

VIDEO EXAMPLES

SECTION 1.5

Example 1 Discuss the continuity of the function at the point where $x = 2$.

a. $f(x) = \dfrac{x^2 - x - 2}{x - 2}$

b. $g(x) = \begin{cases} \dfrac{x^2 - x - 2}{x - 2} & \text{if } x \neq 2 \\ 1 & \text{if } x = 2 \end{cases}$

c. $h(x) = \begin{cases} \dfrac{x^2 - x - 2}{x - 2} & \text{if } x \neq 2 \\ 3 & \text{if } x = 2 \end{cases}$

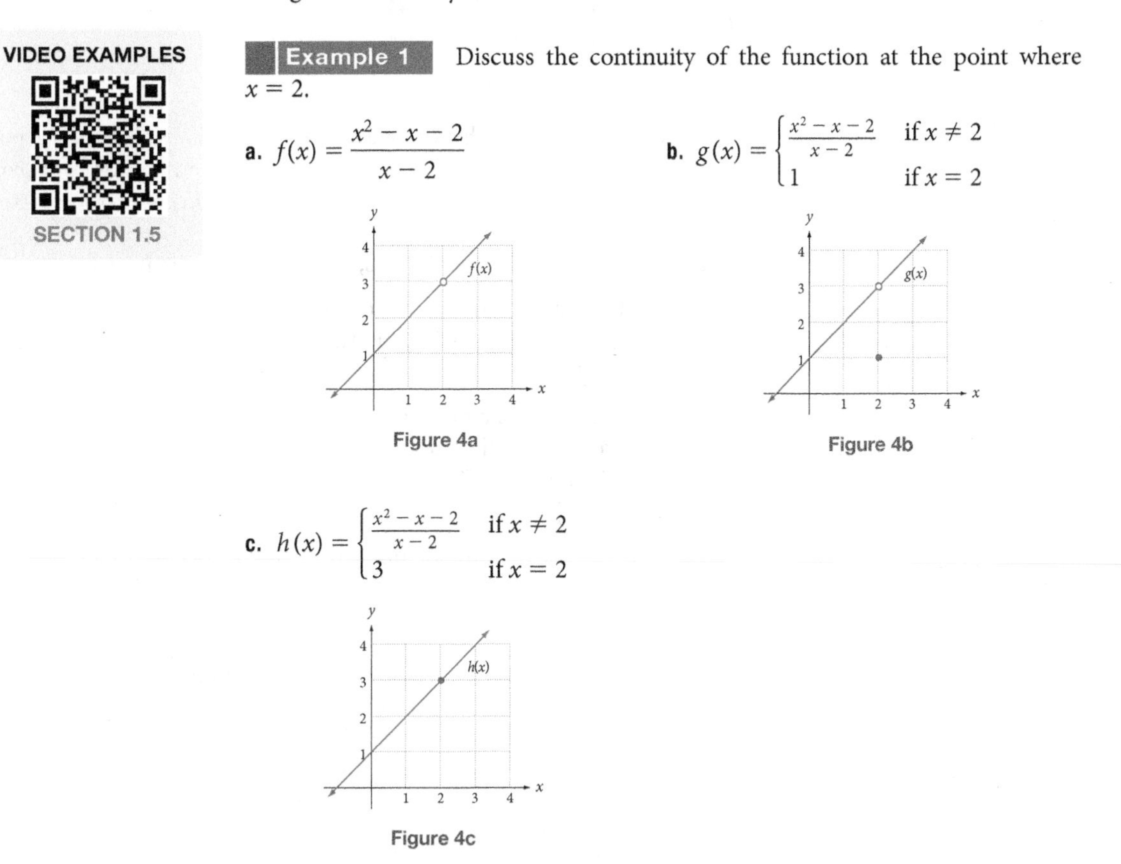

Figure 4a

Figure 4b

Figure 4c

Solution The only one of the three functions that is continuous at $x = 2$ is $h(x)$, the last one. Here are the details.

a. $f(x)$ is not continuous at $x = 2$ because $f(2)$ does not exist. (Continuity Condition 1 is not satisfied.)

b. $g(x)$ is not continuous at $x = 2$ because $g(2) = 1$ which is different from $\lim\limits_{x \to 2} g(x)$.

$$\lim_{x \to 2} g(x) = 3 \neq g(2) = 1$$

(Continuity Condition 3 is not satisfied.)

c. $h(x)$ is continuous at $x = 2$ because $h(2)$ exists and

$$\lim_{x \to 2} h(x) = 3 = h(2)$$

(All 3 Continuity Conditions are satisfied.) ■

Example 2 Discuss the continuity of the function at $x = 4$.

a. $f(x) = \begin{cases} \frac{1}{2}x - 3 & \text{for } x \le 4 \\ -x + 3 & \text{for } x > 4 \end{cases}$

b. $g(x) = \begin{cases} -\frac{1}{2}x + 4 & \text{for } x \le 4 \\ 3x - 9 & \text{for } x > 4 \end{cases}$

c. $h(x) = \begin{cases} 2x - 1 & \text{for } x < 4 \\ \frac{1}{4}x + 5 & \text{for } x > 4 \end{cases}$

Solution

a. $f(4)$ exists, so we proceed to Condition 2.

$$\lim_{x \to 4^-} f(x) = \frac{1}{2}(4) - 3 = -1 \text{ and } \lim_{x \to 4^+} f(x) = -(4) + 3 = -1$$

Since the limits from the left and right are equal, we have $\lim_{x \to 4} f(x) = -1$. Finally, $f(4) = -1$ and $\lim_{x \to 4} f(x) = -1$, so $f(x)$ is continuous at $x = 4$.

b. $g(4)$ exists, so we proceed to Condition 2.

$$\lim_{x \to 4^-} g(x) = -\frac{1}{2}(4) + 4 = 2 \text{ and } \lim_{x \to 4^+} g(x) = 3(4) - 9 = 3$$

Since the limits from the left and right are not equal, $\lim_{x \to 4} g(x)$ does not exist. Therefore, $g(x)$ is not continuous at $x = 4$.

c. Looking closely at the definitions for the piecewise function $h(x)$, we see that $h(4)$ does not exist. Neither "$x < 4$" nor "$x > 4$" includes the value of 4, so the function is not defined there, and is therefore not continuous at that point. ■

Our next example involves the piecewise function we used in the introduction to this section to model the profit for the fundraising dance put on by the college sorority. The interesting point on the graph is at $x = 150$. As you will see, we are unable to find a limit of $P(x)$, as x approaches 150, because as we move up to 150 from the left we approach one number, and as we move down to 150 from the right we get a different number.

Example 3 Determine the continuity of the function at each of the given points.

© smithcjb/iStockPhoto

$$P(x) = \begin{cases} 20x - 100 & 0 \le x \le 150 \\ 20x - 250 & 150 < x \le 350 \end{cases}$$

a. $x = 145$ **b.** $x = 150$ **c.** $x = 155$

Solution Let's look at the close-up graph of our function that we used in Sections 1.3 and 1.4. As you can see from the graph, two of the limits can be found from the graph, and one of the limits does not exist.

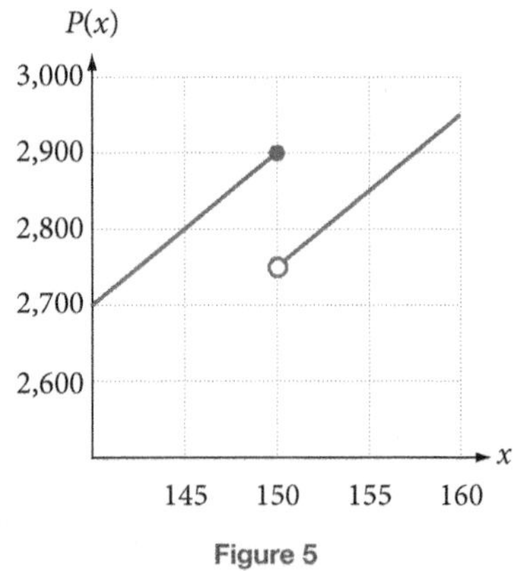

Figure 5

a. The function is continuous at $x = 145$ because $P(145) = 2{,}800 = \lim_{x \to 145} P(x)$.

b. The function is not continuous at $x = 150$ because $\lim_{x \to 150} P(x)$ does not exist.

c. The function is continuous at $x = 155$ because $P(155) = 2{,}850 = \lim_{x \to 155} P(x)$. ■

Example 4 Some people, when placed in extremely stressful situations, exhibit explosive bursts of anger. Such behavior might be described by the function pictured in Figure 6.

The horizontal axis records the amount of stress and the vertical axis, the amount of anger. Make a statement about the continuity of the function $A(s)$ at $s = c$ and give a possible interpretation to the point $s = c$.

Solution This function is discontinuous at $s = c$ since Continuity Condition 2 fails ($\lim_{s \to c} A(s)$ does not exist). However, $\lim_{s \to c-} A(s)$ does exist, and in fact, $\lim_{s \to c-} A(s) = a$. We might say that as this person's stress level approaches c, that is, as $s \to c^-$, he is losing his patience. Then, as the stress level gets very close to c, he is *reaching his limit* of patience. Finally, at $s = c$, he loses his patience completely and exhibits explosive behavior, and his anger level hits 1.

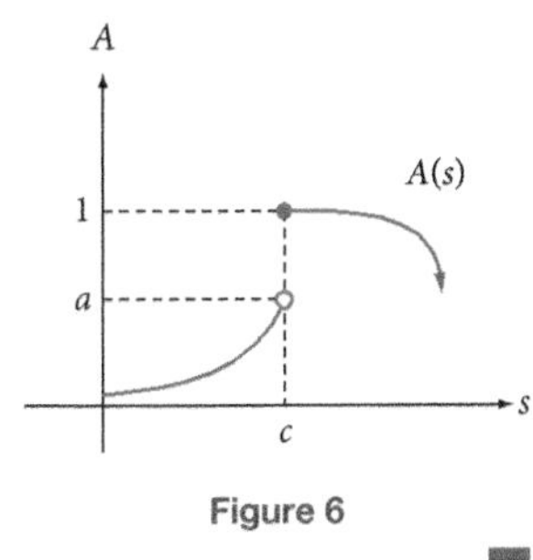

Figure 6

■

> *Note* *Interval Notation:* Recall that the "open interval (a, b)" is simply another way of indicating "$a < x < b$." Similarly, the "closed interval $[a, b]$" is equivalent to "$a \le x \le b$."

Continuity of Polynomials and Rational Functions

1. Every polynomial function is continuous on every open interval (a, b) and on every closed interval $[a, b]$.
2. Every rational function is continuous on every open interval (a, b) and on every closed interval $[a, b]$ except for the points in the interval that make the denominator 0.

These important continuity properties tell us that polynomials will always graph as unbroken curves, and that rational expressions will be broken only at the c values for which the denominator is 0.

Example 5 Discuss the continuity of each function over the real numbers. Use interval notation to show where each is continuous.

a. $f(x) = x^3 + 3x - 1$

b. $g(x) = \dfrac{x + 2}{x^2 - 7x + 12}$

Solution

a. Since $f(x)$ is a polynomial function, it is continuous for all x. It is continuous on the interval $(-\infty, \infty)$.

b. Since $g(x)$ is a rational function, it is continuous for all real numbers x except those that make the denominator 0. Since the denominator factors as $(x - 3)(x - 4)$, we know that $x = 3$ and $x = 4$ are points of discontinuity. The function $g(x)$ is continuous on $(-\infty, 3) \cup (3, 4) \cup (4, \infty)$. ■

Continuity and Idealized Functions

Often in business, in the social sciences, and in the life sciences, the domain of the function of interest consists of integers in some interval rather than all real numbers in that interval. So that problems can be analyzed, it is necessary to *idealize* functions so that they are continuous and, therefore, useful. The domain of such idealized functions is then all real numbers in some interval rather than just the integers in that interval.

For example, a motel might describe the relationship between its occupancy rate and the week of the year by the continuous function pictured in Figure 7a. The function is idealized as continuous. In reality, it is not continuous. The actual function may look more like the function pictured in Figure 7b. Calculus may be applied to the idealized continuous function but not the actual discontinuous function.

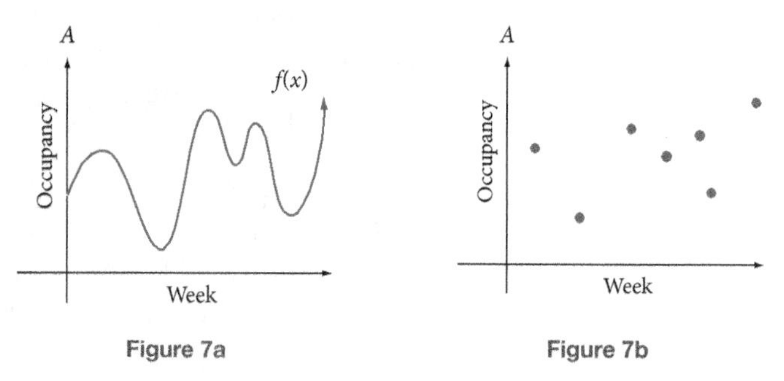

Figure 7a **Figure 7b**

On the other hand, our model for the speed of a dragster is actually very accurate. The actual function would have to be continuous because the dragster has a speed at any instant during the race. And the speed increases smoothly, without any bumps, in a normal race. In this case, the idealized model and the actual model are very close to each other.

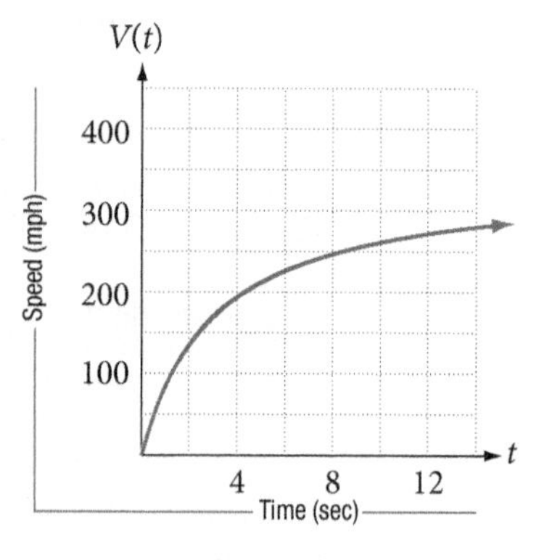

Figure 8

Example 6 The graph below was used previously to model the progress of a person on a diet. The graph is an idealized model of weight loss. Discuss how the actual graph of this diet could differ from the idealized graph.

© esolla/iStockPhoto

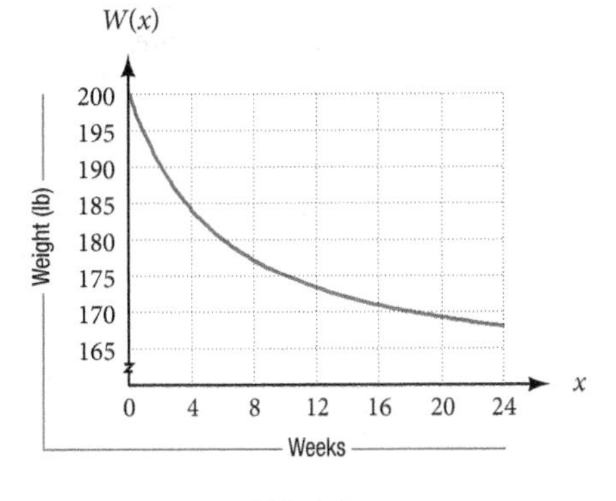

Figure 9

Solution The idealized model is not too far from the actual graph. We know that the actual model is a continuous function because at any time, the person on the diet has an actual weight. Also, we know that people lose more weight at the beginning of a diet than they do later on, so the fact that our graph drops more quickly in the beginning than it does at the end is realistic. Finally, real-life weight loss generally includes some fluctuation. We would not expect 20-pound losses or gains in a day or even a week, but we would expect some bumps in the curve. Therefore, the actual graph could look something like this:

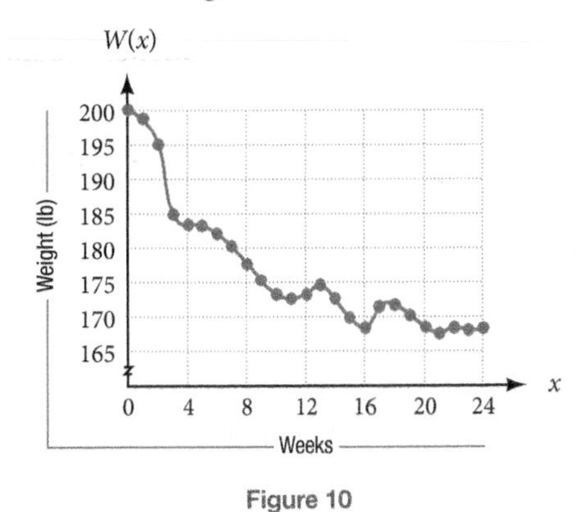

Figure 10

Using Technology 1.5

You can use your graphing calculator to examine the continuity of a function.

Example 7 Graph $f(x) = \dfrac{x^2 - 7x + 12}{x - 3}$ using these windows:

a. $-10 \leq x \leq 10$ and $-3 \leq y \leq 3$

b. $-9.4 \leq x \leq 9.4$ and $-3 \leq y \leq 3$

Solution

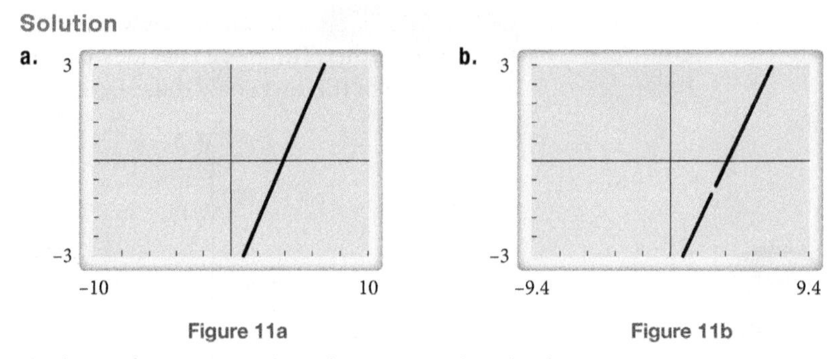

Figure 11a **Figure 11b**

We know from our work in this section that the function

$$f(x) = \frac{x^2 - 7x + 12}{x - 3}$$

is discontinuous at $x = 3$ because Continuity Condition 1 fails; that is, $f(3)$ is undefined. However, Figure 11a appears continuous at $x = 3$. We are able to see the hole in Figure 11b because we chose a window based on the screen width of 94 pixels. Can you see why this helps? ■

Example 8 Graph $g(x) = \dfrac{x + 3}{x - 2}$.

Solution Compare the graphs below. The one on the left was constructed using a TI-83 calculator, while the one on the right was constructed using the TI-84 Plus model.

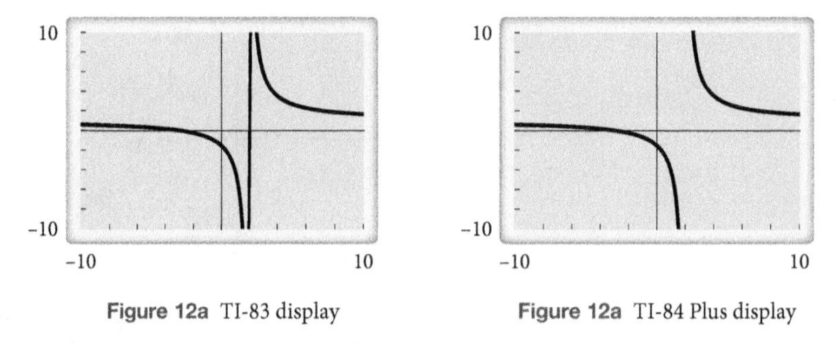

Figure 12a TI-83 display **Figure 12a** TI-84 Plus display

We know from our work in this section that the function $g(x) = \frac{x+3}{x-2}$ is discontinuous at $x = 2$. In fact, there is a vertical asymptote at $x = 2$. The TI-84 Plus accurately displays this discontinuity, but the TI-83 incorrectly connects the two branches. (It may look like an asymptote, but it is simply the calculator trying to connect the curve when it is in Connected mode.) You can fix this problem by placing your calculator in Dot mode.

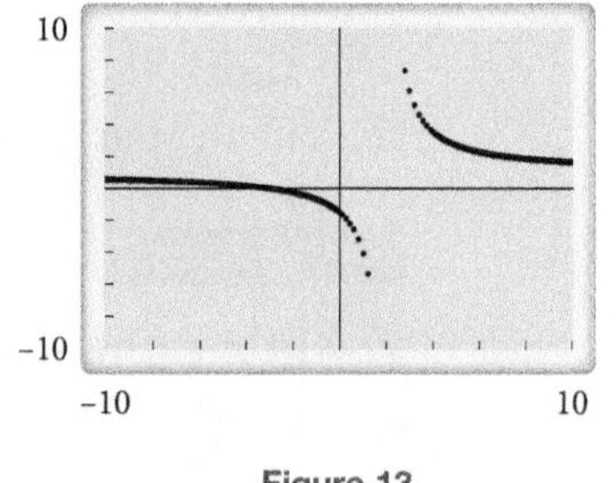

Figure 13
TI-83 Dot mode display

■

Your calculator can construct graphs of discontinuous functions that are defined piecewise. The following example illustrates this.

Example 9 Graph $h(x) = \begin{cases} x^2 - 2x - 2 & \text{if } x \leq 3 \\ \frac{1}{3}x - \frac{7}{3} & \text{if } x > 3 \end{cases}$.

Solution Enter the function into your Y-variables list as follows:

$$Y1=(X^{\wedge}2-2X-2)(X\leq3)+(X/3-7/3)(X>3)$$

Notice the different results below.

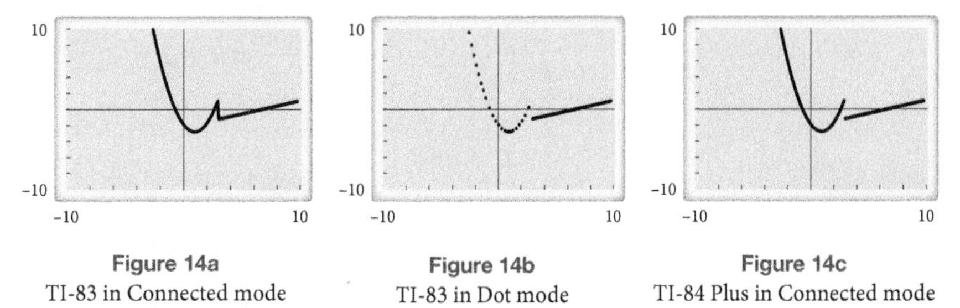

Figure 14a TI-83 in Connected mode

Figure 14b TI-83 in Dot mode

Figure 14c TI-84 Plus in Connected mode

Getting Ready for Class

After reading through the preceding section, respond in your own words and in complete sentences.

A. What is the difference between continuous and discontinuous functions?

B. Explain why, for a function to be continuous at $x = a$, it is not enough for $f(a)$ and $\lim_{x\to a} f(x)$ to exist, but it also must be true that $\lim_{x\to a} f(x) = f(a)$.

C. When discussing the continuity of $f(x)$, where

$$f(x) = \begin{cases} x^2 - 2x - 2 & \text{if } x \leq 3 \\ \frac{1}{3}x - \frac{7}{3} & \text{if } x > 3 \end{cases}$$

it is necessary to consider both one-sided limits. However, when discussing the continuity of $g(x)$, where

$$g(x) = \begin{cases} \frac{x^2 - 7x + 12}{x - 3} & \text{if } x \neq 3 \\ -3 & \text{if } x = 3 \end{cases}$$

it is not necessary to analyze the one-sided limits. Explain the difference between these two cases.

Problem Set 1.5

Skills Practice

For Problems 1-14, discuss the continuity of each function at the specified value of x. Be sure to justify your results, as we did in Examples 1 and 2.

P. 92

1. $f(x) = 3x^2 - 7x + 4$ at $x = -1$

2. $f(x) = -5x^3 + x^2 + 2$ at $x = 2$

3. $f(x) = \dfrac{x^2 - 5x + 6}{x - 2}$ at $x = 3$

4. $f(x) = \dfrac{x^2 - 9x + 8}{x - 1}$ at $x = -2$

5. $f(x) = \dfrac{x^2 - 5x + 6}{x - 2}$ at $x = 2$

6. $f(x) = \dfrac{x^2 - 9x + 8}{x - 1}$ at $x = 1$

7. $f(x) = \begin{cases} 3x + 1 & \text{if } x < 1 \\ x^2 - 2x + 5 & \text{if } x > 1 \end{cases}$ at $x = 1$

8. $f(x) = \begin{cases} x^2 - 4 & \text{if } x \leq 0 \\ \frac{3}{4}x^2 - 3 & \text{if } x > 0 \end{cases}$ at $x = 1$

9. $f(x) = \begin{cases} |x - 3| & \text{if } x < 3 \\ -x^2 + 6x - 8 & \text{if } x \geq 3 \end{cases}$ at $x = 0$

10. $f(x) = \begin{cases} -x^2 - 2 & \text{if } x < -1 \\ 3 & \text{if } x = -1 \\ x^2 + 2x + 3 & \text{if } x > -1 \end{cases}$ at $x = -1$

11. $f(x) = \begin{cases} x^2 + 8x + 13 & \text{if } x < -4 \\ 0 & \text{if } x = -4 \\ -3 & \text{if } x > -4 \end{cases}$ at $x = -4$

12. $f(x) = \begin{cases} \frac{x^2 + 7x + 10}{x + 2} & \text{if } x < -2 \\ 3 & \text{if } x = -2 \\ -x + 1 & \text{if } x > -2 \end{cases}$ at $x = -2$

13. $f(x) = \begin{cases} \frac{2x^2 - 5x - 3}{x - 3} & \text{if } x \neq 3 \\ 7 & \text{if } x = 3 \end{cases}$ at $x = 3$

14. $f(x) = \begin{cases} \frac{x^2 - 4x - 12}{x - 6} & \text{if } x \neq 6 \\ 3 & \text{if } x = 6 \end{cases}$ at $x = 6$

Check Your Understanding

For Problems 15-26, determine if each illustrated function is continuous or discontinuous at the point $x = a$. If the function is continuous, so state. If the function is discontinuous, specify which of the three continuity conditions is the first to fail.

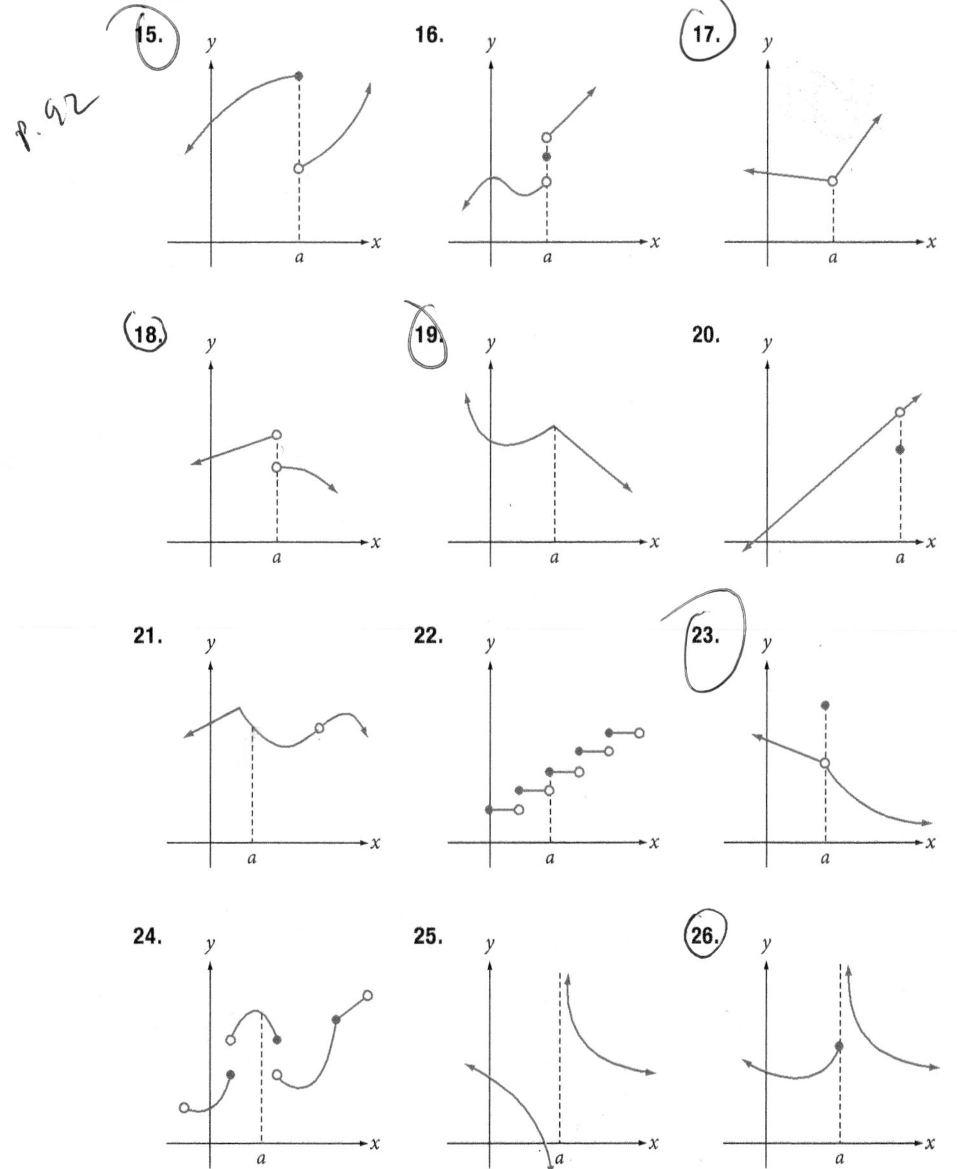

Modeling Practice

27. **Psychology: Learning Curve** Psychologists have established that many types of learning follow a well-established "learning curve." Such a curve is pictured below. The horizontal axis records the amount of time spent studying and the vertical axis records the amount of understanding of a particular concept or technique.

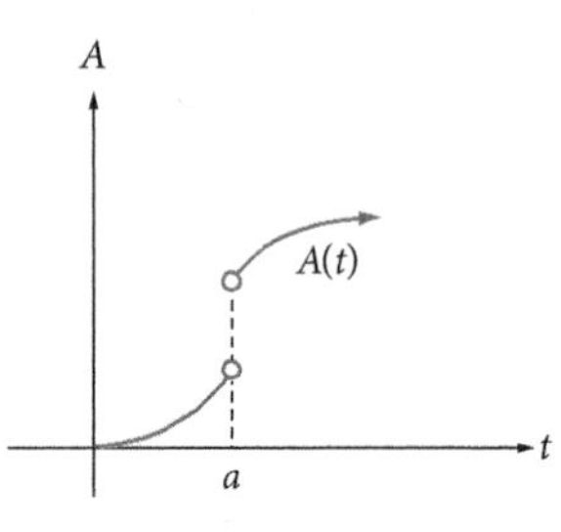

Make a statement about the continuity of the function $A(t)$ at $t = a$, and give a possible interpretation to the point $t = a$.

28. **Economics: Population Growth** When observations are made over long enough periods of time, the growth of many populations can be viewed as a smooth, continuous curve. However, some growth can be disrupted and produce a curve such as that pictured below. The horizontal axis records the amount of time the population has existed and the vertical axis records the number of members in the population.

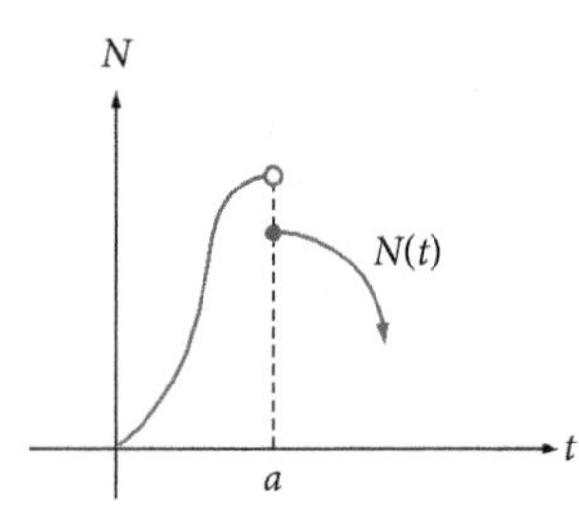

Make a statement about the continuity of the function $N(t)$ at $t = a$ and give a possible interpretation to the point $t = a$.

29. **Economics: Profit** When observations are made over long enough periods of time, the curve that represents the profits of a Wall Street financial brokerage house can be viewed as a smooth, continuous curve. However, certain conditions, market or otherwise, can produce a curve such as that pictured below.

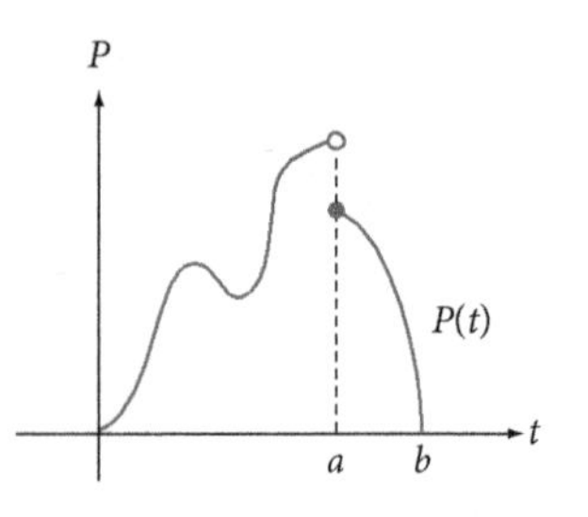

Make a statement about the continuity of the function $P(t)$ at $t = a$ and give possible interpretations to the points $t = a$ *and* $t = b$.

30. **Economics: Demand** When idealized, the function relating time (in months) and the demand for personal computers can be viewed as a smooth, continuous curve. However, certain market conditions can produce a curve such as that pictured below.

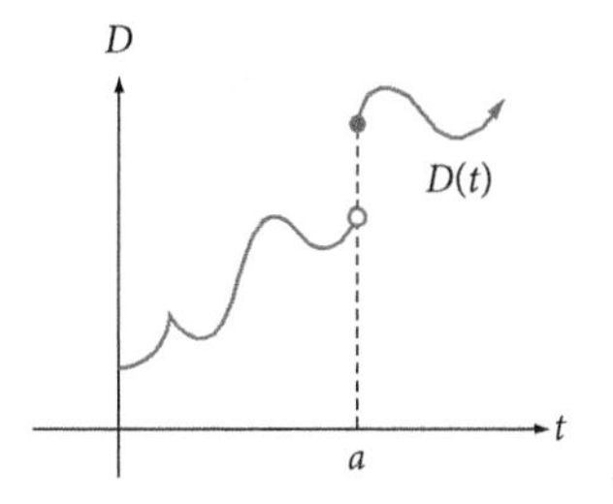

Make a statement about the continuity of the function $D(t)$ at $t = a$ and give a possible interpretation to the point $t = a$.

31. **Ecology: Oil Pollution** When idealized, the function relating time (in months) to the amount (in thousands of gallons) of crude oil polluting a particular stretch of a coastal wildlife preserve can be viewed as a smooth, continuous curve. However, certain phenomena can produce a curve such as that pictured in the following illustration.

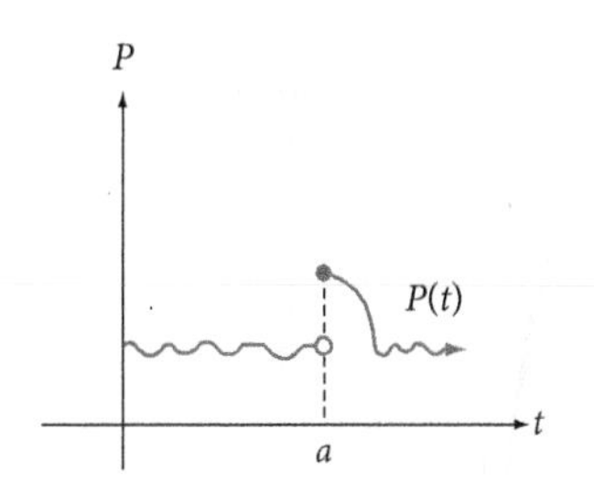

Make a statement about the continuity of the function $P(t)$ at $t = a$ and give a possible interpretation to the point $t = a$.

32. **Business: Parking Fees** When idealized, the function relating time (in semesters or quarters) and a college's or university's parking fee can be viewed as the broken curve pictured below.

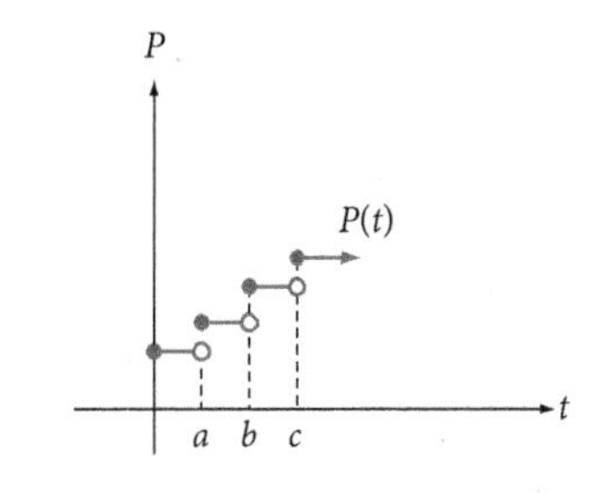

Give a possible interpretation for the discontinuities at the points $t = a$, $t = b$, and $t = c$.

33. Business: Revenue Growth XYZ Textbooks/MathTV is a small publisher of college textbooks (publishing quality material at reasonable prices). Their first year of operations was 2010. At the beginning of that year they had sold 0 books. By the end of the year they had sold 5,000 books. After that, books doubled in sales each year. The idealized function that describes the growth of XYZ Textbooks/MathTV during its first few years of operation is

$$S(t) = \begin{cases} 5{,}000t & 0 \le t < 1 \\ 5{,}000 \cdot 2^{t-1} & t \ge 1 \end{cases}$$

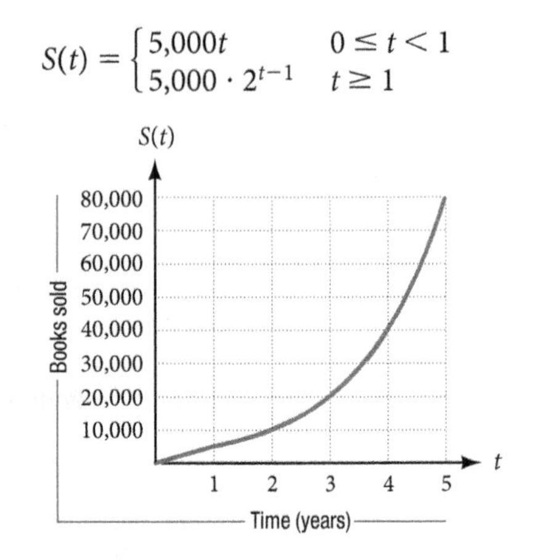

where S is the number of books sold, and t is time in years. The graph of the idealized function is shown above. Discuss how the actual graph could differ from the idealized graph.

© Lajos Repasi/iStockPhoto

34. Health Science: Flu Epidemic Health officials, concerned with a flu epidemic that has hit the north Atlantic coast, have estimated that the number of people ill with flu symptoms can be approximated by the function

$$N(t) = 90t^2 - t^3, \quad 0 \le t \le 70$$

where t is measured in days since the beginning of the epidemic. The graph of this function is shown below. The function and the graph are both idealized models of the situation. Discuss how the actual graph could differ from the idealized graph.

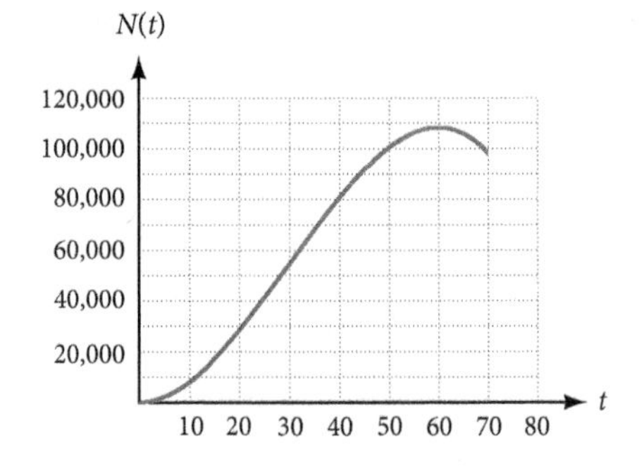

35. Business: Average Cost A producer and distributor of gift cards produces x boxes of cards per day at an average cost $\overline{C}(x)$ (in dollars) of

© Warchi/iStockPhoto

$$\overline{C}(x) = \frac{6x + 8,000}{2x + 300}$$

a. What is the domain for this function?

b. At what value of x is this function discontinuous? Why does the discontinuity not appear on the graph?

Using Technology Exercises

Use your calculator to solve the following problems. Keep in mind that your model may require some modification to get the graph you want.

36. Graph $f(x) = \frac{2x^2 - x - 3}{x + 1}$ and state any discontinuities. (Hint: Set up your graphing window so that $-9.4 \leq x \leq 9.4$.)

37. Graph $f(x) = \frac{x^2 + 9x + 20}{x + 4}$ and state any discontinuities.

38. Graph $f(x) = \frac{x - 5}{x + 3}$ and state any discontinuities.

39. Graph $f(x) = \frac{(x + 4)(x - 3)}{(x + 1)(x + 4)}$ and state any discontinuities.

40. Graph $f(x) = \begin{cases} x^2 - 3x + 4 & \text{if } x \leq -1 \\ 2x + 5 & \text{if } x > -1 \end{cases}$ and state any discontinuities.

41. Graph $f(x) = \begin{cases} x^2 + 2x - 3 & \text{if } x \leq 2 \\ 3x - 1 & \text{if } x > 2 \end{cases}$ and state any discontinuities.

Getting Ready for the Next Section

42. For the function $f(x) = 3\sqrt{x} + 1$, find $f(8) - f(2)$.

43. For the function $f(x) = 5x + 8$, find $f(x + h)$.

44. For the function $P(x) = -x^2 + 13x - 22$, find $\frac{P(6) - P(3)}{6 - 3}$.

45. For the function $f(x) = -x^2 + 13x - 22$, find $\frac{f(x + h) - f(x)}{h}$.

46. For $f(x) = 3x^2 + 7x - 2$, find

a. $f(1)$ **b.** $f(2)$ **c.** $\frac{f(x + h) - f(x)}{h}$

47. For the function $f(x) = -5x^2 + 200x$, find $\lim_{h \to 0} \frac{f(x + h) - f(x)}{h}$.

48. For the function $s(t) = 32t - 16t^2$, find $\lim_{h \to 0} \frac{s(t + h) - s(t)}{h}$.

Average and Instantaneous Rates of Change

A skydiver jumps from an airplane. For the first few seconds, the distance she falls is given by the function

$$f(t) = 16t^2$$

We can use the function to find a number of quantities associated with the fall.

© Hector Mandel/iStockPhoto

Distance If we want to find how far the skydiver has fallen in the first 3 seconds, we find $f(3)$.

$$\text{Distance after 3 seconds} = f(3) = 16 \cdot 3^2 = 144 \text{ ft}$$

Average Speed If we want her average speed from second 1 to second 3, we find the difference quotient.

$$\text{Average speed from 1 to 3 seconds} = \frac{f(3) - f(1)}{3 - 1} = 64 \text{ ft/sec}$$

If we want the average speed from second 1 to some time t after that, we find the difference quotient.

$$\text{Average speed from 1 to } t \text{ seconds} = \frac{f(t) - f(1)}{t - 1} = 16(t + 1) \text{ ft/sec}$$

Instantaneous Speed If we want the instantaneous speed when $t = 1$, we find the following limit.

$$\text{Instantaneous speed at 1 second} = \lim_{t \to 1} \frac{f(t) - f(1)}{t - 1} = 32 \text{ ft/sec}$$

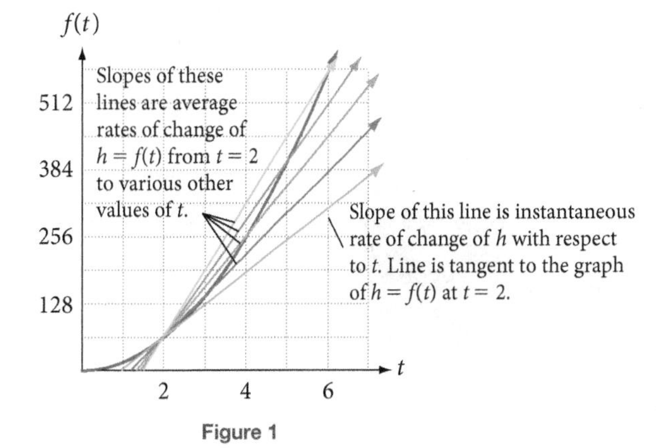

Figure 1

Finding instantaneous rates of change is what we will study in this section. To begin, let's review what we have done previously with average rate of change.

Note The skydiving example that we are using here is an idealized model of the situation because we are neglecting the resistance of air that impedes the fall of the skydiver. The faster the skydiver falls, the more this resistance increases, until the skydiver reaches a terminal velocity, usually from 125 mph to 200 mph, depending on the position and the altitude of the skydiver. For our example, our model works well. For times at $t = 15$ or more, our model would not be as accurate.

Note As this example is being written, skydiver Felix Baumgartner broke the record for terminal velocity, reaching a speed of 833.9 mph, when he jumped from an altitude of 128,100 feet.

Average Rate of Change of a Function

As we mentioned in Section 1.3, a reasonable measure of the rate at which the dependent quantity changes relative to the independent quantity is the **average rate of change**. The following box summarizes what we know so far.

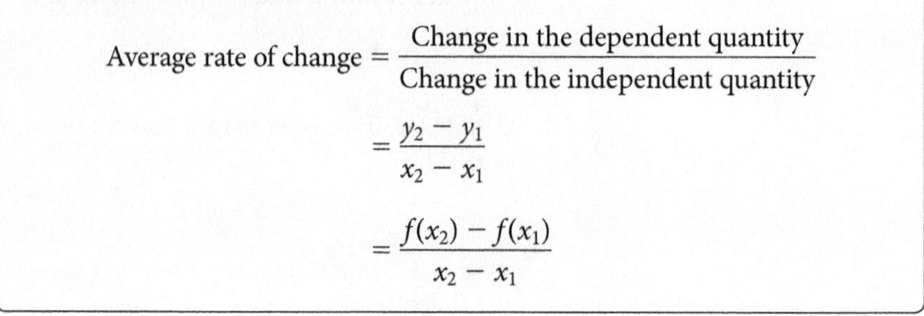

Average Rate of Change of a Function

$$\text{Average rate of change} = \frac{\text{Change in the dependent quantity}}{\text{Change in the independent quantity}}$$

$$= \frac{y_2 - y_1}{x_2 - x_1}$$

$$= \frac{f(x_2) - f(x_1)}{x_2 - x_1}$$

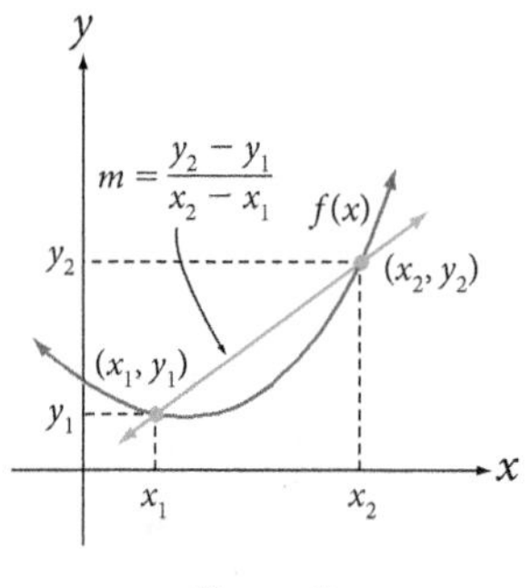

Figure 2

Recall that the average rate of change is the slope of the secant line. Figure 2 illustrates the fact that the slope of the line through the points (x_1, y_1) and (x_2, y_2) is the average rate of change of the function as the input values change from x_1 to x_2. In Section 1.3, we also introduced the term "difference quotient" to represent the average rate of change. We will explore this further in this section.

VIDEO EXAMPLES

SECTION 1.6

Example 1 For the function $f(x) = 3\sqrt{2x} + 1$, find the average rate of change with respect to x on the interval $[2, 8]$.

Solution The problem asks us to find the average rate of change of $f(x)$ as the value of x changes from 2 to 8.

$$\frac{f(x_2) - f(x_1)}{x_2 - x_1} = \frac{f(8) - f(2)}{8 - 2}$$

$$= \frac{13 - 7}{8 - 2}$$

$$= \frac{6}{6}$$

$$= 1$$

The slope of the secant line connecting the points $(2, 7)$ and $(8, 13)$ is 1. Over the interval $[2, 8]$, $f(x)$ changes an average of 1 unit for every 1 unit change in x. ■

Example 2 A college bookstore sells mechanical pencils for which it pays \$1 each. The manager of the bookstore feels that if the selling price is \$$x$ each, she will sell $12 - x$ pencils each week. The bookstore has a fixed cost of \$10. The function relating profit P and selling price x is

$$P(x) = -x^2 + 13x - 22 \quad 2 \leq x \leq 11$$

Find the average rate of change of the profit with respect to the selling price if the selling price changes from \$3 to \$6.

Solution

$$\begin{aligned}\text{Average rate of change} &= \frac{\text{Change in profit}}{\text{Change in price}}\\ &= \frac{P(6) - P(3)}{6 - 3}\\ &= \frac{20 - 8}{3}\\ &= \frac{12}{3}\\ &= 4 \quad \left(= \frac{4}{1}\ \frac{\text{Change in profit}}{\text{Change in price}}\right)\end{aligned}$$

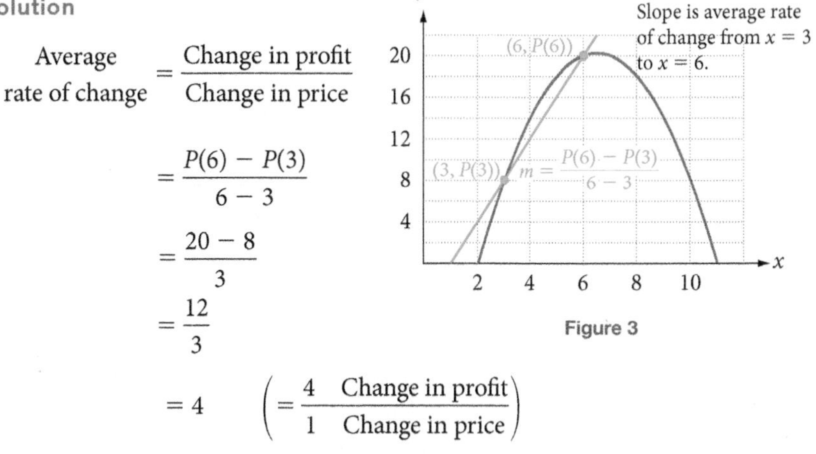

Figure 3

Interpretation
For selling prices between \$3 and \$6, each \$1 increase in the selling price increases the bookstore's weekly profit, on the average, by \$4.

Average Rate of Change (Version 2)

We can write the expression $\frac{f(x_2) - f(x_1)}{x_2 - x_1}$ for the average rate of change in a more useful way. Rather than depending on both the initial value x_1 and the final value x_2, we can express the final value in terms of the initial value. Suppose that x represents the initial value of the independent variable. If we change this value by h units, $x + h$ will represent the new value. If h is positive, then $x + h$ is the final value and represents an increase in the initial value. If h is negative, then $x + h$ is the final value and represents a decrease in the initial value. (If h is negative, $x + h$ is smaller than x.) We can now express the average rate of change formula in a more general and useful way.

$$\begin{aligned}\frac{\text{Final dependent value} - \text{Initial dependent value}}{\text{Final independent value} - \text{Initial independent value}} &= \frac{f(x + h) - f(x)}{(x + h) - x}\\ &= \frac{f(x + h) - f(x)}{h}\end{aligned}$$

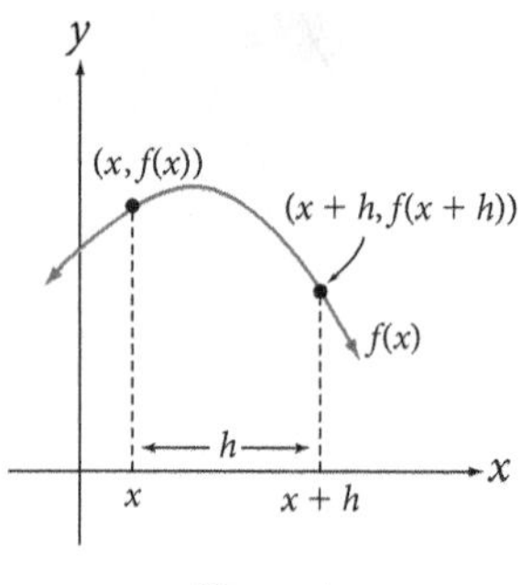

Figure 4

Figure 4 illustrates this notation graphically. We now have a new form for the difference quotient that is much more useful in our calculus studies.

Difference Quotient

$$\text{Average rate of change} = \frac{f(x+h)-f(x)}{h}$$

This form of the difference quotient is convenient because it allows us easy access to the amount of change in the independent variable. We can adjust the change in the independent variable, making it smaller or larger, simply by adjusting the value of h.

Example 3 For the function $f(x) = 5x + 8$, find the difference quotient

$$\frac{f(x+h)-f(x)}{h}$$

Solution

$$\begin{aligned}\frac{f(x+h)-f(x)}{h} &= \frac{5(x+h)+8-(5x+8)}{h}\\ &= \frac{5x+5h+8-5x-8}{h}\\ &= \frac{5h}{h}\\ &= \frac{5h}{h}, \text{ if } h \neq 0\\ &= 5 \qquad \left(= \frac{5}{1}\ \frac{\text{change in } f(x)}{\text{change in } x}\right)\end{aligned}$$

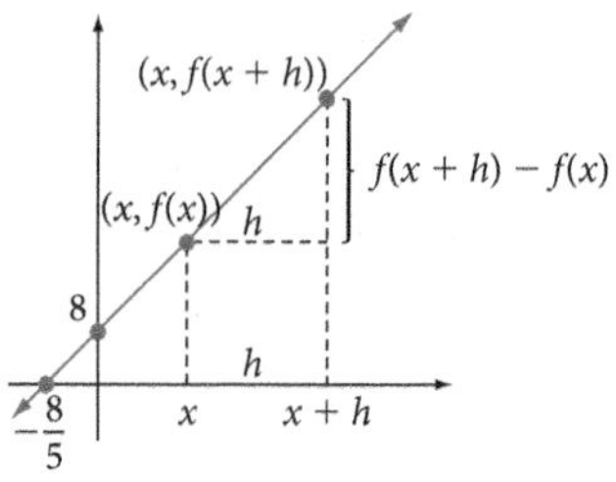

Figure 5

Note Whenever we have a linear function, such as the one in Example 3, the average rate of change will simply be the value of the slope.

Interpretation

The average rate of change of the function $f(x) = 5x + 8$ with respect to x is 5. Notice that this rate of change is constant; it does not depend on x. Regardless of the initial value of x, a 1-unit change in its value produces a 5-unit change in the value of the dependent variable.

Example 4 For the function $f(x) = -x^2 + 13x - 22$, find and simplify the difference quotient $\frac{f(x+h) - f(x)}{h}$. (Notice that this is the function we examined in Example 2.)

Solution

$$\frac{f(x+h) - f(x)}{h} = \frac{-(x+h)^2 + 13(x+h) - 22 - (-x^2 + 13x - 22)}{h}$$

$$= \frac{-(x^2 + 2hx + h^2) + 13x + 13h - 22 + x^2 - 13x + 22}{h}$$

$$= \frac{-x^2 - 2hx - h^2 + 13x + 13h - 22 + x^2 - 13x + 22}{h}$$

$$= \frac{-2hx - h^2 + 13h}{h}$$

$$= \frac{h(-2x - h + 13)}{h}$$

$$= \frac{h(-2x - h + 13)}{h} \quad \text{if } h \neq 0$$

$$= -2x - h + 13$$

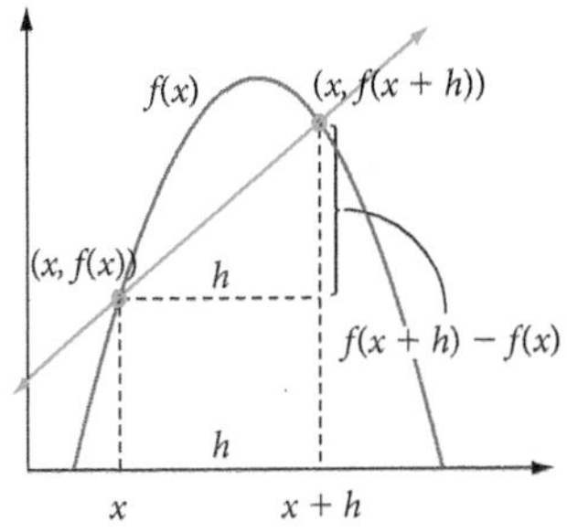

Figure 6

Interpretation

The average rate of change of $f(x) = -x^2 + 13x - 22$ with respect to x is given by $-2x - h + 13$. Unlike the rate of change of Example 3, this rate of change depends on both the initial value of x and the amount of change, h. For example, recalling that x represents the initial dollar selling price of the mechanical pencils and h represents the dollar change in the selling price, then when $x = 3$ and $h = 3$,

$$-2x - h + 13 = -2(3) - 3 + 13 = 4$$

This result agrees with the result we obtained in Example 2.

Instantaneous Rate of Change

We have seen that the average rate of change gives only an expected value for the rate at which the dependent variable, y, is changing with respect to changes in the independent variable, x. The average rate of change describes the change in y in the long term. To see how y is changing in the short term, or even at a particular instant, the change in x should be made small. The smaller the change in x, the more accurately the average rate of change describes the short-term change in y. The difference quotient

$$\lim_{h \to 0} \frac{f(x+h) - f(x)}{h}$$

allows us to easily adjust the change in x. Recall that h represents a change in x. By letting h get smaller and smaller, the average rate of change, as computed by the difference quotient

$$\lim_{h \to 0} \frac{f(x+h) - f(x)}{h}$$

better reflects the short-term change in y. If h is very close to 0, the average rate of change very closely approximates the immediate, or instantaneous, change in y. The idea of examining a function around values closer and closer to x reminds us of the limit process. The instantaneous rate of change of a function is found by finding the limit of the difference quotient of the function as h gets smaller and smaller, that is, as h approaches 0.

Instantaneous Rate of Change

Instantaneous rate of change of $f(x) = \lim_{h \to 0} \frac{f(x+h) - f(x)}{h}$ if this limit exists.

Notice that although there are two variables, x and h, in this limit expression, we are interested only in changes in h. Evaluation of this limit by direct substitution will always result in the indeterminate form $\frac{0}{0}$ (verify this by direct substitution). Therefore, when evaluating this limit, we will always proceed immediately with the algebraic manipulations we used to find limits in Section 1.4.

Example 5 Find the instantaneous rate of change of the function

$f(x) = 3x^2 + 7x - 2$ at

a. $x = 1$ **b.** $x = 2$ **c.** $x = -4$

Solution

Since the instantaneous rate of change $= \lim_{h \to 0} \frac{f(x+h) - f(x)}{h}$, we need to compute this limit, if it exists.

$$\lim_{h \to 0} \frac{f(x+h) - f(x)}{h} = \lim_{h \to 0} \frac{3(x+h)^2 + 7(x+h) - 2 - (3x^2 + 7x - 2)}{h}$$

$$= \lim_{h \to 0} \frac{3(x^2 + 2hx + h^2) + 7x + 7h - 2 - 3x^2 - 7x + 2}{h}$$

$$= \lim_{h \to 0} \frac{3x^2 + 6hx + 3h^2 + 7x + 7h - 2 - 3x^2 - 7x + 2}{h}$$

$$= \lim_{h \to 0} \frac{6hx + 3h^2 + 7h}{h}$$

$$= \lim_{h \to 0} \frac{h(6x + 3h + 7)}{h}$$

$$= \lim_{h \to 0} \frac{h(6x + 3h + 7)}{h} \qquad \text{Since } h \neq 0$$

$$= \lim_{h \to 0} (6x + 3h + 7) \qquad \text{Take the limit by substitution}$$

$$= 6x + 7$$

Thus, the instantaneous rate of change is $6x + 7$. Since the expression involves the variable x, the instantaneous rate of change is itself variable and depends on the initial value selected for x. For example,

a. When $x = 1$, the instantaneous rate of change of y is $6(1) + 7 = 6 + 7 = 13$.

b. When $x = 2$, the instantaneous rate of change of y is $6(2) + 7 = 12 + 7 = 19$.

c. When $x = -4$, the instantaneous rate of change of y is $6(-4) + 7 = -24 + 7 = -17$. ■

We are not quite ready to interpret these values. A short discussion of the geometry associated with the instantaneous rate of change will put us in the right position.

Interpreting the Instantaneous Rate of Change

We saw that, geometrically, the average rate of change is the slope of a secant line. The instantaneous rate of change has its geometric counterpart. By making h smaller and smaller in the difference quotient

$$\frac{f(x + h) - f(x)}{h}$$

we know we get more accurate information about the changes in y.

To find the instantaneous rate of change, we let h approach 0. We want to know what happens to the slope of the secant line as h approaches 0.

As h approaches 0, the point Q moves along the curve toward point P, and the secant line has changing slope. As h gets very close to 0 (or as Q gets very close to P), the changes in the secant lines' slopes are very small. In fact, as h approaches 0, the secant line approaches a fixed line that passes through point P, and the slope of the secant line approaches the slope of this fixed line. Figure 7, along with the sequential graphs in Figure 8, illustrates this fact.

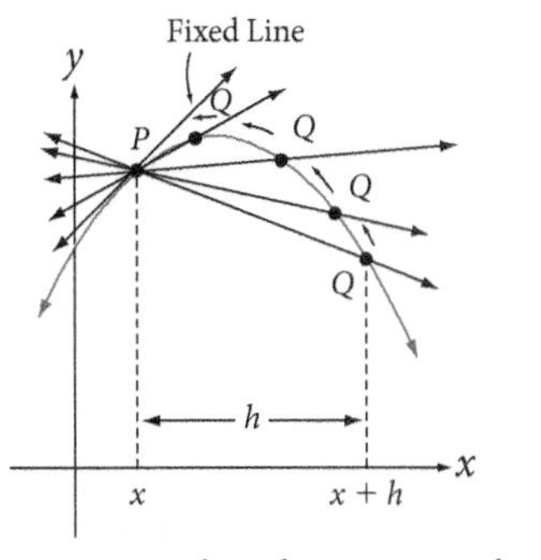

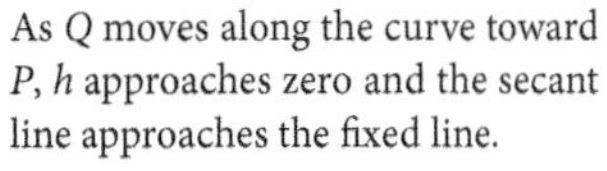
As Q moves along the curve toward P, h approaches zero and the secant line approaches the fixed line.

Figure 7

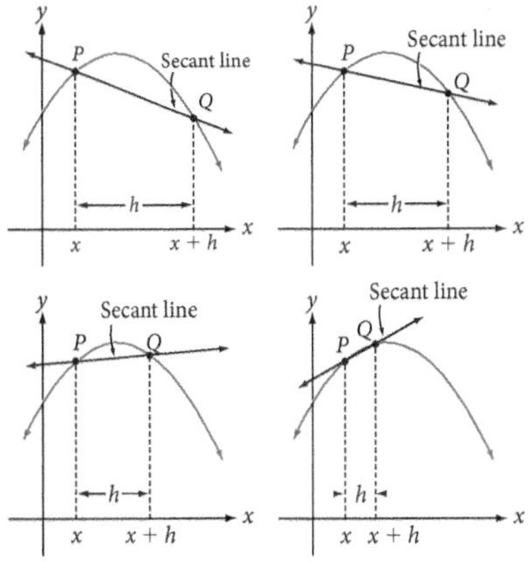

Figure 8

The fixed line passing through the point P in Figure 7 is called the **tangent line** to the curve at the point P. The slope of this tangent line is the limit, as h approaches 0, of the slope of the secant line.

We now know that the slope of the tangent line of a function is the instantaneous rate of change of the function.

The Slope of a Tangent Line

For a function $f(x)$, the instantaneous rate of change of $f(x)$ at x is

$$\lim_{h \to 0} \frac{f(x+h) - f(x)}{h}$$

That is, the instantaneous rate of change of a function is the slope of the tangent line to the graph of the function at $(x, f(x))$.

Example 6 A magazine publisher has determined that the cost C (in dollars) associated with x number of half-page articles is given by the function

$$C(x) = -5x^2 + 200x \quad 0 \le x \le 20$$

a. If the magazine is currently running 12 half-page articles, what is the expected increase in costs if it were to run 13?

b. If the magazine is currently running 15 half-page articles, what is the expected increase in cost if it were to run 16?

Solution

a. Since we are interested in the change in cost with respect to a 1-unit (very small) change in the number of half-page articles, the instantaneous rate of change is an appropriate approximation.

$$\lim_{h \to 0} \frac{f(x+h) - f(x)}{h} = \lim_{h \to 0} \frac{-5(x+h)^2 + 200(x+h) - (-5x^2 + 200x)}{h}$$

$$= \lim_{h \to 0} \frac{-5x^2 - 10hx - 5h^2 + 200x + 200h + 5x^2 - 200x}{h}$$

$$= \lim_{h \to 0} \frac{-10hx - 5h^2 + 200h}{h}$$

$$= \lim_{h \to 0} \frac{h(-10x - 5h + 200)}{h}$$

$$= \lim_{h \to 0} \frac{\not{h}(-10x - 5h + 200)}{\not{h}} \qquad \text{If } h \neq 0$$

$$= \lim_{h \to 0} (-10x - 5h + 200) \qquad \text{Take the limit by substitution}$$

$$= -10x + 200$$

Thus, the instantaneous rate of change $= -10x + 200$. It involves a variable and therefore will produce an approximation as x increases by 1 unit. If $x = 12$,

$$\begin{aligned} -10x + 200 &= -10(12) + 200 \\ &= -120 + 200 \\ &= 80 \end{aligned}$$

Interpretation
If the number of half-page articles is increased by 1, from 12 to 13, the cost will increase by approximately \$80. Note that we can calculate the exact increase directly by evaluating $C(13) - C(12) = \$75$.

b. If $x = 15$,

$$\begin{aligned} -10x + 200 &= -10(15) + 200 \\ &= -150 + 200 \\ &= 50 \end{aligned}$$

Interpretation
If the number of half-page articles is increased by 1, from 15 to 16, the cost will increase by approximately \$50.

Comparing parts a and b, it appears that the rate of cost increase decreases as the number of half-page articles increases. ■

Example 7 Kendra tosses a softball into the air with an underhand motion. The distance of the ball above her hand is given by the function in symbolic form as

$$s(t) = 32t - 16t^2 \quad \text{for} \quad 0 \le t \le 2$$

where s is the height of the ball in feet and t is the time in seconds. Find the instantaneous rate of change of the height at $t = 1/4$ second and $t = 1$ second.

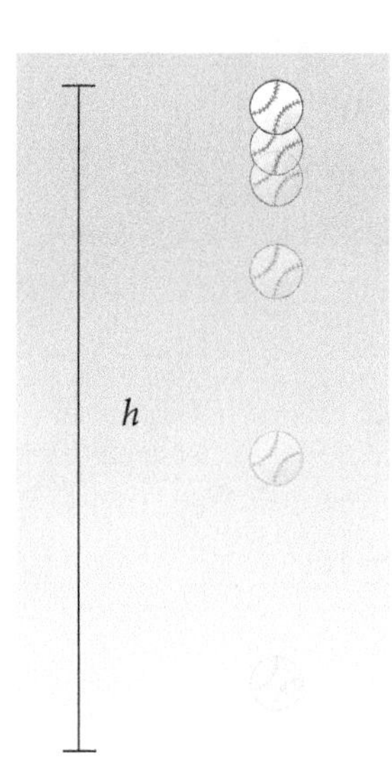

Solution To find the instantaneous rate of change we use our formula:

$$\begin{aligned} \text{Instantaneous rate of change} &= \lim_{h \to 0} \frac{s(t+h) - s(t)}{h} \\ &= \lim_{h \to 0} \frac{[32(t+h) - 16(t+h)^2] - [32t - 16t^2]}{h} \\ &= \lim_{h \to 0} \frac{32t + 32h - 16(t^2 + 2th + h^2) - 32t + 16t^2}{h} \end{aligned}$$

$$= \lim_{h \to 0} \frac{32t + 32h - 16t^2 - 32th - 16h^2 - 32t + 16t^2}{h}$$

$$= \lim_{h \to 0} \frac{32h - 32th - 16h^2}{h}$$

$$= \lim_{h \to 0} (32 - 32t - 16h)$$

$$= 32 - 32t$$

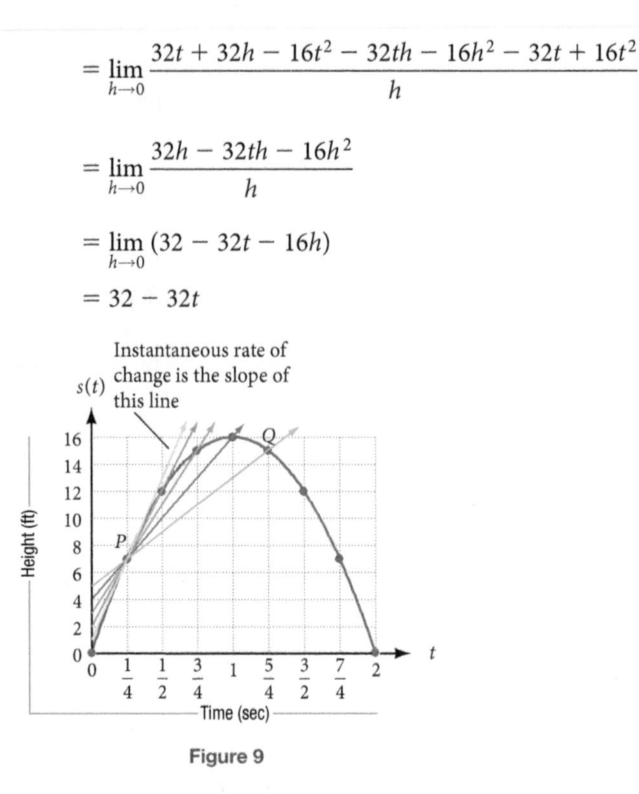

Figure 9

When $t = \frac{1}{4}$, the instantaneous rate of change is $32 - 8 = 24$ ft/sec. When $t = 1$, the instantaneous rate of change is $32 - 32 = 0$ ft/sec. Note that when $t = 1$, the line tangent to the curve at this point is horizontal. ■

Using Technology 1.6

You can use your graphing calculator to find the average rate of change of a function and to approximate the instantaneous rate of change of a function.

Example 8 For the function $f(x) = (5x - 3)^{2/3}$, find

a. the average rate of change as x increases from 16 to 20.

b. the approximate instantaneous rate of change at $x = 16$.

Solution

a. Set up your Y-variable this way:

Y1 = (5X − 3)^(2/3)

Now, in the computation window, enter

(Y1(20) − Y1(16))/(20 − 16)

The result is displayed as 0.7530075664.

b. To approximate the instantaneous rate of change at $x = 16$, you can use $h = 0.001$. Then, in the computation window, enter

$$(\text{Y1}(16.001) - \text{Y1}(16))/(16.001 - 16)$$

The result is displayed as 0.783508665. ■

You can also use your graphing calculator to draw a line tangent to the graph of a function at a particular point.

Example 9 Graph the function $f(x) = x^2 + x - 1$. Then draw the line tangent to the curve at $x = 1$.

Solution Enter the function in your Y-variables list:

$$\text{Y1} = \text{X}^\wedge 2 + \text{X} - 1$$

Use [ZOOM] [6] to set up a standard window and graph the function. Then press [DRAW] [5] ([2nd] [PRGM] [5]) to access the tangent. Enter the X value of 1 and press [ENTER]. The tangent line will be drawn as shown in Figure 10.

Note that the calculator also provides the equation of the tangent line at that point. If you wish to draw tangent lines at other points, simply press [DRAW] [1] ([2nd] [PRGM] [1]) to delete the current tangent line and repeat the steps above with a new value of X.

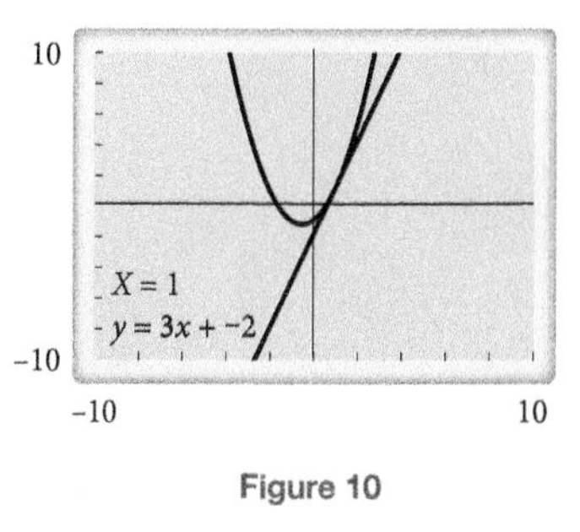

Figure 10

■

Getting Ready for Class

After reading through the preceding section, respond in your own words and in complete sentences.

A. Describe the difference between the average rate of change and the instantaneous rate of change of a function.

B. Explain why we must keep the lim symbol affixed to each expression except the last when computing the instantaneous rate of change of a function using the difference quotient.

C. Explain why the word *approximately* needs to be used when interpreting the instantaneous rate of change of a nonlinear function. (A picture may help your explanation.)

Problem Set 1.6

Skills Practice

For Problems 1-6, find the average rate of change of $f(x)$ with respect to x on each interval.

1. $f(x) = x^2 - 2; \quad [1, 4]$
2. $f(x) = x^2 + 3x - 11; \quad [6, 10]$
3. $f(x) = 4x - 5; \quad [-1, 5]$
4. $f(x) = 2\sqrt{x} - 1; \quad [4, 9]$
5. $f(x) = \dfrac{x^2 - 25}{x - 5}; \quad [6, 15]$
6. $f(x) = \begin{cases} 2x^2 + x - 4 & \text{if } 1 \le x < 3 \\ x - 7 & \text{if } 3 \le x \le 6 \end{cases} \quad [1, 6]$

For Problems 7-12, find and simplify the difference quotient $\frac{f(x+h) - f(x)}{h}$ for each function.

7. $f(x) = 3x - 8$
8. $f(x) = x^2 + 5x - 4$
9. $f(x) = -\sqrt{x} + 2$
10. $f(x) = \dfrac{x + 3}{x - 1}$
11. $f(x) = -\dfrac{1}{x}$
12. $f(x) = 4x^2 - 3$

For Problems 13-20, find the instantaneous rate of change of each function at the specified value of x.

13. $f(x) = x^2 + 4; \quad x = 3$
14. $f(x) = x^2 - 3x + 1; \quad x = 6$
15. $f(x) = 5x^2 + 4x + 2; \quad x = 0$
16. $f(x) = -6x^2 - 3x + 11; \quad x = 1$
17. $f(x) = 4x + 4; \quad x = -2$
18. $f(x) = -6x + 5; \quad x = 5$
19. $f(x) = \dfrac{x + 2}{x - 3}; \quad x = 2$
20. $f(x) = \dfrac{3x + 2}{x + 5}; \quad x = 1$

21. Find $f(20) - f(0)$ if $f(x) = 1{,}200x - x^3$.
22. Find $f(52.70) - f(0)$ if $f(x) = 13.89x$.
23. Find $f(5) - f(2)$ if $f(x) = \dfrac{x^3}{3} - \dfrac{5x^2}{2} + 4x$.
24. Find $f(4) - f(2)$ if $f(x) = -\dfrac{x^3}{3} + \dfrac{5x^2}{2} - 4x$.

Check Your Understanding

25. **Medicine: Drug Concentration** The concentration C (in mg/cc) of a particular drug in a person's bloodstream is related to the time t (in hours) after injection in such a way that as t changes from 0 to 8, C changes from 0.8 to 0.1. Find and interpret the average rate of change of C with respect to t.

26. **Economics: City Redevelopment** The population P (in thousands of people) of a city involved in redevelopment is related to time t (in years) after the beginning of redevelopment in such a way that as t changes from 2 to 10, P changes from 65 to 82. Find and interpret the average rate of change of P with respect to t.

27. **Business: Retail Profit** When a retailer increases the selling price of an item from 75 cents to 80 cents, his weekly profit in sales of that item increases 300 cents. With these changes, the average rate of change of profit with respect to selling price is 60. Interpret the average rate of change of profit with respect to selling price.

28. **Psychology: Reaction Time** A psychologist has determined that for 20-year-old women, the reaction time t (in seconds) to a particular stimulus is 2 seconds, whereas for 23-year-old women, the reaction time to the same stimulus is 8 seconds. The average rate of change of reaction time with respect to age is 2. Interpret the average rate of change of reaction time with respect to age.

29. The relationship between the quantities x and y is expressed by the function $y = f(x)$. The instantaneous rate of change of y with respect to x is $8x - 1$. Find and interpret the instantaneous rate of change of y with respect to x when $x = 4$.

30. The relationship between the quantities x and y is expressed by the function $y = f(x)$. The instantaneous rate of change of y with respect to x is $-3x^2 + 5x - 3$. Find and interpret the instantaneous rate of change of y with respect to x when $x = 1$.

Modeling Practice

For Problems 31 and 32, use the information presented in the table below. The information was excerpted from the business section of the *American Medical News*, February, 2012. It presents the number of jobs in an occupation in the year 2010 and the anticipated number of jobs in that occupation in the year 2020.

Occupation	2010	2020
Health	45,200	62,000
Marketing/Sales	26,800	25,100

31. **Health Science: Physician Assistant** Find and interpret the average rate of change with respect to time of the number of jobs in physician assistant occupations. Round your result to the nearest whole number.

32. **Marketing/Sales: Advertising Associate** Find and interpret the average rate of change with respect to time of the number of jobs in advertising associate occupations. Round your result to the nearest whole number.

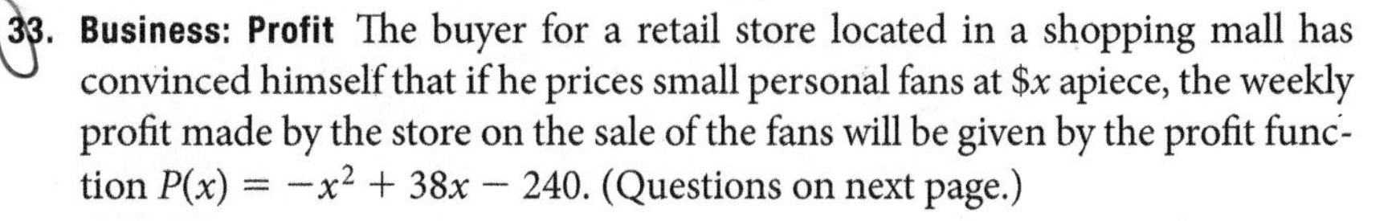

33. **Business: Profit** The buyer for a retail store located in a shopping mall has convinced himself that if he prices small personal fans at $\$x$ apiece, the weekly profit made by the store on the sale of the fans will be given by the profit function $P(x) = -x^2 + 38x - 240$. (Questions on next page.)

a. If the store is currently selling the small personal fans for \$10 each, what is the average rate of change in the weekly profit if it increases the selling price of each fan to \$11?

b. If the store is currently selling the small personal fans for \$20 each, what is the average rate of change in the weekly profit if it increases the selling price of each fan to \$21?

34. Business: Profit The buyer for a retail store located in a shopping mall has convinced herself that if she prices leather wallets at \$$x$ apiece, the weekly profit made by the store on the sale of the wallets will be given by the profit function $P(x) = -x^2 + 100x - 875$.

a. If the store is currently selling the leather wallets for \$40 each, what is the average rate of change in the weekly profit if it increases the selling price of each wallet to \$41?

b. If the store is currently selling the leather wallets for \$55 each, what is the average rate of change in the weekly profit if it increases the selling price of each wallet to \$56?

35. Social Science: Marriage Studies show that there is a relationship between the median age A (in years) of a woman at her first marriage and the time t (in years since 1950). The relationship between A and t is described by the function $A(t) = 0.08t + 19.7$. By how many years would we expect the median age of a woman at her first marriage to change between the years

a. 1975 to 1976? (Hint: $t = 1975 - 1950 = 25$.)

b. 1990 to 1991?

c. 2012 to 2013?

36. Anthropology: Anatomy Anthropologists often use the function

$$h(x) = 2.75x + 71.48$$

to estimate the height h (in centimeters) of a human female. The height h depends on the length x (in centimeters) of the humerus bone (the bone from the shoulder to the elbow). By how much do we expect the projected height of a 400-year-old fossil female to change if more detailed research shows that the length of the remains of her humerus bone is closer to 41 centimeters than to an earlier estimate of 40 centimeters?

37. Physical Science: Drag Racing The following rational function gives the speed $V(t)$, in miles per hour, of a dragster at each second t during a quarter-mile race. Find the instantaneous rate of change in speed when $t = 2$.

© Shaun Lowe Photographic/ iStockPhoto

$$V(t) = \frac{340t}{t + 3}$$

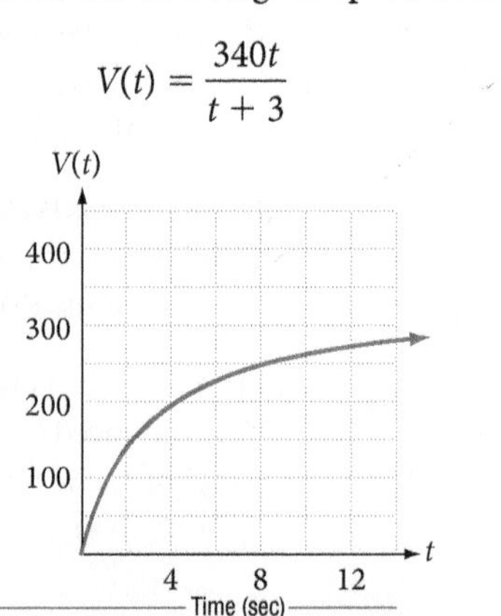

© Warchi/iStockPhoto

38. Business: Average Cost A producer and distributor of gift cards produces x boxes of cards per day at an average cost $\overline{C}(x)$ (in dollars) of

$$\overline{C}(x) = \frac{6x + 8{,}000}{2x + 300}$$

a. Find the average rate of change in the average cost, as production increases from 50 boxes/day to 100 boxes per day.

b. Find the average rate of change in the average cost, as production increases from 80 boxes/day to 100 boxes per day.

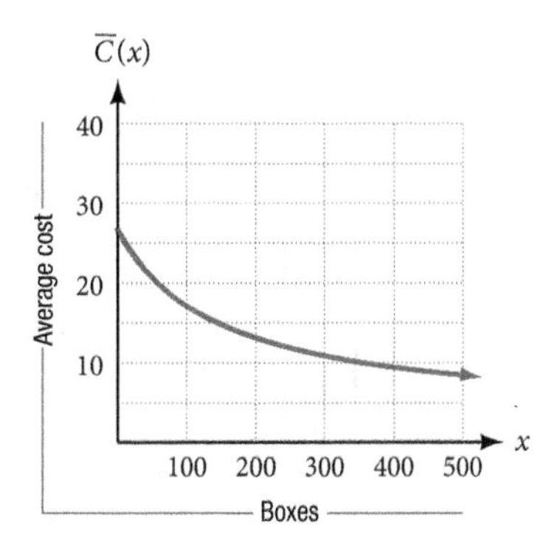

Using Technology Exercises

Use your graphing calculator to find rates of change and to draw tangent lines.

39. For the function $f(x) = (2x + 3)^{3/4}$, find

a. the average rate of change as x increases from 12 to 15.

b. the approximate instantaneous rate of change at $x = 12$. (Use $h = 0.001$ as you did in Example 8.)

40. Graph the function $f(x) = -\frac{1}{2}x^2 - 2x + 3$. Then draw the line tangent to the curve at $x = 2$.

Getting Ready for the Next Section

41. Evaluate $f(5)$ for $f(x) = x^2 + 6x - 2$.

42. Evaluate $\frac{f(x + h) - f(x)}{h}$ for $f(x) = x^2 + 6x - 2$.

43. Evaluate $\lim_{h \to 0} \frac{f(x + h) - f(x)}{h}$ for $f(x) = x^2 + 6x - 2$.

44. Compute $\frac{2}{3}x^{(2/3)-1}$ for $x = 8$.

45. Write the equation of the line having slope 4 and passing through the point $(-1, 1)$.

46. For $f(x) = x^2 - 4x + 3$, find the values of x for which $f(x) = 0$.

47. Compute $f(8)$ for $f(x) = 6x^{-1/3}$.

Spotlight on **Success**

Diane Van Deusen (Napa Valley College)

You may think that all your mathematics instructors started their college math sequence with precalculus or calculus, but that is not always the case. Diane Van Deusen, a full time mathematics instructor at Napa Valley College in Napa, California, started her career in mathematics in elementary algebra. Here is part of her story from her website:

I was not encouraged to attend college after high school, and in fact, had no interest in "more school." Consequently, I didn't end up taking a college class until I was 31 years old! Before returning to and while attending college, I worked locally in the restaurant business as a waitress and bartender and in catering. In fact, I sometimes wait tables a few nights a week during my summer breaks.

When I first came back to school, at Napa Valley College (NVC), I thought I might like to enter the nursing program but soon found out nursing was not for me. As I started working on general education requirements, I took elementary algebra and was surprised to learn that I really loved mathematics, even though I had failed 8th grade algebra! As I continued to appreciate and value my own education, I decided to become a teacher so that I could support other people seeking education goals. After earning my AA degree from NVC, I transferred to Sonoma State where I earned my bachelors degree in mathematics with a concentration in statistics. Finally, I attended Cal State Hayward to earn my master's degree in applied statistics. It took me ten years in all to do this.

Diane's story shows that you can start here and go as far as you want in mathematics. Who knows, you may end up teaching mathematics one day, just like Diane Van Deusen.

Chapter 1 Summary

EXAMPLES

Relations and Functions [1.1]

1. The relation

$$\{(8, 1), (6, 1), (-3, 0)\}$$

is also a function because no ordered pairs have the same first coordinates. The domain is $\{8, 6, -3\}$ and the range is $\{1, 0\}$.

A **function** is a rule that pairs each element in one set, called the **domain**, with exactly one element from a second set, called the **range**.

A **relation** is any set of ordered pairs. The set of all first coordinates is called the **domain** of the relation, and the set of all second coordinates is the **range** of the relation. A function is a relation in which no two different ordered pairs have the same first coordinates.

Vertical Line Test [1.1]

If a vertical line crosses the graph of a relation in more than one place, the relation cannot be a function. If no vertical line can be found that crosses a graph in more than one place, then the graph represents a function.

Function Notation [1.1]

2. If $f(x) = 5x - 3$ then

$$\begin{aligned} f(0) &= 5(0) - 3 \\ &= -3 \\ f(1) &= 5(1) - 3 \\ &= 2 \\ f(-2) &= 5(-2) - 3 \\ &= -13 \\ f(a) &= 5a - 3 \end{aligned}$$

The alternative notation for y is $f(x)$. It is read "f of x" and can be used instead of the variable y when working with functions. The notation y and the notation $f(x)$ are equivalent; that is, $y = f(x)$.

Algebra with Functions [1.2]

3. If $f(x) = 2x + 3$ and $g(x) = x - 5$, then

$$\begin{aligned} (fg)(x) &= f(x)g(x) \\ &= (2x + 3)(x - 5) \\ &= 2x^2 - 7x - 15 \end{aligned}$$

$(f + g)(x) = f(x) + g(x)$ The function $f + g$ is the sum of functions f and g.

$(f - g)(x) = f(x) - g(x)$ The function $f - g$ is the difference of functions f and g.

$(fg)(x) = f(x)g(x)$ The function fg is the product of functions f and g.

$\left(\frac{f}{g}\right)(x) = \frac{f(x)}{g(x)}$ The function $\frac{f}{g}$ is the quotient of functions f and g, where $g(x) \neq 0$.

Composition of Functions [1.2]

4. If $f(x) = 2x + 3$ and $g(x) = x - 5$, then

$$\begin{aligned} (f \circ g)(x) &= f[g(x)] \\ f(x - 5) &= 2(x - 5) + 3 \\ &= 2x - 7 \end{aligned}$$

If f and g are two functions for which the range of each has numbers in common with the domain of the other, then we have the following definitions:

The composition of f with g: $(f \circ g)(x) = f[g(x)]$
The composition of g with f: $(g \circ f)(x) = g[f(x)]$

The Slope of a Line [1.3]

5. The slope of the line through $(1, -1)$ and $(6, 9)$ is

$$m = \frac{9 - (-1)}{6 - 1} = \frac{10}{5} = 2$$

The **slope** of the line containing points (x_1, y_1) and (x_2, y_2) is given by

$$\text{Slope} = m = \frac{\text{Rise}}{\text{Run}} = \frac{y_2 - y_1}{x_2 - x_1}$$

Horizontal lines have 0 slope, and vertical lines have no slope.

Difference Quotients [1.3, 1.6]

6. If $f(x) = x^2$, then

$$\frac{f(x) - f(a)}{x - a} = \frac{x^2 - a^2}{x - a} = \frac{(x + a)(x - a)}{x - a} = x + a$$

Each of the following expressions is called a **difference quotient**.

$$\frac{f(b) - f(a)}{b - a} \qquad \frac{f(x) - f(a)}{x - a} \qquad \frac{f(x_2) - f(x_1)}{x_2 - x_1} \qquad \frac{f(x + h) - f(x)}{h}$$

Each represents the average rate of change of a function from one point to another.

The Slope-Intercept Form of a Line [1.3]

7. The equation of the line with slope 5 and y-intercept 3 is

$$y = 5x + 3$$

The equation of a line with slope m and y-intercept b is given by

$$y = mx + b$$

The Point-Slope Form of a Line [1.3]

8. The equation of the line through (3, 2) with slope -4 is

$$y - 2 = -4(x - 3)$$

which can be simplified to

$$y = -4x + 14$$

The equation of the line through (x_1, y_1) that has slope m can be written as

$$y - y_1 = m(x - x_1)$$

Piecewise-Defined Functions [1.3]

9. The following is a piecewise-defined function.

$$f(x) = \begin{cases} x & \text{if } x \leq 0 \\ x^2 & \text{if } x > 0 \end{cases}$$

A function that is defined by more than one formula for different values in its domain is a piecewise function.

The Limit Notation [1.4]

The limit of $f(x)$, as x approaches c, is L

$$\lim_{x \to c} f(x) = L$$

if we can get as close as we like to L with $f(x)$, by taking x sufficiently close to c.

Limit Theorem [1.4]

10. Given the piecewise function

$$f(x) = \begin{cases} x + 1 & x \leq 2 \\ x^2 - 2x + 3 & x > 2 \end{cases}$$

we see that $\lim_{x \to 2^-} f(x) = 3$ and $\lim_{x \to 2^+} f(x) = 3$. Therefore, $\lim_{x \to 2} f(x)$ exists and $\lim_{x \to 2} f(x) = 3$.

$\lim_{x \to c} f(x) = L$ if and only if $\lim_{x \to c^-} f(x) = \lim_{x \to c^+} f(x) = L$. That is, $\lim_{x \to c} f(x) = L$ if and only if the limits from the left and right both exist and are equal.

Limit Properties Involving Infinity [1.4]

Property 1 $\lim_{x\to\infty}(x^n) = \infty$ (or *does not exist*), where $n > 0$

Property 2 $\lim_{x\to\infty}\left(\frac{1}{x^n}\right) = 0$ where $n > 0$

11. $\lim_{x\to\infty} x^6 = \infty$ (or *does not exist*)

$\lim_{x\to\infty} \frac{1}{x^6} = 0$

Dominant Term Property [1.4]

If $P(x)$ and $Q(x)$ are polynomials, then

$$\lim_{x\to\infty} \frac{P(x)}{Q(x)} = \lim_{x\to\infty} \frac{\text{Dominant term in } P(x)}{\text{Dominant term in } Q(x)}$$

12. $\lim_{x\to\infty} \frac{2x^2 - 4x + 6}{3x^2 + 2x - 9} = \lim_{x\to\infty} \frac{2x^2}{3x^2}$

$= \lim_{x\to\infty} \frac{2}{3} = \frac{2}{3}$

Asymptotes [1.4]

Horizontal Asymptote For the graph of the function $y = f(x)$, a horizontal line that is approached by the graph as the independent variable tends toward infinity is called a horizontal asymptote. That is, if $\lim_{x\to-\infty} f(x) = \lim_{x\to\infty} f(x) = b$, then the horizontal line $y = b$ is a horizontal asymptote for the graph of $y = f(x)$.

Vertical Asymptote If $\lim_{x\to c} f(x) = \infty$ or $\lim_{x\to c} f(x) = -\infty$ as $x \to c^-$ or $x \to c^+$, then $x = c$ is a vertical asymptote for the graph of $y = f(x)$.

13. Since $\lim_{x\to\infty} W(x) = \lim_{x\to-\infty} W(x) = 160$, there is a horizontal asymptote at $y = 160$. Since $\lim_{x\to-6^-} W(x) = -\infty$ and $\lim_{x\to-6^+} W(x) = +\infty$, there is a vertical asymptote at $x = -6$.

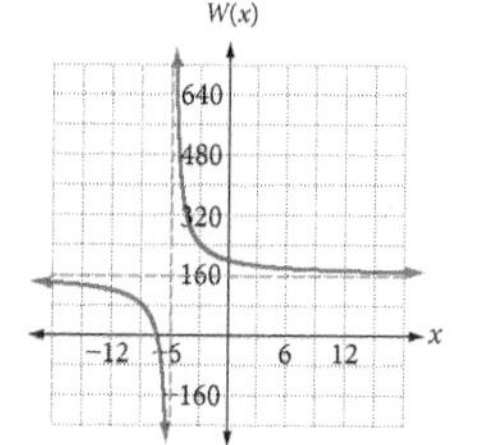

Continuity Conditions [1.5]

If $f(x)$ is a function, then $f(x)$ is continuous at point $x = a$ if all three of the following conditions are satisfied.

1. $f(a)$ is defined.
2. $\lim_{x\to a} f(x)$ exists.
3. $\lim_{x\to a} f(x) = f(a)$.

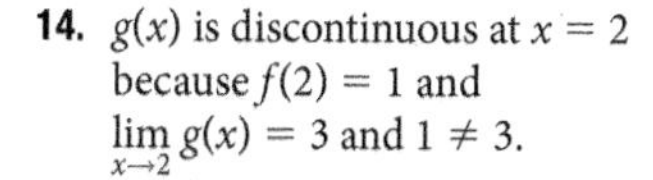

14. $g(x)$ is discontinuous at $x = 2$ because $f(2) = 1$ and $\lim_{x\to 2} g(x) = 3$ and $1 \neq 3$.

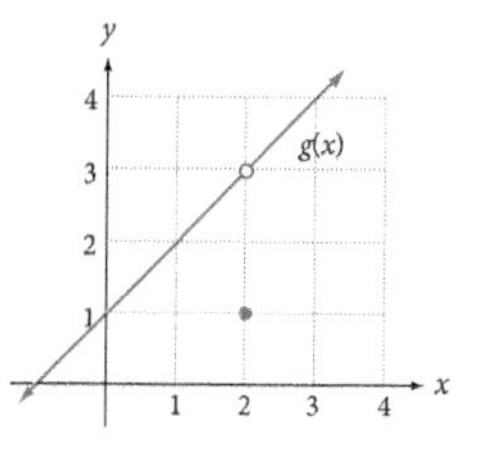

Average Rate of Change [1.6]

The average rate of change of a function is defined as

$$\frac{\text{Change in the dependent quantity}}{\text{Change in the independent quantity}}$$

We calculate the average rate of change using one of the difference quotients shown earlier. The average rate of change is also the slope of a secant line.

15. The average rate of change for $f(x) = x^2 - 3x + 4$ over the interval $[1, 5]$ is

$$\frac{f(5) - f(1)}{5 - 1} = \frac{14 - 2}{4} = 3$$

Instantaneous Rate of Change [1.6]

Instantaneous rate of change of a function $f(x)$ is defined as

$$\lim_{h \to 0} \frac{f(x + h) - f(x)}{h}$$

if this limit exists. The instantaneous rate of change is also the slope of the tangent line at the indicated point.

16. The instantaneous rate of change for $f(x) = x^2 - 3x + 4$ is given by

$$\lim_{h \to 0} \frac{f(x + h) - f(x)}{h} =$$

$$\lim_{h \to 0} \frac{(x + h)^2 - 3(x + h) + 4 - (x^2 - 3x + 4)}{h} =$$

$$\lim_{h \to 0} \frac{2xh + h^2 - 3h}{h} =$$

$$\lim_{h \to 0} 2x + h - 3 = 2x - 3$$

At $x = 1$, the instantaneous rate of change is $2(1) - 3 = -1$.

Chapter 1 Test

1. State the domain for the function. [1.1]

$$h(x) = \frac{1}{x^2 - 9}$$

2. Let $f(x) = x - 2$, $g(x) = 3x + 4$, and $h(x) = 3x^2 - 2x - 8$, and find the following. [1.2]

a. $(fg)(3)$ **b.** $h(3)$ **c.** $f(g(3))$

3. For the following straight line, identify the x-intercept, y-intercept, and slope, and sketch the graph. [1.3]

$$2x + y = 6$$

Find the equation for each line. [1.3]

4. Give the equation of the line through $(-1, 3)$ that has slope $m = 2$.

5. Give the equation of the line through $(-3, 2)$ and $(4, -1)$.

Graph the piecewise-defined function. [1.3]

6. $f(x) = \begin{cases} x - 2 & \text{if } x \leq 2 \\ (x - 2)^2 & \text{if } x > 2 \end{cases}$

7. Use the tables below to determine $\lim_{x \to 13} f(x)$, if it exists. [1.4]

x (from the left)	f(x)
12.5	1.7
12.7	1.9
12.9	1.99
12.99	1.999
12.999	1.9999

x (from the right)	f(x)
13.5	2.2
13.3	2.1
13.1	2.01
13.01	2.001
13.001	2.0001

8. Use the graph to find each limit, if it exists. [1.4]

a. $\lim_{x \to 1} f(x)$

b. $\lim_{x \to 3} f(x)$

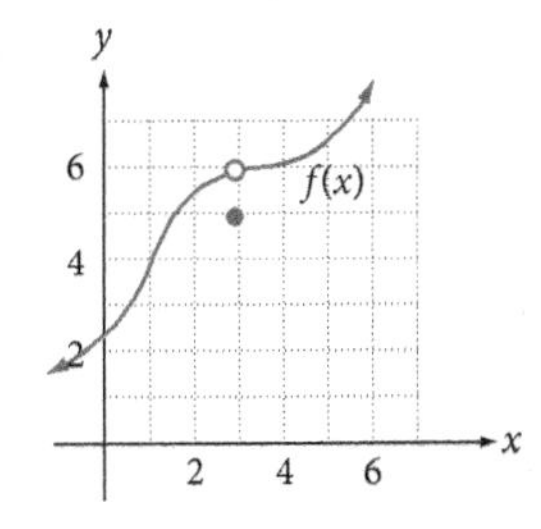

For Problems 9-15, find the indicated limit, if it exists. [1.4]

9. $\lim_{x \to 6} 8$

10. $\lim_{x \to 4} \frac{x^2 - 2x - 8}{x + 1}$

11. $\lim_{x \to 3} \frac{x^2 - 2x - 3}{x - 3}$

12. $\lim_{x \to 0} \frac{3x^3 - 8x}{2x}$

13. $\lim_{x\to\infty} \frac{7x^2 + 3x - 1}{8x^2 + x + 3}$

14. $\lim_{x\to\infty} \frac{x^4 + 2x^3 + x + 1}{x^5 + 3x - 9}$

15. $\lim_{x\to-\infty} \frac{x^2 + 11x + 10}{x + 1}$

For Problems 16 and 17, determine if each illustrated function is discontinuous or continuous at the point $x = a$. If the function is continuous, so state. If the function is discontinuous, specify the first continuity condition that is not satisfied. [1.5]

16. **17.**

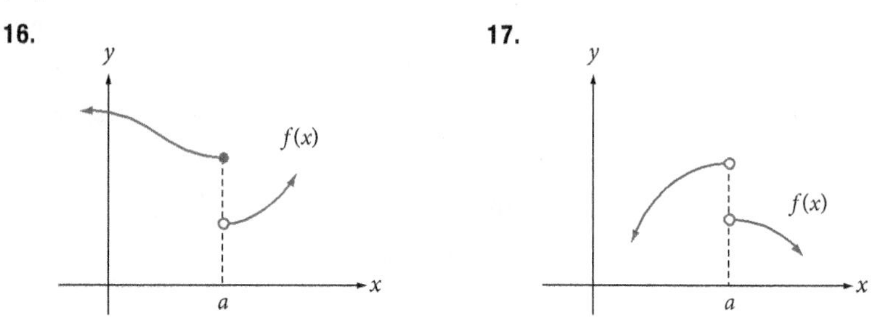

For Problems 18 and 19, discuss the continuity of each function at the specified value of x. If the function is discontinuous, specify the first continuity condition that is not satisfied. [1.5]

18. $f(x) = \frac{x^2 - 5x - 24}{x + 3}$ at $x = 6$

19. $g(x) = \begin{cases} 3x - 1 & \text{if } x \le 1 \\ x^2 - 2x + 5 & \text{if } x > 1 \end{cases}$ at $x = 1$

20. Find the average rate of change of $f(x)$ with respect to x on the specified interval. [1.6]

$$f(x) = \frac{x^2 - 4}{x + 2}; \text{ on } [-1, 1]$$

21. For $f(x) = -3x^2 + x - 2$, find the difference quotient $\frac{f(x + h) - f(x)}{h}$. [1.6]

22. Find the instantaneous rate of change of $f(x)$ at the specified value of x. [1.6]

$$f(x) = 3x^2 - 8x + 1; x = 5$$

23. Find the slope of the tangent line to the graph of the function at the specified point. [1.6]

$$f(x) = \frac{x + 2}{x - 6}; \quad x = 5$$

© Lajos Repasi/iStockPhoto

24. Health Science: Flu Epidemic Health officials, concerned with a flu epidemic that has hit the north Atlantic coast, have estimated that the number of people ill with flu symptoms can be approximated by the function $N(t) = 90t^2 - t^3$, $0 \le t \le 70$, where t is measured in days since the beginning of the epidemic.

a. How many people will have the flu on day 15?

b. What is the average rate of change in the number of people with flu symptoms from day 40 to day 70?

c. What is the instantaneous rate of change in the number of people with flu symptoms on day 40? [1.6]

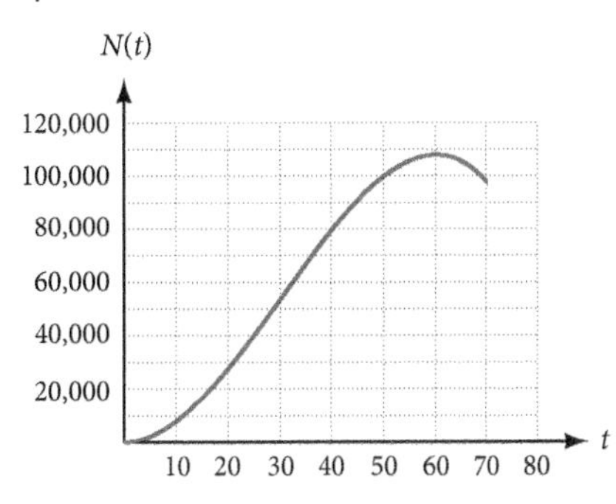

25. Business: Average Cost A producer and distributor of gift cards produces x boxes of cards per day at an average cost $\overline{C}(x)$ (in dollars) of

© Warchi/iStockPhoto

$$\overline{C}(x) = \frac{6x + 8{,}000}{2x + 300}$$

a. What is the domain for this function?

b. Find $\overline{C}(280)$.

c. Find the average rate of change in the average cost, as production increases from 280 boxes/day to 300 boxes/day.

d. Find $\lim_{x \to \infty} \overline{C}(x)$ and explain what your results mean.

e. Find $\lim_{x \to -150} \overline{C}(x)$ and explain what your results mean with respect to the graph of the function $\overline{C}$ if the domain included negative numbers for x. [1.6]

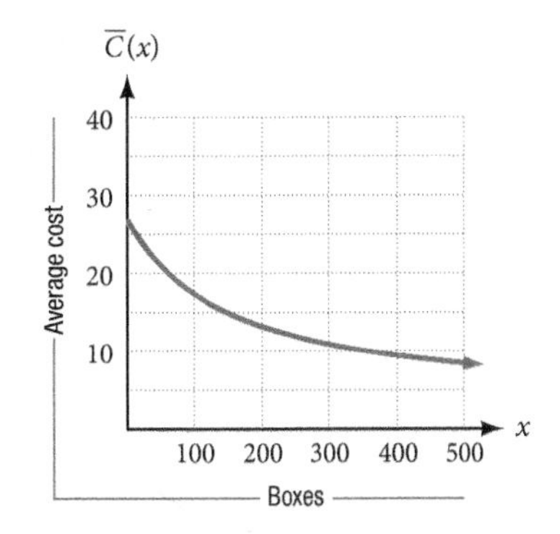

Differentiation: The Language of Change

2

Chapter Outline

© Dean Bertoncelj/iStockPhoto

Note When you see this icon next to an example or problem in this chapter, you will know that we are using the topics in this chapter to model situations in the world around us.

People tend to drink more water on hot days than they do on cold days. Because most drinking water is sold in plastic bottles that can be recycled, as the temperature goes up, there is an increase in the number of plastic bottles recycled. We can model this situation with the following diagram.

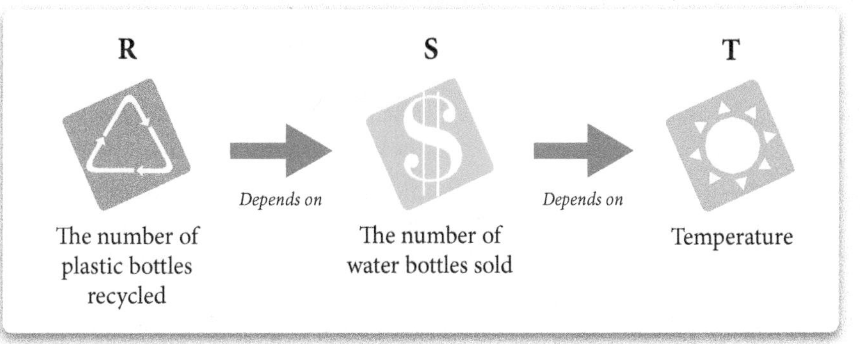

Once we have the function for these quantities, we can differentiate them, giving us some rates of change. Further, we have two types of questions we can answer, one using just algebra, and the other using calculus.

Algebra How many plastic bottles are recycled when the temperature is 85 °F?

Calculus How fast is the number of plastic bottles recycled changing when the temperature is 85 °F?

As you might expect, this situation involves a composite function because the number of plastic bottles recycled depends on the number of plastic bottles sold, which in turn, depends on the temperature. Among the rules we will develop in this chapter is the rule for differentiating a composite function.

Study Skills

If you have successfully completed Chapter 1, then you have made a good start at developing the study skills necessary to succeed in all math classes. Some of the study skills for this chapter are a continuation of the skills from Chapter 1, while others are new to this chapter.

1. **Continue to Set and Keep a Schedule** Sometimes I find students do well in Chapter 1 and then become overconfident. They will begin to put in less time with their homework. Don't do it. Keep to the same schedule.
2. **Increase Effectiveness** You want to become more and more effective with the time you spend on your homework. Increase those activities that are the most beneficial and decrease those that have not given you the results you want.
3. **List Difficult Problems** Begin to make lists of problems that give you the most difficulty. These are the problems that you continue to get wrong. They may also be the key to your success in this course.
4. **Begin to Develop Confidence with Modeling Problems** It seems that the main difference between people who are good at working application problems and those who are not is confidence. People with confidence know that no matter how long it takes them, they will eventually be able to solve the problem. Those without confidence begin by saying to themselves, "I'll never be able to work this problem." If you are in this second category, then instead of telling yourself that you can't do application problems, decide to do whatever it takes to master them. The more application problems you work, the better you will become at them.
5. **Use the Matched Problems Worksheets** These worksheets are available online with each section of the eBook. Each problem on a worksheet is similar to an example from that section. Every time you complete the Matched Problem Worksheet for a section of the book, you know you have worked every type of important problem for that section. If you consistently complete the Matched Problem Worksheets before going to class, you will increase your chances for success in this course significantly.

The Derivative of a Function and Two Interpretations

2.1

© Lajos Repasi/iStockPhoto

When a flu epidemic occurs in a densely populated area, health officials need to know not only how many people have the flu, but how fast the flu is increasing, or decreasing, in the population. If they have a mathematical model that gives the number of people with the flu at any given time, then the derivative of that model will describe how fast the flu is increasing or decreasing in the population.

This section presents the derivative, one of the two main topics of calculus. The derivative is the calculus instrument used for examining change. This section presents two ways of interpreting the derivative of a function: one as the slope of the tangent line to a curve described by the function, and the other as the instantaneous rate of change of the function.

The Derivative of a Function

We have discovered that, for a function $f(x)$, the instantaneous rate of change of f with respect to x, and the slope of the tangent line to $f(x)$, are both given by the function

$$m_{tan} = \lim_{h \to 0} \frac{f(x+h) - f(x)}{h}$$

This new function—the instantaneous rate of change function—that is *derived* from the original function $f(x)$, is called the **derivative of $f(x)$.** A common notation for the derivative of the function $f(x)$ is $f'(x)$ (read "f prime of x"). The process of determining the derivative of a function is called **differentiation**. In a problem in which change is encountered, differentiation should come to mind because, mathematically, *differentiation is the language of change.*

Definition: Derivative

Limit Definition of the Derivative

The **derivative** of the function $f(x)$, denoted $f'(x)$, is

$$f'(x) = \lim_{h \to 0} \frac{f(x+h) - f(x)}{h}$$

provided this limit exists.

From our work in the previous chapter, we know we can interpret the derivative in two ways.

Interpreting the Derivative

The derivative $f'(x)$ of the function $f(x)$

1. is the slope of the tangent line to $f(x)$.
2. is the instantaneous rate of change of f with respect to x.

VIDEO EXAMPLES

SECTION 2.1

Example 1 Find the derivative of the function $f(x) = 5x + 8$.

Solution Applying our new definition, we have

$$
\begin{aligned}
f'(x) &= \lim_{h \to 0} \frac{f(x+h) - f(x)}{h} \\
&= \frac{5(x+h) + 8 - (5x+8)}{h} \\
&= \lim_{h \to 0} \frac{5x + 5h + 8 - 5x - 8}{h} \\
&= \lim_{h \to 0} \frac{5h}{h} \\
&= \lim_{h \to 0} \frac{5h}{h}, \text{ if } h \neq 0 \\
&= 5
\end{aligned}
$$

Interpretation

The instantaneous rate of change of the function $f(x) = 5x + 8$ with respect to x is a constant 5 everywhere, which we expect because our original function is linear.

Notation

There are several notations for the derivative of a function $f(x)$. Depending on the situation, one notation or description may be more useful than another.

Derivative Notations

If $y = f(x)$, then each of the following notations represents the derivative of $f(x)$.

$$f'(x) \qquad y' \qquad \frac{dy}{dx} \qquad \frac{df}{dx} \qquad \frac{d}{dx}f$$

The notation $\frac{dy}{dx}$ is read as *dee y dee x* and was developed by Gottfried Leibniz around 1670. For now, $\frac{dy}{dx}$ is to be considered only a notation, not a fraction. It indicates the derivative of y with respect to x, not the operation dy divided by dx.

Common notations indicating the *value* of the derivative at $x = a$ are

$$f'(a) \qquad y'\big|_{x=a} \qquad \frac{dy}{dx}\bigg|_{x=a} \qquad \frac{df}{dx}\bigg|_{x=a}$$

Example 2 If $f(x) = x^2 + 6x - 2$, find each of the following and interpret each result.

a. $f(5)$ **b.** $f'(x)$ **c.** $f'(5)$

Solution For part a we use the function itself.

a. $f(5) = 5^2 + 6 \cdot 5 - 2 = 25 + 30 - 2 = 53$

Interpretation
When $x = 5$, $y = 53$. When the input is 5, the output is 53.

b. Using the definition of the derivative, we have

$$f'(x) = \lim_{h \to 0} \frac{f(x+h) - f(x)}{h}$$

$$= \lim_{h \to 0} \frac{\overbrace{(x+h)^2 + 6(x+h) - 2}^{f(x+h)} - \overbrace{(x^2 + 6x - 2)}^{f(x)}}{h}$$

$$= \lim_{h \to 0} \frac{x^2 + 2hx + h^2 + 6x + 6h - 2 - x^2 - 6x + 2}{h}$$

$$= \lim_{h \to 0} \frac{2hx + h^2 + 6h}{h}$$

$$= \lim_{h \to 0} \frac{h(2x + h + 6)}{h}$$

$$= \lim_{h \to 0} \frac{\cancel{h}(2x + h + 6)}{\cancel{h}} \quad \text{Since } h \neq 0$$

$$= \lim_{h \to 0} (2x + h + 6) \quad \text{Take the limit by substitution}$$

$$= 2x + 6$$

Thus, $f'(x) = 2x + 6$.

Interpretation
$f'(x) = 2x + 6$ gives the instantaneous rate of change of f with respect to x.

c. To find $f'(5)$, we will substitute 5 for x in $2x + 6$.

$$f'(5) = 2(5) + 6 = 10 + 6 = 16 \quad \text{or} \quad \frac{16}{1}$$

Interpretation
The slope of the line tangent to the graph of $y = f(x) = x^2 + 6x - 2$ is 16 at the point (5, 53).

■

Next, we develop rules for finding derivatives that allow us shortcuts from using the definition of the derivative as we have in Examples 1 and 2. These rules will make our lives easier.

Fundamental Rules of Differentiation

Finding the derivative of a function $f(x)$ using the limit definition

$$f'(x) = \lim_{h \to 0} \frac{f(x+h) - f(x)}{h}$$

can be cumbersome and time-consuming; it is not the most efficient use of our time and energy. By using the limit definition, however, we can prove differentiation rules that are both manageable and efficient. Since we are concerned more with the applications of calculus than the theory on which it is built, we will simply state these rules without proof. Here is our first one.

> **Derivative of a Constant**
> If $y = c$, where c is any real number, then $\frac{dc}{dx} = 0$.

Alternatively, if the function is expressed as $f(x) = c$, then $f'(x) = 0$. This derivative rule states that the derivative of any constant is zero.

This result is understood geometrically by observing that the graph of $y = c$ is a horizontal line passing through c on the y-axis, and that the slope of a horizontal line is 0. Also, an increase of 1 unit in any x-value will produce no change in the y-value. (See Figure 1.)

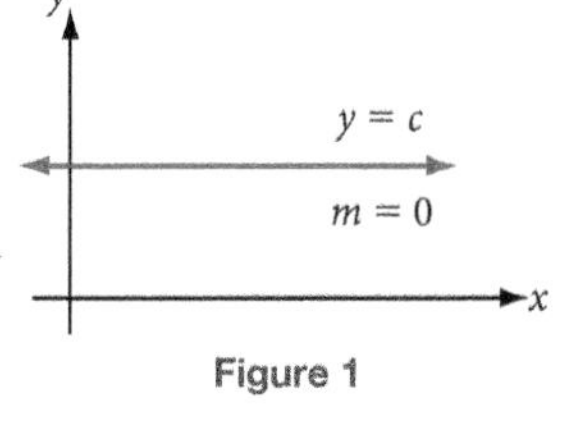

Figure 1

Example 3 Here are two functions we differentiate using our first rule.

a. If $f(x) = 4$, then $f'(x) = 0$.

b. If $f(x) = -16$, find $f'(2)$.

Since $f(x) = -16$, a constant, $f'(x) = 0$ for every value of x. In particular, $f'(2) = 0$.

Our next rule tells us how to differentiate functions of the form $f(x) = x^n$.

Simple Power Rule

> **Simple Power Rule Derivative of a Simple Power**
> If $y = x^n$, where n is any real number, then $\frac{dy}{dx} = n \cdot x^{n-1}$.

Alternatively, if the function were expressed as $f(x) = x^n$, then $f'(x) = n \cdot x^{n-1}$. This derivative rule states that the derivative of x^n is found by multiplying x by the original exponent, then decreasing the exponent on x by 1.

Example 4 Here are some examples that use our simple power rule.

a. If $y = x^2$, then $\frac{dy}{dx} = 2x^{2-1} = 2x^1 = 2x$.

b. If $f(x) = x^5$, then $f'(x) = 5x^{5-1} = 5x^4$.

c. If $y = x = x^1$, then $\frac{dy}{dx} = 1x^{1-1} = 1x^0 = 1 \cdot 1 = 1$.

The problem shown in Example 4c occurs so often that we make a special note of it.

Derivative of $y = x$

$$\text{If } y = x, \text{ then } y' = 1.$$

We can use the power rule to find derivatives of radical functions as well. If we want to find the derivative of $f(x) = \sqrt[3]{x}$, for example, we simply convert it to the exponential form $f(x) = x^{1/3}$ and go from there.

Example 5 If $f(x) = \sqrt[3]{x^2}$, find

a. $f(8)$ b. $f'(8)$

Solution

a. To find $f(8)$, we use the original function: $f(8) = (\sqrt[3]{8})^2 = 2^2 = 4$

b. To find the derivative, we write $\sqrt[3]{x^2}$ as $x^{2/3}$ and apply the power rule for derivatives.

$$f'(x) = \frac{2}{3}x^{(2/3)-1} = \frac{2}{3}x^{-1/3} = \frac{2}{3x^{1/3}}$$

Next, we evaluate the derivative at $x = 8$.

$$f'(8) = \frac{2}{3 \cdot 8^{1/3}} = \frac{2}{3 \cdot 2} = \frac{1}{3}$$

Example 6 If $f(x) = \frac{1}{x^4}$, find

a. $f'(x)$

b. The equation of the line tangent at $(-1, 1)$

Solution

a. We must try to make $f(x)$ match one of our derivative forms. We can get $f(x) = \frac{1}{x^4}$ into the form $f(x) = x^n$ by using the rules of exponents: $f(x) = \frac{1}{x^4} = x^{-4}$. Now, we can differentiate $f(x)$ as follows.

$$f'(x) = -4x^{-4-1} = -4x^{-5} = \frac{-4}{x^5}$$

b. We have the point $(-1, 1)$; all we need is the slope of the tangent line at this point. The slope is

$$f'(-1) = \frac{-4}{(-1)^5} = 4$$

The equation of the tangent line at $(-1, 1)$ is

$$y - y_1 = m(x - x_1)$$

with $x_1 = -1,\ y_1 = 1$, and $m = 4$. Therefore,

$$y - 1 = 4(x + 1)$$
$$y - 1 = 4x + 4$$
$$y = 4x + 5$$

The graph of $y = \dfrac{1}{x^4}$, along with the tangent line at $(-1, 1)$, is shown in Figure 2.

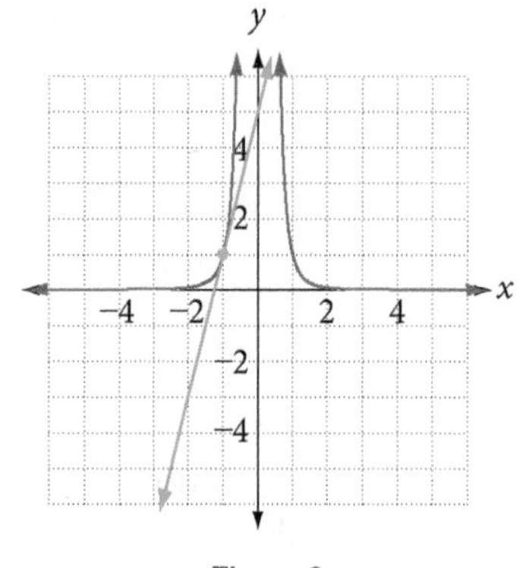

Figure 2

■

Derivative of a Constant Times a Function

If $f(x)$ is a differentiable function and c is a constant, then

$$\frac{d}{dx}cf(x) = c \cdot \frac{d}{dx}f(x).$$

Derivative of a Constant Times a Function

Note When we say a function is differentiable, we mean its derivative exists for each element in its domain.

This rule states that to differentiate a function that is multiplied by a constant, we differentiate the function, then multiply it by the constant.

Example 7 If $f(x) = \dfrac{4}{x^8}$, find $f'(x)$.

Solution We write the function in the more convenient form $f(x) = 4x^{-8}$, then we differentiate.

$$f'(x) = 4 \cdot (-8)x^{-8-1}$$
$$= -32x^{-9}$$
$$= \frac{-32}{x^9} \quad \text{or} \quad -\frac{32}{x^9}$$

■

Example 8 If $y = cx$, where c is a constant, find $\dfrac{dy}{dx}$.

Solution Proceeding as we have in the previous examples, we have

$$\frac{dy}{dx} = c \cdot \frac{d}{dx}x$$
$$= c \cdot 1$$
$$= c$$

■

The result in Example 9 occurs so often that we make a special note of it.

Derivative of $y = cx$

If c is a constant and $y = cx$, then $y' = c$.

Example 9

a. If $y = 8x$, then $y' = 8$.

b. If $f(x) = -14x$, then $f'(x) = -14$.

Next we consider derivatives of sums and differences of functions.

Derivative of a Sum or Difference

Derivative of a Sum or a Difference

If $f(x)$ and $g(x)$ are both differentiable functions, then

$$\frac{d}{dx}[f(x) \pm g(x)] = \frac{d}{dx}f(x) \pm \frac{d}{dx}g(x).$$

This rule states that to differentiate the sum or difference of two functions (that are themselves differentiable), we differentiate each function individually, then add or subtract the resulting derivatives.

In all the examples that follow, we will eliminate some of the intermediate steps.

Example 10 Find $\frac{dy}{dx}$ for $y = 5x^3 + 4x^2 + 9x - 7$.

Solution We differentiate each term separately.

$$\frac{dy}{dx} = \frac{d}{dx}(5x^3) + \frac{d}{dx}(4x^2) + \frac{d}{dx}(9x) - \frac{d}{dx}(7)$$

$$= 15x^2 + 8x + 9 - 0 \quad \text{Differentiate each individual function}$$

$$= 15x^2 + 8x + 9$$

Example 11 If $f(x) = x^2 - 4x + 3$, find

a. the value of x for which $f(x) = 0$.

b. the value of x for which $f'(x) = 0$.

Solution

a. We set $f(x)$ equal to 0 and solve:

$$x^2 - 4x + 3 = 0$$

$$(x - 3)(x - 1) = 0$$

$$x - 3 = 0 \quad \text{or} \quad x - 1 = 0$$

$$x = 3 \quad \text{or} \quad x = 1$$

The function is 0 when $x = 1$ or $x = 3$.

b. First we differentiate term by term:

$$f'(x) = 2x - 4$$

Next, we set this derivative equal to 0:

$$2x - 4 = 0$$
$$x = 2$$

The derivative is 0 when x is 2.

If we know that the slope of the tangent line is 0, then we also know that it is a horizontal line. Figure 3 shows the graph of f, the points where $f(x) = 0$, and the line tangent when $f'(x) = 0$.

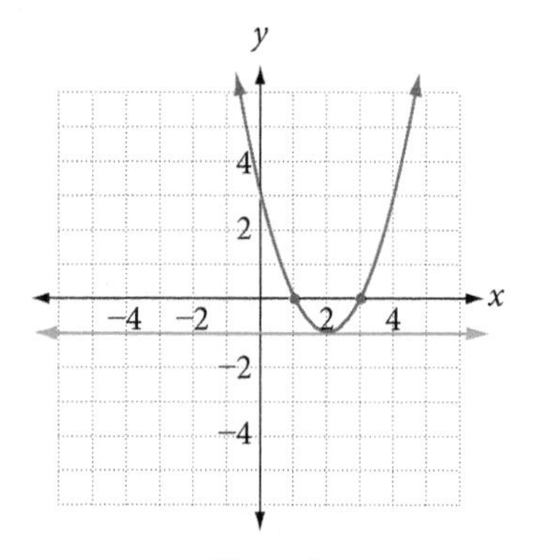

Figure 3

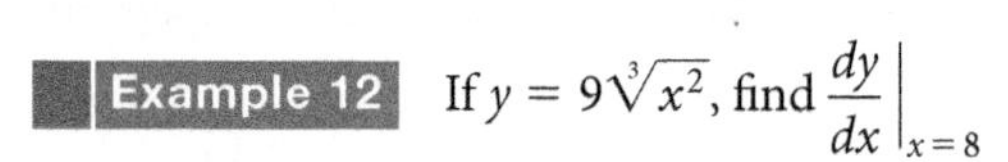

Example 12 If $y = 9\sqrt[3]{x^2}$, find $\left.\frac{dy}{dx}\right|_{x=8}$

Solution We will begin by writing y in the more convenient form $y = 9x^{2/3}$, recalling that the notation $\left.\frac{dy}{dx}\right|_{x=8}$ means to evaluate $\frac{dy}{dx}$ at $x = 8$. Then

$$\frac{dy}{dx} = 9 \cdot \frac{2}{3} x^{(2/3)-1}$$

$$\frac{dy}{dx} = 6x^{-1/3} = \frac{6}{x^{1/3}} \qquad = \frac{6}{\sqrt[3]{x}}$$

Now we evaluate the derivative at $x = 8$.

$$\left.\frac{6}{\sqrt[3]{x}}\right|_{x=8} = \frac{6}{\sqrt[3]{8}}$$
$$= \frac{6}{2} = 3$$

Interpretation

The slope of the line tangent to the curve $y = 9\sqrt[3]{x^2}$ is 3 when x is 8.

Marginal Change

Intervals of length 1 are very important in business applications. For example, a publisher that is currently producing and selling textbooks realizes a profit of $P(5{,}000)$ dollars from the sale of those 5,000 units. If the publisher increases the number of sales by 1, to 5,001 books, the increase in profit is $P(5{,}001) - P(5{,}000)$. But we can approximate this number by finding $P'(5{,}000)$, the derivative evaluated at $x = 5{,}000$. This approximation is called the **marginal profit**. It is an approximation to the actual change in profit brought on by a one-unit increase in sales.

Example 13 If $y = 6x^4$, find $\frac{dy}{dx}$. Then find $\left.\frac{dy}{dx}\right|_{x=3}$ and interpret the result.

Solution

$$\begin{aligned}\frac{dy}{dx} &= 6 \cdot \frac{d}{dx}x^4 \\ &= 6 \cdot 4x^{4-1} \\ &= 6 \cdot 4x^3 \\ &= 24x^3\end{aligned}$$

Our derivative is $y' = 24x^3$. Evaluating this at $x = 3$, we have,

$$\left.\frac{dy}{dx}\right|_{x=3} = 24 \cdot 3^3 = 648$$

Interpretation

If x increases 1 unit in value, from, say, $x = 3$ to $x = 4$, then $f(x)$ will increase in value by approximately 648 units.

More About Marginal Change

Here are two notes that clarify the relationship between the derivative and marginal change.

Note Most business applications will fall into the first category.

1. If the function that is being used to model a relationship does not curve much at the point of interest, the tangent line and the curve will be close together. Consider the graph of $y = \sqrt{x}$ in Figure 4.

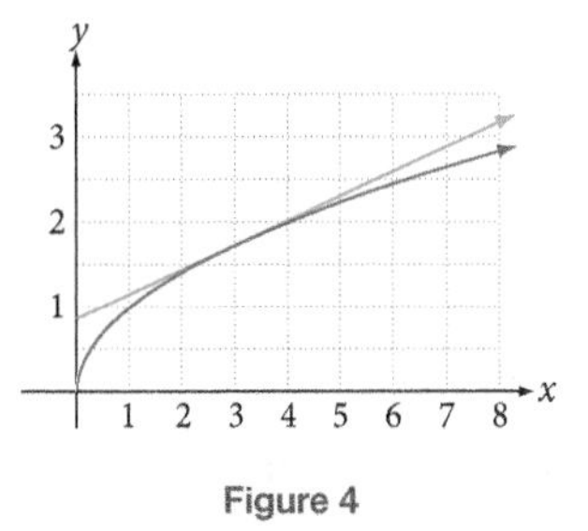

Figure 4

Figure 4 shows that by increasing x by 1-unit, from 3 to 4, there is almost no distance between the curve and the tangent line.

$$\text{Marginal change: } f'(3) = 0.289$$

$$\text{Actual change: } f(4) - f(3) = \sqrt{4} - \sqrt{3} = 0.268$$

The difference between the actual change and marginal change is about 0.02, two-hundredths of a unit.

2. If the function that is being used to model a relationship is not close to linear at the point of interest, the derivative may not, and typically will not, provide a reasonable approximation to an actual change brought on by a 1-unit change in x. Consider the graph of $y = 6x^4$ in Figure 5.

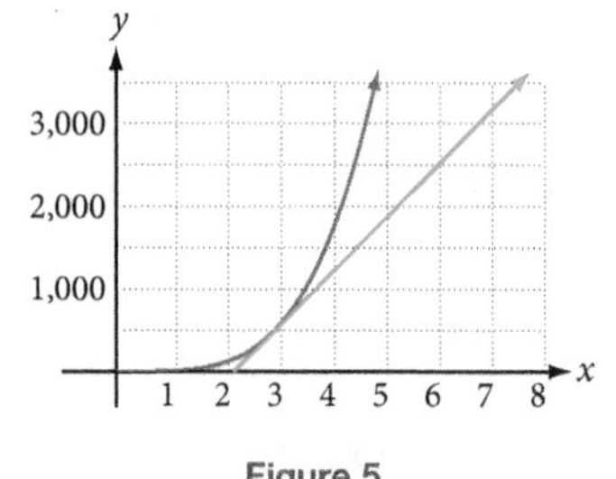

Figure 5

The graph shows that by increasing x by 1-unit, from 3 to 4, there is a significant distance between the curve and the tangent line.

$$\text{Marginal change: } f'(3) = 648$$

$$\text{Actual change: } f(4) - f(3) = 1{,}050$$

The difference between the actual change and marginal change is 402 units.

Example 14 Health officials, concerned with a flu epidemic that has hit the north Atlantic coast, have estimated that the number of people ill with flu symptoms can be approximated by the function $N(t) = 90t^2 - t^3$, $0 \leq t \leq 70$, where t is measured in days since the beginning of the epidemic.

© Lajos Repasi/iStockPhoto

a. How many people have the flu on day 15?

b. At what rate are flu symptoms spreading on day 15 of the epidemic?

c. When are the flu symptoms spreading at the rate of 1,500 people per day?

Solution

a. To find the number of people with the flu on day 15, we find $N(15)$.

$$\begin{aligned} N(15) &= 90(15)^2 - 15^3 \\ &= 16{,}875 \end{aligned}$$

On day 15, there are 16,875 people with the flu.

b. The rate at which the flu symptoms are spreading is given by $N'(t)$, the derivative of $N(t)$. Using the sum rule,

$$N'(t) = 180t - 3t^2 \quad 0 \leq t \leq 70$$

To find the rate of change at $t = 15$, we compute

$$N'(15) = 180(15) - 3(15)^2 = 2{,}025 \text{ more people.}$$

Note If we calculate the value of the original function at 16, we get

$$N(16) = 18{,}944$$

so the derivative gives a very close approximation.

Interpretation

On day 15 of the epidemic, there are 16,875 people ill with flu symptoms. If t increases by 1, from 15 to 16, $N(t)$ increases by approximately 2,025. That is, from day 15 to day 16 of the epidemic, we expect to see an increase of approximately 2,025 cases of flu symptoms. The epidemic is on the rise. So, on day 16 of the epidemic, we expect to see approximately $16{,}875 + 2{,}025 = 18{,}900$ people ill with flu symptoms.

c. Since we are asked the question when, we need to find t. Specifically, we need to find t when $N'(t) = 1{,}500$. We set $N'(t) = 180t - 3t^2$ equal to 1,500 and solve for t.

$$1{,}500 = 180t - 3t^2$$
$$3t^2 - 180t + 1{,}500 = 0$$
$$3(t^2 - 60t + 500) = 0$$
$$3(t - 10)(t - 50) = 0$$

The values 10 and 50 are solutions to this equation.

Interpretation

At both 10 days and 50 days after the beginning of the epidemic, the rate at which people are taken ill with flu symptoms is 1,500 people per day.

■

So far, all the functions we have worked with have derivatives in the places where we were looking for derivatives. But that is not always the case. To end this section, let's look at the conditions under which we cannot find a derivative. That is, we examine the situations in which the derivative does not exist.

Conditions for Nondifferentiability

A function $f(x)$ will not be differentiable at $x = a$ (that is, $f'(x)$ will not exist at $x = a$) if any of the following three conditions exists.

1. $f(x)$ is not continuous at $x = a$. (See Figure 6.)

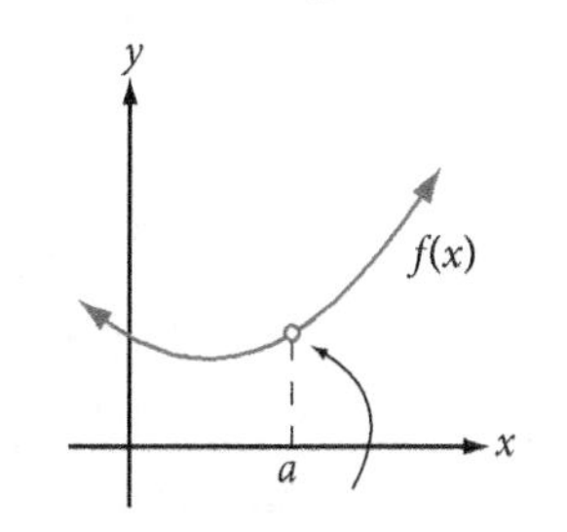

There can be no tangent line here. Therefore, there is no derivative here.

Figure 6

2. $f(x)$ has a corner (Figure 7) at $x = a$.

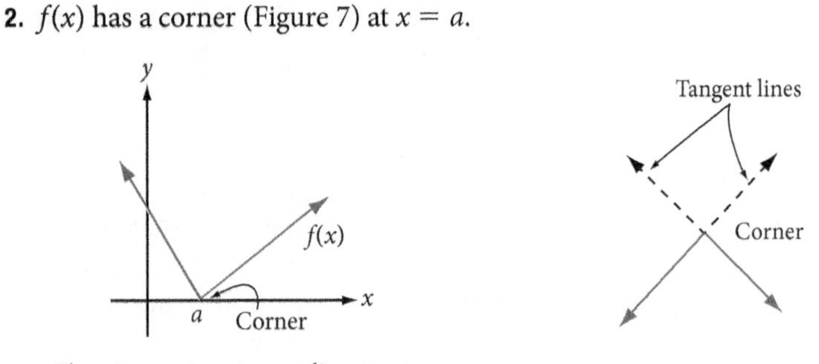

There is no unique tangent line at $y = a$.

Figure 7

3. $f(x)$ has a vertical tangent line at $x = a$. (See Figure 8.) Note that the vertical tangent line in the leftmost graph occurs where there is a *cusp* in the graph.

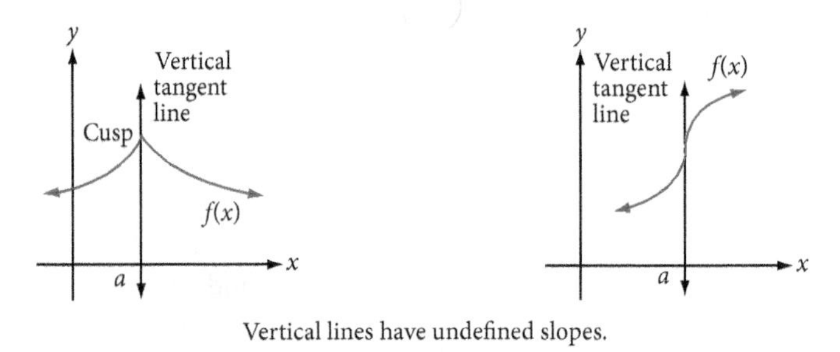

Vertical lines have undefined slopes.

Figure 8

Using Technology 2.1

You can use your graphing calculator to compute the numerical derivative of a function. This numerical derivative will be an approximation to the exact numerical derivative when the derivative is computed symbolically. The reason is that the calculator approximates the slope for the tangent line with the slope of a secant line that is just slightly different from the tangent line.

Example 15 Use your calculator to compute the derivative of

$$f(t) = 90t^2 - t^3$$

Then evaluate it at $t = 15$. This is the function we encountered in Example 14.

Solution Set up the Y-variables like this:

Y1 = 90X^2 − X^3
Y2 = nDeriv(Y1,X,X,0.001)

Note nDeriv(can be found as item 8 under the MATH menu.

Now, in the computation window, enter

Y2(15)

The calculator responds with the value 2024.999999, which you can approximate as 2,025. Thus, as x increases 1 unit, from 15 to 16, the function increases approximately 2,025 units.

You could also compute the value of the derivative at 15 by storing 15 to X and computing Y2 as follows. In the computation window, enter

$15 \rightarrow X$
Y2

This will also produce the value $2024.999999 \approx 2{,}025$. Note that the actual change in the function value can be computed as

Y1(16) − Y1(15)

The actual change is 2,069. ■

Wolfram|Alpha can provide you with both the symbolic and numerical derivative of a function.

Example 16 Use Wolfram|Alpha to find the derivative of the function

$$f(t) = 90t^2 - t^3$$

Then evaluate it at $t = 15$.

Solution Go to www.wolframalpha.com and in the entry field enter

differentiate 90t^2 – t^3

Wolfram|Alpha displays your entry, its interpretation of your entry, and the derivative.

$$-3(t - 60)t$$

This is the same derivative we got in Example 14, but in a different form. Wolfram|Alpha may sometimes generate a form different from the form you get using pencil and paper.

$$-3(t - 60)t = -3t^2 + 180t = 180t - 3t^2$$

To find the value of the derivative when $t = 15$, enter into the Wolfram|Alpha entry field

differentiate 90t^2 – t^3 at t = 15

Wolfram|Alpha responds with 2,025, which agrees with what we computed in Examples 14 and 15. ■

Getting Ready for Class

After reading through the preceding section, respond in your own words and in complete sentences.

A. As if you were explaining to a friend who has not yet taken a calculus course, explain what the derivative of a function is.
B. Describe the conditions under which the derivative of a function gives the exact rate of change of the function or the approximate rate of change of the function.
C. Explain why functions having cusps or corners do not have derivatives at those points.
D. Explain, with a geometric reference, why the derivative of a constant is zero.

Problem Set 2.1

Skills Practice

For Problems 1-6, use the limit definition to find the derivative of each function.

1. $f(x) = x^2 + 2x - 7$
2. $y = 5x^2 - x$
3. $y = 8x + 9$
4. $f(x) = 5x - 8$
5. $f(x) = \dfrac{3}{x}$
6. $y = \dfrac{10}{x}$

For Problems 7-18, use the differentiation rules to find each derivative.

7. $f(x) = 10x^3 - 12x + 6$
8. $f(x) = 4x^4 + 2x^3 + 8x^2 - x - 10$
9. $f(x) = 8x^{5/2}$
10. $y = 9x^{2/3} + x + 1$
11. $y = 6x^{-3/2} + 4x^{1/2} + 8x^{3/2}$
12. $y = 10x^{-5/2} + 2x^{-7/2} - 4x^{-9/2}$
13. $y = \dfrac{1}{x}$
14. $y = \dfrac{5}{x^3} + \dfrac{1}{x} - \dfrac{1}{7}$
15. $f(x) = \dfrac{2}{x^3} + \dfrac{3}{x^2} + \dfrac{3}{x} - \dfrac{1}{4}$
16. $f(x) = \sqrt[3]{x} + \sqrt[4]{x^3} - x - 1$
17. $f(x) = \dfrac{-2}{\sqrt[5]{x^2}} + \dfrac{5x^{4/5}}{\sqrt[5]{x^3}}$
18. $y = 12$

For Problems 19-24, find what is requested.

19. For $f(x) = x^2 + 6x - 4$, find $f(3), f'(3)$, and $f(4) - f(3)$.
20. For $f(x) = 3x^2 + 11x + 1$, find $f(5), f'(5)$, and $f(6) - f(5)$.
21. For $y = 2x^4 + 3x^3 + x + 16$, find $\left.\dfrac{dy}{dx}\right|_{x=0}$
22. For $y = 8x^3 - 8x + 2$, find $\left.\dfrac{dy}{dx}\right|_{x=-1}$
23. Find all the values of x where the slope of the tangent line to
$$f(x) = x^2 + 6x + 1$$
is 0. (Hint: Find $f'(x)$, then set it equal to 0 and solve for x.)
24. Find all the values of x where the slope of the tangent line to
$$f(x) = 4x^2 + 2x + 3$$
is 0.

Paying Attention to Instructions The next two problems are intended to give you practice reading, and paying attention to, the instructions that accompany the problems you are working.

25. If $f(x) = 4x^2 - 16x + 17$, find
    a. $f(1)$.
    b. $f'(1)$.
    c. x so that $f'(x) = 0$.
    d. the equation of the line tangent at $x = 3$.
26. If $g(t) = 2t^3 - 9t^2$, find
    a. $g(2)$.
    b. $g'(2)$.
    c. all t for which $g'(t) = 0$.
    d. the t-intercepts of its graph.

Check Your Understanding

27. Business: Profit from Advertising Suppose, for a particular item, $P(x)$ relates the profit P (in thousands of dollars) to the amount x (in thousands of dollars) spent on advertising. Interpret both $P(60) = 40$ and $\left.\frac{dP}{dx}\right|_{x=60} = 2$.

28. Anatomy: Animal Weight and Tail Thickness Suppose that for a particular animal, the animal's weight w (in pounds) and tail thickness x (in millimeters) are related by the function $w(x)$. Interpret both $w(30) = 20$ and $w'(30) = 0.76$.

29. Psychology: Task Proficiency Psychologists have determined that as a person repeatedly performs a task, he gets more proficient at it. Suppose that t hours are necessary to build N objects. The function $N(t)$ relates t and N. Interpret both $N(150) = 1{,}200$ and $N'(150) = 21.25$.

30. Business: Compact Disc Sales For short periods of time, the number N (in thousands) of compact discs sold by a particular rock group each week is related to the number x of times the group's music video is played on MTV. For a particular type of rock music, N and x are related by the function $N(x)$. Interpret both $N(30) = 74$ and $N'(30) = 7$.

31. Education: Exam Scores An analysis by a group of educational psychologists shows that the average score S (in points) attained by students at a particular high school on a standardized exam is strongly related to the number n of teachers having college degrees in the subject they teach. The research indicates that the function $S(n)$ relates S and n. Interpret both $S(44) = 87$ and $S'(44) = 1.5$.

32. Business: Magazine Circulation A computer magazine publishing company has estimated that its national circulation C (in thousands) depends on the number x of reviews of current computer software it places in each issue. The function $C(x)$ relates C and x. Interpret both $C(12) = 2{,}000$ and $C'(12) = 20$.

33. Ecology: Air Pollution In a particular community in which a large chemical producer is located, under low wind conditions the function $d = f(t)$ relates the total distance d (in miles) a cloud of toxic gas travels and the time t (in hours) since the release of the cloud. Interpret both $f(2) = 7$ and $f'(2) = 3$.

34. Business: Job Applications The director of the personnel office of a large company has determined that the function $N(x)$ approximates the relation between the number N of applicants applying for an accounting position and the amount of dollars x (in hundreds) it advertises as benefits. Interpret both

$$N(42) = 340 \quad \text{and} \quad \left.\frac{dN}{dx}\right|_{x=42} = 66.$$

Modeling Practice

35. Business: Online Advertising Revenue Social networking and video-sharing sites, such as Facebook, dating sites, and YouTube, often include rich media and display advertising placed around user-generated content. Suppose that, for a particular social networking company, the annual revenue from rich media advertisements, in millions of dollars, for the years 2007 through 2012 can be approximated with the model $R(x) = -x^4 + 11x^3 - 39x^2 + 45x$, where x is the number of years from the beginning of 2007.

Source: eMarketer April 2008

a. For this company, what was the revenue from rich media advertisements at the beginning of 2011?

b. At what rate was the revenue changing in the year 2011?

36. Business: Equipment The function

$$W(x) = -0.006x^3 + 0.233x^2 + 1.826x + 19.812$$

approximates the percent of total investment a winery of size x (in cases of wine produced) makes on cellar equipment. Find and interpret:

a. $W(15)$

b. $W'(15)$

Source: Small Winery Investment and Operation Costs, Fickle, Folwell, Ball, and Clary

37. Anatomy: Surface Area and Mass The function $A(m) = 0.11m^{2/3}$ approximates the relationship between the surface area A (in square meters) of a person and his mass m (in kilograms).

a. Find the approximate surface area of a man whose mass is 54 kg; that is, find $A(54)$. (54 kg $\approx$ 120 pounds)

b. Find and interpret $A'(54)$.

Source: www.halls.md

38. Business: Winery Production A vat in a winery has wine flowing into it for fermentation. The amount A (in gallons) of wine in the vat after t minutes is given by the function

$$A(t) = 40t - 30\sqrt{t} \quad 0 \leq t \leq 25$$

a. How much wine is in the vat after 16 minutes?

b. Approximately how many gallons of wine can be expected to flow into the vat between minutes 16 and 17? (Hint: Use $A'(t)$.)

Source: NW Winery Calculators

39. Biology: Heart Rate Biologists have determined that the relationship between heart rate h (in beats per minute) and body weight w (in pounds) is approximated by the function

$$h(w) = \frac{250}{w^{1/4}}$$

a. Find the heart rate for a 16-pound baby.

b. Find the heart rate for a 256-pound man.

c. Find and interpret $h'(w)$ for an 81-pound child. Express $h'(81)$ as a decimal rounded to two decimal places.

Source: Suggested by an article in the *American Journal for Clinical Nutrition*

40. Medicine: Detection of a Tumor For a certain type of tumor, the weight w (in grams) of the tumor t months from the time of detection is given by the function $w(t) = 0.03t^3 + 0.006$. By how much can the weight of the tumor be expected to change in the next month if it has been 4 months since the detection of the tumor?

41. Business: Advertising Revenue The function

$$R(x) = 85 + 7.130\sqrt{x} \quad 0 \le x \le 15{,}000$$

indicates that the monthly revenue R (in thousands of dollars) depends on the amount of dollars x spent on advertising each month.

a. By how much would the monthly revenue be expected to change if the monthly expenditure on advertising were to be raised from its current level of \$6,000 to \$6,001? (Hint: find the marginal revenue.)

b. What is the revenue when the amount spent on advertising is \$6,000?

Medicine: Glucose Tolerance To test for hypoglycemia, physicians often have the patients undergo a glucose tolerance test. For one such test, the amount A (in milligrams) of glucose remaining in one cubic centimeter of blood t hours after ingestion of the glucose is given by the function

$$A(t) = 3.1 + \frac{4.8}{\sqrt{t}} \quad 0 \le t \le 6$$

Source: Suggested by an article in Web MD

42. Find the amount of glucose in the one cubic centimeter of blood 2 hours after ingestion of the glucose.

43. Find the rate at which the concentration of glucose is changing 2 hours after ingestion of the glucose.

Using Technology Exercises

Use your calculator or Wolfram|Alpha to find or compute the derivative of each function.

44. $f(x) = 4x^3 - 5x^2$ at $x = 0.3$

45. $f(x) = x^{3/5} + x^{2/5} - 5x^{1/5}$ at $x = 1.5$

46. $f(x) = 3\sqrt[4]{x^3} + \frac{4}{\sqrt[3]{x^2}}$ at $x = 1$

47. The function $f(x) = 3\sqrt[4]{x^3} + \sqrt[3]{x^2}$ is defined at $x = 0$. What is the calculator or Wolfram|Alpha's response when you try to find $f'(0)$? Why is the response as it is?

48. Biology: Surface Area of Cells The ratio between surface area and volume of biological cells is important to the survival of the cells. To survive, cells must constantly exchange ions, gases, nutrients, and wastes with their environment. The surface area of an object is the summation of the areas of all the exposed sides of the object, and the volume of an object is a measure of how much space the object occupies. In biology, a filament is a box-shaped object. To calculate the surface area of a filament, calculate the surface area of each face of the filament. To calculate the volume of a filament, use volume = length times width times height. Suppose a particular neuronal filament has width x nanometers (nm), height $0.3x$ nm, and length $100x$ nm.

a. Develop the formula for the ratio of the surface area of the filament and its volume.

b. Find and interpret the derivative of the ratio when the width of the filament is 10 nm.

Source: *Neuronal Intermediate Filaments*, Liem

Getting Ready for the Next Section

Multiply.

49. $(5x^2 + 4)(x^3 + 11)$

50. $2.3(7.1x^2 + 210)$

51. $(x^2 - 1)(3x^2 + 2)$

52. $-2x(x^3 + 2x - 1)$

Simplify.

53. $2t(t^2 - 4) + 2t(t^2 + 1)$

54. $2(x - 3) - (2x - 5)$

Solve. Give only real answers.

55. $(t^2 - 4)(t^2 + 1) = 0$

56. $4t^3 - 6t = 0$

57. If $g(x) = \dfrac{x^4 - 5x^2 + 2x - 2}{(x^2 - 1)^2}$, find $g(2)$.

58. If $f(x) = \dfrac{2x - 5}{x - 3}$, find $f(4)$.

59. If $g(x) = \dfrac{-1}{(x - 3)^2}$, find $g(4)$.

60. If $g(x) = -16.33x^2 + 483$, find $g(4)$.

61. Find the equation of the line through $(4, 3)$ with slope -1.

62. Find the equation of the line through $(-4, 1)$ with slope 3.

Find the derivative.

63. $y = 5x^2 + 4$

64. $y = t^2 - 4$

65. $y = 2x - 5$

66. $y = x^3 + 11$

67. $y = t^2 + 1$

68. $y = x - 3$

69. $y = 5x^5 + 4x^3 + 55x^2 + 44$

70. $y = 7.1x^2 + 210$

Differentiating Products and Quotients

2.2

Book publishers carefully consider many numbers before deciding to publish and release a book. They are keenly interested in having a good sense of the profit the book is likely to bring each week of its sales. Not only is the profit number important, publishers wish to have an idea of how the profit is likely to change from one week to the next. Publishers wish to know the *marginal profit*, that is, the profit it realizes in adding one additional week to its number of weeks since release. Marginal profit is one of the topics we will cover in this section.

The differentiation rules we examined in Section 2.1 significantly increased our ability to differentiate functions. The product and quotient rules we will examine in this section will extend our ability to differentiate more sophisticated functions. The rules for the derivatives of products and quotients of functions are not as intuitive as are the rules for the derivatives of sums and differences.

The Product Rule

To differentiate the product of two differentiable functions, use the product rule.

The Product Rule

If $u(x)$ and $v(x)$ are differentiable functions, then

$$\frac{d}{dx}[u(x) \cdot v(x)] = \frac{d}{dx}[u(x)] \cdot v(x) + u(x) \cdot \frac{d}{dx}[v(x)]$$

In the prime notation, the product rule is written as

$$\frac{d}{dx}[u(x) \cdot v(x)] = u'(x) \cdot v(x) + u(x) \cdot v'(x)$$

The product rule states that the derivative of a product is

$$\begin{pmatrix}\text{the derivative}\\ \text{of the first factor}\end{pmatrix}\begin{pmatrix}\text{the second}\\ \text{factor}\end{pmatrix} + \begin{pmatrix}\text{the first}\\ \text{factor}\end{pmatrix}\begin{pmatrix}\text{the derivative}\\ \text{of the second factor}\end{pmatrix}$$

In symbols,

$$\frac{d}{dx}[u(x)v(x)] = \underbrace{u'(x)}_{(\text{first})'} \cdot \underbrace{v(x)}_{(\text{second})} + \underbrace{u(x)}_{(\text{first})} \cdot \underbrace{v'(x)}_{(\text{second})'}$$

VIDEO EXAMPLES

SECTION 2.2

Example 1 Find the derivative of $f(x) = (5x^2 + 4)(x^3 + 11)$.

Solution If we identify $u(x) = 5x^2 + 4$ and $v(x) = x^3 + 11$, we have the product $f(x) = u(x)v(x)$, and

$$\begin{aligned} f'(x) &= \underbrace{u'(x)}_{(\text{first})'} \cdot \underbrace{v(x)}_{(\text{second})} + \underbrace{u(x)}_{(\text{first})} \cdot \underbrace{v'(x)}_{(\text{second})'} \\ &= \underbrace{(10x)}_{(\text{first})'} \cdot \underbrace{(x^3 + 11)}_{(\text{second})} + \underbrace{(5x^2 + 4)}_{(\text{first})} \cdot \underbrace{(3x^2)}_{(\text{second})'} \end{aligned}$$

$$= 10x^4 + 110x + 15x^4 + 12x^2$$

$$= 25x^4 + 12x^2 + 110x$$

Thus, $f'(x) = 25x^4 + 12x^2 + 110x$. ■

Note In addition to illustrating the product rule, Example 1 illustrates two important facts.

1. The derivative of a product could also be found by performing the multiplication, then differentiating term-by-term using the sum rule.

$$f(x) = (5x^2 + 4)(x^3 + 11)$$
$$= 5x^5 + 4x^3 + 55x^2 + 44$$

so that

$$f'(x) = 25x^4 + 12x^2 + 110x$$

as before.

It is not always practical or possible to perform the multiplication. For example, it is impossible to multiply out $(5x + 4)^7(x^3 + 11)^{-2/3}$.

2. A note of caution: the derivative of a product is *not* the product of the individual derivatives. That is,

$$\frac{d}{dx}[u(x)v(x)] \neq u'(x) \cdot v'(x)$$

For example, if

$$f(x) = (5x^2 + 4)(x^3 + 11)$$

then

$$f'(x) \neq (10x)(3x^2)$$

Example 2 For the function $g(t) = (t^2 - 4)(t^2 + 1)$, find

a. the values of t for which $g(t) = 0$.

b. the values of t for which $g'(t) = 0$.

Solution

a. We set $g(t)$ equal to 0, and solve for t.

$$(t^2 - 4)(t^2 + 1) = 0$$
$$(t + 2)(t - 2)(t^2 + 1) = 0$$
$$t + 2 = 0 \quad \text{or} \quad t - 2 = 0 \quad \text{or} \quad t^2 + 1 = 0$$
$$t = -2 \quad \text{or} \quad t = 2 \quad \text{or} \quad \text{no real solution}$$

The function $g(t)$ is 0 when $t = -2$, and when $t = 2$.

b. First we differentiate g with respect to t using the product rule:

$$g'(t) = \frac{d}{dt}(t^2 - 4)(t^2 + 1) + (t^2 - 4)\frac{d}{dt}(t^2 + 1)$$
$$= (2t)(t^2 + 1) + (t^2 - 4)(2t)$$
$$= 2t^3 + 2t + 2t^3 - 8t$$
$$= 4t^3 - 6t$$

Next, we set the derivative equal to 0 and solve for t.

$$4t^3 - 6t = 0$$
$$2t(2t^2 - 3) = 0$$
$$2t = 0 \quad \text{or} \quad 2t^2 - 3 = 0$$
$$t = 0 \quad \text{or} \quad t^2 = \frac{3}{2}$$
$$t = 0 \quad \text{or} \quad t = \pm\sqrt{\frac{3}{2}} = \pm\frac{\sqrt{6}}{2}$$

The derivative is 0 at $t = 0$, $-\frac{\sqrt{6}}{2}$, and $\frac{\sqrt{6}}{2}$. At these values of t the slope of the lines tangent to the graph are 0; those tangent lines are horizontal.

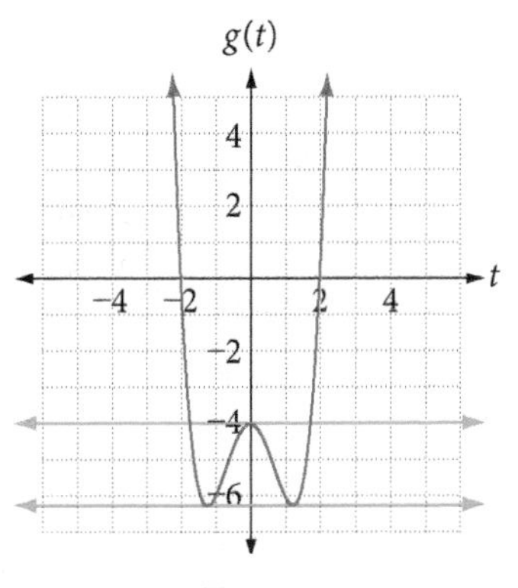

Figure 1

The Quotient Rule

To differentiate the quotient of two differentiable functions, use the quotient rule.

The Quotient Rule

If $u(x)$ and $v(x)$ are differentiable functions and $v(x) \neq 0$, then

$$\frac{d}{dx}\left[\frac{u(x)}{v(x)}\right] = \frac{v(x) \cdot \frac{d}{dx}[u(x)] - u(x) \cdot \frac{d}{dx}[v(x)]}{[v(x)]^2}$$

In the prime notation, the quotient rule is written as

$$\frac{d}{dx}\left[\frac{u(x)}{v(x)}\right] = \frac{v(x) \cdot u'(x) - u(x) \cdot v'(x)}{[v(x)]^2}$$

The quotient rule states that the derivative of a quotient is

$$\frac{d}{dx}\left[\frac{\text{numerator}}{\text{denominator}}\right] = \frac{(\text{denom.}) \cdot (\text{numer.})' - (\text{numer.}) \cdot (\text{denom.})'}{(\text{denom.})^2}$$

or (less technical, but easier to say),

$$\frac{d}{dx}\left[\frac{\text{top}}{\text{bottom}}\right] = \frac{(\text{bottom}) \cdot (\text{top})' - (\text{top}) \cdot (\text{bottom})'}{(\text{bottom})^2}$$

Example 3 Find $f'(2)$ for $f(x) = \frac{x^3 + 2x - 1}{x^2 - 1}$.

Solution We first recall that $f'(2)$ indicates that the derivative of $f(x)$ is to be evaluated at $x = 2$. Thus, initially, we need to find $f'(x)$. Since $f(x)$ is a quotient of functions, we will use the quotient rule. If we identify $u(x) = x^3 + 2x - 1$ and $v(x) = x^2 - 1$, we have the quotient $f(x) = \frac{u(x)}{v(x)}$.

$$f'(x) = \frac{v(x) \cdot u'(x) - u(x) \cdot v'(x)}{[v(x)]^2}$$

or

$$f'(x) = \frac{(\text{bottom}) \cdot (\text{top})' - (\text{top}) \cdot (\text{bottom})'}{(\text{bottom})^2}$$

$$= \frac{\overbrace{(x^2 - 1)}^{(\text{bottom})}\overbrace{(3x^2 + 2)}^{(\text{top})'} - \overbrace{(x^3 + 2x - 1)}^{(\text{top})}\overbrace{(2x)}^{(\text{bottom})'}}{(x^2 - 1)^2}$$

Now we simplify using the distributive property:

$$= \frac{3x^4 - x^2 - 2 - 2x^4 - 4x^2 + 2x}{(x^2 - 1)^2}$$

$$= \frac{x^4 - 5x^2 + 2x - 2}{(x^2 - 1)^2}$$

Thus,

$$f'(x) = \frac{x^4 - 5x^2 + 2x - 2}{(x^2 - 1)^2}$$

Now we can evaluate $f'(x)$ at $x = 2$.

$$f'(2) = \frac{2^4 - 5 \cdot 2^2 + 2 \cdot 2 - 2}{(2^2 - 1)^2}$$

$$= \frac{16 - 20 + 4 - 2}{9}$$

$$= \frac{-2}{9}$$

In addition to illustrating the quotient rule, Example 3 illustrates an important fact.

The derivative of a quotient is *not* the quotient of the individual derivatives of the function. That is,

$$\frac{d}{dx}\left[\frac{u(x)}{v(x)}\right] \neq \frac{u'(x)}{v'(x)}$$

For example, if

$$f(x) = \frac{x^3 + 2x - 1}{x^2 - 1}$$

then

$$f'(x) \neq \frac{3x^2 + 2}{2x}$$

There is a significant difference between the quotient of derivatives, $\frac{f'(x)}{g'(x)}$, and derivative of a quotient, $\left[\frac{f(x)}{g(x)}\right]'$. The quotient rule gives us that

$$\left[\frac{f(x)}{g(x)}\right]' = \frac{g(x) \cdot f'(x) - f(x) \cdot g'(x)}{[g(x)]^2}$$

Comparing expressions, it is hard to imagine that the left and right sides below could ever be equal:

$$\frac{g(x) \cdot f'(x) - f(x) \cdot g'(x)}{[g(x)]^2} \neq \frac{f'(x)}{g'(x)}$$

Thus,

$$\left[\frac{f(x)}{g(x)}\right]' \neq \frac{f'(x)}{g'(x)}$$

Example 4 illustrates this difference with an application of a revenue function.

Example 4 Two quantities can be compared using division. Division indicates how many times more one quantity is than another. For example, $\frac{15}{3} = 5$ indicates that 15 is 5 times as big as 3. Suppose that analyst A uses the revenue model $R_A(t)$ to predict the annual revenue for a company t years from now. For her revenue function, she computes $R_A(3) = 513{,}684$ and $R_A'(3) = 4{,}623$. Analyst B uses the revenue model $R_B(t)$ to predict the annual revenue of the same company also t years from now. For his revenue function, he computes $R_B(3) = 512{,}145$ and $R_B'(3) = 4{,}097$.

a. What information is contained in the quotient

$$\frac{R_A(3)}{R_B(3)} \approx 1.003$$

b. What information is contained in the quotient

$$\frac{R_A'(3)}{R_B'(3)} = \frac{4{,}623}{4{,}097} \approx 1.13$$

c. What information is contained in the quotient

$$\left[\frac{R_A(t)}{R_B(t)}\right]' \approx 0.001 \quad \text{when } t = 3$$

Solution

a. This quotient indicates that 3 years from now, the revenue predicted by analyst A will be about 1.003 times as big as that predicted by analyst B.

b. This quotient indicates that 3 years from now, the revenue predicted by analyst A will be increasing at a rate that is about 1.13 times the rate predicted by analyst B.

c. This means that in year 4, the ratio of A's predicted value to B's predicted value will increase by about 0.001. The ratio in year 3 is 1.003. In year 4, then, we expect the ratio of A's predicted value to B's predicted value to be

$$1.003 + 0.001 = 1.004$$

Example 5 The publisher of a musician's recently released autobiography expects that over the first 13 months after its release, the monthly profit can be approximated with the model

$$P(t) = \frac{500 + 100t - 10t^2}{t^2 + 5}$$

where P is in thousands of dollars and t is in months from the time of release.

a. Derive a model that gives the marginal profit of the book.

b. Specify and interpret the expected marginal profit 5 months after release.

c. Specify and interpret the expected marginal profit 10 months after release.

d. Offer a mathematical argument to demonstrate that from month 1 after release, the monthly profit realized from the sales of the book decreases steadily.

Solution

a. The marginal profit is given by the derivative of the profit function. Applying the quotient rule,

$$P'(t) = \frac{(t^2 + 5) \cdot (100 - 20t) - (500 + 100t - 10t^2) \cdot (2t)}{(t^2 + 5)^2}$$

$$= \frac{-100(t^2 + 11t - 5)}{(t^2 + 5)^2}$$

b. At $t = 5$, $P'(5) = \frac{-100(t^2 + 11t - 5)}{(t^2 + 5)^2} \approx -8.3$. The publisher can expect the profit realized on the sale of the book to decrease by about \$8,300 from month 5 to month 6.

c. At $t = 10$, $P'(10) = \frac{-100(t^2 + 11t - 5)}{(t^2 + 5)^2} \approx -1.9$. The publisher can expect the profit realized on the sale of the book to decrease by about \$1,900 from month 10 to month 11.

d. For all values of $t > 1$, $P'(t) < 0$, demonstrating that the profit function $P(t)$ is always decreasing.

Example 6 Find the equation of the line tangent to the graph of

$$f(x) = \frac{2x - 5}{x - 3}$$

when $x = 4$.

Solution First we use the function to find the value of y that comes from $x = 4$.

$$f(4) = \frac{2 \cdot 4 - 5}{4 - 3}$$

$$= \frac{3}{1} = 3$$

The point on the graph is (4, 3).

Next we find the slope of the tangent line by first finding the derivative.

$$f'(x) = \frac{(x - 3)\frac{d}{dx}(2x - 5) - (2x - 5)\frac{d}{dx}(x - 3)}{(x - 3)^2}$$

$$= \frac{(x - 3)(2) - (2x - 5)(1)}{(x - 3)^2}$$

$$= \frac{2x - 6 - 2x + 5}{(x - 3)^2}$$

$$= \frac{-1}{(x - 3)^2}$$

We find the slope of the tangent line by substituting 4 for x in the derivative.

$$f'(4) = \frac{-1}{(4 - 3)^2} = -1$$

Using the point-slope form of a line $y - y_1 = m(x - x_1)$, we get the equation of the line tangent to $y = \frac{2x - 5}{x - 3}$ when $x = 4$:

$$y - 3 = -1(x - 4)$$

$$y - 3 = -x + 4$$

$$y = -x + 7$$

The graph of the function and the line tangent at $x = 4$ are shown in Figure 2.

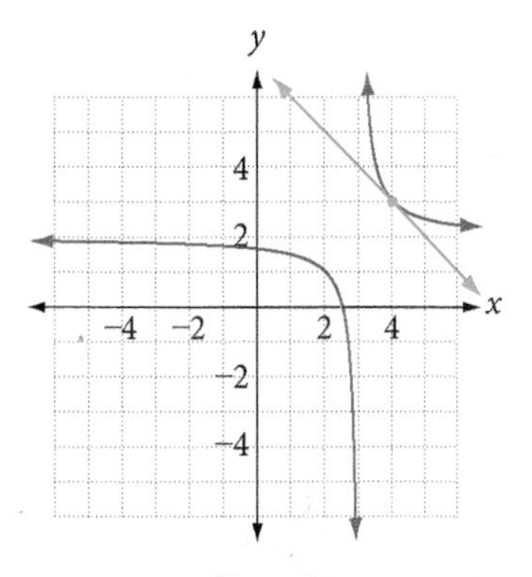

Figure 2

Example 7 The campaign committee for a politician estimates that in an upcoming national election the politician can, by campaigning for x days in a particular county, secure $2.3x$ (in thousands) of votes. However, campaigning costs money and the committee estimates that the cost for campaigning x days will be $7.1x^2 + 210$ dollars. The function

© ABDESIGN/iStockPhoto

$$f(x) = \frac{\text{number of votes}}{\text{cost for those votes}} = \frac{2.3x}{7.1x^2 + 210}$$

produces the number of votes (in thousands) per dollar spent. Find both $f'(4)$ and $f'(6)$.

Solution Since $f(x)$ is a quotient, we'll use the quotient rule. If we identify $u(x) = 2.3x$ and $v(x) = 7.1x^2 + 210$, we have

$$f(x) = \frac{v(x) \cdot u'(x) - u(x) \cdot v'(x)}{[v(x)]^2}$$

or

$$f(x) = \frac{(\text{bottom}) \cdot (\text{top})' - (\text{top}) \cdot (\text{bottom})'}{(\text{bottom})^2}$$

$$= \frac{\overbrace{(7.1x^2 + 210)}^{(\text{bottom})}\overbrace{(2.3)}^{(\text{top})'} - \overbrace{(2.3x)}^{(\text{top})}\overbrace{(14.2x)}^{(\text{bottom})'}}{(7.1x^2 + 210)^2}$$

Now we simplify:

$$= \frac{16.33x^2 + 483 - 32.66x^2}{(7.1x^2 + 210)^2}$$

$$= \frac{-16.33x^2 + 483}{(7.1x^2 + 210)^2}$$

Thus,

$$f'(x) = \frac{-16.33x^2 + 483}{(7.1x^2 + 210)^2}$$

Now, we can find $f'(4)$ and $f'(6)$.

$$f'(4) = \frac{-16.33(4)^2 + 483}{(7.1(4)^2 + 210)^2}$$

$$\approx 0.0021$$ 2.1 votes per dollar spent per day

$$f'(6) = \frac{-16.33(6)^2 + 483}{(7.1(6)^2 + 210)^2}$$

$$\approx -0.0005$$ −0.5 votes per dollar spent per day

Interpretation

If the politician increases the number of campaigning days in the county by 1, from 4 days to 5 days, he will secure approximately 2.1 more votes for each dollar spent. If he increases the number of campaigning days in the county by 1, from 6 days to 7 days, he will lose approximately one-half vote for each dollar spent. His campaign committee must determine the optimal number of campaigning days. If he campaigns fewer days than that optimal number, he will not secure the greatest number of votes per dollar. If he campaigns more days than that optimal number, the efficiency of "buying" votes will go down. The committee must decide when the politician has had just the right amount of exposure. (Section 3.3 is concerned with the process of optimization.)

Using Technology 2.2

You can use Wolfram|Alpha to help you find derivatives of functions that involve products and quotients.

Example 8 Use Wolfram|Alpha to find the derivative of the function

$$f(x) = \frac{x^2 + 3x}{x^2 - 4x}$$

then evaluate it at $x = 5$.

Solution Go to www.wolframalpha.com and in the entry field enter

differentiate (x^2 + 3x) / (x^2 – 4x)

Wolfram|Alpha displays your entry, its interpretation of your entry, and the derivative.

$$-\frac{7}{(x - 4)^2}$$

To find the value of the derivative when $x = 5$, enter into the Wolfram|Alpha entry field

differentiate (x^2 + 3x) / (x^2 - 4x) at x = 5

Wolfram|Alpha responds with -7.

Note Wolfram|Alpha also displays other information about the function, such as the graph and domain and range, but we can ignore those features for now.

Getting Ready for Class

After reading through the preceding section, respond in your own words and in complete sentences.

A. State the product rule for derivatives.

B. Is the derivative of a quotient the same as the quotient of the derivatives?

C. If $g(t) = (t^2 - 4)(t^2 + 1)$, what is $g(2)$?

D. If $g(t) = (t^2 - 4)(t^2 + 1)$, what is $g'(0)$?

Problem Set 2.2

Skills Practice

For Problems 1–20, use the product and/or quotient rule to find each derivative.

1. $f(x) = (2x + 1)(4x - 5)$
2. $f(x) = (3x + 4)(2x + 5)$
3. $f(x) = (x^2 + 4)(x - 2)$
4. $f(x) = (x^2 + 3)(2x + 3)$
5. $f(x) = (x^2 - 1)(x^2 + 2)$
6. $f(x) = (x^2 + 7)(x^2 - 5)$
7. $f(x) = \dfrac{4x + 3}{5x + 2}$
8. $f(x) = \dfrac{2x + 1}{2x - 1}$
9. $f(x) = \dfrac{x - 4}{3x + 2}$
10. $f(x) = \dfrac{x + 6}{4x - 1}$
11. $f(x) = \dfrac{3x^3 - 7x + 1}{x^2 - 4}$
12. $f(x) = \dfrac{6x^2 + 8x - 5}{2x^2 + x + 1}$
13. $f(x) = \dfrac{x^4 - 6}{x^3 + 1}$
14. $f(x) = \dfrac{2x^5 + 9}{x - 7}$
15. $f(x) = \dfrac{(2x - 1)(x^2 + 5)}{7x + 2}$
16. $f(x) = \dfrac{(x - 4)(9x + 14)}{x^2 + 1}$
17. $f(x) = \dfrac{8}{x^2 + 3x + 4}$
18. $f(x) = \dfrac{-6}{x^3 + 5}$
19. $f(x) = \dfrac{\sqrt{x}}{x - 4}$
20. $f(x) = \dfrac{x^{2/3}}{x^{1/3} + x}$

For Problems 21–24, find and interpret the derivative of the given function at the indicated value of the independent variable.

21. $f(x) = (9x + 2)(x - 6), \quad x = 4$
22. $f(x) = (x^2 + x - 1)(5x - 7), \quad x = 2$
23. $g(t) = \dfrac{5t + 4}{8t + 6} \quad t = 10$
24. $h(s) = \dfrac{9s + 1}{9s - 4} \quad s = 1$

Check Your Understanding

For Problems 25-32, a function is given and the derivatives of its parts are given. (For now, do not worry about how some of the derivatives were obtained. You will learn those techniques in a later section.) Place the parts and the derivatives in their proper positions so as to express the derivative of the function. (You need not simplify.)

25. $f(x) = 3x^4(x^3 + 1)$; $3x^4$, $x^3 + 1$, $\dfrac{d}{dx}(3x^4) = 12x^3$, $\dfrac{d}{dx}(x^3 + 1) = 3x^2$
26. $f(x) = 5x(2x - 3)^4$; $5x$, $(2x - 3)^4$, $\dfrac{d}{dx}(5x) = 5$, $\dfrac{d}{dx}(2x - 3)^4 = 8(2x - 3)^3$
27. $f(x) = (3x + 7)\sqrt{5x + 2}$; $3x + 7$, $\sqrt{5x + 2}$,
$\dfrac{d}{dx}(3x + 7) = 3$, $\dfrac{d}{dx}\sqrt{5x + 2} = \dfrac{5}{2\sqrt{5x + 2}}$

28. $f(x) = \sqrt{x^2+6}\sqrt[3]{2x-9}$; $\sqrt{x^2+6}$, $\sqrt[3]{2x-9}$,

$\dfrac{d}{dx}\sqrt{x^2+6} = \dfrac{x}{\sqrt{x^2+6}}$, $\dfrac{d}{dx}\sqrt[3]{2x-9} = \dfrac{2}{3\sqrt[3]{(2x-9)^2}}$

29. $f(x) = \dfrac{5x-7}{8x+1}$; $5x-7$, $8x+1$, $\dfrac{d}{dx}(5x-7)=5$, $\dfrac{d}{dx}(8x+1)=8$

30. $f(x) = \dfrac{9x-4}{2x-5}$; $9x-4$, $2x-5$, $\dfrac{d}{dx}(9x-4)=9$, $\dfrac{d}{dx}(2x-5)=2$

31. $f(x) = (x^2+3)^{1/4}(2x+7)$; $(x^2+3)^{1/4}$, $2x+7$,

$\dfrac{d}{dx}(x^2+3)^{1/4} = \dfrac{1}{4}(x^2+3)^{-3/4}(2x)$, $\dfrac{d}{dx}(2x+7)=2$

32. $f(x) = x^{-2/3}(5x^2+x-4)^{1/6}$; $x^{-2/3}$, $(5x^2+x-4)^{1/6}$,

$\dfrac{d}{dx}x^{-2/3} = \dfrac{-2}{3}x^{-5/3}$, $\dfrac{d}{dx}(5x^2+x-4)^{1/6} = \dfrac{1}{6}(5x^2+x-4)^{-5/6}(10x+1)$

Modeling Practice

33. Psychology: Memorization The function

$$N(t) = \frac{80t}{105t-80}$$

is a form of a function developed by psychologist L.L. Thurstone that relates the number N of facts that a person can remember t hours after memorizing them. Find and interpret $N(1)$, $N(3)$, $N'(1)$, and $N'(3)$.

34. Business: Company Takeovers During the buyout attempt of company X by company Y, the accountants of company Y developed a formula that relates the total assets A (in millions of dollars) company Y can expect from the acquisition of company X t years from the time of takeover. The function is

$$A(t) = \frac{t^{5/4}+1}{3t+6} \quad 0 \le t \le 15$$

Find and interpret $A(4)$ and $A'(4)$.

Source: Suggested by an article in Bloomberg.com

© Shaun Lowe Photographic/ iStockPhoto

35. Physical Science: Drag Racing The following rational function models the speed $V(t)$, in miles per hour, of a dragster t seconds since it starts its quarter-mile run.

$$V(t) = \frac{340t}{t+3}$$

a. What is the speed of the dragster 4 seconds into its run? Round your answer to one decimal place.

b. Write an expression that models the rate at which the speed changes as time changes.

c. How fast is the speed changing 4 seconds into the dragster's run? Round your answer to one decimal place.

d. Interpret the value you obtained in part c.

© Warchi/iStockPhoto

36. Business: Average Cost A producer and distributor of gift cards produces x boxes of cards per day at an average cost $\overline{C}(x)$ (in dollars) of

$$\overline{C}(x) = \frac{6x + 8{,}000}{2x + 300}$$

If the producer is currently producing and distributing 280 boxes of cards each day, what change can she expect in the average daily cost of production if she increases production to 281 boxes?

37. Medicine: Drug Concentration The concentration $C(t)$ (in milligrams per cubic millimeter) of a particular drug in the human bloodstream t hours after ingestion is given by

$$C(t) = \frac{21t}{t^2 + t + 4}$$

What change in the concentration can the prescribing physician expect between the third and fourth hours after ingestion?

38. Economics: Demand for a Product Economists have established that the demand D for an item decreases as the price x increases. The daily number D of tubes of roofing tar that people are willing to buy at x cents each is given by

$$D(x) = \frac{85{,}000}{\sqrt{x} + 10}, \quad 200 \leq x \leq 700$$

If the current price per tube is \$2.50, find and interpret the marginal demand for tubes of roofing tar.

39. Business: Online Advertising Revenue Two niche social networking companies, one for financial advisors and one for accountants, generate revenue from rich media advertisements. The function $F(x) = -3x^2 + 15x$ models the revenue (in millions of dollars) for the financial advisors network for the years 2007 to 2012. The function $A(x) = -x^2 + 9x$ models the revenue (in millions of dollars) for the accountants network for the years 2007 to 2012. The ratio $\frac{F(x)}{A(x)}$ is a measure of the size of $F(x)$ relative to $A(x)$. For example, if $\frac{F(x)}{A(x)} = 3$, then $F(x) = 3 \cdot A(x)$ and you can see that $F(x)$ is three times the size of $A(x)$. That is, the revenue realized by the financial advisors network is three times the revenue realized by the accountants network. (Assume x is the number of years from January 1, 2007.)

a. How does the revenue of the financial advisors network compare to that of the accountants niche in 2011?

b. Write an expression for the rate at which the ratio of revenues changes as the number of years from 2007 changes.

c. At what rate is the ratio changing in the year 2011?

d. Interpret the value you obtained in part c. Writing the value as a decimal number over 1 will help you to do so.

40. Health Science: Dieting The function

$$W(x) = \frac{80(2x + 15)}{x + 6}$$

might be used to model the progress of a person on a particular diet. The quantity $W(x)$ represents the person's weight (in pounds) x weeks from the first day of the diet.

a. Write an expression for the rate of change of weight with respect to x.

b. At what rate is a person's weight changing if the person is 10 weeks into the diet? Five weeks into the diet?

41. Business: Marginal Revenue A manufacturer of ceramic vases has determined that her weekly revenue and cost functions for the manufacture and sale of x vases are $R(x) = 95x - 0.08x^2$ dollars and $C(x) = 1{,}200 + 35x - 0.04x^2$ dollars, respectively. Given that profit equals revenue minus cost,

a. find the marginal revenue, marginal cost, and marginal profit functions.

b. estimate the revenue realized by selling the 601st vase and the 900th vase. What seems to be happening?

c. estimate the cost incurred by producing the 601st vase and the 900th vase. What seems to be happening?

d. estimate the profit realized by producing the 601st vase and the 900th vase. What seems to be happening?

42. Business: Marginal Profit The publisher of a recently released nonfiction book expects that over the first 20 months after its release, the monthly profit can be approximated with the model

$$P(t) = \frac{900 + 600t - 30t^2}{t^2 + 10}$$

where P is in thousands of dollars and t is in months from the time of release.

a. Derive a model that gives the marginal profit of the book.

b. Specify and interpret the expected marginal profit 5 months after release.

c. Specify and interpret the expected marginal profit 10 months after release.

d. Offer a mathematical argument to demonstrate that from month 2 after release, the monthly profit realized from the sales of the book decreases steadily.

43. Business: Marginal Profit The publisher of a recently released biography book expects that over the first 15 months after its release, the monthly profit can be approximated with the model

$$P(t) = \frac{200 + 500t - 10t^2}{t^2 + 10}$$

where P is in thousands of dollars and t is in months from the time of release.

a. Derive a model that gives the marginal profit of the book.

b. Specify and interpret the expected marginal profit 5 months after release.

c. Specify and interpret the expected marginal profit 10 months after release.

d. Offer a mathematical argument to demonstrate that from month 3 after release, the monthly profit realized from the sales of the book decreases steadily.

44. Manufacturing: Productivity A manufacturer's records over the last 80 months show that is productivity (in units produced each day) is

$$P(t) = 0.60t^2 - 100t + 9{,}500$$

in month number t where $t = 0$ represents the beginning of the first month of the eight-year period. Find and interpret the manufacturer's marginal productivity in

a. at the end of the 20th month.

b. at the end of the 60th month.

c. at the end of the 75th month. What seems to be happening?

45. Business: Comparisons of Revenue Predictions Suppose that analyst Star uses the revenue model $R_S(t)$ to predict the revenue of the Solar Car Company t years from now. For her revenue function, she computes $R_S(2) = 84{,}946.90$ and $R'_S(2) = 2{,}548.41$. Analyst Moon uses the revenue model $R_M(t)$ to predict the revenue of the same car company also t years from now. For his revenue function, he computes $R_M(2) = 81{,}616.10$ and $R'_M(2) = 816.16$.

a. What information is contained in the quotient

$$\frac{R_S(2)}{R_M(2)} = \frac{84{,}946.90}{81{,}616.10} \approx 1.04$$

b. What information is contained in the quotient of the derivatives

$$\frac{R'_S(2)}{R'_M(2)} = \frac{2{,}548.41}{816.16} \approx 3.12$$

c. What information is contained in the derivative of the quotient

$$\left[\frac{R_S(t)}{R_M(t)}\right]' \approx 0.02 \quad \text{when } t = 2$$

46. Life Science: Comparison of Drug Concentrations Suppose a researcher measuring the concentration of two drugs in a person's bloodstream, uses the two concentration models $C_1(t)$ for drug 1 and $C_2(t)$ for drug 2 to predict the number of milligrams in the bloodstream t hours from the time the drug is introduced into the bloodstream. For these functions, she computes $C_1(1) = 106.84$, $C_2(1) = 104.081$, $C'_1(1) = 6.37$, and $C'_2(1) = 4.16$.

a. What information is contained in the quotient

$$\frac{C_1(1)}{C_2(1)} = \frac{106.84}{104.081} \approx 1.03$$

b. What information is contained in the quotient of the derivatives

$$\frac{C'_1(1)}{C'_2(1)} = \frac{6.37}{4.16} \approx 1.53$$

c. What information is contained in the derivative of the quotient

$$\left[\frac{C_1(t)}{C_2(t)}\right]' \approx 0.02 \quad \text{when } t = 1$$

Using Technology Exercises

Use Wolfram|Alpha to compute the derivative of each function.

47. $f(x) = \dfrac{x^2 - 5x}{x^2 + 6x}$

48. $f(x) = \dfrac{2x^3 + 1}{2x^3 - 1}$

49. $f(x) = \dfrac{x^{2/3} + x}{x^{2/3} - x}$ when $x = 64$

50. Biology: Population Size Knowing the size of biological populations is important to evolutionary biologists in assessing the health of the population and in identifying processes that shape the evolution of the population. But the census size of the population may not be the best measure. In some populations, there may be a drastic difference between the number of sexually mature individuals and the number of those that are not capable of breeding. In such populations, biologists often use the effective size N_e rather than the census size to measure the size of a population. The formula for effective size is

$$N_e = \frac{4N_f N_m}{N_m + N_f}$$

where N_f and N_m represent the number of breeding females and breeding males, respectively.

a. Suppose a population has x number of breeding males and 400 breeding females. Rewrite the effective population function given above so that it appears as a function $N_e(x)$.

b. Differentiate $N_e(x)$ to find an expression for the rate at which the effective population size is changing for a population.

c. Use the expression you found in part b to find the rate at which the effective population size is changing for a population in which there are 400 breeding females and 20 breeding males.

Source: The Biology Project, Department of Biochemistry and Molecular Biophysics at the University of Arizona

51. When x liters of distilled water are added to 5 liters of a 30% acid solution, the acid solution is diluted to a P%-acid solution according to the function

$$P(x) = \frac{1.5}{x + 5}$$

Compute and interpret both $P(3)$ and $P'(3)$.

52. Social Science: Age at Which Men First Marry The median age at which men marry for the first time can be modeled by the cubic function

$$M(x) = -0.0044x^2 + 0.2001x + 24.7$$

and the median age at which women marry for the first time by

$$W(x) = -0.0040x^2 + 0.2330x + 22.07$$

where in each case, x represents the number of years since January 1, 1980. The ratio of these two functions is $\frac{M(x)}{W(x)}$.

a. What does the quotient of these two functions measure? (A quotient is a ratio. Think about what a ratio represents.)

b. Compute and interpret $\frac{M(x)}{W(x)}$ for the year 1980, that is, when $x = 0$.

c. Compute and interpret $\frac{M(x)}{W(x)}$ for the year 2010, that is, when $x = 30$.

Source: www.census.gov

Getting Ready for the Next Section

53. If $f(x) = \frac{1}{2}x^4 - 4x^2$, find $f(2)$.

54. If $g(x) = 2x^3 - 8x$, find $g(2)$.

Solve.

55. $x^3 - 9x = 0$

56. $3x^2 - 9 = 0$

Differentiate each function.

57. $g(x) = 2x^3 - 8x$

58. $h(x) = 180x^2 - 48x$

59. $f(x) = x^3 - 9x$

60. $y = \frac{1}{x}$

Spotlight on **Success**

Student Instructor Gordon

Math takes time. This fact holds true in the smallest of math problems as much as it does in the most math intensive careers. I see proof in each video I make. My videos get progressively better with each take, though I still make mistakes and find aspects I can improve on with each new video. In order to keep trying to improve in spite of any failures or lack of improvement, something else is needed. For me it is the sense of a specific goal in sight, to help me maintain the desire to put in continued time and effort.

When I decided on the number one university I wanted to attend, I wrote the name of that school in bold block letters on my door, written to remind myself daily of my ultimate goal. Stuck in the back of my head, this end result pushed me little by little to succeed and meet all of the requirements for the university I had in mind. And now I can say I'm at my dream school bringing with me that skill.

I recognize that others may have much more difficult circumstances than my own to endure, with the goal of improving or escaping those circumstances, and I deeply respect that. But that fact demonstrates to me how easy but effective it is, in comparison, to "stay with the problems longer" with a goal in mind of something much more easily realized, like a good grade on a test. I've learned to set goals, small or big, and to stick with them until they are realized.

Higher-Order Derivatives

2.3

Suppose a college student is doing a research project on water hyacinth growth rates in a lake. This student may use a derivative to determine the amount of and evaluate the growth of the plant. The derivative $f'(x)$ gives information about the rate of change of the function $f(x)$. The derivative can tell us if $f(x)$ is increasing or decreasing over an interval. In this section, we examine how to obtain information about $f'(x)$ that will tell us if $f(x)$ is increasing at an increasing or decreasing rate or decreasing at an increasing or decreasing rate.

© Roel Smart/ iStockPhoto

Higher-Order Derivatives

From our experience with differentiation, we know (or may surmise) that the derivative f' of a function f is itself a function. Hence, it may be possible to differentiate f'. The derivative of f' is denoted by f'' and is called the *second derivative* of f. Again, f'' is a function and may be differentiated, giving the *third derivative*, f'''. Continuing this way, we get the fourth derivative of f, fifth derivative of f, sixth derivative of f, and all other **higher-order derivatives of f**.

For convenience, the notation changes slightly for the fourth and higher-order derivatives of f.

f' indicates the first derivative of f.
f'' indicates the second derivative of f.
f''' indicates the third derivative of f.
$f^{(4)}$ indicates the fourth derivative of f.
$f^{(5)}$ indicates the fifth derivative of f.
$f^{(6)}$ indicates the sixth derivative of f.

$\vdots$

$f^{(n)}$ indicates the nth derivative of f.

The $\frac{d}{dx}$ notation is also used to indicate a higher-order derivative. If $y = f(x)$ is a differentiable function, then

$\frac{dy}{dx}$ and $\frac{df}{dx}$ both indicate the first derivative of the function.

$\frac{d^2y}{dx^2}$ and $\frac{d^2f}{dx^2}$ both indicate the second derivative of the function.

$\frac{d^3y}{dx^3}$ and $\frac{d^3f}{dx^3}$ both indicate the third derivative of the function.

$\frac{d^4y}{dx^4}$ and $\frac{d^4f}{dx^4}$ both indicate the fourth derivative of the function.

$\frac{d^5y}{dx^5}$ and $\frac{d^5f}{dx^5}$ both indicate the fifth derivative of the function.

$\vdots$

$\frac{d^ny}{dx^n}$ and $\frac{d^nf}{dx^n}$ both indicate the nth derivative of the function.

VIDEO EXAMPLES

SECTION 2.3

Example 1 Find $f'(x)$, $f''(x)$, $f'''(x)$, and $f^{(4)}(x)$ for

$$f(x) = 3x^5 - 2x^4 + x^2 - 10x + 4.$$

Solution

$f'(x) = 15x^4 - 8x^3 + 2x - 10$ — Now differentiate $f'(x)$ to get $f''(x)$

$f''(x) = 60x^3 - 24x^2 + 2$ — Now differentiate $f''(x)$ to get $f'''(x)$

$f'''(x) = 180x^2 - 48x$ — Now differentiate $f'''(x)$ to get $f^{(4)}(x)$

$f^{(4)}(x) = 360x - 48$

Example 2 For the function $f(x) = \frac{1}{2}x^4 - 4x^2$, find

a. $f(2)$ **b.** $f'(2)$ **c.** $f''(2)$

Solution First, let's find the first and second derivatives.

$$f(x) = \frac{1}{2}x^4 - 4x^2$$

$$f'(x) = 2x^3 - 8x$$

$$f''(x) = 6x^2 - 8$$

Next, we evaluate each function at $x = 2$.

a. $f(2) = \frac{1}{2}(2)^4 - 4(2)^2 = \frac{1}{2} \cdot 16 - 4 \cdot 4 = -8$

b. $f'(2) = 2 \cdot 2^3 - 8 \cdot 2 = 2 \cdot 8 - 8 \cdot 2 = 0$

c. $f''(2) = 6 \cdot 2^2 - 8 = 6 \cdot 4 - 8 = 16$

Example 3 If $g(t) = t^3 - 9t$, find the values of t for which

a. $g(t) = 0$ **b.** $g'(t) = 0$ **c.** $g''(t) = 0$

Solution Here are our first two derivatives:

$$g(t) = t^3 - 9t$$

$$g'(t) = 3t^2 - 9$$

$$g''(t) = 6t$$

Next we set each function equal to 0 and solve for t.

a. If $g(t) = 0$, then $t^3 - 9t = 0$

$$t(t^2 - 9) = 0$$

$$t(t + 3)(t - 3) = 0$$

$$t = 0 \quad \text{or} \quad t = -3 \quad \text{or} \quad t = 3$$

The function is 0 when t is 0, -3, and 3.

b. If $g'(t) = 0$, then $3t^2 - 9 = 0$

$$t^2 = \frac{9}{3} = 3$$

$$t = \pm\sqrt{3}$$

The first derivative is 0 when t is $-\sqrt{3}$ and $\sqrt{3}$.

c. If $g''(t) = 0$, then $6t = 0$

$$t = 0$$

The second derivative is 0 when t is 0.

Example 4 If $y = \frac{1}{x}$ find y', y'', and y'''.

Solution If we write $\frac{1}{x}$ as x^{-1}, our derivatives will be easy to find.

$$y = \frac{1}{x} = x^{-1}$$

$$y' = -1 \cdot x^{-2} = -x^{-2} = \frac{-1}{x^2}$$

$$y'' = -1(-2) \cdot x^{-3} = 2x^{-3} = \frac{2}{x^3}$$

$$y''' = 2(-3) \cdot x^{-4} = -6x^{-4} = \frac{-6}{x^4}$$

Interpreting Higher-Order Derivatives

Just as the first derivative f' of a function f describes the rate of change of f, the second derivative f'' describes the rate of change of f'; that is, f'' provides the rate of change of the rate of change of f. To understand the meaning of the *rate of change of the rate of change*, consider the following example.

Manufacturers experience variations in their flow of revenue. Over some time periods, revenue may increase or decrease. Revenue potentially changes as time changes. If revenue is changing, a question is *how* is it changing? If, at some particular time, revenue is increasing, is it doing so at an increasing rate or a decreasing rate? Is this month's increase higher or lower than last month's increase? Or, if, at some particular time, revenue is decreasing, is it doing so at an increasing rate or a decreasing rate? Is this month's decrease higher or lower than last month's decrease? These questions are answered by the second derivative.

The first and second derivatives of a function describe the trend in the function's behavior over some interval. A function could be

1. Increasing at an increasing rate
2. Increasing, but at a decreasing rate
3. Decreasing at an increasing rate
4. Decreasing, but at a decreasing rate

The signs of the first and second derivatives provide information about the possible trends. You can remember them by thinking of them in terms of the multiplication of signed numbers.

Summarizing, we know that

For multiplication:

1. The product of two numbers with the *same sign* is positive (making us think of increasing).
2. The product of two numbers with *opposite signs* is negative (making us think of *decreasing*).

For derivatives:

1. Two derivatives f' and f'' with the *same sign*, indicate that the function f changes at an *increasing rate.*

   $f' > 0, f'' > 0 \rightarrow f$ is increasing at an increasing rate.
   $f' < 0, f'' < 0 \rightarrow f$ is decreasing at an increasing rate.

2. Two derivatives f' and f'' with *opposite signs*, indicate that the function f changes at a *decreasing rate.*

   $f' > 0, f'' < 0 \rightarrow f$ is increasing at a decreasing rate.

   $f' < 0, f'' > 0 \rightarrow f$ is decreasing at a decreasing rate.

Example 5 Suppose the function $C(t)$ represents the amount of money (in thousands of dollars) a county spends on the local drug war t months from January 1, 2000. Interpret the information provided by $C'(15) = 1.6$ and $C''(15) < 0$.

© Rapid Eye/iStockPhoto

Solution Because the first derivative of C is positive, the cost of the drug war is increasing. Because the second derivative is negative (so that the derivatives have opposite signs), the function $C'(t)$ is decreasing and the function $C(t)$ itself is increasing at a decreasing rate. We make the following interpretation.

Interpretation

Fifteen months from January 1, 2000, although the cost of fighting the drug war is increasing, it is increasing at a decreasing rate. This means that the citizens of this county can expect the cost of fighting the drug war to increase each month, but they can expect each monthly increase to be less than the previous one.

Figure 1 illustrates this curve between $t = 15$ and $t = 16$. Notice that, as t increases, the curve rises, indicating that $C(t)$ is increasing. Notice also that as t increases, the slopes of the tangent lines decrease. This gives the curve its *opening downward appearance*, indicating that the rate of change in C is itself decreasing.

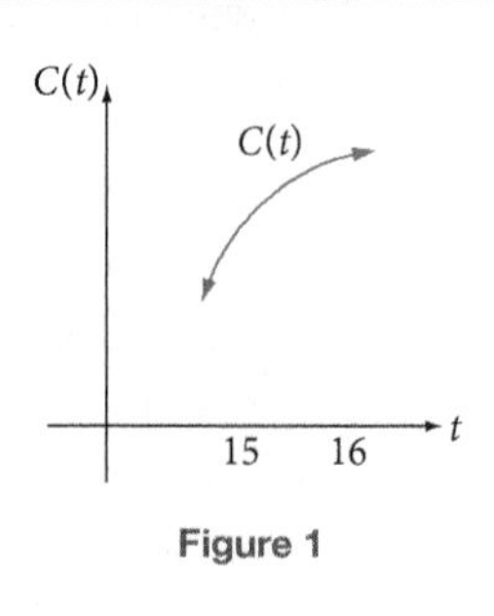

Figure 1

Example 6 Suppose the function $W(x)$ represents the amount of money (in thousands of dollars) a county spends on welfare each month, where x is the number of people (in thousands) applying for and receiving welfare. Interpret the information provided by $W'(16) = -8$ and $W''(16) < 0$.

Solution Because $W'(16)$ is negative, $W(x)$ is decreasing when $x = 16$. Because the second derivative is negative (so that both derivatives have the same sign), $W(x)$ is decreasing at an increasing rate.

Interpretation
At the time when 16,000 people are applying for and receiving welfare, the citizens of this county can not only expect the amount of money spent each month on welfare to decrease, but they can expect each decrease to be greater than the previous decrease.

Figure 2 shows $W(x)$ for $x = 16$ to $x = 17$. Notice that as x increases, the curve falls, indicating that W is decreasing. Also notice that as x increases, the curve bends more sharply downward, indicating that not only is W decreasing, but it is doing so at an increasing rate.

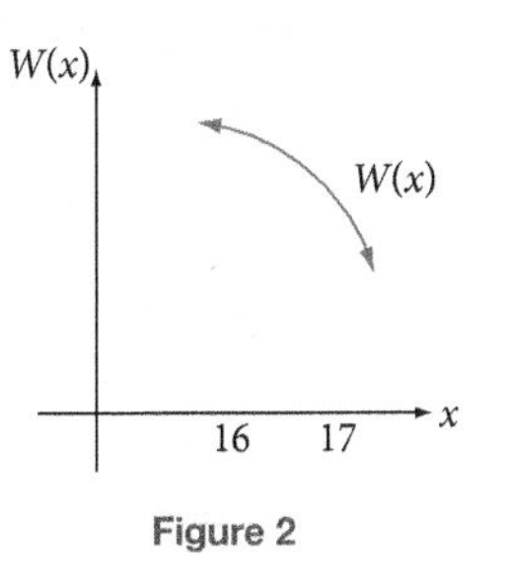

Figure 2

Example 7 Suppose the function $A(t)$ represents the amount (in square meters) of water hyacinth growing wild in a lake t months from the beginning of a university research project at the lake. Interpret the information provided by the first and second derivatives, $A'(25) = -14$ and $A''(25) > 0$.

Solution Because the first derivative is negative, the amount of water hyacinth growing wild in the lake 25 months after the beginning of the research project is decreasing. Because the second derivative is positive (so that the derivatives have opposite signs), $A(t)$ is decreasing at a decreasing rate.

© Roel Smart/ iStockPhoto

Interpretation
Twenty-five months after the beginning of the research project, the amount of water hyacinth in the lake is decreasing, but at a decreasing rate. This means that although the amount of water hyacinth in the lake is decreasing each month, each decrease will be less than the previous decrease.

Figure 3 shows $A(t)$ for $t = 25$ to $t = 26$. Notice that as t increases, the curve falls, indicating that A is decreasing. Also notice that as x increases, the curve is tending to bend upward, indicating that not only is A decreasing, but it is doing so at a decreasing rate.

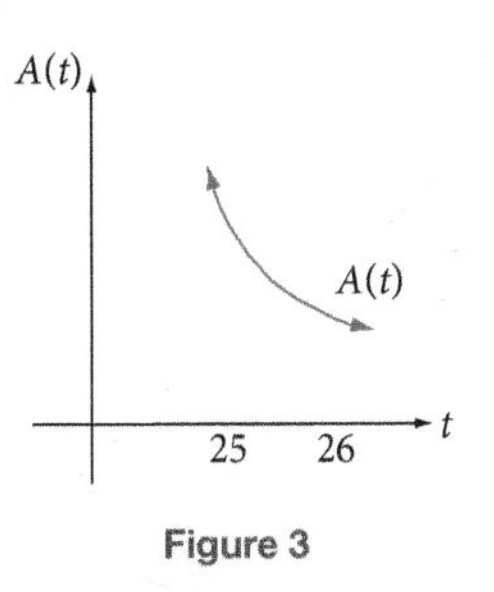

Figure 3

Using Technology 2.3

Wolfram|Alpha can be very helpful in finding and computing higher-order derivatives.

Example 8 Use Wolfram|Alpha to find $f'(x)$, $f''(x)$, and $f'''(3)$ for the function

$$f(x) = \frac{x^2 + 3x}{x^2 - 4x}$$

Solution To obtain the first derivative, go to www.wolframalpha.com and in the entry field enter

d/dx for (x^2 + 3x) / (x^2 – 4x)

Wolfram|Alpha displays your entry, its interpretation of your entry, and the derivative.

$$-\frac{7}{(x-4)^2}$$

To obtain the second and third derivatives, enter

d^2/dx^2 for (x^2 + 3x) / (x^2 – 4x)

and

d^3/dx^3 for (x^2 + 3x) / (x^2 – 4x)

respectively. Wolfram|Alpha displays your entry, its interpretation of your entry, and the derivatives.

$$\frac{14}{(x-4)^3} \quad \text{and} \quad -\frac{42}{(x-4)^4}$$

Getting Ready For Class

After reading through the preceding section, respond in your own words and in complete sentences.

A. How do you find the second derivative of a function?

B. If both $f'(x)$ and $f''(x)$ are positive numbers, what does that tell you about the function itself?

C. If we know that a function is increasing, but at a decreasing rate, what can we say about the first and second derivatives?

D. Will $g'''(t)$ ever be 0?

Problem Set 2.3

Skills Practice

For Problems 1-6, find f', f'', f''', and $f^{(4)}$ for each function.

1. $f(x) = x^6 - 2x^5 + 6x - 4$
2. $f(x) = x^5 - x^3 + x + 6$
3. $f(x) = x^4 + 2x^3 + 11x^{-1}$
4. $f(x) = -3x^4 + x^3 - x^2 + 5x + 4x^{-1}$
5. $f(x) = x^3 + 6x^2 - 4\sqrt{x}$
6. $f(x) = 3x^2 + 8x + 10\sqrt{x}$

For Problems 7-10, find:

7. $f'(x)$ and $f''(x)$ for $f(x) = \dfrac{x-4}{x}$
8. $f'(x)$ and $f''(x)$ for $f(x) = \dfrac{3x-5}{x}$
9. $f'(x)$ and $f''(x)$ for $f(x) = \dfrac{2x+7}{x^3}$
10. $f'(x)$ and $f''(x)$ for $f(x) = \dfrac{x^3+3}{x^2}$

For Problems 11-12, find and interpret $f'(1)$, $f''(1)$.

11. $f(x) = x^2 + 5x - 4$
12. $f(x) = -5x^2 + x - 4$

Paying Attention to Instructions The next two problems are intended to give you practice reading, and paying attention to, the instructions that accompany the problems you are working.

13. If $f(x) = (4x - 1)(2x - 3)$, find
    a. $f(1)$.
    b. $f'(1)$.
    c. $f''(1)$.
    d. the values of x for which $f(x) = 0$.
    e. the values of x for which $f'(x) = 0$.
14. If $f(x) = \dfrac{2x+1}{x-3}$, find
    a. $f(4)$.
    b. $f'(4)$.
    c. $f''(4)$.
    d. the slope of the tangent line when $x = 2$.
    e. the values of x for which $f'(x) = 0$.

Check Your Understanding

15. **Business: Production Costs** Suppose that $C(x)$ represents the cost (in dollars) to a manufacturer for producing x decorative throw pillows. Interpret the information provided by $C'(4{,}500) = 0.65$ and $C''(4{,}500) > 0$.
16. **Economics: Population** Suppose that $P(t)$ represents the population (in thousands) of a particular geographic area in some state t years from 2000. Interpret the information provided by $P'(11) = 1.2$ and $P''(11) > 0$.
17. **Management: Water Supply** Suppose that $A(t)$ represents the amount of water (in thousands of gallons) flowing out of a reservoir each day into adjacent creeks t days from the beginning of the year. Interpret the information provided by $A'(230) = -685$ and $A''(230) < 0$.

18. **Business: Sales Volume** Suppose that $S(r)$ represents a company's monthly sales volume (in millions of dollars) when r (in millions of dollars) is spent on research and development. Interpret the information provided by $S'(1.6) = 5.8$ and $S''(1.6) < 0$.

19. **Business: Marketing** Suppose that $U(c)$ represents the useful life (in hours) of a toy of complexity index c (in a marketer's defined units). Interpret the information provided by $U'(7) = 30$ and $U''(7) < 0$.

20. **Business: Capital Formation** Suppose that $C(t)$ represents the capital formation (in thousands of dollars) of a revenue flow t years after the start-up of a company. Interpret the information provided by $C'(3) = 350$ and $C''(3) > 0$.

21. **Law Enforcement: Drunk Driving** Suppose that $N(t)$ represents the number of alcohol-related traffic accidents in a city t months after a strict new driving-under-the-influence law has taken effect. Interpret the information provided by $N'(15) = -15$ and $N''(15) < 0$.

22. **Business: Advertising** Suppose that $N(t)$ represents the number of people (in thousands) in a city who have heard of a new product being advertised on radio and television t days after the beginning of the advertisement. Interpret the information provided by $N'(21) = 0.6$ and $N''(21) < 0$.

23. **Business: Manufacturing** The global manufacturing value added, in US$ billions for China, can be approximated with the quadratic function

$$G(t) = 1.2t^2 + 64.3t + 1989$$

where t represents the number of years from January 1, 1980. Find both $G'(32)$ and $G''(32)$ to help you make a statement about the behavior of China's global manufacturing value added in the year 2012.

Source: Strategy+Business, Booz & Company

Modeling Practice

For Problems 24-27, determine the signs of the first and second derivatives and make a statement about the trend of the function as it applies to the given situation.

24. **Business: Profits** The function

$$P(x) = \frac{180x}{110 + x^2}$$

relates the weekly profit P (in thousands of dollars) a company makes to the number x (in hundreds) of automobile sheepskin seat covers it sells each week. In this case,

$$P'(x) = \frac{180(110 - x^2)}{(110 + x^2)^2} \quad \text{and}$$

$$P''(x) = \frac{-360x(330 - x^2)}{(110 + x^2)^3}$$

What is the trend at $x = 10$?

25. Medicine: Drug Reaction The reaction function

$$S(x) = \frac{x}{2}\sqrt{260 - x}$$

measures the strength of the average person's reaction to x milligrams of a particular drug. In this case,

$$S'(x) = \frac{520 - 3x}{4\sqrt{260 - x}} \quad \text{and}$$

$$S''(x) = \frac{3x - 1{,}040}{8\sqrt{(260 - x)^3}}$$

What is the trend at $x = 40$?

© Shaun Lowe Photographic/ iStockPhoto

26. Physical Science: Drag Racing The rational function $V(t) = \frac{340t}{t+3}$ models the speed $V(t)$, in miles per hour, of a dragster t seconds after it starts its quarter-mile run. For this function,

$$V'(t) = \frac{1{,}020}{(x + 3)^2} \quad \text{and} \quad V''(t) = -\frac{2{,}040}{(x + 3)^3}$$

What is the trend in the dragster's speed 4 seconds into the run?

27. Biology: Bacterial Toxins The toxins function

$$P(t) = \frac{30t}{t^2 + 5}, \quad t > 1$$

relates the population P (in millions) of bacteria in a culture to the time t (in hours) after a particular toxin is introduced into the culture. In this case,

$$P'(t) = \frac{150 - 30t^2}{(t^2 + 5)^2} \quad \text{and}$$

$$P''(t) = \frac{60t(t^2 - 15)}{(t^2 + 5)^3}$$

What is the trend at $t = 4$?

For Problems 28-32, find the first and second derivatives and make a statement about the trend of the function as it applies to the given situation.

28. Business: Sales Trend The number N of hand-held hair dryers a manufacturer believes it can sell each month at price p (in dollars) is given by the function

$$N(p) = \frac{1{,}100}{p^2}$$

What is the trend when the price of a hair dryer is \$12?

29. Business: Advertising The sales department of a company believes that t days after the end of an advertising campaign, the number N of items it will sell each day is given by the function

$$N(t) = -3t^2 + 14t + 450$$

What is the trend 7 days after the end of the advertising campaign?

30. **Education: Word Processing** The faculty of a business school's word processing department believes that the number N of words a student can correctly input after t weeks of instruction is given by the function

$$N(t) = t^2 + 6t + 10 \quad 0 \leq t \leq 6$$

What is the trend after 4 weeks of instruction?

31. **Business: Online Advertising Revenue** The revenue, in millions of dollars, realized by a social networking site is modeled by the function

$$R(x) = -x^4 + 11x^3 - 39x^2 + 45x$$

where x is the number of years from the beginning of 2007. What is the trend in this site's revenue in the year 2012?

32. **Physical Science: Drag Racing** At the beginning of the textbook, in Section 1.1, we noted that the function

© 20th Century Fox Film Corp.

$$V(t) = \frac{340t}{t + 3} \quad t \geq 0$$

described drag racer Shirley Muldowney's dragster's speed in one of the races in the movie *Heart Like a Wheel.* Because the dragster is moving in essentially a straight line, the formula actually gives the velocity of the dragster at time t. Velocity is speed with a direction. Velocity is the first derivative of a position function. The function $V(t)$ represents Shirley's velocity at time t and it is the first derivative of Shirley's dragster's position function (unknown to us right now).

   a. The second derivative of the position function describes acceleration. When you speed up or slow down in your car, you are experiencing acceleration. Find the acceleration function for Shirley's dragster.

   b. The third derivative of the position function describes the phenomenon called jerk. Imagine being in your car and accelerating up a freeway on-ramp. Just as you are about to enter the rightmost lane you see a car coming up in the lane. To avoid a collision, you press the accelerator to the floor. When you do this, you feel your car suddenly jerk forward. Hence, the word jerk. Find the jerk function for Shirley's dragster.

Using Technology Exercises

33. For the function $f(x) = \dfrac{x^2 - 5x}{x^2 + 6x}$, find $f''(x)$.

34. For the function $f(x) = (x^{1/3} + x^{1/4})(x^{1/5} + x^{1/6})$, find $f''(2)$.

35. Determine the behavior of the function $f(x) = \dfrac{x^{1/3} + x}{x^{1/3} - x}$ when $x = 64$.

36. Biology: Population Size This problem is an extension of Using Technology Problem 50 in Section 2.2. Knowing the size of biological populations is important to evolutionary biologists in assessing the health of the population and in identifying processes that shape the evolution of the population. But the census size of the population may not be the best measure. In some populations, there may be a drastic difference between the number of sexually mature individuals and the number of those that are not capable of breeding. In such populations, biologists often use the effective size N_e rather than the census size to measure the size of a population. The formula for effective size is

$$N_e = \frac{4N_f N_m}{N_m + N_f}$$

where N_f and N_m represent the number of breeding females and breeding males, respectively.

a. Suppose a population has x number of breeding males and 400 breeding females. Rewrite the effective population function given above so that it appears as a function $N_e(x)$.

b. Compute both $N_e'(20)$ and $N_e''(20)$.

c. Use the information you obtained from the derivatives in part b to make a statement about the behavior of the effective population size for a population in which there are 400 breeding females and x number of breeding males at the point in time where there are 20 breeding males.

Source: The Biology Project, Department of Biochemistry and Molecular Biophysics at the University of Arizona

Getting Ready for the Next Section

Simplify.

37. $4(2x - 9) + 6$

38. $3(2x - 9) + (2x - 9)^2$

39. Write $\sqrt[5]{(7x - 8)^3}$ as an equivalent expression that does not contain a radical.

40. Write $4(6 - 2x)^{-5}$ with a positive exponent.

Use a calculator to approximate each expression to the nearest hundredth.

41. $17^{1/3}$

42. $17^{-2/3}$

43. $3{,}363^{3/4}$

44. $3{,}363^{-1/4}$

Differentiate each function.

45. $y = 2x - 9$

46. $y = 6 - 2x$

47. $y = 7x - 8$

48. $y = 3x + 2$

49. $y = 4x^3 + 5x + 2$

50. $y = x^2 + 3x$

51. $y = x^6$

52. $y = x^{3/4}$

53. $y = x^{3/5}$

54. $y = x^{-5}$

Spotlight on **Success**

University of North Alabama

Pride is a personal commitment.
It is an attitude which separates excellence from mediocrity.
—William Blake

Photo courtesy UNA

The University of Northern Alabama places its Pride Rock, a 60-pound granite stone engraved with a lion's paw print, behind the north end zone at all home football games. The rock reminds current Lion players of the proud athletic traditions that has been established at the school, and to take pride in their efforts on the field.

The same idea holds true for your work in your math class. Take pride in it. When you turn in an assignment, it should be accurate and easy for the instructor to read. It shows that you care about your progress in the course and that you take pride in your work. The work that you turn in to your instructor is a reflection of you. As the quote from William Blake indicates, pride is a personal commitment; a decision that you make, yourself. And once you make that commitment to take pride in the work you do in your math class, you have directed yourself toward excellence, and away from mediocrity.

The Chain Rule and General Power Rule

2.4

As we mentioned in the introduction to this chapter, relationships are often such that one quantity depends on another quantity, which in turn depends on yet another quantity. Such relationships are called composite relationships and are modeled by composite functions.

The diagram below shows that the number of aluminum cans recycled depends on the number of soft drinks purchased. Since people consume more soft drinks on hot days, the number of soft drinks purchased depends on the temperature.

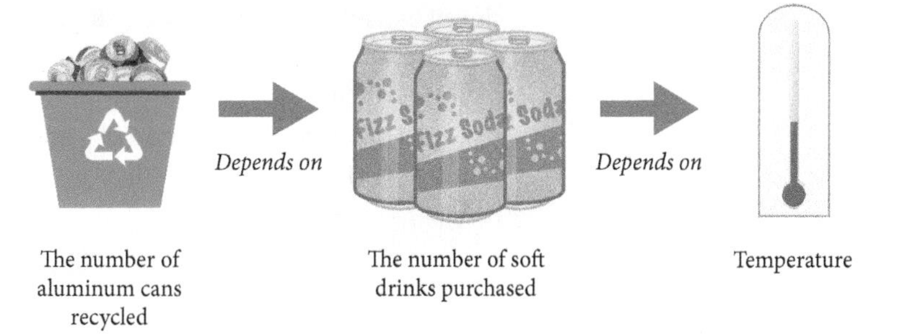

As we progress through this section we will come back to this problem. We will use composite functions, and calculus, to see how a one-degree change in temperature affects the number of aluminum cans that are available for recycling. Let's begin this section with the chain rule, a rule that allows us to differentiate composite functions.

The Chain Rule

Suppose that y is a function of u and that u, in turn, is a function of x. That is, the value of the variable y depends on the value of u and the value of u in turn depends on the value of x. A change in the value of x may produce a change in the value of u and a change in the value of u may produce a change in the value of y. Thus, through a chain of events, a change in the value of x may produce a change in the value of y. Since differentiation is the language of change, these changes can be described using derivative notation.

$\frac{dy}{du}$ represents the ratio of a change in y to a change in u.

$\frac{du}{dx}$ represents the ratio of a change in u to a change in x.

$\frac{dy}{dx}$ represents the ratio of a change in y to a change in x.

We can suggest an expression for $\frac{dy}{dx}$ in terms of $\frac{dy}{du}$ and $\frac{du}{dx}$ by making the following observation. Suppose $\frac{dy}{du} = 7$ and $\frac{du}{dx} = 4$. This means that y changes seven times as fast as u and that u changes four times as fast as x. Consequently, any change in x would produce a four times greater change in u which, in turn, would produce a seven times greater change in y. Since $\underbrace{7}_{\frac{dy}{du}} \cdot \underbrace{4}_{\frac{du}{dx}} = \underbrace{28}_{\frac{dy}{dx}}$, we conclude that through the chain of changes, y changes 28 times as fast as x, and suggest the following formula.

$$\frac{dy}{du} \cdot \frac{du}{dx} = \frac{dy}{dx}$$

More conveniently,

$$\frac{dy}{dx} = \frac{dy}{du} \cdot \frac{du}{dx}$$

The formula indicates that since y is linked to u and u is linked to x, then y is linked to x and is subject to x's changes.

The Chain Rule

If y is a differentiable function of u and u is a differentiable function of x—that is, $y = f(u)$ and $u = g(x)$—then $y = f(u) = f[g(x)]$, and y is a function of x. Also,

$$\frac{dy}{dx} = \frac{dy}{du} \cdot \frac{du}{dx}$$

Alternatively,

$$\frac{dy}{dx} = f'[g(x)] \cdot g'(x)$$

VIDEO EXAMPLES

SECTION 2.4

Example 1 Suppose $y = u^2 + 3u$ and, in turn, $u = 2x - 9$. Find:

a. $\frac{dy}{dx}$ and interpret this result, and

b. $\frac{dy}{dx}$ when $x = 5$ and interpret this result.

Solution

a. Since y is a function of u and u is a function of x, y is a function of x, and we can use the chain rule.

$$\begin{aligned}\frac{dy}{dx} &= \frac{dy}{du} \cdot \frac{du}{dx} \\ &= \frac{d}{du}(u^2 + 3u) \cdot \frac{d}{dx}(2x - 9) \\ &= (2u + 3) \cdot 2 \\ &= 4u + 6\end{aligned}$$

This statement does not reflect what we wish to know, because $4u + 6$ involves u not x. Since we know that $u = 2x - 9$, we replace all occurrences of u with $2x - 9$ and simplify.

$$\begin{aligned}\frac{dy}{dx} &= 4(2x - 9) + 6 \\ &= 8x - 30\end{aligned}$$

Thus, $\frac{dy}{dx} = 8x - 30$.

Interpretation

1. The expression $8x - 30$ gives the slope of the tangent line at any point along the curve once the x-value is specified.
2. The expression $8x - 30$ gives the instantaneous rate of change of y with respect to x.

b. When $x = 5$,

$$\frac{dy}{dx} = 8(5) - 30$$

$$= 40 - 30$$

$$= 10 \qquad \left(\frac{10}{1} \text{ as a slope}\right)$$

Interpretation

When the value of x increases by 1 unit, from 5 to 6, the value of y increases by approximately 10 units.

$$\left(\frac{dy}{dx} = 10 = \frac{10 \leftarrow \text{change in } y}{1 \leftarrow \text{change in } x}\right)$$

■

Notice that we could have found $\frac{dy}{dx}$ for the functions of Example 1 without using the chain rule by defining y as a function of x at the outset.

$$y = u^2 + 3u \quad \text{and} \quad u = 2x - 9$$

We then replace all occurrences of u in $y = u^2 + 3u$ with $2x - 9$ and simplify.

$$y = (2x - 9)^2 + 3(2x - 9)$$

$$= 4x^2 - 36x + 81 + 6x - 27$$

$$= 4x^2 - 30x + 54$$

Then, $\frac{dy}{dx} = 8x - 30$ as before.

Now, realize that if y were defined as $y = u^{15} + 3u$, this technique would involve an outrageous amount of work! It would take an extraordinary amount of effort to expand and simplify $y = (2x - 9)^{15} + 3(2x - 9)$. (This function will be differentiated using the general power rule, a rule that we will examine later in this section.)

The function in Example 2 illustrates how the chain rule can be applied to a business application.

Modeling: Temperature and Recycling

The example below is a follow up to the introduction to this section. It shows how we can use composite functions, and the chain rule, to estimate the number of cans that will be recycled based on a one-degree change in temperature.

Example 2 For temperatures greater than 10 °C, the average daily number of soft drinks sold, s (in hundreds), in a city depends on the temperature, t, as

$$s(t) = 1{,}200 + 23t + 600(t - 10)^{1/3}$$

The number, in hundreds, of soft drink aluminum cans recycled, r, depends on the number of soft drinks sold, s, as

$$r(s) = s^{3/4}$$

a. Give some reasonable numbers for the domain of the function s.

b. How many cans are recycled when the temperature is 27 °C?

c. How does the number of cans recycled change as the temperature changes from 27 °C to 28 °C?

d. Suppose we want to use the Fahrenheit temperature scale, instead of the Celsius scale. How would the problem change?

Solution Here is a diagram of the situation.

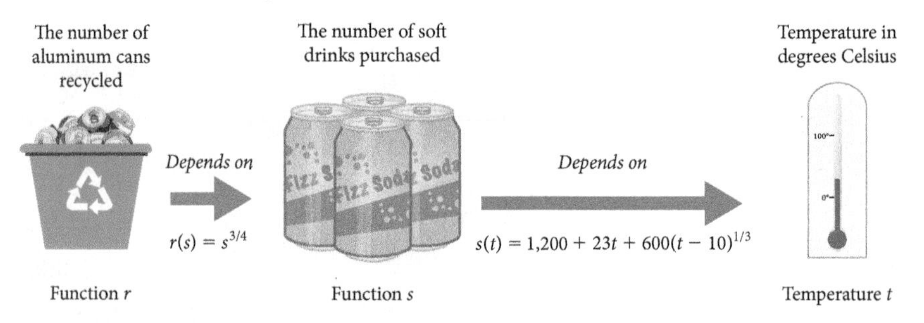

a. People tend to drink more soft drinks on hot days than on cold days. If we restrict the domain of this function so that t ranges from 20 °C (which is 68 °F) to 40 °C (104 °F), we should have a reasonable range of values to input for temperature.

b. To find the number of cans recycled when the temperature is 27 °C, we simply substitute 27 for t and use our formulas.

$$\begin{aligned}\text{When } t = 27, \text{ then } s(27) &= 1{,}200 + 23(27) + 600(27 - 10)^{1/3} \\ &= 1{,}200 + 621 + 600(17)^{1/3} \\ &\approx 1{,}200 + 621 + 600(2.57) \\ &\approx 3{,}364\end{aligned}$$

Since s represents hundreds of cans, the number of cans sold is 336,400.

To find the number of cans recycled, we substitute $s = 3{,}364$ into the formula for the number of cans recycled, giving us

$$r(3{,}364) = 3{,}364^{3/4} \approx 442$$

Since r is in hundreds of cans, the number of cans recycled when the temperature is 27 °C is 44,200.

c. To answer the question we need to find $\frac{dr}{dt}$. We know that r is a function of s and, in turn, s is a function of t. Hence, we can apply the chain rule.

$$\frac{dr}{dt} = \frac{dr}{ds} \cdot \frac{ds}{dt}$$

Since $\frac{dr}{ds} = \frac{3}{4}s^{-1/4}$ and $\frac{ds}{dt} = 23 + 200(t - 10)^{-2/3}$, we have

$$\frac{dr}{dt} = \frac{dr}{ds} \cdot \frac{ds}{dt}$$

$$= \frac{3}{4}s^{-1/4} \cdot \left[23 + 200(t - 10)^{-2/3}\right]$$

But $s(t) = 1{,}200 + 23t + 600(t - 10)^{1/3}$, so we replace all occurrences of s with $1{,}200 + 23t + 600(t - 10)^{1/3}$.

$$\frac{dr}{dt} = \frac{3}{4}\left[1{,}200 + 23t + 600(t - 10)^{1/3}\right]^{-1/4} \cdot \left[23 + 200(t - 10)^{-2/3}\right]$$

At $t = 27$,

$$\frac{dr}{dt} = \frac{3}{4}\left[1{,}200 + 23(27) + 600(27 - 10)^{1/3}\right]^{-1/4} \cdot \left[23 + 200(27 - 10)^{-2/3}\right]$$

$$\approx 5.244$$

Interpretation

If the temperature in the city increases by 1 °, from 27 °C to 28 °C, the number of aluminum cans recycled in this city will increase by approximately 524.

d. Let's use the variable T to represent the temperature in degrees Fahrenheit. If we want to input our temperatures in degrees Fahrenheit, we must add a third function onto the end of the chain that will convert our degrees Fahrenheit (T) to degrees Celsius (t), in the beginning of the problem. Here is a diagram of this situation.

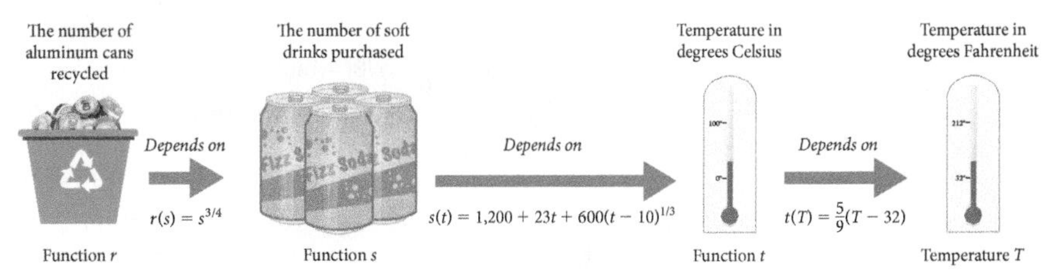

If we want to find the rate of change of the number of cans recycled, with respect to the temperature in degrees Fahrenheit, we would apply the chain rule this way:

$$\frac{dr}{dT} = \frac{dr}{ds} \cdot \frac{ds}{dt} \cdot \frac{dt}{dT}$$

■

The General Power Rule

Our list of differentiation rules includes the chain rule, which allows us to differentiate a function such as $y = (4x - 5)^{12}$. If we consider y as a composite function with $u = 4x - 5$ and $y = u^{12}$, we can apply the chain rule:

$$\begin{aligned} \frac{dy}{dx} &= \frac{dy}{du} \cdot \frac{du}{dx} \\ &= \frac{d}{du}(u^{12}) \cdot \frac{d}{dx}(4x - 5) \\ &= 12u^{11} \cdot 4 \end{aligned}$$

Substituting $4x - 5$ for u, we get

$$\frac{dy}{dx} = 12(4x - 5)^{11} \cdot 4$$

Without simplifying, let's observe the result of this differentiation again, looking for a pattern.

$$\frac{dy}{dx} = \underbrace{12(4x - 5)^{11}}_{\text{derivative of exponential part}} \underbrace{\cdot}_{\text{times}} \underbrace{4}_{\text{derivative of the base}}$$

The pattern we observe is in two parts. Differentiating the exponential part follows the rule for powers:

$$\frac{dy}{du} = \frac{d}{du}(u^n) = nu^{n-1} \rightarrow 12(4x - 5)^{11}$$

and the derivative of the base is $\frac{du}{dx} \rightarrow 4$.

Thus,

$$\frac{dy}{dx} = \underbrace{12(4x - 5)^{11}}_{nu^{n-1}} \cdot \underbrace{4}_{\frac{du}{dx}}$$

From this pattern, we can suggest the following differentiation rule for powers.

The General Power Rule

If y is a differentiable function of u, and u is a differentiable function of x and $y = u^n$, then

$$\frac{dy}{dx} = nu^{n-1} \cdot \frac{du}{dx}$$

or,

if $y = [u(x)]^n$, then $\frac{dy}{dx} = n[u(x)]^{n-1} \cdot \frac{du}{dx}$, where n is any real number.

The general power rule might be remembered as: *the derivative of a power is the derivative of the exponential part times the derivative of the base.* It may be that the derivative of the base requires several differentiation rules.

Example 3 Find the derivative of each of the following functions.

a. $y = (4x^3 + 5x + 2)^6$ **b.** $f(x) = \dfrac{4}{(6 - 2x)^5}$ **c.** $y = \sqrt[5]{(7x - 8)^3}$

Solution

a. This function has the form $y = u^6$, where $u = 4x^3 + 5x + 2$. Using our power rule formula $\dfrac{dy}{dx} = nu^{n-1} \cdot \dfrac{du}{dx}$, where $u = 4x^3 + 5x + 2$,

$$\begin{cases} \text{derivative of exponential part: } n[u]^{n-1} = 6(4x^3 + 5x + 2)^5, \\ \text{derivative of the base: } \dfrac{du}{dx} = \dfrac{d}{dx}(4x^3 + 5x + 2) = 12x^2 + 5 \end{cases}$$

$$\frac{dy}{dx} = 6(4x^3 + 5x + 2)^5(12x^2 + 5)$$

b. $f(x) = \dfrac{4}{(6 - 2x)^5}$

Since this quotient has a constant in the numerator, we will rewrite it as

$$f(x) = 4(6 - 2x)^{-5}$$

From the general power rule, $f'(x) = n[u(x)]^{n-1} \cdot \dfrac{du}{dx}$, where $u = 6 - 2x$,

$$\begin{cases} \text{derivative of exponential part: } n[u(x)]^{n-1} = -5(6 - 2x)^{-6}, \text{ and} \\ \text{derivative of the base: } \dfrac{du}{dx} = \dfrac{d}{dx}(6 - 2x) = -2 \end{cases}$$

$$f'(x) = 4(-5)(6 - 2x)^{-6}(-2)$$

$$= 40(6 - 2x)^{-6} \qquad \text{Multiply 4, } -5\text{, and } -2$$

$$= \frac{40}{(6 - 2x)^6} \qquad \text{Eliminate the negative exponent}$$

c. $y = \sqrt[5]{(7x - 8)^3}$

Since our rules of differentiation involve exponents and not radicals, we will begin by writing y in exponential form.

$$y = (7x - 8)^{3/5}$$

From the general power rule, $nu^{n-1} \cdot \dfrac{du}{dx}$, where $u = 7x - 8$,

$$\begin{cases} \text{derivative of exponential part: } nu^{n-1} = \dfrac{3}{5}(7x - 8)^{-2/5}, \text{ and} \\ \text{derivative of the base: } \dfrac{du}{dx} = \dfrac{d}{dx}(7x - 8) = 7 \end{cases}$$

$$\frac{dy}{dx} = \frac{3}{5}(7x - 8)^{-2/5} \cdot 7$$

$$= \frac{21}{5}(7x - 8)^{-2/5} \qquad \text{Multiply } \frac{3}{5} \text{ and 7}$$

$$= \frac{21}{5(7x - 8)^{2/5}} \qquad \text{Write with positive exponent}$$

$$= \frac{21}{5\sqrt[5]{(7x - 8)^2}} \qquad \text{Convert back to radical notation}$$

Example 4 illustrates how it is sometimes necessary to use several differentiation rules to differentiate a function.

Example 4 Find the derivative of the function $f(x) = \frac{(2x-1)^4}{3x+2}$ at $x = -1$.

Solution Viewing this function globally, we see a quotient. Locally, we see a power. We will begin by using the quotient rule.

$$f'(x) = \frac{(3x+2)\cdot\frac{d}{dx}(2x-1)^4 - (2x-1)^4\cdot\frac{d}{dx}(3x+2)}{(3x+2)^2} \quad \text{Quotient rule}$$

$$= \frac{(3x+2)\cdot 4(2x-1)^3\cdot 2 - (2x-1)^4\cdot 3}{(3x+2)^2} \quad \text{Power rule}$$

$$= \frac{8(3x+2)(2x-1)^3 - 3(2x-1)^4}{(3x+2)^2} \quad \text{Simplify}$$

Now we need to factor the numerator. The smallest exponent on the common factor $2x - 1$ is 3. Thus, the greatest common factor of the expression in the numerator is $(2x - 1)^3$. Continuing to simplify the derivative we have

$$f'(x) = \frac{(2x-1)^3[8(3x+2) - 3(2x-1)]}{(3x+2)^2}$$

$$= \frac{(2x-1)^3[24x+16-6x+3]}{(3x+2)^2}$$

$$= \frac{(2x-1)^3[18x+19]}{(3x+2)^2}$$

When $x = -1$,

$$f'(-1) = \frac{(-27)(1)}{1}$$

$$= \frac{-27}{1} \quad \text{or} \quad -27$$

Example 5 Find the derivative of the function $f(x) = x^5(4x-1)^{1/4}$.

Solution Viewing this function globally, we see a product. Locally, we see powers. We will begin by using the product rule.

$$f'(x) = \frac{d}{dx}x^5\cdot(4x-1)^{1/4} + x^5\cdot\frac{d}{dx}(4x-1)^{1/4}$$

$$= 5x^4(4x-1)^{1/4} + x^5\cdot\frac{1}{4}\cdot(4x-1)^{-3/4}\cdot 4 \quad \text{Simplify}$$

$$= 5x^4(4x-1)^{1/4} + x^5(4x-1)^{-3/4} \quad \text{Simplify}$$

$$= x^4(4x - 1)^{-3/4}[5(4x - 1) + x]$$ Factor

$$= x^4(4x - 1)^{-3/4}[20x - 5 + x]$$ Eliminate the negative exponent

$$= \frac{x^4(21x - 5)}{(4x - 1)^{3/4}}$$

Example 6 Find $f'(x)$ for the function $f(x) = \left(\frac{2x + 5}{8x + 7}\right)^4$.

Solution We see this solution globally as a power. Locally, we see a quotient. To differentiate this function we will use both the power rule and the quotient rule.

$$f'(x) = 4\left(\frac{2x + 5}{8x + 7}\right)^3 \cdot \frac{d}{dx}\left(\frac{2x + 5}{8x + 7}\right)$$

$$= 4\left(\frac{2x + 5}{8x + 7}\right)^3 \cdot \frac{(8x + 7)(2) - (2x + 5)(8)}{(8x + 7)^2}$$

$$= 4\frac{(2x + 5)^3}{(8x + 7)^3} \cdot \frac{16x + 14 - 16x - 40}{(8x + 7)^2}$$

$$= \frac{4(2x + 5)^3(-26)}{(8x + 7)^5}$$

$$= \frac{-104(2x + 5)^3}{(8x + 7)^5}$$

Using Technology 2.4

Wolfram|Alpha can help you to find derivatives of composite functions when calculating by hand would be long and complicated. It is also useful for constructing linear approximations to functions.

Example 7 Compute both the first and second derivatives of the function

$$f(x) = \frac{\sqrt{x^2 - 1}}{(x^2 + 1)^2}$$

at $x = 2$ and use the information you obtain from their signs to comment on the behavior of $f(x)$ at $x = 2$.

Solution In the entry line, type

evaluate d/dx (x^2 - 1)^(1/2) /(x^2 + 1)^2
and d^2/dx^2 (x^2 - 1)^(1/2) /(x^2 + 1)^2 where x = 2

Wolfram|Alpha responds with $\left\{-\frac{14}{125\sqrt{3}}, \frac{179}{1875\sqrt{3}}\right\}$

Since $f'(2) < 0$ and $f''(2) > 0$, we note that at $x = 2$, the function is decreasing at a decreasing rate.

Example 8 Figure 1 shows the graph of the function

$$f(x) = \sqrt{x - 6}$$

and the tangent line to the function at $x = 15$. The equation of the tangent line is $y = \frac{1}{6}x + \frac{1}{2}$. The slope of the tangent line is, of course, the derivative of $f(x)$ at $x = 15$. That is, $m = f'(15)$.

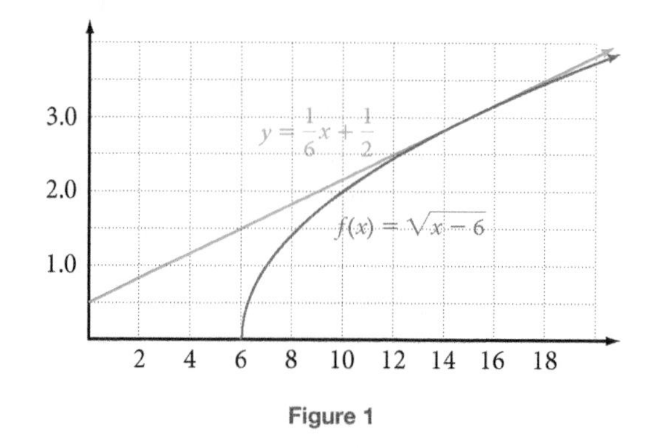

Figure 1

The graph shows that for input values close to 15, the output values produced by both the function and the tangent line function are nearly the same. This means that for x-values near $x = 15, f(x) = \sqrt{x - 6}$ could be approximated with $y = \frac{1}{6}x + \frac{1}{2}$.

Approximating a function with a linear function, $f(x) = mx + b$, is called *linear approximation*. Approximating a function with a simple linear function can be useful if the function is intractable, that is, it is not easily managed or manipulated. We will not see such functions in this text, but an example is the function

$$f(x) = \sqrt[5]{(x^{3/2} - 5x)^4}\, e^{\sqrt{x^2 - 1}}$$

You can use Wolfram|Alpha to help you find the equation of the tangent line to a function at a $(a, f(a))$. The equation of the tangent line can be used to approximate values of $f(x)$ for values of x near $x = a$. Use Wolfram|Alpha to find the equation of the tangent line to the function $f(x) = \sqrt{x - 6}$ at $x = 15$.

Solution The tangent line to the function has the slope-intercept form $y = mx + b$. You can use Wolfram|Alpha to find the slope m and then the equation of the line. Wolfram|Alpha uses the point-slope form, then solves it for y to put the equation in slope-intercept form, $y = mx + b$. You need only supply the value of the derivative and the point (x_1, y_1).

1. To find the y-value of the point (x_1, y_1) as well as the slope of the line at $x = 15$, compute the value of the function itself at $x = 15$ and the derivative at $x = 15$. Go to www.wolframalpha.com and in the entry field enter

   evaluate sqrt(x - 6) and d/dx sqrt(x - 6) where x = 15

   Wolfram|Alpha displays your entry, its interpretation of your entry, and the value of the derivative.

   $$\left\{3, \frac{1}{6}\right\} \text{ meaning that } y_1 = 3 \text{ and } m = f'(15) = \frac{1}{6}$$

2. To get the equation of the tangent line to the curve at $x = 15$, enter into the entry field

$$\text{line with slope } \frac{1}{6} \text{ and point } (15, 3)$$

Wolfram|Alpha responds with the graph of the tangent line but also the equation of the tangent line, $y = \frac{1}{6}x + \frac{1}{2}$.

You can now use the equation of the tangent line to approximate output values of $f(x) = \sqrt{x - 6}$ for values of x near $x = 15$. For example if $x = 14.97$, then $f(14.97) \approx 2.994996$ and $y = 2.995$, a slight difference. However, if $x = 9$, then $f(9) \approx 1.73$ and $y = 2$, and the equation of the tangent line produces a larger error than when the x-value was closer to 15. ■

Getting Ready for Class

After reading through the preceding section, respond in your own words and in complete sentences.

A. What is a composite function?

B. If $y = u^2 + 3u$, what is $\frac{dy}{du}$?

C. If y is a function of u, and u is a function of t, how do you find $\frac{dy}{dt}$?

D. When using the general power rule, when do you differentiate the base of the exponential function?

Problem Set 2.4

Skills Practice

For Problems 1-18, find $\frac{dy}{dx}$.

1. $y = 12u - 7$ and $u = 3x + 1$
2. $y = 16u^2 + u - 3$ and $u = 8x - 5$
3. $y = 10u + 1$ and $u = -x^2 + 3$
4. $y = 8u - 6$ and $u = -x^4 + 3x^2 + x$
5. $y = 3u^2 + 2$ and $u = 5x^2 + 4$
6. $y = 4u^2 + u - 3$ and $u = 3x - 8$
7. $y = u^3 - 6u$ and $u = 5x^2 - 2$
8. $y = u^3 - 8u$ and $u = 3x^2 - 4$
9. $y = (5x^2 + 2x - 6)^3$
10. $y = (6x^3 + x)^4$
11. $y = \sqrt[5]{(x^2 + 7)^3}$
12. $y = \sqrt[6]{(4x^2 + 2x - 1)^5}$
13. $y = (x^2 + 3)^3(2x - 1)^5$
14. $y = (4x + 1)^6(2x^2 + 3x - 4)^8$
15. $y = x^4(5x + 2)^{1/5}$
16. $y = 3x^2(6x - 8)^{1/6}$
17. $y = 2x^7(4x + 1)^{3/4}$
18. $y = 5x^3(8x^2 + 6)^{5/16}$

For Problems 19-30, find $\frac{dy}{dx}$ without using the quotient rule; rather, rewrite each function by using a negative exponent and then use the product rule and the general power rule to find the derivative.

19. $y = \frac{1}{x + 3}$
20. $y = \frac{1}{x - 5}$
21. $y = \frac{8}{(x - 4)^3}$
22. $y = \frac{5}{(x + 3)^2}$
23. $y = \frac{3}{(2x + 7)^4}$
24. $y = \frac{4}{(5x - 11)^6}$
25. $y = \frac{(x^2 + 6)^7}{x - 2}$
26. $y = \frac{(x^2 - 5)^4}{x - 8}$
27. $y = \frac{3x^2 + 1}{(x + 2)^3}$
28. $y = \frac{3x + 4}{(x^2 + 1)^5}$
29. $y = \left(\frac{9x - 4}{3x + 2}\right)^3$
30. $y = \left(\frac{5x - 4}{8x - 1}\right)^5$

31. If $u(x) = 5x^2 + 4$ and $v(x) = x^3 + 11$, find $u(x) \cdot v(x)$.
32. If $u(t) = t^2 - 4$, $v(t) = t^3 + 1$, and $d(t) = 2t$, find $h(t) = u(t) \cdot d(t) + v(t) \cdot d(t)$.
33. If $u(x) = x^3 + 5x + 4$, $v(x) = x^2 - 3$, $f(x) = 3x^2 + 2$, and $g(x) = 2x$, find $h(x) = v(x) \cdot f(x) - u(x) \cdot g(x)$.
34. If $f(x) = x^2 + 3x$ and $g(x) = 2x - 9$, find $(f \circ g)(x)$ and $(g \circ f)(x)$.
35. If $f(u) = u^2 + 3u$ and $u(x) = 2x - 9$, find $(f(u(x))$.
36. If $f(x) = x^6$ and $g(x) = 4x^3 + 5x + 2$, find $(f \circ g)(x)$ and $(g \circ f)(x)$.

Check Your Understanding

For Problems 37-40, state whether or not the chain rule is the appropriate differentiation technique for finding the specified derivative.

37. $y = 8u + 6$ and $u = 2x - 7$; $\dfrac{dy}{dx}$

38. $m = 2x + 16$ and $x = n^2 + 6n + 1$; $\dfrac{dm}{dn}$

39. $y = 4a + 1$ and $a = 3b + 6$; $\dfrac{dy}{da}$

40. $y = 15t^3 + 2t^2 - 8$ and $x = 9t - 1$; $\dfrac{dy}{dt}$

41. Interpret the statement $\dfrac{dy}{dx} = 4$ when $x = 2$.

42. Interpret the statement $\dfrac{dy}{dx} = -6$ when $x = 15$.

43. Business: Jobs A population researcher has found that the population P (in thousands) of a city in a particular year is a function of the number N (in thousands) of the jobs available in that year which is, in turn, a function of the economy E (in billions of dollars) of the state in which that city is located. The researcher has determined that for a particular year, $E = 204$, $\frac{dP}{dN} = 8$, and $\frac{dN}{dE} = 0.6$. Find $\frac{dP}{dE}$ and interpret its value for $E = 204$.

44. Economics: Educational Endowments A researcher has found that the number N of endowments made to a university in a particular year is a function of the size P (in hundreds of full-time students) of the university and that the size of the university is a function of the number J of papers published in major journals by the faculty. The researcher has determined that for a particular year, $J = 78$, $\frac{dN}{dP} = 4$, and $\frac{dP}{dJ} = 0.7$. Find $\frac{dN}{dJ}$ and interpret its value for $J = 78$.

Modeling Practice

45. Economics: Government Spending The rate of government spending (in billions of dollars) in a country is related to the country's unemployment by the function

$$g(u) = 193 + u^{6/5}$$

where g is the level of government spending and u is the percentage unemployment rate.

The percentage unemployment rate, in turn, is related to the inflation rate as

$$u(i) = 80(i + 10)^{-1}$$

where i is the inflation rate. What change in spending can the government expect if the inflation rate rises from 7% to 8%?

46. Business: Demand and Cost An empirical relationship between the number of computers sold in a city and the median age of the people living in the city is given by the function

$$N(a) = -a^2 + 60a + 160$$

where N is the number of computers sold and a is the median age of the people.

The value of the software packages sold in the city depends on the number of computers sold as

$$V(N) = 100N + 500N^{3/4}$$

where V is the value of the software packages sold.

a. Write a symbolic description that shows how the value of the software packages sold changes with the median age of the people in the city.

b. How will the value change if the median age increases from 30 years to 31 years?

47. Business: Demand and Cost Demand for a product depends on the price of the product as

$$d(p) = 420p^{-2}$$

where d is the number of units demanded and p is the price of the product in dollars. The cost per unit sold depends on the number produced as

$$c(d) = 20 + 10d^{-1/3}$$

where c is the cost per unit in dollars.

a. Write a symbolic description that shows how the cost of producing a unit changes with the sales price.

b. How does the cost of production change as the price increases from \$50 to \$51 per unit?

48. Economics: Money Supply The money supply is a function of interest rate as

$$m(i) = 360 - 84i^{3/5}$$

where m is the money supply and i is the interest rate.

The general price level is a function of the money supply as

$$p(m) = 10\sqrt{m}$$

where p is the general price level.

a. Write a symbolic description of how the general price level changes with the interest rate.

b. What change in prices can be expected if interest rates rise from 8% to 9%?

Source: Suggested by the article *Money Supply* at Bloomberg.econoday.com

49. Business: Production The distribution manager of a national supplier of paper note pads knows that the number N (in thousands) of pads to be packaged is related to the price P (in dollars) of a pad. That relationship is expressed by the function $N(P) = -P^2 + 4P + 12, 2 \le P \le 6$. The price of each pad is, in turn, related to the cost C (in dollars) of a unit of paper. That relationship is expressed by the function $P(C) = 1.5C + 0.5$. The cost of paper is currently \$1.70 per unit but is expected to increase by one dollar next month. The distribution manager needs to prepare the packaging facility for a change in the number of pads to be packaged.

a. Write a symbolic description of how the number of pads to be packaged changes with the cost of paper.

b. What approximate change in the number of pads to be packaged should the manager expect?

© Microgen/iStockPhoto

50. **Ecology: Lake Toxicity** The amount A (in hundreds of gallons) of toxic material entering a lake is related to the time t (in months) an industrial factory near one of the lake's inlet rivers has been operating. The relationship is expressed by the function

$$A(t) = (0.8t^{1/5} + 2)^4$$

At what rate is the toxic material in the lake changing after 32 months?

Source: Suggested by an article in pesticideinfo.com

51. **Business: Jobs** The marketing department of a pool supply company has determined that the demand D (in thousands of tablets) for its pool chlorination tablets is related to the price x (in cents) by the function

$$D(x) = 400\left(\frac{8}{2x - 25}\right)^{1/4} \quad 15 \leq x \leq 100$$

Currently, the price of a tablet is \$0.40. What can the marketing department expect of the demand for the tablet if the price of a tablet is raised to \$0.41?

52. **Physical Science: Surface Area of a Sphere** A balloon in the shape of a sphere is being inflated with air by a machine that pumps in air at an increasing rate. The formula for the surface area of a sphere is $S(r) = 4\pi r^2$, where r is the radius of the balloon in centimeters. The radius of the balloon depends on how much air is in the balloon and air is being pumped in according to the formula

$$r(t) = \left(\frac{1}{20}t + 1\right)^{5/2}$$

where t is the number of seconds after the air pumping machine is turned on and starts pumping air into the balloon.

a. Write a formula that gives the surface area of the balloon in terms of time, t.

b. Write a formula that gives the rate at which the surface area changes as time changes.

c. How fast is the surface area of the balloon changing 1 second after the machine starts pumping air into the balloon? Approximate π with 3.14 and round your answer to the nearest one unit.

Using Technology Exercises

53. Find $f'(x)$ for $f(x) = (3x^2 + 4)^5$.

54. Comment on the behavior of the function $f(x) = \dfrac{(x^{1/3} + x)^{1/2}}{(x^{1/2} + x)^{1/3}}$ when $x = 2$.

55. The function $f(x) = (x - 7)^{2/3} + 25$ is defined at $x = 7$. Explain why Wolfram|Alpha is not able to find the equation of the tangent line to the function at $x = 7$.

Getting Ready for the Next Section

56. Which of the following functions is solved explicitly for y?

$$x^2 + y^2 = 4 \qquad y = \frac{2x - 5}{x^2 - 5}$$

57. If $x^2 + y^2 = 4$, solve for y.

58. If $x = 1$ and $y = 2$, find the value of $\dfrac{-15x^4}{8y^3 + 1}$.

59. **a.** Find the area of a circle with a radius of 600 feet.

b. Evaluate $2\pi r \dfrac{dr}{dt}$ when $r = 600$ and $\dfrac{dr}{dt} = 2$. Use 3.14 for π.

60. **a.** Write the expression below using a rational exponent, rather than a radical:

$$\frac{20{,}000}{\sqrt[3]{2p^2 - 5}} + 350$$

b. Find $\dfrac{dx}{dp}$ for $x = \dfrac{20{,}000}{\sqrt[3]{2p^2 - 5}} + 350$.

Implicit Differentiation

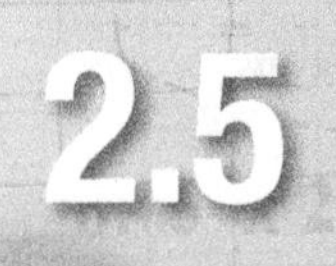

2.5

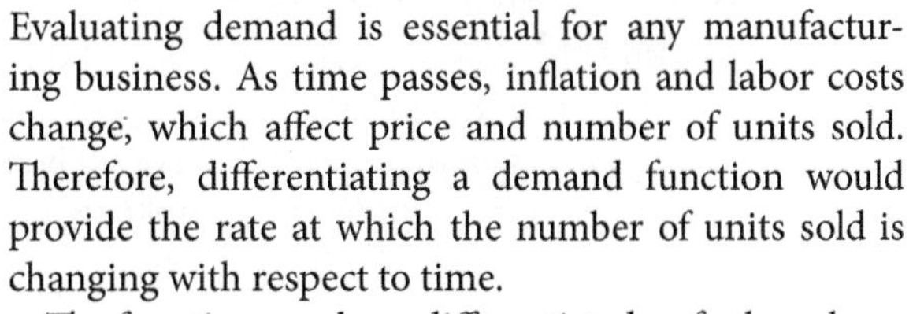

Evaluating demand is essential for any manufacturing business. As time passes, inflation and labor costs change, which affect price and number of units sold. Therefore, differentiating a demand function would provide the rate at which the number of units sold is changing with respect to time.

The functions we have differentiated so far have been of the form

$$f(x) = (\text{some expression in } x)$$

That is, all instances of the input variable are on one and only one side of the equal sign. In this section we examine implicit differentiation, a technique for finding the derivative of a function when the function is not solved for y; that is, when the function is not of the form $y = (\text{some expression in } x)$.

Explicit and Implicit Functions

Equations that are expressed as $y = f(x)$ directly convey the relationship between y and x. Since y is expressed explicitly as a function of x, the form $y = f(x)$ conveys y as an **explicit function** of x. All the functions we have worked with so far have been explicit functions. Some examples are

$$y = 5x - 6 \qquad y = 8x^2 + 7x - 11$$

$$y = \frac{2x - 5}{x^2 - 6} \qquad y = (x - 4)^3(3x + 5)^2$$

Some equations do not directly convey the relationship between y and x, but only imply it. Equations such as

$$xy = 3 \qquad x^2 + y^2 = 4$$

$$y^2 = x \qquad x^2 + 3xy + y^2 = 5$$

indirectly convey the relationship between y and x. Equations that only imply that y is a function of x are said to convey y as an **implicit function** of x.

Each of the above implicit functions can be converted to explicit functions by solving them for y. For example, for $x^2 + y^2 = 4$,

$$x^2 + y^2 = 4$$

$$y^2 = 4 - x^2$$

$$y = \pm\sqrt{4 - x^2}$$

so that $y = \sqrt{4 - x^2}$ and $y = -\sqrt{4 - x^2}$. We could now use the general power rule to find $\frac{dy}{dx}$.

Quite often, however, it is impractical, or even impossible, to convert an implicit function of x to an explicit function of x by solving for y. For example, it is impractical to try to solve the function

$$x^5 + 3x^2y^3 + y^5 = 2$$

for y in terms for x. But we can find $\frac{dy}{dx}$ (or y') for such a function using a method called implicit differentiation.

Implicit Differentiation

Implicit differentiation is the method of finding $\frac{dy}{dx}$ (or y') when y is an implicit function of x. It is a direct application of the general power rule, and we can gain insight to it by studying the following example.

Suppose y is an implicit function of x. Since y is not conveyed directly as a function of x, its form, in terms of x, is unknown. Let $f(x)$ represent this unknown form. Then, with $y = f(x)$, $y^n = [f(x)]^n$, and

$$\begin{aligned} \frac{d}{dx} y^n &= \frac{d}{dx}[f(x)]^n \\ &= n \cdot [f(x)]^{n-1} \cdot f'(x) && \text{By the general power rule} \\ &= ny^{n-1}y' && \text{Since } f'(x) = y' \end{aligned}$$

Derivative of an Implicit Function

If y is an implicit function of x, then

$$\frac{d}{dx} y^n = ny^{n-1}y'$$

Caution You might think of this rule as "the derivative of the outside times the derivative of the inside."

It is very common for students to find the derivative of y^n, with respect to x, as ny^{n-1}. This is *not* correct. The expression ny^{n-1} represents $\frac{d}{dy}y^n$, the derivative with respect to y, not $\frac{d}{dx}y^n$, the derivative with respect to x.

Suppose $y^3 - 5x^2 = 8$. We say that y is an implicit function of x. Since y is not conveyed directly in terms of x, its form, in terms of x, is unknown. Let $f(x)$ represent this unknown form. Then, with $y = f(x)$, we see that

$$y^3 - 5x^2 = 8 \text{ is equivalent to } [f(x)]^3 - 5x^2 = 8$$

Now differentiate each side with respect to x, using the general power rule.

$$\frac{d}{dx}[f(x)]^3 - \frac{d}{dx}[5x^2] = \frac{d}{dx}[8]$$

$$3 \cdot [f(x)]^2 \cdot f'(x) - 10x = 0$$

Now replace $f(x)$ with y and $f'(x)$ with y', and solve for y'.

$$\begin{aligned} 3y^2y' - 10x &= 0 \\ 3y^2y' &= 10x \\ y' &= \frac{10x}{3y^2} \end{aligned}$$

Notice that y' depends on both x and y.

In the examples that follow, we will omit the step of replacing y with its unknown form $f(x)$, and directly use the fact that

$$\frac{d}{dx}[y^n] = ny^{n-1}y'$$

VIDEO EXAMPLES

SECTION 2.5

Example 1 Find y' for $3x^5 + 2y^4 + y = 37$ at the point $(1, 2)$.

Solution Since y is an implicit function of x, we need to differentiate implicitly.

$$\frac{d}{dx}[3x^5] + \frac{d}{dx}[2y^4] + \frac{d}{dx}y = \frac{d}{dx}[37]$$

$$15x^4 + 8y^3y' + y' = 0 \qquad \text{Isolate } y'$$

$$8y^3y' + y' = -15x^4 \qquad \text{Factor out } y'$$

$$y'(8y^3 + 1) = -15x^4 \qquad \text{Divide by } 8y^3 + 1$$

$$y' = \frac{-15x^4}{8y^3 + 1}$$

Now evaluate y' at $x = 1$ and $y = 2$.

$$y'\Big|_{x=1,\, y=2} = \frac{-15x^4}{8y^3 + 1}\Big|_{x=1,\, y=2}$$

$$y' = \frac{-15(1)^4}{8(2)^3 + 1}$$

$$y' = \frac{-15(1)}{64 + 1}$$

$$y' = \frac{-15}{65}$$

$$y' = \frac{-3}{13}$$

Example 2 Find y' for $x^2y^3 - 3x + 4y = 10$.

Solution We see that y is an implicit function of x, so we use the method of implicit differentiation. We differentiate term-by-term, and since the first term is a product, we use the product rule.

$$\frac{d}{dx}[x^2y^3] - \frac{d}{dx}[3x] + \frac{d}{dx}[4y] = \frac{d}{dx}[10]$$

$$\frac{d}{dx}[x^2] \cdot y^3 + x^2 \cdot \frac{d}{dx}[y^3] - \frac{d}{dx}[3x] + \frac{d}{dx}[4y] = \frac{d}{dx}[10]$$

$$2x \cdot y^3 + x^2 \cdot 3y^2y' - 3 + 4y' = 0$$

$$2xy^3 + 3x^2y^2y' - 3 + 4y' = 0$$

$$3x^2y^2y' + 4y' = 3 - 2xy^3$$

$$y'(3x^2y^2 + 4) = 3 - 2xy^3$$

$$y' = \frac{3 - 2xy^3}{3x^2y^2 + 4}$$

Related Rates

Suppose the dependent variable y is related to the independent variable x by some function $y = f(x)$. Suppose also that y and x are both related to a third variable t (usually time in applications). Then $\frac{dy}{dt}$ and $\frac{dx}{dt}$, respectively, express the rates of change of y and x with respect to t. If the equation relating y and x is differentiated implicitly with respect to t, an equation involving $\frac{dy}{dt}$ and $\frac{dx}{dt}$ will result, and that equation will relate the rates of change $\frac{dy}{dt}$ and $\frac{dx}{dt}$. Problems in which rates of change are related to other rates of change are called **related rate problems.**

Related rate problems are solved using implicit differentiation.

Example 3 Suppose both y and x are differentiable functions of t and that the relationship between y and x is expressed by the equation $4x^3 + 3y^5 = 960$. Find and interpret $\frac{dy}{dt}$ when $\frac{dx}{dt} = 4$, $x = 6$, and $y = 2$.

Solution Since both y and x are functions of t, we need to differentiate them using implicit differentiation.

$$\frac{d}{dt}[4x^3] = 12x^2\frac{dx}{dt}, \quad \frac{d}{dt}[3y^5] = 15y^4\frac{dy}{dt}, \quad \frac{d}{dt}[960] = 0$$

Then, differentiating both sides of $4x^3 + 3y^5 = 960$, we get

$$\frac{d}{dt}(4x^3 + 3y^5) = \frac{d}{dt}960$$

$$\frac{d}{dt}[4x^3] + \frac{d}{dt}[3y^5] = \frac{d}{dt}[960]$$

$$12x^2\frac{dx}{dt} + 15y^4\frac{dy}{dt} = 0$$

Now we solve for $\frac{dy}{dt}$.

$$15y^4\frac{dy}{dt} = -12x^2\frac{dx}{dt}$$

$$\frac{dy}{dt} = \frac{-12x^2\frac{dx}{dt}}{15y^4}$$

$$= \frac{-4x^2\frac{dx}{dt}}{5y^4}$$

Now we can evaluate $\frac{dy}{dt}$ when $\frac{dx}{dt} = 4$, $x = 6$, and $y = 2$.

$$\frac{dy}{dt} = \frac{-4(6)^2(4)}{5(2)^4}$$

$$= \frac{-576}{80}$$

$$= -7.2$$

> **Interpretation**
>
> If t is increased by 1 unit, when $\frac{dx}{dt} = 4$, $x = 6$, and $y = 2$, then the value of y can be expected to decrease by approximately 7.2 units.

Example 4 Oil is leaking in a circular shape from a tanker in such a way that the radius of the circular spill is increasing at the rate of 2 feet per hour. To manage the spill, the oil company needs to know the rate at which the area covered by oil is increasing. Find the rate at which the area is increasing at the time that the radius of the spill is 600 feet. (Approximate π by 3.14.)

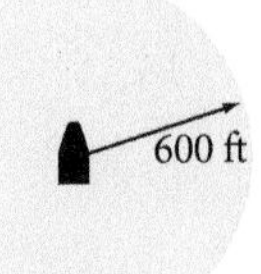

Solution When it is possible, drawings are very helpful in related rate problems. They make relationships more visual and promote better access to the equation. When we see the word *rate*, we think *derivative*. We wish to find $\frac{dA}{dt}$ when $\frac{dr}{dt} = 2$ and $r = 600$. (Note: $\frac{dr}{dt} = 2$, since the radius of the spill is increasing at the rate [which means *derivative*] of 2 feet per hour.)

The equation relating A and r is $A = \pi r^2$ (the formula for the area of a circle). Since we are interested in the rate of change with respect to time t, we differentiate each side of the equation with respect to t.

$$A = \pi r^2$$

$$\frac{d}{dt}(A) = \frac{d}{dt}(\pi r^2)$$

$$\frac{dA}{dt} = \pi \frac{d}{dt}(r^2) \qquad \text{Now use the chain rule}$$

$$\frac{dA}{dt} = \pi \cdot 2r\frac{dr}{dt} \qquad \text{Substitute } \pi = 3.14,\ r = 600 \text{ ft, and } \frac{dr}{dt} = 2 \text{ ft/hr}$$

$$\frac{dA}{dt} \approx (3.14) \cdot 2(600 \text{ ft})(2 \text{ ft/hr})$$

$$\frac{dA}{dt} \approx 7{,}536 \text{ square feet/hour}$$

> **Interpretation**
>
> In the next hour, the area of the oil spill can be expected to increase by approximately 7,536 square feet.

Example 5 A manufacturer of precision instruments has determined that when the price of an instrument is p (in hundreds of dollars), it will sell x of those instruments each month. The demand function relating p and x is

$$x = \frac{20{,}000}{\sqrt[3]{2p^2 - 5}} + 350$$

Due to inflation and changing labor costs, both p and x depend on time t (in months). Find the rate at which the number of instruments sold is changing with respect to time, when the price of an instrument is \$400 and is changing at the rate of one dollar per month.

Solution We see the word *rate* and think *derivative*. We need to find $\frac{dx}{dt}$ when $\frac{dp}{dt} = 0.01$ and $p = 4$. (Remember, p is in hundreds so \$400 means that $p = 4$ and \$1 means that $\frac{dp}{dt} = \frac{1}{100} = 0.01$.)

We differentiate each side of $x = \frac{20{,}000}{\sqrt[3]{2p^2 - 5}} + 350$ with respect to t. Since the numerator is a constant, we will rewrite the equation in the more convenient form $x = 20{,}000(2p^2 - 5)^{-1/3} + 350$. Then

$$\frac{dx}{dt} = 20{,}000 \cdot \frac{d}{dt}(2p^2 - 5)^{-1/3} + \frac{d}{dt}(350)$$

$$= 20{,}000\left[\frac{-1}{3}(2p^2 - 5)^{-4/3} \cdot 4p \cdot \frac{dp}{dt}\right] + 0$$

$$= \frac{-80{,}000(2p^2 - 5)^{-4/3}p}{3}\frac{dp}{dt}$$

$$= \frac{-80{,}000p}{3(2p^2 - 5)^{4/3}}\frac{dp}{dt}$$

$$= \frac{-80{,}000p}{3\sqrt[3]{(2p^2 - 5)^4}}\frac{dp}{dt}$$

Then, when $p = 4$ and $\frac{dp}{dt} = 0.01$,

$$\frac{dx}{dt} = \frac{-80{,}000(4)}{3\sqrt[3]{(2 \cdot 4^2 - 5)^4}} \cdot (0.01)$$

$$= \frac{-3{,}200}{3(\sqrt[3]{27})^4}$$

$$\approx -13.17$$

$$\approx -13$$

Interpretation

At the time when the price of a precision instrument is \$400 and increasing at the rate of \$1 per month, the company can expect to sell about 13 fewer instruments in the next month.

■

Using Technology 2.5

Do you have an implicitly defined function you wish to differentiate or a related rate problem you wish to solve? Wolfram|Alpha may be able to help you.

Example 6 Use Wolfram|Alpha to help you find $\frac{dy}{dx}$ for the implicitly defined function $y^3 - 5x^2 = 8$.

Solution Go to www.wolframalpha.com and in the entry field enter

find dy/dx of y^3 – 5x^2 = 8

Wolfram|Alpha displays your entry, its interpretation of your entry, and the value of the derivative.

$$y'(x) = \frac{10x}{3y^2}$$

Example 7 Solve the related rate problem of Example 5. We wish to differentiate the function

$$x = \frac{20{,}000}{\sqrt[3]{2p^2 - 5}} + 350$$

with respect to t and then evaluate the derivative when $p = 4$ and $\frac{dp}{dt} = 0.01$.

Solution Since both p and x depend on time t, we need to specify that. In the entry line, type

d/dt x(t) = 20,000/(2p(t)^2 - 5)^(1/3) + 350

Wolfram|Alpha responds with

$$\frac{dp}{dt} = -\frac{80{,}000p(t)p'(t)}{3(2p(t)^2 - 5)^{4/3}}$$

Now, in the entry field, enter the expression for $\frac{dp}{dt}$ with the values for $p(t)$ and $p'(t)$. Use $p = 4$ and $p'(t) = 0.01$. Take care to match left and right parentheses.

evaluate (–80,000(4)(0.01))/(3(2*4^2 - 5)^(4/3))

Wolfram|Alpha responds with

$$-13.1687\ldots \approx -13$$

which agrees with our work in Example 5.

Getting Ready for Class

After reading through the preceding section, respond in your own words and in complete sentences.

A. Give an example of a function in which y is given explicitly in terms of x.

B. Give an example of a function in which y is an implicit function of x.

C. How would you find $\frac{dy}{dx}$ for $x^2 + y^2 = 9$?

D. What is a related rate problem?

Problem Set 2.5

Skills Practice

For Problems 1-15, find y' for each function.

1. $3y^4 + 2x^3 = 4$
2. $5y^2 - 11x^2 = 9$
3. $3y^3 - 2y^2 + 5x^4 - x = 1$
4. $3x^4 + 6y^4 - 5y^3 = 5x^4 - 2x$
5. $(y + 6)^8 = 4x^2 + x - 4$
6. $(3y^2 + 4)^4 - 3x^5 + 5 = 0$
7. $(2y^3 + 5x^2)^3 = 6x^2 + 11$
8. $5xy^2 + y^3 = 4$
9. $6x^2 + 4x^2y^4 = 7x - 1$
10. $6x - 4x^5y^3 + y^5 = 1$
11. $5x^2 - 3x^2y^2 - y^3 = 2x$
12. $\sqrt[4]{2y + 3} = 9x + 4$
13. $\sqrt[5]{(y + 1)^2} = 1 - 4x$
14. $3x + 1 = 10(y^5 - 2)$

For Problems 15-20, find y' at the specified point.

15. $x^2 + y^2 = 25$, at $(0, 5)$
16. $y^2 + y^3 = 12x + 12$, at $(2, 3)$
17. $4y^3 + 3x^4 + y = 53$, at $(-2, 1)$
18. $\dfrac{1}{x + y} = \dfrac{1}{5}$, at $(6, -1)$
19. $x^{-3} + y^{-3} = -\dfrac{35}{216}$, at $(-2, -3)$
20. $(xy)^{2/3} = 9$, at $(3, 9)$

Paying Attention to Instructions The next two problems are intended to give you practice reading, and paying attention to, the instructions that accompany the problems you are working.

21. For the circle $x^2 + y^2 = 25$,
    a. find y when $x = 4$.
    b. find the slope of the tangent line where $x = 4$.
    c. find the points at which $\frac{dy}{dx} = 0$.
    d. If the radius starts increasing at a constant rate of 2 cm/sec, how fast is the area increasing when $r = 6$ cm?
22. For the equation $x^2y = 4$,
    a. find x when y is 1.
    b. find $\frac{dy}{dx}$ when y is 1.
    c. find the equation of the tangent line where $x = 2$.
    d. If $\frac{dx}{dt} = 3$, find $\frac{dy}{dt}$ when $x = 2$.

Check Your Understanding

In Problems 23-26, y is expressed implicitly as a function of x. The function is then differentiated and solved for y'. However, each differentiation process contains at least one error. Specify the first step at which the error occurs and correct it. (The step numbers are shown in parentheses.)

23. $4y^3 - 8x^2 + y = 0$

$12y^2 - 16x + y' = 0 \quad ...(1)$

$y' = 16x - 12y^2 \quad ...(2)$

24. $-6y^4 + 3y^2 + 5x^2 + 3y = 9$

$-24y^3 + 6y + 10x + 3y' = 0 \quad ...(1)$

$3y' = 24y^3 - 6y - 10x \quad ...(2)$

$y' = \dfrac{24y^3 - 6y - 10x}{3} \quad ...(3)$

25. $(2y^3 - 5x^2)^4 + 6y = 0$

$4(2y^3 - 5x^2)^3 + 6y' = 0 \quad ...(1)$

$6y' = -4(2y^3 - 5x^2)^3 \quad ...(2)$

$y' = \dfrac{-4(2y^3 - 5x^2)^3}{6} \quad ...(3)$

26. $y - 8x = (3x + 2y)^2$

$y' - 8x = 2(3x + 2y)(3 + 2) \quad ...(1)$

$y' - 8x = 10(3x + 2y) \quad ...(2)$

$y' = 10(3x + 2y) + 8 \quad ...(3)$

Modeling Practice

27. Business: Price and Demand The number x (in thousands) of cat flea collars demanded each year when the price of a collar is p dollars is expressed by the function $x^3 + 250p^2 = 18{,}000$. The collars are currently selling for \$4 each and the annual number of sales is 24,101. Find the approximate decrease in sales of the collar if the price of each collar is raised by \$1.

28. Business: Price and Demand Suppose that for a particular item the equation $8p + 2xp + 5x^2 = 400$, relates the price p (in dollars) and the demand x (in thousands of units) for the item. Find the approximate change in sales (demand) if the selling price is increased one dollar from \$11 to \$12.

Problems 29-32 are related rate problems.

29. Manufacturing: Computer Costs The cost C (in dollars) of manufacturing x number of high-quality computer laser printers is

$$C(x) = 15x^{4/3} + 15x^{2/3} + 600{,}000$$

Currently, the level of production is 1,728 printers and that level is increasing at the rate of 350 printers each month. Find the rate at which the cost is increasing each month.

30. Manufacturing: Air Quality and Automobiles The air quality monitoring department of a city has established a relationship between the quality of air and the number of automobiles driven in the city. The air quality is measured in pollution index units, I, $0 \leq I \leq 100$, and the number of automobiles driven in the city, x, is measured in thousands. The relationship is expressed by the function

$$I(x) = 30 + (5x - 277)^{3/5}$$

Currently, there are 75,000 automobiles being driven in the city ($x = 75$) and that number is increasing at the rate of 4,500 per year. By how many points can the people at the air quality monitoring department expect the pollution index to increase in the next year?

Source: Suggested by an article in
The Reference in Scientific Document Supply,

31. **Business: Revenue** The monthly revenue R (in dollars) of a telephone polling service is related to the number x of completed responses by the function

$$R(x) = -12{,}000 + 25\sqrt{3.5x^2 + 25x} \quad 0 \le x \le 1{,}500.$$

If the number of completed responses is increasing at the rate of 10 forms per month, find the rate at which the monthly revenue is changing when $x = 750$.

32. **Economics: Foreign Aid** The amount of economic aid E (in millions of dollars) a country receives from the United States is related to the amount of military aid x (in millions of dollars) it receives by the function

$$E(x) = 10 + \sqrt{5x^3 + x^2} \quad 0 \le x \le 25.$$

If military aid is decreasing at the rate of 4 million dollars per year, find the rate at which the economic aid is changing when the military aid is \$18 million a year.

Using Technology Exercises

33. Find y' for the implicitly defined function $2y^3 + 3x^2 = 25$.

34. Find y' for the implicitly defined function $y^{1/2} + x^{1/2} = 30^{1/2}$.

35. Find and interpret y' for $x^3 + 3y^3 + y = 92$ at the point $(2, 3)$.

36. Find the equation of the tangent line to the circle

$$x^2 + y^2 = 1$$

at the point $(-\frac{4}{5}, \frac{3}{5})$.

37. When there is no change in heat, the pressure and the volume of a gas change according to the function $PV^\gamma = C$, where P represents the pressure of the gas, V represents the volume of the gas, γ (the lower case Greek letter *gamma*) is a constant that depends on the gas under consideration, and C is a constant. For air, $\gamma = 1.4$. Suppose that at a particular time, the pressure on 500 in^3 of air is 30 lb/in^2 and the volume of air is decreasing at the rate of 10 in^3/sec. What is the rate of change of the pressure? (In Wolfram|Alpha, use the English letter a in place of the Greek letter γ. The left side of the equation is a product, $P \cdot V^\gamma$, so Wolfram|Alpha uses the product rule. Be sure to tell Wolfram|Alpha that both P and V are functions of t.)

Source: UNSW, School of Physics, Sydney, Australia

38. **Business: Revenue** The monthly revenue R (in dollars) realized by a chain of nursery stores is related to the number x of house plants it can secure and sell each month by the function

$$R(x) = 35{,}000 + \sqrt{\frac{x^4 + 800}{3x + 650}} \quad 0 \le x \le 5{,}000$$

If the number of house plants being secured and sold is increasing at the rate of 120 plants each month, at what rate is the revenue changing when the number of plants the chain is securing and selling is 1,500?

39. Manufacturing: Cost of Production The manufacturer of reinforced cardboard packing boxes has determined that its monthly cost C (in dollars) of producing such boxes is related to the price p (in dollars) it has to pay for a unit of cardboard by the function

$$C(p) = 45{,}000 + 18{,}000\left(\frac{250p + 20}{80p + 90}\right)^{3/4}$$

If the current price of a unit of cardboard is \$0.40 but is increasing at the rate of \$0.02 per month, at what rate is the cost of producing the boxes changing?

Getting Ready for the Next Section

Find the intercepts for each function.

40. $y = x^{3/5} - 1$

41. $f(x) = \dfrac{5x + 2}{3x - 4}$

42. Find the y-intercept for $y = x^3 - 3x^2 + 4$.

43. Find $\lim\limits_{x \to \infty} \dfrac{5x + 2}{3x - 4}$.

Solve.

44. $3x^2 - 6x < 0$

45. $6x - 6 > 0$

Find the first and second derivatives for each function.

46. $f(x) = x^{3/5} - 1$

47. $f(x) = \dfrac{5x + 2}{3x - 4}$

Chapter 2 Summary

EXAMPLES

1. If $f(x) = 3x + 2$, then

$f'(x) = \lim_{h \to 0} \frac{f(x+h) - f(x)}{h}$

$= \lim_{h \to 0} \frac{3(x+h) + 2 - (3x+2)}{h}$

$= \lim_{h \to 0} \frac{3x + 3h + 2 - 3x - 2}{h}$

$= \lim_{h \to 0} \frac{3h}{h}$ Since $h \neq 0$

$= \lim_{h \to 0} 3$ The limit of a constant is the constant

$= 3$

Limit Definition of the Derivative [2.1]

The **derivative** of the function $f(x)$, denoted $f'(x)$, is

$$f'(x) = \lim_{h \to 0} \frac{f(x+h) - f(x)}{h},$$

provided this limit exists. This is one of the most important definitions in the course. Be sure you have it memorized.

2. If $f(x) = x^2 + 2x$, then $f'(x) = 2x + 2$ and $f'(-3) = 2(-3) + 2 = -4$. So -4 is the slope of the tangent line when x is -3.

Interpreting the Derivative [2.1]

The derivative $f'(x)$ of the function $f(x)$

a. is the slope of the tangent line to $f(x)$.

b. is the instantaneous rate of change of f with respect to x.

3. **a.** If $y = 3x + 2$, then $y' = 3$ which gives a constant rate of change for every x.

b. If $y = x^2 - 1$, then $y' = 2x$, which is a variable rate of change.

Exact and Approximate Rates of Change [2.1]

a. If the rate of change remains constant as x increases one unit from $x = a$ to $x = a + 1$, the derivative, $f'(x)$, produces the exact rate of change. (The expression for $f'(x)$ will involve no variable.)

b. If the rate of change varies as x increases one unit from $x = a$ to $x = a + 1$, the derivative, $f'(x)$, produces the approximate rate of change. (The expression for $f'(x)$ will involve variables.)

4. If $R(t) = t^3 - 50t + 150$, $0 \leq t \leq 10$, represents a company's expected revenue for the next 10 weeks, then

$R'(t) = 3t^2 - 50$, and

$R'(2) = 3 \cdot 2^2 - 50 = -38$

represents the marginal revenue at week 2.

Marginal Change [2.1]

The change in the output variable that is produced by a 1-unit change in the input variable is called the **marginal** change in the output.

5. If $y = g(x) = x^4$, then

$y' = g'(x) = \frac{dy}{dx} = \frac{dg}{dx} = 4x^3$

Derivative Notations [2.1]

If $y = f(x)$, then each of the following notations represents the derivative of $f(x)$.

$$f'(x) \qquad y' \qquad \frac{dy}{dx} \qquad \frac{df}{dx} \qquad \frac{d}{dx}f$$

Derivative of a Constant [2.1]

6. If $y = 5$, then $y' = 0$.

If $y = c$, where c is any real number, then $\frac{dc}{dx} = 0$.

Derivative of a Simple Power [2.1]

7. If $y = x^6$, then $y' = 6x^5$

If $y = x^n$, where n is any real number, then $\frac{dy}{dx} = n \cdot x^{n-1}$.

Derivative of a Constant Times a Function [2.1]

8. If $f(x) = 8x^3$ then $f'(x) = 8 \cdot 3x^2 = 24x^2$

If $f(x)$ is a differentiable function, then $\frac{d}{dx}cf(x) = c \cdot \frac{d}{dx}f(x)$.

The Derivative of a Sum or a Difference [2.1]

9. If $y = 3x^2 + 8x^3$, then

$$\frac{dy}{dx} = \frac{d}{dx}3x^2 + \frac{d}{dx}8x^3$$

$$= 6x + 24x^2$$

If $f(x)$ and $g(x)$ are both differentiable functions, then

$$\frac{d}{dx}[f(x) \pm g(x)] = \frac{d}{dx}f(x) \pm \frac{d}{dx}g(x)$$

The Product Rule [2.2]

10. If $y = x^3(3x^2 + 4)$, then

$$\overset{u' \cdot v \;+\; u \cdot v'}{y' = (3x^2)(3x^2 + 4) + (x^3)(6x)}$$

$$= 9x^4 + 12x^2 + 6x^4$$

$$= 15x^4 + 12x^2$$

If $u(x)$ and $v(x)$ are differentiable functions, then

$$\frac{d}{dx}[u(x) \cdot v(x)] = \frac{d}{dx}[u(x)] \cdot v(x) + u(x) \cdot \frac{d}{dx}[v(x)]$$

The Quotient Rule [2.2]

11. If $f(x) = \frac{x^2}{2x - 3}$, then

$$f'(x) = \frac{\overset{v \cdot u' - u \cdot v'}{(2x - 3)(2x) - x^2(2)}}{(2x - 3)^2}$$

$$= \frac{4x^2 - 6x - 2x^2}{(2x - 3)^2}$$

$$= \frac{2x^2 - 6x}{(2x - 3)^2}$$

If $u(x)$ and $v(x)$ are differentiable functions and $v(x) \neq 0$, then

$$\frac{d}{dx}\left[\frac{u(x)}{v(x)}\right] = \frac{v(x) \cdot \frac{d}{dx}[u(x)] - u(x) \cdot \frac{d}{dx}[v(x)]}{[v(x)]^2}$$

Higher-Order Derivatives [2.3]

12. If $y = \frac{1}{x^2}$, then

$y' = -\frac{2}{x^3}$

$y'' = \frac{6}{x^4}$

$y''' = -\frac{24}{x^5}$

$y^{(4)} = \frac{120}{x^6}$

If $y = f(x)$ is a differentiable function, then

$y', f'(x), \frac{dy}{dx}$, and $\frac{df}{dx}$ all indicate the first derivative of the function.

$y'', f''(x), \frac{d^2y}{dx^2}$, and $\frac{d^2f}{dx^2}$ all indicate the second derivative of the function.

$y''', f'''(x), \frac{d^3y}{dx^3}$, and $\frac{d^3f}{dx^3}$ all indicate the third derivative of the function.

$y^{(4)}, f^{(4)}(x), \frac{d^4y}{dx^4}$, and $\frac{d^4f}{dx^4}$ all indicate the fourth derivative of the function.

The Chain Rule [2.4]

13. If $y = u^2 + 3u$ and $u = 3x - 1$, then

$$\begin{aligned}\frac{dy}{dx} &= \frac{dy}{du} \cdot \frac{du}{dx} \\ &= (2u + 3)(3) \\ &= 6u + 9 \\ &= 6(3x - 1) + 9 \\ &= 18x - 6 + 9 \\ &= 18x + 3\end{aligned}$$

If y is a differentiable function of u and u is a differentiable function of x—that is, $y = f(u)$ and $u = g(x)$—then $y = f(u) = f[g(x)]$, and y is a function of x. Also,

$$\frac{dy}{dx} = \frac{dy}{du} \cdot \frac{du}{dx}$$

Alternatively,

$$\frac{dy}{dx} = f'[g(x)] \cdot g'(x)$$

The General Power Rule [2.4]

14. If $y = (6x - 1)^4$, then

$$\begin{aligned}\frac{dy}{dx} &= 4(6x - 1)^3(6) \\ &= 24(6x - 1)^3\end{aligned}$$

If y is a differentiable function of u, and u is a differentiable function of x and $y = u^n$, then

$$\frac{dy}{dx} = nu^{n-1} \cdot \frac{du}{dx}$$

or,

if $y = [u(x)]^n$, then $\frac{dy}{dx} = n[u(x)]^{n-1} \cdot \frac{du}{dx}$, where n is any real number.

Derivative of an Implicit Function [2.5]

If y is an implicit function of x, then $\dfrac{d}{dx}y^n = ny^{n-1}y'$.

15. If $x^2 + y^2 = 9$, then

$$\frac{d}{dx}(x^2 + y^2) = \frac{d}{dx}(9)$$

$$2x + 2y\frac{dy}{dx} = 0$$

$$\frac{dy}{dx} = -\frac{x}{y}$$

Related Rates [2.5]

If $y = f(x)$ is such that both x and y are also functions of time t, then we can differentiate both sides of $y = f(x)$ with respect to t. The result will be a function that gives the relationship between the two rates of change, $\frac{dx}{dt}$ and $\frac{dy}{dt}$.

16. If $x^2 + y^2 = 9$, then

$$\frac{d}{dt}(x^2 + y^2) = \frac{d}{dt}(9)$$

$$2x\frac{dx}{dt} + 2y\frac{dy}{dt} = 0$$

$$x\frac{dx}{dt} = -y\frac{dy}{dt}$$

Chapter 2 Test

For Problems 1-10, find the derivative of each function. [2.1, 2.2, 2.4]

1. $f(x) = x^3 + 5x + 4$

2. $f(x) = 8$

3. $f(x) = (5 - 3x)^{4/3}$

4. $f(x) = 3x - 4x^2 + x^3$

5. $f(x) = (2x + 1)(3x - 8)$

6. $f(x) = 3x^2(5x + 4)^3$

7. $f(x) = (x - 4)^2(x + 1)^3$

8. $f(x) = \dfrac{2}{\sqrt[3]{3x + 4}}$

9. $f(x) = \dfrac{x + 1}{x + 3}$

10. $f(x) = \dfrac{(x + 2)^3}{(x - 4)^2}$

11. Create a three-term polynomial function $f(x)$ for which $f^{(4)}(x) = 0$. [2.3]

For Problems 12 and 13, find $\dfrac{dy}{dx}$. [2.4]

12. $y = 3u + 4, u = x + 1$

13. $y = u^2 + 3, u = 2x + 3$

14. Find y' for $x^2 + 3y^3 + 4 = 0$. [2.5]

15. Find y' for $5x - y^3 + x^2y^2 = 10$. [2.5]

16. For $f(x) = 4x^3 + 3x^2 - x$, find
  a. $f'(x)$.
  b. $f''(x)$.
  c. $f'''(x)$. [2.3]

17. For $f(x) = (x - 3)^3$, find
  a. $f'(x)$.
  b. $f''(x)$. [2.3, 2.4]

18. If $f(x) = 3x^2 - x$, find
  a. $f(2)$.
  b. $f'(2)$. [2.1]

19. If $f(x) = x^3(x^2 + 1)$, find the slope of the line tangent at $x = 0$. [2.2]

20. If $f(x) = 7x^2 + 21x$, find the value of x for which $f'(x) = 0$. [2.1]

21. Find the equation of the line tangent to $y = (x - 3)^2 + 2$ where $x = 4$. [2.4]

Modeling Practice

22. Manufacturing: Container Strength A manufacturer of plastic drinking cups has established that the function $d = f(t)$ relates the breaking strength (in pounds per square inch) of a plastic cup to the thickness t (in millimeters) of the cup. Interpret $f(18) = 10$ and $f'(18) = 0.8$. [2.1]

23. Manufacturing: Grain Size The mechanical properties of steel are affected by grain size. Reduction in grain size improves yield strength but has a significant effect on the ductile/brittle transition temperature. Suppose the function $D(s)$ relates the ferrite grain size in units of $(d/\mu m)^{1/2}$ and the ductile/brittle transition temperature in degrees Celsius. Interpret the information provided by

$$D(8) = 50, D'(8) > 0, \text{ and } D''(8) < 0 \qquad [2.2]$$

Source: Metallurgy for Dummies, The Metallurgy's 810g for beginners

24. The Human Cannonball David Smith, Jr., The Bullet, was shot from a cannon at the Washington County Fair in Oregon. He reached a height of 70 feet before landing in a net 160 feet from the cannon. The shape of his path is the parabola

$$f(x) = -\frac{7}{640}x^2 + \frac{7}{4}x$$

a. Find the derivative of $f(x)$.

b. Evaluate the derivative at $x = 80$. [2.1]

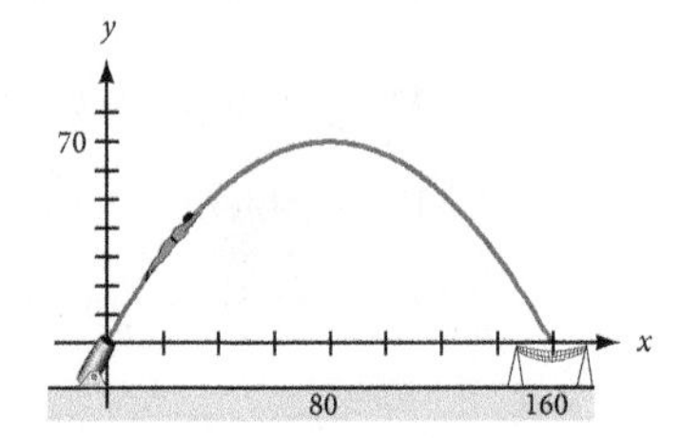

25. Physical Science: Medicine The drug vancomycin is dripping from an IV bottle onto the surface of a table and is forming a circular disk. The radius of the disk is increasing at the rate of 2 cm per minute. Find the rate at which the area of the disk is increasing at the time when the radius is 4 cm. [2.5]

26. Business: Inventory Costs The manager of a large department store has determined that the equation

$$C(x) = \frac{203{,}000}{x} + 2x$$

relates the inventory cost C (in dollars) to the number of packages x of men's socks it stocks. The manager is currently ordering 317 packages of socks. What change in the inventory costs can the manager expect if she changes her order to 318 packages? [2.1]

Source: Suggested by an article in Scribd

27. Business: Marginal Profit The publisher of a fiction book expects a total profit from a recently published book that is expected to sell for many years to be P (in thousands of dollars). If t represents the number of years since the release of the book, P and t are related by the function

$$P(t) = \frac{850t^2}{t^2 + 20}$$

Find the publisher's marginal profit on this book at year 10 and then at year 11.

3 Applying the Derivative

© Kovalchuk Oleksandr/Shutterstock

Chapter Outline

Note When you see this icon next to an example or problem in this chapter, you will know that we are using the topics in this chapter to model situations in the world around us.

In 2011, extensive flooding in Thailand severely disrupted the worldwide supply of computer hard drives. Nearly overnight, hard drive prices began to climb affecting the cost of everything from server farms to laptop computers to digital video recorders that use hard drives to save television shows.

In an attempt to keep hard drives in stock, retailers raised prices and put limits on the number of hard drives customers could purchase. As the price increased, the number of sales went down. In normal circumstances, knowing the relationship between the price of an item and the number of sales that will occur at that price, is the key to maximizing profit.

For example, Sally owns an online business and is trying to figure out how best to price her inventory of 3-terabyte hard drives to maximize her profits. Currently, Sally sells 20 hard drives (that cost the store \$200 each) a week for \$400 each. If she drops the price to \$390, she estimates that she will sell 22 hard drives a week. As you will see in this chapter, this information can lead to the profit equation

$$P(r) = 4{,}000 + 200r - 20r^2$$

where r is the number of \$2 reductions in price. Differentiating the profit function and setting the result to zero

$$P'(r) = 0$$

will lead us to her maximum profit. By applying the derivative to the function that describes her hard drive sales, Sally can find out what this optimum price is. In this chapter you'll learn to interpret the derivative as it applies to a variety of real-world scenarios like Sally's.

When determining how to price her hard drives, Sally focuses on finding a price point that maximizes her profits. This process is called *optimization*, and it is one of the topics we cover in this chapter.

Study Skills

The study skills for this chapter are about attitude. They are points of view that point toward success.

1. **Be Focused, Not Distracted** We have students who begin their assignments by asking themselves, "Why am I taking this class?" If you are asking yourself similar questions, you are distracting yourself from doing the things that will produce the results you want in this course. Don't dwell on questions and evaluations of the class that can be used as excuses for not doing well. If you want to succeed in this course, focus your energy and efforts toward success, rather than distracting yourself from your goals.
2. **Be Resilient** Don't let setbacks keep you from your goals. You want to put yourself on the road to becoming a person who can succeed in this class, or any class in college. Failing a test or quiz, or having a difficult time on some topics, is normal. No one goes through college without some setbacks. Don't let a temporary disappointment keep you from succeeding in this course. A low grade on a test or quiz is simply a signal that you need to reevaluate your study habits.
3. **Intend to Succeed** We have a few students who simply go through the motions of studying without intending to master the material. It is more important to them to look like they are studying than to actually study. You need to study with the intention of being successful in the course. Intend to master the material, no matter what it takes.

The First Derivative and the Behavior of Functions

3.1

Functions are important mathematical instruments because they chronicle the behavior of physical or theoretical phenomena. They allow us to analyze the past, current, and future behavior of the phenomena. For example, the graph below shows the behavior of the concentration of a medication in a patient's system t hours after it is administered.

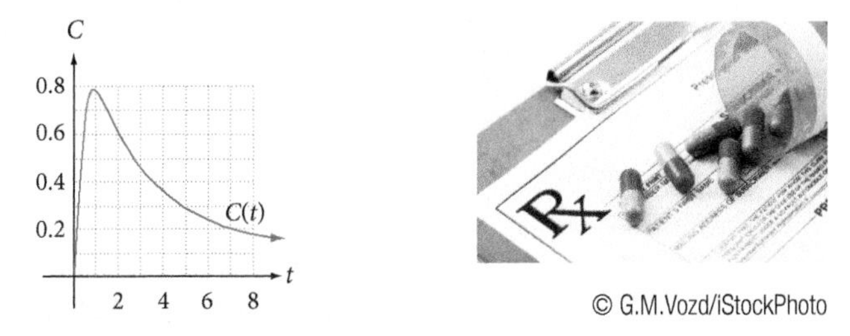

© G.M.Vozd/iStockPhoto

Of the various forms of a function (tables of values, sets of ordered pairs, expressions, and graphs), the graphical form is particularly appealing because it can reveal information that may not be evident from the other forms. Graphs of functions can be generated by computers or calculators by inputting the ordered pairs or the expression that defines the function.

When examined from left to right, the graph of a function can exhibit properties that allow us to describe the behavior of the function. For example, the graph of the function can be

1. rising, falling, or staying constant
2. opening upward or downward
3. attaining a highest value or a lowest value relative to nearby values
4. attaining a highest value or lowest value relative to the entire domain of the function
5. approaching asymptotes; or
6. intersecting one or both of the coordinate axes.

Intervals on Which a Function Is Increasing or Decreasing

A function $f(x)$ is **increasing** on an interval (a, b) if the graph of the function rises through (a, b). If a curve rises through (a, b), its tangent lines will also rise and will, therefore, have positive slopes. The graphs of Figure 1 illustrate functions that increase through an interval (a, b).

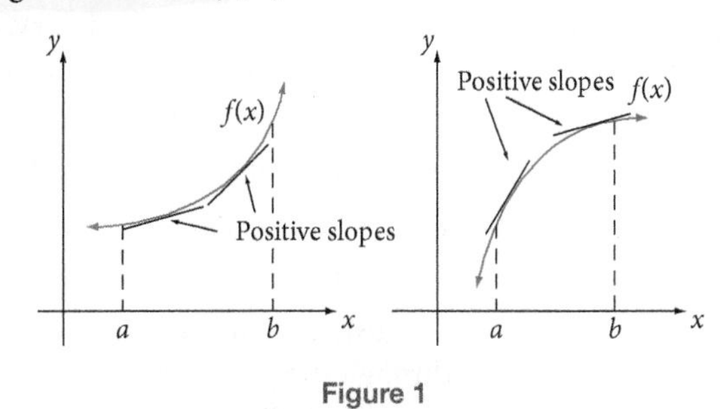

Figure 1

A function $f(x)$ is **decreasing** on an interval (a, b) if the graph of the function falls through (a, b). If a curve falls through (a, b), its tangent lines will also fall and will, therefore, have negative slopes. The graphs of Figure 2 illustrate functions that decrease through an interval (a, b).

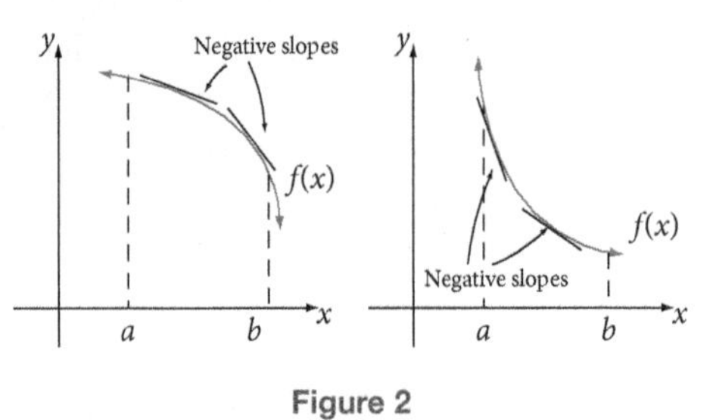

Figure 2

Since the slope of a tangent line to a curve is the derivative of the function corresponding to the curve, the increasing and decreasing behavior of the function can be described in terms of the signs of the first derivative of the function.

The First Derivative and the Increasing/Decreasing Behavior of a Function
Suppose that $f(x)$ is a continuous, differentiable function defined on the interval (a, b). Then

1. If $f'(x) > 0$ for every value of x in (a, b), then $f(x)$ is increasing on (a, b). Conversely, if $f(x)$ is increasing on (a, b), then $f'(x) > 0$ for every value of x in (a, b).
2. If $f'(x) < 0$ for every value of x in (a, b), then $f(x)$ is decreasing on (a, b). Conversely, if $f(x)$ is decreasing on (a, b), then $f'(x) < 0$ for every value of x in (a, b).
3. If $f'(x) = 0$ for every value of x in (a, b), then $f(x)$ is neither increasing nor decreasing on (a, b), but is constant on (a, b). Conversely, if $f(x)$ is constant on (a, b), then $f'(x) = 0$ for every value of x in (a, b).

VIDEO EXAMPLES

SECTION 3.1

Example 1 The graph illustrated in Figure 3 displays the relationship between time t (in years from the beginning of the year 2000) and the number N of townhouse sales in the southeastern part of the United States in the period from 2000 to 2010. Discuss the behavior of $N(t)$ in terms of increasing and decreasing.

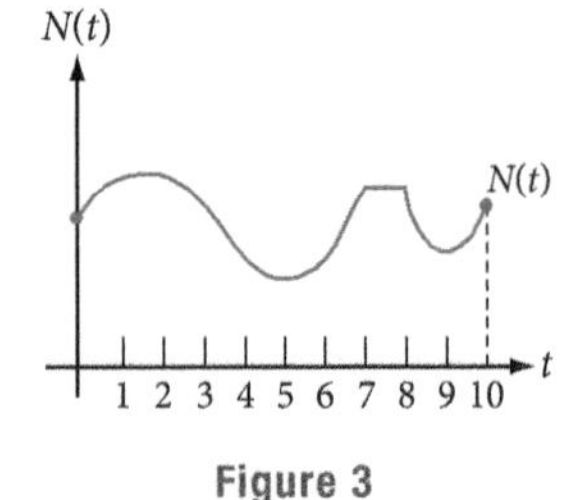

Figure 3

Solution

- $N(t)$ rises through the intervals $(0, 2)$, $(5, 7)$, and $(9, 10)$ so that $N(t)$ is increasing and $N'(t) > 0$ through each of these intervals.
- $N(t)$ falls through the intervals $(2, 5)$ and $(8, 9)$ so that $N(t)$ is decreasing and $N'(t) < 0$ through each of these intervals.
- $N(t)$ neither rises nor falls through $(7, 8)$ so that $N(t)$ is constant and $N'(t) = 0$ through $(7, 8)$.

Thus, we can conclude that the number of townhouse sales was increasing through the years 2000-2002, 2005-2007, and 2009-2010, decreasing through the years 2002-2005 and 2008-2009, and constant through the years 2007-2008. ■

Example 1 illustrates that graphs allow us to make estimates rather than obtain exact results. But very often in applied problems, estimates are perfectly acceptable. When exact results are desired, an analytic approach is needed. Using information obtained from the first derivative of a function can be the analytic approach needed to find the intervals upon which a function increases or decreases.

Sign charts are useful for determining and displaying the intervals upon which a function increases or decreases. A sign chart is a number line picture on which the signs of the derivative of a function are displayed. Those signs indicate if the function is increasing or decreasing at a particular location.

In constructing sign charts, we use the fact that the graph of a continuous function possesses no holes, gaps, or breaks. A continuous function (in this case, the first derivative function) can change signs from positive to negative or from negative to positive only if it equals zero or is undefined at some point. See Figure 4.

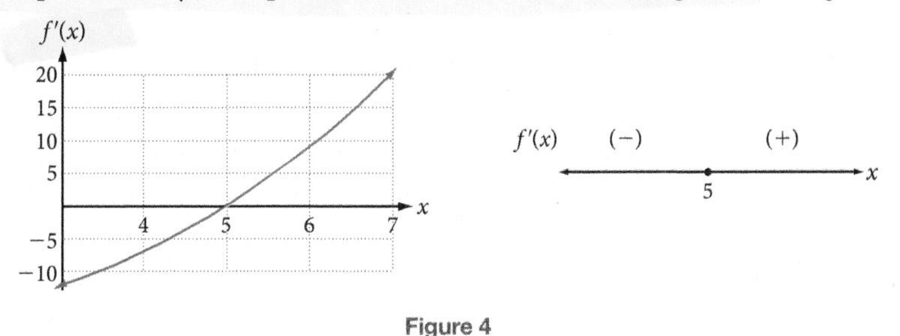

Figure 4

Figure 4 shows the graph of the derivative of a continuous function $f(x)$. Notice that the graph of the derivative function changes signs as it passes through the zero at $x = 5$. To the left of $x = 5$, the graph of $f'(x)$ lies below the x-axis so that it is negative $(-)$. This tells us that the function $f(x)$ itself is decreasing to the left of $x = 5$. To the right of $x = 5$, the graph of $f'(x)$ lies above the x-axis so that it is positive $(+)$. This tells us that the function $f(x)$ itself is increasing to the right of $x = 5$. Since the number 5 is critical to the accurate description of the behavior of $f(x)$, it is called a critical value of $f(x)$.

Critical Values of a Function

If a function $f(x)$ is defined at $x = c$, and if either $f'(c) = 0$ or $f'(c)$ is undefined, then the number c is called a **critical value** of $f(x)$. Conversely, if the number c is a critical value of $f(x)$, then either $f'(c) = 0$ or $f'(c)$ is undefined.

The Procedure for Constructing a Sign Chart for $f'(x)$

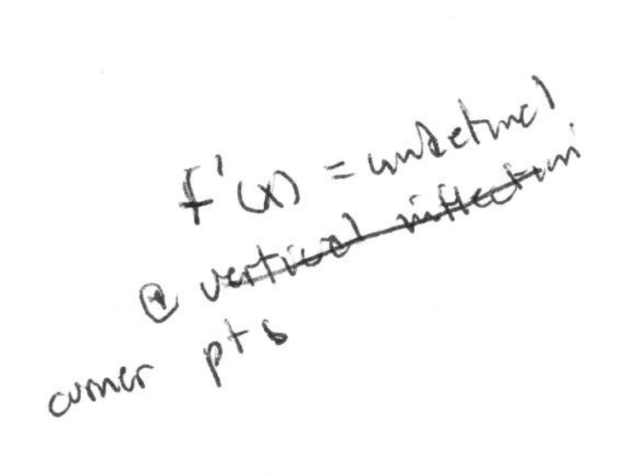

1. Find all the critical values of $f(x)$. That is, find all values of x for which either $f'(x) = 0$ or $f'(x)$ is undefined. If $f'(x)$ is a fraction, $f'(x)$ will be zero only when the numerator equals zero. It will be undefined only when the denominator equals zero.
2. Draw a number line and place points at the critical values. These points divide the line into the open intervals upon which the function increases or decreases.
3. In each interval, choose a test number a and determine the sign of $f'(x)$ at $x = a$.
   a. If $f'(a) > 0$, then $f'(x) > 0$ throughout the interval and the function $f(x)$ is increasing on the interval.
   b. If $f'(a) < 0$, then $f'(x) < 0$ throughout the interval and the function $f(x)$ is decreasing on the interval.

Example 2 Find the interval on which $f(x) = x^2 - 4x$ is increasing and the interval upon which it is decreasing.

Solution

1. Find the critical values of $f(x)$ by setting $f'(x)$ equal to zero and solving for x. Since $f(x)$ is a polynomial function, it and all its derivatives are never undefined.

$$f'(x) = 2x - 4$$

and

$$f'(x) = 0, \qquad \text{when}$$
$$2x - 4 = 0$$
$$2(x - 2) = 0$$
$$x = 2$$

The number 2 is the only critical value of the function $f(x)$.

2. Draw the number line and place a point at $x = 2$. (See the number line in Figure 5 below.) This identifies the two open intervals $(-\infty, 2)$ and $(2, \infty)$. These are the only intervals upon which $f(x)$ could be increasing or decreasing.

3. In each interval, choose a test point a and compute the sign of $f'(x)$ at $x = a$.

   a. In the interval $(-\infty, 2)$, choose a value to test, say $x = 0$. Then

   $$f'(0) = 2(0 - 2) = 2(-2) = -4$$

   so that $f'(x) < 0$ throughout this interval. We conclude that $f(x)$ is decreasing on $(-\infty, 2)$.

   b. In the interval $(2, \infty)$, choose a value to test, say $x = 3$. Then

   $$f'(3) = 2(3 - 2) = 2(1) = 2$$

   so that $f'(x) > 0$ throughout this interval. We conclude that $f(x)$ is increasing on $(2, \infty)$.

   Figure 5 shows the sign chart for and graph of $f(x) = x^2 - 4x$. The sign chart does indeed reflect the behavior shown in the graph.

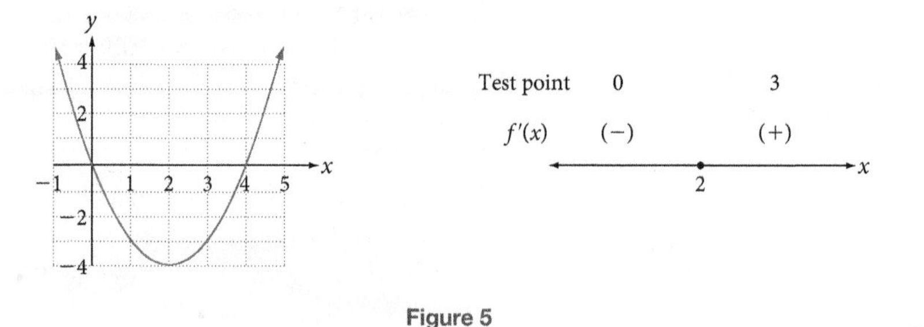

Figure 5

Example 3 Discuss the first derivative of $g(x) = 4$ and its relationship to its graph.

Solution $g'(x) = 0$ so that $g(x)$ is constant. The function values are always 4. As the values of x increase, the function values neither increase nor decrease. The graph in Figure 6 reflects this constant behavior.

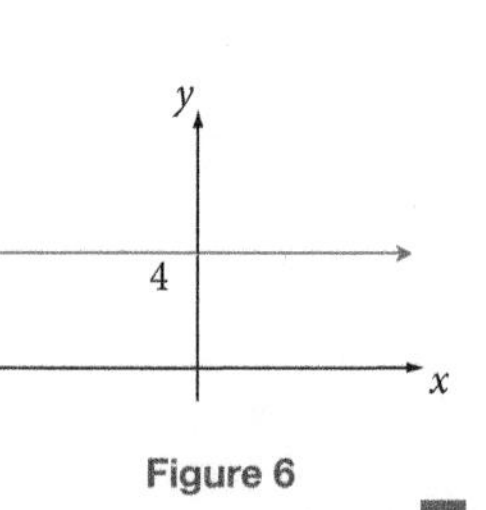

Figure 6

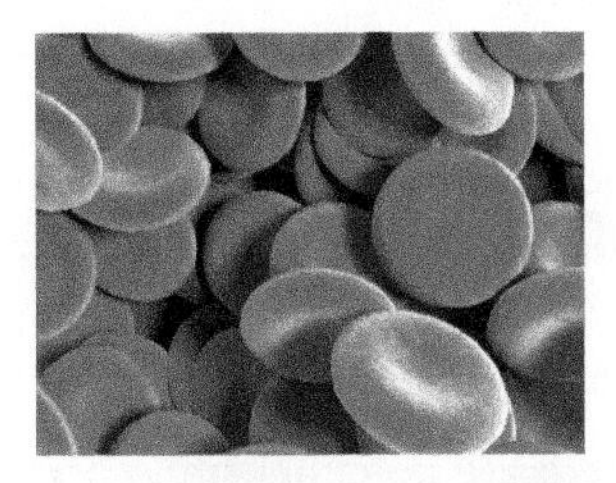

Relative Maxima and Minima

The function illustrated graphically in Figure 7 describes the relationship between time t (in minutes) and the concentration C (in milligrams per deciliter) of sugar in a person's blood when a highly concentrated sugar solution is ingested and then, 30 minutes later, an experimental drug that sharply reduces the concentration of sugar is injected. Notice that for values in a small interval around $t = 30$, it is $t = 30$ that produces the largest function value (the highest sugar concentration). That is, $t = 30$ produces a maximum value relative to the other t values near $t = 30$. Thus, the point $(30, C(30))$ is the **relative maximum** because, relative to points on the curve near $(30, C(30))$, the function has the largest value at that point.

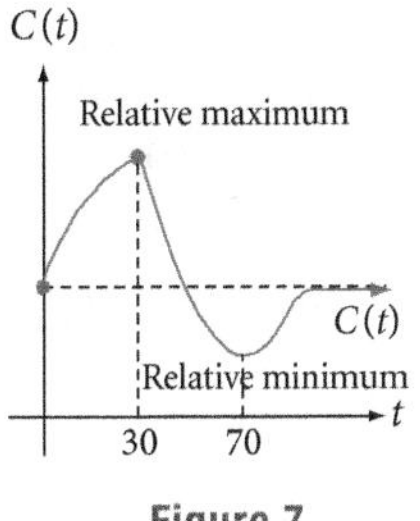

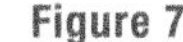

Figure 7

We can visualize this by adding more detail to Figure 7, which we have done in Figure 8. The leftmost graph of Figure 8 illustrates this idea. The t value 30 is called a **critical value** since it is critical to the accurate description of the behavior of the function. Notice that at this critical value, the derivative $C'(30)$ is undefined (since the function has a corner here).

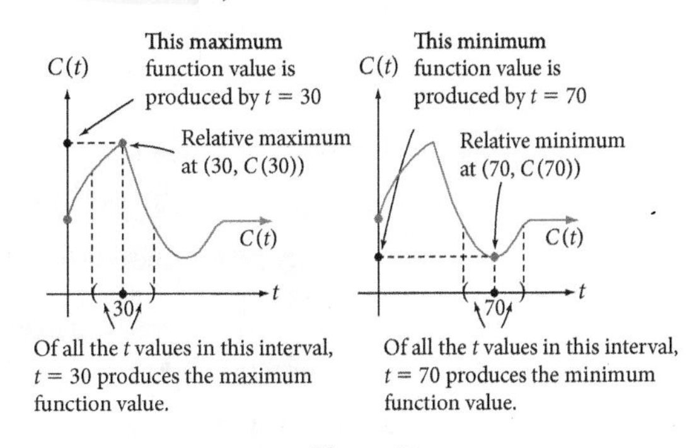

Figure 8

Notice that for values in a small interval around $t = 70$, it is $t = 70$ that produces the smallest function value (the lowest sugar concentration). That is, $t = 70$ produces a minimum value relative to the other t values near $t = 70$. Thus, the point $(70, C(70))$ is the **relative minimum** because, *relative* to points on the curve near $(70, C(70))$, the function has the smallest value at that point. The rightmost graph of Figure 8 illustrates this idea.

As before, the t value 70 is called a critical value since it is critical to the accurate description of the behavior of the function. Notice that at this critical value, the derivative $C'(x)$ equals zero.

These observations suggest the following conclusions.

Relative Extrema and Critical Values

Critical values can, but do not necessarily, produce relative maxima and relative minima. The term **relative extrema** is used to describe points that can be relative maxima or minima.

Critical values for which the first derivative is zero are potential smooth relative maximum or minimum points. Critical values for which the first derivative is undefined are potential cusp or corner relative maximum or minimum points.

Figure 9 shows the graph of a function having three points at which horizontal tangents occur. At each point, $x = a$, $x = b$, and $x = c$, the derivative is zero. That is, $f'(a) = 0$, $f'(b) = 0$, and $f'(c) = 0$. The point $x = a$ produces a relative minimum, the point $x = c$ produces a relative maximum, but the point $x = b$ produces neither a relative maximum nor a relative minimum. We can refer to such a point as a "terrace point."

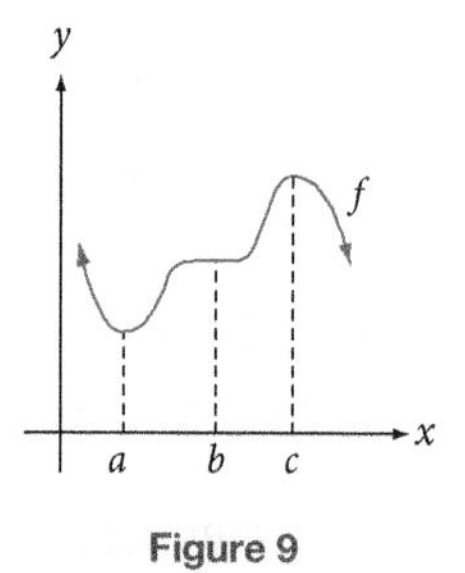

Figure 9

The next observation is known as the **first derivative test** for relative extrema.

The First Derivative Test

If c is a critical value of the function $f(x)$, and

1. $f(x)$ is increasing immediately to the left of c and decreasing immediately to the right of c, then c produces a relative maximum. In terms of the first derivative, if $f'(x) > 0$ to the immediate left of c and $f'(x) < 0$ to the immediate right of c, then c produces a relative maximum.
2. $f(x)$ is decreasing immediately to the left of c and increasing immediately to the right of c, then c produces a relative minimum. In terms of the first derivative, if $f'(x) < 0$ to the immediate left of c and $f'(x) > 0$ to the immediate right of c, then c produces a relative minimum.

Figure 10 visually summarizes the first derivative test.

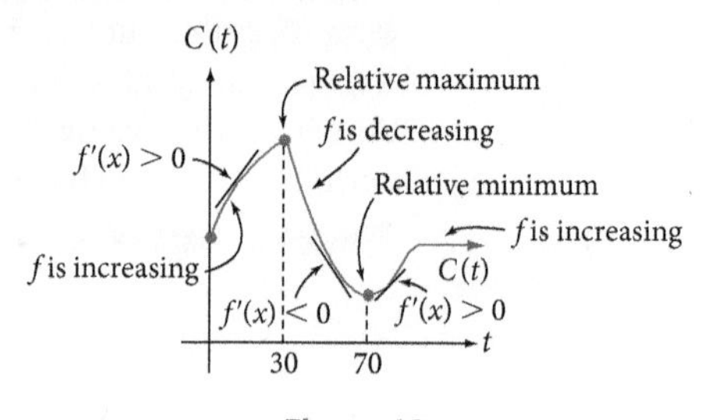

Figure 10

Example 4 Find the relative extrema for

$$f(x) = x^4 - 8x^2.$$

Solution We illustrate two methods.

Method 1 We use the graphical method to obtain estimates for the relative extrema.

We observe the graph of $f(x) = x^4 - 8x^2$, and as best we can, estimate the points at which the relative extrema occur. Figure 11 shows the graph of $f(x) = x^4 - 8x^2$.

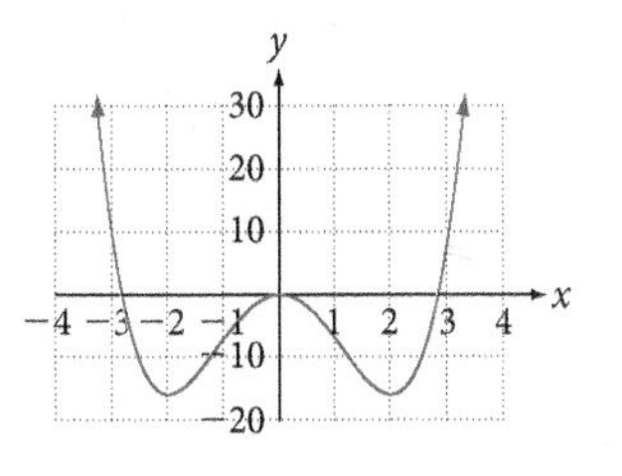

Figure 11

We can see from the graph that there are (at least) two relative minima and one relative maximum. The relative minima appear to occur at $(-2, -16)$ and $(2, -16)$. The relative maximum appears to occur right at the origin $(0, 0)$.

Are we sure we have all the relative extrema? How do we know that as the curve moves farther to the left or to the right, the curve does not turn around and produce another relative maximum? There are algebraic properties of functions that tell us we are seeing the global behavior of this curve, that it does not turn around at points farther out. But if we are not familiar with those properties, the graphical method may not be our best choice.

Method 2 We use the analytic method to obtain exact results for the relative extrema.

The plan is to find the derivative of $f(x) = x^4 - 8x^2$, determine the critical values, and construct a sign chart from which we can note behavior. The derivative is $f'(x) = 4x^3 - 16x$. Since $f'(x)$ is a polynomial function and polynomials are always smooth, continuous curves (no holes, gaps, or breaks), the derivative is never undefined. Critical values occur only at x-values for which $f'(x) = 0$.

$$f'(x) = 0, \qquad \text{when}$$

$$4x^3 - 16x = 0$$

$$4x(x^2 - 4) = 0$$

$$4x(x + 2)(x - 2) = 0$$

$$x = 0, -2, 2$$

We have three critical values, $x = -2, x = 0, x = 2$. This function has three potential smooth relative extreme points and no cusp/corner extreme points. We construct a sign chart using these three values. (See Figure 12.)

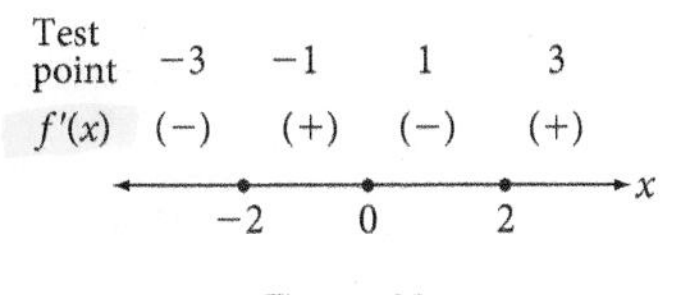

Figure 12

Employing the first derivative test gives us the relative maxima and relative minima. We observe sign changes at $x = -2$, from negative to positive, at $x = 0$, from positive to negative, and at $x = 2$, from negative to positive. We summarize the behavior of

$$f(x) = x^4 - 8x^2$$

in a summary table. (See Table 1.)

Interval/Value	$f'(x)$	$f(x)$	Behavior of $f(x)$
$(-\infty, -2)$	$(-)$		$f(x)$ is decreasing
-2	0	-16	$(-2, -16)$ is a relative minimum
$(-2, 0)$	$(+)$		$f(x)$ is increasing
0	0	0	$(0, 0)$ is a relative maximum
$(0, 2)$	$(-)$		$f(x)$ is decreasing
2	0	-16	$(2, -16)$ is a relative minimum
$(2, \infty)$	$(+)$		$f(x)$ is increasing

Table 1

Absolute Maxima and Minima

Often when working with functions that describe physical or theoretical phenomena, the primary objective is to find the absolute maximum or the absolute minimum value of the function. For example, the function graphed in Figure 13 shows that on the interval $[a, b]$, there are two relative maxima, one at $x = r$ and another at $x = t$; one relative minimum at $x = s$; an absolute maximum at $x = t$; and an absolute minimum at the endpoint $x = a$. Notice that an absolute extreme point can occur at an endpoint or at a relative extreme point.

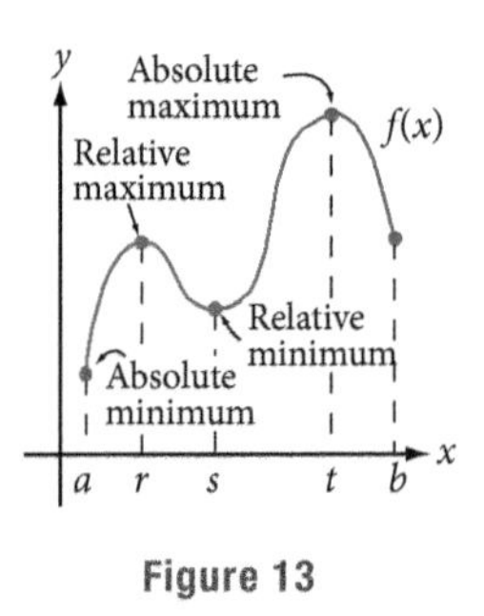

Figure 13

The **absolute maximum** of a function is the y-value of the highest point on the graph, and the **absolute minimum** is the y-value of the lowest point on the graph.

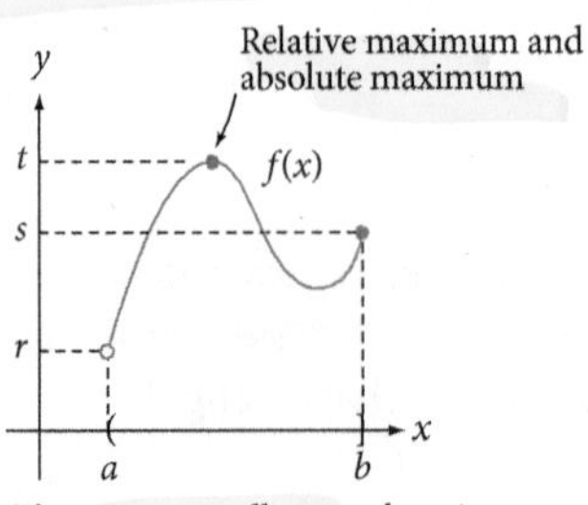

There is no smallest y value. As x approaches a from the right, y only approaches r from above. As x gets closer to a, y gets closer to r. Thus, there is no smallest y value and, hence, no absolute minimum.

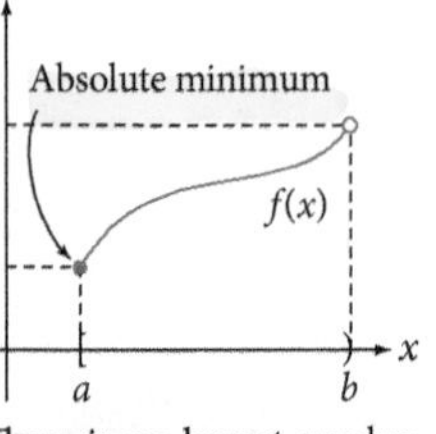

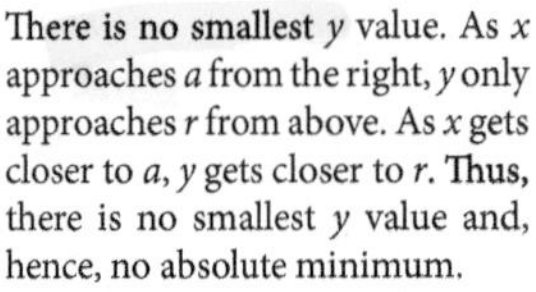

There is no largest y value. As x approaches b from the left, y only approaches s from below. As x gets closer to b, y gets closer to s. Thus, there is no largest y value and, hence, no absolute maximum.

Figure 14

Not all continuous functions have an absolute maximum or absolute minimum. However, if a function is continuous on the closed interval $[a, b]$, it will have both an absolute maximum and an absolute minimum. (Try to draw a graph of a continuous function over a closed interval $[a, b]$ that does not have one or the other.) Figure 14 and Figure 15 show some of the possibilities. A continuous function that is defined over an interval that is not closed, such as $(a, b]$, $[a, b)$, or (a, b), may or may not have an absolute maximum or absolute minimum (or both). A function will not have an absolute maximum (or an absolute minimum) if the largest value (or smallest value) occurs adjacent to an open endpoint since we are not able to identify the value. The rightmost illustration of Figure 15 shows a function that is defined on an open interval (a, b) that has both types of absolute extrema. Absolute extrema are found by finding the y-values of all the relative extrema and the y-values of the endpoints, then comparing to determine which is the largest and which is the smallest.

Note The leftmost graph in Figure 15 shows a function with a relative minimum and relative maximum, but no absolute maximum or absolute minimum.

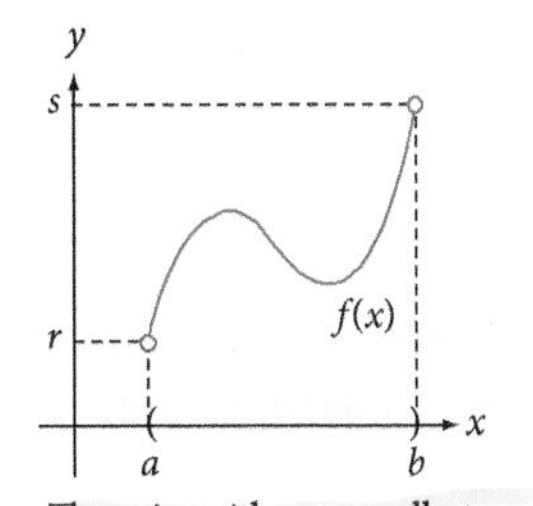

There is neither a smallest nor a largest y value. As x approaches a from the right, y only approaches r. As x approaches b from the left, y only approaches s. Thus, there is neither a smallest nor a largest y value and, hence, no absolute minimum or maximum.

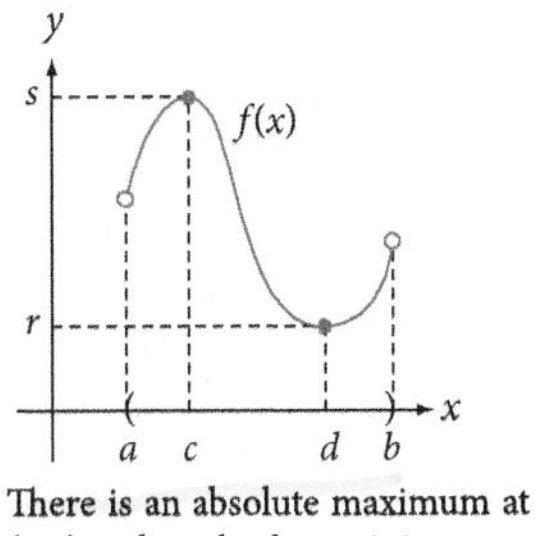

There is an absolute maximum at (c, s) and an absolute minimum at (d, r).

Figure 15

Example 5 Construct a sign chart for the graph shown in Figure 16 and use it to construct a summary table for the function.

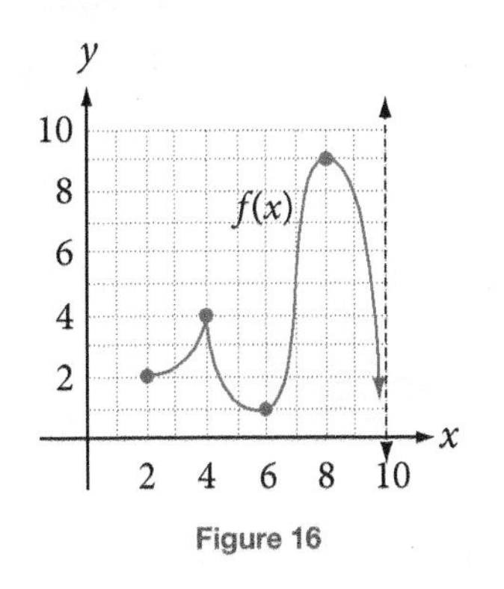

Figure 16

Solution

Interval/Value	$f'(x)$	$f(x)$	Behavior of $f(x)$
2		2	
(2, 4)	(+)		$f(x)$ is increasing
4	Undefined	4	(4, 4) is a relative maximum
(4, 6)	(−)		$f(x)$ is decreasing
6	0	1	(6, 1) is a relative minimum
(6, 8)	(+)		$f(x)$ is increasing
8	0	9	(8, 9) is an absolute maximum
(8, 10)	(−)		$f(x)$ is decreasing
			$f(x)$ approaches the vertical asymptote at $x = 10$.

Table 2

Example 6 illustrates how we can use mathematics to rigorously scrutinize claims made by a drug company about one of its new drugs.

Modeling: Testing a Claim

Example 6 A drug manufacturer makes three claims about a drug it has developed for women weighing between 90 and 120 pounds. The manufacturer bases its claims on the drug concentration model

$$C(t) = \frac{7t}{5t^2 + 20}$$

1. Two hours after the drug is administered, the concentration will be at its maximum value.
2. From two hours on, the concentration will only decrease.
3. Not only will the concentration decrease, it will decrease to 0.

Assuming the drug concentration model to be true, should the manufacturer's claims be accepted or rejected?

Solution For convenience, the graph of $C(t)$ appears in Figure 17. The manufacturer is making a claim about a function's maximum value and its decreasing behavior. Such a claim involves the first derivative.

Begin by finding the derivative of the function.

$$C'(t) = \frac{(5t^2 + 20) \cdot 7 - 7t \cdot 10t}{(5t^2 + 20)^2}$$

$$= \frac{-35(t^2 - 4)}{(5t^2 + 20)^2}$$

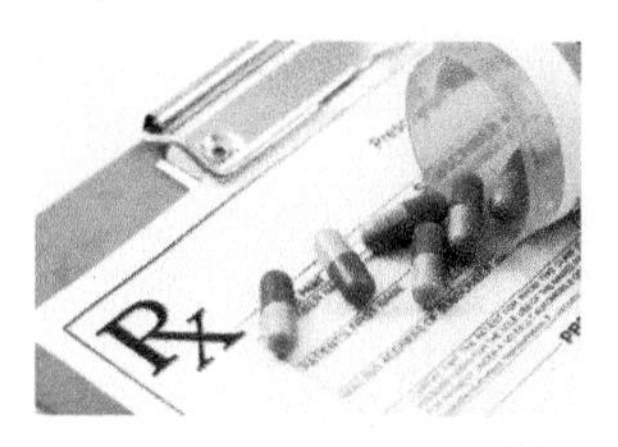

© G.M.Vozd/iStockPhoto

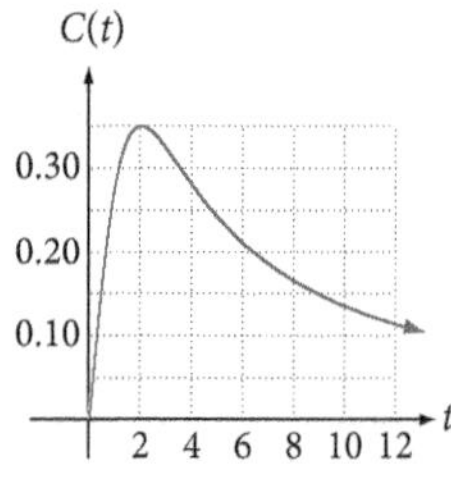

Figure 17

Next, we find the critical values. Since the denominator can never be 0, the only possible critical values will occur when the numerator is 0.

$$t^2 - 4 = 0$$
$$(t + 2)(t - 2) = 0$$
$$t = -2, 2$$

Since t represents time, the only critical value is 2.

Both the first derivative test and the graph show 2 to produce a maximum concentration, just as the manufacturer claims. Accept this claim.

For values of t greater than 2,

$$C'(t) = \frac{-35(t^2 - 4)}{(5t^2 + 20)^2} = \frac{(-)(+)}{(+)} = (-)$$

so that the concentration is always decreasing. Accept this claim.

Furthermore, since

$$\lim_{t \to \infty} C(t) = \lim_{t \to \infty} \frac{7t}{5t^2 + 20} = 0$$

the concentration decreases to 0. Accept this claim.

Using the derivative, we are able to substantiate this manufacturer's claims. ■

Example 7 Find, if they exist, the absolute maximum and minimum of the function $f(x) = (x - 4)^{2/7} + 3$ on the interval $[0, 6]$.

Solution The plan is to find the derivative of

$$f(x) = (x - 4)^{2/7} + 3$$

to determine the critical values, and construct a sign chart. The derivative is

$$f'(x) = \frac{2}{7(x - 4)^{5/7}}$$

Critical values occur at x-values for which $f'(x) = 0$ or for which $f'(x)$ is undefined.

$f'(x)$ is a fraction and fractions are zero only when the numerator is zero. This numerator is 2, and therefore, never zero. This function has no smooth extrema.

$f'(x)$ is undefined when the denominator is zero. This denominator is zero when $x = 4$. Be alert! Is 4 in the domain of the original function? Critical values are values that are in the domain of the original function. Is this function defined at $x = 4$? Yes, it is. $f(4) = (4 - 4)^{2/7} + 3 = 0^{2/7} + 3 = 3$. This function has a potential cusp/corner maximum or minimum.

We have only one critical value, $x = 4$. We construct a sign chart using this value. (See Figure 18.)

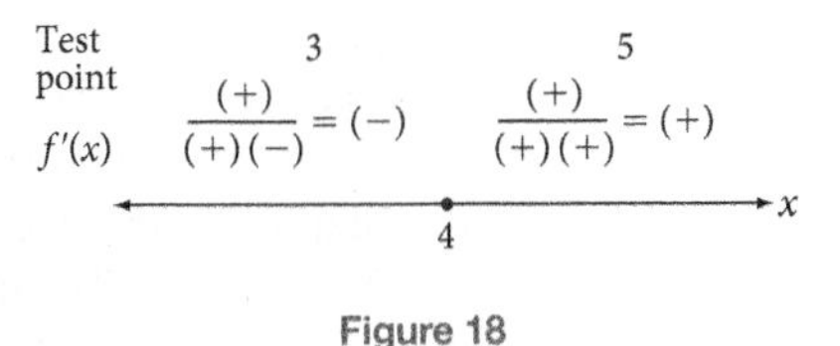

Figure 18

Applying the first derivative test shows us the relative maxima and relative minima. We observe sign changes at $x = 4$, negative to positive, indicating there is a relative minimum at $x = 4$. Since we are looking for the absolute maximum and absolute minimum, we evaluate the original function at the endpoints of the interval $[0, 6]$ as well as at the critical value $x = 4$.

$$f(0) = (0 - 4)^{2/7} + 3 = (-4)^{2/7} + 3 \approx 4.486$$

$$f(4) = (4 - 4)^{2/7} + 3 = (0)^{2/7} + 3 = 3$$

$$f(6) = (6 - 4)^{2/7} + 3 = (2)^{2/7} + 3 \approx 4.219$$

The smallest value is 3. Therefore, the absolute minimum is 3 and occurs at the point $(4, 3)$.

The largest value is 4.486. Therefore, the absolute maximum is 4.486 and occurs at the point $(0, 4.486)$.

The graph of $f(x) = (x - 4)^{2/7} + 3$ appears in Figure 19. It provides visual evidence for our analytic conclusion. Notice the cusp at $x = 4$.

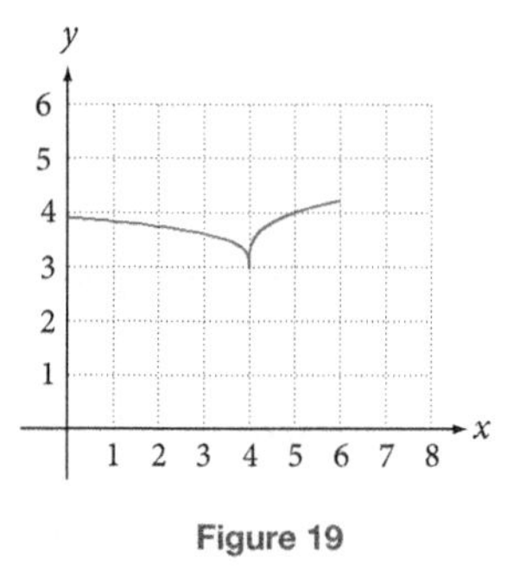

Figure 19

Example 8 Find, if they exist, all relative and absolute extrema of the function

$$f(x) = \frac{1}{x - 4}$$

Solution The plan is to find the derivative of

$$f(x) = \frac{1}{x - 4}$$

to determine the critical values, and construct a sign chart. The derivative is

$$f'(x) = -\frac{1}{(x - 4)^2}$$

$f'(x)$ is a fraction and fractions are zero only when the numerator is zero. This numerator is 1, and therefore, never zero. This function has no smooth maxima or minima.

$f'(x)$ is undefined when the denominator is zero. This denominator is zero when $x = 4$. Be alert! Is 4 in the domain of the original function? Critical values are values that are in the domain of the original function. Is this function defined at $x = 4$? No, it is not. The number 4 produces 0 in the denominator of the original function. This function has no potential cusp/corner maxima or minima.

This function has no critical values, and therefore, no relative extrema. Also, because there are no endpoints to test, the function has no absolute extrema.

> *Note* Although there are no extrema for this function, we can still use the first derivative to determine intervals on which the function is increasing or decreasing. Since the first derivative is always negative, this function is decreasing wherever it is defined; that is, $(-\infty, 4)$ and $(4, \infty)$.

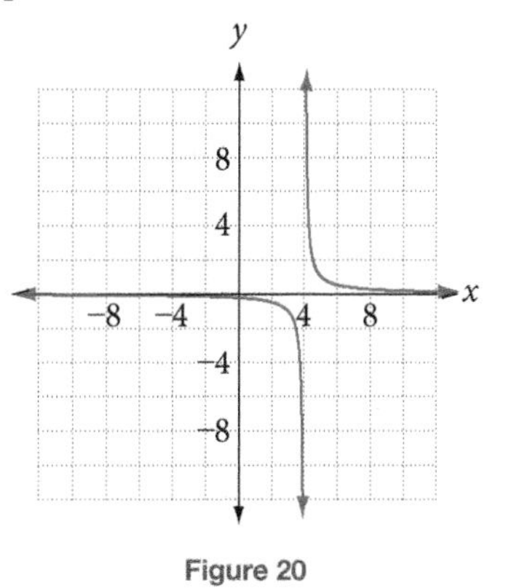

Figure 20

Using Technology 3.1

You can use Wolfram|Alpha to help you find the critical values and even the relative and absolute extrema of a function.

Example 9 Use Wolfram|Alpha to find any relative maxima and minima of the function

$$f(x) = \frac{1}{4}x^4 - 2x^3 - \frac{1}{2}x^2 + 30x.$$

Solution Online, go to www.wolframalpha.com. In the entry field, enter

find the max and min of (1/4) x^4 - 2x^3 - (1/2) x^2 + 30x

Wolfram|Alpha returns all the critical values, labels them as max or min, and draws the graph so you can see them as well as the intervals upon which the function increases or decreases. Notice that Wolfram|Alpha uses the term "global" where we use "absolute" and "local" where we use "relative."

Global minimum:

$$\min\left\{\frac{x^4}{4} - 2x^3 - \frac{x^2}{2} + 30x\right\} = -42 \text{ at } x = -2$$

Local maximum:

$$\max\left\{\frac{x^4}{4} - 2x^3 - \frac{x^2}{2} + 30x\right\} = \frac{207}{4} \text{ at } x = 3$$

Local minimum:

$$\min\left\{\frac{x^4}{4} - 2x^3 - \frac{x^2}{2} + 30x\right\} = \frac{175}{4} \text{ at } x = 5$$

Wolfram Alpha LLC. 2012. Wolfram|Alpha
http://www.wolframalpha.com/
(access September 25, 2012)

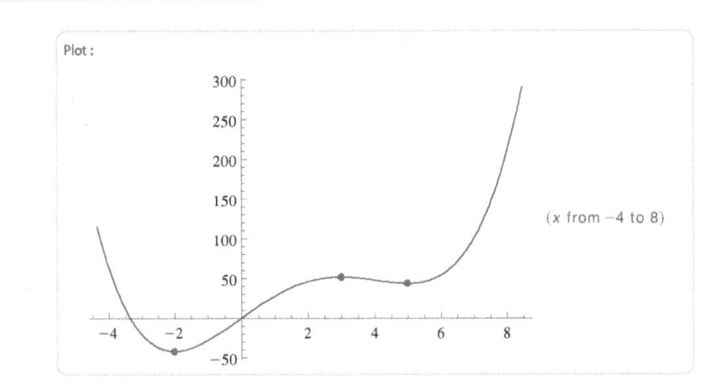

Wolfram Alpha LLC. 2012. Wolfram|Alpha
http://www.wolframalpha.com/
(access September 25, 2012)

From the output, we see that relative minima occur at the points $(-2, -42)$ and $\left(5, \frac{175}{4}\right)$, and that a relative maximum occurs at the point $\left(3, \frac{207}{4}\right)$. Note that the point $(-2, -42)$ is also an absolute minimum and that the graph does not have an absolute maximum. The graph of $f(x)$ shows the change in the sign of the derivative just to the left and right of these points. Looking at the graph, we can see the changes from increasing to decreasing or from decreasing to increasing around these points.

■

Example 10 Use Wolfram|Alpha to help you find the critical values of the function $f(x) = 0.03x^3 - 2x + 3$.

Solution Critical values occur at values of x that make the first derivative 0. We can have Wolfram|Alpha find the derivative, then set it equal to 0 to find these values. Into the entry field enter

solve d/dx (0.03x^3 - 2x + 3) = 0

Wolfram|Alpha returns the two critical values

$$x = \pm\frac{10\sqrt{2}}{3} \approx \pm 4.7140$$

■

Getting Ready for Class

After reading through the preceding section, respond in your own words and in complete sentences.

A. What is a relative maximum?
B. How do you find the critical values of a function, if any exist?
C. What is the first derivative test?
D. What is an absolute minimum?

Problem Set 3.1

Skills Practice

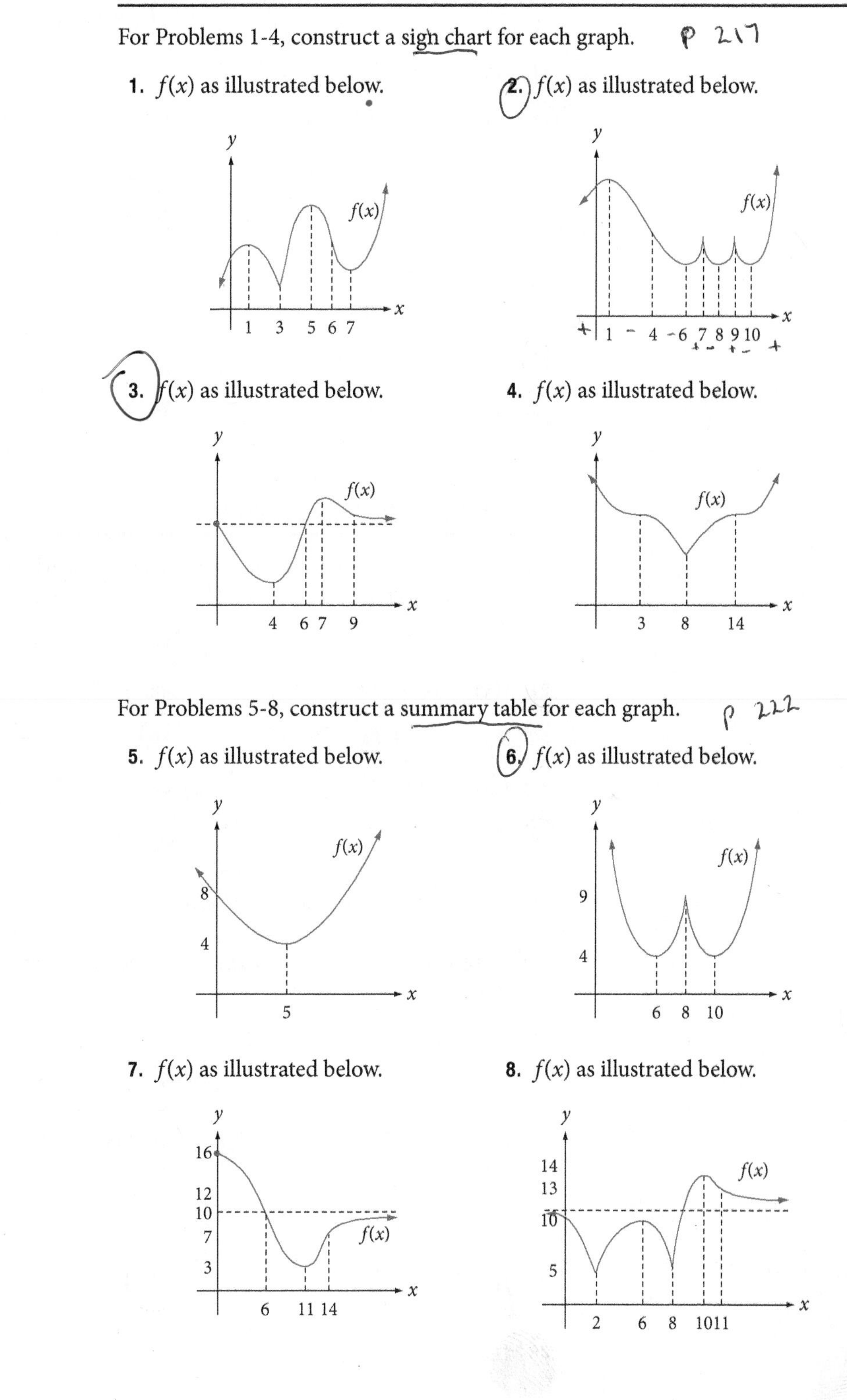

For Problems 1-4, construct a sign chart for each graph.

1. $f(x)$ as illustrated below.

2. $f(x)$ as illustrated below.

3. $f(x)$ as illustrated below.

4. $f(x)$ as illustrated below.

For Problems 5-8, construct a summary table for each graph.

5. $f(x)$ as illustrated below.

6. $f(x)$ as illustrated below.

7. $f(x)$ as illustrated below.

8. $f(x)$ as illustrated below.

Find, if any exist, the critical values of each function.

9. $f(x) = 3x + 2$
10. $f(x) = x - 12$
11. $f(x) = 2x^2 - 3x$
12. $f(x) = 4x^2 + 6x$
13. $f(x) = \frac{1}{4}x^4 + x^3 - 2x^2$
14. $f(x) = x^4 + 4x^3 + 4x^2$
15. $f(x) = x^4 + 4x^3 + 5$
16. $f(x) = x^5 - 5x^4 - 7$
17. $f(x) = \frac{1}{x - 5}$
18. $f(x) = \frac{3}{x + 7}$
19. $f(x) = \frac{4x}{x - 3}$
20. $f(x) = \frac{5x}{2x - 1}$
21. $f(x) = \frac{x^2}{x + 2}$
22. $f(x) = \frac{3x^2}{x^2 + 2}$
23. $f(x) = x^{3/5} + 9$
24. $f(x) = x^{1/7} - 4$

Check Your Understanding

For Problems 25-35, determine the intervals upon which the given function is increasing or decreasing.

25. $f(x) = 8x - 3$
26. $f(x) = -5x + 12$
27. $f(x) = 3x^3 - 12x^2$
28. $f(x) = 2x^3 + 15x^2$
29. $f(x) = 2x^4 + 8x^3$
30. $f(x) = x^4 + 8x^3$
31. $f(x) = x^{1/3} + 4$
32. $f(x) = 5 - 9x^{2/3}$
33. $f(x) = (x - 6)^{2/5} - 3$
34. $f(x) = \sqrt{x - 4}$
35. $f(x) = \sqrt{2x - 10}$

For Problems 36-46, find, if any, the relative maxima and relative minima of each function.

36. $f(x) = x^2 - 6x$
37. $f(x) = x^2 + 12x$
38. $f(x) = 8x - x^2$
39. $f(x) = 32x - 4x^2$
40. $f(x) = x^3 - 9x^2$
41. $f(x) = x^5 - \frac{15}{2}x^4 + 10$
42. $f(x) = x + \frac{1}{x} + 5$
43. $f(x) = x + \frac{4}{x}$
44. $f(x) = x^2 - \frac{16}{x^2}$
45. $f(x) = \frac{1}{x - 7}$
46. $f(x) = \frac{1}{x - 4}$

For Problems 47-50, find, if they exist, the absolute maxima and absolute minima of each function on the given interval.

47. $f(x) = x^3 - 4x^2 - 3x + 1$ on $[0, 5]$

48. $f(x) = x^3 - 3x^2 - 7$ on $[0, 5]$

49. $f(x) = \dfrac{2}{x - 1}$ on $[3, \infty)$

50. $f(x) = \dfrac{x}{x - 4}$ on $[0, 4)$

Modeling Practice

For Problems 51-58, a situation is given along with the symbolic form and graph form of a function that describes the situation. As best you can (meaning you may have to approximate), use the graph to determine the critical value that maximizes or minimizes the function and then discuss the behavior of the function for values on both sides of the critical value.

51. Business: Manufacturing Costs A manufacturer of lightweight, durable containers is contracted to make right cylindrical containers that hold 20π in^3 (cubic inches) of liquid. The material used to construct the side of the can costs \$3 per in^2, and the material used to construct the top and bottom of the can costs \$4 per in^2. What radius will minimize the cost of the can? The function relating the cost C (in dollars) and the radius r (in inches) is

$$C(r) = 8\pi r^2 + \frac{377}{r}$$

and its graph, for $r > 0$, appears below.

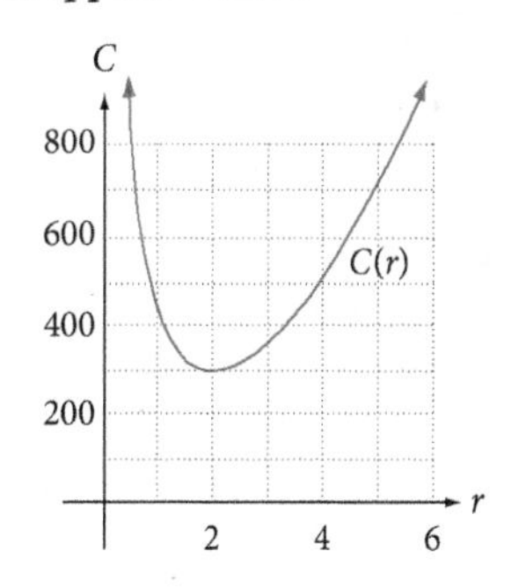

52. **Business: Manufacturing** A box with a square base and an open top is to be made from a square piece of cardboard by cutting out four squares of equal size from the corners and folding up the sides. What size should the corner cuts be so that the volume of the box is as large as possible? The function relating the volume V (in cubic inches) and the size x (in inches) of the corner cuts is $V(x) = 4x^3 - 60x^2 + 225x$, and its graph appears below.

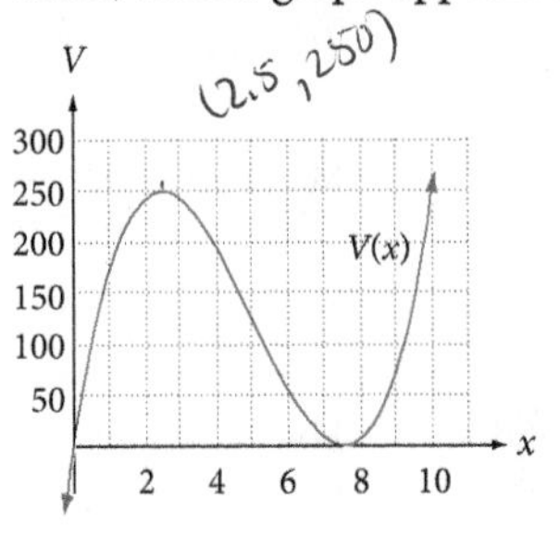

53. **Business: Manufacturing** A manufacturing company that uses toxic chemicals for cleaning precision parts plans to use 600 feet of fencing to enclose a rectangular region to store the chemical containers. One of the boundaries of the region will be a wall of a building and will, therefore, not need to be fenced. The function relating the area A (in square feet) to the width x (in feet) of the rectangular region is $A(x) = 600x - 2x^2$, and its graph appears below. What width will maximize the area?

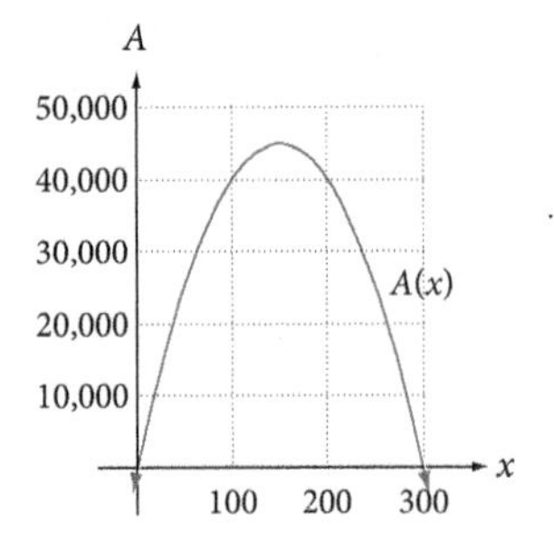

54. **Health Science: Flu Epidemic** A city's health officials believe that the number N of people contracting the flu t days from the beginning of an epidemic is predicted by the function $N(t) = -2t^2 + 180t + 40$. When is the number of people contracting the flu greatest? The graph of $N(t)$ appears below.

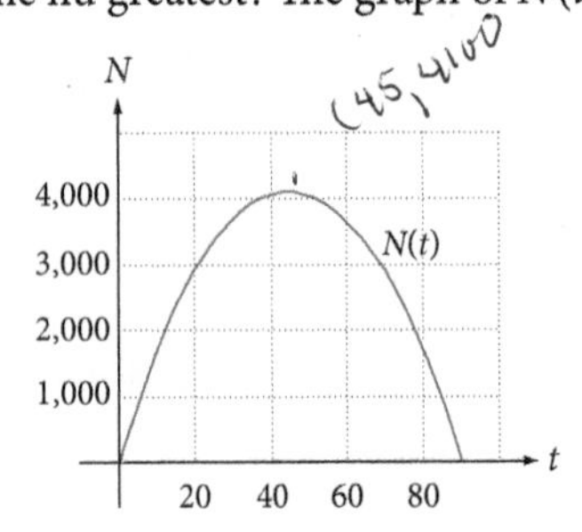

55. Business: Publishing Costs A publishing company has directed its printer to print pages that have one-half-inch margins on the sides and three-fourth-inch margins on the top and bottom and that contain 70 square inches of print. The function that relates the area A (in square inches) of a page and the width w (in inches) of a page is

$$A(w) = (w - 1)\left(\frac{140 - 3w}{2w}\right)$$

At what width is the area of the page maximized? The graph of $A(w)$ appears below.

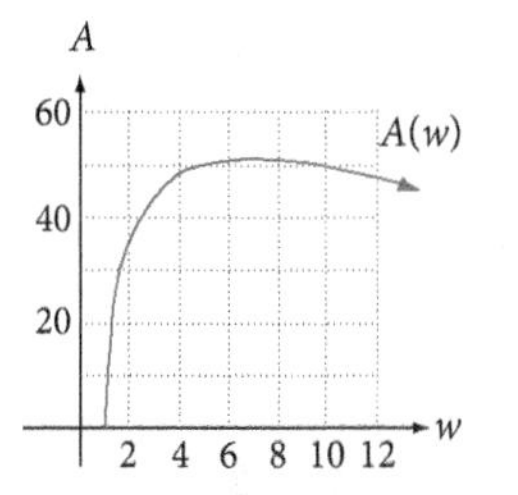

56. Business: Construction Costs In the following figure, a power station is located on a straight coastline at location B. A cable is to be laid that will connect the power station to a laboratory located at P on an island 2,000 feet from point A. The cost of laying the cable on land is \$2 per foot, and under water the cost is \$7 per foot. In the first figure, if $x = 0$, the cable will be laid from point P directly to point A, then from point A along the coastline to point B. If $x = 7{,}500$, then the cable will be laid completely under water from point P to point B. If x is between 0 and 7,500, the cable will be laid from B along the coastline to point C and then at an angle to point P. What value of x will minimize the cost of laying the cable?

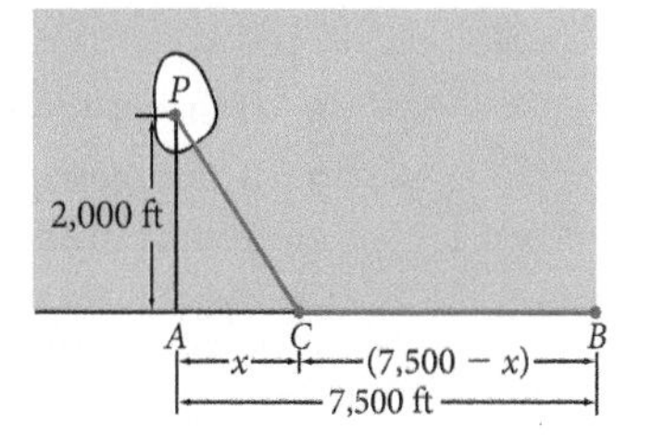

The function relating the cost C (in dollars) to the distance x (in feet) is $C(x) = 15{,}000 - 2x + 7\sqrt{x^2 + 2{,}000^2}$. The graph of $C(x)$ appears below.

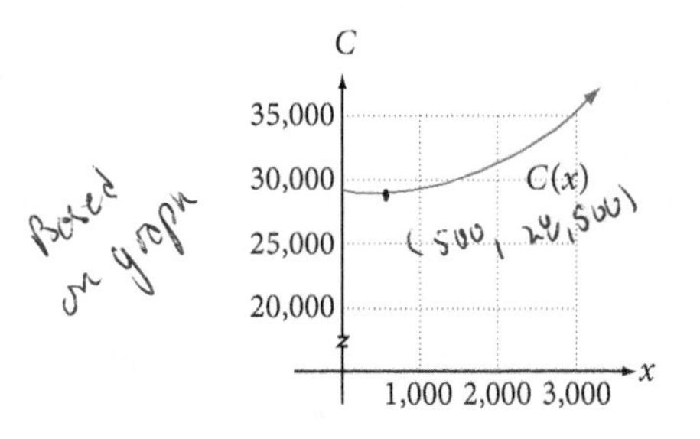

57. Political Science: Candidate Recognition A political candidate's campaign officials believe that t weeks after the beginning of a campaign, the percentage P of the electorate favorably recognizing the candidate's name is given by the function

$$P(t) = \frac{46t}{t^2 + 15} + 0.39$$

Find the value of t that maximizes $P(t)$. The graph of $P(t)$ appears below.

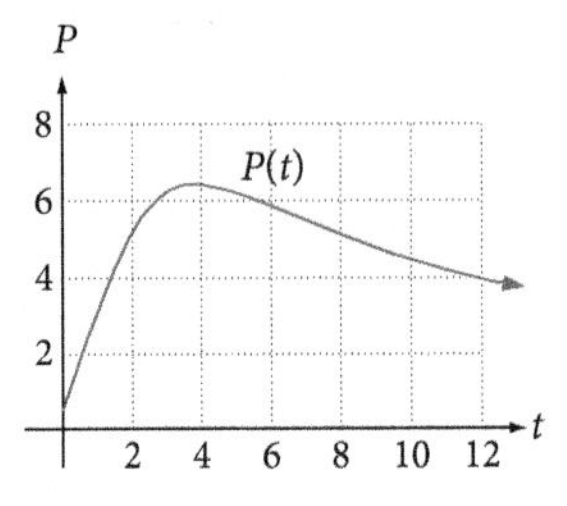

58. Business: Online Advertising Revenue In the year 2010, analysts at a new economics professional's social networking site modeled the revenue (in thousands of dollars) the site could realize over the next x years with the function

$$R(x) = -x^4 + 72x^2$$

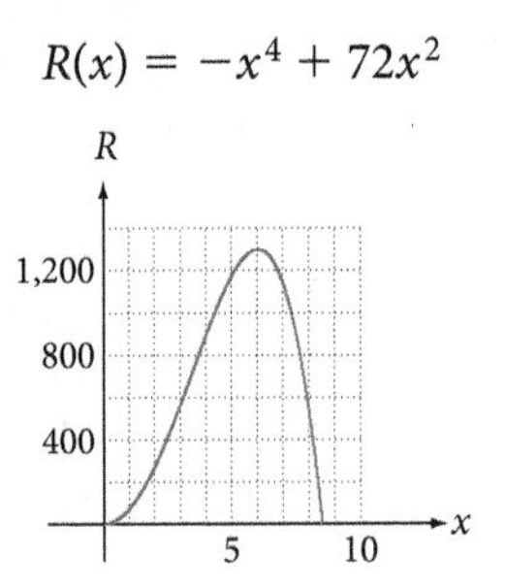

a. If this model is correct, in which year will the revenue be maximum?

b. If the management of the site does not change the way they do business, how many years from 2010 will the revenue fall to zero?

For Problems 59-64, use the first derivative test to maximize or minimize each function.

© Shaun Lowe Photographic/ iStockPhoto

59. Physical Science: Drag Racing The following rational function models the speed $V(t)$, in miles per hour, of a dragster t seconds since it starts its quarter-mile run:

$$V(t) = \frac{340t}{t + 3}$$

a. Respond to the claim that during the first 4 seconds of the dragster's run, it attains a maximum speed then begins to slow down. Is the claim true or false? Support your response mathematically.

b. What is the absolute maximum speed of the dragster on the interval $[0, 4]$ and at what time in the first 4 seconds of its run does it occur? Round your maximum speed to two decimal places.

60. Health Science: Drug Concentration The concentration $C(t)$ (in milligrams per cubic millimeter) of a particular drug in the human bloodstream t hours after the drug is ingested is modeled by the function

$$C(t) = \frac{21t}{t^2 + t + 4}$$

a. How many hours after the drug is ingested is the concentration at its maximum?

b. Provide a mathematical argument to demonstrate that once the concentration attains its maximum value, it will only decrease to 0.

61. Manufacturing: Minimizing Production Costs Power over Ethernet (PoE) systems pass electrical power and data safely on Ethernet cabling. The function

$$C(x) = \frac{144 + x^2}{180x}$$

relates the weekly cost (in thousands of dollars) to a company to the number x (in hundreds) of PoE systems it manufactures and sells each week.

a. Determine the number of systems the manufacturer must produce to minimize its cost.

b. Provide a mathematical argument to demonstrate that once the manufacturer produces the minimizing number of systems, the cost will only increase.

Source: Fluke Networks

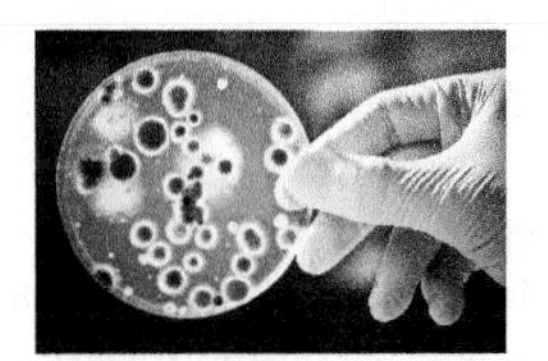

© Alexander Raths /iStockPhoto

62. Biology: Bacteria Population The function

$$P(t) = \frac{30t}{t^2 + 16} \qquad t > 1$$

relates the population P (in millions) of bacteria in a culture to the time t (in days) after a particular slow-acting toxin is introduced into the culture.

a. Determine the number of days after the introduction of the toxin that the population of bacteria will be at its maximum.

b. Provide a mathematical argument to demonstrate that once the population is at its maximum value, it will only decrease to 0.

63. Health Science: Drug Reaction The function

$$S(x) = \frac{1}{2}x\sqrt{300 - x} \qquad 0 \le x \le 300$$

measures the strength of a typical person's reaction to x milligrams of a particular drug.

a. Determine the number of milligrams of the drug that will maximize a person's reaction to it.

b. Provide a mathematical argument to demonstrate that once the reaction to the drug is at its maximum value, it will only decrease.

64. Health Science: Dieting The function

$$W(x) = \frac{80(2x + 15)}{x + 6}$$

might be used to model the progress of a person on a particular diet. The quantity $W(x)$ represents the person's weight (in pounds) x weeks from the first day of the diet.

a. Respond to the claim that during the first 20 weeks of the program, the typical 200-pound person will see an initial decrease in weight, but then attain some minimum value and then begin to experience an increase in weight. Support your response mathematically.

b. What is the absolute minimum weight a 200-pound person can expect to see if he or she stays on the program for 20 weeks and at how many weeks into the program does that weight occur? Round your minimum weight to the nearest one pound.

For Problems 65-68, use the first derivative of the given function to provide a mathematical argument that will substantiate the given claim.

65. Business: Stock Investment Stocks offer investors a cash dividend and/or the potential for capital gains. To compute their expected return, R, investors can use the formula

$$R(p) = \frac{d + c}{p}$$

where d is the expected dividend value, c is the expected capital gains value, and p is the price per share. All values are in dollars. Suppose an investor decides on values for d and c. That is, d and c are considered to be constants and the price of a stock, p, is variable. It may seem like common sense that when the price of a stock increases, the expected return always decreases. Show, by means of a mathematical argument, that this common sense idea is true.

Source: Money-Zine.com

66. Physical Science: Double Stars An interesting astronomical phenomenon is that of the double star. A double star is a system of two stars revolving around each other. Astronomers can use telescopes to view double stars. Since stars are so far away, double stars often appear as a single source of light in the sky. For a telescope to show the two individual stars in a double star system, it must have a minimum magnification. The magnification needed to split a double star is given by the model

$$M(d) = \frac{1}{d}$$

where d is the angular separation of the double star and M is the magnification required of the telescope to separate the stars. Make a mathematical argument to support the claim that as the angular separation of two stars in a double star system increases, the required magnification decreases.

Source: Saguaro Astronomy Club

67. Manufacturing: Modulus of Rupture In manufacturing, it is important to know the mechanical properties of materials. The modulus of rupture is the maximum strength to which a particular material can be subjected before if breaks. When working with wooden beams, the model

$$M(d) = \frac{R}{bd^2}$$

relates the modulus of rupture, M, in pounds per cubic inch, R is the rupture stress constant (in pounds), and b and d, the width and depth (in inches) of the beam, respectively. Determine if the modulus of rupture for a wooden beam having a rupture stress constant of 20 pounds and width of 4 inches, is increasing or decreasing as the depth of the beam increases.

U.S. Department of Agriculture, Forest Service

68. Health Science: Differential Pressure in Human Lungs Differential pressure is the difference between two pressures. Creating a differential pressure can drive a physical process such as breathing. It is the diaphragm of the human body that, by pulling up then down on the lungs, creates a differential pressure across the lungs creating air flow in and out of the lungs. The function

$$K(d) = \frac{1.59923P\rho d^4}{W^2}$$

relates the differential pressure, K, to the change in pressure P (measured in bars), the pipe diameter d (in millimeters), the density of the fluid passing through the pipe (air in the case of human lungs) ρ (in kg/m^3) and the mass flow rate W (in kg/hr).

Make a mathematical argument to support the claim that, with the change in pressure, the density of the fluid, and the mass flow rate kept constant, the differential pressure increases as the diameter of the pipe increases.

Source: CR4, The Engineer's Place for News and Discussion

Using Technology Exercises

Use Wolfram|Alpha to solve each of the following problems..

69. Find any maxima and minima of the function

$$f(x) = x^3 - 4x + 5$$

70. Find the critical values of the function

$$f(x) = \frac{x^3 - 25x}{x(x + 5)}$$

Explain how your knowledge of elementary algebra gives you the answer to this question.

71. Health Science: Drug Concentration The concentration $C(t)$ (in mg/cm^3) of a drug in a patient's bloodstream t hours after the drug is administered is approximated by the function

$$C(t) = \frac{0.02t}{t^2 + 6t + 9}$$

Find the maximum concentration and the time at which it occurs.

72. Find the absolute maximum and absolute minimum values of the function

$$f(x) = 2x(x - 6)^{2/3}$$

on the interval $[1, 8]$. Round to the nearest whole numbers.

Getting Ready for the Next Section

73. Find $f'(x)$ for $f(x) = -x^3 + 9x^2$.

74. Find $f'(x)$ for $f(x) = x^{3/5} - 1$ and write your answer so that only positive exponents appear.

75. Find $f'(x)$ for $f(x) = \dfrac{3}{5x^{2/5}}$.

76. Solve the equation $25x^{7/5} = 0$.

77. Find $f'(x)$ for $f(x) = \dfrac{5x + 2}{3x - 4}$.

78. Find $f'(x)$ for $f(x) = \dfrac{-26}{(3x - 4)^2}$.

79. For which values of x, if any, is $f(x) = \dfrac{-26}{(3x - 4)^2}$ undefined?

80. For which values of x, if any, is $f'(x) = \dfrac{156}{(3x - 4)^3}$ undefined?

81. Find $f''(x)$ for the function $f(x) = x^3 + \dfrac{9}{2}x^2 - 12x + 11$.

82. Find $\displaystyle\lim_{x \to \infty} \frac{5x + 2}{3x - 4}$.

The Second Derivative and the Behavior of Functions

The graph in Figure 1 is of a function that represents the relationship between the amount A (in millions of gallons) of reclaimed water used by a county each year and the time t (in years) since 1940. By evaluating the graph and using the first and second derivatives of the function, we would be able to describe the function's behavior.

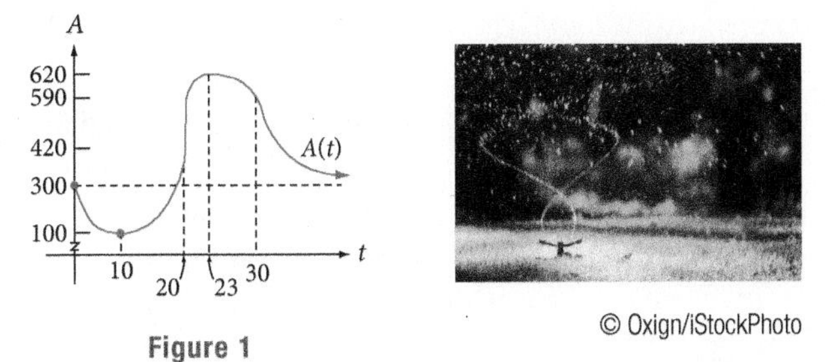

Figure 1

The first derivative of a function is useful in helping us describe some practical behaviors of functions. Using the first derivative, we can determine, if any exist, the critical values of the function, and from those, we can locate

1. the intervals upon which a function increases or decreases, and
2. the relative minimum and relative maximum points, if they exist, of the function.

The second derivative of a function is also useful in helping us describe some practical features of functions. Using the second derivative, we can determine, if any exist, the hypercritical values of the functions, and from those, we can locate

1. the intervals upon which the function is concave upward or downward, and
2. the points of inflection, if they exist, of the function.

Using both the first and second derivative together can provide us with valuable information about the behavior of a function. In this section, we see

1. how to use the second derivative to extract information contained in the symbolic or graphical description of a function, and
2. how to use both the first and second derivatives together to give a practical and accurate description of a function's behavior.

Concavity and Points of Inflection

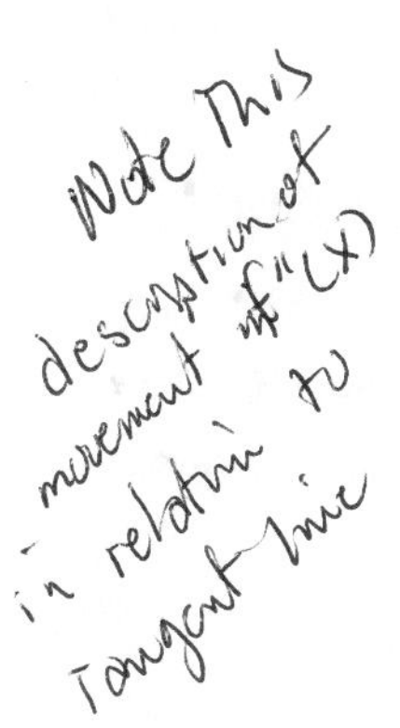

If the graph of a function opens upward at $x = c$ — that is, the curve is above the tangent line at $x = c$ — the function is said to be **concave upward** at $x = c$. For the curve to open upward, the slopes of the tangent lines must get larger. The rate at which the slopes change is described by the second derivative; thus, if $f''(x) > 0$, then the curve is concave upward. If the graph of a function opens downward at $x = c$ — that is, the curve is below the tangent line at $x = c$ — the function is said to be **concave downward** at $x = c$. For the curve to open downward, the slopes of the tangent lines must get smaller; thus, $f''(x) < 0$. Figure 2 illustrates a function that is concave upward at $x = 4$ and concave downward at $x = 8$. Notice that both $x = 4$ and $x = 8$ are critical values since $f'(4) = 0$, and $f'(8) = 0$.

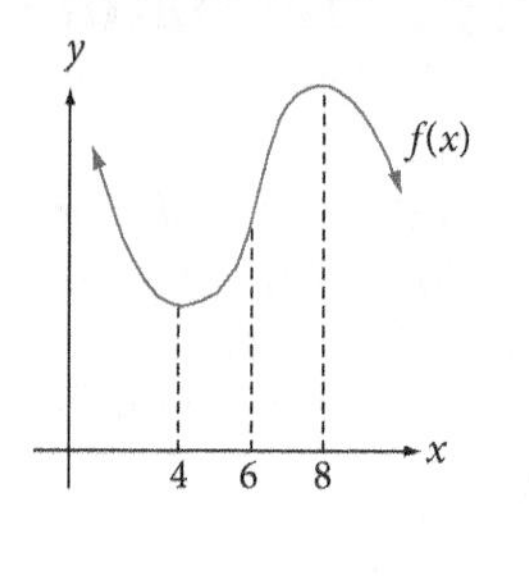

$f'(x)$ (−) (+) (+) (−)

$f''(x)$ (+) 4 (+) 6 (−) 8 (−) x

Figure 2

The Second Derivative and the Concavity of a Function

Suppose that $f(x)$ is a continuous, twice-differentiable function defined on the interval (a, b). Then

1. If $f''(x) > 0$ for every value of x in (a, b), then $f(x)$ is concave upward on (a, b). Conversely, if $f(x)$ is concave upward on (a, b), then $f''(x) > 0$ for every value of x in (a, b).
2. If $f''(x) < 0$ for every value of x in (a, b), then $f(x)$ is concave downward on (a, b). Conversely, if $f(x)$ is concave downward on (a, b), then $f''(x) < 0$ for every value of x in (a, b).

At $x = 6$ in Figure 2, there is a change in concavity; the graph of $f(x)$ changes from concave upward to concave downward. Points at which there is a change of concavity are called **points of inflection**. Since the function changes concavity at a point of inflection, the second derivative must change signs at that point (from positive to negative or from negative to positive). This means that at a point of inflection, the second derivative must either equal zero or be undefined; that is, if $x = c$ produces a point of inflection, then either $f''(c) = 0$, or $f''(c)$ is undefined. Values that produce possible points of inflection are called **hypercritical values**. In Figure 2, $x = 6$ is a hypercritical value and $(6, f(6))$ is a point of inflection.

Hypercritical Values of a Function

If a function $f(x)$ is defined at $x = c$, and if either $f''(c) = 0$ or $f''(c)$ is undefined, then the number c is called a *hypercritical value* of $f(x)$. Conversely, if the number c is a hypercritical value of $f(x)$, then either $f''(c) = 0$ or $f''(c)$ is undefined.

The First and Second Derivatives Together

We can use information provided by both the first and second derivatives together to describe how a function behaves over some interval. For example, over an interval, a function could

1. Increase at an increasing rate. Figure 3a illustrates this behavior.
2. Increase at a decreasing rate. Figure 3b illustrates this behavior.
3. Decrease at a decreasing rate. Figure 3c illustrates this behavior.
4. Decrease at an increasing rate. Figure 3d illustrates this behavior.

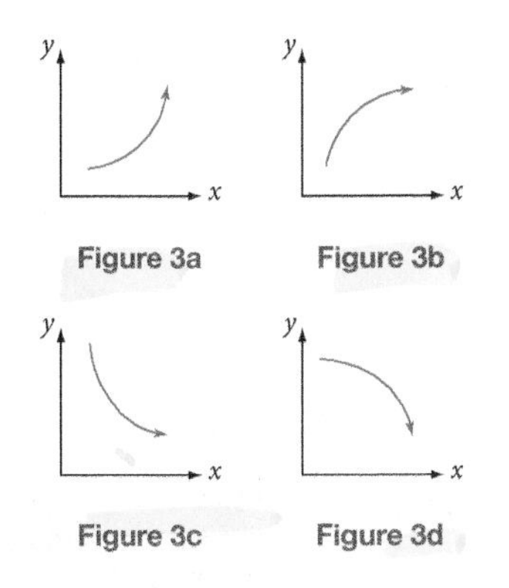

Figure 3a Figure 3b

Figure 3c Figure 3d

These four behaviors are directly related to the signs of the first and second derivatives.

1. The graph of a function that is increasing at an increasing rate shows that the function is increasing and concave upward, that is, both $f'(x) > 0$ and $f''(x) > 0$.

   Remember it this way: $(+)(+) = (+)$, and $(+)$ reminds us of *increase*.

$$\underbrace{(+)(+) = (+)}_{\text{Increasing at an increasing rate}}$$

2. The graph of a function that is increasing at a decreasing rate shows that the function is increasing but concave downward, that is, $f'(x) > 0$ but $f''(x) < 0$.

$$\underbrace{(+)(-) = (-)}_{\text{Increasing at a decreasing rate}}$$

3. The graph of a function that is decreasing at a decreasing rate shows that the function is decreasing but concave upward, that is, $f'(x) < 0$ but $f''(x) > 0$.

$$\underbrace{(-)(+) = (-)}_{\text{Decreasing at a decreasing rate}}$$

4. The graph of a function that is decreasing at an increasing rate shows that the function is decreasing and concave downward, that is, $f'(x) < 0$ and $f''(x) < 0$.

$$\underbrace{(-)(-) = (+)}_{\text{Decreasing at an increasing rate}}$$

We will now illustrate this process with a variety of different functions. Make note of the common systematic process that is used regardless of the function.

VIDEO EXAMPLES

SECTION 3.2

Example 1 An efficiency study conducted on the day shift of a manufacturer showed that the number of units of a product produced by a typical employee t hours after 8:00 AM could be modeled by the function

$$f(t) = -t^3 + 9t^2 \quad 0 \le t \le 8$$

In the context of this problem, describe the behavior of this function.

Solution For $f(t) = -t^3 + 9t^2, 0 \le t \le 8, f'(t) = -3t^2 + 18t$, and $f''(t) = -6t + 18$. The critical values are $t = 0$ and $t = 6$ and the hypercritical value is $t = 3$. The sign chart appears in Figure 4.

f'		(+)		(+)		(−)	
f''	0	(+)	3	(−)	6	(−)	t

Figure 4

From the sign chart, we see that

> On (0, 3), the sign analysis shows $(+)(+) = (+)$ which tells us the function increases at an increasing rate.
>
> On (3, 6), the sign analysis shows $(+)(-) = (-)$ which tells us the function increases but at a decreasing rate.
>
> On (6, 8), the sign analysis shows $(-)(-) = (+)$ which tells us the function decreases at an increasing rate.

We conclude that from 8:00 AM to 11:00 AM, efficiency is increasing at an increasing rate. From 11:00 AM to 2:00 PM, efficiency is increasing but at a decreasing rate. Then from 2:00 PM, efficiency is decreasing at an increasing rate. The maximum efficiency occurs at 2:00 PM, 6 hours after the start of the shift, and is 108 units of product per hour.

A point of inflection where the concavity changes from upward to downward is often called the point of diminishing returns. The Point of Diminishing Returns is the point at which the addition of another factor produces less output than did the addition of the previous factor. It is the point at which time and energy generate a minimal increase in output. Figure 5 illustrates a point of diminishing returns for a function $P(a)$ where a is the amount of money (in thousands of dollars) spent on advertising and P is the profit (in thousands of dollars) realized from the advertised product. As a increases from \$0 to \$90,000, the profit increases at an increasing rate. After \$90,000 has been spent on advertising, the profit still increases but progressively more slowly. It now takes more advertising money to realize the same increase in profit than it previously did.

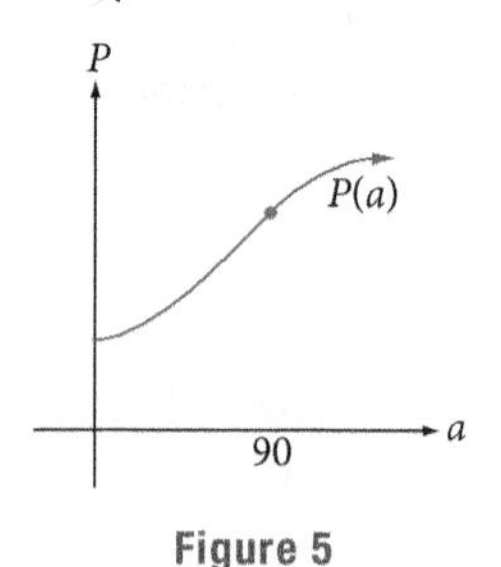

Figure 5

Example 2 Describe the behavior of the function $f(x) = x^{3/5} - 1$.

Solution First we note that the domain of $f(x)$ is the set of all real numbers. Next, we find the first and second derivatives.

$$f'(x) = \frac{3}{5x^{2/5}} \quad \text{and} \quad f''(x) = \frac{-6}{25x^{7/5}}$$

Then, to find the critical and hypercritical values, we check both conditions.

1. To find where $f'(x) = 0$ we need to solve $\frac{3}{5x^{2/5}} = 0$ for x. But a fraction is 0 *only* when the numerator is zero, and $3 \neq 0$. Thus, there are no critical values from $f'(x) = 0$. Likewise, to find where $f''(x) = 0$ we need to solve $\frac{-6}{25x^{7/5}} = 0$ for x. Again, a fraction is 0 *only* when the numerator is zero and $-6 \neq 0$. Thus, there are no hypercritical values from $f''(x) = 0$.
2. To find where $f'(x)$ is undefined we note that a fraction is undefined *only* when the denominator is 0. Thus, we set the denominator equal to 0 and solve for x.

$$5x^{2/5} = 0 \quad \text{so that} \quad x = 0$$

$f'(x)$ is undefined when $x = 0$, and 0 is in the domain of $f(x)$. Thus, 0 is a critical value of $f(x)$ and, therefore, a possible extreme point.

Where is $f''(x)$ undefined? Again, we set the denominator equal to 0 and solve for x.

$$25x^{7/5} = 0 \quad \text{so that} \quad x = 0$$

$f''(x)$ is undefined when $x = 0$, and 0 is in the domain of $f(x)$. Thus, 0 is a hypercritical value of $f(x)$ and, therefore, a possible point of inflection.

This means that $x = 0$ is both a critical value and a hypercritical value. We can now construct a sign chart (illustrated in Figure 6) for $f(x)$. Using the sign chart, we can construct the following summary table. From the summary table (see Table 1) we can construct a sketch of $f(x)$ and describe its behavior (as shown in Figure 6).

Note:
incr = increasing
decr = decreasing
cu = concave up
cd = concave down
und = undefined

Interval/Value	$f'(x)$	$f''(x)$	$f(x)$	Behavior of $f(x)$
$(-\infty, 0)$	+	+	incr/cu	incr at an incr rate
0	und	und	-1	pt of infl at $(0, -1)$
$(0, +\infty)$	+	−	incr/cd	incr at a decr rate

Table 1

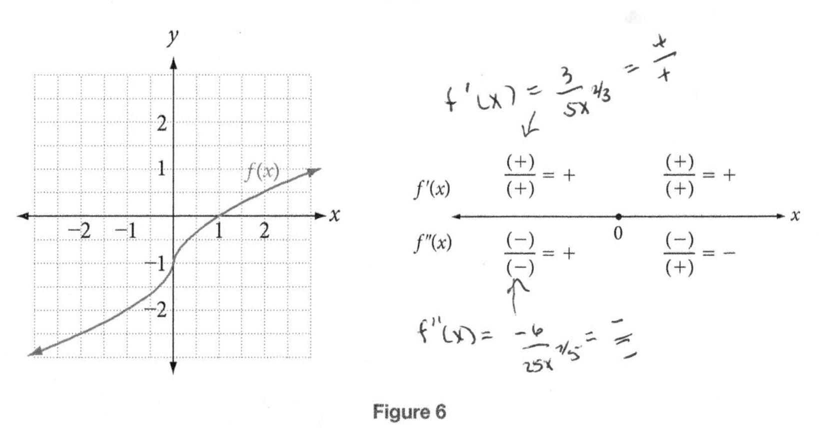

Figure 6

Summary of behavior The function $f(x) = x^{3/5} - 1$ increases at an increasing rate when x is less than 0. When $x = 0$, $f(x) = -1$ and the function experiences a point of inflection (which alerts us to a coming change in concavity—that is, the speed of the increase is going to change). Then $f(x)$ increases at a decreasing rate when x is greater than 0. ■

Example 3 Summarize the behavior of $f(x) = x^3 - 3x^2 + 4$ and sketch its graph.

Solution Since $f(x)$ is a polynomial, its domain is the set of all real numbers. The first and second derivatives are

$$f'(x) = 3x^2 - 6x \quad \text{and} \quad f''(x) = 6x - 6$$

To find the critical and hypercritical values, we check both conditions.

1. Where is $f'(x) = 0$? We need to solve $3x^2 - 6x = 0$ for x. Factoring produces $3x(x-2) = 0$. So, $x = 0, 2$.
   Where is $f''(x) = 0$? We need to solve $6x - 6 = 0$ for x. Factoring produces $6(x-1) = 0$. So, $x = 1$.
2. Where are $f'(x)$ and $f''(x)$ undefined? Since both $f'(x)$ and $f''(x)$ are polynomials, they are never undefined.

Thus, $x = 0, 2$ are critical values and, therefore, possible relative extrema, and $x = 1$ is a hypercritical value and, therefore, a possible point of inflection. We construct the sign chart illustrated in Figure 7a. Using the sign chart, we construct the summary table for $f(x)$.

$f'(x) = 3x(x-2)$

$f''(x) = 6x-6 = 6(x-1)$

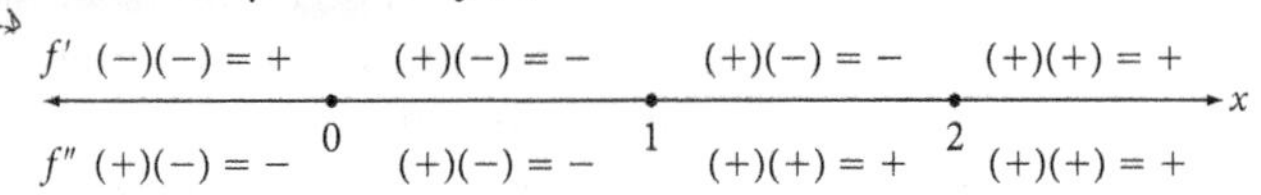

	$(-\infty, 0)$	0	$(0, 1)$	1	$(1, 2)$	2	$(2, +\infty)$
f'	$(-)(-) = +$		$(+)(-) = -$		$(+)(-) = -$		$(+)(+) = +$
f''	$(+)(-) = -$		$(+)(-) = -$		$(+)(+) = +$		$(+)(+) = +$

Figure 7a

Interval/Value	$f'(x)$	$f''(x)$	$f(x)$	Behavior of $f(x)$
$(-\infty, 0)$	+	−	incr/cd	incr at a decr rate
0	0	−	(0, 4)	rel max at (0, 4)
(0, 1)	−	−	decr/cd	decr at an incr rate
1	−	0	(1, 2)	pt of infl at (1, 2)
(1, 2)	−	+	decr/cu	decr at a decr rate
2	0	+	(2, 0)	rel min at (2, 0)
$(2, +\infty)$	+	+	incr/cu	incr at an incr rate

Table 2

Using the information from the summary table, we construct a sketch of $f(x)$ as shown in Figure 7b.

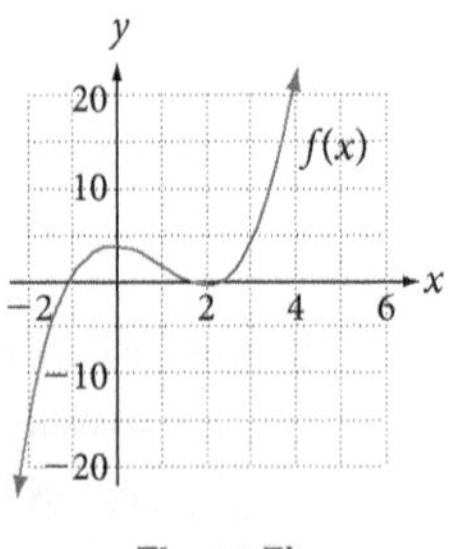

Figure 7b

Summary of behavior As x increases to 0, $f(x)$ increases but at a decreasing rate. At $x = 0$, $f(x) = 4$ and the function attains a relative maximum. As x increases to 1, $f(x)$ decreases at an increasing rate. At $x = 1$, the function experiences a point of inflection (which alerts us to a coming change in concavity—that is, the speed of the decrease is going to change). As x increases to 2, $f(x)$ decreases but does so now at a decreasing rate (the rate of decrease is slowing). At $x = 2$, $f(x) = 0$ and the function attains a relative minimum. Then, as x increases without bound, $f(x)$ increases at an increasing rate. ■

Example 4 Summarize the behavior of $f(x) = \frac{5x+2}{3x-4}$, and sketch its graph.

Solution The rational expression $\frac{5x+2}{3x-4}$ is undefined when $3x - 4 = 0$; that is, when $x = \frac{4}{3}$. Since $\frac{4}{3}$ is not in the domain of $f(x)$, it cannot be a critical or hypercritical value. The first and second derivatives are

$$f'(x) = \frac{-26}{(3x-4)^2} \quad \text{and} \quad f''(x) = \frac{156}{(3x-4)^3}$$

We search for critical and hypercritical values by checking the two conditions.

1. Since fractions are 0 only when their numerators are zero, $f'(x)$ and $f''(x)$ are never 0.
2. Since fractions are undefined only when their denominators are 0, $f'(x)$ and $f''(x)$ are undefined when $x = \frac{4}{3}$. But $\frac{4}{3}$ is not in the domain of the original function $f(x)$ so it cannot be a critical or hypercritical value. We will, however, plot a point on the sign chart at $x = \frac{4}{3}$ because of its importance in the description of the behavior of $f(x)$.

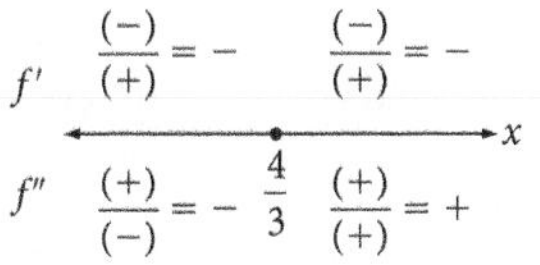

f': $\frac{(-)}{(+)} = -$ | $\frac{(-)}{(+)} = -$

x-axis point: $\frac{4}{3}$

f'': $\frac{(+)}{(-)} = -$ | $\frac{(+)}{(+)} = +$

Figure 8

Thus, $f(x)$ has no critical or hypercritical values and, therefore, no relative extreme points or points of inflection. The sign chart for $f(x)$ appears in Figure 7. Before constructing the summary table for $f(x)$, we notice that

$$\lim_{x\to\infty} \frac{5x+2}{3x-4} = \frac{5}{3}$$

This means that $f(x)$ has a horizontal asymptote at $y = \frac{5}{3}$. (When working with fractions, we always check for horizontal asymptotes.)

Now, using the sign chart, we construct the summary table for $f(x)$. Using the information in the summary table, we construct the graph of $f(x)$ shown in Figure 9.

Interval/Value	$f'(x)$	$f''(x)$	$f(x)$	Behavior of $f(x)$
$\left(-\infty, \frac{4}{3}\right)$	$-$	$-$	decr/cd	decr at an incr rate
$x = \frac{4}{3}$	und	und	und	Vertical Asymptote
$\left(\frac{4}{3}, +\infty\right)$	$-$	$+$	decr/cu	decr at a decr rate
$y = \frac{5}{3}$				Horizontal Asymptote

Table 3

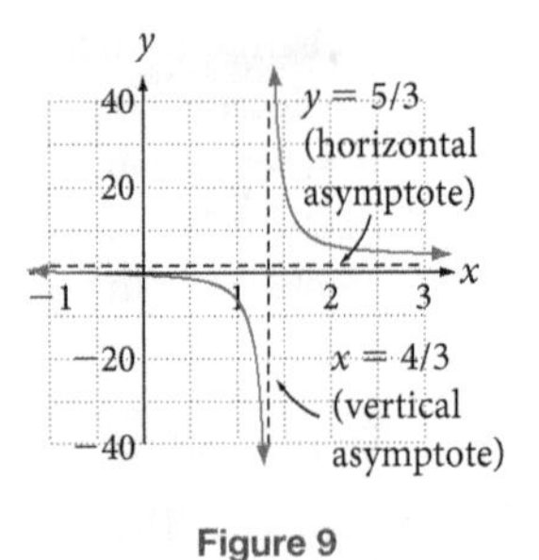

Figure 9

Summary of behavior When x is very small (far to the left of 0), $f(x)$ is very near, but always less than $\frac{5}{3}$. As x increases to $\frac{4}{3}$, $f(x)$ decreases at an increasing rate. In fact, as x approaches $\frac{4}{3}$, $f(x)$ gets unboundedly small. At $x = \frac{4}{3}$, $f(x)$ is undefined. In fact, $x = \frac{4}{3}$ is a vertical asymptote. When x is just beyond $\frac{4}{3}$, $f(x)$ is unboundedly large. As x increases beyond $\frac{4}{3}$, $f(x)$ decreases at a decreasing rate and approaches, but is always just greater than, $\frac{5}{3}$. ■

We noted in Section 3.1 that a powerful feature of mathematics is that it can provide us with the ability to rigorously validate or invalidate arguments and claims. The next example demonstrates how the derivative can be used to scrutinize an economic forecast made by an analyst about the future of a start-up social media site.

Modeling: Testing Economic Forecasts

Example 5 In the year 2011, a group of musicians launched a surf music social website, a site on which people who enjoy listening to instrumental surf guitar could communicate and interact with each other. The site receives revenue from organizations that place advertisements on the site's pages. The musicians constructed the revenue function

$$R(x) = -\frac{1}{4}x^4 + 200x^2$$

to model the revenue $R(x)$ the site should realize from online advertising x months after the beginning of 2011.

A financial analyst claims that under this model, the site will realize its maximum revenue 30 months after the beginning of 2011 and then from that time on the revenue will decrease slowly to 0. Assuming the revenue model to be true, should the website's managers accept or reject this analyst's forecast?

Solution The analyst makes statements about a function's maximum value and its decreasing-at-a-decreasing-rate behavior. Such a claim involves the first and second derivatives.

Begin by finding the first and second derivative of $R(x)$ and the critical values.

$$R'(x) = -x^3 + 400x \quad \text{and} \quad R''(x) = -3x^2 + 400$$

$$R'(x) = 0 \quad \text{when} \quad -x^3 + 400x = 0$$

Solving this equation, we get

$$-x(x^2 - 400) = 0$$
$$-x(x + 20)(x - 20) = 0$$
$$x = 0, -20, 20$$

The values 0 and -20 make no sense in the context of the problem, so the only critical value is 20. We can make a sign chart to see where the function is increasing or decreasing.

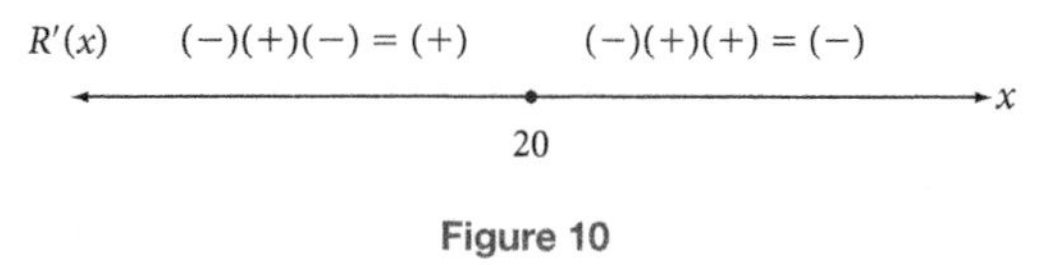

Figure 10

As the sign chart shows, at $x = 20$, $R(x)$ changes from increasing to decreasing, indicating a maximum occurs at that point. This means the media site will see its maximum revenue 20 months after the beginning of 2011, contradicting the analyst's claim of 30 months.

Also for all values of $x > 20$, $R'(x) < 0$ and $R''(x) < 0$. These results mean that for all months after the maximum revenue is realized, the revenue function decreases at an increasing rate. This, again, contradicts the analyst's forecast that the site's revenue will decrease slowly to 0. The revenue will decrease quickly to 0. The website's managers should reject both the analyst's forecasts. ■

In Section 3.1, we used the first derivative test to determine if a critical value of a function f produced a relative maximum or a relative minimum. Because a function is concave upward at a point if its second derivative is positive at that point, it makes sense that if $x = c$ is a critical value of the function, then it must produce a relative minimum. (See Figure 11.) Likewise, because a function is concave downward at a point if its second derivative is negative at that point, it makes sense that if $x = c$ is a critical value of the function, then it must produce a relative maximum. See Figure 11.

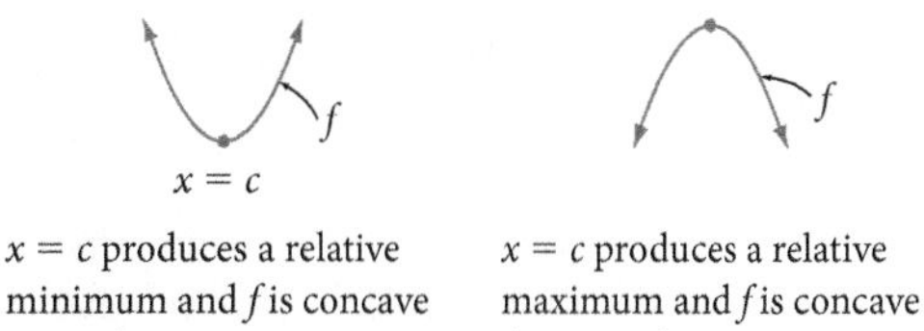

$x = c$ produces a relative minimum and f is concave upward at $x = c$

$x = c$ produces a relative maximum and f is concave downward at $x = c$

Figure 11

We summarize these facts in the Second Derivative Test.

The Second Derivative Test

If a function $f(x)$ is such that both $f'(x)$ and $f''(x)$ exist (can be computed), and the number c is a critical value of f, then

1. If $f''(c) > 0$, then f has a relative minimum at $x = c$.
2. If $f''(c) < 0$, then f has a relative maximum at $x = c$.
3. If either $f''(c) = 0$ or $f''(c)$ is undefined, the test fails and cannot be used to obtain information about the relative extrema of f at $x = c$.

Example 6 Use the second derivative test to find all the relative extrema of the function

$$f(x) = x^3 + \frac{9}{2}x^2 - 12x + 11$$

Solution The plan is to compute both $f'(x)$ and $f''(x)$, set $f'(x)$ equal to zero to find the critical values, then apply the second derivative test to classify them as relative maximum or relative minimum.

$$f(x) = x^3 + \frac{9}{2}x^2 - 12x + 11$$

$$\begin{aligned} f'(x) &= 3x^2 + 9x - 12 \\ &= 3(x^2 + 3x - 4) \\ &= 3(x + 4)(x - 1) \\ f''(x) &= 6x + 9 \end{aligned}$$

To find the critical values, set $f'(x)$ equal to zero and solve for x.

$$\begin{aligned} f'(x) &= 0 \quad \text{when} \\ 3(x + 4)(x - 1) &= 0, \quad \text{and} \\ x &= -4, 1 \end{aligned}$$

There are two critical values to test, $x = -4$ and $x = 1$. Substitute these values into the second derivative to determine the sign. $f''(-4) = 6(-4) + 9 = -15 < 0$ so that $x = -4$ produces a relative maximum. Since $f''(-4) < 0$ and $f(-4) = 67$, $f(x)$ has a relative maximum at $(-4, 67)$. $f''(1) = 6(1) + 9 = 15 > 0$ so that $x = 1$ produces a relative minimum. Since $f''(1) > 0$ and $f(1) = \frac{9}{2}$, $f(x)$ has a relative minimum at $\left(1, \frac{9}{2}\right)$. ■

Using Technology 3.2

You can use Wolfram|Alpha to help you find the hypercritical values and even the points of inflection of a function.

Example 7 Use Wolfram|Alpha to find any points of inflection of the function

$$f(x) = \frac{1}{4}x^4 - 2x^3 - \frac{1}{2}x^2 + 30x$$

Solution Online, go to www.wolframalpha.com. In the entry field, enter

find the points of inflection of (1/4) x^4 - 2x^3 - (1/2) x^2 + 30x

Wolfram|Alpha returns all the hypercritical values and draws the graph so you can see if sign changes occur.

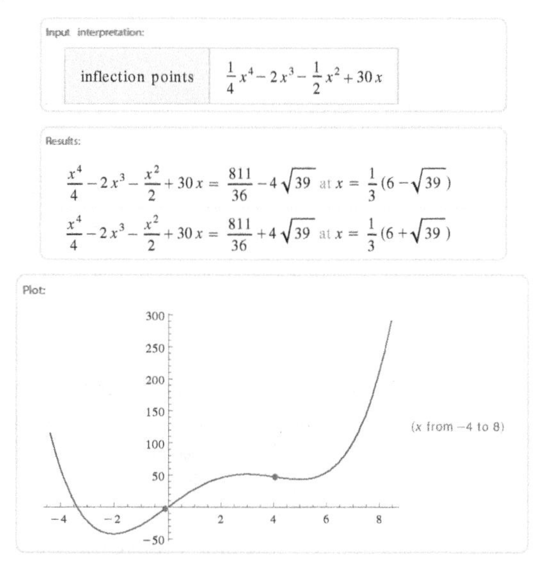

Wolfram Alpha LLC. 2012. Wolfram|Alpha
http://www.wolframalpha.com/
(access September 25, 2012)

The Wolfram|Alpha output indicates two points of inflection, one at

$$\left(\frac{1}{3}(6 - \sqrt{39}), \frac{811}{36} - 4\sqrt{39}\right)$$

and one at

$$\left(\frac{1}{3}(6 + \sqrt{39}), \frac{811}{36} + 4\sqrt{39}\right)$$

The graph shows the change in concavity at these points. ■

Example 8 Use Wolfram|Alpha to find the hypercritical values of the function $f(x) = 0.03x^3 - 2x + 3$.

Solution Hypercritical values occur at values of x that make the second derivative 0. We can have Wolfram|Alpha find the second derivative, then set it equal to 0 to find these values.

Into the entry field enter

solve d^2/dx^2 (0.03x^3 - 2x + 3) = 0

Wolfram|Alpha returns the one hypercritical value $x = 0$.

It is instructive to observe the graphs of both f and f'' on the same coordinate system as that allows us to see the how the concavity of the function relates to the sign of the second derivative.

Into the entry field enter

graph (0.03x^3 - 2x + 3) and d^2/dx^2 (0.03x^3 - 2x + 3)

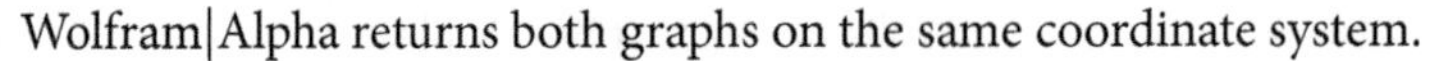

Wolfram|Alpha returns both graphs on the same coordinate system.

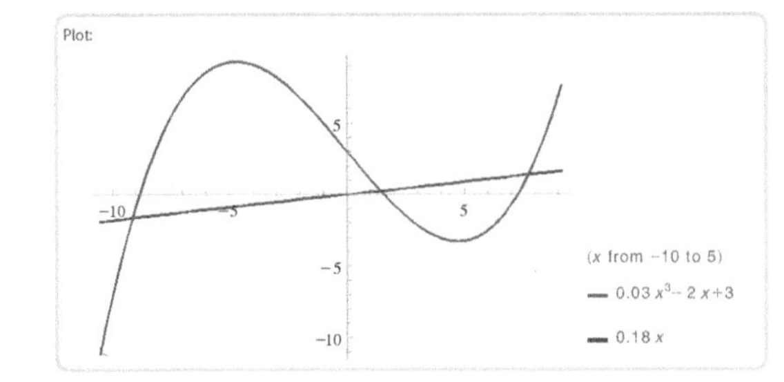

Wolfram Alpha LLC. 2012. Wolfram|Alpha
http://www.wolframalpha.com/
(access September 25, 2012)

The original function is cubic and is represented by the curve. The second derivative of the function is linear and is represented by the line.

The line is below the x-axis to the left of zero, meaning the second derivative is negative when $x < 0$. Notice that the curve, too, is concave downward when $x < 0$. It makes visual sense that $f'' < 0 \leftrightarrow f$ is concave downward.

The line is above the x-axis to the right of zero, meaning the second derivative is positive when $x > 0$. Notice that the curve, too, is concave upward when $x > 0$. It makes visual sense that $f'' > 0 \leftrightarrow f$ is concave upward.

Observing the graph, we can only guess that the point of inflection is at $x = 0$. We cannot be sure that it is not a bit to the left or right of zero. Using calculus provides conclusive evidence that the only point of inflection of this function is exactly at $x = 0$. ■

Getting Ready for Class

After reading through the preceding section, respond in your own words and in complete sentences.

A. The first derivative gives us information about the increasing/decreasing behavior of a function. About which characteristics of a function does the second derivative provide information?

B. From which derivative do critical values come? From which do hypercritical values come?

C. In terms of the first and second derivatives, explain how to describe a function that increases at a decreasing rate.

D. What is the sign, positive or negative, of the second derivative at a relative minimum of a function? At a relative maximum?

Problem Set 3.2

Skills Practice

For Problems 1-9, construct a sketch of the graph of a function f that has the given properties.

1. f passes through the points $(0, 0)$, $(8, 0)$, has a relative minimum at $(4, -16)$, and is everywhere concave upward.

2. f passes through the points $(0, 0)$, $(4, 0)$, has a relative maximum at $(2, 12)$, and is everywhere concave downward.

3. f passes through the point $(3, 2)$, which is a relative maximum, and $f''(x) < 0$ everywhere.

4. f passes through the point $(4, 4)$, which is a relative minimum, and $f''(x) > 0$ everywhere.

5. f is such that

$$f'(5) = 0$$
$$f'(x) < 0 \text{ on } (-\infty, 5)$$
$$f'(x) > 0 \text{ on } (5, \infty)$$
$$f''(x) > 0 \text{ on } (-\infty, \infty)$$

6. f is such that

$$f(0) = 0$$
$$f'(-3) = 0 \text{ and } f'(4) = 0$$
$$f'(x) < 0 \text{ on } (-3, 0) \text{ and } (0, 4)$$
$$f'(x) > 0 \text{ on } (-\infty, -3) \text{ and } (4, \infty)$$
$$f''(x) < 0 \text{ on } (-\infty, 0)$$
$$f''(x) > 0 \text{ on } (0, \infty)$$

7. f is such that

$$f(4) \text{ is undefined}$$
$$f'(4) \text{ is undefined}$$
$$f''(4) \text{ is undefined}$$
$$f(0) = 0$$
$$f'(x) < 0 \text{ on } (-\infty, 4) \text{ and } (4, \infty)$$
$$f'(x) > 0 \text{ nowhere}$$
$$f''(x) < 0 \text{ on } (-\infty, 4)$$
$$f''(x) > 0 \text{ on } (4, \infty)$$

8. f is such that

$f(0) = 0$

$f'(0) = 0$

$f''(0) = 0$

When $x < 0$, f increases at a decreasing rate

When $x > 0$, f increases at an increasing rate

9. f is such that

$f(0) = 6$ and $f(5) = 6$

$(0, 6)$ and $(5, 6)$ are both relative minima

$f'(0)$ *is* undefined

$f''(x) < 0$ on $(-\infty, 0), (0, 5), (5, \infty)$

$f''(x)$ is never positive

10. A function f possesses two critical values, $x = -2$ and $x = 3$. The function's second derivative is $f''(x) = 8x - 4$. Determine if the critical values produce relative maxima or relative minima.

11. A function f possesses three critical values, $x = -3$, 0 and $x = 3$. The function's second derivative is $f''(x) = 3(x^2 - 3)$. Determine if the critical values produce relative maxima or relative minima.

For Problems 12-20, find all the hypercritical values of the function.

12. $f(x) = \frac{2}{3}x^3 - 4x^2 + 6$

13. $f(x) = \frac{5}{6}x^3 - 2x^2 + 4$

14. $f(x) = \frac{1}{5}x^6 - \frac{3}{2}x^4 - 12x^2$

15. $f(x) = \frac{1}{4}x^4 - 2x^3 + x - 5$

16. $f(x) = \frac{1}{x^2 + 2}$

17. $f(x) = x^{2/3}$

18. $f(x) = x^{1/5}$

19. $f(x) = \frac{x - 1}{x + 1}$

20. $f(x) = \frac{x - 2}{x + 3}$

21. The function $f(x) = 2x^3 + 4$ has $f'(x) = 6x^2$, and $f''(x) = 12x$. Determine the intervals, if any, upon which $f(x)$ is concave upward or concave downward.

22. The function

$$f(x) = \frac{7}{6}x^3 - 7x^2 + 2$$

has $f'(x) = \frac{7}{2}x^2 - 14x$, and $f''(x) = 7x - 14$. Determine the intervals, if any, upon which $f(x)$ is concave upward or concave downward.

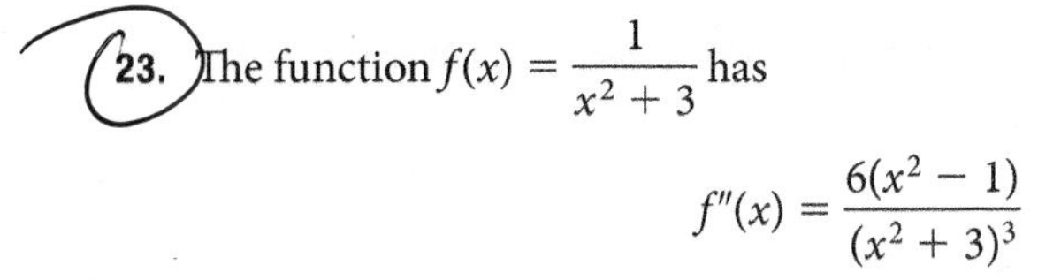

23. The function $f(x) = \dfrac{1}{x^2 + 3}$ has

$$f''(x) = \frac{6(x^2 - 1)}{(x^2 + 3)^3}$$

Determine the intervals, if any, upon which $f(x)$ is concave upward or concave downward.

24. The function $f(x) = \dfrac{x}{x^2 + 3}$ has

$$f''(x) = \frac{2x(x^2 - 9)}{(x^2 + 3)^3}$$

Determine the intervals, if any, upon which $f(x)$ is concave upward or concave downward.

25. The function $f(x) = \dfrac{x - 1}{x + 1}$ has

$$f''(x) = \frac{-4}{(x + 1)^3}$$

Determine the intervals, if any, upon which $f(x)$ is concave upward or concave downward.

26. The function $f(x) = \dfrac{x^2 - 1}{x + 2}$ has

$$f''(x) = \frac{6}{(x + 2)^3}$$

Determine the intervals, if any, upon which $f(x)$ is concave upward or concave downward.

Check Your Understanding

For Problems 27-35, find the intervals where the function is concave upward or concave downward.

27. $f(x) = x^2 - 4x + 3$

28. $f(x) = 5x^2 - 5x + 3$

29. $f(x) = x^3 - 3x^2 - 2$

30. $f(x) = x^4 + 4x^3$

31. $f(x) = (1 - x)^3$

32. $f(x) = (x + 1)^3$

33. $f(x) = (x + 1)^{1/3}$

34. $f(x) = \dfrac{x^2 - 4}{x}$

35. $f(x) = \dfrac{10}{x^2 + 3}$

36. Business: Advertising Revenue A company uses the model

$$R(x) = \frac{600x^2 - x^3}{20{,}000} \qquad 0 \le x \le 400$$

to relate its revenue, R, in thousands of dollars, to the amount of money, x, in thousands of dollars, it spends on advertising. Find the point of diminishing returns for this model.

37. Business: Advertising Revenue A company uses the model

$$R(x) = 10x^2 - \frac{2}{3}x^3 \quad 0 \le x \le 10$$

to relate its revenue, R, in millions of dollars, to the amount of money, x, in millions of dollars, it spends on advertising. Find the point of diminishing returns for this model.

38. Manufacturing: Employee Efficiency An efficiency study conducted on the day shift of a manufacturer showed that the number of units of a product produced by a typical employee t hours after 8:00 AM could be modeled by the function

$$f(x) = \frac{-4}{3}x^3 + 8x^2 + 22x \quad 0 \le x \le 8$$

Find the point of diminishing returns for this model.

39. Business: Advertising Revenue The owners of a photo framing store relates the amount of money they spend on advertising to the number of people who see their advertisement with the function

$$R(x) = \frac{900x^2 - x^3}{30{,}000} \quad 0 \le x \le 900$$

where R is in individuals and x is in dollars. Find the point of diminishing returns for this model.

40. Business: Inventory Costs Retailers are concerned about inventory costs. A manufacturer models its inventory costs by the function

$$C(x) = 2x + \frac{1{,}600}{x}$$

where x represents the lot size of an order.

a. Determine the critical value of this function.

b. Use the second derivative to show that this function is always concave upward.

c. Use the second derivative test to determine the lot size that minimizes the inventory cost.

For Problems 41-46, specify all hypercritical values and the intervals upon which the function $f(x)$ is concave upward and upon which it is concave downward.

41. $f(x)$ as illustrated below.

42. $f(x)$ as illustrated below.

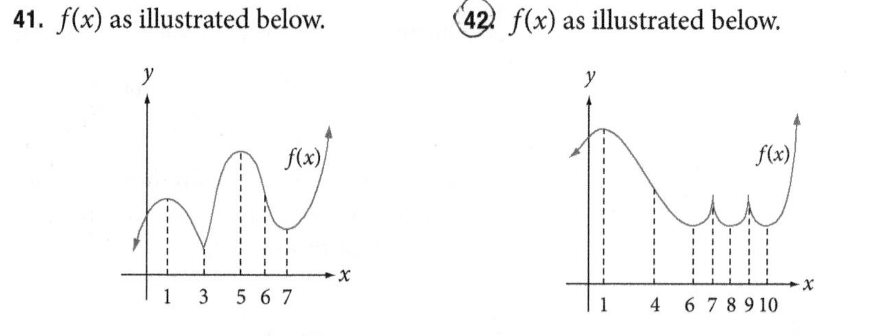

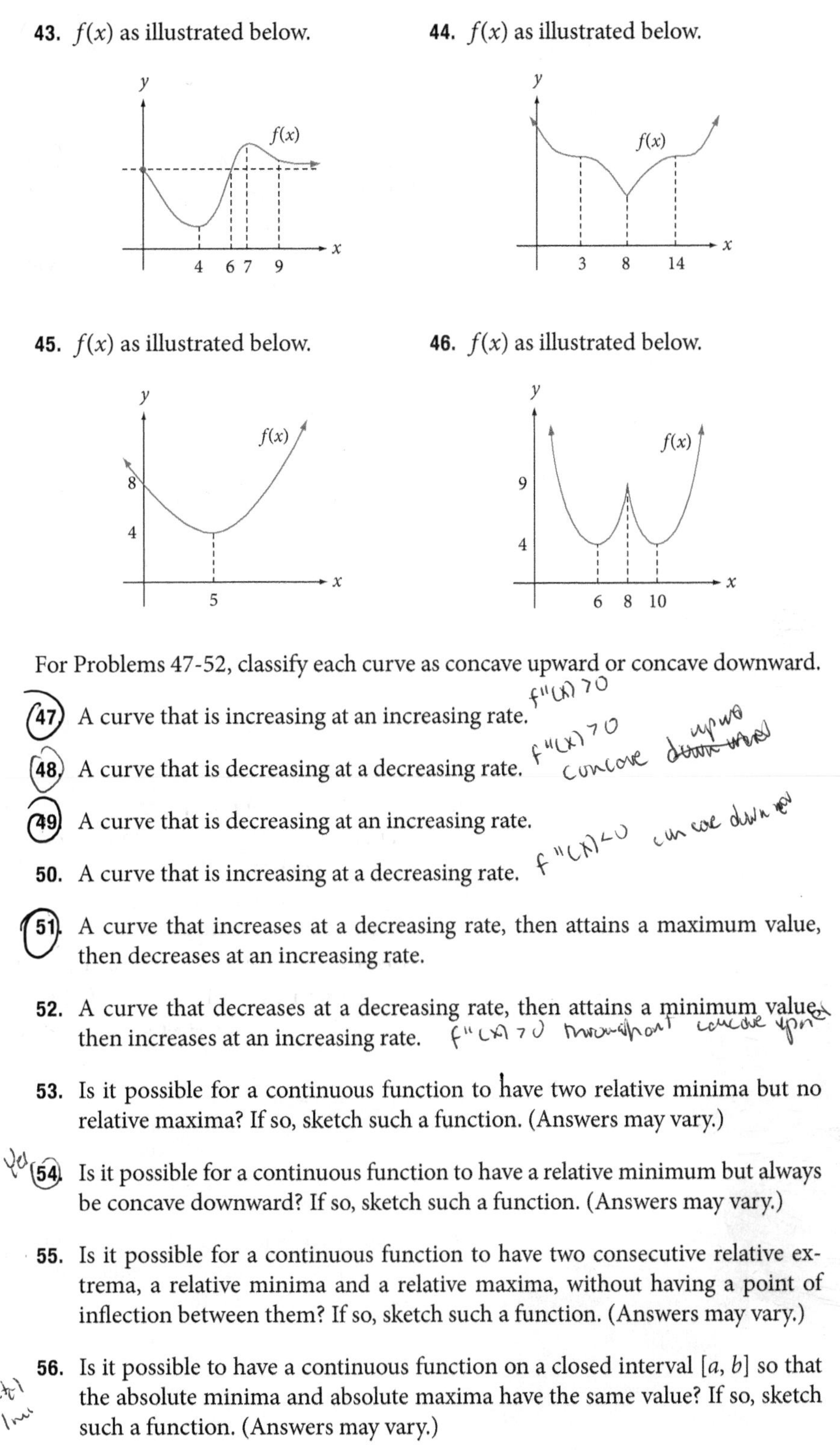

43. $f(x)$ as illustrated below.

44. $f(x)$ as illustrated below.

45. $f(x)$ as illustrated below.

46. $f(x)$ as illustrated below.

For Problems 47-52, classify each curve as concave upward or concave downward.

47. A curve that is increasing at an increasing rate.

48. A curve that is decreasing at a decreasing rate.

49. A curve that is decreasing at an increasing rate.

50. A curve that is increasing at a decreasing rate.

51. A curve that increases at a decreasing rate, then attains a maximum value, then decreases at an increasing rate.

52. A curve that decreases at a decreasing rate, then attains a minimum value, then increases at an increasing rate.

53. Is it possible for a continuous function to have two relative minima but no relative maxima? If so, sketch such a function. (Answers may vary.)

54. Is it possible for a continuous function to have a relative minimum but always be concave downward? If so, sketch such a function. (Answers may vary.)

55. Is it possible for a continuous function to have two consecutive relative extrema, a relative minima and a relative maxima, without having a point of inflection between them? If so, sketch such a function. (Answers may vary.)

56. Is it possible to have a continuous function on a closed interval $[a, b]$ so that the absolute minima and absolute maxima have the same value? If so, sketch such a function. (Answers may vary.)

Modeling Practice

For Problems 57-60, describe the behavior of the given function. (You need not construct a summary table.)

57. **Business: Average Cost** The graph illustrated below displays the relationship between the average cost A (in dollars) of a particular commodity and the number q of the commodity produced.

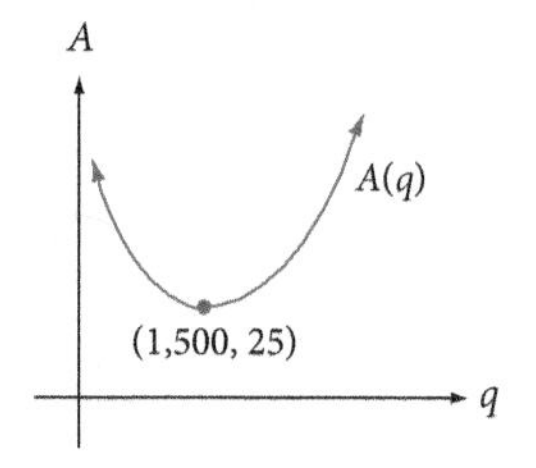

58. **Psychology: Excitement Level** The graph illustrated below exhibits the relationship between the average excitement level A (in appropriate units) and the amount of time t (in days) since the occurrence of a local championship sports game.

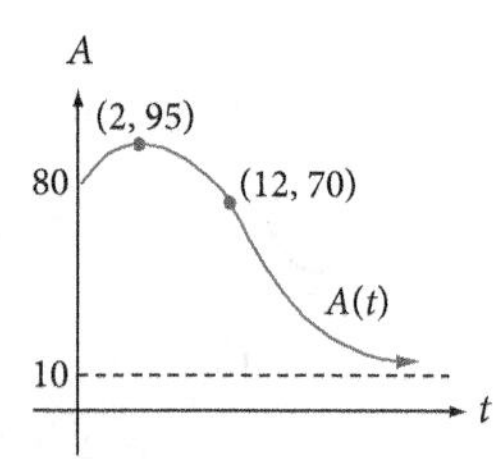

59. **Economics: University Tuition** The graph illustrated below exhibits the relationship between the increase I (in percent) in tuition at a state university and the average enrollment E (in thousands) in community colleges near the state university.

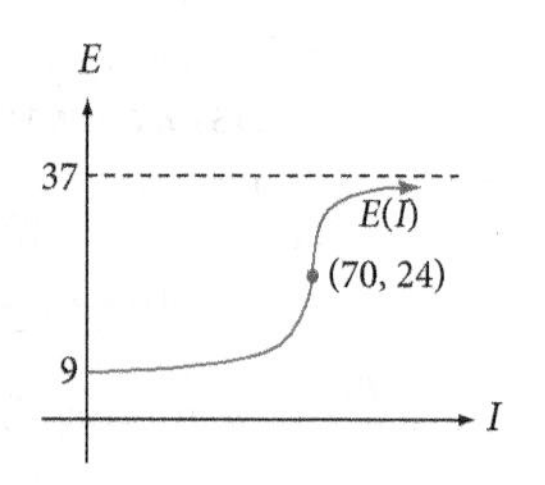

60. **Economics: Use of Water** The graph illustrated below displays the relationship between the amount A (in millions of gallons) of reclaimed water used by a county each year and the time t (in years) since 1940.

© Oxign/iStockPhoto

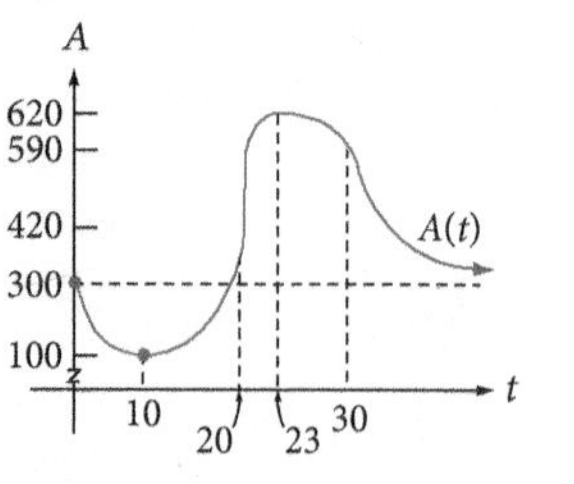

© Shaun Lowe Photographic/ iStockPhoto

61. Physical Science: Drag Racing The rational function

$$V(t) = \frac{340t}{t + 3}$$

models the speed $V(t)$, in miles per hour, of a dragster t seconds since it starts its quarter-mile run. Respond to the claim that during the first 4 seconds of the dragster's run, its speed is increasing at an increasing rate. Assuming the speed model to be true, do you accept or reject the claim? Support your response mathematically.

62. Manufacturing: Costs of Production A manufacturer of precision medical instruments uses the function

$$C(x) = \frac{400x + 1{,}000}{x + 1}$$

to model the cost C (in dollars) of producing x number of instruments. A company analyst claims that if the company increases its production up through 100 units, its cost will decrease but do so at a decreasing rate. Assuming the cost model to be true, do you accept or reject the claim? Support your response mathematically.

63. Health Science: Drug Reaction The function

$$S(x) = \frac{1}{2}x\sqrt{300 - x} \quad 0 \leq x \leq 300$$

measures the strength of a typical person's reaction to x milligrams of a particular drug. The manufacturer of the drug makes three claims:

a. That as the dosage increases from 0 mg to 200 mg, the reaction to the drug will increase but at a decreasing rate.

b. That at a dose of 200 mg, the typical person will experience the maximum reaction to the drug.

c. That as the dosage increases from 200 mg to 300 mg, the reaction will decrease at a decreasing rate.

Assuming that the strength function accurately models the typical person's reaction to the drug, do you accept or reject each claim?

64. Physical Science: Double Stars An interesting astronomical phenomenon is that of the double star. A double star is a system of two stars revolving around each other. Astronomers can use telescopes to view double stars. Since stars are so far away, double stars often appear as a single source of light in the sky. For a telescope to show the two individual stars in a double star system, it must have a minimum magnification. The magnification needed to split a double star is given by the model

$$M(d) = \frac{1}{d}$$

where d is the angular separation of the double star and M is the magnification required of the telescope to separate the stars.

Make a mathematical argument to support the claim that as the angular separation of two stars in a double star system increases, the required magnification decreases and does so at a decreasing rate.

Source: Saguaro Astronomy Club

65. Manufacturing: Minimizing Production Costs Power over Ethernet (PoE) systems pass electrical power and data safely on Ethernet cabling. The function

$$C(x) = \frac{144 + x^2}{180x}$$

relates the weekly cost (in thousands of dollars) to a company to the number x (in hundreds) of PoE systems it manufactures and sells each week.

Provide a mathematical argument to demonstrate that once the manufacturer produces more than 1,200 ($x = 12$) systems, the cost of producing systems will increase but at an increasing rate.

Source: Fluke Networks

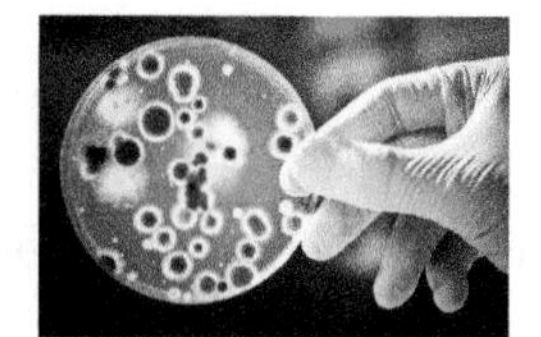

© Alexander Raths /iStockPhoto

66. Biology: Bacteria Population The function

$$P(t) = \frac{30t}{t^2 + 16} \qquad t > 1$$

relates the population P (in millions) of bacteria in a culture to the time t (in days) after a particular slow-acting toxin is introduced into the culture. The manufacturer of the toxin claims

a. that 4 days after the introduction of the toxin the population of bacteria will be at its maximum.

b. that from 4 to 6 days, the population will decrease at an increasing rate.

c. that from 7 days on, the population will decrease but at a decreasing rate. Assuming the population model is true, do you accept or reject the manufacturer's claims?

For Problems 67-70, sketch the graph of a function that matches the summarized behavior.

67. Mathematics: Probability The probability of many types of events is described by a curve that increases at an increasing rate as x increases from negative infinity to -1. At $x = -1$, $f(x) \approx 0.24$ and the function experiences a point of inflection; in particular, this point of inflection is a point of diminishing returns. As x increases beyond -1 to 0, the function continues to increase but at a decreasing rate. When $x = 0$, $f(x) \approx 0.4$ and the function attains a relative maximum. As x then increases to $+1$, $f(x)$ decreases at an increasing rate. At $x = +1$, $f(x) \approx 0.24$ and the function experiences a point of inflection. As x increases from $+1$ to $+\infty$, the function continues to decrease but at a decreasing rate. The function is always positive and the x-axis is a horizontal asymptote.

68. Economics: Demand for a Commodity The demand D (in single units) for a particular commodity is related to the price p (in dollars) of the commodity. For a particular commodity, $D = 0$ when $p = 0$. Then as the price increases from \$0 to \$250, the demand increases but at a decreasing rate. The demand attains a maximum of 1,400 units when the price is \$250. As the price increases from \$250 to \$400, the demand for the commodity decreases at an increasing rate. When $p = 400$, $D = 350$ and experiences a point of inflection. When the price increases beyond \$400 per unit, the demand for it continues to decrease but now at a decreasing rate. The function is always positive and the x-axis is a horizontal asymptote.

69. Manufacturing: Newton's Law of Cooling Newton's Law of Cooling describes how a heated object cools when placed into a cooler environment. A particular object has temperature $T = 98\ °C$ when it is first taken from an oven. It then cools according to Newton's Law of Cooling. The temperature of the object decreases at a decreasing rate and approaches 0 as time goes on. In fact, the temperature of the object is only about 18 °C after 60 minutes.

70. Business: Investment When an amount of \$2,000 is invested at 7% compounded continuously, the amount A accumulated grows at an increasing rate as time t (in years) goes by. In fact, after 30 years, \$16,332 will have accumulated.

Using Technology Exercises

71. Use Wolfram|Alpha to find, if any exist, the hypercritical values of the function $f(x) = x^3 - 4x^2 + 5$.

72. Use Wolfram|Alpha to find any points of inflection of the function

$$f(x) = \frac{1}{9}x^4 - \frac{5}{2}x^3 + 4x + 25$$

73. Manufacturing: Training A manufacturer believes a trainee's skill level, L, is related to the number of days, t, spent training by the function

$$L(t) = 0.05(t - 4)^3 + 0.5t + 1.4$$

For the first 4 days of training, the trainee's skill level is increasing at a decreasing rate. Describe the behavior of the trainee's skill level from day 4 through day 7.

74. Use Wolfram|Alpha to approximate the critical values of

$$f(x) = x^5 - x^3$$

Then use the second derivative test to determine which one produces the relative maximum and which one the relative minimum. Specify those maximum and minimum values.

Getting Ready for the Next Section

75. The length of a rectangle is 20 feet longer than its width. Write an expression that gives the perimeter of the rectangle in terms of one variable, the width.

76. Multiply $(80 - 2w)w$.

77. What is the critical value of the function $A'(w) = 80 - 4w$?

78. The function $f(x) = x^3 - 13x^2 + 40x$ has two critical values. Find them.

79. Subtract $200(20 + 2r)$ from $(400 - 10r)(20 + 2r)$.

80. Find $P(5)$ for $P(r) = 4{,}000 + 200r - 20r^2$.

81. Find $V'(h)$ for $V(h) = 160h - 52h^2 + 4h^3$.

82. Find, if it exists, $\lim_{t\to\infty} \frac{7t}{t^3 + 18}$.

83. Solve the equation $\frac{0.21 - 0.03t}{(t + 7)^3} = 0$.

84. Solve the equation $14(9 - t^3) = 0$.

Applications of the Derivative: Optimization

3.3

© Kovalchuk Oleksandr/Shutterstock

As we mentioned in the introduction to this chapter, derivatives can be used to determine the optimum price for a product; that is, the price that produces the most profit for a business. We looked at the example of Sally, who owns an online business that sells hard drives. By applying a derivative to a function that relates the price of hard drives, the number sold, the cost of the hard drive to the retailer, and the number of hard drives purchased by the retailer, Sally can determine how best to price her product. These sorts of calculations are done on a constant basis by big Internet retailers as they seek to maximize their profits.

The process of determining the maximum or minimum values of a function is called **optimization**. Using the differentiation techniques discussed in Sections 3.1 and 3.2, we will see how the derivative is applied to optimize functions that model practical applications from business, economics, medicine, and other fields.

A Strategy for Solving Optimization Problems

We will now consider how differential calculus can be used to solve optimization problems. These are problems in which we are interested in maximizing or minimizing a particular quantity. For example, we may be interested in knowing the sales price of an item that will maximize the profit on that item, or how many days after an insecticide is applied to a population of insects, the population will be at its minimum.

A strategy that is helpful in solving optimization problems follows.

1. Draw a figure when one is appropriate.
2. Assign a variable to each quantity mentioned in the problem.
3. Select the quantity that is to be optimized and construct a function that relates it to some or all of the other quantities.
4. Since all the rules of differentiation deal with functions of only one variable, the function must involve only one variable. If the function constructed in step 3 involves more than one variable, use the information contained in the problem to eliminate variables until you have a function of only one variable.
5. Use differentiation to find the critical values of the function and apply the test for absolute extrema. Disregard all answers that are not relevant to the situation.

Applying the Strategy: Examples

We will illustrate the optimization strategy with five examples. We will show the first example with great detail and then include less detail where appropriate in the remaining examples.

VIDEO EXAMPLES

SECTION 3.3

Example 1 A rectangular region along the bank of a straight section of a river is to be enclosed with 80 feet of fencing. Only three sides need to be fenced since the river will bound the other side. What are the dimensions that will produce the greatest enclosed area?

Solution We will apply the optimization strategy.

1. We will construct a figure since one is appropriate. (See Figure 1.)

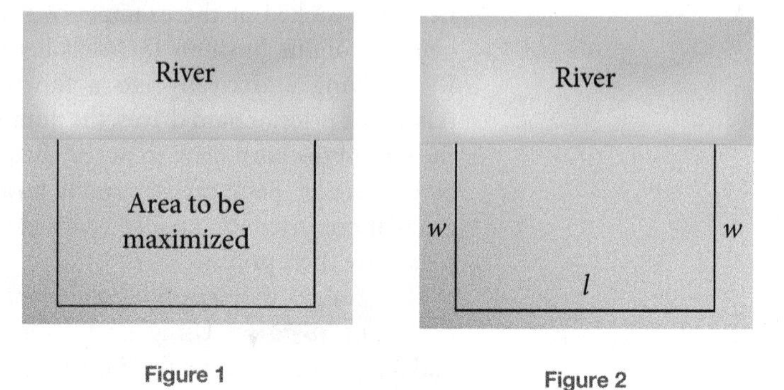

Figure 1 Figure 2

2. The quantities mentioned in this problem are the area, the dimensions of the region, and the amount of fencing available. We will label each with a variable and then affix the appropriate labels to our picture, as shown in Figure 2.
   $A =$ the area of the rectangular region
   $l =$ the length of the rectangular region
   $w =$ the width of the rectangular region
   $P =$ the perimeter of the rectangular region (that is, the amount of fencing available)

3. The quantity to be maximized is the area, so we need to construct an area function. Since the region is rectangular, we will use the formula for the area of a rectangle.

   Maximize: $A(l, w) = lw$.
   (Remember how functions are read: $A(l, w)$ means that the area A depends on both the length l and the width w.)

4. This function involves two variables, l and w; we need to eliminate one. To do so, we will use the information contained in the problem to express l in terms of w or w in terms of l. The problem restricts the amount of fencing to 80 ft, which means that the *perimeter*, P, of the fenced region is 80 ft. The perimeter of this shape is the sum of the lengths of the three fenced sides, so

$$P = w + l + w = 80$$
$$= l + 2w = 80$$

   Thus, $l + 2w = 80$, and we can solve for either l or w. (We can avoid fractions by solving for l. Avoiding fractions may not always be possible.)

$$l + 2w = 80$$
$$l = 80 - 2w$$

Now, replacing l in the function $A(l, w) = lw$ with $80 - 2w$ gives us the function of the one variable w.

$$A(w) = (80 - 2w)w$$
$$A(w) = 80w - 2w^2$$

A negative area is not acceptable; thus, solving $A(w) \geq 0$ (by sign charting) gives us $0 \leq w \leq 40$. Therefore, $A(w)$ is defined over a closed interval $[0, 40]$.

5. To find the critical values and then the absolute extreme values, we first find the first derivative.

$$A'(w) = 80 - 4w$$

a. We then ask: Where is $A'(w) = 0$?

$$A'(w) = 0$$
$$80 - 4w = 0$$
$$-4w = -80$$
$$w = 20$$

So, $A'(w) = 0$ when $w = 20$.

b. We then ask: Where is $A'(w)$ undefined? $A'(w)$ is never undefined since it is a polynomial function.

Thus, the only critical value is $w = 20$. Since we are interested only in the absolute extrema, we only need to evaluate $A(w)$ at 0, 20, and 40, to see which produces the greatest output (area).

$$A(0) = 0,$$
$$A(20) = 80(20) - 2(20)^2 = 800,$$
$$A(40) = 80(40) - 2(40)^2 = 0$$

Now, we conclude that 20 produces the absolute maximum value of the function. We can also use $l = 80 - 2w$ to find that $l = 80 - 2(20) = 40$.

Interpretation

The dimensions that produce the maximum enclosed area for 80 feet of fencing are a length of 40 ft and a width of 20 ft. The graph of $A(w) = 80 - 2w^2$ appears in Figure 3 and reinforces the fact that the absolute maximum area occurs when the width is 20 feet.

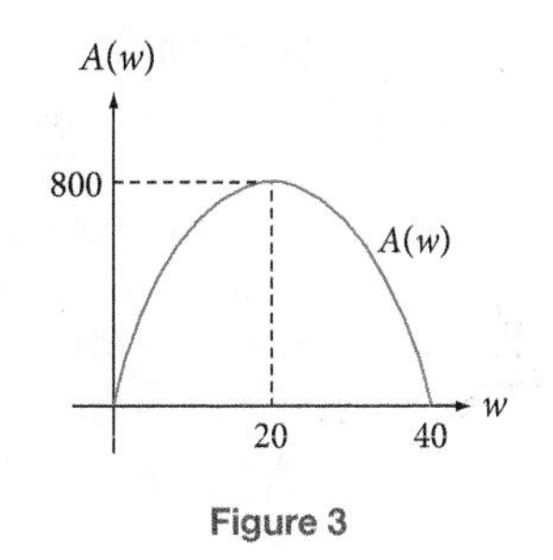

Figure 3

Example 2 The concentration, C, of a drug in the bloodstream t hours after it has been administered is approximated (in mg/cm^3) by the function

$$C(t) = \frac{7t}{t^3 + 18} \qquad t \geq 0$$

When is the concentration the greatest?

Solution In this problem, we do not need to draw a picture, nor do we need to construct a function. From the outset, we have a function of one variable and we can use differentiation to find the critical values (strategy item 5).

$$\begin{aligned} C'(t) &= \frac{(t^3 + 18) \cdot 7 - 7t(3t^2)}{(t^3 + 18)^2} \\ &= \frac{7[(t^3 + 18) - 3t^3]}{(t^3 + 18)^2} \\ &= \frac{7(18 - 2t^3)}{(t^3 + 18)^2} \\ &= \frac{14(9 - t^3)}{(t^3 + 18)^2} \end{aligned}$$

a. Where is the derivative 0? Recalling that a fraction is 0 only when the numerator is 0 gives us

$$\begin{aligned} 14(9 - t^3) &= 0 \\ 9 - t^3 &= 0 \\ t^3 &= 9 \\ t &= \sqrt[3]{9} \end{aligned}$$

b. Where is the derivative undefined? A fraction is undefined when the denominator is 0. But since the denominator is squared and $t \geq 0$, the denominator is never 0, so the fraction is never undefined. Thus, $C(t)$ has the single critical value $\sqrt[3]{9}$.

We need to evaluate $C(t)$ at 0 and $\sqrt[3]{9}$ to determine which produces the absolute maximum. Since $C(0) = 0$, $\sqrt[3]{9}$ must produce the absolute maximum of this function. (As the graph in Figure 4 shows, that maximum is approximately 0.5.)

Interpretation

The concentration of the drug is maximum $\sqrt[3]{9} \approx 2.08$ hours after the drug is administered. Figure 4 displays the graph of $C(t)$. Does this match your expectations for this type of situation? The concentration begins at 0 when the drug is administered. It then climbs quickly to reach its maximum, then falls off slowly as the body metabolizes it out of its system. After enough time has gone by, the concentration is again essentially 0. Notice that this last piece of information is given by

$$\lim_{t \to \infty} \frac{7t}{t^3 + 18}$$

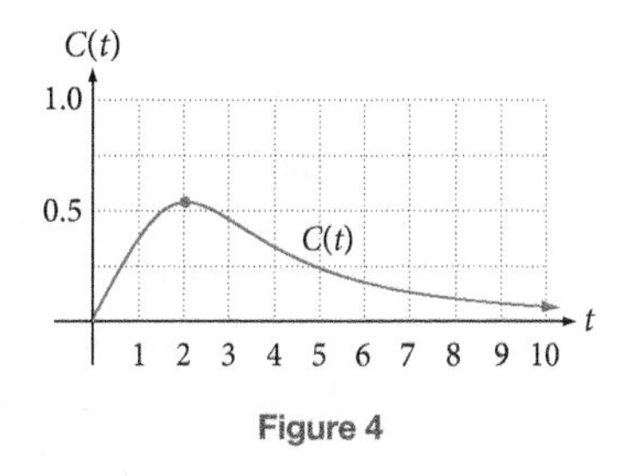

Figure 4

Example 3 Sally owns an online business and wants to know how to price her inventory of 3-terabyte hard drives in order to maximize the profit she makes from their sale. Currently, Sally sells 20 hard drives a week for $400 each (which cost the store $200 each). She estimates that for each $10 reduction in price, she can sell two more drives each week. Find the price of each drive that will produce the maximum profit for her business.

Solution

1. A picture is not appropriate here so we will not construct one.
2. We will introduce some variables to represent the quantities mentioned in the problem.
   r = the number of $10 reductions in price
   p = the price of each hard drive after a reduction
   n = the number of hard drives purchased and sold
   R = the revenue produced by all the sale of n hard drives
   C = the cost to the store of n hard drives
   P = the profit realized by the sale of n hard drives
3. We are asked to find the price that maximizes the profit, so we need to construct a profit function. Since profit equals revenue minus cost, we have

   $$P = R - C$$

   But this is a function of two variables. Since revenue is price times the number sold, and cost is the individual cost times the number purchased,

   $$R = \underbrace{(400 - 10r)}_{\text{Price}} \cdot \underbrace{n}_{\text{Number sold}} \quad \text{and} \quad C = \underbrace{200}_{\text{Cost}} \cdot \underbrace{n}_{\text{Number purchased}}$$

4. That is, $R = (400 - 10r)n$ and $C = 200n$. The expression for R comes from the fact that the sales price is $400 minus the $10 for every reduction; thus, $400 − $10r$. But also, the number of drives sold n is 20 plus 2 times the number of $10 price reductions, that is, $n = 20 + 2r$. Hence,

   $$P(r) = (400 - 10r)(20 + 2r) - 200(20 + 2r)$$
   $$= 4{,}000 + 200r - 20r^2$$

   and now the profit, P, depends only on the number, r, of $10 price reductions.

5. $P'(r) = 200 - 40r$.
   The critical value is $r = 5$, and the first derivative test shows this to produce a relative maximum.

Interpretation
The best price per drive is $\$400 - 10(5) = \$350$. This would produce sales of $n = 20 + 2(5) = 30$ drives per week, and a maximum profit (per week) of

$$P(5) = 4{,}000 + 200(5) - 20(5)^2 = \$4{,}500$$

Example 4 A rectangular box with an open top is to be made from a 10-in.-by-16-in. piece of cardboard by removing small squares of equal size from the corners and folding up the remaining flaps. What should be the size of the squares cut from the corners so that the box will have the largest possible volume?

Solution

1. A figure is appropriate and very helpful. Figure 5 shows the construction of the box from a flat piece of cardboard.

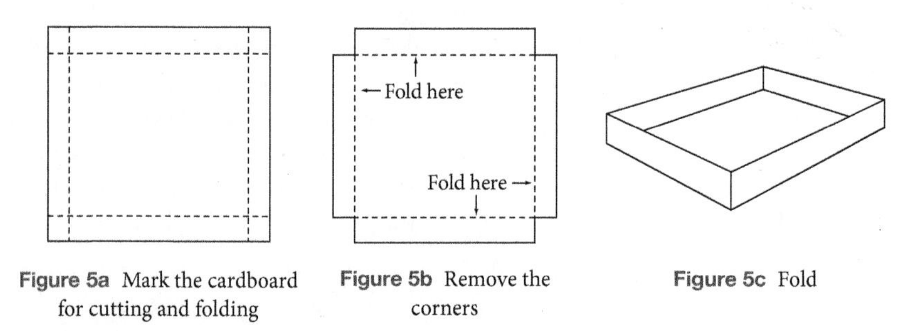

Figure 5a Mark the cardboard for cutting and folding **Figure 5b** Remove the corners **Figure 5c** Fold

2. The quantities mentioned or implied in this problem are volume, length, width, and height. We will label each with a variable and then affix the labels to our picture. Since the cut determines the height of the box which, in turn, determines both the length and width of the box, we will let
   h = the length of the side of the square that is cut from the flat piece of cardboard. We label h on the figure to help establish w and l.
   l = the length of the box $= 16 - 2h$.
   w = the width of the box $= 10 - 2h$.
   V = the volume of the box.

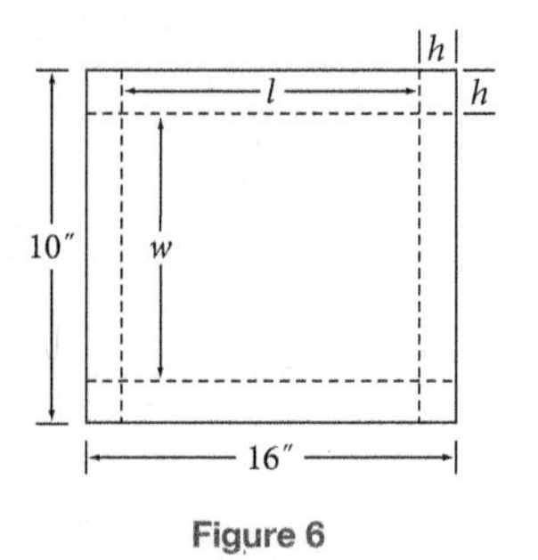

Figure 6

3. The quantity to be maximized is the volume, so we need to construct a volume function. Since the region is a rectangular box, we will use the formula for the volume of a rectangle box. Maximize: $V(l, w, h) = lwh$

4. Use the information in Figure 6 to rewrite the volume equation in just one variable. Since $l = 16 - 2h$ and $w = 10 - 2h$, we have

$$V = (16 - 2h)(10 - 2h)h = 160h - 52h^2 + 4h^3$$

5. $V'(h) = 160 - 104h + 12h^2$

   a. Where is the derivative equal to 0?

$$160 - 104h + 12h^2 = 0$$
$$40 - 26h + 3h^2 = 0$$
$$(3h - 20)(h - 2) = 0$$

   so that $h = \frac{20}{3}$ or 2.

   b. Where is the derivative undefined? Never (since $V'(h)$ is a polynomial function).

Thus, the critical values are $h = \frac{20}{3}$ and 2. However, since $0 < h < 5$ (why?), only $h = 2$ needs to be tested. Figure 7 shows the sign chart for $V'(h)$ and indicates that 2 produces a relative maximum.

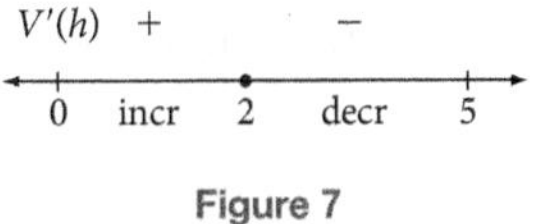

Figure 7

Interpretation

To maximize the volume of the box, the cutout squares should be 2 inches by 2 inches.

Example 5 A chemical company claims that 6 days after its new insecticide is applied to a one-acre field infested with a population of a certain insect, the population will be minimal. The company bases its claim on its belief that the function

$$P(t) = \frac{2[13 + (t - 5)^2]}{t + 1} + 15 \quad 0 \le t \le 20$$

accurately predicts the population size P (in ten thousands) of the insect t days after application. Assuming the population model is accurate, should the company's claim be accepted or rejected?

Solution A figure is not appropriate here. The company is making a claim about a function's maximum value. Mathematically, this is a statement about the derivative of the function. Since we are given the function, we will differentiate it.

$$P'(t) = \frac{2(t + 8)(t - 6)}{(t + 1)^2}$$

Now, $P'(t) = 0$ when $t = -8$ and 6 and is undefined at $t = -1$. But $t = -8$ and -1 make no sense in terms of the physical situation, so we disregard them. The only critical value is $t = 6$. Figure 8 shows the sign chart for $P'(t)$ and indicates that 6 produces a relative minimum.

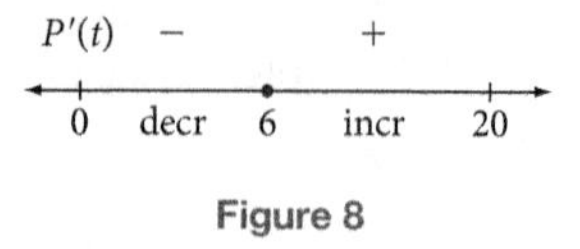

Figure 8

Interpretation
Six days after the insecticide is applied, the population of insects is at a minimum. Assuming the population model is correct, the company's claim should be accepted.

Using Technology 3.3

Many functions that model applied phenomena are data-generated functions and therefore well suited to the approximation techniques of the graphing calculator.

Example 6 Repeat Example 5 using your graphing calculator.

Solution On your calculator, enter the function

Y1=(2(13+(X-5)^2))/(X+1)+15

Next, on your graphing calculator, activate the minimum value operation and locate the function's minimum which is (6, 19).

Getting Ready for Class

After reading through the preceding section, respond in your own words and in complete sentences.

A. Describe the strategy you would use to solve Problem 13. Do not solve the problem; simply describe the strategy as if you were explaining to another person how to go about solving the problem.

B. To optimize a function using calculus requires that you find the derivative of the function to be optimized. What are the two questions you ask yourself about the derivative that help you to find the critical values?

C. Maximum and minimum points on the graph of a function can be smooth or can be cusps or corners. What is it about the derivative of the function that distinguishes the two?

Problem Set 3.3

Skills Practice

1. **Health Science: Drug Concentration** The concentration of a drug in the bloodstream $C(t)$ at any time t, in minutes, is described by the equation

$$C(t) = \frac{100t}{t^2 + 16}$$

where $t = 0$ corresponds to the time at which the drug was swallowed. Determine how long it takes the drug to reach its maximum concentration.

2. **Health Science: Flu Outbreak** According to a model developed by a public health group, the number of people $N(t)$, in hundreds, who will be ill with the Asian flu at any time t, in days, next flu season is described by the equation

$$N(t) = 90 + \frac{9}{4}t - \frac{1}{40}t^2 \quad 0 \le t \le 120$$

where $t = 0$ corresponds to the beginning of December. Find the date when the flu will have reached its peak and state the number of people who will have the flu on that date.

3. **Business: Heating Costs** A homeowner wishes to insulate his 1,500-square-foot attic. The total cost $C(r)$ for r inches of insulation and the heating costs for that amount of insulation over the next 10 years is given by

$$C(r) = 120r + \frac{4{,}320}{r}$$

How many inches of insulation should be placed in the attic if the total cost is to be minimized?

Source: CPS Energy

4. **Business: Maximization of Space** A rancher has 200 feet of fencing to enclose two adjacent rectangular corrals, as shown in the following figure. The equation describing the enclosed area is

$$A(x) = 2x\left(\frac{200}{3} - \frac{4}{3}x\right)$$

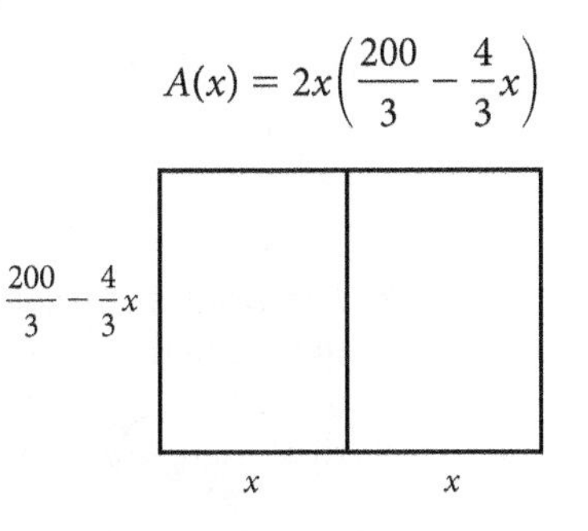

To enclose the maximum area, what should be the dimensions of each corral?

5. **Business: Revenue** Suppose a baby food company has determined that its total revenue R for its food is given by

$$R = -x^3 + 63x^2 + 1{,}200x$$

where R is measured in dollars and x is the number of units (in thousands) produced. What production level will yield a maximum revenue?

6. **Wildlife Management: Deer Population** The game commission in a certain state introduces 500 deer into some newly acquired land. The population $N(t)$ of the herd is given by

$$N(t) = \frac{500 + 500t}{1 + 0.04t^2}$$

where t is time in years. Determine the number of years for the deer population to become a maximum. What is likely to happen to this deer population in the long run?

Check Your Understanding

7. **Economics: Population** The following figure illustrates the relationship between time t (in years) and the number of people P (in thousands) in a given area over a 20-year period. During which year is the population at its maximum?

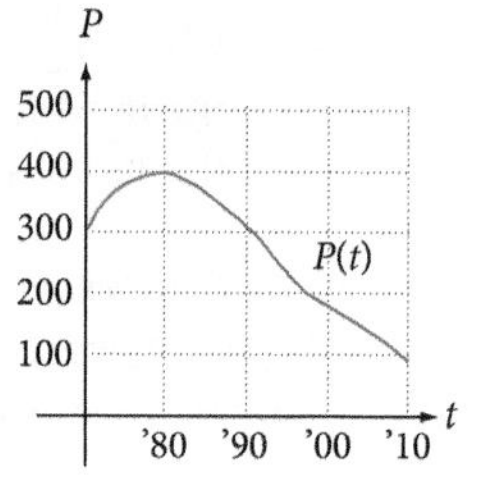

8. **Medicine: Drug Concentration** The following figure shows the concentration levels C (in mg) of a certain drug in the bloodstream over the time t (in hours) since consumption. What is the maximum concentration of the drug and when does it reach this level?

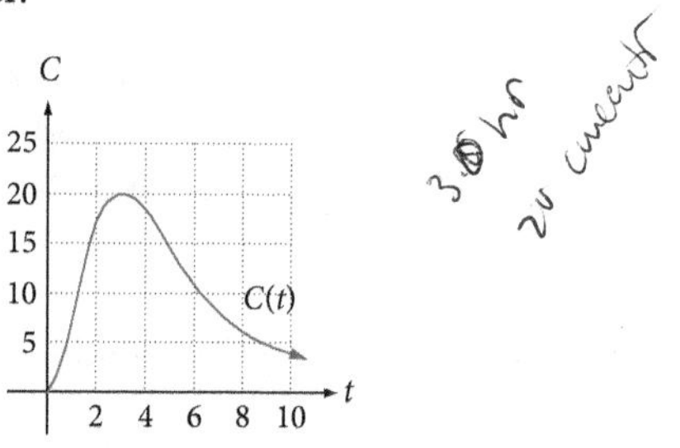

9. **Mathematics: Triangles** The following figure shows the relationship between the hypotenuse of a right triangle H and the length of one leg L when the sum of the two legs is always 20 units. What are the lengths of the two legs of the right triangle with a minimum hypotenuse length?

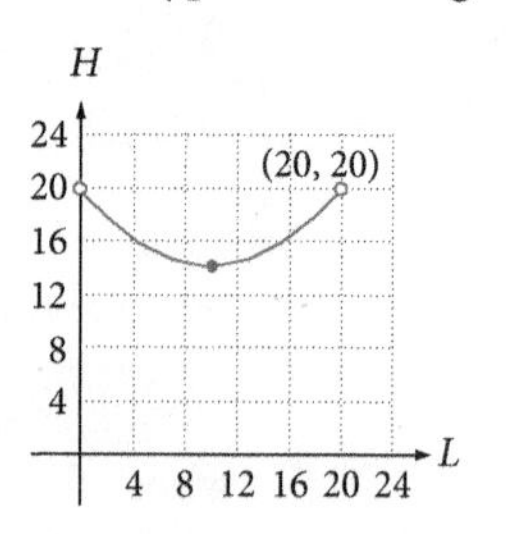

10. **Business: Operation Costs** The following figure illustrates the cost C, in dollars, of operating a truck at speed, V, in miles per hour. What is the most economical speed at which to operate the truck?

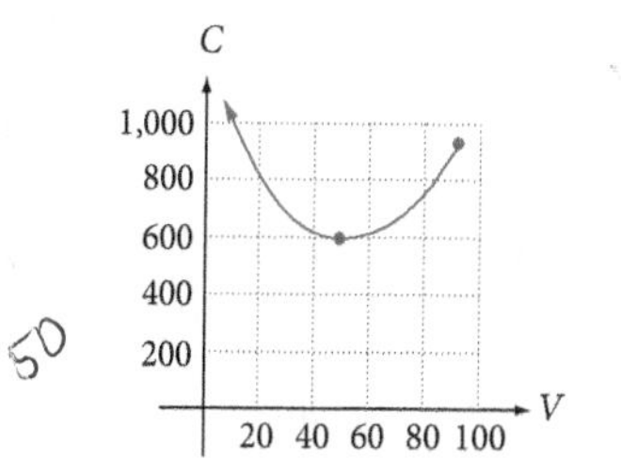

11. **Business: Profit** The following figure shows the profit P, in thousands of dollars, a magazine publisher can make if it sells n, in hundreds, magazines. How many magazines should be sold to maximize profit?

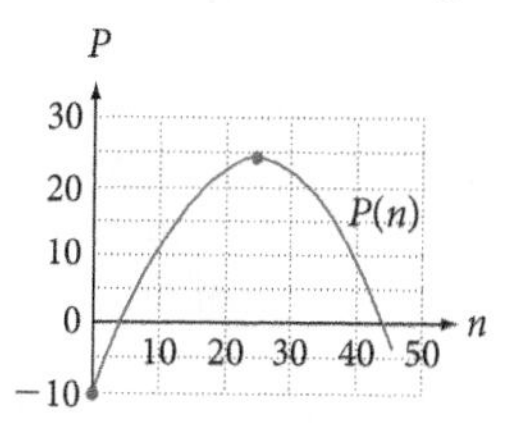

Modeling Practice

12. **Economics: Taxes** A state legislature is considering a bill that would impose a meals tax of t cents per dollar spent on food in restaurants. Economists estimate the tax will cause a decline in the amount of money $A(t)$, in millions of dollars, spent by the public on food in restaurants. This is given by

$$A(t) = 216 - 2t^2 \quad 0 \leq t \leq 8$$

Graph $A(t)$ and determine the value of t that maximizes $A(t)$.

13. **Business: Maximization of Space** A farmer has 1,200 feet of fence and wishes to build two identical rectangular enclosures, as in the following figure. What should be the dimensions of each enclosure if the total area is to be a maximum?

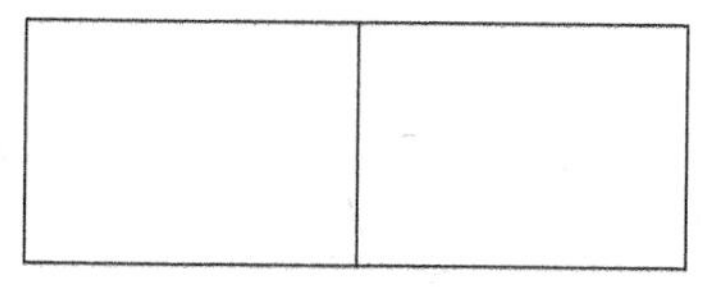

14. **Business: Space Utilization** The owner of a ranch has 3,000 yards of fencing material with which to enclose a rectangular piece of grazing land along a straight portion of a river. If fencing is not required along the river, what are the dimensions of the pasture having the largest area? What is the largest area?

15. **Business: Hotel Revenue** A 210-room hotel is filled when the room rate is \$50 per day. For each \$1 increase in the rate, three fewer rooms are rented. Find the room rate that maximizes daily revenue.

16. Business: Airline Revenue A charter flight club charges its members $250 per year. But for each new member above a membership of 65, the charge for all members is reduced by $2 each. What number of members leads to a maximum revenue?

17. Business: Agricultural Yield An orchard presently has 30 trees per acre. Each tree yields 480 peaches. It is determined that for each additional tree that is planted per acre, the yield will be reduced by 12 peaches per tree. How many trees should be added per acre to maximize the yield?

18. Business: Demand and Revenue Demand for an electric fan is related to its selling price p (in dollars) by the equation

$$n = 2{,}880 - 90p$$

where n is the number of fans that can be sold per month at a price p. Find the selling price that will maximize the revenue received.

19. Business: Packaging Postal regulations specify that a parcel sent by parcel post may have a combined length and girth (distance around) not exceeding 100 inches. If a rectangular package has a square cross section, find the dimensions of the package with the largest volume that may be sent through the mail.

20. Business: Packaging Costs The More Beef Company requires its corned beef hash containers to have a capacity of 64 cubic inches. The containers are in the shape of right circular cylinders. Find the radius and height of the container that can be made at a minimum cost, if the tin alloy for the side and bottom costs 4 cents per square inch and the aluminum for the pull-off lid costs 2 cents per square inch.

21. Business: Maximum Revenue A tool company determines that it can achieve 500 daily rentals of jackhammers per year at a daily rental fee of $30. For each $1 increase in rental price, 10 fewer jackhammers will be rented. What rental price maximizes revenue?

22. Business: Motel Revenue A motel finds that it can rent 200 rooms per day if it charges $80 per room. For each $5 increase in rental rate, 10 fewer rooms will be rented per day. What room rate maximizes revenue?

Using Technology Exercises

23. Find the value of x that minimizes the function

$$f(x) = x^{5/3} - 5x^{2/3} + 3$$

on the interval $[0, 5]$ and specify the value of x at which it occurs.

24. The concentration, C, of a drug in the bloodstream t hours after it has been administered is approximated by the function

$$C(t) = \frac{12t}{t^{3.2} + 24} \qquad t \geq 0$$

When is the concentration the greatest and what is that concentration?

25. A chemical company determines that the population P (in ten thousands) of a certain insect t days after its insecticide is applied to a one-acre field in which the insects reside is approximated by the function

$$P(t) = \frac{3[15 + (t - 8)^{2.3}]}{t^{1.1} + 1.1} + 20 \quad 0 \le t \le 30$$

Determine the number of days after an application of the insecticide at which the population of insects will be at a minimum. What is the minimum population? Round the number of days to the nearest day and the number of insects to the nearest one thousand.

26. A drug manufacturer claims that the effectiveness of a new drug it has produced is related to the time after the drug was introduced into the bloodstream by the function

$$E(t) = -0.04t^{3.1} + 0.081t^{1.9} + 0.39t$$

where $E(t)$ represents the effectiveness of the drug on a scale of 0 to 1 and t represents the number of hours since introduction, $0 \le t \le 5$. After how many hours since introduction will the drug be at its maximum effectiveness? What is that effectiveness? Round the number of hours to the nearest tenth and the effectiveness to the nearest hundredth.

27. When learning a new subject, it is often the case that a person's interest level L increases near the beginning of learning but then declines after having spent some time with the material. Suppose for a 30-minute period, the function

$$L(t) = \frac{105t}{7.5 + 0.25t^{1.7}}$$

approximates a person's level of interest on a scale of 0-to-75, 75 being highest interest. The variable t represents the number of minutes a person spends learning the new material. According to this model, how long will it take a person to reach maximum interest level in this material? Round the number of minutes to the nearest one minute.

Source: Theories of Learning, Oxford Brookes University

Getting Ready for the Next Section

28. If $R(x) = x^3 + 4x^2 + 160x$,

a. compute $R(70)$

b. find $R'(x)$

29. If $C(x) = -0.035x^2 + 40x + 25$,

a. find $C'(x)$

b. find $C'(600)$

30. Subtract $0.02x^3 - 0.5x^2 + 10x + 150$ from $-1.1x^2 + 41.5x$.

31. Find the derivative of the function

$$P(x) = -0.02x^3 - 0.6x^2 + 31.5x - 150$$

32. Multiply $14 - 0.002x$ by x.

33. Multiply $32 \cdot \frac{x}{2}$

34. Solve the equation $16 - \frac{160{,}000}{x^2} = 0$. Approximate your answer to two decimal places.

35. Convert the expression $\frac{4{,}500}{\sqrt[3]{p^2}}$ to one that has a rational exponent rather than a radical sign and so that no denominator appears.

36. Find $\frac{dq}{dp}$ for $q = \frac{97{,}533}{\sqrt[5]{p^{24}}}$

37. Simplify $500 \text{ units} \cdot \frac{\$27}{\text{unit}}$. Pay attention to the units.

Applications of the Derivative in Business and Economics

3.4

© Marcus Clackson/iStockPhoto

In business and economic applications, effects of changes over wide intervals rather than effects of a change over an interval of length 1 are more frequently of interest. Analysis over wider intervals is defined as incremental analysis and it is best understood by having a basic understanding of marginal analysis. In this section we do not study incremental analysis. That subject is more appropriately studied in managerial courses. However, we lay the foundation for the study of incremental analysis by examining in more detail the concepts of marginal revenue, marginal cost, and marginal profit. We also continue our study of profit maximization and cost minimization, while introducing the new topics of inventory cost, the economic order quantity, and elasticity of demand.

Marginal Analysis

In business and economics, the word *marginal* is equivalent to the phrase *rate of change*, or more precisely, the *instantaneous rate of change*. The marginal concept in business and economics is the derivative concept in mathematics.

1. **Marginal revenue** is the approximate change in the revenue associated with a 1-unit change in the number of units sold. It is the derivative of the revenue function. Revenue functions are commonly denoted with the letter R. The marginal revenue is commonly represented with the letters MR. Symbolically, $R' = MR$.
2. **Marginal cost** is the approximate change in the cost associated with a 1-unit change in the number of units produced. It is the derivative of the cost function. Cost functions are commonly denoted with the letter C. The marginal cost is commonly represented with the letters MC. Symbolically, $C' = MC$.
3. **Marginal profit** is the approximate change in the profit associated with a 1-unit change in the number of units sold. It is the derivative of the profit function. Profit functions are commonly denoted with the letter P. The marginal profit is commonly represented with the letters MP. Symbolically, $P' = MP$.

VIDEO EXAMPLES

SECTION 3.4

Example 1 The revenue realized by a company on the sale of x units of its product is given by the revenue function $R(x) = x^3 + 4x^2 + 160x$. Compute and interpret both $R(70)$ and $R'(70)$.

Solution Since we are asked for information about $R'(x)$, we compute it first.

$$R'(x) = 3x^2 + 8x + 160$$

a. Substituting 70 for x in $R(x)$, we get $R(70) = 373{,}800$. This means that when the company sells 70 units of its product, it realizes a revenue of \$373,800.

b. Substituting 70 for x in $R'(x)$, we get $R'(70) = 15{,}420$. This means that if the company increases the number of units it sells by 1, from 70 to 71, its revenue will increase by approximately \$15,420. (Notice that $R'(x)$ and $R'(70)$ could be denoted $MR(x)$ and $MR(70)$, respectively.)

Example 2 The cost to a company to produce x units of a product is given by the cost function $C(x) = -0.035x^2 + 40x + 25$. Compute and interpret both $C(600)$ and $C'(600)$.

Solution Since we are asked for information about $C'(x)$, we compute it first.

$$C'(x) = -0.070x + 40$$

a. Substituting 600 for x in $C(x)$, we get $C(600) = 11{,}425$. This means that when the company produces 600 units of its product, it does so at a cost of \$11,425.

b. Substituting 600 for x in $C'(x)$, we get $C'(600) = -2$. This means that if the company increases the number of units it produces by 1, from 600 to 601, the cost to do so will decrease by approximately \$2. (Notice that $C'(x)$ and $C'(600)$ could be denoted $MC(x)$ and $MC(600)$, respectively.) ■

Example 3 The cost to a company to produce x units of a product is given by the cost function $C(x) = 0.02x^3 - 0.5x^2 + 10x + 150$. The revenue the company expects to realize from the sale of x units of the product is given by the revenue function $R(x) = -1.1x^2 + 41.5x$. If $P(x)$ represents the profit realized on the sale of x units of the product, compute and interpret both $P(20)$ and $P'(20)$.

Solution Since we are asked for information about $P'(x)$, we compute it first. To do so requires that we find the profit function $P(x)$. In business and economics, profit is defined to be the difference between the revenue and the cost. That is,

$$P(x) = R(x) - C(x)$$

In this case,

$$\begin{aligned} P(x) &= -1.1x^2 + 41.5x - (0.02x^3 - 0.5x^2 + 10x + 150) \\ &= -1.1x^2 + 41.5x - 0.02x^3 + 0.5x^2 - 10x - 150 \\ &= -0.02x^3 - 0.6x^2 + 31.5x - 150 \end{aligned}$$

Differentiating, we get $P'(x) = -0.06x^2 - 1.2x + 31.5$.

a. Substituting 20 for x in $P(x)$, we get $P(20) = 80$. This means that when the company produces and sells 20 units of its product, it realizes a profit of \$80.

b. Substituting 20 for x in $P'(x)$, we get $P'(20) = -16.50$. This means that if the company increases the number of units it produces and sells by 1, from 20 to 21, the profit it realizes will decrease by approximately \$16.50. (Notice that $P'(x)$ and $P'(20)$ could be denoted $MP(x)$ and $MP(20)$, respectively.) ■

Profit, Revenue, and Cost Optimization

Businesses wish to optimize their profits, revenues, and costs. We can use the optimization techniques of Section 3.3 to compute these values and the input values that produce them.

Example 4 A company estimates that it can sell 2,000 units each week of its product if it prices each unit at \$10. However, its weekly number of sales will increase by 50 units for each \$0.10 decrease in price. The company has fixed costs of \$800. The cost to make each unit is \$0.80. Find the level of production that maximizes the company's profit if the company must produce and sell between and including 2,000 and 3,500 units.

Solution We use the fact that profit = revenue − cost. To do so requires that we find both the revenue and cost functions. The revenue realized on the sale of a product is found by multiplying the number of units sold by the sales price. Letting x represent the number of units sold and p represent the sales price, $\boldsymbol{R(x) = x \cdot p}$.

But the sales price need not be constant; it can change. As it decreases, the number of units sold increases above 2,000. The sales price and the level of production are related. We need to find an expression that relates the number of units sold to the sales price. Let $p(x)$ represent the sales price at production level x.

The base production level is 2,000 units and the base price is \$10. We can represent price decreases with the expression $10 - p(x)$. The number of units sold above 2,000 is 50 times the number of \$0.10 price decreases. That number is found by finding the difference between the original \$10 price and the new price, $p(x)$, and then determining how many \$0.10 decreases are in that difference. Symbolically, this is

$$\frac{10 - p(x)}{0.10}$$

Then, the quantity sold is

$$x = 2{,}000 + 50 \cdot \frac{10 - p(x)}{0.10}$$

Solving this equation for $p(x)$ gives us the sales price when the number of units sold is x,

$$p(x) = 14 - 0.002x$$

Thus, the revenue function is

$$\begin{aligned} R(x) &= x \cdot p(x) \\ &= x(14 - 0.002x) \\ &= 14x - 0.002x^2 \end{aligned}$$

The cost $C(x)$ of producing x units is the sum of the fixed cost \$800 and the variable cost of \$0.80 per unit, or \$0.80x. Thus,

$$C(x) = 800 + 0.80x$$

Now, we can find the profit function, which is the revenue function minus the cost function.

$$\begin{aligned} P(x) &= R(x) - C(x) \\ &= 14x - 0.002x^2 - (800 + 0.80x) \\ &= 14x - 0.002x^2 - 800 - 0.80x \\ &= -0.002x^2 + 13.20x - 800 \end{aligned}$$

To maximize this function, we compute the derivative, find the values of x for which $P'(x)$ is 0 or undefined, and test for an absolute maximum on the interval [2,000, 3,500].

$$P'(x) = -0.004x + 13.2$$

$P'(x) = 0$ when $x = 3{,}300$. Since $P(x)$ is a polynomial function, it is never undefined. Since 3,300 is in the interval [2,000, 3,500], it is the only critical point. We construct a table to find the absolute maximum using the critical point and the two endpoints. As the table shows, the maximum profit is \$20,980, and it occurs when the level of production is 3,300 units each week.

x	$P(x)$
2,000	17,600
3,300	20,980
3,500	20,900

Table 1

Inventory Cost and the Economic Order Quantity

For retailers to do well, they must be careful about the number of units of a product they have in their inventory over some particular time period. The business can control its inventory in either of two ways. One way is to place one order that will cover the entire anticipated demand. This method is attractive since the company is likely to have enough units in stock to meet anticipated demand. The company will incur only one ordering cost, but it may also incur **carrying costs** such as storage and insurance costs as well as costs due to the possibility of obsolescence, spoilage, or breakage. Carrying costs are sometimes called *holding costs.*

Another way is to place a series of smaller orders throughout the time period. This method is attractive since it will keep carrying costs down. However, this method will make the company susceptible to reordering costs as well the possibility of not having enough units in stock to meet demand.

The **total inventory cost** is the sum of the carrying costs and the order/reorder costs. The problem for a business, then, is to determine the **lot size**—that is, the number of units in an order—that will minimize its total inventory cost. This is a minimization problem for which we can use the optimization techniques of Section 3.3. To do so, we must get a total inventory cost function.

We will analyze this problem by making three assumptions: that the sales of the product are made at a relatively uniform rate over the time period under consideration; that the lot size, x, of each reorder is the same; and that as each inventory in stock falls to 0, another order immediately arrives. We will also make the following representations:

1. $C(x)$ = the total inventory cost for lot size x.
2. $H(x)$ = the carrying cost for lot size x.
3. $R(x)$ = the reordering cost for lot size x.

Since both carrying cost and reorder cost depend on the lot size x, they are functions of x. Therefore, the total inventory cost is a function of x. In symbols, the total inventory cost function is

$$C(x) = H(x) + R(x)$$

and our goal is to minimize it.

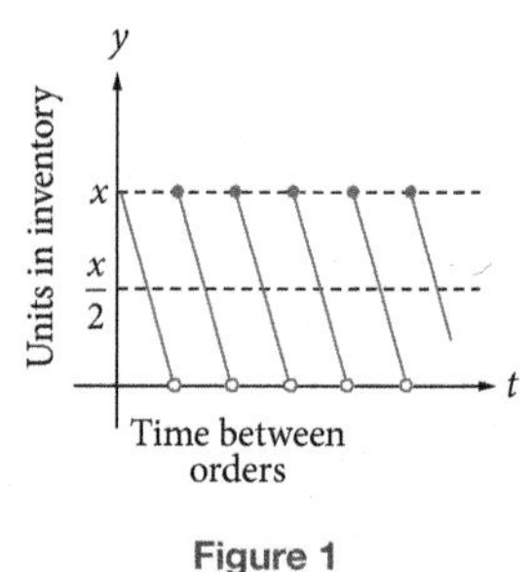

Figure 1

Now, at any time, the largest inventory in stock is x.

Since the units sell at a uniform rate, the average inventory during the time period is $\frac{x}{2}$. Figure 1 illustrates these ideas. To find explicit expressions for $H(x)$ and $R(x)$, we reason as follows.

$$\text{Carrying Costs} = (\text{holding cost per unit}) \cdot (\text{average no. of units})$$

$$H(x) = (\text{holding cost per unit}) \cdot \frac{x}{2}$$

$$\text{Reordering Costs} = (\text{cost per order}) \cdot \left(\frac{\text{no. of units sold during time period}}{\text{lot size}}\right)$$

$$R(x) = (\text{cost per order}) \cdot \left(\frac{\text{no. of units sold during time period}}{x}\right)$$

Then, assembling these expressions together,

$$C(x) = H(x) + R(x)$$

Example 5 A company anticipates selling 3,200 units of its product at a uniform rate over the next year. Each time the company places an order for x units, it is charged a flat fee of \$50. Carrying costs are \$32 per unit per year. How many times should the company reorder each year and what should be the lot size of each order to minimize inventory cost? What is the minimum inventory cost?

Solution Since the company anticipates selling 3,200 units, it can place anywhere from 1 order for 3,200 units to 3,200 orders of 1 unit each. We need to find the number of orders that will minimize inventory costs. To do so, we need an inventory cost function that is valid on the interval [1, 3,200]. We know that the general form of the inventory function is

$$C(x) = H(x) + R(x), \quad \text{where}$$

$$H(x) = (\text{carrying cost per unit}) \cdot \frac{x}{2} \quad \text{and}$$

$$R(x) = (\text{cost per order}) \cdot \left(\frac{\text{number of units sold during time period}}{x}\right)$$

In this case, $H(x) = 32 \cdot \frac{x}{2} = 16x$ and $R(x) = (50) \cdot \frac{3{,}200}{x} = \frac{160{,}000}{x}$

so that $C(x) = 16x + \frac{160{,}000}{x}$, $1 \leq x \leq 3{,}200$.

Now, $C'(x) = 16 - \frac{160{,}000}{x^2}$ and $C''(x) = \frac{320{,}000}{x^3}$

Minimize $C(x)$ by computing $C'(x)$ and determining the x-values for which it is 0 or undefined.

Where is $C'(x) = 0$?

$$16 - \frac{160{,}000}{x^2} = 0$$

$$16x^2 - 160{,}000 = 0$$

$$16(x^2 - 10{,}000) = 0$$

$$16(x + 100)(x - 100) = 0$$

$$x = -100 \text{ or } 100$$

We can reject -100 since it is not inside the interval [1, 3,200]. We can also reject 0 (the value of x that makes $C'(x)$ undefined) for the same reason.

The only critical number is 100. The second derivative test assures us that 100 produces a relative minimum of $C(x)$:

$$C''(100) = \frac{320{,}000}{100^3} > 0 \text{ for every } x \text{ in } [1, 3{,}200]$$

Therefore, to minimize the inventory cost, the company should place $\frac{3{,}200}{100} = 32$ orders each of lot size 100. The minimum cost will be

$$C(100) = 16 \cdot 100 + \frac{160{,}000}{100} = \$3{,}200$$

The graph illustrates that (100, 3,200) is an absolute minimum of $C(x)$.

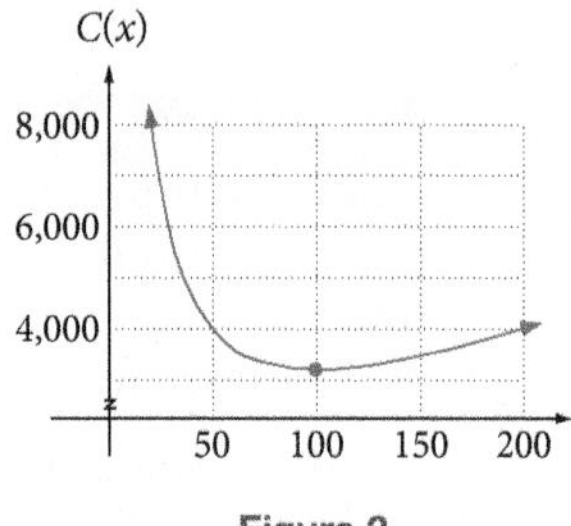

Figure 2

The Economic Order Quantity (EOQ)

The economic order quantity (EOQ) is the order quantity that minimizes total holding and ordering costs for the year. Even if the assumptions of relatively uniform sales, fixed item holding costs, and fixed item ordering and reordering costs don't all hold exactly, the EOQ provides a good estimate of the order quantity that minimizes the total inventory cost.

The basic EOQ formula is

$$C(x) = H(x) + R(x) + P$$

which is the total inventory cost function along with the inventory purchase cost per year function, P.

$$P = (\text{price per unit})(\text{number of units to be purchased in time period})$$

Notice that since P is just the price of a unit times the number of units purchased, it is a constant.

To minimize $C(x)$, find its critical values.

$$C(x) = H(x) + R(x) + P$$

$$= \begin{pmatrix}\text{carrying cost}\\ \text{per unit}\end{pmatrix} \cdot \frac{x}{2} + \begin{pmatrix}\text{cost per}\\ \text{order}\end{pmatrix} \cdot \frac{\begin{matrix}\text{number of units sold}\\ \text{during time period}\end{matrix}}{x} + P$$

$$= \begin{pmatrix}\text{carrying cost}\\ \text{per unit}\end{pmatrix} \cdot \frac{x}{2} + \begin{pmatrix}\text{cost per}\\ \text{order}\end{pmatrix} \cdot \begin{pmatrix}\text{number of units sold}\\ \text{during time period}\end{pmatrix} \cdot x^{-1} + P$$

Then,

$$C'(x) = \begin{pmatrix}\text{carrying cost}\\ \text{per unit}\end{pmatrix} \cdot \frac{1}{2} - 1 \cdot \begin{pmatrix}\text{cost per}\\ \text{order}\end{pmatrix} \cdot \begin{pmatrix}\text{number of units sold}\\ \text{during time period}\end{pmatrix} \cdot x^{-2} + 0$$

$$C'(x) = \begin{pmatrix}\text{carrying cost}\\ \text{per unit}\end{pmatrix} \cdot \frac{1}{2} - \begin{pmatrix}\text{cost per}\\ \text{order}\end{pmatrix} \cdot \frac{\begin{matrix}\text{number of units sold}\\ \text{during time period}\end{matrix}}{x^2}$$

$C'(x) = 0$ when

$$(\text{carrying cost per unit}) \cdot \frac{1}{2} = (\text{cost per order}) \cdot \frac{\text{units sold during time period}}{x^2}$$

Solve for x by multiplying each side of the equation by the least common denominator, $2x^2$.

$$(\text{carrying cost per unit}) \cdot x^2 = 2 \cdot (\text{cost per order}) \cdot (\text{units sold during time period})$$

$$x^2 = \frac{2 \cdot (\text{cost per order}) \cdot (\text{units sold during time period})}{(\text{carrying cost per unit})}$$

EOQ = $$x = \sqrt{\frac{2 \cdot (\text{cost per order}) \cdot (\text{units sold during time period})}{(\text{carrying cost per unit})}}$$

This number is called the EOQ, the Economic Order Quantity.

Example 6 The company in Example 5 anticipated selling 3,200 units of its product at a uniform rate over the next year. Each time the company placed an order for x units, it is to be charged a flat fee of \$50. Carrying costs are \$32 per unit per year. We found that to minimize its inventory cost, the company should place 32 orders of size 100 each.

But now that we have developed it, we can use the formula for EOQ to get the lot size that will minimize inventory cost.

$$\begin{aligned} x &= \sqrt{\frac{2 \cdot 50 \cdot 3{,}200}{32}} \\ &= \sqrt{10{,}000} \\ &= 100 \end{aligned}$$

Elasticity of Demand

There are economic theories about the ways in which households and businesses are likely to respond to financial changes. If two variables, x and y, are related and a change in x causes almost no change in y, then y is not very sensitive to changes in x. If, however, a change in x causes a large change in y, then y is very sensitive to changes in x.

Elasticity

When a change in one variable causes a response in another variable, **elasticity** is a measure of the size of the response. Elasticity measures how sensitive the value of one variable is to the value of another. If a percent change in the input variable causes a relatively *large* change in the output variable, the relationship between the variables is called **elastic.** If a percent change in the input variable causes a relatively *small* change in the output variable, the relationship between the variables is called **inelastic.**

For example, imagine that two companies, Company A and Company B, each increase the price of one of the products its sells by 1%. As price increases, demand tends to decrease so both Company A and Company B experience a decrease in their number of sales. But suppose Company A's 1% price increase deters only a few customers so that its revenue stays the same as that before the price increase. In this case, the percent change in demand is not very sensitive to a change in price. Now suppose Company B's 1% price increase deters many customers so that its revenue is substantially less than that before the price increase. In this case, the percent change in demand is very sensitive to a change in price.

When the variables that are related are demand for a commodity and the price of the commodity, we speak of the **elasticity of demand**. Elasticity of demand is denoted with the lowercase Greek letter η (pronounced *eta*) and is computed by dividing the percent change in the quantity demanded by the percent change in the price. Symbolically, the percent change in q is represented by $\frac{\Delta q}{q}$, and the percent change in p is represented by $\frac{\Delta p}{p}$. Now, since η *is* computed by dividing the percent change in the quantity demanded by the percent change in the price,

$$\eta = \frac{\Delta q}{q} \div \frac{\Delta p}{p}$$

$$= \frac{\Delta q}{q} \cdot \frac{p}{\Delta p}$$

$$= \frac{p}{q} \cdot \frac{\Delta q}{\Delta p}$$

If the price p is a continuous function of the demand q, then

$$\lim_{\Delta p \to 0} \frac{\Delta q}{\Delta p} = \frac{dq}{dp}$$

Then, the **point elasticity of demand**, which is denoted ε (pronounced *epsilon*), is the instantaneous rate of change of the η). That is,

$$\varepsilon = \lim_{\Delta p \to 0} \eta$$

$$\varepsilon = \lim_{\Delta p \to 0} \frac{p}{q} \cdot \frac{\Delta q}{\Delta p}$$

$$\varepsilon = \frac{p}{q} \cdot \lim_{\Delta p \to 0} \frac{\Delta q}{\Delta p}$$

$$\varepsilon = \frac{p}{q} \cdot \frac{dq}{dp}$$

Notice that p and q are always positive (representing price and demand). The derivative $\frac{dq}{dp}$, however, is almost always negative, since it represents the rate of change (or slope) of the demand function, and demand is usually a decreasing function of price. (See Figure 3.) This means that the product $\frac{p}{q} \cdot \frac{dq}{dp}$ will be negative. Since positive numbers are generally more user-friendly, economists insert a negative sign in the definition, making epsilon a number that is positive (or zero).

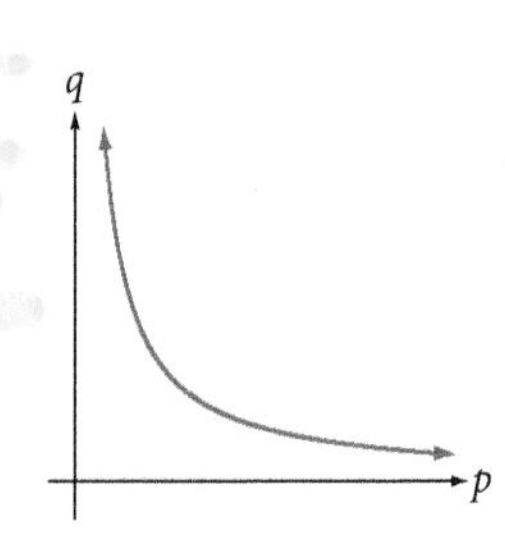

Figure 3

Point Elasticity of Demand

The **point elasticity of demand**, ε, is defined mathematically as

$$\varepsilon = -\frac{p}{q} \cdot \frac{dq}{dp}$$

where q is the original quantity demanded when the original price per unit is p. Point elasticity measures the elasticity at a particular point and is interpreted as the percent change in the demand for a 1% change in the price.

Example 7 A company estimates that the weekly sales q of its product is related to the product's price p by the function

$$q = \frac{4{,}500}{\sqrt[3]{p^2}}$$

where p is in dollars. Currently, each unit of the product is selling for \$27. Determine the point elasticity of demand of this product.

Solution We use the point elasticity of demand formula $\varepsilon = -\frac{p}{q} \cdot \frac{dq}{dp}$ with $p = 27$. To use the formula, however, we need the values of q and $\frac{dq}{dp}$.

First, when $p = 27$, $q = \dfrac{4{,}500}{\sqrt[3]{27^2}} = 500$. Second,

$$\begin{aligned}\frac{dq}{dp} &= \frac{d}{dp}\left[\frac{4{,}500}{\sqrt[3]{p^2}}\right] \\ &= \frac{d}{dp}\left[\frac{4{,}500}{p^{2/3}}\right] \\ &= \frac{d}{dp}\left[4{,}500 \cdot p^{-2/3}\right] \\ &= -3{,}000p^{-5/3} \\ &= \frac{-3{,}000}{p^{5/3}}\end{aligned}$$

When $p = 27$, $\dfrac{dq}{dp} = \dfrac{-3{,}000}{27^{5/3}} \approx -12.35$. Now, we have

$$\begin{aligned}\varepsilon &= -\frac{p}{q} \cdot \frac{dq}{dp} \\ &\approx -\frac{27}{500} \cdot (-12.35) \\ &\approx 0.67 \qquad \left(\frac{0.67}{1} = \frac{0.67\%}{1\%}\right)\end{aligned}$$

Thus $\varepsilon \approx 0.67$.

Interpretation
If the price of each unit of this product is raised by 1%, from \$27 to \$27.27, the demand for the product will *decrease* by approximately 0.67%, from 500 units to approximately 497 units (a relatively small decrease). According to the definition of elasticity we gave previously, this would constitute an *inelastic* relationship between price and demand.

Relating Point Elasticity to Revenues

Point elasticity and revenues are related to each other, and it is possible to derive and state that relationship symbolically. To do so, recall that

$$\text{revenue} = (\text{quantity sold}) \cdot (\text{price per unit})$$

Letting q represent either the quantity sold or the quantity purchased, and p the price per unit, and keeping in mind that the quantity sold or purchased is a function $q(p)$ of the price, we have

$$\begin{aligned} R(p) &= q \cdot p \\ &= q(p) \cdot p \end{aligned}$$

To observe how changes in the price can affect the revenue, we differentiate the product function $R(p) = q(p) \cdot p$ with respect to p.

$$\begin{aligned} R'(p) &= q'(p) \cdot p + q(p) \cdot 1 \\ &= q'(p) \cdot p + q(p) \\ &= q'(p) \cdot p + q \end{aligned}$$

We would like to write this derivative so that it involves ε, the point elasticity. We notice that the derivative involves p, as does the formula for ε. We will solve the formula relating p and ε for p. Since $\varepsilon = -\frac{p}{q} \cdot \frac{dq}{dp}$ and $\frac{dq}{dp} = q'(p)$, solving for p produces

$$p = \frac{-\varepsilon q}{q'(p)}$$

Now, we can rewrite $R'(p) = q'(p) \cdot p + q$ so that it involves ε.

$$\begin{aligned} R'(p) &= q'(p) \cdot p + q \\ R'(p) &= q'(p) \cdot \left(-\frac{\varepsilon q}{q'(p)}\right) + q \\ R'(p) &= -\varepsilon q + q \\ R'(p) &= q(1 - \varepsilon) \end{aligned}$$

Since q represents some number of units, it is always positive. Thus,

1. $R'(p)$ is negative when the quantity $(1 - \varepsilon) < 0$. This will happen when $\varepsilon > 1$.
2. $R'(p)$ is 0 when the quantity $(1 - \varepsilon) = 0$. This will happen when $\varepsilon = 1$.
3. $R'(p)$ is positive when the quantity $(1 - \varepsilon) > 0$. Solving this inequality for ε gives $\varepsilon < 1$. However, since ε is never negative, $0 < \varepsilon < 1$.

This leads us to three interesting cases involving values of ε.

Elastic, Unit Elastic, and Inelastic Demand

1. If $\varepsilon > 1$, then $R'(p) < 0$ means that the total revenue falls with price increases and rises with price decreases. Demand is **elastic.**
2. If $\varepsilon = 1$, then $R'(p) = 0$ means that the total revenue is unaffected by changes in price. Demand is **unit elastic.**
3. If $0 < \varepsilon < 1$, then $R'(p) > 0$ means that the total revenue rises with price increases and falls with price decreases. Demand is **inelastic.**

Example 8

a. In Example 7, we computed ε to be ≈ 0.67. Since $0 < \varepsilon < 1$, demand is inelastic, meaning that a 1% increase in the price of the product, from \$27 to $\$27 + \$27(0.01) = \$27.27$, produces a 0.67% decrease in the demand, but an increase in revenue. Revenue rises because a price increase produces a relatively small decrease in demand. (A decrease from 500 units to $500 + 500(-0.0067) \approx 497$ units is relatively small.) In fact, revenue increases from

$$500 \text{ units} \cdot \frac{\$27}{\text{unit}} = \$13{,}500$$

to

$$497 \text{ units} \cdot \frac{\$27.27}{\text{unit}} = \$13{,}553.19$$

b. Suppose that in Example 7 the quantities q and p were related by the function

$$q = \frac{97{,}533}{\sqrt[5]{p^{24}}}$$

Now with an initial price of \$3 per unit, ε computes to 4.8, a number greater than 1. In this case demand is elastic, meaning that a 1% increase in the price of the product, from \$3 to \$3.03, produces a 4.8% decrease in demand (from 500 units to $500 + 500(-0.048) = 476$ units) and a decrease in revenue from

$$500 \text{ units} \cdot \frac{\$3}{\text{unit}} = \$1{,}500$$

to

$$476 \text{ units} \cdot \frac{\$3.03}{\text{unit}} = \$1{,}442.28$$

Revenue falls because a price increase produces a relatively large decrease in demand. (A decrease from 500 units to 476 units is relatively large.)

c. Finally, suppose that in Example 7 the quantities q and p were related by the function

$$q = \frac{62{,}500}{p}$$

Now with an initial price of $125 per unit, ε computes to 1. In this case demand is unit elastic, meaning that a 1% increase in the price of the product, from $125 to $126.25, produces no change in revenue. Revenue remains at $62,500. Revenue is unaffected because a price increase is offset by a decrease in demand: a $1.25 increase in price is offset by a 5-unit decrease in demand. ■

Using Technology 3.4

Wolfram|Alpha may be able to help you find the point elasticity of demand and how it might affect revenue.

Example 9 Use Wolfram|Alpha to help you find the point elasticity of demand for a product if the demand function for the product is $q = 2{,}500p - p^3$ and each unit of the product is currently selling for $30. Is the demand for this product elastic, unit elastic, or inelastic? How would a 1% increase in price affect revenue?

Solution To compute the point elasticity of demand for this product you need the values of p, q, and $\frac{dq}{dp}$ when $p = 30$.

1. Find q when $p = 30$. Go to www.wolframalpha.com and in the entry field enter

   2500p - p^3 where p = 30

   Wolfram|Alpha responds with

   48,000

   When the price of each unit of the product is $30, the demand is 48,000 units.

2. Find $\frac{dq}{dp}$ when $p = 30$. In the Wolfram|Alpha entry line, type

   d/dp 2500p - p^3 where p = 30

   Wolfram|Alpha responds with

   -200

3. Since

$$\varepsilon = -\frac{p}{q} \cdot \frac{dq}{dp} = \left(-\frac{30}{48{,}000}\right)(-200) = 0.125$$

making $0 < \varepsilon < 1$, the demand for the product is inelastic.

If the price of each unit of this product is raised by 1%, from $30 to $30.30, the demand for the product will decrease by about 0.125% from 48,000 units to about $48{,}000 + (-0.00125)48{,}000 = 47{,}940$ units. Since the demand is inelastic, a 1% increase in price produces a decrease in demand but an increase in revenue. At the original $30 price, the revenue is

$$48{,}000 \text{ units} \cdot \$30/\text{unit} = \$1{,}440{,}000$$

At the new $30.30 price, the revenue is

$$47{,}940 \text{ units} \cdot \$30.30/\text{unit} = \$1{,}452{,}582$$

A 1% increase in price will produce a $12,582 increase in revenue. ■

Getting Ready for Class

After reading through the preceding section, respond in your own words and in complete sentences.

A. When demand is elastic, what effect will a price increase have on revenue? What effect will a price decrease have on revenue?

B. Describe the relationship that exists between marginal revenue and marginal cost at the point where the marginal profit is zero.

C. As we analyzed the inventory control problem, we made an assumption that produced the value $\frac{x}{2}$, where x was the lot size. If that assumption were changed, could we still use $\frac{x}{2}$ in the computations?

Problem Set 3.4

Skills Practice

1. For a particular commodity, the demand function is $q = 140 - 10p$.
   a. Find ε when $p = 10$.
   b. Is demand elastic, inelastic, or unit elastic?

2. For a particular commodity, the demand function is $q = -\frac{2}{5}p + 10$.
   a. Find ε when $p = 20$.
   b. Is demand elastic, inelastic, or unit elastic?

3. For a particular commodity, the demand function is $q = \frac{5{,}000}{\sqrt[5]{p^2}}$.
   a. Find ε when $p = 32$.
   b. Is demand elastic, inelastic, or unit elastic?

4. For a particular commodity, the demand function is $q = \frac{16{,}000}{\sqrt[3]{p^2}}$.
   a. Find ε when $p = 64$.
   b. Is demand elastic, inelastic, or unit elastic?

5. For a particular commodity, the demand function is $q = \frac{1}{4}(400 - p^2)$.
   a. Find ε when $p = 10$.
   b. Is demand elastic, inelastic, or unit elastic?

6. For a particular commodity, the demand function is $q = -\frac{12}{5}p + 1{,}014$.
   a. Find ε when $p = 360$.
   b. Is demand elastic, inelastic, or unit elastic?

7. For a particular commodity, the demand function is $q = \frac{864}{\sqrt{p^3}}$.
   a. Find ε when $p = 36$.
   b. Is demand elastic, inelastic, or unit elastic?

8. For a particular commodity, the demand function is $q = \frac{120{,}000}{p}$.
   a. Find ε when $p = 10$.
   b. Is demand elastic, inelastic, or unit elastic?

9. Suppose for a particular commodity, $\varepsilon = 0.5$ and that currently 40,000 units are selling for \$8 each. How does a 1% increase in price affect the revenue?

10. Suppose for a particular commodity, $\varepsilon = 1.6$ and that currently 800,000 units are selling for \$12 each. How does a 1% increase in price affect the revenue?

11. Suppose for a particular commodity, $\varepsilon = 1.6$ and that currently 150,000 units are selling for \$20 each. How does a 1% increase in price affect the revenue?

12. Suppose for a particular commodity, $\varepsilon = 2.3$ and that currently 8,500 units are selling for \$350 each. How does a 1% increase in price affect the revenue?

Check Your Understanding

13. **Business: Marginal Cost** For a manufacturer, $C(x)$ represents the dollar cost of producing x units of product. Interpret both $C(250) = 12{,}000$ and $C'(250) = -85$.

14. **Business: Marginal Revenue** For a company, $R(x)$ represents the revenue, in thousands of dollars, realized from the expenditure of x thousands of dollars for advertising. Interpret both $R(6) = 125$ and $R'(6) = 16$.

15. **Business: Marginal Profit** For a company, $P(x)$ represents the profit, in thousands of dollars, realized from the sale of a product that sells for x dollars per unit. Interpret both $P(55) = 16$ and $P'(55) = -4$.

16. **Business: Marginal Profit** For a retail store, $P(x)$ represents the dollar profit realized from the sale of a product when it devotes x square feet to the display of the product. Interpret both $P(30) = 4{,}500$ and $P'(30) = 150$.

17. **Business: Inventory Costs** The inventory cost C, in dollars, to a company is related to the lot size of inventory orders x by the function $C(x)$. Interpret both $C(200) = 1{,}500$ and $C'(200) = 300$.

18. **Business: Inventory Costs** The inventory cost C, in dollars, to a company is related to the lot size of inventory orders x by the function $C(x)$. Interpret both $C(2{,}600) = 3{,}500$ and $C'(2{,}600) = 700$.

19. **Business: Elasticity of Demand** A company determines that the point elasticity of demand for one of its products is 2.6. Is this demand elastic, unit elastic, or inelastic? If the price of the product is increased by 1%, will the company's revenue increase, decrease, or remain the same?

20. **Business: Elasticity of Demand** A company determines that the point elasticity of demand for one of its products is 0.22. Is this demand elastic, unit elastic, or inelastic? If the price of the product is increased by 1%, will the company's revenue increase, decrease, or remain the same?

21. **Business: Elasticity of Demand** A company determines that the point elasticity of demand for one of its products is 1. Is this demand elastic, unit elastic, or inelastic? If the price of the product is increased by 1%, will the company's revenue increase, decrease, or remain the same?

Modeling Practice

22. **Business: Marginal Cost** A manufacturer believes that the function

$$C(x) = 0.15x^2 - 18x + 960$$

where $0 \leq x \leq 130$, approximates the dollar cost of producing x units of a product. Compute the marginal cost of producing 35 units. Interpret this value, as well as $C(35)$. What level of production will ~~maximize~~ minimize the manufacturer's costs? What is that cost?

23. **Business: Marginal Cost** A manufacturer believes that the cost function

$$C(x) = \frac{3}{2}x^2 + 45x + 720$$

approximates the dollar cost of producing x units of a product. The manufacturer believes it cannot make a profit when the marginal cost goes beyond \$210. What is the most units the manufacturer can produce and still make a profit? What is the total cost at this level of production?

24. **Business: Marginal Revenue** The revenue function realized by a company by the sale of x units of a product is thought to be $R(x) = -0.0125x^2 + 112.5x$, where $0 \leq x \leq 9{,}000$. Determine the interval(s) on which the revenue is increasing/decreasing. At what level of production is the revenue maximum? What is the total revenue at that level?

25. **Business: Marginal Profit** A retail company estimates that if it spends x thousands of dollars on advertising during the year, it will realize a profit of $P(x)$ dollars, where $P(x) = -0.25x^2 + 80x + 1{,}400$, where $0 \leq x \leq 336$. What is the company's marginal profit at the \$100,000 and \$260,000 advertising levels? What advertising expenditure would you recommend to this company?

26. **Business: Marginal Profit** A company estimates that it can sell 5,000 units each week of its product if it prices each unit at \$20. However, its weekly number of sales will increase by 100 units for each \$0.10 decrease in price. The company has fixed costs of \$1,000. The cost to make each unit is \$1.60. Find the level of production that maximizes the company's profit if the company must produce and sell between and including 5,000 and 20,000 units. What is the maximum profit?

27. **Business: Marginal Profit** A company estimates that it can sell 3,000 units each month of its product if it prices each unit at \$75. However, its monthly number of sales will increase by 20 units for each \$0.25 decrease in price. The company has fixed costs of \$350. The cost to make each unit is \$4.20. Find the level of production that maximizes the company's profit if the company must produce and sell between and including 1,000 and 7,000 units. What is the maximum profit?

28. **Business: Inventory Costs** An auto parts retailer anticipates selling 960 brake pad sets at a uniform rate over the next year. Each time the retailer places an order for x units of brake pad sets, it is charged a flat fee of \$15. Carrying costs are \$8 per unit per year. How many times should the retailer reorder each year and what should be the lot size to minimize inventory costs? What is the minimum inventory cost?

Source: BendixBrakes

29. **Business: Inventory Costs** A retailer anticipates selling 300 fuel and water separators at a uniform rate over the next year. Each time the retailer places an order for x units of separators, it is charged a flat fee of \$15. Carrying costs are \$10 per unit per year. How many times should the retailer reorder each year and what should be the lot size to minimize inventory costs? What is the minimum inventory cost?

Source: WIX Filters

30. **Business: Inventory Costs** KP Engineering anticipates selling 9,000 Cavern Stairwell Controlled Lighting fixtures at a uniform rate over the next year. Each time KP Engineering places an order for x units of fixtures, it is charged a flat fee of \$400. Carrying costs are \$20 per unit per year. The company buyer determines that inventory costs can be minimized at \$10,000 by placing 15 orders of 600 fixtures each. Is the buyer's conclusion accurate? If not, can you offer a conclusion that better reflects the true cost?

31. **Business: Economic Order Quantity** S & K Printing anticipates using 40,000 reams of paper at a uniform rate over the next year. Each time S & K places an order for x units of reams of paper, it is charged a flat fee of $200. Carrying costs are $4 per unit per year. Nick, the company buyer, determines that inventory costs can be minimized at $8,000 by placing 20 orders of 2,000 reams each. Is Nick's conclusion accurate? If not, can you offer a conclusion that better reflects the true cost?

32. **Business: Economic Order Quantity** In Problem 28, you noted that an auto parts retailer anticipates selling 960 brake pad sets at a uniform rate over the next year. Each time the retailer places an order for x units of brake pad sets, it is charged a flat fee of $15. Carrying costs are $8 per unit per year. Use the formula for the EOQ to verify the result you obtained in Problem 28.

33. **Business: Economic Order Quantity** In Problem 29, you noted that a retailer anticipates selling 300 fuel and water separators at a uniform rate over the next year. Each time the retailer places an order for x units of separators, it is charged a flat fee of $15. Carrying costs are $10 per unit per year. Use the formula for the EOQ to verify the result you obtained in Problem 29.

34. **Business: Elasticity of Demand** A company estimates that the weekly sales q of its product is related to the product's price p by the function

$$q = \frac{2{,}160}{\sqrt[5]{p^3}}$$

where p is in dollars. Currently, each unit of the product is selling for $32. Determine the point elasticity of demand of this product, state whether the demand for this product is elastic, inelastic, or unit elastic, and interpret the meaning of this number in terms of the company's revenue.

35. **Business: Elasticity of Demand** A company estimates that the weekly sales q of its product is related to the product's price p by the function

$$q = \frac{14{,}625}{\sqrt[4]{p^{11}}}$$

where p is in dollars. Currently, each unit of the product is selling for $4. Determine the point elasticity of demand of this product, state whether the demand for this product is elastic, inelastic, or unit elastic, and interpret the meaning of this number in terms of the company's revenue.

36. **Business: Elasticity of Demand** Suppose that a company is currently selling 700 units of a product at $16 per unit. Suppose also that the point elasticity of demand for this product is 2.6. By how much, if any, does the revenue increase or decrease if the price is increased by 1%?

37. **Business: Elasticity of Demand** Suppose that a company is currently selling 35,000 units of a product at $160 per unit. Suppose also that the point elasticity of demand for this product is 8.8. By how much, if any, does the revenue increase or decrease if the price is increased by 1%?

38. **Business: Elasticity of Demand** Suppose that a company is currently selling 55,000 units of a product at \$38 per unit. Suppose also that the point elasticity of demand for this product is 0.39. By how much, if any, does the revenue increase or decrease if the price is increased by 1%?

Using Technology Exercises

39. Find the elasticity of demand for a product if the demand function for the product is $q = 4{,}500p - 2p^3$ and each unit of the product is currently selling for \$30. Is the demand for this product elastic, unit elastic, or inelastic?

40. Find the elasticity of demand for a product if the demand function for the product is

$$q = \frac{(300p + 8)^{5/4}}{(p + 8)^{3/2}}$$

and each unit of the product is currently selling for \$50. Is the demand for this product elastic, unit elastic, or inelastic?

41. Find the elasticity of demand for a product if, currently, each unit of the product is selling for \$8 and the demand function is $q = 2{,}160p^{1.1}$. Is the demand for this product elastic, unit elastic, or inelastic?

42. In an effort to increase revenue, a state's automobile registration agency is considering increasing the fee for vanity license plates by 1%, from \$100 per year to \$101 per year. The price elasticity of demand for vanity plates is 1.6. Currently, there are 85,000 people with vanity license plates in the state. What will be the effect on revenue if the increase in fee is instituted?

43. The administrator of a county's museum is considering increasing the price of an annual pass by 1%, from \$50 to \$50.50. At the current price, the museum sells 2000 annual passes and the point of elasticity is 0.73. What effect would the increase in price have on the museum's annual pass revenue?

Getting Ready for the Next Section

44. Power functions have the form x^b, where b is a constant. Give an example of a power function.

45. Compute both 2^{-3} and 2^3.

46. To one decimal place, find the value of e^{2t} when $t = 3$.

47. To the nearest whole number, find the value of $5e^{0.06931t}$ when $t = 120$.

48. To two decimal places, find the value of $35 \cdot e^{-0.11(p-1)}$ when $p = 6$.

49. To two decimal places, find the value of $2{,}000\left(1 + \frac{0.08}{4}\right)^{4t}$ when $t = 15$.

Spotlight on **Success**

Student Instructor Octabio

The best thing about the future is that it comes one day at a time.
—Abraham Lincoln

For my family, education was always the way to go. Education would move us ahead, but the path through education was not always clear. My parents had immigrated to this country and had not had the opportunity to continue in education. Luckily though, with the help of school counselors and the A.V.I.D. (Advancement Via Individual Determination) program in our school district, my older sister and brother were able to get into some of their top colleges. Later, with A.V.I.D. and the guidance of my siblings I was able to take the right courses and was lucky enough to be accepted at my dream university.

Math has been my favorite subject ever since I can remember. When I got to Calculus, however, I struggled more than I had with previous levels of math. This struggle initially stopped me from enjoying the class, but as my understanding grew, I became more and more interested in seeing how things connected. The connection I found most astounding was that between acceleration, velocity, and distance covered by an object. If we have a formula for one of these, we can make a simple connection with the other two. Also, the relationship between area underneath a curve and how this relates to the value of the integral blew my mind. This was just a little taste though as to how Calculus can be connected directly to many areas.

Chapter 3 Summary

EXAMPLES

Function Behavior Noted from the First Derivative [3.1]

Using the first derivative, we can determine, if any exist, the critical values of the function, and from those, we can locate

1. The intervals upon which a function increases or decreases, and
2. The relative minimum and relative maximum points, if they exist, of the function.

1. For the function

$$f(x) = \frac{1}{3}x^3 - 6x^2 + 32x$$

$$f'(x) = x^2 - 12x + 32$$
$$= (x - 4)(x - 8)$$

The critical values are $x = 4$ and $x = 8$.

f' $(-)(-) = (+)$ $(+)(-) = (-)$ $(+)(+) = (+)$

incr 4 decr 8 incr x

The function increases on $(-\infty, 4)$ and $(8, \infty)$ and decreases on $(4, 8)$.

Increasing/Decreasing Functions [3.1]

A function $f(x)$ is increasing on an interval (a, b) if the graph $f(x)$ rises through (a, b).

A function $f(x)$ is decreasing on an interval (a, b) if the graph $f(x)$ falls through (a, b).

If $f'(x) > 0$ for every value of x in (a, b), then $f(x)$ increases through (a, b). Conversely, if $f(x)$ increases through (a, b), then $f'(x) > 0$ for every value of x in (a, b).

If $f'(x) < 0$ for every value of x in (a, b), then $f(x)$ decreases through (a, b). Conversely, if $f(x)$ decreases through (a, b), then $f'(x) < 0$ for every value of x in (a, b).

If $f'(x) = 0$ for every value of x in (a, b), then $f(x)$ is neither increasing nor decreasing on (a, b), but is constant on (a, b). Conversely, if $f(x)$ is constant on (a, b), then $f'(x) = 0$ for every value of x in (a, b).

2. The function

$$f(x) = x^2 - 8x + 19$$

decreases on $(-\infty, 4)$ and increases on $(4, \infty)$.

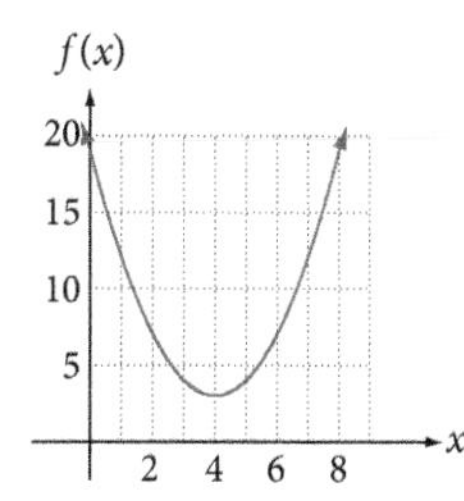

Critical Values [3.1]

Critical values of a function f are values for which f is defined but for which either $f'(x) = 0$ or $f'(x)$ is undefined. The values are called critical values because they are critical to the accurate description of the function f.

3. For the function

$$f(x) = x^2 - 8x + 19$$

$f'(x) = 2x - 8 = 2(x - 4)$. Since $f'(x) = 0$ when $x = 4$, the number 4 is a critical value of $f(x)$.

For the function

$$f(x) = (x - 5)^{2/3}$$

$f'(x) = \frac{2}{3(x - 5)^{1/3}}$. Since $f'(x)$ is undefined when $x = 5$, the number 5 is a critical value of $f(x)$.

Relative Maxima and Relative Minima [3.1]

A point $(a, f(a))$ is called a relative maximum if for all values of x in a small neighborhood about $x = a$, the output value $f(a)$ is the largest.

A point $(a, f(a))$ is called a relative minimum if for all values of x in a small neighborhood about $x = a$, the output value $f(a)$ is the smallest.

Relative maxima and relative minima describe a function's local behavior, that is, the behavior of the function near or at particular input values.

4. For the function

$$f(x) = -x^2 + 10x - 21$$

the point (5, 4) is a relative maximum because for all the x-values in a small neighborhood near $x = 5$, the output value $f(5) = 4$ is the largest. For any x-value near $x = 5$, the output values are smaller than 4.

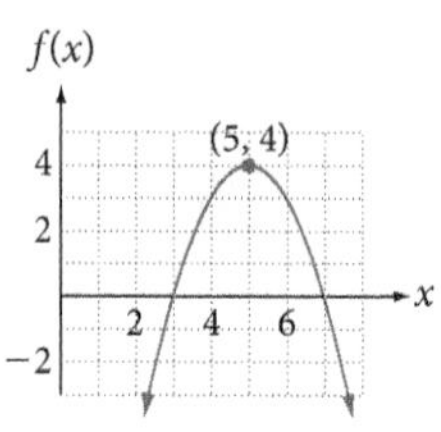

For the function

$$f(x) = x^2 - 6x + 11$$

the point (3, 2) is a relative minimum because for all the x-values in a small neighborhood near $x = 3$, the output value $f(3) = 2$ is the smallest. For any x-value near $x = 3$, the output values are larger than 2.

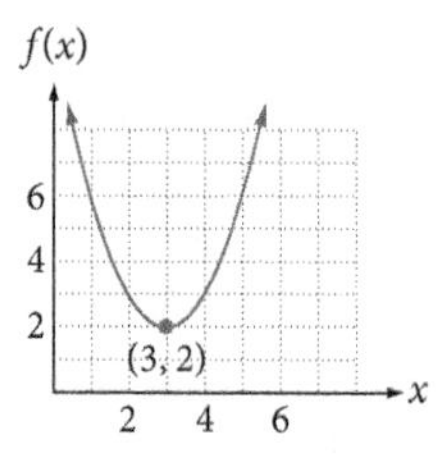

First Derivative Test [3.1]

If c is a critical value of the function $f(x)$, and

1. If $f(x)$ is increasing immediately to the left of c and decreasing immediately to the right of c, then the critical value c produces a relative maximum. That is if $f'(x) > 0$ to the left of c and $f'(x) < 0$ to the right of c, then the critical value c produces a relative maximum.

2. If $f(x)$ is decreasing immediately to the left of c and increasing immediately to the right of c, then the critical value c produces a relative minimum. That is if $f'(x) < 0$ to the left of c and $f'(x) > 0$ to the right of c, then the critical value c produces a relative minimum.

5. $f(x)$

(5, 4)

$f'(x) > 0$ for every x here

$f'(x) < 0$ for every x here

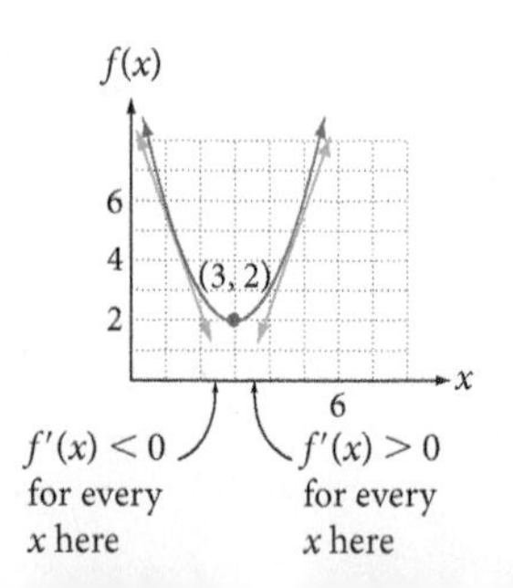

Absolute Extrema [3.1]

Relative maxima and relative minima relate to a function's local behavior, that is, its behavior near or at particular input values. Absolute maxima and absolute minima relate to a function's global behavior, that is, the function's behavior over its entire domain. Points that are absolute maxima or absolute minima are also referred to as global maxima or global minima.

6. On the interval [0, 6], the function

$$f(x) = \frac{1}{3}x^3 - \frac{7}{2}x^2 + 10x - 5$$

has an absolute minimum at $(0, -5)$ and an absolute maximum at $\left(2, \frac{11}{3}\right)$. The point $\left(2, \frac{11}{3}\right)$ is also a relative maximum.

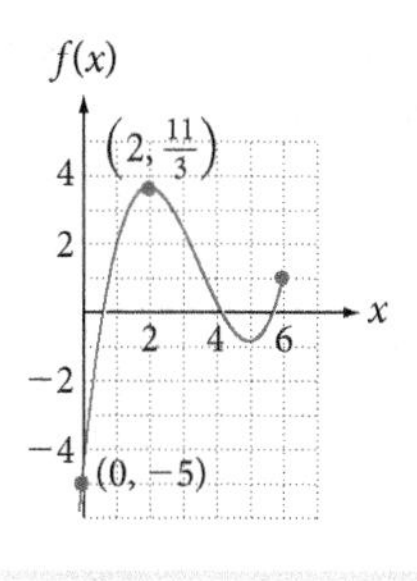

Function Behavior Noted from the Second Derivative [3.2]

Using the second derivative, we can determine, if any exist, the hypercritical values of the function, and from those, we can locate

1. The intervals upon which the function is concave upward or downward, and
2. The points of inflection, if they exist, of the function.

7. For the function

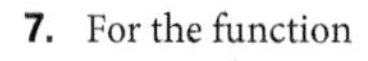

$$f(x) = \frac{1}{3}x^3 - 6x^2 + 32x$$

$$f''(x) = 2x - 12 = 2(x - 6)$$

The hypercritical value is $x = 6$.

$f''(x)$ (−) (+)

Concave downward | Concave upward

The function is concave downward on $(-\infty, 6)$ and concave upward on $(6, \infty)$.

Concavity and Points of Inflection [3.2]

A function is concave upward on an interval (a, b) if its second derivative is positive through the interval. That is, a function $f(x)$ is concave upward on the interval (a, b) if $f''(x) > 0$. Curves that are concave upward open upward.

A function is concave downward on an interval (a, b) if its second derivative is negative through the interval. That is, a function $f(x)$ is concave downward on the interval (a, b) if $f''(x) < 0$. Curves that are concave downward open downward.

A point on a curve where the concavity changes is called a point of inflection and these occur at values for which $f''(x) = 0$ or $f''(x)$ is undefined. Input values that produce points of inflection are called hypercritical values.

8. The function

$$f(x) = x^3 - 15x^2 + 66x - 80$$

is concave upward on $(5, \infty)$ because

$$f''(x) = 6x - 30 = 6(x - 5) > 0$$

when $x > 5$. It is concave downward on $(-\infty, 5)$ because

$$f''(x) = 6x - 30 = 6(x - 5) < 0$$

when $x < 5$. Since

$$f''(x) = 6x - 30 = 6(x - 5) = 0$$

when $x = 5$, the number 5 is a hypercritical value of $f(x)$ and because there is a change in concavity from downward to upward at $x = 5$, the point $(5, 0)$ is a point of inflection.

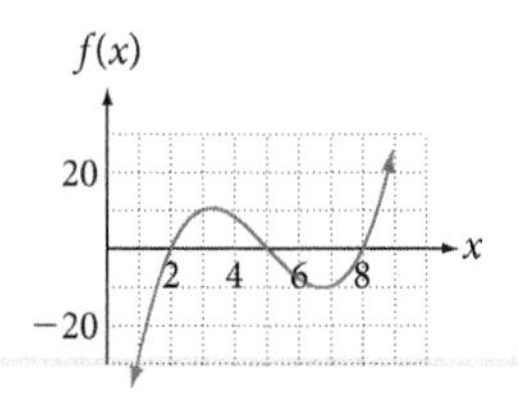

The First and Second Derivatives Together [3.2]

We can use information provided by both the first and second derivatives together to describe how a function behaves over some interval. For example, over an interval, a function could

1. increase at an increasing rate. Figure 1a illustrates this behavior.
2. increase at a decreasing rate. Figure 1b illustrates this behavior.
3. decrease at a decreasing rate. Figure 1c illustrates this behavior.
4. decrease at an increasing rate. Figure 1d illustrates this behavior.

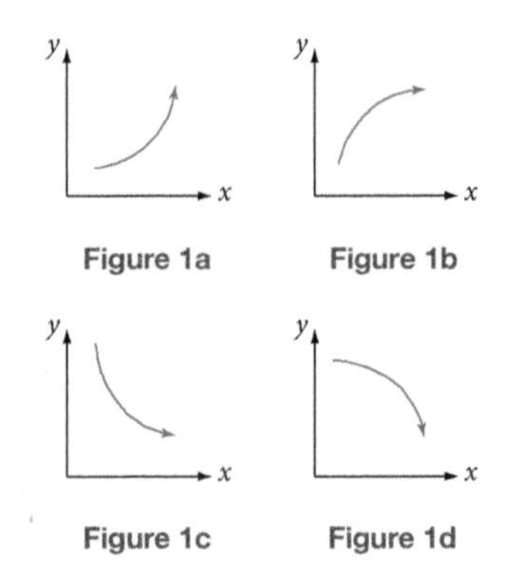

The Second Derivative Test [3.2]

If a function $f(x)$ is such that both $f'(x)$ and $f''(x)$ exist (can be computed), and the number c is a critical value of f, then

1. If $f''(c) > 0$, then f has a relative minimum at $x = c$.
2. If $f''(c) < 0$, then f has a relative maximum at $x = c$.
3. If either $f''(c) = 0$ or $f''(c)$ is undefined, the test fails and cannot be used to obtain information about the relative extrema of f at $x = c$.

Profit Optimization [3.4]

Use the fact that profit = revenue − cost to assist you in determining the production level that maximizes a company's profit. Find revenue and cost functions using the following equation,

$R(x) = x \cdot p$

The Total Inventory Cost [3.4]

9. A company anticipates selling 5,500 units of its product at a uniform rate over the next year. When the company places an order for x units, it is charged a flat fee of $40. Carrying costs are $26 per unit per year. How many times should the company reorder each year and what should be the lot size of each order to minimize inventory cost? What is the minimum inventory cost?

$$C(x) = H(x) + R(x)$$
$$= 26 \cdot \frac{x}{2} + (40) \cdot \frac{5{,}500}{x}$$
$$= 13x + \frac{220{,}000}{x}$$

Then

$$C'(x) = 13 - \frac{220{,}000}{x^2}$$

and $C'(x) = 0$ when $x \approx 130$.

To minimize inventory costs, the company should place $\frac{5{,}500}{130} \approx 42$ orders each of lot size 130.

The **total inventory cost** is the sum of the carrying costs and the order/reorder costs. Businesses want to determine the number of units in an order (lot size) that will minimize the total inventory cost function.

We make three assumptions when working with inventory problems.

1. The sales of the product are made at a relatively uniform rate over the time period under consideration;
2. the lot size, x, of each reorder is the same; and
3. as each inventory in stock falls to 0, another order immediately arrives.

We use the following representations:

1. $C(x)$ = the total inventory cost for lot size x
2. $H(x)$ = the carrying cost for lot size x
3. $R(x)$ = the reordering cost for lot size x

The total inventory cost function is

$$C(x) = H(x) + R(x)$$

$$\text{Carrying Costs} = (\text{holding cost per unit}) \cdot (\text{average no. of units})$$

$$H(x) = (\text{holding cost per unit}) \cdot \frac{x}{2}$$

$$\text{Reordering Costs} = (\text{cost per order}) \cdot \left(\frac{\text{no. of units sold during time period}}{\text{lot size}}\right)$$

$$R(x) = (\text{cost per order}) \cdot \left(\frac{\text{no. of units sold during time period}}{x}\right)$$

Then, assembling these expressions together,

$$C(x) = H(x) + R(x)$$

The Economic Order Quantity [3.4]

10. For Example 9 on the previous page,

$$x = \sqrt{\frac{2 \cdot 40 \cdot 5{,}500}{26}} \approx 130$$

The economic order quantity (EOQ) is the order quantity that minimizes total holding and ordering costs for the year. The basic EOQ formula is

$$C(x) = H(x) + R(x) + P$$

which is the total inventory cost function along with the inventory purchase cost per year function, P.

$$P = \text{(price per unit)(number of units to be purchased in time period)}$$

$C(x)$ is minimum when

$$x = \sqrt{\frac{2 \cdot \text{(cost per order)} \cdot \text{(units sold during time period)}}{\text{(carrying cost per unit)}}}$$

Point Elasticity of Demand [3.4]

The **point elasticity of demand,** ε, is defined mathematically as

$$\varepsilon = -\frac{p}{q} \cdot \frac{dq}{dp}$$

where q is the original quantity demanded when the original price per unit is p. Point elasticity measures the elasticity at a particular point and is interpreted as the percent change in the demand for a 1% change in the price.

Elastic, Unit Elastic, and Inelastic Demand [3.4]

1. If $\varepsilon > 1$, then $R'(p) < 0$ means that the total revenue falls with price increases and rises with price decreases. Demand is **elastic.**
2. If $\varepsilon = 1$, then $R'(p) = 0$ means that the total revenue is unaffected by changes in price. Demand is **unit elastic.**
3. If $0 < \varepsilon < 1$, then $R'(p) > 0$ means that the total revenue rises with price increases and falls with price decreases. Demand is **inelastic.**

Chapter 3 Test

1. Specify all the critical values, all the intervals on which the function $f(x)$ is increasing, and all the intervals on which the function is decreasing.

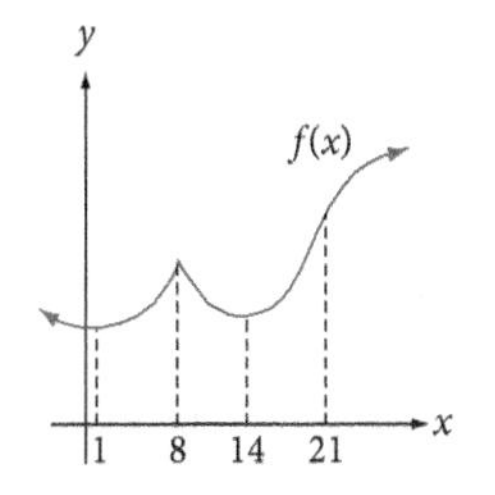

2. Specify all the hypercritical values, all the intervals on which the function $f(x)$ (as illustrated in the previous problem) is concave upward, and all the intervals on which the function is concave downward.

3. Specify any relative and absolute extrema of $f(x)$ as illustrated below.

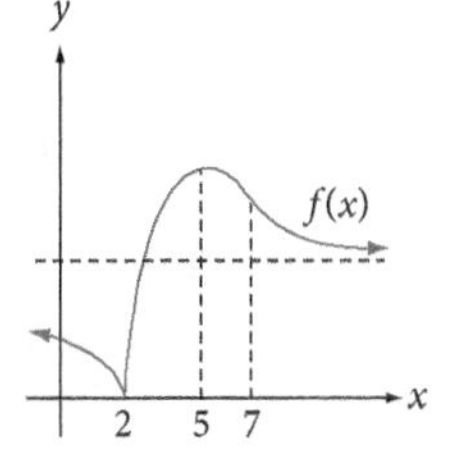

For Problems 4-5, classify each curve as concave upward or concave downward.

4. A curve that is increasing at a decreasing rate.

5. A curve that is decreasing at an increasing rate.

6. Construct sign charts for the first and second derivatives for the graph below.

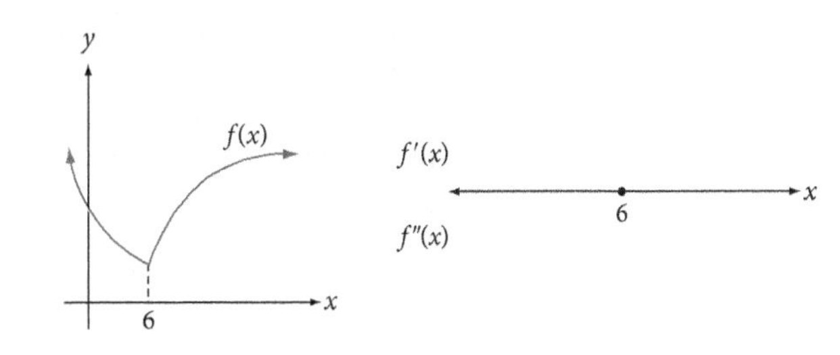

7. **Business: Packaging** A box with no top is to be made from a square piece of cardboard by cutting out squares of equal size from each corner and folding up the sides. Approximately how long should the cut be so that the volume of the box is a maximum? What is the approximate maximum volume? The function relating the length of the cut x to the volume of the box V is

$$V(x) = 625x - 100x^2 + 4x^3$$

The graph of $V(x)$ appears in the figure on the following page.

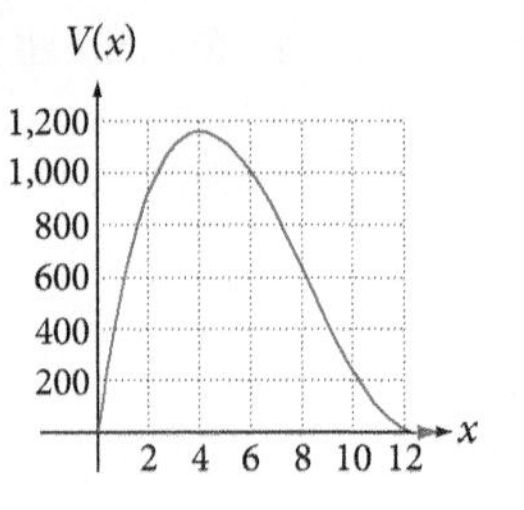

8. Construct a summary table for the graph.

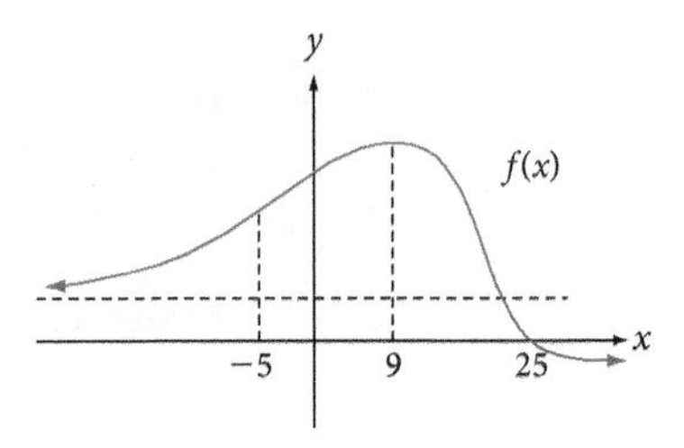

9. Economics: Investment The graph illustrated below displays the relationship between the amount in an account A when an initial investment of \$25,000 is compounded continuously and the number t of years for which the investment is made. In terms of the time and amount of money in the account, describe the behavior of this function.

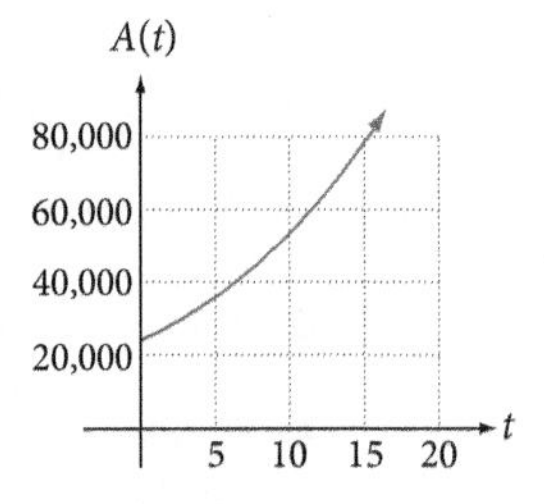

10. Medicine: Antibody Production The graph illustrated below displays the relationship between time t, in hours, after the injection of a medication into the body and the rate of production r, in units per hour, of antibodies. In terms of the time and rate of production of antibodies, describe the behavior of this function.

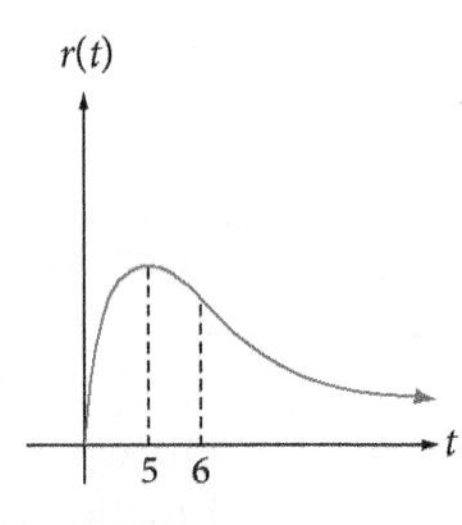

11. Find, if any exist, the critical values of $f(x) = 12x^2 - 17x - 5$.

12. Find, if any exist, the hypercritical values of $f(x) = \frac{1}{12}x^4 - \frac{7}{6}x^3 + 5x - 8$.

13. A function f possesses three critical values, $x = 0, -2$, and 2. The function's second derivative is $f''(x) = 4(x^2 - 2)$. Determine if the critical values produce relative maxima, relative minima, or neither.

14. Determine the intervals upon which $f(x) = x^3 - 5x^2 - 36x + 4$ is increasing or decreasing.

15. Find, if any exist, the relative extrema of the function

$$f(x) = x^4 - 8x^3 - 32x^2 + 20.$$

16. Find, if they exist, the absolute maximum and absolute minimum of the function $f(x) = \frac{5}{x-4}$ on $[5, \infty)$.

17. Determine the intervals upon which $f(x) = (2x + 2)^3$ is concave upward or concave downward.

18. A wholesale store can sell 40 ceramic filters each week when it prices them at \$100 each. The manager estimates that for each \$5 reduction in price, he can sell two more filters. The filters cost the store \$60 each. Find the best price and the quantity that will maximize the store's profit. What is that profit?

19. The inventory cost C, in dollars, to a company is related to the lot size of orders x by the function $C(x)$. Interpret both $C(370) = 5{,}000$ and $C'(370) = 200$.

20. Suppose a company is currently selling 1,000 units of a product at \$40 per unit. Suppose also, that the point elasticity of demand for this product is 0.74. If the price per unit is increased by 1%, will the company's revenue increase or decrease, and by how much?

21. The total monthly cost, in dollars, incurred by the Austenitic Stainless Steel manufacturing company is modeled by

$$C(x) = -0.003x^2 + 5x + 12{,}000 \quad 0 \le x \le 2{,}400$$

a. What is the actual cost of producing the 401st unit and the 2,001st unit?

b. What is the marginal cost for $x = 400$ and $x = 2{,}000$?

22. **Economics: Marginal Utility** Economists define the **marginal utility** of a commodity as the additional satisfaction or benefit (utility) that a consumer derives from buying an additional unit of a commodity or service. The total utility of a commodity typically increases as more of the commodity is consumed, but the marginal utility usually decreases with each additional increase in the consumption of the commodity. For example, suppose the commodity is a chocolate bar. After eating one chocolate bar, both your total utility and marginal utility are probably high. But, as you eat more and more bars, the pleasure you get from each additional bar will likely be less than the pleasure you got from eating the previous bar. The marginal utility of the chocolate bar is decreasing. Typically, for any commodity, total utility increases and marginal utility decreases. If $U(x)$ represents utility of a commodity when x units of it are consumed, make a statement about the signs of the derivatives $U'(x)$ and $U''(x)$.

Source: Encyclopedia Britannica

23. A rectangular page of paper is to contain 24 square inches of photographs. The top and bottom margins are to be 1 inch each and the left and right side margins are to be 1½ inches each. What should be the dimensions of the page so that the area of the page is a minimum?

24. A company anticipates selling 15,000 units of its product at a uniform rate over the next year. Each time the company places an order for x units, it is to be charged a flat fee of \$500. Carrying costs are \$15 per unit per year. Use the EOQ formula to find the lot size that will minimize the inventory cost.

The Natural Exponential and Logarithmic Functions

4

Chapter Outline

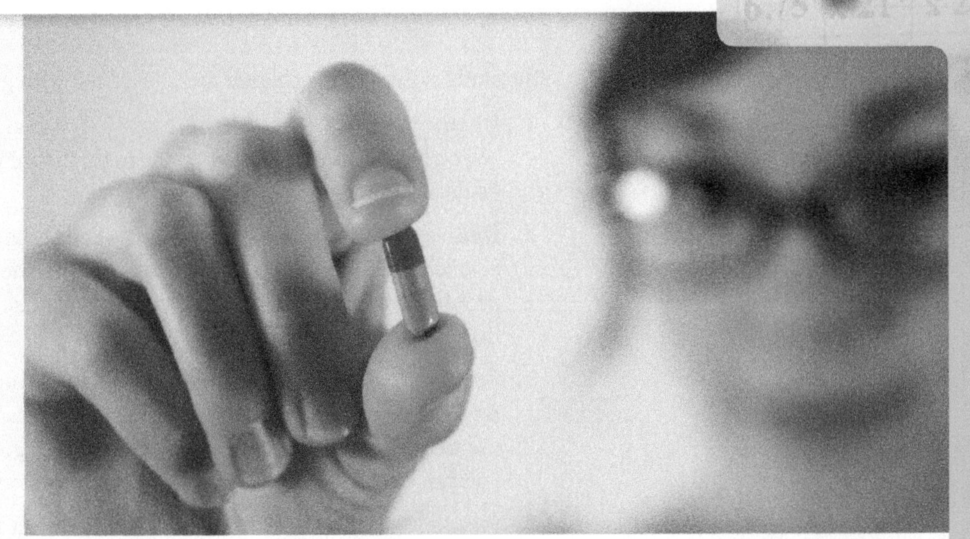

© LoriCaryn/iStockPhoto

When they are introduced into the bloodstream, drugs decay and, over time, eventually are metabolized out of the body. Some drugs decay in specific, predictable ways. You can imagine that once a person has taken a drug, it could be important to know how much of the drug is in the bloodstream after a particular amount of time.

The following table and diagram show how the concentration of a popular antidepressant changes over time once the patient stops taking it. In this particular case, the concentration in the patient's system is 80 ng/mL (nanograms per milliliter) when the patient stops taking the antidepressant, and the half-life of the antidepressant is 5 days.

Note When you see this icon next to an example or problem in this chapter, you will know that we are using the topics in this chapter to model situations in the world around us.

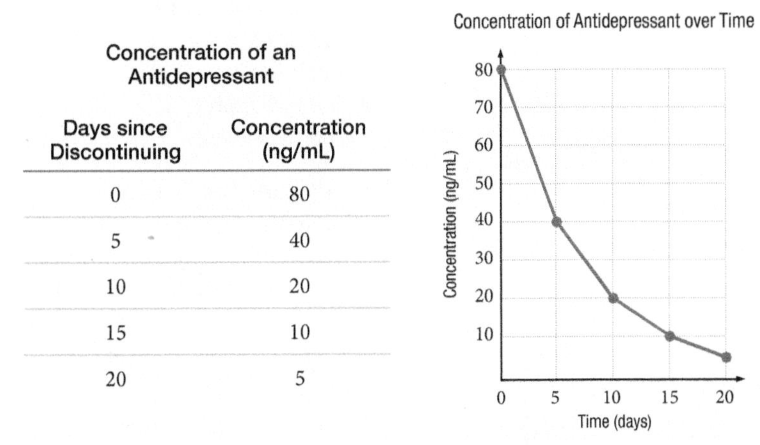

Concentration of an Antidepressant

Days since Discontinuing	Concentration (ng/mL)
0	80
5	40
10	20
15	10
20	5

The half-life of a medication is an indicator of how quickly the medication is eliminated from a person's system: Medications with a long half-life are eliminated slowly, whereas those with a short half-life are eliminated more quickly. Half-life is the key to constructing the preceding table and graph. When you are finished with this chapter, you will be able to go beyond a table and graph and use the half-life of a medication to construct a mathematical model for its decay.

Study Skills

The study skills for this chapter are concerned with getting ready to take an exam.

1. **Getting Ready to Take an Exam** Try to arrange your daily study habits so you have little studying to do the night before your next exam. The next two goals will help you achieve goal number 1.
2. **Review with the Exam in Mind** You should review material that will be covered on the next exam every day. Your review should consist of working problems. Preferably, the problems you work should be problems from your list of difficult problems.
3. **Continue to List Difficult Problems** You should continue to list and rework the problems that give you the most difficulty. It is this list that you will use to study for the next exam. Your goal is to go into the next exam knowing you can successfully work any problem from your list of hard problems.
4. **Pay Attention to Instructions** Taking a test is different from doing homework. When you take a test, the problems will be mixed up. When you do your homework, you usually work a number of similar problems. Sometimes students who do well on their homework become confused when they see the same problems on a test, because they have not paid attention to the instructions on their homework. For example, suppose you see the equation $y = 3x - 2$ on your next test. By itself, the equation is simply a statement. There isn't anything to do unless the equation is accompanied by instructions. Each of the following is a valid instruction with respect to the equation $y = 3x - 2$ and the result of applying the instructions will be different in each case:

   - Find x when y is 10.
   - Solve for x.
   - Graph the equation.
   - Find the intercepts.
   - Find the slope.

   There are many things to do with the equation. If you train yourself to pay attention to the instructions that accompany a problem as you work through the assigned problems, you will not find yourself confused about what to do with a problem when you see it on a test.

The Exponential Functions

4.1

Imagine that at a particular time, a kitchen counter has on it 5 bacteria and that this type of bacteria doubles in number every 10 minutes. How many bacteria could be predicted to be on the counter 2 hours later? Practical problems such as this require a model different from the polynomial models with which we are familiar. In this section we investigate a new type of function, the exponential function, a function that models a great many practical problems.

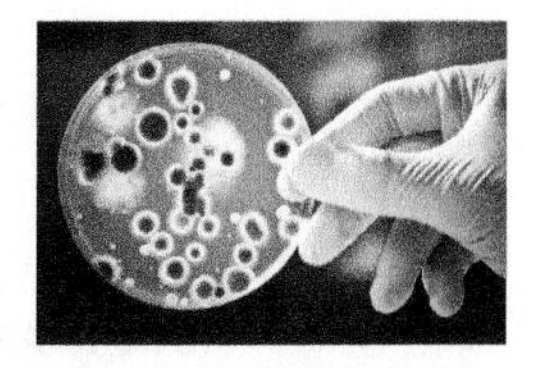

To this point in our study of calculus, we have worked with only **algebraic functions**; that is, functions that are created using the algebraic operations of addition, subtraction, multiplication, division, and powers. The functions

$$f(x) = 3x^2 + 5x - 2, \quad f(x) = \frac{x^2 + 5}{x^3 - 4}, \quad f(x) = \left(\frac{2x + 1}{x + 5}\right)^{2/3}$$

are examples of algebraic functions. In particular, the power function $f(x) = x^b$, where x is a variable and b is a rational number, is an algebraic function. In a power function, the base is a variable and the power is a constant.

We now turn our attention to another type of function that is extremely important in both pure and applied mathematics. This function is a non-algebraic function called the **exponential function.**

Algebraic Functions Compared with Exponential Functions

We begin by defining the exponential function, and we then distinguish it from the power function.

Exponential Function

The function $f(x) = b^x$, where $b > 0$, $b \neq 1$, and x is a variable, is the **exponential function**.

Note that the power function $f(x) = x^b$ and the exponential function $f(x) = b^x$ are two entirely distinct functions.

Power Function	Exponential Function
$f(x) = x^b$	$f(x) = b^x$
Form: $(\text{variable})^{(\text{constant})}$	Form: $(\text{constant})^{(\text{variable})}$

It is also important to note the restriction on the base b of the exponential function. The base cannot be just any constant. It must be greater than 0 but not equal to 1. (The reason for this restriction will be made evident at the end of this section.)

Graphs of Exponential Functions

The exponential function is used to describe relationships in learning theory, population growth and decay, finance and economics, medicine, archaeology, ecology, psychology, sociology, chemistry, biology, physics, mathematics, and other fields.

Because the exponential function is so useful and occurs so often, it is important to have a good "feel" for its behavior. To examine the behavior of the exponential function, we will examine the various forms of its graph. We will begin by looking at two specific examples,

$$f(x) = 2^x \quad \text{and} \quad g(x) = \left(\tfrac{1}{2}\right)^x$$

Example 1 Use tables to graph $f(x) = 2^x$ and $g(x) = \left(\frac{1}{2}\right)^x$.

Solution To construct the graphs of $f(x) = 2^x$ and $g(x) = \left(\frac{1}{2}\right)^x$, we build tables of values as in Table 1, plot the corresponding points, then connect those points with a smooth, continuous curve. To connect the points with a smooth, continuous curve, we will assume (correctly and without proof) that b^x is defined for all real number exponents and not just the rational exponents developed in elementary and intermediate algebra courses.

x	$f(x) = 2^x$	$g(x) = \left(\frac{1}{2}\right)^x$
−3	1/8	8
−2	1/4	4
−1	1/2	2
0	1	1
1	2	1/2
2	4	1/4
3	8	1/8

Table 1

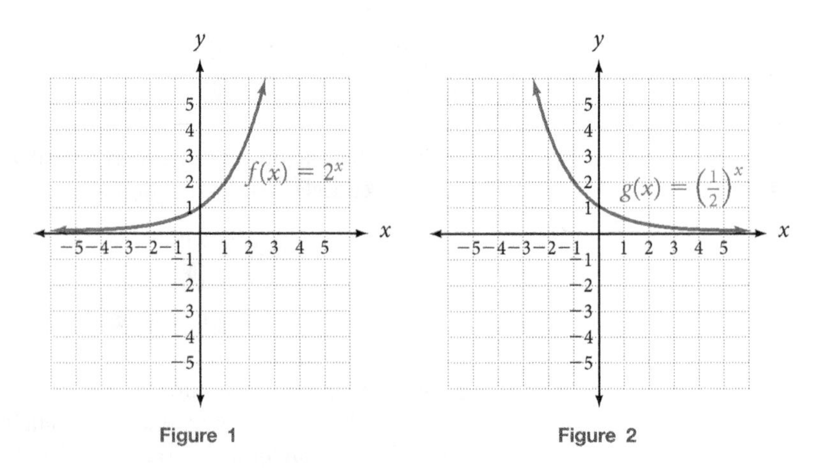

Figure 1 **Figure 2**

In Figure 1, we see that the base is greater than 1 and the function increases as x increases. In Figure 2, we see that the base is between 0 and 1 and that the function decreases as x increases. These graphs illustrate the general forms of the exponential function.

Behaviors of Exponential Functions

If the base is greater than 1 ($b > 1$), the output variable increases as the input variable increases, or geometrically, the curve rises as we look left to right.

Since the function $f(x) = b^x$, $b > 1$, *increases* as x increases, this function is referred to as a **growth** function. (See Figure 3.)

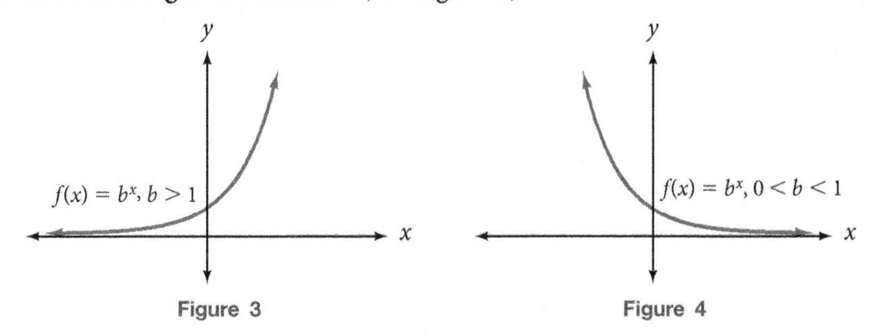

Figure 3

Figure 4

If the base is between 0 and 1 ($0 < b < 1$), the output variable decreases as the input variable increases, or geometrically, the curve falls as we look left to right.

Since the function $f(x) = b^x$, $0 < b < 1$, *decreases* as x increases, this function is referred to as a **decay** function. (See Figure 4.)

Before examining growth and decay functions in more detail, we introduce a number that is very important in the study of growth and decay processes.

The Number *e*

When describing relationships using the exponential function $f(x) = b^x$, a base that occurs remarkably often is the number represented by the lowercase letter e. The number e is an irrational number and is defined precisely in the following way.

The Number *e*

$$e = \lim_{x \to \infty}\left(1 + \frac{1}{x}\right)^x$$

The expression $\left(1 + \frac{1}{x}\right)^x$, or forms of it, often occurs when describing relationships in the fields mentioned earlier. Since exponential functions with a base of e can describe natural phenomena so well, we refer to them as natural exponential functions. We shall see the natural exponential function $f(x) = e^x$ (or a form of it) in applications throughout this chapter.

We can calculate the value of e to any precision we want by computing the expression $\left(1 + \frac{1}{x}\right)^x$ for large enough values of x. To 15 decimal places the value of e is 2.718281828459045.... Since the number e is irrational, it never ends and it contains no repeating block of digits. When calculating in applications with e, we commonly approximate its value to be 2.718 (although most calculators have the value built into them). Table 2 and Figure 5 on the following page will help convince us that as x grows larger and larger, the expression $\left(1 + \frac{1}{x}\right)^x$ approaches e. (Notice that as x gets bigger and bigger, $\left(1 + \frac{1}{x}\right)^x$ gets closer and closer to a limit that, to four significant digits, is 2.718.)

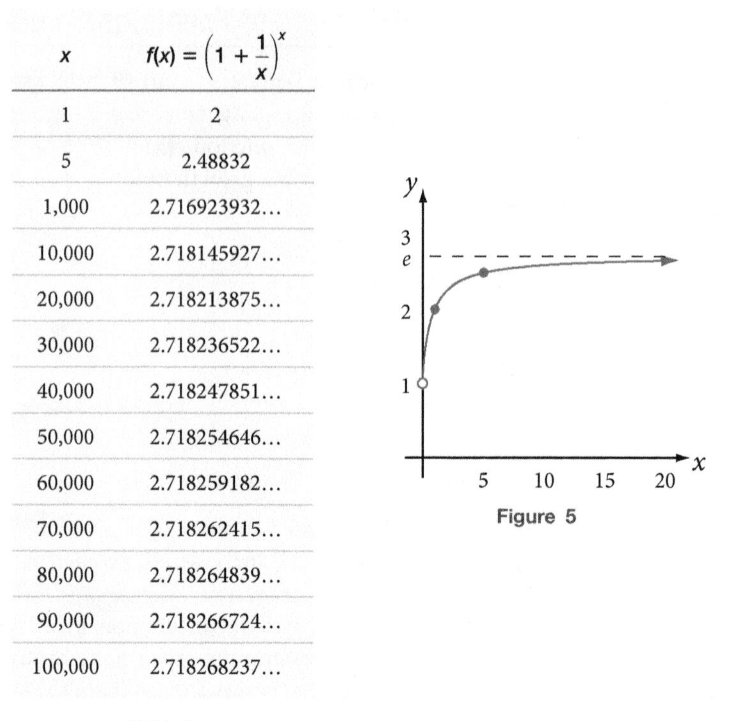

Figure 5

x	$f(x) = \left(1 + \frac{1}{x}\right)^x$
1	2
5	2.48832
1,000	2.716923932...
10,000	2.718145927...
20,000	2.718213875...
30,000	2.718236522...
40,000	2.718247851...
50,000	2.718254646...
60,000	2.718259182...
70,000	2.718262415...
80,000	2.718264839...
90,000	2.718266724...
100,000	2.718268237...

Table 2

Graphs of the General Natural Exponential Function

With the number e as the base, we have the very important and useful general exponential function $f(x) = A_0e^{kx}$.

The forms commonly used to predict growth or decay in business and the life and social sciences use a positive number k as the coefficient of the variable t for time. To predict growth, the form is

$$f(t) = A_0e^{kt}$$

and to predict decay, the form is

$$f(t) = A_0e^{-kt}$$

The number A_0 represents the amount of a quantity present at the initial observation; that is, at $t = 0$, where t represents time. Notice that at $t = 0$,

$$f(0) = A_0e^{k \cdot 0} = A_0e^0 = A_0 \cdot 1 = A_0$$

The number k is the **growth constant**. It is a measure of how frequently the output of the function grows by a factor of e. Exponential functions are functions that have rates of change proportional to their current values.

We can get a feel for exponential growth functions by examining the behavior of $f(t) = A_0e^{kt}$ for a specific value of A_0 and various values of the growth constant k. For convenience, we will take $A_0 = 1$. (Any other number would do. The only effect A_0 has on the graph is the vertical position of the curve. It does not affect the shape of the curve. Taking $A_0 = 1$ allows us to sketch the graph near the origin.)

Example 2 Graph each curve: $f(t) = e^{0.3t}$; $f(t) = e^{1t}$; $f(t) = e^{2t}$.

Solution We begin by making a table, plotting points, and connecting them with a smooth curve.

t	$e^{0.3t}$	e^{1t}	e^{2t}
−2	0.5	0.1	0.02
−1	0.7	0.4	0.1
0	1	1	1
1	1.3	2.7	7.4
2	1.8	7.4	54.6
3	2.5	20.1	403.4

Table 3

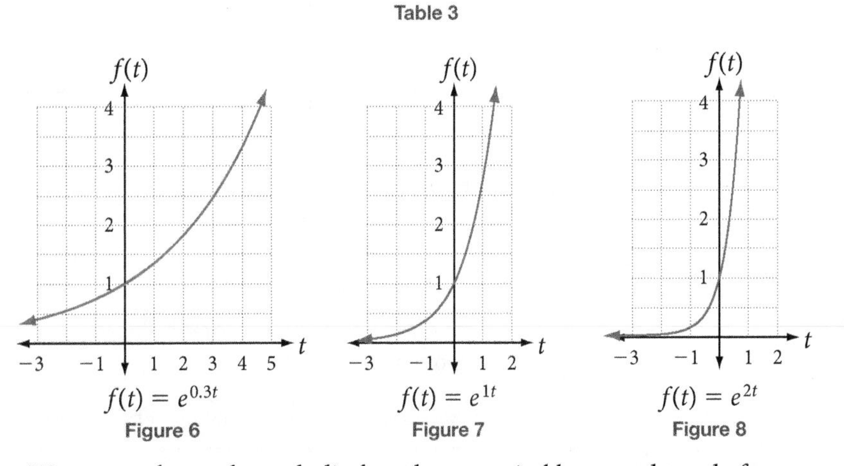

$f(t) = e^{0.3t}$ Figure 6

$f(t) = e^{1t}$ Figure 7

$f(t) = e^{2t}$ Figure 8

We can see that each graph displays the recognizable upward trend of an exponential growth function. We also notice, looking at these graphs in order from Figure 6 to Figure 8, that as the value of the growth constant k increases, the curve bends more sharply upward. This indicates that for small values of k, growth is increasing relatively slowly and that for large values of k, growth is increasing relatively quickly.

Next, to successfully examine the function $f(t) = A_0e^{-kt}$, we make an important observation. Since $e \approx 2.718 > 1$, it is true that $0 < \frac{1}{e} < 1$. Now, e^{-k} can be written as $e^{-1 \cdot k}$. Then $e^{-k} = e^{-1 \cdot k} = (e^{-1})^k = \left(\frac{1}{e}\right)^k$. Since $0 < \frac{1}{e} < 1$, $\left(\frac{1}{e}\right)^k$ is also between 0 and 1. (Recall that earlier we showed that exponential functions with such bases represented decay.) So, if e^{-k} is used as the base of an exponential function, it will describe a decay function.

Now let's examine the behavior of $f(t) = A_0e^{-kt}$, an exponential decay function.

Example 3 Graph $f(t) = e^{-0.3t}$; $f(t) = e^{-1t}$; $f(t) = e^{-2t}$.

Solution We make a table, plot points, and sketch the graph by drawing a smooth curve through the points.

t	$e^{-0.3t}$	e^{-1t}	e^{-2t}
−2	1.8	7.4	54.6
−1	1.3	2.7	7.4
0	1	1	1
1	0.7	0.4	0.1
2	0.5	0.1	0.02

Table 4

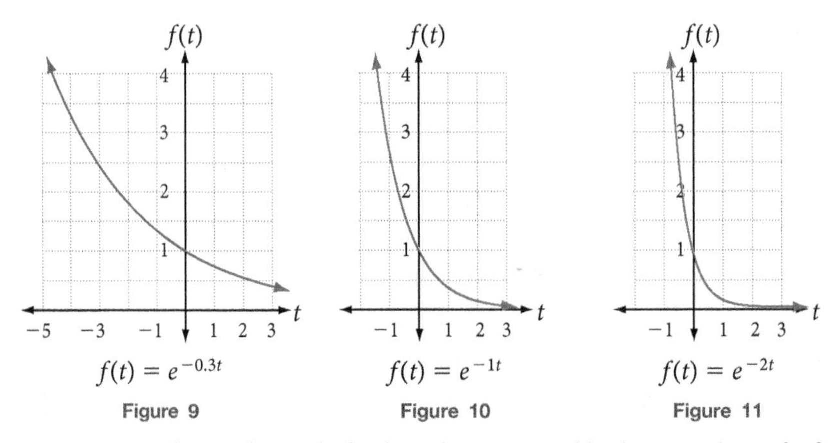

$f(t) = e^{-0.3t}$ Figure 9 $f(t) = e^{-1t}$ Figure 10 $f(t) = e^{-2t}$ Figure 11

We can see that each graph displays the recognizable downward trend of an exponential decay function. ■

We now make the following general observations about exponential functions.

Exponential Functions

The exponential function $f(x) = b^x$ and specifically the natural exponential functions

$$f(x) = A_0 e^{kx} \quad \text{and} \quad g(x) = A_0 e^{-kx} \quad (k > 0)$$

have the following properties.

1. The domains are all real numbers, which means that $f(x)$ and $g(x)$ can be calculated for any value of x.
2. The range of each function is all positive numbers. ($f(x)$ and $g(x)$ are never 0 or negative.)
3. Both functions are continuous.
4. Both functions pass through $(0, A_0)$.
5. $f(x)$ increases as x increases. $g(x)$ decreases as x increases.

Applications of the Exponential Function

The following examples demonstrate how exponential functions (both natural, with base e, and otherwise) can be used to obtain information about real-world phenomena.

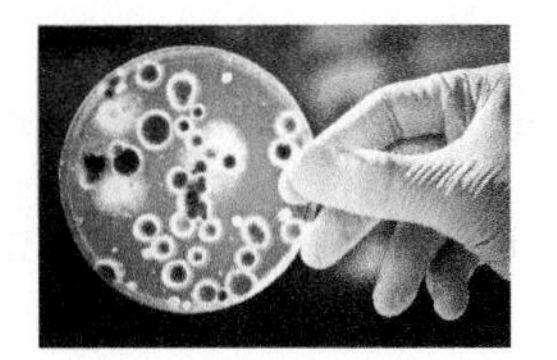
© Alexander Raths/iStockPhoto

Example 4 Imagine that at a particular time, a kitchen counter has on it 5 bacteria and that this type of bacteria doubles in number every 10 minutes. The exponential function $A(t) = 5e^{0.06931t}$ is a good model for predicting the number of bacteria, $A(t)$, in the population t minutes from some particular start time. How many bacteria could be predicted to be on the counter 2 hours later?

Solution We are asked to find the number of bacteria, $A(t)$, in the population after 2 hours have passed since there were 5 bacteria present. Since the variable t in the model is given in minutes, we start by converting 2 hours to minutes.

$$2 \text{ hours} = 2 \text{ hours} \times \frac{60 \text{ minutes}}{1 \text{ hour}}$$

$$= 2 \times 60 \text{ minutes}$$

$$= 120 \text{ minutes}$$

Now, we substitute 120 for t in the model $A(t) = 5e^{0.06931t}$ and compute to find $A(120)$.

$$A(120) = 5e^{0.06931(120)}$$

$$\approx 20{,}468$$

Interpretation

If there are initially 5 bacteria on the counter and the population of bacteria doubles every 10 minutes, then after 2 hours, there will be about 20,468 bacteria on the counter.

Example 5 The sociologists Stephan and Mishler found that the exponential function

$$N(p) = N_1e^{-0.11(p-1)} \quad 1 \le p \le 10$$

is produced when members of a discussion group of 10 people are ranked according to the number of times each participated. N_1 represents the number of times the first-ranked person participated, and $N(p)$ represents the number of times the pth-ranked person participated. If, in a discussion group of 10 people, the first-ranked person participated 35 times, how many times did the sixth-ranked person participate?

© Alex Nikada/iStockPhoto

Solution We are asked to find the number of times the sixth-ranked person participated. This means that we need to find the value of $N(6)$. Since the first-ranked person participated 35 times, we substitute 35 for N_1. Since we are interested in the number of times the sixth-ranked person participated, we substitute 6 for p. This produces the computation

$$N(6) = 35e^{-0.11(6-1)}$$

Using a calculator, we find that

$$N(6) = 35 \cdot e^{-0.11(6-1)}$$

$$= 35 \cdot e^{-0.11(5)}$$

$$= 35 \cdot e^{-0.55}$$

$$\approx 35 \cdot 0.5769498$$

$$\approx 20.193243$$

Interpretation
We conclude that if the first-ranked person participated 35 times, the sixth-ranked person participated approximately 20 times.
Notice that this function is a decay function, which seems reasonable in this situation. We would expect that as p increases, $N(p)$ will decrease.

© Kyoungil Jeon/iStockPhoto

Example 6 When a particular amount of money P, called the principal, is invested at the interest rate r and is compounded n times a year, the amount A accumulated after t years is

$$A(t) = P\left(1 + \frac{r}{n}\right)^{nt}$$

Determine the amount of money accumulated after 15 years if \$2,000 is invested in an account that pays 8% interest compounded quarterly.

Solution The initial investment is \$2,000, so we replace P with 2,000. Since the interest rate is 8%, we substitute 0.08 for r. Since interest is compounded quarterly, meaning four times a year, we substitute 4 for n. Finally, since the investment is made for 15 years, we substitute 15 for t.

$$\begin{aligned} A(15) &= 2{,}000\left(1 + \frac{0.08}{4}\right)^{4\cdot 15} \\ &= 2{,}000(1 + 0.02)^{60} \\ &= 2{,}000(1.02)^{60} \\ &\approx 2{,}000(3.2810308) \\ &\approx 6{,}562.0616 \end{aligned}$$

Interpretation
We conclude that after 15 years, \$2,000 will grow to approximately \$6,562.06 if it is invested at 8% interest compounded quarterly.

The last example we will examine involves a variation of the natural exponential growth function called the **logistic growth function**. The logistic growth function is used to describe *restricted growth*; that is, growth that is restricted to some upper bound by factors imposed by some particular conditions. It is used, for example, by biologists to predict or describe the growth of a population in which environmental factors can restrict that growth, and by sociologists to describe the spread of a rumor. It has the form

$$f(t) = \frac{M}{1 + Ae^{-kt}}$$

The numbers M, k, and A are positive constants, k is the growth constant, and t represents the amount of time the process has been going on. M is the maximum level the output of the particular function can reach. The graph of the logistic function

$$f(t) = \frac{5}{1 + 15e^{-0.8t}}$$

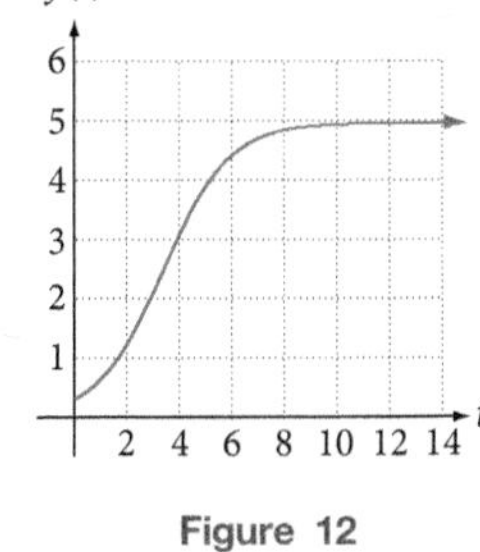

Figure 12

is illustrated in Figure 12. Notice that for small values of t, the logistic function closely resembles the exponential growth function. However, as the values of t get larger, the curve levels off to the maximum value prescribed in the numerator. This leveling-off effect reflects restrictions on the growth of the subject being studied, such as a population, spread of a rumor, or automobile sales.

Example 7 Let us model the number of automobile sales in a developing country as

© Maria Pavlova/iStockPhoto

$$N(t) = \frac{156{,}000}{1 + 5.6e^{-0.8(t-2010)}} \qquad t \geq 2010$$

where N is the number of automobiles sold in one year and t is the year.

a. How many cars were sold in 2010, the first year of observation?

b. How many cars are projected to be sold in the year 2015?

c. If the model holds well into the future, at what level will the sale of cars stabilize?

Solution

a. Since t represents the year, the value of t at the beginning of the observation period is 2010. Then

$$N(t) = \frac{156{,}000}{1 + 5.6e^{-0.8(t-2010)}} \text{ and}$$

$$\begin{aligned} N(2010) &= \frac{156{,}000}{1 + 5.6e^{-0.8(2010-2010)}} \\ &= \frac{156{,}000}{1 + 5.6e^{-0.8(0)}} \\ &= \frac{156{,}000}{1 + 5.6e^{0}} \\ &= \frac{156{,}000}{1 + 5.6 \cdot 1} \\ &= \frac{156{,}000}{1 + 5.6} \\ &\approx 23{,}636 \end{aligned}$$

In the year 2010, about 23,636 cars were sold.

b. In the year 2015, the value of t is 2015 and

$$N(t) = \frac{156{,}000}{1 + 5.6e^{-0.8(t-2010)}}$$

$$N(2015) = \frac{156{,}000}{1 + 5.6e^{-0.8(2015-2010)}}$$

$$= \frac{156{,}000}{1 + 5.6e^{-0.8(5)}}$$

$$\approx 141{,}488$$

In the year 2015, the model estimates that about 141,488 cars will be sold.

c. If the trend continues, t will increase without bound, which leads us to taking the limit of the function as t approaches infinity.

$$\lim_{t\to\infty} \frac{156{,}000}{1 + 5.6e^{-0.8(t-2010)}} = \lim_{t\to\infty} \frac{156{,}000}{1+0}$$

$$= 156{,}000$$

We conclude that if the trend continues, car sales can be expected to increase to and stabilize at about 156,000 cars per year.

Notice that the 156,000 is the numerator, which in the logistic growth function represents the maximum output level the function attains. ■

We end this section by seeing, graphically, why the base b is restricted to being positive but not equal to 1. The fact that the base b is restricted so that it is greater than 0 but not equal to 1 is responsible for the basic shape of the exponential curve. If b were allowed to be negative or 1, exponential functions would not always exhibit the recognizable shape they now possess.

As examples, the functions $f(x) = (-2)^x$ and $f(x) = 1^x$ are graphed in Figures 13 and 14, respectively. Notice that the points we have chosen to show on the graph of $f(x) = (-2)^x$ have not been joined with a smooth curve. Since the expression $(-2)^x$ is not defined for certain exponents (such as 1/2, 1/4, 1/6, ..., which represent even roots), the graph has infinitely many pieces. The function $f(x) = 1^x$ is actually the constant function $f(x) = 1$, since $1^x = 1$ for all values of x. The graphs of constant functions, as we know, are horizontal lines. Notice that neither graph exhibits the growth or decay structure apparent when $b > 0$ and $b \neq 1$.

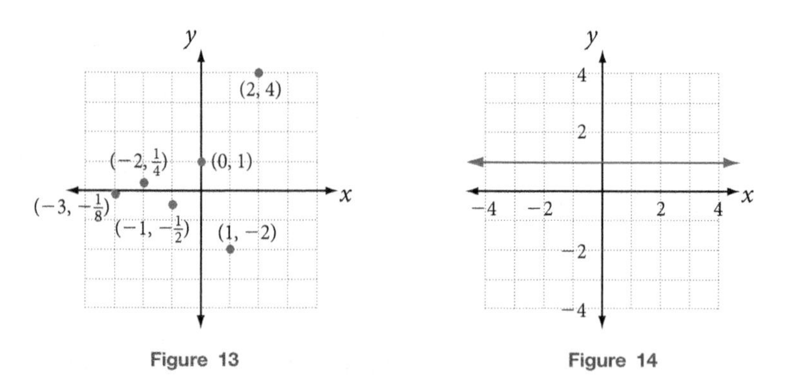

Figure 13

Figure 14

Using Technology 4.1

In Example 1, we constructed the graphs of the exponential growth function $f(x) = 2^x$ and the exponential decay function $f(x) = \left(\frac{1}{2}\right)^x$ by building a table of values, plotting the corresponding points, then connecting the points with a smooth, continuous curve. Graphs of functions can be constructed more accurately and efficiently using technology.

Example 8 Use Wolfram|Alpha to construct the graph of both functions

$$f(x) = 2^x \quad \text{and} \quad f(x) = \left(\frac{1}{2}\right)^x$$

on the same coordinate system from $x = -2$ to $x = 2$.

Solution Online, go to www.wolframalpha.com. In the entry field, enter

Graph 2^x and (1/2)^x from x = -2 to x = 2

You will see the graph of both functions on the same coordinate system. Here is the graph:

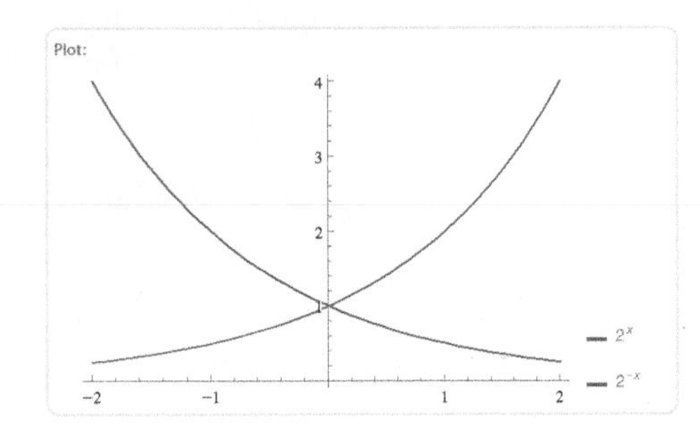

Wolfram Alpha LLC. 2012. Wolfram|Alpha
http://www.wolframalpha.com/
(access November 2, 2012)

Example 9 Constructing the graph of a logistic function by hand can be challenging. Use Wolfram|Alpha to help you construct the graph of the logistic function

$$f(t) = \frac{10}{1 + 10e^{-0.8t}}$$

on the interval $[0, 10]$.

Solution Go to www.wolframalpha.com and in the entry field type

graph 10 / (1 + 10e^(-0.8t)) from t = 0 to 10

Wolfram|Alpha responds with the graph.

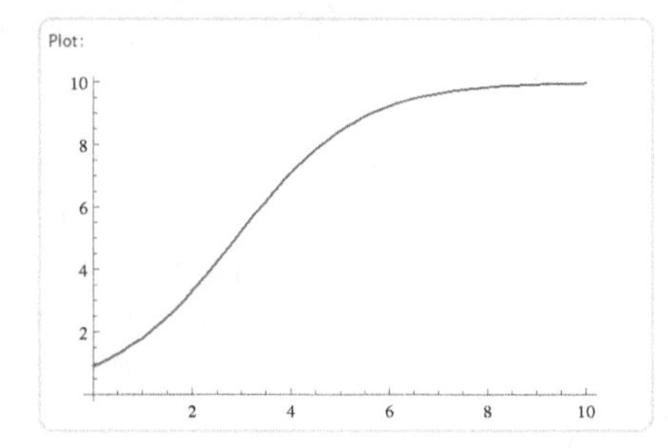

Wolfram Alpha LLC. 2012. Wolfram|Alpha
http://www.wolframalpha.com/
(access November 2, 2012)

Getting Ready for Class

After reading through the preceding section, respond in your own words and in complete sentences.

A. What is an exponential function and how is it different from a power function?
B. Give a definition of the number *e*.
C. What is a growth function?
D. When is an exponential function an increasing function?

Problem Set 4.1

Skills Practice

For Problems 1-11, use a calculator to compute the value of each function. Identify each exponential function as a growth or decay function. Round to three decimal places.

1. Find $f(3)$ for $f(x) = 10^x$.
2. Find $f(2)$ for $f(x) = 10^{-2x}$.
3. Find $f(6)$ for $f(x) = (0.99)^{40x}$.
4. Find $f(2)$ for $f(x) = (1.06)^{50x}$.
5. Find $f(5)$ for $f(x) = (1.35)^{3x+4}$.
6. Find $f(55)$ for $f(x) = (1.01)^{10x+120}$.
7. Find $f(3)$ for $f(x) = \left(\frac{1}{2}\right)^{x+1}$.
8. Find $f(4)$ for $f(x) = e^x$.
9. Find $f(2)$ for $f(x) = e^{-x}$.
10. Find $f(6)$ for $f(x) = 2{,}000e^{0.04x}$.
11. Find $f(105)$ for $f(x) = 3{,}560e^{0.06x}$.

For Problems 12-16, use a calculator to compute the value of each function. Round to three decimal places.

12. Find $f(25)$ for $f(x) = 1 - e^{0.11x}$.
13. Find $f(0.10)$ for $f(x) = 1 - e^{-0.3x}$.
14. Find $f(5)$ for $f(x) = 15{,}000 + 20{,}000\left(\frac{3}{8}\right)^{0.75x}$.
15. Find $f(3)$ for $f(x) = 100 - 60\left(\frac{1}{4}\right)^{0.95x}$.
16. Find $f(20)$ for $f(t) = 5{,}500{,}000 \cdot 3^{t/45}$.

For Problems 17 and 18, use the corresponding figure and, as best you can, match each function with its graph (A, B, C, D).

17. a. $f(x) = e^{2x/3}$ D
    b. $f(x) = e^x$
    c. $f(x) = e^{4x/5}$
    d. $f(x) = e^{2x}$ A

18. a. $f(x) = e^{-x}$
    b. $f(x) = e^{-3x/5}$
    c. $f(x) = e^{-x/3}$
    d. $f(x) = e^{-2.4x}$

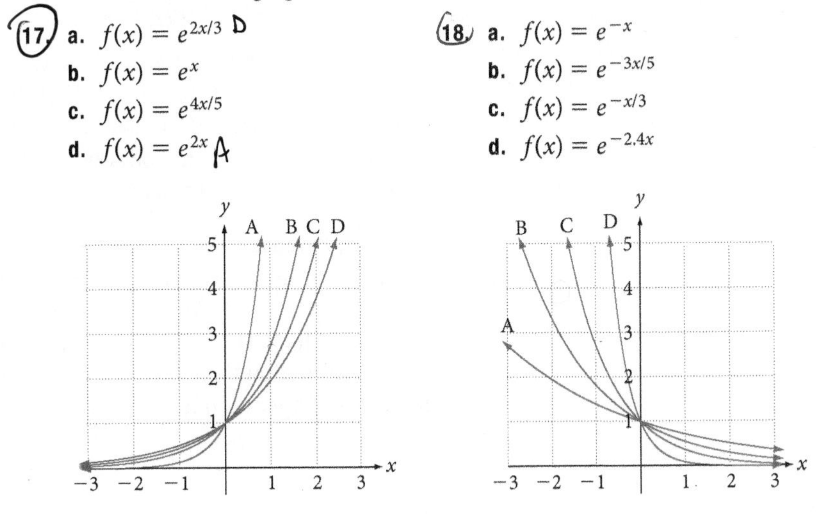

Check Your Understanding

For Problems 19-29, identify each function as algebraic or exponential.

19. $f(x) = 5x^3 - 2x + 11$

20. $f(x) = (x + 4)^3(x - 5)^2$

21. $f(x) = 5^x + 12$

22. $f(x) = x^2(x + 1)^8$

23. $f(x) = \left(\dfrac{x+3}{x-1}\right)^{-2}$

24. $f(x) = 12^{x-1} + x$

25. $f(x) = 8^{3x}$

26. $f(x) = 4x^e$

27. $f(x) = e^{-2x+1}$

28. $f(x) = (5 + e)^{3x}$

29. $f(x) = 2x^{-3} - 2ex^2$

30. A power function has the form $(\text{variable})^{(\text{constant})}$. Specify the form of the exponential function.

31. Exponential functions are given names according to whether the base is greater than 1 or between 0 and 1. Specify these names.

32. Give a common approximation value for the number e.

Modeling Practice

33. Business: Investment in Art An investor and collector of art is considering buying a painting by a well-established artist for \$4,500. The gallery owner tells the investor that in 5 years the painting will be worth \$10,000 and that in 10 years the painting will be worth \$25,000. As her advisor, the investor asks you to evaluate the gallery owner's assessment of the painting's future value. Using data you have collected from previous sales of the artist, you assign to this new painting the function

$$V(t) = 4{,}500 \cdot 2^{t/5} \quad t \geq 0$$

where t is the number of years from the time the piece is purchased and V is its value in dollars at that time. What do you tell the investor? What is your estimate of the value of the painting in 5 years and in 10 years? Is the gallery owner's assessment of the painting's future values accurate or has the owner under or over estimated the painting's future value?

34. Business: Revenue Growth XYZ Textbooks is a small publisher of college textbooks (publishing quality material at reasonable prices). Their first year of operations was 2010. At the beginning of that year they had sold 0 books. By the end of the year they had sold 5,000 books. After that, sales doubled each year. The function that describes the growth of XYZ Textbooks during its first few years of operation is

$$S(t) = \begin{cases} 5{,}000t & 0 \leq t < 1 \\ 5{,}000 \cdot 2^{t-1} & t \geq 1 \end{cases}$$

As a financial advisor to XYZ, what would you tell these publishers to expect in terms of number of sales in 2015 (5 years from 2010)? Would you tell them to expect sales to increase consistently or decrease consistently after 2015? How do you justify your assessment about the number of sales?

35. Life Science: Bacterial Growth Imagine that at a particular time, a kitchen counter has on it 2 bacteria and that this type of bacteria doubles in number every 5 minutes. The exponential function $A(t) = 2e^{0.13863t}$ is a good model for predicting the number of bacteria, $A(t)$, in the population t minutes from some particular start time. How many bacteria could be predicted to be on the counter

a. 1 hour later? (Use 60 minutes for 1 hour.)

b. 2 hours later?

Source: suggested by the article *Bacteria and Foodborne Illness*, National Digestive Diseases, Information Clearinghouse

36. Psychology: Forgetting Curves In his book *Memory, A Contribution to Experimental Psychology*, published in 1885, psychologist Hermann Ebbinghaus presented a *forgetting curve* that describes the relationship between the percentage P of words remembered from a list or story and the time t (in weeks) since memorization took place. The curve is modeled by the exponential decay function

$$P(t) = (100 - a)e^{-kt} + a$$

The numbers a and k are positive constants, and $0 < a < 100$. Suppose that for a particular person under particular conditions,

$$P(t) = (100 - 10)e^{-0.75t} + 10$$

Find

a. $P(0)$ **b.** $P(1)$ **c.** $P(2)$ **d.** $P(7)$ **e.** $P(20)$

Source: based on material in the text *Essentials of Human Memory*, Baddeley

37. Life Science: Bacterial Decay The number, N, of bacteria present in a culture at any time t (in hours) is given by the exponential decay function

$$N(t) = 15{,}000 \cdot 3^{-0.05t}$$

Find

a. $N(0)$ **b.** $N(1)$ **c.** $N(5)$ **d.** $N(24)$

For Problems 38 and 39, use the following information. When a particular amount of money, P, called the principal, is invested at the interest rate, r, and is compounded n times a year, the amount A of money accumulated after t years is

$$A(t) = P\left(1 + \frac{r}{n}\right)^{nt}$$

38. Business: Investment Find the amount of money accumulated in 5 years if \$12,000 is invested at 8% interest and compounded

a. annually **b.** quarterly

c. monthly **d.** daily (assume 365 days in a year)

39. Business: Investment Find the amount of money accumulated in 5 years if \$500 is invested at 11% interest and compounded

a. annually **b.** quarterly

c. monthly **d.** daily (assume 365 days in a year)

For Problems 40 and 41, use the following information. We know that when a particular amount of money P, called the principal, is invested at the interest rate r and is compounded n times a year, the amount A of money accumulated after t years is

$$A(t) = P\left(1 + \frac{r}{n}\right)^{nt}$$

If the number n of compoundings increases without bound, that is, if compounding takes place continuously, the above formula reduces to $A(t) = Pe^{rt}$.

© Kyoungil Jeon/iStockPhoto

40. Business: Investment Find the amount of money accumulated if \$2,000 is invested for 3 years at 7% interest and is compounded

a. quarterly

b. monthly

c. daily (assume 365 days in a year) and

d. continuously.

e. Compare the values you find in parts a-d. Are they close? (They should be.) Since there is such a small difference in the amount of money a bank pays out in interest between compounding quarterly and compounding continuously, one might think that by advertising continuous compounding, a bank would have customers flocking to invest their money, and could, therefore, make quite a bit of money from this advertising ploy. But, although bankers are aware of the continuous compounding formula (from their college calculus course), they rarely, if ever, offer continuous compounding. Can you think of a reason why not?

41. Business: Investment Find the amount of money accumulated if \$1,600 is invested for 5 years at 9% interest and is compounded

a. quarterly

b. monthly

c. daily (assume 365 days in a year), and

d. continuously.

e. Compare the values you get in parts a-d. Are they close? (They should be.)

42. Economics: Oil Consumption The world's oil consumption (in billions of tonnes) can be modeled with the logistic equation

$$C(t) = \frac{4{,}000}{1 + 373e^{-0.089t}}$$

where t is the number of years since the year 1900.

a. Estimate the world's oil consumption in the year 1900.

b. Estimate the world's oil consumption in the year 2010.

c. Estimate the expected world consumption of oil in the year 2020.

d. What is the maximum level possible for world oil consumption?

Source: Energy Bulletin

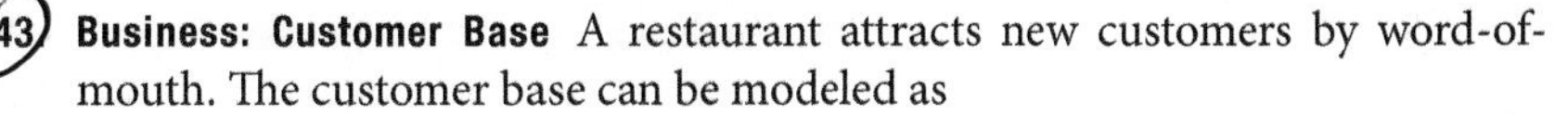

43. Business: Customer Base A restaurant attracts new customers by word-of-mouth. The customer base can be modeled as

$$C(t) = \frac{240}{1 + e^{-0.2t}}$$

where C is the customer base and t is measured in days since the opening of the restaurant.

a. What was the level of the customer base when the restaurant opened?

b. What is the maximum attainable level for the customer base?

44. Business: Demand for a Product Demand for a product depends on the price as

$$D(p) = 500 + \frac{1{,}000}{1 + e^{-0.05p}}$$

where D is the quantity demanded and p is the price in dollars. How many units of the product will be sold when the price is \$50?

45. Business: Number of Telephones The number of telephones per capita in a country depends on the median yearly income of the people living in the country as

$$T(i) = \frac{0.9}{1 + 20e^{-i/400}}$$

where T is the number of telephones per capita and i is the median income.

a. What will be the number of telephones per capita if the median income is \$400? (Round to 1 decimal point.)

b. What will be the number of telephones per capita if the median income is \$4,000? (Round to 1 decimal point.)

46. Marketing: Sales The marketing department of a toy company has determined that t weeks after an advertising campaign ends, the number N of the advertised toys it sells each month decreases according to the exponential decay function $N(t) = 2{,}700e^{-0.12t}$. Find how many of the advertised toys the company will sell

a. immediately at the close of the advertising campaign.

b. one week after the end of the campaign.

c. five weeks after the end of the campaign.

d. 12 weeks after the end of the campaign.

47. Health Science: Drug Concentration When a drug is intravenously administered to a patient on a continuous basis, the amount A (in milligrams) of the drug in the patient's bloodstream t minutes after injection of the drug is given by

$$A(t) = \frac{1}{k}(a - Ce^{-kt})$$

In this case, the growth constant k is called the absorption constant. The number a is the amount of the drug added to the bloodstream each minute, and the number C is a constant that depends on the values of a and k. For the case when $a = 2$ mg, $C = 35$, and $k = 0.20$, find the amount of the drug in the patient's bloodstream

a. 15 minutes after injection.

b. 60 minutes after injection.

48. Medicine: Spread of Disease Suppose that in a large city, 21,000 people are susceptible to a particular flu virus. When that flu epidemic does break out, the city's health agency determines that t weeks after the beginning of the outbreak the number of cases of flu is given by the logistic function

$$N(t) = \frac{21{,}000}{1 + 165e^{-0.85t}}$$

a. How many people had the flu virus when the health agency began recording?

b. How many people had the flu virus 5 weeks after the outbreak?

c. If the epidemic continues, how many people can be expected to contract the flu?

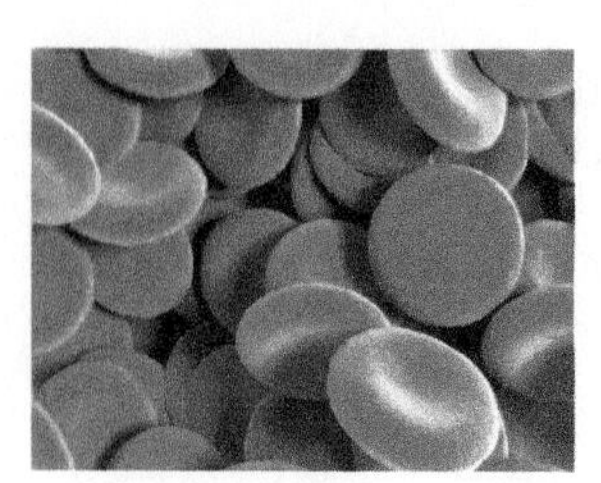

© Andrew Dorey/iStockPhoto

49. Health Science: Contraction of Disease The health agency of a city has determined that the number N (in thousands) of people contracting a particular disease t weeks after the outbreak of the disease is described by the logistic function

$$N(t) = \frac{45}{1 + 38e^{-0.095t}}$$

How many people have contracted the disease

a. when the health agency makes its first recording?

b. 3 weeks after the outbreak?

c. 10 weeks after the outbreak?

d. 26 weeks after the outbreak?

e. in the long run?

Source: suggested by the article
T cell senescence and contraction of T cell repertoire diversity in patients with chronic obstructive pulmonary disease,
2009 British Society for Immunology

50. Business: Advertising When information is diffused through the mass media to a population of size A, the number N of people hearing the news by time t (in days) is given by the function $N(t) = A(1 - e^{-kt})$. A clothing company, through frequent daily announcements on television, is advertising, for a three-month period, a new line of clothes. If the total television audience for that three-month period is estimated to be 350,000 people, how many people can be expected to hear about the new clothing line (assuming $k = 0.059$)

a. 10 days after the advertisements begin?

b. at the end of the three-month period? (A month consists of a 30 day period.)

Using Technology Exercises

Use Wolfram|Alpha to construct the graphs.

51. Construct the graph of

$$f(x) = \left(1 + \frac{1}{x}\right)^x$$

from $x = 0$ to $x = 1{,}000{,}000$. Use the graph to describe the behavior of the function. Include in your description what happens to $f(x)$ as $x \to \infty$.

52. On the same coordinate system, construct the graphs of

$$f(x) = 3^x \text{ and } f(x) = \left(\frac{1}{3}\right)^x$$

from $x = -2$ to $x = 2$.

53. On the same coordinate system, construct the graphs of the functions

$$f(x) = 10e^x, f(x) = 30e^x, f(x) = 60e^x$$

from $x = -2$ to $x = 2$. What effect does the coefficient of e^x have on the graph of the function?

54. Psychology: Memory In his book *Memory, A Contribution to Experimental Psychology*, published in 1885, psychologist Hermann Ebbinghaus presented a *forgetting curve* that describes the relationship between the percentage P of words remembered from a list or story and the time t (in weeks) since memorization took place. The curve is modeled by the exponential decay function

$$P(t) = (100 - a)e^{-kt} + a$$

The numbers a and k are positive constants, and $0 < a < 100$. Suppose that for a particular person under particular conditions,

$$P(t) = (100 - 10)e^{-0.05t} + 10$$

Construct the graph of $P(t)$ over the first 100-week period, use it to describe the behavior of the function, and use it to estimate $\lim_{t \to \infty} P(t)$. Be careful to enclose the exponent on e in parentheses.

55. Social Science: Rumors A person starts a rumor and t hours later N people have heard it. Suppose N and t are related by the logistic function

$$N(t) = \frac{600}{1 + 400e^{-0.8t}}$$

Construct the graph of this function and use it to describe its behavior over a 24-hour time period. Your description should include a statement about the maximum number of people who can hear the rumor. Does this number appear in the function? Be careful to enclose the entire denominator in parentheses as well as the exponent on e.

56. Health Science: Heart Failure Medication Cardiac glycosides are a class of medications used to treat heart failure. For patients on maintenance doses for the cardiac glycoside digoxin, the serum level, C, of the drug at time t hours after it has been measured to show a level of 1.6 ng/mL is modeled by the exponential decay function $C(t) = 1.6e^{-0.01444t}$ as long as no further drug is administered. Use this model to determine the serum level of this drug after

a. 2 hours.

b. 24 hours.

57. Psychology: Short-Term Recognition Memory Psychological recognition studies indicate that the relationship between the mean proportion, P, of "yes" responses and the time in seconds since an item was last viewed by a subject can be modeled with the exponential decay function

$$P(t) = 0.0150 + 0.985e^{-0.0086t}$$

Use your calculator or Wolfram|Alpha to determine what proportion of "yes" responses would reasonably be expected

a. 200 seconds after the first viewing.

b. 1,000 seconds after the first viewing.

Source: *Exponential Decay of Episodic Traces in Short Term Recognition Memory* Boneau & Daily

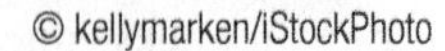

© kellymarken/iStockPhoto

58. Social Science/Linguistics: Predicting Language Change Scientific studies, including language-related research, indicate that the changes made in a language over time can be modeled by a logistic function. In the year 2000, a researcher collected data from Yamagata, a northeast region of Japan, in which a distinct local dialect was often observed. He studied the degree to which people of different ages replaced the equivalent of the English word "sneaked" with the equivalent of "snuck." From his data, he produced the logistic function

$$N(t) = \frac{99.21}{1 + 0.009e^{0.0798t}}$$

where t represents the ages of people and N represents the degree, in percent of people using the new word, to which the language had changed. Determine the degree to which people who are about

a. 65 years old make changes to their language.

b. 25 years old make changes to their language.

c. How does this logistic function differ from the logistic growth functions we encountered previously?

Source: R. Köhler, *Issues in Quantitative Linguistics*

Getting Ready for the Next Section

Appendix A.2 at the back of the book contains a review of logarithms. The problems below show what you are expected to know about logarithms to be successful in the next section.

59. In one word, describe a logarithm.

60. What is the base of the natural logarithm?

61. Convert $y = e^{5x}$ to its natural logarithmic form.

62. Convert $2.6 = \ln(x + 5)$ to its corresponding exponential form.

63. Construct a sketch of the graph of $y = \ln x$.

64. Specify the domain of the natural logarithm function.

65. Expand the expression $\ln \frac{x^3y^2}{z^4}$ into three individual logarithmic expressions in which each variable has an exponent of 1.

66. Compress the expression $4 \ln x - 2 \ln y - 5 \ln z$ into one logarithmic expression.

The Natural Logarithm Function

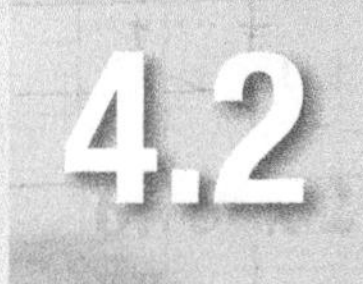

Suppose a hospital spends \$635,000 on a piece of equipment that depreciates at the rate of 27.4% each year. You could imagine the hospital wanting to know when the value of the piece of equipment has depreciated to \$63,500. The value of the piece of equipment at any time after purchase is modeled by the exponential function $A(t) = 635{,}000e^{-0.32t}$. Given a value, $A(t)$, it is possible to solve for the exponent, t. But to do so requires the use of a new function, the natural logarithm function.

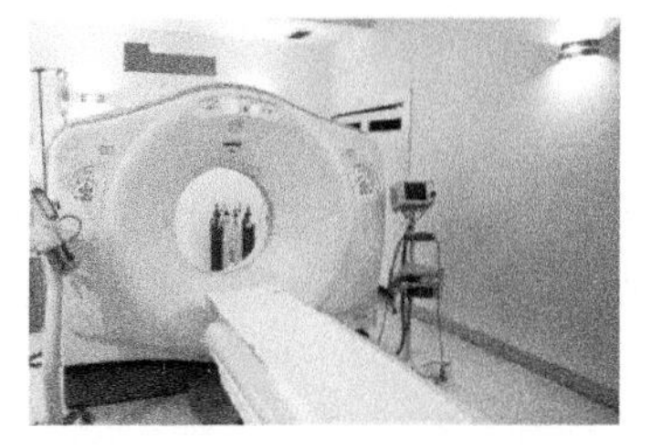

© Baran Özdemir/iStockPhoto

In this section we introduce the natural logarithm function and examine its close relationship to the natural exponential function. A review of general logarithms is provided in Appendix A.2. There you will find the general definition and properties of all logarithms. This section includes the information provided in Appendix A.2, but specific to the base e, and will also extend your understanding of the logarithmic function in general.

Logarithms Are Exponents

In the exponential function $y = A_0e^{kx}$ that we studied in Section 4.1, we were given a value for the independent variable x and computed the corresponding value of the dependent variable y. For example, when working with wounds to the skin, medical researchers have found that the function $A = A_0e^{-0.11t}$ relates the area A (in square centimeters) of unhealed skin to the number t of days since the skin received the wound. A_0 represents the initial area of the wound. Suppose that someone's skin receives a wound of 2 square centimeters in area. The number of square centimeters of unhealed skin remaining after 8 days is found by substituting 8 for t and 2 for A_0 in the equation $A = A_0e^{-0.11t}$.

$$\begin{aligned} A &= A_0e^{-0.11t} \\ &= 2e^{-0.11(8)} \\ &= 2e^{-0.88} \\ &\approx 2(0.414783) \\ &\approx 0.83 \end{aligned}$$

We conclude that after 8 days there remains approximately 0.83 square centimeters of unhealed skin.

An equally important question is posed by turning the previous question around and asking, "For a skin wound of 2 square centimeters, how many days will have to pass before only about 0.83 square centimeters of unhealed skin remain?" With the question posed this way, we need to find t in the equation $0.83 = 2e^{-0.11t}$.

The variable whose value we wish to determine is in the exponent. However, none of the standard algebraic operations can be used to isolate it and therefore solve for it. But mathematicians have developed a technique for finding the unknown value of an exponent and we can try to understand it by considering the exponential equation $a = e^t$.

The problem posed above comes down to this: If we know the value of a in the equation $a = e^t$, what value of the exponent t will produce that value of a? The exponent t used in this way is given a special name, the **natural logarithm.**

The Natural Logarithm (First Form)
In the equation $a = e^t$, the number t that when used as an exponent on the base e produces the number a, is called the **natural logarithm** of a.

The words *natural logarithm* are abbreviated with the symbol ln. Then, if $a = e^t$, we have

$$t \text{ is the natural logarithm of } a$$

or, in symbols,

$$t = \log_e a = \ln a$$

Thus, the definition of the natural logarithm can be restated as follows.

The Natural Logarithm (Final Form)
$t = \ln a \quad \text{if and only if} \quad a = e^t$

The number a is called the **argument of the natural logarithm.**

Here we make four important notes:

Note 1 Since t is a logarithm and an exponent, *logarithms are exponents.*

Note 2 $t = \ln a$ if and only if $a = e^t$.

Note 3 We know that e^t is always a positive number. Thus, in $a = e^t$ the number a is *always* positive. Since $a = e^t$ if and only if $t = \ln a$, the argument of the natural logarithm *must always* be a positive number.

Note 4 $t = \ln a$ expresses t as a *function* of a since for each value of a, $\ln a$ produces exactly one value of t.

The following examples illustrate the relationship between exponential form and logarithmic form, as indicated in notes 1 and 2.

VIDEO EXAMPLES

SECTION 4.2

Example 1 Convert each exponential equation to the corresponding logarithmic equation.

a. $y = e^{2x}$ **b.** $a = e^{4x+3}$ **c.** $20 = e^{-0.3x}$

Solution

a. We know the logarithm is the exponent, and the exponent is $2x$. Therefore, we know the logarithm is $2x$. The base is e so we use the natural logarithm. The exponential equation $y = e^{2x}$ corresponds to the logarithmic equation $\ln y = 2x$.

b. We know the logarithm is the exponent, and the exponent is $4x + 3$. Therefore, we know the logarithm is $4x + 3$. The base is e so we use the natural logarithm. The exponential equation $a = e^{4x+3}$ corresponds to the logarithmic equation $\ln a = 4x + 3$.

c. We know the logarithm is the exponent, and the exponent is $-0.3x$. Therefore, we know the logarithm is $-0.3x$. The base is e so we use the natural logarithm. The exponential equation $20 = e^{-0.3x}$ corresponds to the logarithmic equation $\ln 20 = -0.3x$. ■

Example 2 Convert each logarithmic equation to the corresponding exponential equation.

a. $5 = \ln(7x)$ b. $x + 5 = \ln(y - 3)$ c. $-0.01 = \ln(x + 8)$

Solution

a. We know the logarithm is the exponent, and the logarithm is 5. Since we are using the natural logarithm, the base is e. Exponents go on bases, so the 5 becomes the exponent of e. Then, $5 = \ln(7x)$ converts to $e^5 = 7x$.

b. We know the logarithm is the exponent, and the logarithm is $x + 5$. Since we are using the natural logarithm, the base is e. Exponents go on bases, so the $x + 5$ becomes the exponent of e. Then, $x + 5 = \ln(y - 3)$ converts to $e^{x+5} = y - 3$.

c. We know the logarithm is the exponent, and the logarithm is -0.01. Since we are using the natural logarithm, the base is e. Exponents go on bases, so the -0.01 becomes the exponent of e. Then, $-0.01 = \ln(x + 8)$ converts to $e^{-0.01} = x + 8$. ■

Example 3 illustrates two important and frequently occurring natural logarithm values. Notice that in evaluating these expressions, we use the fact that "logarithms are exponents."

Example 3 Evaluate each of the following:

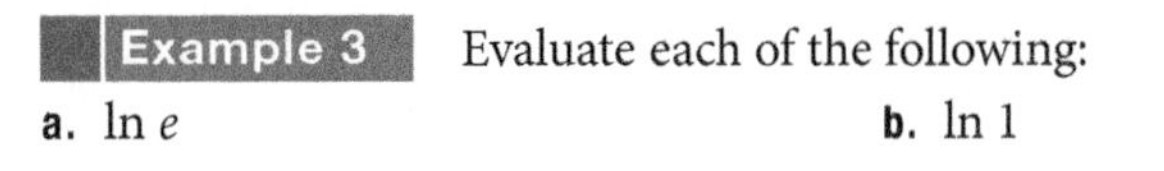

a. $\ln e$ b. $\ln 1$

Solution

a. For the moment, we will assign $\ln e$ the value x. That is, we will let

$$x = \ln e$$

Note 1 on the previous page states that logarithms are exponents. Thus, $x = \ln e$ indicates that x is the logarithm, so x is the exponent. Thus, $x = \ln e$ means that $e = e^x$, and this statement is true (by matching exponents on e) only if $x = 1$. Thus,

$$\ln e = 1$$

b. For the moment, we will assign $\ln 1$ the value x. That is, we will let

$$x = \ln 1$$

$x = \ln 1$ means that $1 = e^x$, and this statement is true only when $x = 0$ (since $e^0 = 1$). Thus,

$$\ln 1 = 0$$

■

Comment The two expressions evaluated in Example 3 make up two of the basic properties of the natural logarithm.

Properties of the Natural Logarithm

The function $y = \ln x$ possesses several properties that are very useful when solving equations that involve natural logarithms. We will now list six of them.

Properties of the Natural Logarithm

1. $\ln e = 1$
2. $\ln 1 = 0$
3. $\ln e^x = x$
4. $\ln (xy) = \ln x + \ln y$ (Multiplication becomes addition.)
5. $\ln \left(\frac{x}{y}\right) = \ln x - \ln y$ (Division becomes subtraction.)
6. $\ln x^r = r \cdot \ln x$ (Exponentiation becomes multiplication.)

Example 4 Use the properties of the natural logarithm to expand each logarithmic expression.

a. $\ln (3x)$ **b.** $\ln \left(\dfrac{8x}{x+4}\right)$ **c.** $\ln x^6$ **d.** $12{,}000 \ln (xy^5)$

Solution

a. $\ln (3x) = \ln 3 + \ln x$ Property 4

$\approx 1.0986 + \ln x$ Using a calculator

b. $\ln \left(\dfrac{8x}{x+4}\right) = \ln (8x) - \ln (x+4)$ Property 5

$= \ln 8 + \ln x - \ln (x+4)$ Property 4

$\approx 2.0794 + \ln x - \ln (x+4)$ Using a calculator

c. $\ln x^6 = 6 \ln x$ Property 6

d. $12{,}000 \ln (xy^5) = 12{,}000[\ln x + \ln y^5]$ Property 4

$= 12{,}000[\ln x + 5 \ln y]$ Property 6

Notice that $\ln (xy^5) \neq 5 \ln (xy)$ since $5 \ln (xy) = \ln [(xy)^5]$. The exponent 5 in the example is not associated with the *entire* argument of the natural logarithm, and therefore property 6 does not apply in the first step. ■

Example 5 Use the properties of the natural logarithm to write each logarithmic expression as an expression with a single logarithm.

a. $\ln x + 6 \ln y$ **b.** $\ln a - \ln b - \ln c$

Solution

a. $\ln x + 6 \ln y = \ln x + \ln y^6$ Property 6

$= \ln (xy^6)$ Property 4

b. $\ln a - \ln b - \ln c = (\ln a - \ln b) - \ln c$

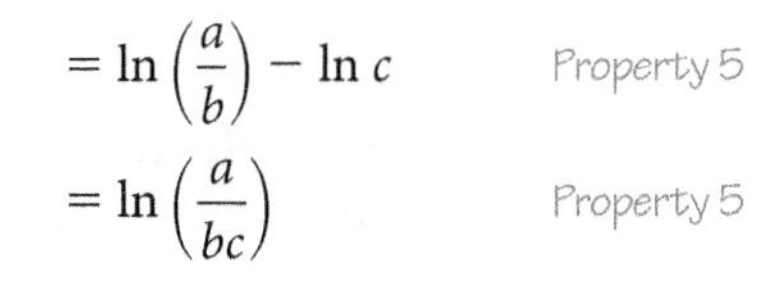

$$= \ln\left(\frac{a}{b}\right) - \ln c \qquad \text{Property 5}$$

$$= \ln\left(\frac{a}{bc}\right) \qquad \text{Property 5}$$

The Reversal Effect of the Natural Exponential and Natural Logarithm Functions on Each Other

An important and useful feature of the natural exponential and natural logarithmic functions is the effect they have on each other. This effect is illustrated in Figure 1 and Figure 2. Figure 1 indicates that $e^{\ln x} = x$ and Figure 2 indicates that $\ln e^x = x$. Since the natural exponential and natural logarithm functions are inverses of each other, composing them acts to reverse the effect of one on the other.

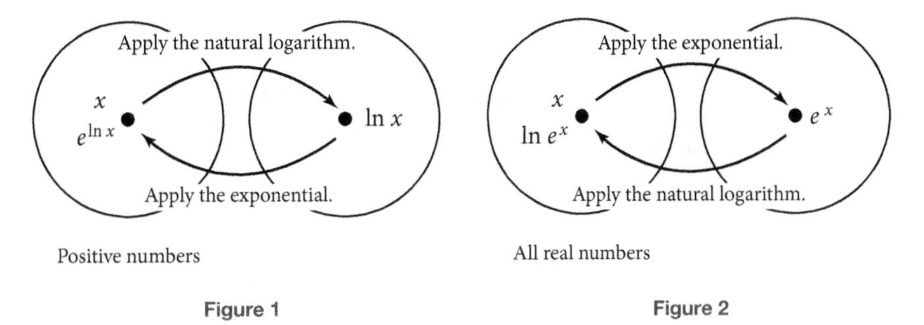

Figure 1 Figure 2

For example, if $x = 8$, then

1. Apply the natural logarithm function to 8, getting ln 8.
2. Now apply the exponential function to ln 8, getting $e^{\ln 8}$.
3. The result is 8; that is, $e^{\ln 8} = 8$.

Vice versa,

1. Apply the exponential function to 8, getting e^8.
2. Now apply the natural logarithm to e^8, getting $\ln e^8$.
3. The result is 8; that is, $\ln e^8 = 8$.

If you try these calculations on your calculator, you may start with 8, but you may get back to a number such as 7.9982753. The discrepancy can be explained by the fact that some calculators round when working with the natural logarithm and exponential functions.

To be sure that $e^{\ln x} = x$, note that $\ln x$ equals some number. For the moment, we will call it y; that is, we will let $y = \ln x$. This means that $x = e^y$. Then,

$$e^{\ln x} = e^y = x$$

The reversal effect is precisely what allows us to solve exponential equations using logarithms, as the next three examples show.

Example 6 To four decimal places, approximate the solution to the equation $e^{2x+5} = 12$.

Solution Exponential equations are commonly solved by taking the natural logarithm of each side. This serves to reverse the effect of the exponential and allows us to isolate the variable and thus solve for it.

$$e^{2x+5} = 12$$
$$\ln e^{2x+5} = \ln 12$$
$$2x + 5 = \ln 12 \qquad \text{Property 3}$$
$$2x = \ln 12 - 5$$
$$x = \frac{\ln 12 - 5}{2}$$
$$x \approx -1.2575 \qquad \text{Round to four decimal places}$$

Example 7 The radioactive isotope uranium-235, ^{235}U, decays according to the exponential decay model $A = A_0e^{-kt}$, where A_0 is the amount of ^{235}U initially present, A is the amount present after t millions of years, and k is the decay rate of ^{235}U.

a. The half-life of ^{235}U is about 700 million years. Use this information to calculate the value of the decay constant k for this model.

b. Use the model with the newly calculated value of k to determine how many years it would take for a particular quantity of ^{235}U to decay to 90% of its original amount. ($0.90A_0$ represents 90% of A_0.)

Solution

a. We wish to find k when the new amount of ^{235}U is $\frac{1}{2}$ of the initial amount (since the information we have is that the *half*-life is 700 million years). That is, we wish to find k when $A = \frac{1}{2}A_0$ and $t = 700$.

$$A = A_0e^{-kt}$$
$$\frac{1}{2}A_0 = A_0e^{-k(700)} \qquad \text{Divide by } A_0$$
$$\frac{1}{2} = e^{-k(700)} \qquad \text{Take the ln of each side}$$
$$\ln\frac{1}{2} = \ln e^{-k(700)} = -700k \qquad \text{Divide by } -700$$
$$k = \frac{\ln\frac{1}{2}}{-700} \approx 0.00099$$

b. The decay model for ^{235}U is now $A = A_0e^{-0.00099t}$. Our next step is to replace A with $0.90A_0$:

$$0.90A_0 = A_0e^{-0.00099t} \qquad \text{Divide each side by } A_0$$
$$0.90 = e^{-0.00099t} \qquad \text{Take the natural log of each side}$$

$$\ln(0.90) = \ln e^{-0.00099t}$$

$$\ln(0.90) = -0.00099t \quad \text{Divide each side by } -0.00099$$

$$\frac{\ln(0.90)}{-0.00099} = t$$

$$t \approx 106.4$$

We conclude that after approximately 106.4 million years, a quantity of ^{235}U will have decayed so that only 90% remains. ■

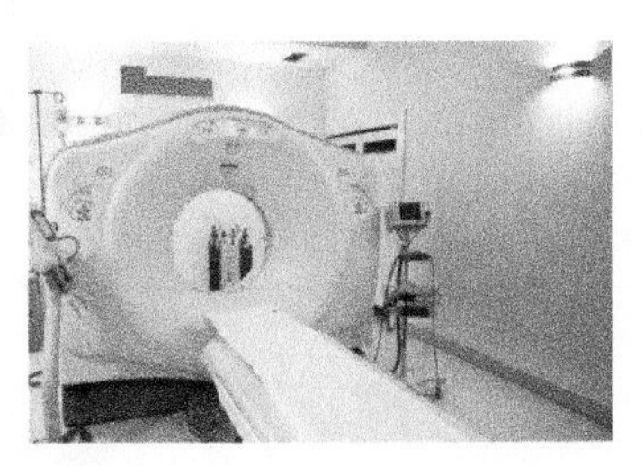

© Baran Özdemir/iStockPhoto

Example 8 A piece of equipment depreciates according to the function $A = A_0e^{-0.32t}$, where t represents the number of years since its purchase. A hospital has purchased a piece of capital equipment for $635,000. It will replace that piece of equipment when its value is 10% of its purchase price. To the nearest year, how many years will the hospital own the equipment?

Solution Since the initial value of the equipment is $635,000, we substitute 635,000 for A_0 in the exponential function $A = A_0e^{-0.32t}$. We can also substitute 63,500 for A since that value represents 10% of the original value. These substitutions produce the equation

$$63{,}500 = 635{,}000e^{-0.32t}$$

We will solve this equation for t using the method followed in Example 7.

$$63{,}500 = 635{,}000e^{-0.32t}$$

$$\frac{63{,}500}{635{,}000} = e^{-0.32t}$$

$$\ln(0.1) = \ln e^{-0.32t}$$

$$-2.302585093 \approx -0.32t$$

$$\frac{-2.302585093}{-0.32} \approx t$$

$$t \approx 7$$

Thus, the hospital will own the machine for approximately 7 years. ■

The Graph of the Natural Logarithm Function

We have just examined the reversal effect that the natural exponential and natural logarithm functions have on each other. To generalize this idea, imagine that the points $(a_1, b_1), (a_2, b_2), \ldots, (a_n, b_n)$ are points on the natural exponential curve $f(x) = e^x$. Reversing the order of the components of each ordered pair results in the set of points $(b_1, a_1), (b_2, a_2), \ldots, (b_n, a_n)$. The graph of these points produces the natural logarithm curve. The natural exponential function $f(x) = e^x$ and the natural logarithm function $f(x) = \ln x$ are illustrated in Figure 3.

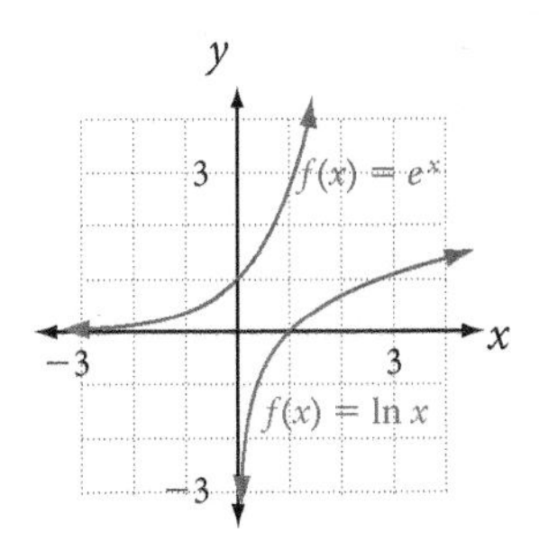

Figure 3

Mathematically, the natural exponential and natural logarithm functions are said to be **inverses** of each other.

The Natural Logarithm Function

Notice the following important features of $f(x) = \ln x$.

1. As x increases, $f(x) = \ln x$ increases but does so at a decreasing rate.
2. For $0 < x < 1$, $\ln x < 0$; that is, for input values of x strictly between 0 and 1, the output values of $\ln x$ are negative.
3. For $x > 1$, $\ln x > 0$; that is, for input values of x strictly greater than 1, the output values of $\ln x$ are positive.
4. $\ln (1) = 0$
5. The domain of the natural logarithm function is the set of all positive numbers. That is, only positive numbers are allowed as input values. The graph in Figure 3 shows that $\ln x$ is not defined for negative numbers or zero.

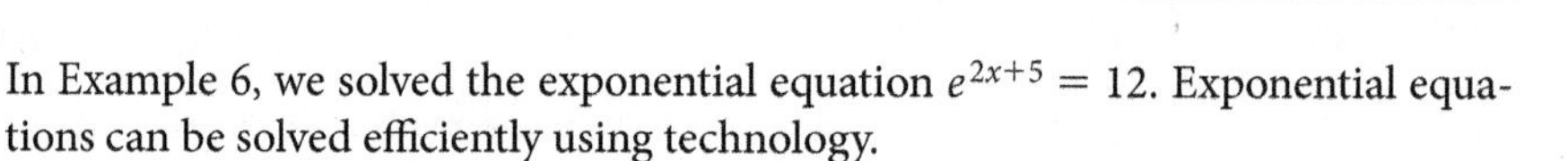

Using Technology 4.2

In Example 6, we solved the exponential equation $e^{2x+5} = 12$. Exponential equations can be solved efficiently using technology.

Example 9 Use Wolfram|Alpha to solve the exponential equation

$$e^{2x+5} = 12.$$

Solution Online, go to www.wolframalpha.com. In the entry field, enter

Solve e^(2x+5) = 12

Wolfram|Alpha solves the equation. ■

Getting Ready for Class

After reading through the preceding section, respond in your own words and in complete sentences.

A. Give a definition for $y = \ln x$.
B. What is $\ln e$?
C. How are $f(x) = e^x$ and $g(x) = \ln x$ related?
D. What is the first step in solving $e^{2x+5} = 12$?

Problem Set 4.2

Skills Practice

For Problems 1-6, express each equation in logarithmic form.

1. $m = e^{3n}$
2. $a = e^{6x}$
3. $y = e^{3x-5}$
4. $x = e^{-3k+8}$
5. $4 = e^{3k+1}$
6. $8.262 = e^{4x-10}$

For Problems 7-12, express each equation in exponential form.

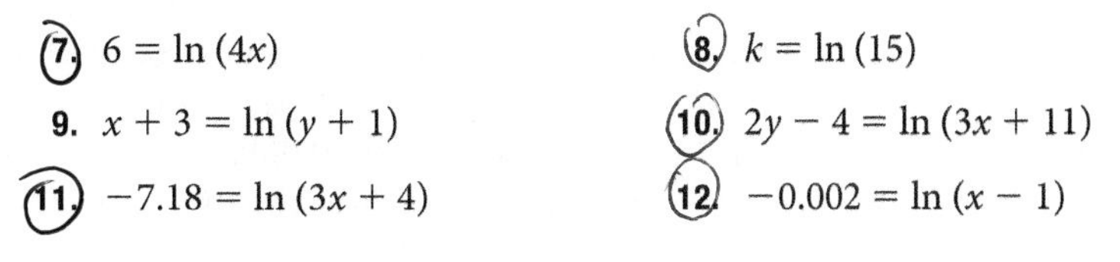

7. $6 = \ln(4x)$
8. $k = \ln(15)$
9. $x + 3 = \ln(y + 1)$
10. $2y - 4 = \ln(3x + 11)$
11. $-7.18 = \ln(3x + 4)$
12. $-0.002 = \ln(x - 1)$

For Problems 13-18, expand each logarithmic expression so that each variable appears with exponent 1.

13. $\ln 5x$
14. $\ln(x^3y^4)$
15. $\ln \dfrac{5}{y^2}$
16. $\ln \dfrac{x^2y^5}{z^4w^3}$
17. $\ln \dfrac{10x}{x-6}$
18. $500 \ln(x^2y)$

For Problems 19-24, write each logarithmic expression as an expression with a single logarithm.

19. $\ln 8 + \ln x$
20. $4 \ln x + 2 \ln y$
21. $\ln x - 3 \ln y - 5 \ln z$
22. $\ln(x + 4) - \ln(x - 4)$
23. $\ln(x + 2) - \ln(x + 7)$
24. $\ln(x + 3) + \ln(x - 3)$

For Problems 25-30, solve each exponential equation. If the solution is not an integer, round it to four decimal places.

25. $e^{3x+7} = 12$
26. $e^{-5x+2} = 70$
27. $31e^{8x-6} = 178$
28. $-62e^{3x-10} = -1{,}085$
29. $e^{-0.03t} = 0.4724$
30. $185e^{-1.6t} = 1.5233$

Check Your Understanding

In each of Problems 31-40, an exponential function is given and a question is posed that makes direct use of that exponential relationship. Pose a new question in such a way that a logarithmic relationship must be used. You do not need to make any computations, and answers may vary. (Refer to the discussion at the beginning of this section.)

31. Life Science: Healing of Skin Wounds The natural exponential function

$$A(t) = 5e^{-0.076t}$$

can, under certain conditions, be used to relate the area $A(t)$ (in square centimeters) of unhealed skin to the number of days t that have passed since the skin received a wound of size 5 square centimeters. Find the number of square centimeters of unhealed skin 20 days after the occurrence of the wound.

Source: *The Role of Sutures on Wound Healing* Infection Control Today, September 1, 2001

32. Business: Revenue Growth An investor and collector of art is considering buying a painting by a well-established artist for \$70,000. The gallery owner believes the function

$$V(t) = 70{,}000e^{0.003t} \quad t \geq 0$$

accurately predicts the value of this artist's work t months from the time of purchase. Estimate the value of this painting 10 months from the time of purchase.

33. Physics: Atmospheric Pressure The function $P(h) = 14.7e^{-0.21h}$ relates the atmospheric pressure P (in pounds per square inch) to the altitude h (in miles above sea level). Find the atmospheric pressure 3.5 miles above sea level.

Source: Pressure with Height WW2010 University of Illinois

34. Business: Investment If \$4,000 is invested at 9% interest compounded continuously, the function $A(t) = 4{,}000e^{0.09t}$ relates the amount A (in dollars) accumulated after t years. Find how much money is accumulated in such an account after 6.8 years.

© Classix/iStockPhoto

35. Economics: City Population The function $P(t) = 25.5e^{0.012t}$ relates the population P (in millions) of Sri Lanka to the time t (in years from 2010). Estimate the population of Sri Lanka in the year 2016.

Source: 2011 CIA World Factbook and other sources

36. Business: Value of an Investment The function $PV(t) = 85{,}000e^{-t^{2/3}}$ relates the present value PV (in dollars) of an investment of \$85,000 to the time t (in years from now) the money has been invested. Find the present value of an \$85,000 investment 23 years from now.

37. Business: Task Completion The function $N(t) = 125 - 125e^{-0.08t}$ relates the number N of tasks a person can complete in an 8-hour working day to the number of days t since the person began performing the tasks. How many tasks can a person who has been working at the job for 21 days be expected to complete in an 8-hour working day?

38. Psychology: Memory Psychologists believe that the function

$$f(t) = A(1 - e^{-kt})$$

relates the number of symbols f that a person can memorize to a particular time period t (in minutes). If $A = 30$ and $k = 0.08$, how many symbols can a person be expected to memorize in 15 minutes?

39. Life Science: Epidemics A city's health agency can approximate the number of people in the city who have a particular disease t days after the outbreak of the disease with the function $N(t) = Ce^{0.03t}$, where C is the number of people who initially have the disease. If, initially, 1500 people have the disease, about how many people are expected to have the disease 15 days after the outbreak of disease?

40. Economics: Price Stability Economists interested in price stability have established that the function $p(t) = (p_0 - p_e)e^{k(a-A)t} + p_e$ relates the current price of a commodity to the time t in months since it was initially priced at p_0 dollars. p_e represents the equilibrium price of the commodity, and the numbers a and A are constants that depend on the supply and demand of the item. If the commodity was originally priced at \$40 and the equilibrium price is \$32 and $k = 0.11$, $a = 4$ and $A = 5$, what is the expected price of the item 14 months after its initial pricing?

Source: Electronic Journal of Qualitative Theory of Differential Equations

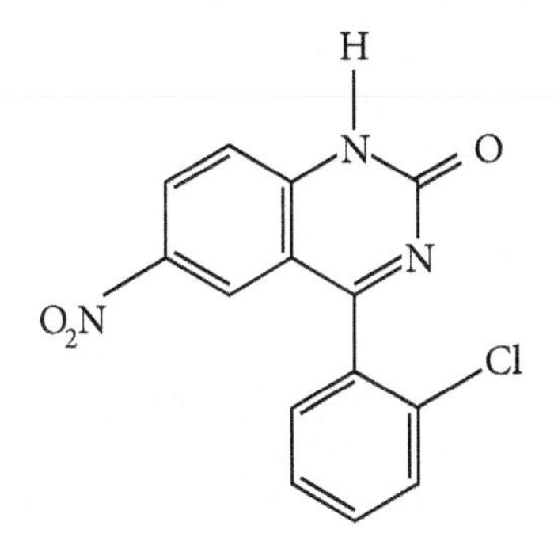

41. Health Science: Drug Half-life The drug clonazepam is marketed by Roche under the trade name Klonopin. It is a benzodiazepine drug having anxiolytic, anticonvulsant, muscle relaxant, and hypnotic properties and has an advertised estimated half-life of between 30 and 40 hours. The half-life of a drug is the amount of time it takes the drug to be reduced to one-half its original amount. Drugs decay in the bloodstream according to the exponential decay model $A = A_0e^{-kt}$, where A_0 is the amount of the drug initially present, A is the amount present t hours after being administered, and k is the decay constant. The exponential decay models for the 30-hour and 40-hour half-life estimates are, respectively, $A = A_0e^{-0.0231t}$ and $A = A_0e^{-0.0173t}$. Just by observing the models, how do you tell which one corresponds to the 30-hour half-life estimate?

Source: Druginfonet.com

42. Health Science: Drug Half-life AndroGel 1.62% is a drug used to treat adult males who have low or no testosterone. It is marketed by Abbott Laboratories and has an advertised estimated half-life of between 10 and 100 hours. The half-life of a drug is the amount of time it takes the drug to be reduced to one-half its original amount. Drugs decay in the bloodstream according to the exponential decay model $A = A_0e^{-kt}$, where A_0 is the amount of the drug initially present, A is the amount present t hours after being administered, and k is the decay constant. The exponential decay models for the 10-hour and 100-hour half-life estimates are, respectively, $A = A_0e^{-0.06931t}$ and $A = A_0e^{-0.00693t}$. Just by observing the models, how do you tell which one corresponds to the 10-hour half-life estimate?

Source: www.drugs.com

Modeling Practice

43. **Economics: Market Value of Property** The market value MV of a piece of city property is related to the time t (in years from now) by the function

$$MV(t) = 240{,}000e^{0.2t^{2/5}}$$

What is the expected market value of the piece of property 15 years from now?

44. **Physical Science: Newton's Law of Cooling** Newton's Law of Cooling relates the temperature T of an object taken from an environment at a temperature of 300 °F to an environment of 70 °F to the time t (in minutes) that the object has been in the cooler environment by the function $T = 70 + 230e^{-0.19018t}$. What is the temperature of such an object 20 minutes after it is placed into the 70 °F environment? (Round to the nearest degree.)

45. **Physical Science: Newton's Law of Cooling** A form of Newton's Law of Cooling is used in police investigations to determine time of death. The exponential function

$$T = T_e + (T_0 - T_e)e^{-kt}$$

relates the victim's current body temperature, T, to his body temperature at the time of death, T_0, the assumed constant temperature of the surrounding environment, T_e, and the number of hours, t, since death. Suppose a police investigator arrives on the scene 7 hours after the victim, Mr. X, has died. If the investigator finds the surrounding air temperature is 68 °F, and assumes that the victim's body temperature at the time of his death is 98.6 °F, and after checking with the police department's supervising mathematician finds that $k = 0.133543$, what is the victim's current body temperature? Round to the nearest one degree.

Source: The Biology Project, BioMath, The University of Arizona

46. **Marketing: Advertising Campaigns** The marketing department of a company that produces animal-shaped fruit snacks has determined that t weeks after an advertising campaign ends, the number N of boxes of such snacks it sells each month decreases according to the function $N(t) = 14{,}000e^{-0.23t}$. How many boxes of animal-shaped fruit snacks can the company expect to sell 7 weeks after the close of the advertising campaign?

47. **Physical Science: Atmospheric Pressure** The function $P(h) = 14.7e^{-0.21h}$ relates the atmospheric pressure P (in pounds per square inch) to the altitude h (in miles above sea level). Find the altitude above sea level at which the atmospheric pressure measures 6.5 pounds per square inch.

48. **Business: Investment** If $18,500 is invested at 9% interest compounded continuously, the function $A = 18{,}500e^{0.09t}$ relates the amount A (in dollars) accumulated after t years. Approximately how long will it take to accumulate $27,051.20?

49. **Economics: City Population** The function $P(t) = 36e^{0.015t}$ relates the population P (in thousands) of a growing city to the time t (in years from the present). Approximately how many years from now will the population of this city be 119,175? (Remember, P is in thousands.)

50. Marketing: Advertising The marketing department of a company that produces and sells inexpensive but durable dinnerware has determined that t weeks after an advertising campaign ends, the number N of dinnerware sets it sells throughout the country each month decreases according to the function $N = 14{,}800e^{-0.23t}$.

a. How many sets of dinnerware does the company expect to sell immediately at the close of an advertising campaign?

b. Approximately how many weeks after the end of an advertising campaign will the company be selling 4,687 dinnerware sets?

51. Manufacturing: Assembly Time Industrial psychologists have determined that the function $N(t) = 125 - 125e^{-0.08t}$ relates the number N of assembly tasks a person can complete each 8-hour working day to the number of days t since the person began doing the tasks. Approximately how many days after starting the particular task will the worker be completing 114 tasks in an 8-hour working day?

52. Political Science: Growth of Bureaucracy The function $N(t) = 14e^{0.019t}$ describes the growth of a county's bureaucracy. N is the size (in thousands) of the bureaucracy, and t is the time (in years) from 2000. Approximately how many years from 2000 will the size of this county's bureaucracy be 20,500? (Remember, N is measured in thousands.)

53. Medicine: Drug Concentration When a drug is intravenously administered to a patient on a continuous basis, the amount A (in milligrams) of the drug in the patient's bloodstream t minutes after injection of the drug is given by

$$A(t) = \frac{1}{0.52}(50 - 50e^{-0.12t})$$

Approximately how much time will have to pass after injection of the drug so that 55 milligrams of the drug will be in the patient's bloodstream?

54. Ecology: Light Intensity in Water For bodies of water that are relatively clear, light intensity is reduced according to the function $I = I_0e^{-kd}$, where I is the intensity of the light d feet below the surface of the water. One of the clearest saltwater bodies of water is the Sargasso Sea off the West Indies ($k = 0.00942$). At what depth is the light intensity reduced to 60% of that at the surface?

Source: *Adaptation and spectral tuning in divergent marine proteorhodopsins,* ISME Journal (2007)

© Michael Utech/iStockPhoto

55. Physical Science: Radioactive Decay The radioactive isotope Strontium-90, which is a by-product of the fission of uranium and plutonium in nuclear reactors, decays according to the formula $A = A_0e^{-0.0248t}$, where A_0 is the amount of Strontium-90 initially present, and A is the amount present after t years. Approximately how many years will have to pass before only 50% of a particular quantity of Strontium-90 remains? Your answer will be the half-life of Strontium-90.

Source: U. S. Environmental Protection Agency Radiation Protection

56. Business: Depreciation of Equipment XYZ Textbooks spends $25,000 on a piece of equipment that depreciates at the rate of 15% each year. The exponential function $A(t) = 25{,}000e^{-0.1625t}$ approximates the value, $A(t)$, of the piece of equipment t years after the piece is purchased. How long will it take this piece of equipment to depreciate to $5,000? Round your result to the nearest year.

Source: PDR.net

57. **Life Science: Drug Concentration** In many cases, the drug Lexapro has a half-life of 30 hours. The exponential function $A(t) = 10e^{-0.0231t}$ can be used to approximate the amount of Lexapro in a person's bloodstream t hours after 10 milligrams of it is ingested.
    a. If a person takes a prescribed dose of 10 milligrams at 6:00 AM, how many milligrams will be in his bloodstream at noon? Round your result to one decimal place.
    b. If a person takes a prescribed dose of 10 milligrams at 6:00 AM, in how many hours will the amount have decayed to about 6.25 milligrams? Round your result to the nearest hour.

58. **Life Science: Effect of Valium** When in the bloodstream, the drug Valium decays according to the function $A = A_0e^{-0.0144t}$ where t is in minutes. Approximately how many minutes after taking a prescribed dose of 5 milligrams of Valium will there be less than 2 milligrams in the bloodstream?

    Source: PDR.net

59. **Economics: Doubling Time** The amount A of dollars accumulated after t years through an investment of A_0 dollars at r% interest compounded continuously can be determined from the function $A = A_0e^{rt}$. The *doubling time* of an investment is the time required for an investment to double in value. Find the doubling time for an investment made at
    a. 6% interest compounded continuously.
    b. 8% interest compounded continuously.
    c. 10% interest compounded continuously.

60. **Life Science: Healing of Wounded Skin** When human skin receives a wound of surface area A_0, the exponential function $A = A_0e^{-0.11t}$ relates the amount A (in square centimeters) of unhealed skin to the number of days t since the wound was received. Suppose a person suffers a skin wound measuring 1.24 cm^2. How many days must pass before only 0.5 cm^2 of unhealed skin remains? (This is similar to the questions we posed at the beginning of this section.)

    Source: The Role of Sutures in Wound Healing
    Infection Control Today September 1, 2001

© Jeremy Edwards/iStockPhoto

61. **Economics: Mutual Funds** A mutual fund expects annual returns to be modeled by the function $A = A_0e^{0.08t}$. How much would an investor have to invest now to have $30,000 available in 10 years?

62. **Business: Market Share** A company believes its sales are modeled by the function $A = A_0e^{0.22t}$ where t is in years. The total value of the market is 100 million dollars. The company currently has 2% of the market. How long before the company can expect to have 20% of the market?

63. **Economics: Credit** The function $A = A_0e^{0.195t}$ models the amount an individual owes a credit company t years after some initial debt. If a person currently owes $2,300 and makes no payments, how much time will pass before that person owes $10,000?

64. **Health Science: Drug Half-life** The half-life of common serum aspirin is estimated to be between 15 and 20 minutes. The half-life of a drug is the amount of time it takes the drug to be reduced to one-half its original amount. Drugs decay in the bloodstream according to the exponential decay model $A = A_0e^{-kt}$, where A_0 is the amount of the drug initially present, A is the amount present t minutes after being administered, and k is the decay constant. The exponential decay models for the 15-minute and 20-minute half-life estimates are, respectively, $A = A_0e^{-0.0462t}$ and $A = A_0e^{-0.0346t}$.
   a. Just by observing the models, how do you tell which one corresponds to the 15-minute half-life estimate?
   b. Compute the quotient $\frac{A_0e^{-0.0346t}}{A_0e^{-0.0462t}}$ at $t = 60$ minutes to determine how many times more aspirin is in a person's bloodstream under the 20-minute half-life estimate than under the 15-minute estimate 60 minutes after taking the aspirin.

Using Technology Exercises

Use Wolfram|Alpha to solve each problem.

65. Approximate the solution to $450e^{3x-5} = 1{,}400$ by rounding your solution to two decimal places.

66. Approximate the solution to $16e^{-0.08x} + 182 = 245$ by rounding your solution to two decimal places.

67. **Business: Equipment Depreciation** A piece of equipment depreciates according to the function $A = A_0e^{-0.32t}$. XYZ Textbooks company has purchased a piece of capital equipment for \$15,000. The company will replace that piece of equipment when its value is 5% of its purchase price. How many years will XYZ own this piece of equipment?

68. **Physical Science: Atmospheric Pressure** The function $P(h) = 14.7e^{-0.21h}$ relates the atmospheric pressure P (in pounds per square inch) to the altitude (in miles above sea level). At how many miles above sea level is the atmospheric pressure 90% of that at sea level? (Note that 14.7 pounds per square inch is the atmospheric pressure P at sea level.)

69. **Economics: Market Value of Property** The market value, V, of a piece of city property is related to the time t (in years from now) by the function

$$V(t) = 240{,}000e^{0.2t^{2/5}}$$

If real estate trends continue as they are now, how many years from now will it be before the expected market value of this piece of property doubles? Round your answer to the nearest year. Construct the graph of this function. Does the graph exhibit the standard shape of either the exponential growth or exponential decay function?

Getting Ready for the Next Section

Simplify.

70. $\ln e$

71. e^1

72. $3 \ln e$

73. $3 \ln e + 1$

Differentiate.

74. $y = 7x^3$

75. $y = 8x^2$

76. $y = (x^2 - 4)^3$

77. $y = 5x^2 + 2x - 7$

Find decimal approximations (to the nearest thousandth) for

78. $7e^3$

79. $7e^2$

80. $\ln 2$

81. $\ln (1.02)$

82. Use the product rule to implicitly differentiate the expression $3x^4y^2$ for y with respect to x.

83. Use the properties of exponents to expand $\ln (xy^2)$.

Differentiating the Natural Logarithm Function

4.3

In many fields of study we are often interested in the rate at which the output variable y changes as the input variable x changes. Of course, when we think *change*, we think *differentiation* (since differentiation is the language of change). In each of the functions introduced in the preceding sections—$y = e^x$ and $y = \ln x$—the output variable y is related to the input variable x by a non-algebraic function, the natural exponential and natural logarithm, respectively. Since our current collection of rules of differentiation apply only to algebraic functions, we will need new rules to find $\frac{d}{dx}[e^x]$ and $\frac{d}{dx}[\ln x]$ and, more generally, $\frac{d}{dx}[e^{g(x)}]$ and $\frac{d}{dx}[\ln [g(x)]]$. Since this is an *applied* calculus text, we will go directly to the formulas and examples of their application. In this section we consider the differentiation of the natural logarithm function.

Differentiating the Natural Logarithm Function

The Derivative of the Natural Logarithm Function

$$\frac{d}{dx}[\ln x] = \frac{1}{x}$$

VIDEO EXAMPLES

SECTION 4.3

Example 1 For the function $f(x) = 7x^3 \ln x$, find

a. $f'(x)$

b. an equation of the line tangent at $x = e$.

Solution

a. Looking at $f(x)$ globally, we see a product. Thus, we will differentiate this function using the product rule.

$$f'(x) = \underbrace{21x^2}_{\text{(first)}'} \cdot \underbrace{\ln x}_{\text{(second)}} + \underbrace{7x^3}_{\text{(first)}} \cdot \underbrace{\frac{1}{x}}_{\text{(second)}'}$$

$$= 21x^2 \ln x + 7x^2$$

$$= 7x^2(3 \ln x + 1)$$

b. When $x = e$, $f(e) = 7e^3 \ln e$

$$= 7e^3$$

$$\approx 140.6$$

We have the point $(e, 140.6)$, which gives us $x_1 = e$ and $y_1 = 140.6$. All we need is the slope of the tangent at this point.

The slope is given by $f'(e)$. Using $f'(x) = 7x^2(3 \ln x + 1)$,

$$f'(e) = 7e^2(3 \ln e + 1)$$

$$\approx 206.9$$

To find the equation of the tangent line at $(e, 140.6)$, we use

$$y - y_1 = m(x - x_1)$$

with $m = 206.9$, $x_1 = e$, $y_1 = 140.6$

$$y - y_1 = m(x - x_1)$$
$$y - 140.6 = 206.9(x - e)$$
$$y - 140.6 = 206.9x - 206.9 \cdot e$$
$$y = 206.9x + 140.6 - 206.9 \cdot e$$
$$y = 206.9x - 421.8$$

■

To differentiate $y = \ln[u]$, where $u = g(x)$, such as $y = \ln(3x^2 + 2x - 5)$, we use the chain rule

$$\frac{dy}{dx} = \frac{dy}{du} \cdot \frac{du}{dx}$$
$$\frac{dy}{dx} = \frac{1}{u} \cdot \frac{du}{dx}$$
$$\frac{dy}{dx}\left[\ln[g(x)]\right] = \frac{1}{g(x)} \cdot g'(x)$$

We have developed the differentiation rule for the generalized natural logarithm function.

Chain Rule for Logarithmic Functions

$$\frac{dy}{dx}\left[\ln[g(x)]\right] = \frac{1}{g(x)} \cdot g'(x)$$

This differentiation rule states that the derivative of the generalized natural logarithm function $f(x) = \ln[g(x)]$ is

(1 over the argument) *times* (the derivative of the argument)

That is,

$$\frac{1}{\text{argument}} \quad \textit{times} \quad (\text{argument})'$$

Example 2 Find $f'(x)$ for $f(x) = \ln(5x^2 + 2x - 7)$.

Solution

$$f'(x) = \overbrace{\frac{1}{5x^2 + 2x - 7}}^{\text{1 over the argument}} \cdot \overbrace{\frac{d}{dx}[5x^2 + 2x - 7]}^{(\text{argument})'}$$
$$= \frac{1}{5x^2 + 2x - 7} \cdot (10x + 2)$$
$$= \frac{10x + 2}{5x^2 + 2x - 7}$$

■

Example 3 Find $f'(x)$ for $f(x) = \ln(x^2 - 4)^3$.

Solution We will illustrate two ways of differentiating this function, the first using a logarithm property, $\ln x^r = r \cdot \ln x$, and the second working from its global form as a natural logarithm.

First, we will use the logarithm property to rewrite $f(x) = \ln(x^2 - 4)^3$ as $f(x) = 3\ln(x^2 - 4)$. (Recall that $\ln(A + B) \neq \ln A + \ln B$; do not mistakenly write $\ln(x^2 - 4)$ as $\ln x^2 - \ln 4$.) Then

$$\begin{aligned} f'(x) &= 3\frac{d}{dx}[\ln(x^2 - 4)] \\ &= 3 \cdot \frac{1}{(x^2 - 4)} \cdot 2x \\ &= \frac{6x}{x^2 - 4} \end{aligned}$$

Now we will differentiate this function using an approach that does not use the property of logarithms. Globally, this function is a natural logarithm function, and locally, the argument of the logarithm is a power function. Thus, we will be sure to differentiate the argument using the generalized power rule.

$$\begin{aligned} f'(x) &= \frac{1}{(x^2 - 4)^3} \cdot \frac{d}{dx}[(x^2 - 4)^3] \\ &= \frac{1}{(x^2 - 4)^3} \cdot 3 \cdot (x^2 - 4)^2 \cdot 2x \\ &= \frac{6x(x^2 - 4)^2}{(x^2 - 4)^3} \\ &= \frac{6x}{x^2 - 4} \end{aligned}$$

Notice that the two methods produce the same results. In this case, the use of a logarithm property saved us from some algebraic manipulations. ■

Example 4 Find $f'(x)$ for $f(x) = [\ln(3x)]^4$.

Solution Looking at this function globally, we see a power. More locally, it is a natural logarithm function. (Recall that $[\ln a]^n \neq \ln a^n$). We will begin differentiating this function using the general power rule.

$$\begin{aligned} f'(x) &= 4 \cdot [\ln(3x)]^3 \cdot \frac{d}{dx}[\ln(3x)] \\ &= 4 \cdot [\ln(3x)]^3 \cdot \frac{1}{3x} \cdot \frac{d}{dx}[3x] \\ &= 4 \cdot [\ln(3x)]^3 \cdot \frac{1}{3x} \cdot 3 \\ &= \frac{4[\ln(3x)]^3}{x} \end{aligned}$$

■

Example 5 Find $f'(x)$ for $f(x) = \dfrac{6}{\ln(8x^2)}$.

Solution Looking at this function globally, we see a quotient. However, since the numerator is a constant, we will write the function in the more convenient form $f(x) = 6[\ln(8x^2)]^{-1}$. Now we will use the general power rule.

$$\begin{aligned} f'(x) &= (-1)(6)[\ln(8x^2)]^{-2} \cdot \frac{d}{dx}[\ln(8x^2)] \\ &= -6[\ln(8x^2)]^{-2} \cdot \frac{1}{8x^2} \cdot \frac{d}{dx}[8x^2] \\ &= -6[\ln(8x^2)]^{-2} \cdot \frac{1}{8x^2} \cdot 16x \\ &= \frac{-12}{x[\ln(8x^2)]^2} \end{aligned}$$

Example 6 Use implicit differentiation to find y' for $3x^4y^2 + \ln(xy^2) = 6$.

Solution We can use the logarithm properties $\ln(xy) = \ln x + \ln y$ and $\ln x^r = r \cdot \ln x$ to simplify the term $\ln(xy^2)$ to $\ln x + 2\ln y$. This helps us avoid the product rule for this part of the expression. Thus, the original equation becomes

$$3x^4y^2 + \ln x + 2\ln y = 6$$

Since this function defines y implicitly as a function of x, we will use implicit differentiation.

$$12x^3 \cdot y^2 + 3x^4 \cdot 2yy' + \frac{1}{x} + 2 \cdot \frac{1}{y} \cdot y' = 0$$

$$12x^3y^2 + 6x^4yy' + \frac{1}{x} + \frac{2}{y}y' = 0$$

Eliminate the fractions by multiplying each side by the lowest common denominator, xy.

$$12x^4y^3 + 6x^5y^2y' + y + 2xy' = 0$$

Now, isolate y'.

$$\begin{aligned} 6x^5y^2y' + 2xy' &= -12x^4y^3 - y \\ y'(6x^5y^2 + 2x) &= -12x^4y^3 - y \\ y' &= \frac{-12x^4y^3 - y}{6x^5y^2 + 2x} \end{aligned}$$

Using Technology 4.3

In Example 1, we differentiated the product $f(x) = 7x^3 \ln x$. Functions that involve logarithms can often be differentiated efficiently using technology.

Example 7 Use Wolfram|Alpha to find $f'(x)$ for $f(x) = 7x^3 \ln x$.

Solution Online, go to www.wolframalpha.com. In the entry field, enter

differentiate 7x^3ln(x)

Wolfram|Alpha returns the derivative.

Derivative:

$$\frac{d}{dx}\left(7\,x^3 \log(x)\right) = 7\,x^2\,(3\log(x) + 1)$$

Wolfram Alpha LLC. 2012. Wolfram|Alpha
http://www.wolframalpha.com/
(access November 6, 2012)

Notice that Wolfram|Alpha used log (x) to represent ln (x) with a note in the bottom right corner to explain the notation. ■

Example 8 Use Wolfram|Alpha to find $f''(x)$ for $f(x) = \ln(x^2 - 4)$.

Solution Online, go to www.wolframalpha.com. In the entry field, enter

Second derivative of ln(x^2 - 4)

Wolfram|Alpha returns the solution.

Derivative :

$$\frac{d^2}{dx^2}(\log(x^2 - 4)) = -\frac{2\,(x^2 + 4)}{(x^2 - 4)^2}$$

Wolfram Alpha LLC. 2012. Wolfram|Alpha
http://www.wolframalpha.com/
(access December 19, 2012)

■

Example 9 Use Wolfram|Alpha to find y' for $3x^4y^2 + \ln(xy^2) = 6$.

Solution Wolfram|Alpha can differentiate a function defined implicitly. Online, go to www.wolframalpha.com. In the entry field, enter

find dy/dx of 3x^4y^2 + ln(xy^2) = 6

Wolfram|Alpha returns the derivative

Result:

$$y'(x) = -\frac{y\,(12\,x^4\,y^2 + 1)}{2\,x\,(3\,x^4\,y^2 + 1)}$$

Getting Ready for Class

After reading through the preceding section, respond in your own words and in complete sentences.

A. How do you differentiate $y = \ln x$?

B. If $f(x) = \ln(5x^2 + 2x - 7)$, is the derivative

$$f'(x) = \frac{1}{5x^2 + 2x - 7}?$$

C. What is the first step in differentiating $y = [\ln x]^4$?

D. Explain how there are two ways to approach finding the derivative for $y = \ln(x^2 - 4)^3$.

Problem Set 4.3

Skills Practice

For Problems 1-18, find the derivative of each function.

1. $f(x) = \ln(2x - 7)$
2. $f(x) = \ln(5x^2 + 3x)$
3. $f(x) = 8\ln(4x^2)$
4. $f(x) = e^2 \ln(5x^3)$
5. $f(x) = x^2 \ln(3x)$
6. $f(x) = 4x \ln x$
7. $f(x) = \ln(2x + 8)^5$
8. $f(x) = \ln(x^2 + 3x)^2$
9. $f(w) = [\ln(w + 4)]^3$
10. $f(r) = 5[\ln(2r - 1)]^6$
11. $f(t) = \ln\sqrt{3t + 5}$
12. $f(c) = \ln\sqrt[3]{7c + 7}$
13. $f(u) = \ln(\ln u)$
14. $f(v) = \ln[(\ln v)^2]$
15. $f(t) = t^2[\ln(t^2)]^2$
16. $f(s) = \dfrac{\ln 6}{5\ln(2s + 1)}$
17. $f(x) = \ln\dfrac{x}{a}$, where a is a constant and $a \neq 0$.
18. $f(x) = \ln\dfrac{ax}{b}$, where both a and b are constants, $a \neq 0$, and $b \neq 0$.

Paying Attention to Instructions The next two problems are intended to give you practice reading, and paying attention to, the instructions that accompany the problems you are working.

19. If $f(x) = 4x^3 \ln x$, find
    a. $f(e)$. Round to the nearest whole number.
    b. $f'(e)$. Round to the nearest whole number.
    c. x so that $f'(x) = 0$. Leave your answer in terms of e.
    d. Using the results of your computations for $f(e)$ and $f'(e)$, find the equation of the line tangent to the curve $f(x)$ at the point $(e, f(e))$. Round decimals to the nearest whole number.

20. If $g(x) = 8\ln(4x)$, find
    a. $g\left(\frac{1}{4}\right)$. What point is this on the graph of $g(x)$?
    b. $g(e)$. Round to the nearest whole number.
    c. $g'(e)$. Round to the nearest whole number.
    d. Using the results of your computations for $g(e)$ and $g'(e)$, find the equation of the line tangent to the curve $g(x)$ at the point $(e, g(e))$. Round decimals to the nearest whole number.

For Problems 21-24, use implicit differentiation to find y'.

21. $3x^3 + \ln(3xy^2) = 2$
22. $4x + \ln(x^2y^4) = 10$
23. $2xy^2 + \ln(xy) = 1$
24. $3x^3y^4 - \ln(5x^3y) = 5$

For Problems 25-29, find and interpret each derivative.

25. Find $f'(3)$ for $f(x) = \ln(x^2 + 2x)$.

26. Find $f'(0)$ for $f(x) = \ln(5x + 1)^2$.

27. Find $f'(1)$ for $f(x) = 3x^2 - 4\ln(x^2)$.

28. Find $f'(6)$ for $f(x) = 3 \ln (4x) - 2 \ln (3x)$.

29. Find $f'(1)$ for $f(x) = \dfrac{\ln (4x)}{\ln (3x)}$.

Check Your Understanding

For Problems 30-39, a function is given along with a claimed derivative of the function. State if the indicated derivative is *correct* or *incorrect*.

30. $f(x) = \ln (x^3)$

$f'(x) = \dfrac{1}{x^3}$

31. $f(x) = \ln (4x^2 + 7x)$

$f'(x) = \dfrac{1}{4x^2 + 7x}$

32. $f(x) = \ln (3x^2 + 5x - 4)$

$f'(x) = \dfrac{1}{6x + 5}$

33. $f(x) = \ln (9x - 1)$

$f'(x) = \dfrac{1}{9x - 1} \cdot \dfrac{1}{9}$

34. $f(x) = 18 \ln (x)$

$f'(x) = \dfrac{18}{x}$

35. $f(x) = 4x^2 \ln 6$

$f'(x) = 8x \ln 6$

36. $f(x) = 10x \ln 8$

$f'(x) = 10$

37. $f(x) = 5 \ln (x^3)$

$f'(x) = \dfrac{1}{5x \cdot x^3} = \dfrac{1}{5x^4}$

38. $f(x) = 12 \ln 2$

$f'(x) = 0$

39. $f(x) = \ln (5x^5)$

$f'(x) = \dfrac{1}{x}$

40. Describe the strategy you would use in differentiating the function

$$f(x) = \ln (5x + 3)$$

41. Describe the strategy you would use in differentiating the function

$$f(x) = [\ln (6x^2)]^5$$

42. Explain how the derivative formula for $f(x) = \ln x$ is a specific case of the generalized differentiation formula for $f(x) = \ln [g(x)]$.

Modeling Practice

43. Manufacturing: Cost of Production The daily cost C (in dollars) for a manufacturer to produce x electronic video components is given by

$$C(x) = 1{,}000 \ln \sqrt[3]{3x^2 - 700}$$

If the manufacturer is currently producing 60 components each day, by how much can the cost be expected to change if production is increased by one component each day to 61 components each day?

44. Physical Science: Radioactive Decay At the start of an observation, there are 100 grams of a radioactive element present. The number of years t that must pass before N grams remain is given by the natural logarithm equation

$$t = 7{,}549.5 - 1{,}639 \ln N$$

Find $\frac{dt}{dN}$ when $N = 20$.

45. Biology: Blood Flow If the velocity of blood in an artery is too high, the flow of blood will become turbulent. The Reynolds number, R, is a measure of blood flow and is determined by the radius r of the aorta and positive constants a and b that relate to the density and viscosity of the blood. The Reynolds number related to blood flow is

$$R = a \ln r - br$$

Determine the rate at which the Reynolds number changes as the radius of the aorta changes.

Source: suggested from material in *Friction factor correlations for laminar, transition and turbulent flow in smooth pipes,* Joseph & Yang

46. Business: Textbook Publishing The senior editor for the life sciences at a textbook publishing company has determined that the equation

$$N(x) = 1.26 + 1.2 \ln (1.1 + x^3)$$

relates the expected number N (in thousands) of copies of an introductory biology textbook sold in a 1-year period to the number x (in thousands) of complimentary copies sent to professors throughout the country. Find and interpret the rate at which N changes as x changes when $x = 6$. (Remember that x is in thousands.)

47. Economics: Price of Art A San Francisco art gallery is currently showing the work of a new artist. The gallery owner believes the function

$$T(V) = 1.44 \ln \left(\frac{V}{A_0}\right) \quad V \text{ and } A_0 > 0$$

accurately relates the value V (in dollars) of this artist's pieces to the number of years T until the piece reaches that value. The purchase price of the painting is $\$A_0$. The gallery owner makes the claim that from the original price of purchase, the time required for each piece of work to reach a higher value will increase, but will do so at a decreasing rate. Assuming the value/time model is accurate, should you accept or reject the owner's claim? Provide mathematical evidence to support your conclusion.

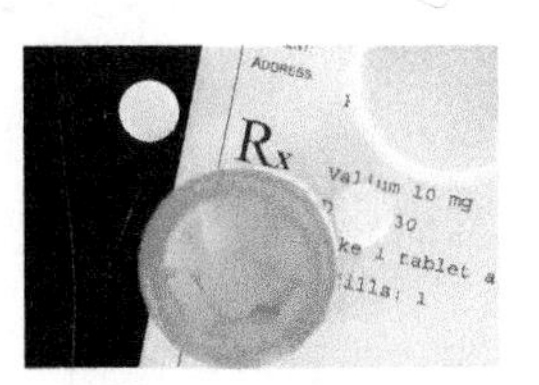

© Michael Flippo/iStockPhoto

48. Life Science: Effect of Valium When the drug Valium is in the bloodstream, the time it takes (in hours) for an initial dose of A_0 milligrams to decay to A milligrams is modeled by the function

$$T(A) = -69.25 \ln \left(\frac{A}{A_0}\right) \quad A \text{ and } A_0 > 0$$

A pharmacist makes the claim that once an initial dose of A_0 milligrams of Valium is introduced into the bloodstream, the amount of Valium in the bloodstream decreases, but does so at a decreasing rate. Assuming the value/time model is accurate, should you accept or reject the pharmacist's claim? Provide mathematical evidence to support your conclusion.

Using Technology Exercises

49. Find $f'(x)$ for $f(x) = 5x^2 \ln(x^2)$.

50. Find $f'(x)$ for $f(x) = \ln(2x)\ln(3x)$.

51. Find $f''(x)$ for $f(x) = x \ln(x)$.

52. **Economics: Value of Goods** In researching a country's industries, a financial analyst concluded that the value of all goods produced by a particular industry in the country is approximated by the function

$$V(t) = 925\left[\frac{\ln(0.075t + 0.085)}{0.75t + 0.85}\right]$$

where V is in millions of dollars and t is the number of years since 1945. Use the derivative of this function to determine by how much the value of all goods produced by this industry will be expected to decrease in the year 2015.

53. Find y' for $x^2y^3 - \ln(x^3y^3) = 10$.

Getting Ready for the Next Section

Simplify

54. e^0

55. e^1

Find the decimal approximations for

56. $e^{-0.15}$

57. $e^{-0.27}$

58. $\ln\frac{10}{20}$

59. $e^{-0.0578(5)}$

Differentiate.

60. $y = 8x^3$

61. $y = 2x^2 + 5$

62. $y = \ln 4x$

63. $y = 3x^2 + 5x$

64. Factor $24x^2e^x + 8x^3e^x$.

65. Find the equation of the line through (0, 0) that has a slope of 7.

Differentiating the Natural Exponential Function

4.4

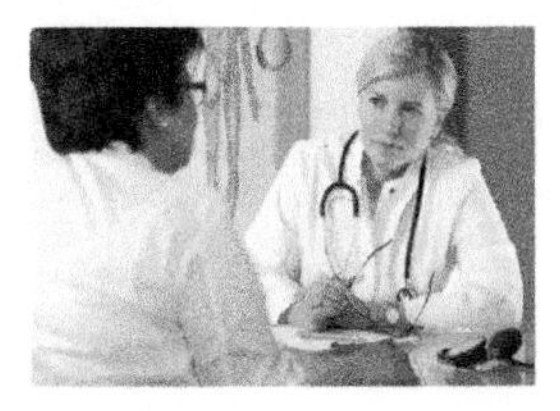

© Alexander Raths/iStockPhoto

The drug Valium is often prescribed to treat anxiety, insomnia, seizures, restless leg syndrome, and alcohol withdrawal. As most drugs do, it decays exponentially when it enters the bloodstream. Because its decay is exponential and its half-life is known, the amount of Valium in the bloodstream at any time can be modeled with an exponential decay function. Algebra can be used to find the amount of Valium in the bloodstream at any time, and calculus can be used to find the rate at which the amount is decreasing at any time.

In this section we differentiate the natural exponential function $f(x) = e^x$ and the generalized natural exponential function $f(x) = e^{g(x)}$. We derive the differentiation formula for $f(x) = e^x$ and examine the special, and often occurring, case for $f(x) = e^{ax}$.

Differentiating the Natural Exponential Function

The differentiation rule for the exponential function $f(x) = e^x$ is quickly derived by noting that

$$\ln e^x = x$$

To get $\frac{d}{dx}[e^x]$, we differentiate each side of $\ln e^x = x$ with respect to x.

$$\frac{d}{dx}[\ln e^x] = \frac{d}{dx}[x]$$

$$\frac{1}{e^x} \cdot \frac{d}{dx}[e^x] = 1$$

We then solve for $\frac{d}{dx}[e^x]$ by multiplying each side by e^x.

$$\frac{d}{dx}[e^x] = 1 \cdot e^x = e^x$$

This means that e^x is its own derivative!

The Derivative of the Natural Exponential Function

$$\frac{d}{dx}[e^x] = e^x$$

VIDEO EXAMPLES

SECTION 4.4

Example 1 Find $f'(x)$ for $f(x) = 8x^3e^x$.

Solution Looking at this function globally, we see a product. We will differentiate this function using the product rule.

$$f(x) = 8x^3e^x$$

$$f'(x) = \underbrace{24x^2}_{(\text{first})'} \cdot \underbrace{e^x}_{(\text{second})} + \underbrace{8x^3}_{(\text{first})} \cdot \underbrace{e^x}_{(\text{second})'}$$

$$= 24x^2e^x + 8x^3e^x \qquad \text{Factor out } 8x^2e^x.$$

$$= 8x^2e^x(3 + x)$$

■

Example 2 Find the equation of the line tangent to the curve $f(x) = 7xe^x$ at the point $(0, 0)$. Round decimals to two places.

Solution We have the point $(0, 0)$, which gives us $x_1 = 0$ and $y_1 = 0$. All we need is the slope of the tangent at this point. The slope at $(0, 0)$ is given by $f'(0)$. Differentiating $f(x) = 7xe^x$, we get

$$f'(x) = 7 \cdot e^x + 7x \cdot e^x$$
$$= 7e^x(x + 1)$$

Then

$$f'(0) = 7e^0(0 + 1)$$
$$= 7 \cdot 1 \cdot 1$$
$$= 7$$

To find the equation of the tangent line at $(0, 0)$, we use $y - y_1 = m(x - x_1)$ with $m = 7, x_1 = 0, y_1 = 0$.

$$y - y_1 = m(x - x_1)$$
$$y - 0 = 7(x - 0)$$
$$y = 7x$$

The graph of $f(x) = 7xe^x$, along with the tangent line at $(0, 0)$, is shown in Figure 1.

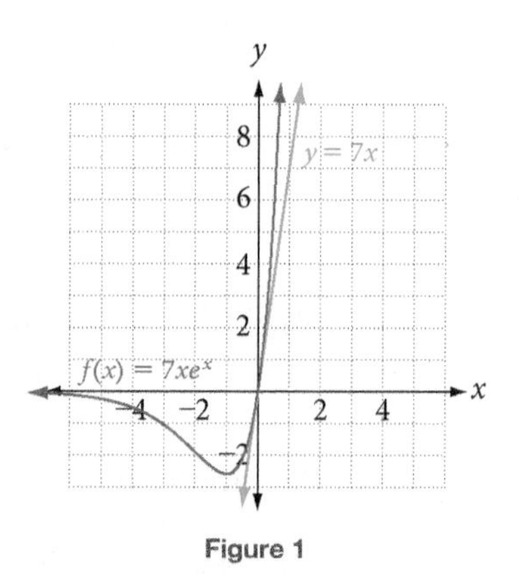

Figure 1

Differentiating the Generalized Natural Exponential Function

The differentiation rule for $y = e^{g(x)}$ is derived using the chain rule with $u = g(x)$.

$$\frac{dy}{dx} = \frac{dy}{du} \cdot \frac{du}{dx}$$
$$= \frac{d}{du}[e^u] \cdot \frac{du}{dx}$$
$$= e^u \cdot \frac{du}{dx}$$
$$= e^{g(x)} \cdot g'(x)$$

The Chain Rule for Exponential Functions

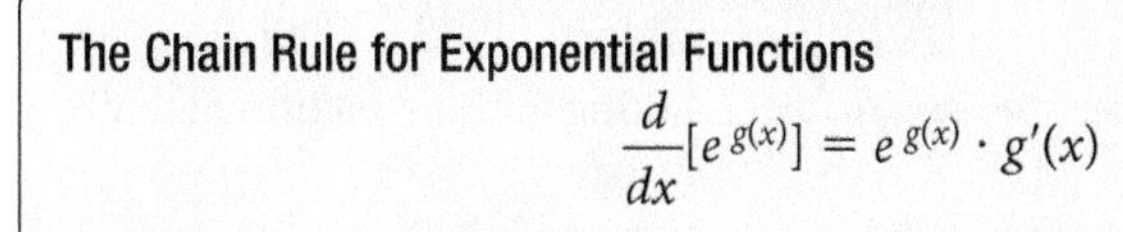

$$\frac{d}{dx}[e^{g(x)}] = e^{g(x)} \cdot g'(x)$$

This differentiation rule states that the derivative of $f(x) = e^{g(x)}$ is

(the function itself) *times* (the derivative of the exponent)

Example 3 Find $f'(x)$ for $f(x) = e^{3x^2+5x}$.

Solution Globally, this function is an exponential function. So,

$$\begin{aligned} f'(x) &= e^{3x^2+5x} \cdot \frac{d}{dx}[3x^2 + 5x] \\ &= \underbrace{e^{3x^2+5x}}_{\text{(function itself)}} \cdot \underbrace{(6x+5)}_{\text{(exponent)}'} \\ &= (6x+5)e^{3x^2+5x} \end{aligned}$$

Example 4 Find $f'(3)$ for $f(x) = \ln(2x + e^{-0.05x})$. Round the result to 4 decimal places.

Solution Globally, this function is a natural logarithm function and we will need the differentiation rule

$$\frac{1}{\text{(argument)}} \quad \textit{times} \quad \text{(argument)}'$$

$$f'(x) = \frac{1}{2x + e^{-0.05x}} \cdot \frac{d}{dx}[2x + e^{-0.05x}]$$

Now we differentiate the second factor.

$$\begin{aligned} f'(x) &= \frac{1}{2x + e^{-0.05x}} \cdot [2 + e^{-0.05x} \cdot (-0.05)] \\ &= \frac{1}{2x + e^{-0.05x}} \cdot [2 - 0.05e^{-0.05x}] \\ &= \frac{2 - 0.05e^{-0.05x}}{2x + e^{-0.05x}} \end{aligned}$$

Then, substituting 3 for x,

$$\begin{aligned} f'(3) &= \frac{2 - 0.05e^{-0.05 \cdot 3}}{2(3) + e^{-0.05 \cdot 3}} \\ &\approx 0.2852 \end{aligned}$$

Example 5 Find $f''(x)$ for $f(x) = e^{2x^2+5}$.

Solution Looking at this function globally, we see an exponential function. So,

$$f'(x) = e^{2x^2+5} \cdot \frac{d}{dx}[2x^2 + 5]$$
$$= e^{2x^2+5} \cdot 4x$$
$$= 4xe^{2x^2+5}$$

We have found the first derivative, so we differentiate $f'(x) = 4xe^{2x^2+5}$ to obtain $f''(x)$. Looking at $f'(x) = 4xe^{2x^2+5}$ globally, we see a product. We will use the product rule to obtain $f''(x)$.

$$f''(x) = \underbrace{4}_{(\text{first})'} \cdot \underbrace{e^{2x^2+5}}_{(\text{second})} + \underbrace{4x}_{(\text{first})} \cdot \underbrace{e^{2x^2+5} \cdot 4x}_{(\text{second})'}$$

$$f''(x) = 4e^{2x^2+5} + 16x^2e^{2x^2+5} \quad \text{Factor out } 4e^{2x^2+5}.$$

$$f''(x) = 4e^{2x^2+5}(1 + 4x^2)$$

Example 6 Find $f'(x)$ for $f(x) = 6e^{-7x} \ln 4x$.

Solution Looking at this function globally, we see a product. We'll differentiate this function using the product rule.

$$f'(x) = \underbrace{6e^{-7x} \cdot -7}_{(\text{first})'} \cdot \underbrace{\ln 4x}_{(\text{second})} + \underbrace{6e^{-7x}}_{(\text{first})} \cdot \underbrace{\frac{1}{4x} \cdot 4}_{(\text{second})'}$$

$$f'(x) = -42e^{-7x} \cdot \ln 4x + 6e^{-7x} \cdot \frac{1}{x}$$
$$= -6e^{-7x}\left(7 \ln 4x - \frac{1}{x}\right)$$

Example 7 Find $f'(0)$ for $f(x) = \dfrac{e^x - e^{-x}}{e^x + e^{-x}}$.

Solution Looking at this function globally, we see a quotient. So we will begin differentiating $f(x)$ using the quotient rule.

$$f'(x) = \frac{(e^x + e^{-x}) \cdot \frac{d}{dx}[e^x - e^{-x}] - (e^x - e^{-x}) \cdot \frac{d}{dx}[e^x + e^{-x}]}{(e^x + e^{-x})^2}$$

$$= \frac{(e^x + e^{-x})(e^x - e^{-x}(-1)) - (e^x - e^{-x})(e^x + e^{-x}(-1))}{(e^x + e^{-x})^2}$$

$$= \frac{(e^x + e^{-x})(e^x + e^{-x}) - (e^x - e^{-x})(e^x - e^{-x})}{(e^x + e^{-x})^2}$$

$$= \frac{e^{2x} + e^0 + e^0 + e^{-2x} - (e^{2x} - e^0 - e^0 + e^{-2x})}{(e^x + e^{-x})^2}$$

$$= \frac{e^{2x} + 1 + 1 + e^{-2x} - (e^{2x} - 1 - 1 + e^{-2x})}{(e^x + e^{-x})^2}$$

$$= \frac{e^{2x} + 2 + e^{-2x} - (e^{2x} - 2 + e^{-2x})}{(e^x + e^{-x})^2}$$

$$= \frac{e^{2x} + 2 + e^{-2x} - e^{2x} + 2 - e^{-2x})}{(e^x + e^{-x})^2}$$

$$= \frac{4}{(e^x + e^{-x})^2}$$

Now we substitute 0 for x and compute to find $f'(0)$.

$$f'(0) = \frac{4}{(e^0 + e^{-0})^2}$$

$$f'(0) = \frac{4}{(1 + 1)^2}$$

$$f'(0) = \frac{4}{4}$$

$$f'(0) = 1$$

Interpretation

If x increases 1 unit in value, from 0 to 1, $f(x) = \frac{e^x - e^{-x}}{e^x + e^{-x}}$ will increase in value by approximately 1 unit. (Remember, we need to say approximately because the derivative involves a variable.)

■

Functions of the form $f(x) = e^{ax}$, where a is a real number, occur frequently and it is convenient to simply write the derivative without writing the individual steps leading up to it. We illustrate those steps by differentiating $f(x) = e^{ax}$.

$$f(x) = e^{ax}$$

$$f'(x) = \underbrace{e^{ax}}_{\text{(function itself)}} \cdot \underbrace{a}_{\text{(exponent)}'}$$

$$= ae^{ax}$$

We have developed the following general rule.

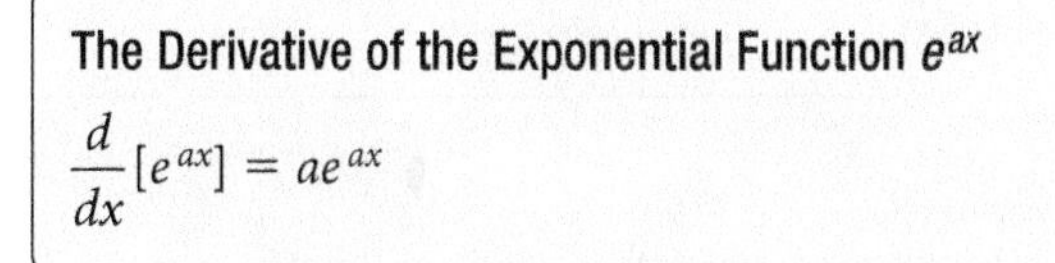

The Derivative of the Exponential Function e^{ax}

$$\frac{d}{dx}[e^{ax}] = ae^{ax}$$

This differentiation rule states that to differentiate e^{ax}, simply multiply e^{ax} by a, the coefficient of the exponent.

Example 8

a. $\frac{d}{dx}[e^{4x}] = 4e^{4x}$

b. $\frac{d}{dx}[e^{-6x}] = -6e^{-6x}$

c. $\frac{d}{dx}[5e^{-0.2x}] = (-0.2)5e^{-0.2x} = -1e^{-0.2x} = -e^{-0.2x}$

■

Example 9 The resale value R (in dollars) of a particular type of laser cutting tool t years after its initial purchase is approximated by

$$R(t) = 450{,}000e^{-0.27t}$$

Find and interpret the rate at which the resale value is changing

a. What is the decay rate of the resale value?

b. 1 year after the initial purchase, and

c. 5 years after the initial purchase.

Solution We need to find and interpret $R'(1)$, and $R'(5)$. We will begin by finding the general form of the derivative, $R'(t)$.

$$R(t) = 450{,}000e^{-0.27t}$$
$$R'(t) = (-0.27)(450{,}000)e^{-0.27t}$$
$$= -121{,}500e^{-0.27t}$$

Thus, $R'(t) = -121{,}500e^{-0.27t}$.

a. From reading the general form of the natural exponential function $A = A_0e^{-k}$ the decay rate is $-0.27 = -27\%$.

b. $R'(1) = -121{,}500e^{-0.27(1)}$

$\approx -92{,}750.61$

Interpretation
If the number of years since the initial purchase of the laser cutting tool increases by 1, from 1 to 2, the resale value of the tool can be expected to decrease by approximately $92,750.61.

c. $R'(5) = -121{,}500e^{-0.27(5)}$

$\approx -31{,}497.69$

Interpretation
If the number of years since the initial purchase of the laser cutting tool increases by 1, from 5 to 6, the resale value of the tool can be expected to decrease by approximately $31,497.69. That is, at the end of year 6, the resale value of the tool can be expected to be about $31,497.69 less than the resale value at the end of year 5.

■

Example 10 If a person takes a prescribed dose of 5 milligrams of Valium, the amount of Valium in a person's bloodstream at any time can be modeled with the exponential decay function $A(t) = 5e^{-0.0144t}$ where t is in hours.

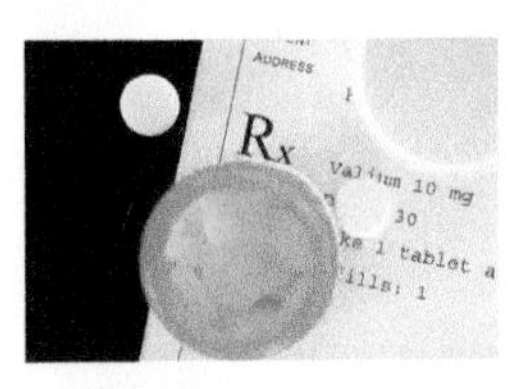

© Michael Flippo/iStockPhoto

a. How much Valium remains in the person's bloodstream 24 hours after taking a 5-mg dose? Round to the nearest tenth milligram.

b. How long will it take 5 mg to decay to 2.5 mg in a person's bloodstream?

c. At what rate is the amount of Valium in a person's bloodstream decaying 5 hours after a 5-mg dose is taken? Round the rate to three decimal places.

Solution

a. We are asked to find $A(t)$ when $t = 24$. Substitute 24 for t in the model $A(t) = 5e^{-0.0144t}$ and compute.

$$\begin{aligned} A(t) &= 5e^{-0.0144t} \\ &= 5e^{-0.0144(24)} \\ &\approx 3.5 \end{aligned}$$

Interpretation
Twenty-four hours after taking a 5-mg dose of Valium, about 3.5 mg of Valium will remain in the bloodstream.

b. We are asked to find t when $A(t) = 2.5$. Substitute 2.5 for $A(t)$ in the model $A(t) = 5e^{-0.0144t}$ and use logarithms to solve for t.

$$\begin{aligned} 2.5 &= 5e^{-0.0144t} \\ \frac{2.5}{5} &= e^{-0.0144t} \\ \ln\frac{2.5}{5} &= \ln e^{-0.0144t} \\ \ln\frac{2.5}{5} &= -0.0144t \\ t &= \frac{\ln\frac{2.5}{5}}{-0.0144} \\ t &\approx 48 \end{aligned}$$

Interpretation
About 48 hours after taking a 5-mg dose of Valium, there will be 2.5 mg of Valium remaining in the bloodstream.

c. We are asked to find the rate at which $A(t)$ is changing when $t = 5$. The word rate tells us we need the derivative, $A'(t)$, of $A(t)$ with respect to t.

$$\begin{aligned} A(t) &= 5e^{-0.0144t} \\ A'(t) &= 5 \cdot \frac{d}{dt}e^{-0.0144t} \\ &= 5(-0.0144)e^{-0.0144t} \\ &= -0.072e^{-0.0144t} \end{aligned}$$

When $t = 5$,

$$\begin{aligned} A'(5) &= -0.072e^{-0.0144(5)} \\ &\approx -0.067 \end{aligned}$$

Interpretation
Five hours after taking a 5-mg dose of Valium, the amount of Valium in the person's bloodstream is decreasing at the rate of about 0.067 mg/hr. That is, in the next hour, hour 6, we can expect about 0.067 mg of Valium to decay out of the bloodstream.

Using Technology 4.4

You can use your graphing calculator to approximate both the first and second derivatives of a function involving a natural exponential function.

Example 11 Approximate the solutions to $R(5)$, $R'(5)$, and $R''(5)$ for the function $R(t) = 450{,}000e^{-0.27t}$.

Solution In the graphing editor, enter

Y1=450000e^(−0.27X)

Y4=nDeriv(Y1,X,X,0.001)

Y5=nDeriv(Y4,X,X,0.001)

In the computation window, enter

Y1(5) to evaluate R(5)

Y4(5) to evaluate R′(5)

Y5(5) to evaluate R″(5)

The results to the nearest tenth are

Y1(5)=116658.1

Y4(5)=-31497.7

Y5(5)=8504.4

You can access Y1, Y4, and Y5 by pressing VARS, then moving the cursor right to [Y-VARS] and selecting the function number you wish to use.

Getting Ready for Class

After reading through the preceding section, respond in your own words and in complete sentences.

A. What is the derivative of $y = e^x$?
B. What is the derivative of $y = e^{2x}$?
C. Describe how the differentiation processes of the functions $f(x) = \ln[g(x)]$ and $f(x) = e^{g(x)}$ are similar.
D. Describe the strategy you would use in differentiating the function $f(x) = e^{5x^2+4}$. Do not perform the differentiation; just describe how you would do it.

Problem Set 4.4

Skills Practice

For Problems 1-20, find the derivative of each function.

1. $f(x) = e^{3x+6}$

2. $f(x) = e^{-4x+6}$

3. $f(x) = e^{6x}$

4. $f(x) = e^{8x}$

5. $f(x) = e^{5x^2+4}$

6. $f(x) = e^{2x^3-x-7}$

7. $f(x) = 9e^{3x+1}$

8. $f(x) = 4x^2 + 5e^{x^2+1}$

9. $f(x) = 5x^3 + 6x - 2e^{-3x}$

10. $f(x) = 6x^3 2e^{2x+1}$

11. $f(x) = 5x^2 e^{x^2+x}$

12. $f(x) = e^{\sqrt{5x+2}}$

13. $f(x) = 3e^{\sqrt[4]{x^2-3x}}$

14. $f(x) = e^{3x} \ln(3x)$

15. $f(x) = 2e^{-x} \ln(x^2 + 2x)$

16. $f(x) = e^{e^{2x}}$

17. $f(x) = 1{,}200 + 50e^{-0.05x}$

18. $f(x) = 3x - 2e^{-0.04x}$

19. $f(x) = \dfrac{e^{3x}}{e^x + e^{2x}}$

20. $f(x) = \dfrac{1 + e^{-x}}{1 - e^{-x}}$

For Problems 21-24, use implicit differentiation to find y'.

21. $2x^3 + e^{3xy} = 120$

22. $7x^2y + e^{4xy^2} = 1$

23. $e^{xy} + \ln(xy) = 0$

24. $e^{y^2} - y^{e^2} = x^2$

For Problems 25 and 26, find $f''(x)$.

25. $f(x) = e^{x^2+4}$

26. $f(x) = e^{-x} + e^x$

For Problems 27-30, find and interpret each derivative.

27. Find $f'(1)$ for $f(x) = 4e^{x^2}$.

28. Find $f'(3)$ for $f(x) = -0.06e^{2x+5}$.

29. Find $f'(40)$ for $f(x) = 25{,}000 + 25{,}000e^{-0.065x}$.

30. Find $f'(1)$ for $f(x) = e^x \ln x$.

Paying Attention to Instructions The next two problems are intended to give you practice reading, and paying attention to, the instructions that accompany the problems you are working.

31. If $f(x) = 4xe^x$, find

a. $f(0)$. Round to the nearest whole number.

b. $f'(0)$. Round to the nearest whole number.

c. x so that $f'(x) = 0$. Keep in mind that e^x is never 0.

d. Using the results of your computations for $f(0)$ and $f'(0)$, find the equation of the line tangent to the curve $f(x)$ at the point $(0, f(0))$.

32. If $f(x) = (e^x)^2$, find

a. $f(0)$

b. $f'(0)$

c. x so that $f'(x) = 1$. Round to 4 decimal places.

d. Using the results of your computations for $f(0)$ and $f'(0)$, find the equation of the line tangent to the curve $f(x)$ at the point $(0, f(0))$.

Check Your Understanding

For Problems 33-42, a function is given along with a claimed derivative of the function. State if the indicated derivative is *correct* or *incorrect*.

33. $f(x) = e^{4x^2+5}$

$f'(x) = 8x$

34. $f(x) = e^{3x^5+x-1}$

$f'(x) = 15x^4 + 1$

35. $f(x) = e^{x^2-6}$

$f'(x) = e^{x^2-6}(x^2 - 6)$

36. $f(x) = e^{-0.03x}$

$f'(x) = e^{-0.03x}(-0.03)$

37. $f(x) = 8e^{2x}$

$f'(x) = 8e^{2x} \cdot 2 = 16e^{2x}$

38. $f(x) = e^{x^2+4} \ln(4x)$

$f'(x) = e^{x^2+4} \cdot 2x \cdot \dfrac{1}{4x} = \dfrac{e^{x^2+4}}{2}$

39. $f(x) = 2xe^x$

$f'(x) = 2e^x + 2xe^x$

40. $f(x) = \ln(e^{6x})$

$f'(x) = \dfrac{1}{e^{6x}} \cdot e^{6x} \cdot 6 = 6$

41. $f(x) = e^{\ln x}$

$f'(x) = \dfrac{1}{e^{\ln x}} \cdot \dfrac{1}{x} = \dfrac{1}{xe^{\ln x}}$

42. $f(x) = \ln x + e^x$

$f'(x) = \dfrac{1}{x} + \dfrac{1}{e^x}$

Modeling Practice

43. Business: Advertising The management of a large chain of frozen yogurt stores in New England believes that t days after the end of an advertising campaign, the volume V of sales is approximated by

$$V(t) = 65{,}000 + 65{,}000e^{-0.46t}$$

At what rate is the sales volume changing

a. 1 day after the end of the advertising campaign?

b. 6 days after the end of the advertising campaign?

44. Manufacturing: Consumption A producer of tetraethylorthosilicate (TEOS), a coating for silicon computer wafers, believes, from an extensive study, that the national consumption C (in appropriate units) of TEOS over the next t years is approximated by

$$C(t) = 14.5e^{-1.2t} + 0.14t^2 \quad 0 \le t \le 8$$

The study was prepared when $t = 0$. Find the rate at which the national consumption of TEOS is changing with respect to time

a. 2 years after the preparation of the study.

b. 5 years after the preparation of the study.

Source: Aerogel.org

45. Life Science: Flu Epidemic The total number N of people in a Washington state community contracting a flu virus t days after the outbreak of an epidemic is approximated by

$$N(t) = \frac{14{,}500}{1 + 65e^{-0.4t}}$$

How many more people can be expected to contract the flu virus between days 4 and 5 of the epidemic?

46. Psychology: Technical Proofreading A psychologist believes the formula

$$N(t) = 85 - 65e^{-0.35t}$$

relates the number N of pages of a technical textbook an average person can proofread and the number of consecutive days t that person has been reading such books. By how many pages can an average person be expected to increase his or her proofreading ability from

a. day 4 to day 5?

b. day 10 to day 11?

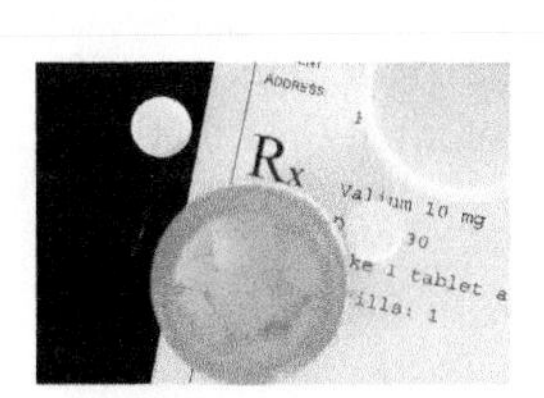

© Michael Flippo/iStockPhoto

47. Life Science: Drug Concentration If a person takes a prescribed dose of 5 milligrams of Valium, the amount of Valium in a person's bloodstream at any time t, in hours, can be modeled with the exponential decay function

$$A(t) = 5e^{-0.0144t}$$

a. How much Valium remains in the person's bloodstream 5 hours after taking a 5-mg dose? Round to one decimal place.

b. How long will it take 5 mg of Valium to decay to 1.25 mg in a person's bloodstream? Round the number of hours to one decimal place.

c. At what rate is the amount of Valium in a person's bloodstream decaying 1 hour after a 5-mg dose is taken? Round the rate to two decimal places.

48. Life Science: Drug Concentration If a person takes a dose of 350 milligrams of aspirin, the amount of aspirin in a person's bloodstream at any time t, in hours, can be modeled with the exponential decay function

$$A(t) = 350e^{-0.0346t}$$

a. How much aspirin remains in the person's bloodstream 5 hours after taking a 350-mg dose? Round to one decimal place.

b. How long will it take 350 mg of aspirin to decay to 2.5 mg in a person's bloodstream? Round the number of hours to one decimal place.

c. At what rate is the amount of aspirin in a person's bloodstream decaying 1 hour after a 350-mg dose is taken? Round the rate to two decimal places.

Source: IPCS INCHEM

49. Physical Science: Radioactive Decay A 120-gram block of radioactive carbon-14, ^{14}C, decays according to the function

$$A(t) = 120e^{-0.00012t}$$

where t is measured in years. By how many grams can the block of ^{14}C be expected to decrease from year 10 to year 11?

Source: WolframAlpha

50. Life Science: Drug Concentration When a particular drug is intravenously administered to a patient on a continuous basis, the amount A (in milligrams) of the drug in the patient's bloodstream t minutes after injection is approximated by

$$A(t) = 175 - 175e^{-0.2t}$$

At what rate is the amount of the drug in the patient's bloodstream changing after 10 minutes?

51. Manufacturing: Fuel Purification In the United States, multi-product pipelines carry fuels such as gasoline, diesel, and jet fuels. The lowest grade of fuel, diesel, typically contains surfactants, chemical compounds that are attracted to surfaces and interfaces and can disrupt the coalescing process of the fuel. To produce clean, dry fuel, filters containing coalescer clay cartridges are often used. Fuel is passed through the cartridges, and the clay treaters remove surfactants and allow the coalescers to perform their function. Normally a string of three 18"-long cartridges is used for receipt of jet fuel that comes from multi-product pipelines. The more clay the fuel passes through, the more surfactants that get removed. A particular type of cartridge removes 15% of surfactants for every one-foot of clay. For this type of filter, the function

$$P(f) = 100e^{-0.163f}$$

calculates the percent P of surfactants that remain in the fuel after the fuel passes through f feet of clay.

a. What percent of surfactants remain in a quantity of fuel after it has passed through 2 feet of clay? Round to one decimal place.

b. Write the expression that models the rate at which the percent of surfactants that remain in the fuel changes as the number of feet of clay increases.

c. At what rate is the percentage of surfactants that remain in a quantity of fuel changing when 2 feet of clay are present? Interpret this result.

Source: Velcon Filters

52. Economics: The Value of Art A gallery in Farmers Branch, Texas, believes the exponential model

$$V_{FB}(t) = 3{,}500e^{0.050t}$$

accurately predicts the value (in dollars) of a particular artist's work over the next 6 years. A gallery in nearby Dallas believes the exponential model

$$V_D(t) = 3{,}500e^{0.057t}$$

accurately predicts the value of that artist's work over the next 6 years. The variable t represents the number of years from now.

a. What is the value both models place on this artist's work right now?

b. Reading from the general form of the natural exponential function

$$A = A_0e^{kt}$$

specify each gallery's estimate of the growth rate for the value of this artist's work.

c. Two quantities can be compared using subtraction. Subtraction indicates how much more one quantity is than another. For example, $15 - 3 = 12$ indicates that 15 is 12 units more than 3. Compute

$$\frac{d}{dt}[V_D(t) - V_{FB}(t)]$$

when $t = 6$ and explain what it means.

d. Two quantities can be compared using division. Division indicates how many times more one quantity is than another. For example $\frac{15}{3} = 5$ indicates that 15 is 5 times as big as 3. Compute

$$\frac{V'_D(t)}{V'_{FB}(t)}$$

when $t = 6$ and explain what it means.

53. Business: Online Advertising Revenue Mr. Azar, a financial analyst, believes the exponential model

$$R_A(t) = 600{,}000e^{-0.012t}$$

accurately predicts a social media's online advertising revenue over the next 5 years. Mr. Hielo, also a financial analyst, believes the exponential model

$$R_H(t) = 600{,}000e^{-0.018t}$$

accurately predicts the social media's online advertising revenue over the next 5 years. The variable t represents the number of years from now.

a. What is the value both models place on this company's online advertising revenue right now?

b. Reading from the general form of the natural exponential function

$$A = A_0e^{kt}$$

specify each analyst's estimate of the decay rate of revenue.

c. Two quantities can be compared using subtraction. Subtraction indicates how much more one quantity is than another. For example, $15 - 3 = 12$ indicates that 15 is 12 units more than 3. Compute

$$\frac{d}{dt}[R'_A(t) - R'_H(t)]$$

when $t = 5$ and explain what it means.

d. Two quantities can be compared using division. Division indicates how many times more one quantity is than another. For example, $\frac{15}{3} = 5$ indicates that 15 is 5 times as big as 3. Compute

$$\frac{R'_H(t)}{R'_A(t)}$$

when $t = 5$ and explain what it means.

54. Business: Revenue Comparisons Permeable pavers are pavers that present a solid surface but still allow drainage and migration of water into the soil beneath them. Suppose that analyst A uses the revenue model

$$R_A(t) = 3{,}000e^{0.07t}$$

to predict the revenue (in thousands of dollars) of Louisiana Paver Company t months from now. Analyst B uses the revenue model

$$R_B(t) = 3{,}000e^{0.065t}$$

to predict the revenue of the same company also t months from now.

a. What information is contained in the quotient

$$\frac{R_A(6)}{R_B(6)} \approx \frac{4{,}565.88}{4{,}430.94} \approx 1.030$$

b. What information is contained in the quotient of the derivatives

$$\frac{R'_A(6)}{R'_B(6)} \approx \frac{319.612}{288.011} \approx 1.11$$

c. What information is contained in the derivative of the quotient

$$\left[\frac{R_A(t)}{R_B(t)}\right]' \approx 0.005$$

when $t = 6$?

Photo Courtesy of the U.S. Air Force

55. Business: Revenue Comparisons A helical antenna is an antenna that is constructed using a conducting wire wound in the form of a helix. They are typically small in size and are for mobile and portable high-frequency communications. To predict its profit over the next twelve months, the Helix Antenna Company has two independent analysts make profit projections. Suppose that analyst A uses the profit model

$$P_A(t) = 30e^{0.048t}$$

to predict the profit (in thousands of dollars) of the company t months from now. Analyst B uses the profit model

$$P_B(t) = 30e^{0.044t}$$

to predict the profit of the company also t months from now.

a. What information is contained in the quotient

$$\frac{P_A(5)}{P_B(5)} \approx \frac{38.1375}{37.3823} \approx 1.02$$

b. What information is contained in the quotient of the derivatives

$$\frac{P'_A(5)}{P'_B(5)} \approx \frac{1.8306}{1.6448} \approx 1.11$$

c. What information is contained in the derivative of the quotient

$$\left[\frac{P_A(t)}{P_B(t)}\right]' \approx 0.004$$

when $t = 5$?

56. Business: Revenue Comparisons To predict its profit over the next 100 days, the Circle Cylinder Company has two independent analysts make profit projections. Suppose that analyst A uses the profit model

$$P_A(t) = 1{,}000e^{0.025t}$$

to predict the profit (in dollars) of the company t days from now. Analyst B uses the profit model

$$P_B(t) = 1{,}000e^{0.020t}$$

to predict the profit of the company also t days from now.

a. What information is contained in the quotient

$$\frac{P_A(25)}{P_B(25)}$$

b. What information is contained in the quotient

$$\frac{P'_A(25)}{P'_B(25)}$$

c. What information is contained in the derivative of the quotient

$$\left[\frac{P_A(t)}{P_B(t)}\right]' \approx 0.0057$$

when $t = 25$?

57. Business: Revenue Comparisons To predict its revenue over the next 10 weeks, the Exponential Logarithm Company has two independent analysts make revenue projections. Suppose that analyst A uses the revenue model

$$R_A(t) = 10e^{0.020t}$$

to predict the revenue (in thousands of dollars) of the company t weeks from now. Analyst B uses the revenue model

$$R_B(t) = 10e^{0.017t}$$

to predict the revenue of the company also t weeks from now.

a. What information is contained in the quotient

$$\frac{R_A(8)}{R_B(8)}$$

Are the two predictions close or are they significantly different?

b. What information is contained in the quotient

$$\frac{R'_A(8)}{R'_B(8)}$$

c. What information is contained in the derivative of the quotient

$$\left[\frac{R_A(t)}{R_B(t)}\right]' \approx 0.003$$

when $t = 8$?

58. Health Science: Drug Rate-of-Change The drug clonazepam is marketed by Roche under the trade name Klonopin and is sometimes used to treat bipolar disorder. One exponential decay model used for measuring the amount of Klonopin in the bloodstream t hours after an initial dose of A_0 mg is administered is $A = A_0e^{-0.0231t}$. If an adult patient is currently taking a dose of 1 mg/day, at what rate is the amount of Klonopin in the bloodstream decreasing 12 hours after it is administered?

Source: PDR.net

59. Health Science: Drug Rate-of-Change AndroGel 1.62% is a drug used to treat adult males who have low or no testosterone and is marketed by Abbott Laboratories. One model for predicting the amount of the drug in the bloodstream after an initial dose of A_0 mg is administered is $A = A_0e^{-0.06931t}$. According to androgel.com, the recommended starting dose is 40.5 mg/day. If a typical adult patient is taking the target dose of 40.5 mg/day, at what rate (to two decimal places) is the amount of AndroGel 1.62% in the bloodstream decreasing 12 hours after it is administered?

Source: www.androgel.com

Using Technology Exercises

Use your calculator to approximate each value to two decimal places.

60. Find $f(3), f'(3)$, and $f''(3)$ for the function $f(x) = 25e^{-0.02x}$.

61. Find $f(0), f'(0)$, and $f''(0)$ for the function $f(x) = 6e^{-0.5x+1} + 50$.

62. A manufacturer believes that the function $N(t) = 3{,}250 + 1{,}425e^{-0.15t}$ models the number of worker hours needed to build the nth unit of a product. Suppose that in writing the function, the "+" sign between 3,250 and 1,425 is mistakenly written as a "−" sign. Compute $N(100)$, $N(200)$, and $N(300)$ for both the "+" sign and the "−" sign forms of the function and use the results to describe what effect the mistake has on the estimations of the required number of worker hours.

63. Compute the derivative of the function

$$f(x) = e^{-0.05x}(x + 50)$$

for the three values $x = -30.1$, $x = -30$, and $x = -29.9$.

a. Use the results of these computations to make a statement about the orientation of the tangent line to the curve at $x = -30$. Is the tangent rising, falling, horizontal, or vertical at $x = -30$?

b. Does the function appear to reach a minimum at $x = -30$? A maximum? Neither a minimum nor a maximum?

64. Manufacturing: Cost of Production The total cost C (in dollars) to a manufacturer to produce x units of a product is given by

$$C(x) = 4{,}000 + 235xe^{x/575}$$

Determine the signs of both $C'(240)$ and $C''(240)$ and use those results to make a statement about the behavior of the cost to the manufacturer when 240 units of the product are produced. Your statement should be something like "When 240 units of the product are being produced, the cost is (increasing/decreasing) at an (increasing/decreasing) rate."

Getting Ready for the Next Section

65. Convert the fraction $\frac{1}{x^4}$ to an expression in which no denominator appears.

66. Find the derivative of $f(x) = \frac{3x^{5/3}}{5}$.

67. Rewrite the expression $\frac{x^{-3}}{-3}$ with positive exponents only.

68. Divide and leave answer with positive exponents only:

$$\frac{x^4 + 3x^3 + 5}{x}$$

69. Find the value of C in $f(x) = x^3 - 2x^2 + 2x + C$ when $f(x) = 4$ and $x = 1$.

70. Find the value of C in $C(x) = -0.02x^2 + 26x + C$ when $C(x) = 6{,}555$ and $x = 150$.

Spotlight on **Success**

Student Instructor CJ

We are what we repeatedly do.
Excellence, then, is not an act, but a habit.
—Aristotle

Something that has worked for me in college, in addition to completing the assigned homework, is working on some extra problems from each section. Working on these extra problems is a great habit to get into because it helps further your understanding of the material, and you see the many different types of problems that can arise. If you have completed every problem that your book offers, and you still don't feel confident that you have a full grasp of the material, look for more problems. Many problems can be found online or in other books. Your professors may even have some problems that they would suggest doing for extra practice. The biggest benefit to working all the problems in the course's assigned textbook is that often teachers will choose problems either straight from the book or ones similar to problems that were not assigned for tests. Doing this will ensure that you do your best in all your classes.

Chapter 4 Summary

EXAMPLES

1. $f(x) = 5^x$
   $g(x) = e^x$
   $h(x) = \left(\frac{3}{4}\right)^x$

2. Domain of $f(x) = A_0e^{kx}$ is $(-\infty, \infty)$
   Range of $f(x) = A_0e^{kx}$ is $(0, \infty)$

The Exponential Function [4.1]

The function $f(x) = b^x$, where $b > 0$, $b \neq 1$, and x is a variable, is the **exponential function.**

The Number e [4.1]

A useful base when representing natural phenomena using an exponential function is the number represented by the lowercase letter e. This number is irrational (the "..." indicates the number goes on forever) and is defined mathematically as

$$e = \lim_{x \to \infty}\left(1 + \frac{1}{x}\right)^x = 2.718281828459045\ldots$$

The Graphs of the Exponential Growth and Exponential Decay Functions [4.1]

a. $f(x) = b^x$, $b > 1$ The exponential growth function

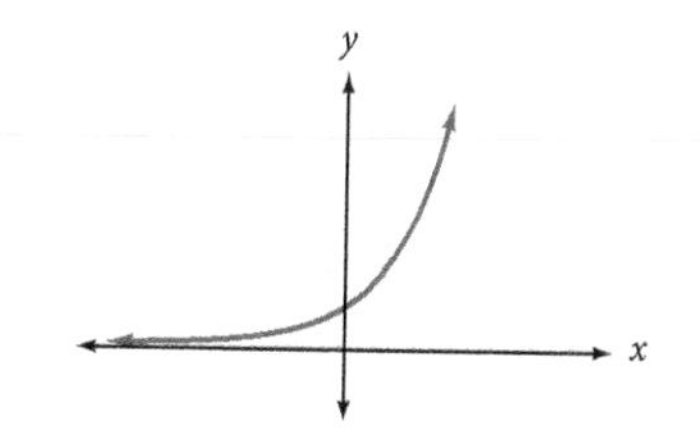

b. $f(x) = b^x$, $0 < b < 1$ The exponential decay function

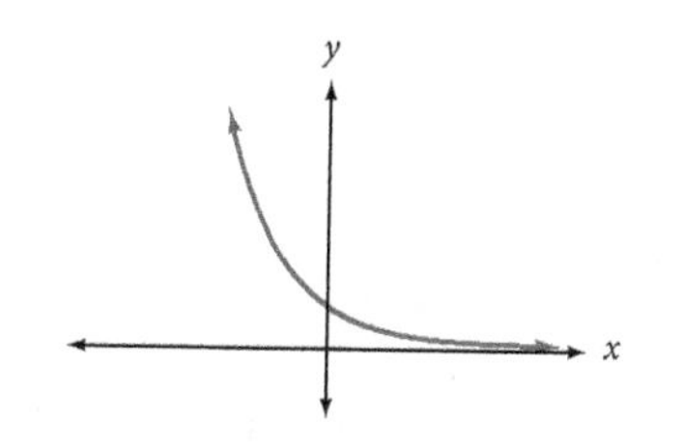

Features of the Natural Exponential Functions [4.1]

1. The domain of the exponential growth function, $f(x) = A_0e^{kx}$, and the exponential decay function, $f(x) = A_0e^{-kx}$, where k is a positive constant, is all real numbers. Both functions are computable for all values of x.
2. The range of both the exponential growth and exponential decay functions is all positive real numbers. Both functions never compute to zero or a negative number. Their graphs lie entirely above the x-axis.

3. Both exponential functions are smooth and continuous for all values of x.
4. Both the natural exponential growth function, $f(x) = A_0e^{kx}$, and the natural exponential decay function, $f(x) = A_0e^{-kx}$, pass through the point A_0 on the y-axis.
5. $f(x) = A_0e^{kx}$ increases as x increases. $f(x) = A_0e^{-kx}$ decreases as x increases.

The Logistic Growth Function [4.1]

Given M, A, and k are positive constants, the logistic growth function $f(t) = \frac{M}{1 + Ae^{-kt}}$ models restricted growth. The value M is the maximum level the output of the particular function can reach. Its graph has the form shown below.

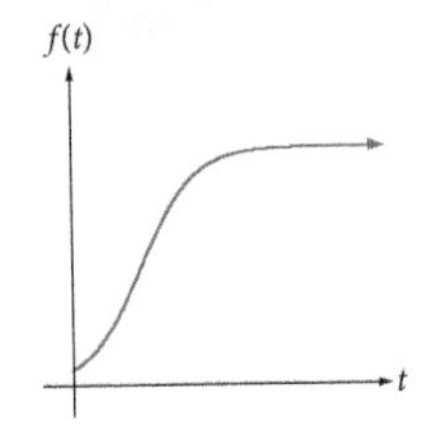

Logarithms Are Exponents [4.2]

3. $4 = \ln x$ means $x = e^4$

$0 = \ln 1$ since $1 = e^0$

$1 = \ln e$ since $e = e^1$

In the equation $a = e^t$, the number t, that when used as an exponent on the base e produces the number a, is called the natural logarithm of the number a. Notice that the number t is the exponent. It is the natural logarithm and an exponent. Logarithms are exponents.

$$t = \ln a \quad \text{if and only if} \quad a = e^t$$

Properties of the Natural Logarithm Function [4.2]

4. $\ln \frac{x^2y^3}{z^4w^5} =$

$2 \ln x + 3 \ln y - 4 \ln z - 5 \ln w$

5. $\ln (3xy) = \ln 3 + \ln x + \ln y$

$\ln \frac{3x}{yz} = \ln 3 + \ln x - \ln y - \ln z$

$\ln x^4 = 4 \ln x$

6. $\ln (2e^5) = \ln 2 + 5$

1. $\ln e = 1$
2. $\ln 1 = 0$
3. $\ln e^x = x$
4. $\ln (xy) = \ln x + \ln y$ — Multiplication becomes addition
5. $\ln \left(\frac{x}{y}\right) = \ln x - \ln y$ — Division becomes subtraction
6. $\ln x^r = r \ln x$ — Exponentiation becomes multiplication

The Reversal Effect of the Natural Exponential and Natural Logarithm Functions on Each Other [4.2]

7. Solve: $e^{2x-3} = 15$

$\ln e^{2x-3} = \ln 15$

$2x - 3 = \ln 15$

$x \approx 2.8540$

The natural logarithm function reverses the effect of the natural exponential function, and the exponential function reverses the effect of the natural logarithm function.

$$\ln e^x = x \quad \text{and} \quad e^{\ln x} = x$$

The Graph of the Natural Logarithm Function [4.2]

If the points (a_1, b_1), (a_2, b_2), … , (a_n, b_n) are points on the natural exponential curve $f(x) = e^x$, reversing the order of the components of each ordered pair results in the set of points (b_1, a_1), (b_2, a_2), … , (b_n, a_n). The graph of these points produces the natural logarithm curve. Mathematically, the exponential and natural logarithm functions are said to be inverses of each other.

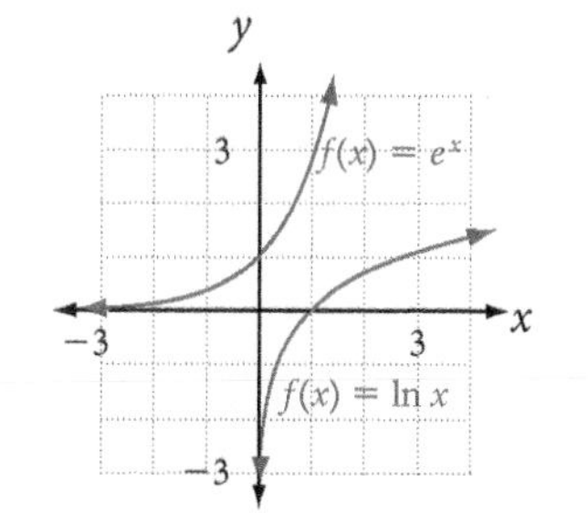

The Derivative of the Natural Logarithm Function [4.3]

8. Find $f'(x)$ for $f(x) = 5x^4 \ln x$

$\frac{d}{dx}[5x^4 \ln x]$

$= \frac{d}{dx}[5x^4] \cdot \ln x + 5x^4 \cdot \frac{d}{dx}\ln x$

$= 20x^3 \cdot \ln x + 5x^4 \cdot \frac{1}{x}$

$= 20x^3 \cdot \ln x + 5x^3$

$= 5x^3(4 \ln x + 1)$

$$\frac{d}{dx}[\ln x] = \frac{1}{x}$$

The Derivative of the Generalized Natural Logarithm Function [4.3]

$$\frac{d}{dx}[\ln [g(x)]] = \frac{1}{g(x)} \cdot g'(x)$$

9. Find $f'(x)$ for $f(x) = \ln [5x^4]$

$$\frac{d}{dx}[\ln (5x^4)] =$$

$$= \frac{1}{5x^4} \cdot \frac{d}{dx}[5x^4]$$

$$= \frac{1}{5x^4} \cdot 20x^3$$

$$= \frac{4}{x}$$

The Derivative of the Natural Exponential Function [4.4]

$$\frac{d}{dx}[e^x] = e^x$$

10. Find $f'(x)$ for $f(x) = 5x^4e^x$

$$\frac{d}{dx}[5x^4e^x] =$$

$$= \frac{d}{dx}[5x^4] \cdot e^x + 5x^4 \cdot \frac{d}{dx}e^x$$

$$= 20x^3 \cdot e^x + 5x^4 \cdot e^x$$

$$= 5x^3e^x(4 + x)$$

The Derivative of the Generalized Natural Exponential Function [4.4]

$$\frac{d}{dx}[e^{g(x)}] = e^{g(x)} \cdot g'(x)$$

11. $\frac{d}{dx}[e^{5x^4}] = e^{5x^4} \cdot \frac{d}{dx}5x^4$

$$= e^{5x^4} \cdot 20x^3$$

$$= 20x^3e^{5x^4}$$

The Derivative of $f(x) = e^{ax}$ [4.4]

$$\frac{d}{dx}[e^{ax}] = ae^{ax}$$

12. $\frac{d}{dx}[e^{5x}] = 5e^{5x}$

$$\frac{d}{dx}[e^{-0.3x}] = -0.3e^{-0.3x}$$

Chapter 4 Test

1. Identify each function as exponential or algebraic. [4.1]

   a. $f(x) = x^3 + 7$

   b. $f(x) = 3e^{4x+1}$

2. Use the properties of logarithms to [4.2]

   a. expand $\ln\left(\dfrac{x^3y^4}{z^2}\right)$.

   b. compress $5 \ln x - 3 \ln y - 4 \ln z$.

Convert each exponential equation to its corresponding logarithmic equation. [4.2]

3. $5 = e^{4x}$

4. $k = e^{3y-1}$

Convert each logarithm equation to its corresponding exponential equation. [4.2]

5. $5 = \ln(3x)$

6. $-2 = \ln(x + 6)$

Solve each exponential equation. Round your result to 4 decimal places. [4.2]

7. $e^{2x+1} = 19$

8. $2{,}500 = 1{,}500 + 50e^{-0.6x}$

9. Find the equation of the line tangent to the curve $f(x) = e^{x^2-3x}$ at the point $(3, 1)$. [4.4]

Find the derivative of each function. [4.3, 4.4]

10. $f(x) = e^{-3x}$

11. $f(x) = e^{5x+4}$

12. $f(x) = x^4e^{6x}$

13. $f(x) = \ln(8x)$

14. $f(x) = \ln(3x - 2)^3$

15. $f(x) = [e^{5x-6}]^4$

16. $f(x) = 4\ln(7x + 7) + 10e^{7x+7}$

17. $f(x) = \dfrac{e^{3x} + 1}{x}$

18. Use implicit differentiation to find y' for $4x^3 + e^y = 1$. [4.4]

Find $f''(x)$ for each function. [4.3, 4.4]

19. $f(x) = e^{3x}$

20. $f(x) = x \ln x$

Find and interpret each value. [4.3, 4.4]

21. $f'(1)$ for $f(x) = 3e^{-x}$

22. $f'(11)$ for $f(x) = \ln(2x - 21)$

For Problems 23-26, solve. [4.2, 4.4]

23. Industrial psychologists have determined that the function

$$N(t) = 40 - 40e^{-0.07t}$$

relates the number, N, of assembly tasks an average person can complete each 8-hour working day to the number of days t the person has been performing the task. Approximately how many days after starting a particular task will the worker be completing 25 tasks in an 8-hour working day? Round to the nearest day.

24. When a drug is intravenously administered to a patient on a continuous basis, the amount A (in milligrams) of the drug in the patient's bloodstream t days after injection is modeled by the function $A(t) = 500 - 150e^{0.20t}$. Approximately how much time will have to pass so that 175 milligrams of the drug will be in the patient's bloodstream? Round to the nearest whole number of days.

25. The function $P(t) = 220e^{0.08t}$ relates the population, P, (in thousands of people) of a city to the time, t, in years from now. How will the population of the city be changing 4 years from now?

26. The amount of plutonium-239, Pu-239, in a sample that initially contains 10 grams of the substance is modeled by the exponential function $A(t) = 10e^{-0.00002876t}$, where t is the number of years from now. How will the amount of Pu-239 in the sample be changing 1,000 years from now?

Source: Cutnell, John D. and Johnson, Kenneth W. Physics, 3rd edition. New York: Wiley, 1995, 1013.

Integration: The Language of Accumulation

5

Chapter Outline

© Dave Hauge/iStockPhoto

In response to global concerns about rising oil prices and the impact of greenhouse gas emissions, farms in the U.S. are producing agricultural crops for energy production. The USDA predicts that by the end of 2016, the U.S. will produce 12 billion gallons of ethanol from corn and 700 million gallons from soybean oil.

At the same time farms are producing corn and soybeans for the production of energy, an increase in quality of life associated with economic growth in developing countries is increasing the demand for food.

By collecting and analyzing data, it is possible to construct functions that model worldwide demand for both energy and food. Once we have the functions, we can use them to answer two types of questions, one using only algebra, and another using integral calculus.

Note When you see this icon next to an example or problem in this chapter, you will know that we are using the topics in this chapter to model situations in the world around us.

Algebra What will be the worldwide demand for food and for energy in the year 2016?

Calculus What will be the total accumulated worldwide demand for food and for energy for the years 2016 through 2020?

In this section, we will work with integrals that represent accumulation, such as the social utility of consumption defined by

$$\int_0^a D(x)\,dx$$

This integral gives the accumulated value the public places on its consumption and can be used to answer the above question.

Study Skills

The study skills for this chapter cover the way you approach new situations in mathematics. The first study skill is a point of view you hold about your natural instincts for what does and doesn't work in mathematics. The second study skill gives you a way of testing your instincts.

1. **Don't Let Your Intuition Fool You** As you become more experienced and more successful in mathematics you will be able to trust your mathematical intuition. For now, though, it can get in the way of your success. For example, if you ask some students to "subtract 3 from -5" they will answer -2 or 2. Both answers are incorrect, even though they may seem intuitively true. Likewise, some students will expand $(a + b)^2$ and arrive at $a^2 + b^2$, which is incorrect. In both cases, intuition leads directly to the wrong answer.
2. **Test Properties of Which You Are Unsure** From time to time, you will be in a situation where you would like to apply a property or rule, but you are not sure it is true. You can always test a property or statement by substituting numbers for variables. For instance, I always have students that rewrite $(x + 3)^2$ as $x^2 + 9$, thinking that the two expressions are equivalent. The fact that the two expressions are not equivalent becomes obvious when we substitute 10 for x in each one.

   When $x = 10$, the expression $(x + 3)^2$ is $(10 + 3)^2 = 13^2 = 169$

   When $x = 10$, the expression $x^2 + 9$ is $10^2 + 9 = 100 + 9 = 109$

   When you test the equivalence of expressions by substituting numbers for the variable, make it easy on yourself by choosing numbers that are easy to work with, such as 10. Don't try to verify the equivalence of expressions by substituting 0, 1, or 2 for the variable, as using these numbers will occasionally give you false results.

It is not good practice to trust your intuition or instincts in every new situation in calculus. If you have any doubt about the generalizations you are making, test them by replacing variables with numbers and simplifying.

Antidifferentiation and the Indefinite Integral

5.1

© Patrick Heagney/iStockPhoto

Differentiation, the language of change, provides us with the capability of determining the rate at which a quantity is changing when we know how large the quantity is at some particular point in time. For example, if $C(x)$ is the cost function associated with the production of x units of a product, $C'(725) = -2.20$ indicates that if the number of units produced is increased by 1, from 725 to 726, the cost per unit will decrease by approximately \$2.20. In applied situations, it is common to observe not the direct relationship between the quantities, but rather the relationship of change. The function observed is the derivative function.

Now, given the derivative function, suppose we would like to find the function that describes the direct relationship between the quantities. The function we have is the derivative of the function we want. To get the function we want, we need to reverse the process of differentiation. It is this process we investigate in this section.

The Antiderivative and the Indefinite Integral

The process of reversing a differentiation is called **antidifferentiation** or **integration**. The function obtained from this process is called the **antiderivative** or **integral** of the derivative function.

> **The Antiderivative or Integral**
> Suppose that $F(x)$ and $f(x)$ are two functions such that $f(x)$ is the derivative of $F(x)$; that is, $F'(x) = f(x)$. Then $F(x)$ is an **antiderivative** or **integral** of $f(x)$.

For example (and you should verify these), if

1. $f(x) = 4x - 5$, then $F(x)$ could be defined by $2x^2 - 5x$.
2. $f(x) = 4x - 5$, then $F(x)$ could be defined by $2x^2 - 5x + 3$.
3. $f(x) = 4x - 5$, then $F(x)$ could be defined by $2x^2 - 5x - 1$.
4. $f(x) = 4x - 5$, then $F(x)$ could be defined by $2x^2 - 5x - 4$.
5. $f(x) = 4x - 5$, then $F(x)$ could be defined by $2x^2 - 5x +$ some constant.

The first four functions are graphed in Figure 1.

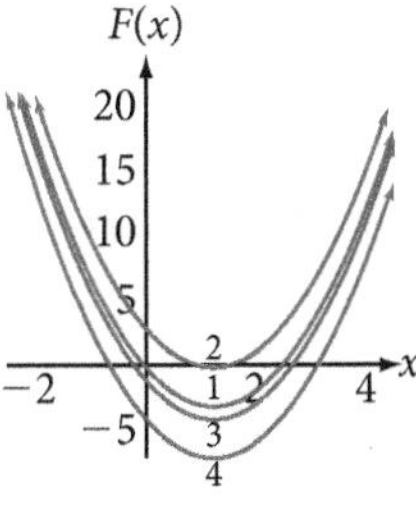

Figure 1

Notice that each of the antiderivative functions $F(x)$ above produces the same derivative function $f(x) = 4x - 5$. This means that when the derivative process is reversed, a unique function $F(x)$ is not produced. Rather a *family* of antiderivative functions consisting of infinitely many members is produced. Each member function of the family, however, differs only by a constant.

Suppose that both $F(x)$ and $G(x)$ are antiderivatives of $f(x)$; that is, suppose that $F'(x) = f(x)$ and $G'(x) = f(x)$. We wish to show that the difference of $F(x)$ and $G(x)$ is a constant. Since the difference of two functions is again a function, let

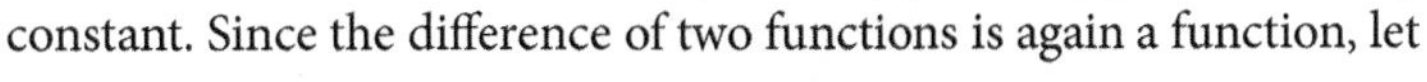
$$H(x) = F(x) - G(x)$$

Then

$$H'(x) = F'(x) - G'(x)$$
$$= f(x) - f(x) \qquad \text{By our supposition}$$
$$= 0$$

But only the constant function has 0 as its derivative. Thus, if C represents some constant,

$$H(x) = C$$

or, since $H(x) = F(x) - G(x)$,

$$F(x) - G(x) = C$$

which means that

$$F(x) = G(x) + C$$

and we have shown that two antiderivatives of the same function differ only by a constant and, thus, that a derivative function has infinitely many antiderivatives.

The Indefinite Integral

Just as $\frac{d}{dx}f(x)$ or $f'(x)$ symbolizes the derivative of $f(x)$ with respect to x, $\int f(x)\,dx$ symbolizes the antiderivative, or integral of $f(x)$, with respect to x. The symbol $\int dx$ indicates the family of antiderivative functions, and the function we wish to find the antiderivative of must be placed between the $\int$ and the dx.

Indefinite Integral

Suppose that $F(x)$ is any one of the infinitely many members of the family of antiderivatives of $f(x)$. Then,

$$\int f(x)\,dx = F(x) + C$$

where c is an arbitrary constant. The function $F(x) + C$ is called the **antiderivative** or **indefinite integral** of $f(x)$. The number C is called the **constant of integration**, the function $f(x)$ is called the **integrand**, and dx is called the **differential**.

In the notation $\int f(x)\,dx$, the differential dx plays the same role it does in the notation $\frac{d}{dx}$. In $\frac{d}{dx}$, dx indicates that the derivative is to be taken with respect to the independent variable x. In $\int f(x)\,dx$, dx indicates that the antiderivative is to be taken with respect to the independent variable x; likewise, in $\int f(y)\,dy$, dy indicates that the antiderivative is to be taken with respect to the independent variable y.

Integration Formulas

The following integration formulas are constructed by reversing the differentiation formulas we have already studied. (You should notice that each one can be proved simply by taking the derivative of the indefinite integral and showing that the integrand results.)

To integrate a constant function, we affix the independent variable to the constant and add an arbitrary constant.

The Integral of a Constant
If k is any constant, then

$$\int k\, dx = kx + C$$

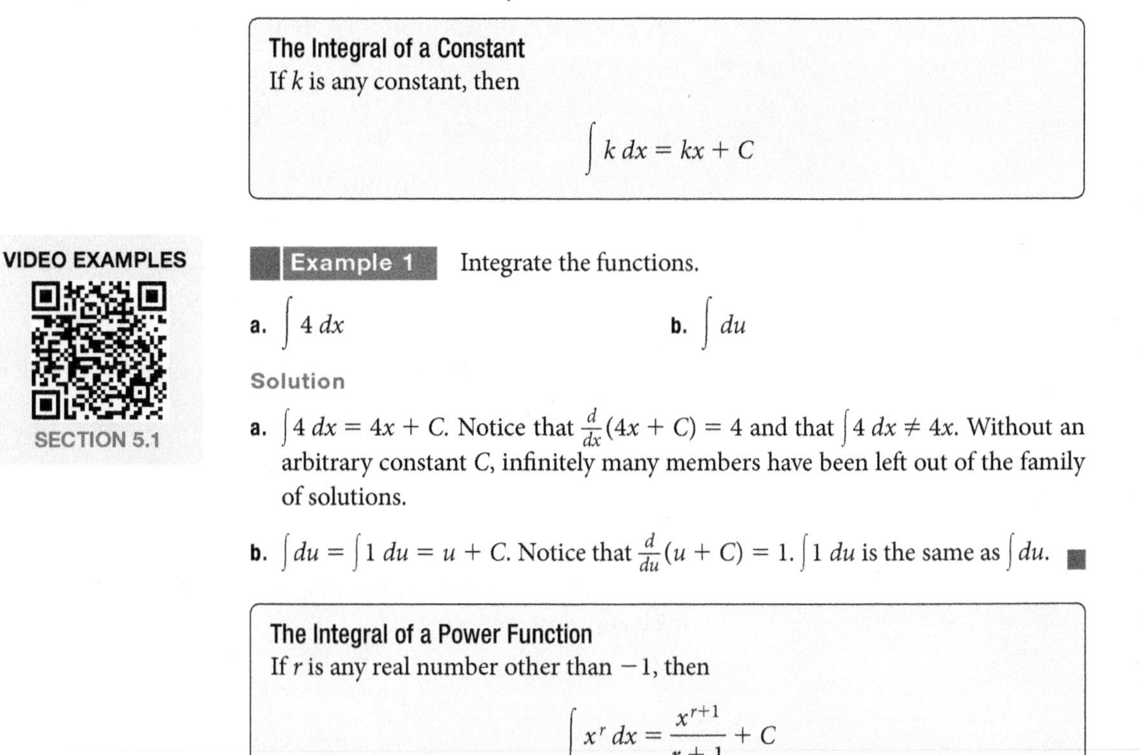

VIDEO EXAMPLES

SECTION 5.1

Example 1 Integrate the functions.

a. $\int 4\, dx$ **b.** $\int du$

Solution

a. $\int 4\, dx = 4x + C$. Notice that $\frac{d}{dx}(4x + C) = 4$ and that $\int 4\, dx \neq 4x$. Without an arbitrary constant C, infinitely many members have been left out of the family of solutions.

b. $\int du = \int 1\, du = u + C$. Notice that $\frac{d}{du}(u + C) = 1$. $\int 1\, du$ is the same as $\int du$. ■

The Integral of a Power Function
If r is any real number other than -1, then

$$\int x^r\, dx = \frac{x^{r+1}}{r+1} + C$$

To integrate a power function, we increase the exponent by 1, divide the resulting power by the new exponent, and add an arbitrary constant.

Example 2 Find each integral.

a. $\int x^3\, dx$ **b.** $\int x^{2/3}\, dx$ **c.** $\int \frac{1}{x^4}\, dx$

Solution

a.

$$\int x^3\, dx = \frac{x^{3+1}}{3+1} + C$$

$$= \frac{x^4}{4} + C$$

Notice that $\frac{d}{dx}\left[\frac{x^4}{4} + C\right] = \frac{4x^3}{4} + 0 = x^3$, the integrand.

b.

$$\int x^{2/3}\, dx = \frac{x^{(2/3)+1}}{\frac{2}{3}+1} + C$$

$$= \frac{x^{5/3}}{\frac{5}{3}} + C$$

$$= \frac{3x^{5/3}}{5} + C$$

c. This integral does not match our formula for the integration of a power function. We can, however, algebraically manipulate the integrand $\frac{1}{x^4}$ so that it does. Since $\frac{1}{x^4} = x^{-4}$, we will express $\int \frac{1}{x^4}\,dx$ as $\int x^{-4}\,dx$. Now,

$$\begin{aligned}\int \frac{1}{x^4}\,dx &= \int x^{-4}\,dx \\ &= \frac{x^{-4+1}}{-4+1} + C \\ &= \frac{x^{-3}}{-3} + C\end{aligned}$$

Since we began with positive exponents, we will express the result using positive exponents. Since $\frac{x^{-3}}{-3} = \frac{1}{-3x^3} = \frac{-1}{3x^3}$,

$$\int \frac{1}{x^4}\,dx = \frac{-1}{3x^3} + C$$

■

The Integral of a Constant Times a Function

If k is any constant, then

$$\int kf(x)\,dx = k\int f(x)\,dx$$

To integrate a constant times a function, we integrate the function, then multiply that result by the constant, then add an arbitrary constant. This rule is analogous to the differentiation rule

$$\frac{d}{dx}[k \cdot f(x)] = k \cdot \frac{d}{dx}[f(x)]$$

Example 3 Find $\int 6x^2\,dx$

Solution

$$\begin{aligned}\int 6x^2\,dx &= 6\int x^2\,dx \\ &= 6 \cdot \frac{x^3}{3} + C \\ &= 2x^3 + C\end{aligned}$$

■

The Integral of a Sum or Difference

For any two functions $f(x)$ and $g(x)$,

$$\int [f(x) \pm g(x)]\,dx = \int f(x)\,dx \pm \int g(x)\,dx$$

To integrate a sum or difference of two functions, we integrate each function individually, then add or subtract those results. That is, we integrate term by term. This rule is analogous to the derivative rule

$$\frac{d}{dx}[f(x) \pm g(x)] = \frac{d}{dx}f(x) \pm \frac{d}{dx}g(x)$$

Example 4 Integrate

a. $\int (4x^2 + 8x - 7)\,dx$ b. $\int (5x^{2/3} - 4)\,dx$

Solution

a.
$$\begin{aligned}\int (4x^2 + 8x - 7)\,dx &= \int 4x^2\,dx + \int 8x\,dx - \int 7\,dx \\ &= 4\int x^2\,dx + 8\int x\,dx - \int 7\,dx \\ &= 4 \cdot \frac{x^3}{3} + C_1 + 8 \cdot \frac{x^2}{2} + C_2 - (7x + C_3) \\ &= \frac{4x^3}{3} + 4x^2 - 7x + C_1 + C_2 - C_3\end{aligned}$$

Each of the three individual integrals produces a constant of integration. We have denoted these by C_1 (produced from the first integral), C_2 (produced from the second integral), and C_3 (produced from the third integral). Since the sum of three (or any number of) constants is again a constant, we will denote $C_1 + C_2 - C_3$ by C. That is, $C_1 + C_2 - C_3 = C$. Thus,

$$\int (4x^2 + 8x - 7)\,dx = \frac{4x^3}{3} + 4x^2 - 7x + C$$

Since the sum of a collection of constants is always another constant, it is convenient to omit the constants that are produced in the intermediate integrations and to simply add an arbitrary constant at the end of the process.

b.
$$\begin{aligned}\int (5x^{2/3} - 4)\,dx &= 5 \cdot \frac{x^{2/3+1}}{\frac{2}{3} + 1} - 4x + C \\ &= 5 \cdot \frac{x^{5/3}}{\frac{5}{3}} - 4x + C \\ &= 5 \cdot \frac{3x^{5/3}}{5} - 4x + C \\ &= 3x^{5/3} - 4x + C\end{aligned}$$

The Integral of the Exponential Function

$$\int e^x\,dx = e^x + C$$

This should be no surprise since the derivative of the exponential function is the exponential function itself.

Example 5 Find $\int (3e^x - 5x)\, dx$

Solution

$$\int (3e^x - 5x)\, dx = \int 3e^x\, dx - \int 5x\, dx$$
$$= 3e^x - \frac{5x^2}{2} + C$$

■

The Integral of the Reciprocal Function

$$\int \frac{1}{x}\, dx = \ln|x| + C$$

Notice that this formula gives us a way to integrate x^{-1}, the case that was excluded in our formula for $\int x^r\, dx$. If $r = -1$, then

$$\int x^r\, dx = \int x^{-1}\, dx = \int \frac{1}{x}\, dx = \ln|x| + C$$

The absolute value bars are necessary because when

$$x > 0, \frac{d}{dx}[\ln x] = \frac{1}{x}$$

and when

$$x < 0, \frac{d}{dx}[\ln(-x)] = \frac{1}{-x} \cdot \frac{d}{dx}[-x]$$
$$= \frac{1}{-x} \cdot (-1)$$
$$= \frac{1}{x}$$

Thus, both $f(x) = \ln x$ and $f(x) = \ln(-x)$ have the same derivative. But, because logarithms are defined only for positive numbers, we need to use the absolute value bars to ensure that the resulting integral is defined.

Example 6

a. Find $\int \frac{7}{x}\, dx$

b. Suppose it is known that $x > 0$. Find $\int \left(6e^x + \frac{4}{x} - 1\right)$.

c. Suppose it is known that $x > 0$. Find $\int \frac{x^4 + 3x^3 + 5}{x}$.

Solution

a.

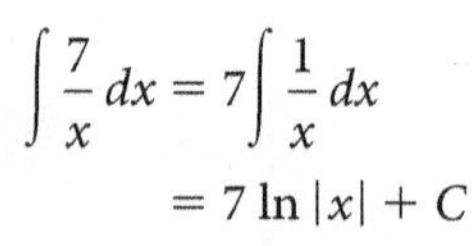

$$\int \frac{7}{x}\, dx = 7\int \frac{1}{x}\, dx$$
$$= 7\ln|x| + C$$

b.
$$\int \left(6e^x + \frac{4}{x} - 1\right) dx = 6e^x + 4\ln|x| - x + C$$
$$= 6e^x + 4\ln x - x + C$$

We omit the absolute value bars here since $x > 0$.

c.
$$\int \frac{x^4 + 3x^3 + 5}{x}\, dx = \int \left(\frac{x^4}{x} + \frac{3x^3}{x} + \frac{5}{x}\right) dx$$
$$= \int \left(x^3 + 3x^2 + \frac{5}{x}\right) dx$$
$$= \frac{x^4}{4} + \frac{3x^3}{3} + 5\ln|x| + C$$
$$= \frac{x^4}{4} + x^3 + 5\ln x + C$$

Once again, we omit the absolute value bars here since $x > 0$. ■

Indefinite Integrals with Initial Conditions

As we noted at the beginning of the section, when we are given the rate at which a quantity changes and we wish to know how large the quantity is at any time, we need to integrate (antidifferentiate) the derivative function. However, the integral produces an arbitrary constant C. To eliminate C and determine precisely the required individual member of the family of antiderivatives that describes the relationship between the input and output values, we must also know some **initial condition**. An initial condition is some specific information about an input value and its corresponding output value. Example 7 illustrates.

Example 7 Find $f(x)$ if $f'(x) = 3x^2 - 4x + 2$ and $f(1) = 4$.

Solution

$$f(x) = \int (3x^2 - 4x + 2)\, dx$$
$$= 3 \cdot \frac{x^3}{3} - 4 \cdot \frac{x^2}{2} + 2x + C$$
$$= x^3 - 2x^2 + 2x + C$$

Now, $f(x) = x^3 - 2x^2 + 2x + C$, and we are given the initial condition that $f(1) = 4$. Substituting into $f(x) = x^3 - 2x^2 + 2x + C$ produces

$$f(1) = (1)^3 - 2(1)^2 + 2(1) + C$$
$$4 = 1 - 2 + 2 + C$$
$$4 = 1 + C$$
$$C = 3$$

Thus, $f(x) = x^3 - 2x^2 + 2x + 3$. ■

Example 8 Let's now consider the marginal cost function

$$C'(x) = -0.04x + 26$$

where x represents the level of production, and $C'(x)$, the marginal cost, and try to determine the cost function, $C(x)$, from which it was derived. If at the production level of 150 items, the cost is known to be \$6,555,

a. compute $C'(530)$.

b. integrate $C'(x)$ to find the cost function $C(x)$.

c. compute the integral $C(530)$.

d. interpret both the derivative $C'(530)$ and the integral $C(530)$.

Solution

a. $C'(530) = -0.04(530) + 26 = 4.8$.

b. Since we are given the marginal cost function (which is the derivative of the cost function), we need to integrate it to find the cost function.

$$\begin{aligned} C(x) &= \int (-0.04x + 26)\,dx \\ &= \frac{-0.04x^2}{2} + 26x + C \\ &= -0.02x^2 + 26x + C \end{aligned}$$

Thus, the family of cost functions is $C(x) = -0.02x^2 + 26x + C$.

To find the member of the family that describes our particular situation (cost = \$6,555 when the level of production is 150 units, or $C(150) = 6{,}555$), we need to find the value of C. We substitute 6,555 for $C(x)$ and 150 for x in $C(x) = -0.02x^2 + 26x + C$ and solve for C.

$$\begin{aligned} 6{,}555 &= -0.02(150)^2 + 26(150) + C \\ C &= 3{,}105 \end{aligned}$$

Therefore, the cost function is $C(x) = -0.02x^2 + 26x + 3{,}105$.

c. $C(530) = -0.02(530)^2 + 26(530) + 3{,}105 = 11{,}267$.

d. $C'(530) = 4.8$ means that the cost of producing the 531st unit is approximately \$4.80. $C(530) = 11{,}267$ means that the cost of producing 530 units is \$11,267.

Using Technology 5.1

The process of reversing a differentiation is antidifferentiation. You can use Wolfram|Alpha to find antiderivatives of functions.

Example 9 Find the antiderivative of the function $f(x) = x^5$. That is, find a function $F(x)$ for which $F'(x) = f(x)$.

Solution Go online to www.wolframalpha.com. Into the entry field enter either of the following two instructions:

find the antiderivative of x^5 or integrate x^5

In either case, Wolfram|Alpha returns

Indefinite integral:

$$\int x^5\,dx = \frac{x^6}{6} + \text{constant}$$

Wolfram Alpha LLC. 2012. Wolfram|Alpha
http://www.wolframalpha.com/
(access October 24, 2012)

Example 10 Find the particular antiderivative, $f(x)$, of

$$f'(x) = -0.04x + 26$$

given that $f(x) = 6{,}555$ when $x = 150$.

Solution Just as in Example 8, we need to integrate and use the initial condition to find the particular constant of integration. Wolfram|Alpha can perform both the integrating and the solving for the constant for us. Into the entry field enter

solve y'(x) = -0.04x + 26, y(150) = 6555

Wolfram|Alpha antidifferentiates $y'(x)$ and uses the initial condition to solve for C, getting the particular antiderivative we are looking for.

y(x) = -0.02x^2 + 26x + 3105

Getting Ready for Class

After reading through the preceding section, respond in your own words and in complete sentences.

A. What are the two names given to the process of reversing a differentiation?

B. What are the two names given to the expression $F(x) + c$ in the expression $\int f(x)\,dx = F(x) + c$?

C. What is the restriction placed on the value of r in the formula for the integral of a power function, $\int x^r\,dx$?

D. Some indefinite integrals come with additional information that allows you to replace the constant of integration c with a specific value. What is the name given to that piece of additional information?

Problem Set 5.1

Skills Practice

For Problems 1-18, evaluate each indefinite integral.

1. $\int 6x^5\,dx$

2. $\int x^{2/3}\,dx$

3. $\int \frac{1}{x^4}\,dx$

4. $\int \frac{-2}{x^{2/3}}\,dx$

5. $\int \frac{8}{\sqrt[5]{x^2}}\,dx$

6. $\int \frac{55}{x}\,dx$

7. $\int 6e^x\,dx$

8. $\int (x^4 + 5x^3 + 2)\,dx$

9. $\int \left(2x^3 + \frac{3}{x^2} + 1\right)dx$

10. $\int \left(\frac{4}{x^4} - \frac{3}{x^2}\right)dx$

11. $\int (x^{-3} + x^{-2} - x^{-1})\,dx$

12. $\int \left(3e^x + \frac{3}{x} + x^3 + 3\right)dx$

13. $\int (0.03 + 0.12x^{-1/2})\,dx$

14. $\int (3x^2 - 6x - 7)\,dx$, where $f(x) = 10$ when $x = 2$

15. $\int (5x^4 - 4x^3 + 8x)\,dx$, where $f(x) = 9$ when $x = 1$

16. $\int \frac{1}{x^{2/3}}\,dx$, where $f(x) = 1$ when $x = 8$

17. $\int \left(3e^x + \frac{4}{x}\right)dx$, where $f(x) = 5e$ when $x = 1$

18. $\int (x^2 + 2e^x + 4\sqrt[3]{x})\,dx$, where $f(x) = 1$ when $x = 0$

Check Your Understanding

19. **Business: Manufacturing** A manufacturer has determined that, because of the seasonal nature of the business, the rate (in pairs per day) at which the number, N, of men's casual pants it produces and sells per day is modeled by the function $N'(t)$, where t is the number of days from January 1. Interpret both the derivative $N'(50) = 30$ and the integral $N(50) = 550$.

20. **Business: Manufacturing** A manufacturer of modeling clay has determined that if the clay is left uncovered in the open air, it will harden. The rate at which the clay hardens is related to the number, t, of minutes the clay has been in the open air after being taken out of its protective wrapping by the function $H'(t)$. The hardness of the clay, H, is in the manufacturer's own hardness units. Interpret both the derivative $H'(20) = 0.3$ and the integral $H(20) = 15$.

21. **Business: Manufacturing** A manufacturer has determined that the rate at which the number, N, of units of her product that fail is related to the speed, s, (in units per minute) at which the units are produced by the function $N'(s)$. Interpret both the derivative $N'(170) = 0.008$ and the integral $N(170) = 3$.

22. **Business: Utilization of a Product** The rate, in hours per week, at which a leisure product is utilized is related to the number of hours, h, of leisure time each week the user has available by the function $T'(h)$. Interpret both $T'(16) = 0.5$ and its integral $T(16) = 4$.

23. **Political Science: Opinion Polls** A political scientist believes that the rate at which a political candidate's favorable rating (in percentage points) in weekly polls changes is related to the number, n, of negative advertisements sponsored each week by the candidate's opponent by the function $F'(n)$. Interpret both $F'(10) = -0.005$ and its integral $F(10) = 0.45$.

24. **Economics: Individual Wealth** The rate at which the wealth (in thousands of dollars) of an individual changes each week is related to the time, t, (in weeks) since the beginning of a series of investments made by the individual by the function $W'(t)$. Interpret both the derivative $W'(30) = -12.25$ and its integral $W(30) = 865$.

25. **City Management: Traffic** On weekdays, the rate (in cars per minute) at which cars enter a boulevard near a city's business park is related to the number of minutes from 5:00 AM by the function $N'(m)$. Interpret both the derivative $N'(120) = 3$ and its integral $N(120) = 12$.

For Problems 26-29, verify, by differentiation, that the function $F(x)$ is an antiderivative of the function $f(x)$.

26. $F(x) = \frac{x^5}{5} + \frac{5x^3}{3} + \frac{7x^2}{2} - x + 6$, where $f(x) = x^4 + 5x^2 + 7x - 1$

27. $F(x) = \frac{3x^5}{5} - \frac{5x^4}{4} + 2x^2 - 8$, where $f(x) = 3x^4 - 5x^3 + 4x$

28. $F(x) = \frac{1}{3}e^{3x} - e^{-x} + x^2 + 4$, where $f(x) = e^{3x} + e^{-x} + 2x$

29. $F(x) = (x^2 + 4)^3(x^2 - 5)^4 + 8$, where $f(x) = 2x(x^2 + 4)^2(x^2 - 5)^3(7x^2 + 1)$

For Problems 30-39, determine if the given integral has been evaluated correctly. If it has, write "correctly evaluated." If it has not, write "not correctly evaluated," and then write the correct evaluation.

30. $\int e^{6x}\,dx = \frac{e^{6x+1}}{6x+1} + C$

31. $\int x^{5/2}\,dx = \frac{2}{7}x^{7/2} + C$

32. $\int \sqrt[3]{x}\,dx = \frac{3}{4}x^{4/3}$

33. $\int (16x^3 + 9x^2)\,dx = 48x^2 + 18x + C$

34. $\int (3x^2 + x^{-1})\,dx = x^3 - x^{-3} + C$

35. $\int (4e^{4x} + x^{-1})\,dx = e^{5x} - x^{-2} + C$

36. $\int (e^{3x} - e^{-3x})\, dx = 3e^{3x} + 3e^{-3x} + C$

37. $\int \left(e^{2x} + \frac{3}{x} + 3\right) dx = \frac{1}{2}e^{2x} + \ln x^3 + 3C$

38. $\int e^{-5x} \ln 4\, dx = \frac{-\ln 4}{5} e^{-5x} + C$

39. $\int \frac{e}{x}\, dx = e \ln|x| + C$

Modeling Practice

40. **Business: Marginal Cost** A company's marginal cost, C', of producing x units of a product is given by the function $C'(x) = -0.104x + 82.12$. At the production level of 600 units, the cost is known to be \$28,752.00.
   a. Find the cost function.
   b. Find the cost of producing 500 items.
   c. Find and interpret both $C'(950)$ and its integral $C(950)$.

41. **Economics: Population of a City** The growth rate of a city t years after 2013 is given by the function $P'(t) = 240\sqrt[5]{t} + 170$. In 2013, the population of the city is 8,400.
   a. Find the population function.
   b. Find the expected population of the city in 2014.
   c. Find and interpret both $P'(6)$ and its integral $P(6)$.

42. **Ecology: Environmental Pollution** A city's environmental commission estimates that the amount A (in parts per million) of carbon monoxide in the summer daytime air will, without new protective measures, increase at the rate of $A'(t) = 0.0027t^2 + 0.04t + 0.08$ parts per million t years from now. Currently, the average daytime level of carbon monoxide in the city's air is 3 parts per million. If no air quality protective measures are taken,
   a. At what rate will the level of carbon monoxide be changing 5 years from now?
   b. What is the expected daytime level of carbon monoxide in this city 5 years from now?

43. **Life Science: Tumor Growth** A medical study has shown that during the first 20 days of radiation therapy, a particular type of malignant tumor decreases at a rate according to the function $M'(t) = -0.018t^2 + 0.01t$, where M represents the mass (in grams) of the tumor and t is the time (in days) since the beginning of the radiation treatment. If a tumor of this particular type has a mass of 200 grams just prior to the start of radiation treatment,
   a. At what rate will the tumor be decreasing in 10 days?
   b. in 15 days?
   c. What will be the mass of the tumor in 10 days?
   d. in 15 days?

44. Business: Sales A manufacturer has determined that the growth rate of sales S (in units) of a newly developed product should be approximated by

$$S'(t) = \frac{2{,}000}{\sqrt[3]{t}}$$

where t is the number of years from now. Assuming there are no sales at the introduction of the product to the market,

a. How many units of the product will have been sold 5 years from now?

b. At what rate will the number of sales be changing after 5 years?

c. Using the result from part b, estimate how many units of the product will have been sold 6 years from now.

d. Calculating as you did in part a and without using the results of part b, how many units of the product will have been sold 6 years from now?

45. Biology: Human Surface Area As a person's mass m (in kilograms) changes, his or her surface area, A, (in square meters) changes according to the function

$$A'(m) = \frac{0.073}{\sqrt[3]{m}}$$

(A reasonable initial condition here is that a person with no mass has no surface area.)

a. Find the surface area of a person with a mass of 64 kg (64 kg $\approx$ 141 pounds).

b. Find the rate at which the surface area of a 64-kg person is changing.

c. Interpret $A'(64)$.

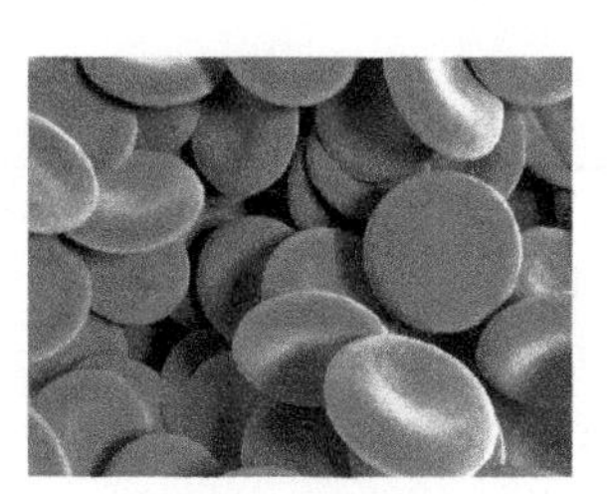

© Andrew Dorey/iStockPhoto

46. Life Science: Drug Concentration For a particular glucose tolerance test for hypoglycemia, physicians believe that the amount, A, (in milligrams) of glucose remaining in the blood t hours after the ingestion of the glucose changes according to the function

$$A'(t) = \frac{2.4}{\sqrt{t}}$$

If for this particular test, 7.9 milligrams of glucose are in the bloodstream 1 hour after the glucose is ingested,

a. At what rate is the glucose in the bloodstream changing after 4 hours?

b. What is the total amount of glucose in the bloodstream after 4 hours?

c. Interpret $A'(4)$.

47. Business: Inventory Costs The inventory cost, C, (in dollars) to a retail shoe chain for ordering boxes of men's socks is changing at the rate

$$C'(x) = 20 - \frac{18{,}000}{x^2}$$

The chain's inventory cost is minimized when they order 15 times per year in lots of size 30. If they order in lots of size 50, the inventory cost is \$1,360. Find and interpret both $C'(30)$ and its integral $C(30)$.

48. Marketing: Advertising and Sales The management of a chain of frozen yogurt stores believes that t days after the end of an advertising campaign, the rate at which the volume V (in dollars) of sales is changing is approximated by $V'(t) = -29{,}900e^{-0.46t}$. On the day the advertising campaign ends ($t = 0$), the sales volume is \$130,000. Find and interpret both $V'(5)$ and its integral $V(5)$.

49. Ecology: National Consumption of Toxic Chemicals A government-prepared study shows that the rate at which the national consumption of a particular toxic chemical is changing is given by the function

$$C'(t) = -17.4e^{-1.2t} + 0.28t$$

where C represents the national consumption of the chemical (in millions of barrels), and t represents the number of years after the study. At the time the study was released, the national consumption of the chemical was 14.5 consumption units. Find and interpret both $C'(2)$ and its integral $C(2)$.

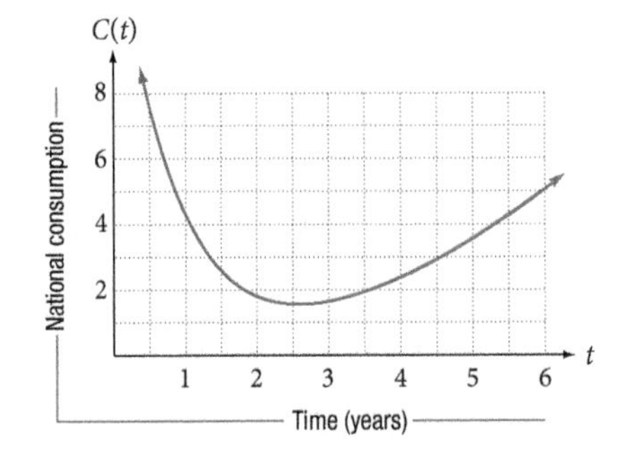

Using Technology Exercises

50. Find the antiderivative of $f(x) = 16x^3 - 18x^2 - 10x + 12$.

51. Specify the coefficient of the 3rd-degree term of the antiderivative of

$$f(x) = (x - 2)^3 + (3x + 3)^4.$$

52. Find the particular antiderivative, $f(x)$, of $f'(x) = \frac{3}{x} + 6x^2$, given that $f(x) = 5$ when $x = 1$.

53. You can check to see if your antiderivative of a function is correct by differentiating the function. If you both integrate and differentiate correctly, the derivative of the antiderivative of a function $f(x)$ will be $f(x)$. Create a Wolfram|Alpha command that checks the integration of $f(x) = 5x^3 + 2x$.

54. Business: Inventory Cost The inventory cost C (in dollars) of stocking x units of a product is changing at the rate

$$C'(x) = \frac{-68{,}500}{x^{2.2}} - 3$$

Find and interpret both $C(300)$ and $C'(300)$. Assume it costs \$2,000 to store 50 units of the product.

Getting Ready for the Next Section

55. Find $\frac{du}{dx}$ for $u = 6x^2 + 4x + 1$.

56. Find $\frac{du}{dx}$ for $u = 3x^2 + 5$.

57. Multiply both sides of $\frac{du}{dx} = 12x + 4$ by dx to solve for du.

58. Multiply both sides of $\frac{du}{dx} = 6x$ by dx to solve for du.

59. Find $\frac{d}{dx}\left[\frac{1}{6}\ln(3x^2 + 5) + C\right]$

60. Find $\frac{d}{dx}[e^{3x+7} + C]$

61. For $u = 5x^2 - 2$, find $\frac{du}{dx}$, du, and dx.

62. For $u = 9x^2 + 2x + 7$, find $\frac{du}{dx}$, du, and $(9x + 1)\,dx$.

For Problems 63 and 64, evaluate.

63. $\frac{1}{6}\int \frac{1}{u}\,du$

64. $\frac{1}{32}\int \frac{1}{\sqrt[3]{u}}\,du$

65. Find $\frac{du}{dx}$ for $u = \frac{2(x + 3)^{5/2}}{5} - 2(x + 3)^{3/2} + C$.

66. Evaluate $\int (u - 3)\sqrt{u}\,du$.

Spotlight on **Success**

Student Instructor Aaron

Sometimes you have to take a step back
in order to get a running start forward.
—Anonymous

As a high school senior I was encouraged to go to college immediately after graduating. I earned good grades in high school and I knew that I would have a pretty good group of schools to pick from. Even though I felt like "more school" was not quite what I wanted, the counselors had so much faith and had done this process so many times that it was almost too easy to get the applications out. I sent out applications to schools I knew I could get into and a "dream school."

One night in my email inbox there was a letter of acceptance from my dream school. There was just one problem with getting into this school. It was going to be difficult and I still had senioritis. Going into my first quarter of college was as exciting and difficult as I knew it would be. But after my first quarter I could see that this was not the time for me to be here. I was interested in the subject matter but I could not find my motivating purpose like I had in high school. Instead of dropping out completely, I decided a community college would be a good way for me to stay on track. Without necessarily knowing my direction, I could take the general education classes and get those out of the way while figuring out exactly what and where I felt a good place for me to be.

Now I know what I want to go to school for and the next time I walk onto a four year campus it will be on my terms with my reasons for being there driving me to succeed. I encourage everyone to continue school after high school, even if you have no clue as to what you want to study. There are always stepping stones, like community colleges, that can help you get a clearer picture of what you want to strive for.

Integration by Substitution

5.2

Consider the following function that represents the level of pollution, P, (in parts per million) over a city that changes as the temperature, T, (in degrees Fahrenheit) changes.

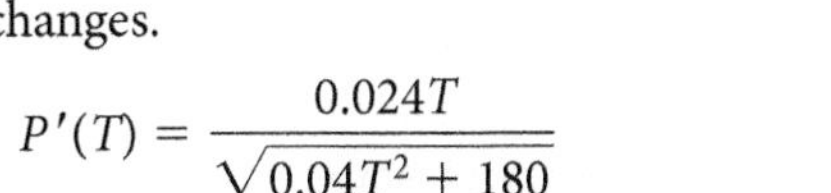

$$P'(T) = \frac{0.024T}{\sqrt{0.04T^2 + 180}}$$

© Kzenon/iStockPhoto

The integral form of this function would be in a form that is different from any of the basic integral forms we have discussed. We will revisit this function later in the problem set.

In this section, we introduce a process that produces an antiderivative of some functions that do not match any of the basic forms discussed in Section 5.1. For example, the integral

$$\int (5x^4 + 2x^2 + 7)\, dx$$

can be evaluated directly because its form matches one of the basic integral forms presented in Section 5.1. However, the integral

$$\int \frac{12x + 4}{6x^2 + 4x + 1}\, dx$$

does not match any of the basic integral forms and therefore cannot by evaluated directly. It can, however, be evaluated indirectly by making a substitution that transforms the integral to a known basic form.

Integration by Substitution

The method of integration called **integration by substitution** involves substituting a variable (any variable other than the original can be used; we often use u) for some part of the integrand and then expressing the rest of the integrand in terms of that variable. This strategy is outlined in the flow diagram in Figure 1.

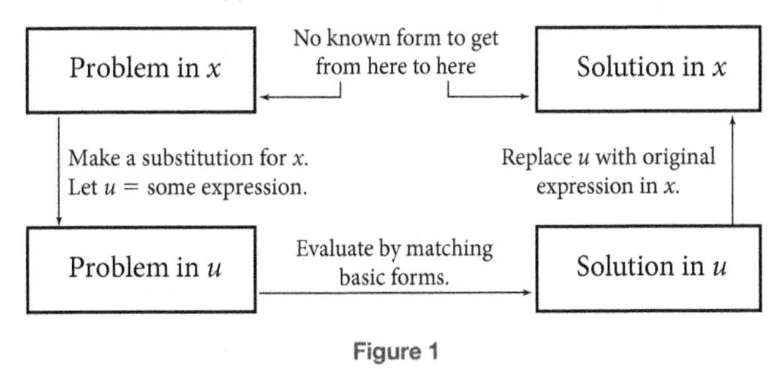

Figure 1

The diagram shows that we want to get from a problem in x to a solution in x. But because we have no form we can match the integral to, we cannot take this route directly. Instead, we must take another route: We make a substitution that transforms the original integrand to a new integrand that does match one of the basic forms. A beneficial substitution often involves a choice of u for which $\frac{du}{dx}$ is a constant multiple of one of the factors in the integrand. We evaluate that integral, and then resubstitute to get to our destination, a solution in x. We will illustrate the technique in Example 1.

VIDEO EXAMPLES

SECTION 5.2

Example 1 Evaluate $\int \frac{12x + 4}{6x^2 + 4x + 1}\, dx$.

Solution We imagine ourselves in the box at the top left of the flow diagram of Figure 1. We have a problem in x with no known basic integral form that will allow us direct passage to a solution in x. We need to take an indirect route via a substitution. In this case, we will let the variable u represent the expression $6x^2 + 4x + 1$. (As you work through the example, you will see why this is a good choice.)

When making a substitution into an integral involving only one variable, it is very important to make a *complete substitution* so that there is only one variable present at all times. This requirement is the reason why not all integrals can be evaluated by substitution.

With $u = 6x^2 + 4x + 1$, we must still replace $(12x + 4)\, dx$ to make the substitution complete. We account for this remaining part of the integrand, $(12x + 4)\, dx$, with the differential of u. Since

$$u = 6x^2 + 4x + 1$$

$$\frac{du}{dx} = 12x + 4$$

so that the differential of u is

$$du = (12x + 4)\, dx$$

Thus, the remaining part of the integrand $(12x + 4)\, dx$ is precisely du when

$$u = 6x^2 + 4x + 1$$

Now we can write a complete substitution:

$$\int \frac{12x + 4}{6x^2 + 4x + 1}\, dx = \int \frac{du}{u}$$

We are now located in the box at the lower left of the flow diagram; we have the problem in u. Furthermore, we can match this form to a basic integral form to get a solution in u.

$$\int \frac{du}{u} = \ln|u| + C$$

With this solution, we arrive in the box at the lower right of the flow diagram and can use our original substitution to resubstitute to get a solution in x. Since $u = 6x^2 + 4x + 1$, we can replace u in $\ln|u| + C$ with $6x^2 + 4x + 1$, and write

$$\int \frac{12x + 4}{6x^2 + 4x + 1}\, dx = \ln|6x^2 + 4x + 1| + C$$ ■

Sometimes, after applying the differential, some adjustments are necessary to get the precise expression that remains in the integrand. Example 2 illustrates this process.

Example 2 Evaluate $\int \frac{x}{3x^2 + 5}\, dx$.

Solution We let $u = 3x^2 + 5$. The remaining part of the integrand is $x\, dx$. If integration by substitution is going to work, we should be able to get $x\, dx$ from the differential of u. We let

$$u = 3x^2 + 5$$

so that

$$\frac{du}{dx} = 6x$$

giving us the differential

$$du = 6x\,dx$$

But we want $x\,dx$, not $6x\,dx$. We need to make an adjustment, so we eliminate the 6 by dividing both sides of $du = 6x\,dx$ by 6.

$$du = 6x\,dx$$

$$\frac{du}{6} = \frac{6x\,dx}{6}$$

so that

$$x\,dx = \frac{1}{6}\,du$$

Thus,

$$\begin{aligned}
\int \frac{x}{3x^2 + 5}\,dx &= \int \frac{\frac{1}{6}\,du}{u} \\
&= \frac{1}{6}\int \frac{du}{u} \\
&= \frac{1}{6}\ln|u| + C \\
&= \frac{1}{6}\ln|3x^2 + 5| + C \\
&= \frac{1}{6}\ln(3x^2 + 5) + C \qquad \text{(since } 3x^2 + 5 \text{ is always positive)}
\end{aligned}$$

You should be able to prove that this result is correct by differentiating

$$\frac{1}{6}\ln(3x^2 + 5) + C$$

to get

$$\frac{x}{3x^2 + 5}$$

■

A Strategy for Substitution

How do we know which part of an integrand to let u represent? Here are some suggestions that *often* work, but *not always*.

Suggestions for Substitutions

If the integrand involves

1. a quotient, let u = (the denominator).
2. an exponential expression, let u = (the exponent).
3. a power or a radical, let u = (the base or the radicand).

Example 3 Evaluate $\int 3e^{3x+7}\,dx$.

Solution First, we will rewrite the integral as $3\int e^{3x+7}\,dx$. Part of the integrand is an exponential so we will try letting u represent the exponent $3x + 7$. The remaining part of the integrand that involves the variable is dx. If integration by substitution is going to work, we should be able to get dx from the differential of u. We let

$$u = 3x + 7$$

so that

$$\frac{du}{dx} = 3$$

giving us the differential

$$du = 3\,dx$$

Therefore,

$$3\,dx = du$$

Thus,

$$\begin{aligned} 3\int e^{3x+7}\,dx &= \int e^{3x+7}\cdot 3\,dx \\ &= \int e^{u}\,du && \text{Which matches a basic integral form} \\ &= e^{u} + C && \text{Replace } u \text{ with } 3x+7 \\ &= e^{3x+7} + C \end{aligned}$$

Example 4 Evaluate $\int 4xe^{5x^2-2}\,dx$.

Solution First, we will rewrite the integral as $4\int xe^{5x^2-2}\,dx$. The integrand is an exponential so we will try letting u represent the exponent $5x^2 - 2$. The remaining part of the integrand is $x\,dx$. If integration by substitution is going to work, we should be able to get $x\,dx$ from the differential of u. We let

$$u = 5x^2 - 2$$

so that

$$\frac{du}{dx} = 10x$$

giving us the differential

$$du = 10x\,dx$$

Therefore,

$$x\,dx = \frac{1}{10}\,du$$

Thus,

$$\begin{aligned} 4\int e^{5x^2-2}x\,dx &= 4\int e^{u}\cdot\frac{1}{10}\,du \\ &= \frac{2}{5}\int e^{u}\,du && \text{Which matches a basic integral form} \\ &= \frac{2}{5}e^{u} + C && \text{Replace } u \text{ with } 5x^2-2 \\ &= \frac{2}{5}e^{5x^2-2} + C \end{aligned}$$

Example 5 Evaluate $\int 10(9x + 1)(9x^2 + 2x + 7)^8\, dx$.

Solution First, we will rewrite the integral as $10\int(9x + 1)(9x^2 + 2x + 7)^8\, dx$. Part of the integrand involves a power so we will try letting u represent the base of the power $9x^2 + 2x + 7$. The remaining part is $(9x + 1)\, dx$. If integration by substitution is going to work, we should be able to get $(9x + 1)\, dx$ from the differential of u. We let

$$u = 9x^2 + 2x + 7$$

so that

$$\frac{du}{dx} = 18x + 2$$

giving us the differential

$$du = (18x + 2)\, dx$$

which when factored gives

$$du = 2(9x + 1)\, dx$$

We then solve for $(9x + 1)\, dx$.

$$(9x + 1)\, dx = \frac{1}{2}\, du$$

Thus,

$$\begin{aligned} 10\int (9x^2 + 2x + 7)^8(9x + 1)\, dx &= 10\int u^8 \cdot \frac{1}{2}\, du \\ &= 5\int u^8\, du \\ &= 5 \cdot \frac{u^9}{9} + C \\ &= \frac{5(9x^2 + 2x + 7)^9}{9} + C \end{aligned}$$

■

Example 6 Evaluate $\int \frac{x^3}{\sqrt[3]{8x^4 - 5}}\, dx$.

Solution Part of the integrand involves a radical so we will try letting u represent the radicand $8x^4 - 5$. The remaining part is $x^3\, dx$. If integration by substitution is going to work, we should be able to get $x^3\, dx$ from the differential of u. We let

$$u = 8x^4 - 5$$

so that

$$\frac{du}{dx} = 32x^3$$

giving us the differential

$$du = 32x^3\, dx$$

Therefore,

$$x^3\, dx = \frac{1}{32}\, du$$

Thus,

$$\int \frac{x^3}{\sqrt[3]{8x^4 - 5}}\,dx = \int \frac{1}{\sqrt[3]{u}} \cdot \frac{1}{32}\,du$$

$$= \frac{1}{32}\int \frac{1}{\sqrt[3]{u}}\,du$$

$$= \frac{1}{32} \cdot \int \frac{1}{u^{1/3}}\,du$$

$$= \frac{1}{32} \cdot \int u^{-1/3}\,du$$

$$= \frac{1}{32} \cdot \frac{u^{-1/3+1}}{-\frac{1}{3}+1} + C$$

$$= \frac{1}{32} \cdot \frac{u^{2/3}}{\frac{2}{3}} + C$$

$$= \frac{1}{32} \cdot \frac{3u^{2/3}}{2} + C$$

$$= \frac{3}{64} \cdot u^{2/3} + C$$

$$= \frac{3}{64} \cdot (8x^4 - 5)^{2/3} + C$$

■

Example 7 Evaluate $\int x\sqrt{x+3}\,dx$.

Solution Part of the integrand is a radical so we will try letting u represent the radicand $x + 3$. The remaining part is $x\,dx$. If integration by substitution is going to work, we should be able to get $x\,dx$ from the differential of u. We let

$$u = x + 3$$

so that

$$\frac{du}{dx} = 1$$

giving us the differential

$$du = 1\,dx$$

or

$$du = dx$$

But this only gives us dx. We still need to get the remaining x in terms of u. We need an adjustment that is different from the adjustments we made in the previous examples. By our original substitution, $u = x + 3$, so that by subtracting 3 from each side we get $x = u - 3$, and then we have x in terms of u. To summarize, we now have:

$$u = x + 3$$

$$x = u - 3$$

$$dx = du$$

Thus,

$$\begin{aligned}
\int x\sqrt{x+3}\,dx &= \int (u-3)\sqrt{u}\,du \\
&= \int (u-3)u^{1/2}\,du && \text{Multiply} \\
&= \int (u^{3/2} - 3u^{1/2})\,du \\
&= \int u^{3/2}\,du - 3\int u^{1/2}\,du \\
&= \frac{u^{5/2}}{\frac{5}{2}} - 3\frac{u^{3/2}}{\frac{3}{2}} + C \\
&= \frac{2u^{5/2}}{5} - 2u^{3/2} + C && \text{Replace } u \text{ with } x+3 \\
&= \frac{2(x+3)^{5/2}}{5} - 2(x+3)^{3/2} + C
\end{aligned}$$

Integrating the Exponential Function

As you may recall we noted that the function $f(x) = e^{ax}$ occurs often in applications and that its derivative, which follows directly from the chain rule, can be obtained quickly by multiplying the expression e^{ax} by a, the coefficient of x. That is, if $f(x) = e^{ax}$, then $f'(x) = ae^{ax}$.

Similarly, since the integral of $f(x) = e^{ax}$ occurs often, it is convenient to have a special rule specifically for it so that we can avoid making a substitution every time we come across it. Therefore, to integrate $f(x) = e^{ax}$, we simply divide the expression e^{ax} by a, the coefficient of x. That is,

The Integral of $f(x) = e^{ax}$

If $f(x) = e^{ax}$, then

$$\int e^{ax}\,dx = \frac{e^{ax}}{a} + C$$

You will be asked to verify this result by substitution in Problem 27.

Example 8

a. $\int e^{6x}\,dx$ **b.** $\int e^{\frac{2}{3}x}\,dx$

Solution

a.
$$\int e^{6x}\,dx = \frac{e^{6x}}{6} + C$$

b.
$$\int e^{\frac{2}{3}x}\,dx = \frac{e^{\frac{2}{3}x}}{\frac{2}{3}} + C = \frac{3e^{\frac{2}{3}x}}{2} + C$$

Using Technology 5.2

You can use Wolfram|Alpha to solve problems requiring integration by substitution.

Example 9 Use integration by substitution to evaluate

$$\int \frac{10x + 3}{5x^2 + 3x - 2}\, dx$$

Soution This integration requires substitution. Doing this by hand, we would let $u = 5x^2 + 3x - 2$. Wolfram|Alpha can do the substitution and evaluation for us.

Go online to www.wolframalpha.com. Into the entry field enter

integrate (10x + 3)/(5x^2 + 3x - 2)

Wolfram|Alpha returns

Indefinite integral:

$$\int \frac{3 + 10x}{-2 + 3x + 5x^2}\, dx = \log(5x^2 + 3x - 2) + \text{constant}$$

Wolfram Alpha LLC. 2012. Wolfram|Alpha
http://www.wolframalpha.com/
(access October 24, 2012)

■

Getting Ready for Class

After reading through the preceding section, respond in your own words and in complete sentences.

A. Sometimes you can evaluate an indefinite integral directly because its form matches one of the basic integral forms. If the integral's form does not match one of the basic forms, what is the name of another method you might use?

B. Integration by substitution involves substituting the variable u for some expression in the original integral. What might you suggest letting u equal if the original integral involves
- **a.** a quotient?
- **b.** an exponential expression?
- **c.** a power or a radical expression?

C. The exponential expression e^{ax} shows up often in differentiation and integration. What is $\frac{d}{dx}e^{ax}$ and $\int e^{ax}\, dx$?

Problem Set 5.2

Skills Practice

1. $\int (x+2)^4\,dx$

2. $\int (x-8)^7\,dx$

3. $\int (5x+1)^4\,dx$

4. $\int (6x+3)^5\,dx$

5. $\int 3x^2(x^3+2)^5\,dx$

6. $\int 20x^3(5x^4+6)^3\,dx$

7. $\int (2x+4)\sqrt{x^2+4x}\,dx$

8. $\int (6x-6)\sqrt{3x^2-6x}\,dx$

9. $\int (60x^2-32)\sqrt{5x^3-8x}\,dx$

10. $\int (4x^3+3)(2x^4+6x)^{1/3}\,dx$

11. $\int \frac{4x}{2x^2+1}\,dx$

12. $\int \frac{5}{5x+4}\,dx$

13. $\int \frac{e^{4x}}{e^{4x}+2}\,dx$

14. $\int \frac{15x^2e^{5x^3}}{e^{5x^3}+2}\,dx$

15. $\int \frac{12x^2+5}{4x^3+5x-6}\,dx$

16. $\int \frac{10x^3+4x}{10x^4+8x^2-1}\,dx$

17. $\int \frac{\ln x+7}{x}\,dx$

18. $\int \frac{\ln(5x)+5}{x}\,dx$

19. $\int \frac{5x}{5x-4}\,dx$

20. $\int (x-7)(x+7)^4\,dx$

21. $\int e^{-0.02x}\,dx$

22. $\int e^{-0.1x}\,dx$

23. $\int 45e^{0.15x}\,dx$

24. $\int \frac{e^{1/x}}{x^2}\,dx$

25. $\int (e^{3x}+12e^{-3x})\,dx$

26. $\int \left(e^{5x-2}+\frac{5}{x}\right)dx$

27. Show by substitution that $\int e^{ax}\,dx = \frac{e^{ax}}{a} + C$.

Check Your Understanding

28. Replace the letter A in the integral $\int A(6x^3+4)^5\,dx$ so that the integral evaluates to $\frac{1}{6}(6x^3+4)^6 + C$.

29. Replace the letter A in the integral $\int A(5x^2+4)^2\,dx$ so that the integral evaluates to $\frac{1}{3}(5x^2+4)^3 + C$.

30. Replace the letter A in the integral $\int Ae^{4x^2}\,dx$ so that the integral evaluates to $\frac{1}{2}e^{4x^2} + C$.

31. Replace the letter A in the integral $\int Ae^{4x^3+x^2}\,dx$ so that the integral evaluates to $\frac{1}{2}e^{4x^3+x^2} + C$.

32. Replace the letter A in the integral $\int \frac{A}{6x^2+8x-1}\,dx$ so that the integral evaluates to $\frac{1}{4}\ln(6x^2+8x-1)^3 + C$.

33. Replace the letter A in the integral

$$\int \frac{A}{\sqrt{x^2+2x}}\,dx$$

so that the integral evaluates to

$$10\sqrt{x^2+2x} + C$$

34. Replace the letter A in the integral $\int Ae^{5x}\,dx$ so that the integral evaluates to $e^{5x} + C$.

35. Replace the letter A in the integral $\int A(\ln x^2)^5\,dx$ so that the integral evaluates to $\frac{1}{6}(\ln x^2)^6 + C$.

Modeling Practice

36. Manufacturing: Assembly Efficiency A worker having no experience with a particular process can assemble 60 units of a product each week. After training and t weeks of experience, a worker can assemble the units at the rate of

$$N'(t) = 450e^{-3t}$$

units per week.

a. Specify the function that describes the worker's weekly output.

b. Find the number of units a worker having 5 weeks of experience can assemble.

c. What is the maximum number any worker can be expected to assemble per week? (Hint: When you verbalize part c to yourself, it should spark the thought of a *limiting value*. You can answer this question by taking the proper limit.)

37. Life Science: Antibody Production A medical researcher believes that the human immune system produces antibodies to a particular vaccine at the rate of

$$N'(t) = \frac{3{,}000}{10t+15}$$

thousand per day. Assume that when no vaccine is administered, the immune system produces no antibodies. That is, assume that when $t = 0$, $N(t) = 0$.

a. Specify the function that describes the antibody production in terms of t.

b. Find the number of antibodies that are produced 5 days after vaccination.

38. Psychology: Proofreading Efficiency The number of pages N of technical matter a proofreader can read each day depends on the number of consecutive days t the proofreader has been reading such material. A psychologist believes that the number of pages changes according to the function

$$N'(t) = 24.5e^{-0.35t}$$

After reading such material for 10 consecutive days, a particular proofreader can read 50 pages of material.

a. Specify the function that describes the number of pages this proofreader can read in terms of t.

b. Find the number of pages of technical material that can be proofread by this reader after having read for 5 consecutive days.

c. What is the upper limit, if one exists, to the number of pages that can be proofread by this proofreader each day?

39. Business: Resale Value The resale value, R, (in dollars) of a laser optical device t months from its time of purchase is decreasing at the rate of

$$R'(t) = -1{,}580e^{-0.004t}$$

dollars each month. Ten months after the time of purchase, the device has a resale value of \$370,000.

a. Specify the function that describes the resale value of the device in terms of t.

b. Find the resale value of the device 24 months from its time of purchase.

c. What is the resale value of this device in the long run?

40. Ecology: Air Pollution The level of pollution P (in parts per million) over a city changes as the temperature T (in degrees Fahrenheit) changes according to the function

$$P'(T) = \frac{0.024T}{\sqrt{0.04T^2 + 180}}$$

When the temperature is 0 °F, the level of pollution is 4 parts per million.

a. Specify the function that describes the level of pollution in terms of the temperature.

b. Find the level of pollution when the temperature is 25 °F.

41. Marketing: Advertising and Sales A company has determined that the number N of daily sales of one of its products t days after an advertising campaign decreases according to

$$N'(t) = -6{,}875e^{-0.32t}$$

If on the day that the advertising campaign ends ($t = 0$), the number of daily sales is 55,000,

a. Specify the function that describes the number of daily sales in terms of the number of days after the advertising campaign.

b. Find the number of daily sales 10 days after the end of the campaign.

c. What is the expected number of daily sales long after the advertising campaign is over?

42. **Manufacturing: Worker Production Level** Data kept by the human resources department of a manufacturing company indicates that t hours after beginning work, the production P (in units completed) of an average worker changes at the rate

$$P'(t) = \frac{0.2(t+2)}{\sqrt{t^2+4t}}$$

If 2 hours after beginning work, an average worker can complete 15 units,

a. Specify the function that describes the number of units an average worker can complete in terms of the number of hours after beginning work.

b. Find the number of units an average worker can complete 6 hours after beginning work.

c. Approximate the number of units an average worker can complete from hour 3 to hour 4.

43. **Economics: Population of a City** A city's population, P, is expected to grow at the rate

$$P'(t) = \frac{126e^{18t}}{1+e^{18t}}$$

where t is the number of months from the present. Presently, the population is 28,000.

a. Specify the function that describes the population of the city in terms of the number of months from the present.

b. Find the population of the city 10 months from now.

c. Approximate the increase in population from month 5 to month 6.

44. **Health Science: Drug Reaction** The function

$$S'(x) = \frac{3}{4} \cdot \frac{200-x}{\sqrt{300-x}} \qquad 0 \le x \le 300$$

measures the rate at which the strength (in drug strength units) of a typical person's reaction to a particular drug changes as the dose x of the drug changes. When the dose of the drug is 0, the strength of the reaction to it is 0 strength units.

a. Specify the function $S(x)$ that measures the strength of a typical person's reaction to x milligrams of a particular drug. (Hint: use substitution and the technique used in Example 7.)

b. Find the strength of reaction to a dose of 200 mg of the drug.

45. **Business: Revenue** The function

$$R'(x) = \frac{1}{3} \cdot \frac{200-3x}{\sqrt{100-x}} \qquad 0 \le x \le 100$$

measures the rate at which a start-up company's revenue, $R(x)$, changes over its first 100 weeks of operation. $R(x)$ is in thousands of dollars and x is in weeks after startup. The revenue is 0 at start-up.

a. Specify the function $R(x)$ that measures the company's revenue x weeks after start-up.

b. Find the company's revenue 51 weeks into its operation.

46. Health Science: Dieting The function

$$W'(x) = \frac{-240}{(x+6)^2}$$

measures the rate at which a person on a particular diet loses weight (in pounds) x weeks from the first day of the diet. At the start of the diet, the person weighs 200 pounds.

a. Specify the function that measures the weight of the person x weeks from the first day of the diet.

b. Find the weight of the person after 10 weeks on the diet.

© Shaun Lowe Photographic/ iStockPhoto

47. Physical Science: Drag Racing The function

$$V'(t) = \frac{1{,}020}{(t+3)^2}$$

models the rate at which the speed $V(t)$, in miles per hour, of a dragster changes t seconds since it starts a quarter-mile run. At the very beginning of the run, the dragster's speed is 0.

a. Specify the function that models the speed of the dragster t seconds into its quarter-mile run.

b. Find the speed of the dragster 2 seconds into its run.

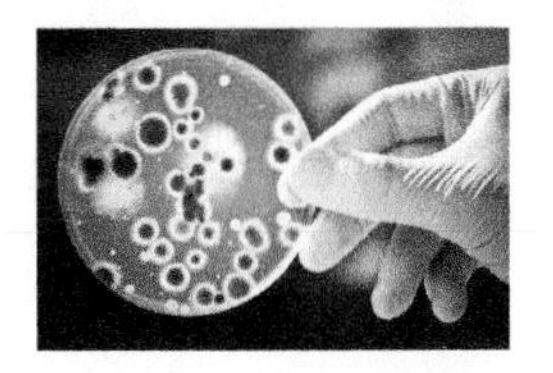

© Alexander Raths /iStockPhoto

48. Biology: Bacteria Population The function

$$P'(t) = \frac{-30(t^2-16)}{(t^2+16)^2} \qquad t > 1$$

measures the rate at which the size $P(t)$ (in millions) of bacteria in a culture changes as the number of days t increases after a slow-acting toxin is introduced into the culture. Two days after the toxin is introduced, there are 3 million bacteria in the culture. (Remember, t is in days and P is in millions.)

a. Specify the function that models the number of bacteria in the culture t days after the toxin is introduced.

b. About how many millions of bacteria are in the culture 20 days after the toxin is introduced?

Using Technology Exercises

49. Find $\int 3x^2(x^3-7)^5\,dx$ using both pencil-and-paper and Wolfram|Alpha. Compare the results. Wolfram|Alpha can perform the substitution and evaluation for us.

50. Use Wolfram|Alpha to find $\int \frac{\ln^4(x)}{x}\,dx$. Before you use the technology, what do you think is the appropriate substitution?

51. Use Wolfram|Alpha to find $\int \frac{\sqrt{4+5\ln x}}{x}\,dx$. Before you use the technology, what do you think is the appropriate substitution?

52. Use Wolfram|Alpha to find $\int \frac{x+2}{(x+4)^5}\,dx$. Before you use the technology, what do you think is the appropriate substitution?

53. In this section, we developed a general formula for $\int e^{ax}\,dx$. Use Wolfram|Alpha to find a general formula for $\int e^{ax+b}\,dx$. Before you use the technology, what do you think is the appropriate substitution?

Getting Ready for the Next Section

54. Evaluate $\int 48e^{1.2t}\,dt$.

55. Compute $T(7) - T(3)$ for $T(t) = 40e^{1.2t} + C$

56. Evaluate $\int x^2\,dx + \int 3x\,dx - \int 2\,dx$.

57. Use substitution to evaluate $\int \frac{1}{x-3}\,dx$.

58. Evaluate $\ln|x - 3| + C$ first for $x = 10$ then for $x = 4$. Subtract the second evaluation ($x = 4$) from the first ($x = 10$).

59. Evaluate $\int \frac{1}{u}\,du$. Then compute the result first for $u = 7$ then for $u = 1$. Subtract the second evaluation ($u = 1$) from the first ($u = 7$).

The Definite Integral

As we have seen so well, differentiation is the language of change, and as we experienced in Sections 5.1 and 5.2, indefinite integration is antidifferentiation. In this section, we will see that another form of integration, called **definite integration**, is the language of accumulation.

Developing the Definite Integral

© pictafolio/iStockPhoto

Suppose that since 2000, the sales of the Dewtex Publishing Company have been growing according to the function

$$S(t) = 48e^{1.2t}$$

where $S(t)$ is the *rate* (think derivative) of sales (in dollars) t years from 2000. What is the total amount of sales of Dewtex for the years 2003 through 2007? That is, how much money did Dewtex accumulate from sales from 2003 to 2007?

This is an accumulation problem since the total amount of sales from 2003 to 2007 is just the amount of dollars that have accumulated from sales over this interval of time. We will let $T(t)$ represent the total accumulated sales. Then the accumulated sales from 2003 to 2007 is found by evaluating $T(7) - T(3)$. (2007 is 7 years from 2000 and 2003 is 3 years from 2000 so that $T(7)$ represents the accumulated sales from 2000 to 2007, $T(3)$ represents the accumulated sales from 2000 to 2003, and $T(7) - T(3)$ represents the required difference.)

But since $S(t)$ is the *rate of change* of $T(t)$, $T(t)$ is an integral of $S(t)$ and

$$\begin{aligned} T(t) &= \int S(t)\,dt \\ &= \int 48e^{1.2t}\,dt \\ &= \frac{48e^{1.2t}}{1.2} + C \\ &= 40e^{1.2t} + C \end{aligned}$$

Thus,

$$T(t) = 40e^{1.2t} + C$$

and

$$\begin{aligned} T(7) &= 40e^{1.2(7)} + C \\ &\approx 177{,}882.67 + C \end{aligned}$$

and

$$\begin{aligned} T(3) &= 40e^{1.2(3)} + C \\ &\approx 1{,}463.93 + C \end{aligned}$$

so that

$$\begin{aligned} T(7) - T(3) &= 177{,}882.67 + C - (1{,}463.93 + C) \\ &= 177{,}882.67 + C - 1{,}463.93 - C \\ &= 176{,}418.74 \end{aligned}$$

Thus, between 2000 and 2007, Dewtex accumulated \$176,418.74 from sales.

Notice that in each integration, the constant of integration C appeared, but cancels out in the end. When using integration to compute an accumulation, we omit the constant of integration since the final evaluation is independent of the choice of C (it may be any value) and C always cancels out.

Let's analyze what we have done.

1. We wished to find the total amount, $F(x)$, accumulated by a function $f(x)$ over some interval $[a, b]$. (In our example, $f(x)$ was $S(t)$, $a = 3$, and $b = 7$.)
2. Although the function $F(x)$ was not originally known, its derivative $f(x)$ was, and we were able to determine the family to which the function $F(x)$ belonged by integrating the given derivative function. In writing the function $F(x)$, we omit the constant of integration, C.
3. The total accumulation was the difference between $F(b)$ and $F(a)$; that is, $F(b) - F(a)$.

It is common to use $F(x)\Big|_a^b$ to indicate the evaluation $F(b) - F(a)$.

The Fundamental Theorem of Calculus

We now present the Fundamental Theorem of Calculus, a theorem that summarizes this evaluation process.

The Fundamental Theorem of Calculus

If $f(x)$ is a continuous function on an interval $[a, b]$, where $a < b$, and if $F(x)$ is an antiderivative of $f(x)$, then the **definite integral** of $f(x)$ from a to b is

$$\int_a^b f(x)\,dx = F(x)\Big|_a^b = F(b) - F(a)$$

and represents the total amount accumulated by $F(x)$ as x increases from a to b. The number a is called the **lower limit of integration** and the number b the **upper limit of integration**.

The Difference Between Indefinite and Definite Integrals

It is worthwhile noting the difference between the indefinite integral

$$\int f(x)\,dx$$

and the definite integral

$$\int_a^b f(x)\,dx$$

1. The indefinite integral $\int f(x)\,dx$ *is a function of* x and represents a family of antiderivatives of $f(x)$. For example,

$$\int x^3\,dx = \frac{x^4}{4} + C, \quad \text{which is a function of } x$$

2. The definite integral $\int_a^b f(x)\,dx$ is an *accumulator* and represents a *real number*, a constant that records the total amount accumulated by a particular member of a family of antiderivatives. For example,

$$\begin{aligned}\int_1^3 x^3\,dx &= \frac{x^4}{4}\bigg|_1^3\\ &= \frac{3^4}{4} - \frac{1^4}{4}\\ &= \frac{81}{4} - \frac{1}{4}\\ &= \frac{80}{4}\\ &= 20, \qquad \text{which is a real number}\end{aligned}$$

Properties of the Definite Integral

The definite integral has the following properties.

1. $\int_a^b kf(x)\,dx = k\int_a^b f(x)\,dx$, where k is a constant.
2. $\int_a^a f(x)\,dx = 0$ No accumulation without change.
3. $\int_a^b f(x)\,dx = -\int_b^a f(x)\,dx$
4. $\int_a^b [f(x) \pm g(x)]\,dx = \int_a^b f(x)\,dx \pm \int_a^b g(x)\,dx$
5. $\int_a^b f(x)\,dx = \int_a^c f(x)\,dx + \int_c^b f(x)\,dx$, where $a \le c \le b$.

VIDEO EXAMPLES

SECTION 5.3

Example 1 Evaluate $\int_3^5 (x^2 + 3x - 2)\,dx$.

Solution

$$\begin{aligned}\int_3^5 (x^2 + 3x - 2)\,dx &= \int_3^5 x^2\,dx + \int_3^5 3x\,dx - \int_3^5 2\,dx\\ &= \frac{x^3}{3}\bigg|_3^5 + \frac{3x^2}{2}\bigg|_3^5 - 2x\bigg|_3^5\\ &= \left(\frac{5^3}{3} - \frac{3^3}{3}\right) + \left(\frac{3\cdot 5^2}{2} - \frac{3\cdot 3^2}{2}\right) - (2\cdot 5 - 2\cdot 3)\\ &= \frac{125}{3} - \frac{27}{3} + \frac{75}{2} - \frac{27}{2} - 10 + 6\\ &= \frac{158}{3}\end{aligned}$$

Interpretation

As x increases from 3 to 5, the function $F(x) = \frac{x^3}{3} + \frac{3x^2}{2} - 2x$ accumulates $\frac{158}{3}$ units.

The Definite Integral and Substitution

Sometimes we must use the substitution method to evaluate a definite integral. Examples 2 and 3 show two different ways. In Example 2, we show the most common and simplest method, replacing the limits of the original integral with new ones that are appropriate for the new variable from the substitution. The new definite integral can be evaluated without returning to the original variable.

Example 2 Evaluate $\int_4^{10} \frac{1}{x-3}\,dx$.

Solution Since the integrand does not match one of the basic integral forms, we will try the substitution method. Since the integrand is a quotient, we will try letting u represent the denominator, $x - 3$.

We let $u = x - 3$. The remaining part of the integrand is $1 \cdot dx = dx$. If integration by substitution is going to work, we should be able to get dx from the differential of u. We let

$$u = x - 3$$

so that

$$\frac{du}{dx} = 1$$

giving us the differential

$$du = dx$$

so that

$$dx = du$$

which is precisely the remaining part of the integrand.

Now we change the limits of integration. (These are the new steps!) Since $u = x - 3$, when $x = 4$, $u = 4 - 3 = 1$, and when $x = 10$, $u = 10 - 3 = 7$. Since all the variables are in terms of u,

$$\begin{aligned}\int_4^{10} \frac{1}{x-3}\,dx &= \int_1^7 \frac{1}{u}\,du \\ &= \ln|u|\Big|_1^7 \\ &= \ln|7| - \ln|1| \\ &= \ln 7 - 0 \qquad \text{(Since } \ln 1 = 0\text{)} \\ &= \ln 7\end{aligned}$$

In Example 3, we will show how this same integral can be evaluated by a process that keeps the limits of integration in terms of x, transforms to u, and then resubstitutes back to x and evaluates. (Note the greater number of steps.)

Example 3 Evaluate $\displaystyle\int_4^{10} \frac{1}{x-3}\,dx$.

Solution We let $u = x - 3$. Then $du = dx$. We need to be careful about what we write at this point. The limits of integration are in terms of x and the integrand is in terms of u. To keep this fact clearly established, we write

$$\int_{x=4}^{x=10} \frac{1}{u}\,du$$

This integrand now matches one of the basic integral forms.

$$\begin{aligned}
\int_{x=4}^{x=10} \frac{1}{u}\,du &= \ln|u|\,\Big|_{x=4}^{x=10} && \text{Now we resubstitute and evaluate}\\
&= \ln|x-3|\,\Big|_{x=4}^{x=10}\\
&= \ln|10-3| - \ln|4-3|\\
&= \ln 7 - \ln 1\\
&= \ln 7 - 0 && \text{(Since } \ln 1 = 0\text{)}\\
&= \ln 7
\end{aligned}$$

Thus, $\displaystyle\int_4^{10} \frac{1}{x-3}\,dx = \ln 7 \approx 1.9459$. ■

Note that the method used in Example 2 is preferred over that used in Example 3.

Using Technology 5.3

You can use your graphing calculator in several ways to approximate a definite integral. The following entries show how to approximate $\int_a^b f(x)\,dx$.

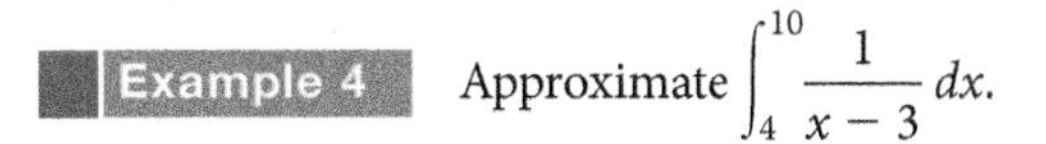

Example 4 Approximate $\displaystyle\int_4^{10} \frac{1}{x-3}\,dx$.

Solution In the function editor enter

Y1 = 1/(x – 3)

Y6 = fnInt(Y1, X, A, B)

The command fnInt is found in the Math editor. The syntax for the fnInt command is fnInt(function, variable of integration, lower limit, upper limit). Use the [STO▸] key to store the values of the lower and upper limit. In the computation (home) window, enter

4[STO▸] A

10[STO▸] B

Then go to [VARS] → [Y-VARS], Select Function and Y6. Press [ENTER]. You should get approximately 1.9459. That is,

$$\int_4^{10} \frac{1}{x-3}\,dx \approx 1.9459$$ ■

Getting Ready for Class

After reading through the preceding section, respond in your own words and in complete sentences.

A. Write the Fundamental Theorem of Calculus.

B. In the definite integral $\int_a^b f(x)\,dx$, what is the name given to the values a and b?

C. Explain the difference between an indefinite and definite integral.

D. Which operation, differentiation or integration, is the language of change? Of accumulation?

Problem Set 5.3

Skills Practice

For Problems 1-38, evaluate each definite integral.

1. $\int_0^2 (3x^2 + 4x - 1)\, dx$

2. $\int_0^4 (6x^2 - 8x - 2)\, dx$

3. $\int_0^4 \left(\frac{3}{2}x^2 - 10x + 8\right) dx$

4. $\int_0^6 \left(\frac{6}{5}x^2 - 2x + 10\right) dx$

5. $\int_2^5 (4x^2 + 4x + 1)\, dx$

6. $\int_3^4 (9x^2 + 2x + 3)\, dx$

7. $\int_5^{10} (5x^2 + x)\, dx$

8. $\int_6^{10} (x^2 + 3x)\, dx$

9. $\int_0^1 (4x^3 - 3x^2)\, dx$

10. $\int_0^1 (5x^4 - 12x^2)\, dx$

11. $\int_0^4 4e^{2x}\, dx$

12. $\int_0^1 3e^{3x}\, dx$

13. $\int_0^2 10e^{-5x}\, dx$

14. $\int_0^6 2e^{-2x}\, dx$

15. $\int_0^{\ln e} e^{5x}\, dx$

16. $\int_0^{\ln e} e^{3x}\, dx$

17. $\int_0^{\ln e} -8e^{-2x}\, dx$

18. $\int_0^{2 \ln e} -6e^{-6x}\, dx$

19. $\int_1^e \frac{3}{x}\, dx$

20. $\int_1^e \frac{5}{x}\, dx$

21. $\int_1^2 \frac{3}{x^2}\, dx$

22. $\int_1^3 \frac{5}{x^2}\, dx$

23. $\int_0^5 7\, dx$

24. $\int_0^6 10\, dx$

25. $\int_4^5 \frac{2}{x - 3}\, dx$

26. $\int_2^5 \frac{1}{x - 1}\, dx$

27. $\int_2^3 \frac{2x}{x^2 - 3}\, dx$

28. $\int_0^2 \frac{3x^2}{x^3 + 1}\, dx$

29. $\int_1^2 \frac{12x + 2}{6x^2 + 2x}\, dx$

30. $\int_1^3 \frac{3x^2 + 1}{x^3 + x}\, dx$

31. $\int_0^5 \frac{6x}{x^2 + 6}\, dx$

32. $\int_0^1 \frac{18x}{3x^2 - 7}\, dx$

33. $\int_0^1 (6x + 1)e^{3x^2 + x}\, dx$

34. $\int_0^1 (x - 2)e^{x^2 - 4x}\, dx$

35. $\int_{\ln 3}^{\ln 6} 3e^x\, dx$

36. $\int_{\ln 2}^{\ln 3} -4e^{-x}\, dx$

37. $\int_2^4 \frac{4}{x \ln^2 x}\, dx$

38. $\int_e^{e^2} \frac{1}{x \ln x}\, dx$

Check Your Understanding

In each of Problems 39-41, an error occurs in the evaluation or interpretation of the definite integral. Specify the step number at which the error occurs, then fix it.

39. Evaluate $\int_2^6 (3x - 4)\,dx$.

Step 1: $\int_2^6 (3x-4)\,dx = \left(\frac{3x^2}{2} - 4x\right)\Big|_2^6$

Step 2: $= \left(\frac{3(2)^2}{2} - 4(2)\right) - \left(\frac{3(6)^2}{2} - 4(6)\right)$

Step 3: $= (6 - 8) - (54 - 24)$

Step 4: $= -2 - 30$

Step 5: $= -32$

40. Evaluate $\int_1^4 (6x - 5)\,dx$.

Step 1: $\int_1^4 (6x-5)\,dx = (3x^2 - 5x)\Big|_1^4$

Step 2: $= (3 \cdot 4^2) - (5 \cdot 1)$

Step 3: $= (3 \cdot 16) - 5$

Step 4: $= 48 - 5$

Step 5: $= 43$

41. Evaluate $\int_2^3 3e^{3x}\,dx$.

Step 1: $\int_2^3 3e^{3x}\,dx = 3 \cdot 3e^{3x}\Big|_2^3$

Step 2: $= 9e^{3x}\Big|_2^3$

Step 3: $= (9e^{3(3)}) - (9e^{3(2)})$

Step 4: $= 9e^9 - 9e^6$

For Problems 42-55, interpret each definite integral.

© Kyoungil Jeon/iStockPhoto

42. Business: Investment If $A(t)$ represents the rate at which the amount of money in an investment is changing from month a to month b, interpret

$$\int_a^b A(t)\,dt$$

43. Business: Advertising If $S(t)$ represents the rate at which a company spends money on advertising from month a to month b, interpret

$$\int_a^b S(t)\,dt$$

44. **Economics: Population** If $P(t)$ represents the rate at which the population of a city is changing between years a and b, interpret

$$\int_a^b P(t)\,dt$$

45. **Economics: Gasoline Consumption** If $G(t)$ represents the rate at which the number of gallons of gasoline are pumped from a local gas station between days a and b, interpret

$$\int_a^b G(t)\,dt$$

46. **Social Science: Spread of News** If $S(t)$ represents the rate at which a population hears news between days a and b, interpret

$$\int_a^b S(t)\,dt$$

47. **Ecology: Toxic Leaks** If $A(t)$ represents the rate at which a toxic chemical is leaked into a city's water supply from hour a to hour b, interpret

$$\int_a^b A(t)\,dt$$

48. **Business: Cost of Production** If $C(x)$ represents the rate at which the cost of producing x items changes as the number of items produced increases from a to b, interpret

$$\int_a^b C(x)\,dx$$

49. **Economics: Personal Wealth** If $W(t)$ represents the rate at which a person's wealth changes as she ages from a years to b years, interpret

$$\int_a^b W(t)\,dt$$

50. **Life Science: Evaporation of a Solution** If $E(T)$ represents the rate at which a solution evaporates between temperatures a and b, interpret

$$\int_a^b E(T)\,dt$$

51. **Economics: Income Streams** If $I(t)$represents the rate at which an income stream is increasing from year a to year b, interpret

$$\int_a^b I(t)\,dt$$

52. **Business: Service Rate Increases** If $U(f)$ represents the rate at which a utility company raises the price of its service to the community relative to fuel prices a and b, $a < b$, interpret

$$\int_a^b U(f)\,df$$

53. **Business: Depreciation** If $D(t)$ represents the rate at which the depreciation of a commodity changes over the period from month a to b, interpret

$$\int_a^b D(t)\,dt$$

54. **Business: Air Travel** If $L(t)$ represents how the decrease in the length of time t a company spends on air travel is changing between years a and b, interpret

$$\int_a^b L(t)\,dt$$

55. **Mathematics: Probability** If $P(t)$ represents the change in the increase in the probability that a particular event occurs between times a and b, interpret

$$\int_a^b P(t)\,dt$$

Modeling Practice

56. **Manufacturing: Filling Containers** A robotic machine fills containers with an acid solution at the rate of $80 + 2t$ milliliters (mL) per second, where t is in seconds and $0 \leq t \leq 60$. How many ml are put into a container in 60 seconds? Evaluate your answer to a whole number.

57. **Business: Projecting the Marginal Cost of Production** The marginal cost of the Xenon Lighting company in producing x number of Xenon lamps is modeled by the function $C'(x) = 100 - 0.6x$. What would be the additional cost to the company if it were to increase its production of lamps from 100 to 120?

58. **Health Science: Delivery of Medication through an IV** An IV is calibrated to deliver $50 + t$ mL per minute of a medication to a patient over a 30-minute period. The variable t represents the number of minutes from the start of the infusion of the medication. How many mL of the medication are delivered
   a. in the first 10 minutes?
   b. in the 30-minute period?
   Evaluate your answer to a whole number.

Source: medic215.com

59. **Ecology: Extraction of Contaminated Water** A company claims it can extract N gallons of contaminated water per day from a deep well at the rate modeled by

$$N(t) = 6t^4 - 720t^3 + 21{,}600t^2$$

where t is the number of days since the extraction begins. How many gallons can the company extract
   a. in the first 10 days of operation?
   b. in 60 days of operation?
   Evaluate your answer to a whole number.

60. Political Science: Projecting the Total Number of Lobbyists Companies in industries such as defense, pharmaceutical and health insurance, utilities, oil and gas, manufacturing, transportation, and securities and investments, as well as labor unions and gun rights groups, spend billions of dollars each year lobbying Congress and federal agencies to secure some desired action. For the years 1998-2012, the number of unique, registered lobbyists who have actively lobbied Congress or a federal agency can be approximated with the 4th-degree polynomial model

$$N(t) = -0.40t^4 + 6.6t^3 - 30t^2 + 440t + 11{,}234 \quad 0 \le t \le 15$$

where t represents the number of years since the beginning of the year 1998.

a. How many lobbyists were there during this 15-year period?

b. If this trend continues, how many lobbyists can be expected from 2012 to 2015?

Source: http://www.opensecrets.org/lobby/

61. Business: Total Online Advertising Revenue Social networking and video-sharing sites, such as Facebook, dating sites, and YouTube, often include rich media and display advertising placed around user-generated content. Suppose that, for a particular social networking company, the annual revenue from rich media advertisements, in millions of dollars, for the five years 2007 through 2011 can be approximated with the model

$$R(x) = -x^4 + 11x^3 - 39x^2 + 45x$$

where x is the number of years from the beginning of 2007. What was the total revenue for this company in the five-year period?

Source: eMarketer April 2008

62. Manufacturing: Flow of a Solvent into Ink In the manufacture of printing ink, solvents are used to suspend or dissolve the pigments that give the ink its color. Suppose that during its manufacture, a process moves a solvent into a vat containing pigments, not at a constant rate, but rather at a rate modeled by the function

$$R(t) = \frac{4t}{\sqrt{2t^2 + 1}}$$

liters per minute, where t is the number of minutes since the start of the process. How many liters of solvent flow into the vat in the first 10 minutes of the process? Round your answer to two decimal places.

Source: Cornell Center for Materials Research

63. Physical Science: Speed of a Dragster At the introduction to Chapter 1, we noted that the function

$$V(t) = \frac{340t}{t + 3} \quad t \ge 0$$

modeled dragster driver Shirley Muldowney's speed, in feet per second, in one of her races in the movie *Heart Like a Wheel.* The antiderivative of the speed function gives the distance traveled through a period of time. To one decimal place, how far did Shirley travel in her dragster in the first 4 seconds of her race?

64. Manufacturing: Projecting Total Cost of Production Power over Ethernet (PoE) systems pass electrical power and data safely on Ethernet cabling. The function

$$C(x) = \frac{144 + x^2}{180x}$$

relates the weekly cost (in thousands of dollars) a company incurs to the number x (in hundreds) of PoE systems it manufactures and sells each week.

a. Show that for production runs of between 100 and 1,000 units, the weekly cost is always decreasing. (x is in hundreds, so, $x = 1$ to 10.)

b. Find the total cost to the company if it increases its production from 100 units to 1,000 units per week.

65. Manufacturing: Projecting Total Cost of Production A manufacturer of precision medical instruments uses the function

$$C(x) = \frac{400x + 1{,}000}{x + 1}$$

to model the cost C (in dollars) of producing x number of instruments.

a. Show that as the company increases its production of instruments from 0 to 100, its cost per instrument will only decrease.

b. What is the least cost per instrument the company can expect? Support your claim with an appropriate mathematical statement.

c. What is the total cost to the company as it increases its production from 0 to 100 instruments?

66. Economics: Investment in Technology A company's financial analyst is advising the management that an investment in a new technology will save it money at the rate of

$$S(t) = 4{,}000(1 - 0.07e^{0.07t}) \quad \text{for } 0 \le t \le 24$$

dollars per month t months after the purchase of the technology.

a. The analyst advises the management that as the number of months from the time of purchase increases from 0 to 24, the amount of money the company will save will increase but will do so more slowly as the number of months go by. As a manager, would you accept or reject this analyst's claim?

b. The analyst advises the management that if it purchases the new technology, it will save, through the next 24 months, just over $100,000. As a manager, would you accept or reject the analyst's claim?

67. Manufacturing: Projecting Total Revenue A company's financial analyst advising the management of a manufacturer of ceramic "green" non-stick frying pans that if it produces and sells x pans per month over the next year, its monthly revenue, in dollars, can be approximated by the model

$$R(x) = 800 - 480e^{-0.002x} \quad \text{for } 0 \le x \le 1{,}000$$

a. The analyst advises the management that as it increases the number of pans it produces and sells each month from 0 to 1,000, its monthly revenue will increase but that increase will slow with the production and sale of each additional pan. As a manager, would you accept or reject this analyst's claim?

b. The analyst advises the management that if it produces and sells 1,000 pans in the next year, it will realize a total annual revenue of over $500,000. As a manager, would you accept or reject the analyst's claim?

68. Business: Projecting Total Sales The sales manager of a company that sells charcoal-filtered water bottles tells the company's president that with an aggressive advertising campaign, the number of sales could grow at the rate of

$$N'(t) = 1{,}000 + 0.62e^{0.031t} \quad \text{for } 0 \le t \le 28$$

bottles per week, t weeks from the beginning of the campaign. The president asks the manager two questions. How would you, if you were the sales manager, answer each one?

a. Will the number of sales always be increasing throughout the 28-week period or will they increase at the beginning of the campaign, then decrease near the end?

b. Do you think we can increase our number of sales from our current number of about 1,000 each week to about 15,000 per week?

69. Business: Projecting Total Cost of Maintenance As machines get older, the cost of maintaining them tends to increase. Suppose for a particular machine, the rate at which the maintenance cost is increasing is approximated by the function

$$C'(t) = (12t + 32)\sqrt{1.5t^2 + 8t} \quad \text{for } 0 \le t \le 10$$

where C is the maintenance cost in thousands of dollars and t is the number of years since purchase. The company will sell or scrap the machine in 10 years and buy a new one if the cost to maintain it in its last year of service is projected to exceed $250,000.

a. What is the total expected cost to maintain this machine over its first year of service? (To the nearest thousand dollars.)

b. What is the total expected cost to maintain this machine in its last year of service? Should the company keep or scrap the machine? (To the nearest thousand dollars.)

c. How many times more expensive is it to maintain the machine in its last year of service than in its first year of service?

70. Business: Analytical Decision Making A company determines that it can make a profit from extracting natural gas from a field if the field can produce a total of at least 20 million cubic feet of gas over the next 5 years. A recent geological survey projects that t years from the time extraction begins, the amount of gas, A, in millions of cubic feet this field can produce each year is modeled by

$$A(t) = \frac{36t}{1.5t^2 + 6} \quad \text{for } 0 \le t \le 5$$

Geologists have confidence in the model for 5 years from the time extraction begins.

a. What is the total amount of natural gas this field is expected to produce over the next 5 years and does it appear the company can make a profit by extracting the natural gas contained in the field?

b. How many years after extraction begins, will the maximum amount of gas be produced from the field? What is that amount?

Using Technology Exercises

71. Use your calculator to approximate

$$\int_0^{30} \frac{400x}{0.3x^2 + 200}\,dx$$

72. Use your calculator to try to approximate $\int_0^5 \frac{1}{x-2}\,dx$. What goes wrong? Explain why the error occurred.

73. A manufacturer has collected data to support his belief that the function

$$N(t) = 355 - 145e^{-0.03t}$$

models the rate at which an average worker can assemble N units per month of a product after t weeks of experience. Find and interpret both $N(5)$ and

$$\int_0^5 (355 - 145e^{-0.03t})\,dt$$

Although the variable here is t, use the variable x in your calculator entries.

74. The income from a chain of convenience stores can flow into a company at a yearly rate given by the function $I(t) = 140{,}000e^{1.5345t}$ dollars, where t represents the number of years since the company's inception. Find and interpret

$$\int_0^2 I(t)\,dt$$

75. Economists use the Gini index, G, to measure how a country's money is distributed among its population. As the value of G approaches 0, the money becomes more evenly distributed to the population. The Gini index is given by

$$G(x) = 1 - \int_0^1 f(x)\,dx$$

Suppose that for country A, $f(x) = x^{0.82}$, and for country B, $f(x) = x^{0.670}$. Find the Gini index for each country and specify the country for which the money supply is more evenly distributed amongst its population.

Source: The CIA World Factbook

Getting Ready for the Next Section

76. Compute the sum $13.5 + 9.3 + 5.81 + 6.54$.

77. Find $\int_1^6 (0.3x^3 - 1.8x^2 + 15)\,dx$.

For Problems 78-80, evaluate the definite integral.

78. $\int_2^5 (x^2 - 5x + 4)\,dx$

79. $\int_2^4 [-(x^2 - 5x + 4)]\,dx + \int_4^5 (x^2 - 5x + 4)\,dx$

80. $\int_5^6 0.20e^{-0.20t}\,dt$

The Definite Integral and Area 5.4

The graph below represents a demand function that a company may use to price its product. The shaded portion of the graph shows the social utility of consumption, which is the accumulated value the public places on its consumption of the product per unit of time. Since the social utility of consumption is an accumulated value, it is defined by an integral.

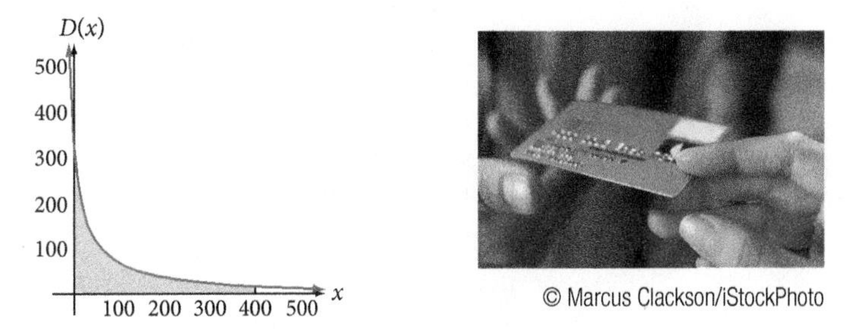

© Marcus Clackson/iStockPhoto

In this section, we will investigate the relationship between area and accumulation, such as that shown in the above graph. We will also tie together the definite integral and the area under a curve.

Area

Before we start computing areas, we need to have some understanding of what area is. Intuitively, the **area** of a region is a *measure* of the size of the region; it is something that can be defined. For example,

1. The area of a rectangle having width x and height h is the product of the width and height. (See Figure 1a.)
2. The area of a triangle having base b and height h is one-half the product of the base and the height. (See Figure 1b.)
3. The area of a circle having radius r is the product of π and the square of the radius. (See Figure 1c.)

In each case, we use nonnegative real numbers to indicate the size of a planar (flat surface) region.

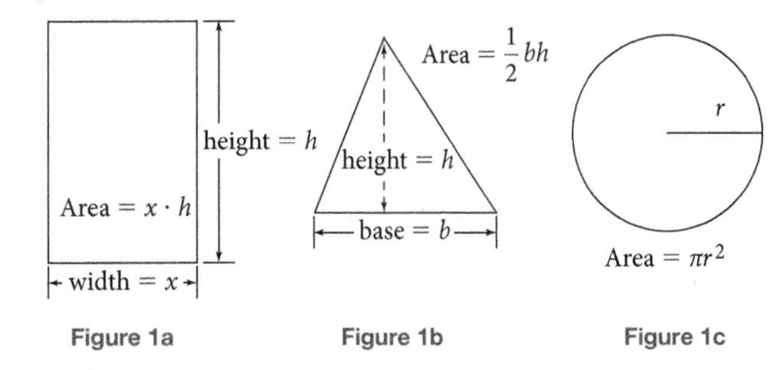

Figure 1a Figure 1b Figure 1c

Approximating Area

As you will see, many applied problems can be solved by relating their characteristics to area. To do so requires that we define the area of more general planar regions. For our purposes, we will begin by defining the area of planar regions that lie in a rectangular coordinate system, above the input axis, below some specified nonnegative function, and between two vertical lines. Figure 2 shows such a region. We do something that is common in mathematics: We define the area in terms of something with which we are already familiar. In particular, we define the area of this planar region in terms of rectangles.

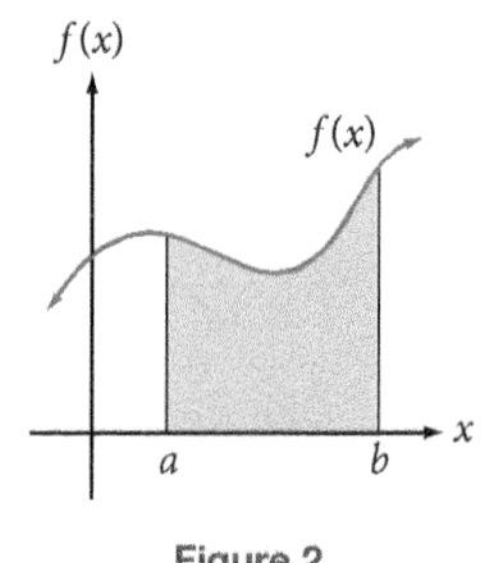

Figure 2

We can use one rectangle, having as its width the entire interval $b - a$, and as its height $f(a)$, the output (height) at the left-hand endpoint of the interval $[a, b]$, as a crude approximation to the number we wish to assign as the area of this region. The left side of Figure 3 shows this approximation. Using one rectangle, the approximation of the area is $f(a)(b - a)$. This is certainly a poor approximation, and we can do better.

Using more rectangles will give us a better approximation. If we subdivide the interval $[a, b]$ into, say, four subintervals of equal width (for convenience), $\frac{b-a}{4}$, and on each subinterval construct a rectangle with height taken as the functional value at the left-hand endpoint of the subinterval, we will get the approximation illustrated in the right side of Figure 3.

Note In Figure 3, we created rectangles with heights coming from the left-most endpoints of the subintervals. We could have also created rectangles with heights coming off the right-most endpoints of the subintervals, or the midpoints of the subintervals, or any point at all in the subintervals.

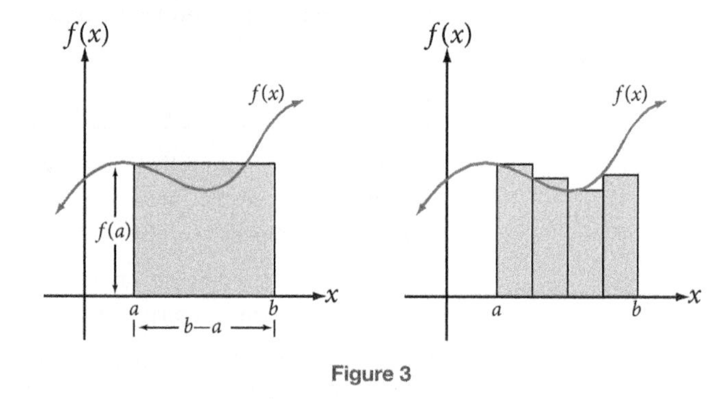

Figure 3

So that we can compute the area of these four rectangles, we introduce the following notation (which is illustrated in Figure 4).

- Let x_0 represent the left-most endpoint of the 1st subinterval. That is, $x_0 = a$.
- Let x_1 represent the left-most endpoint of the 2nd subinterval.
- Let x_2 represent the left-most endpoint of the 3rd subinterval.
- Let x_3 represent the left-most endpoint of the 4th subinterval.

With this notation,

- $f(x_0)$ represents the height of the 1st rectangle.
- $f(x_1)$ represents the height of the 2nd rectangle.
- $f(x_2)$ represents the height of the 3rd rectangle.
- $f(x_3)$ represents the height of the 4th rectangle.

We will also let Δx represent the width of each rectangle. Then

$$\Delta x = \frac{b-a}{4}$$

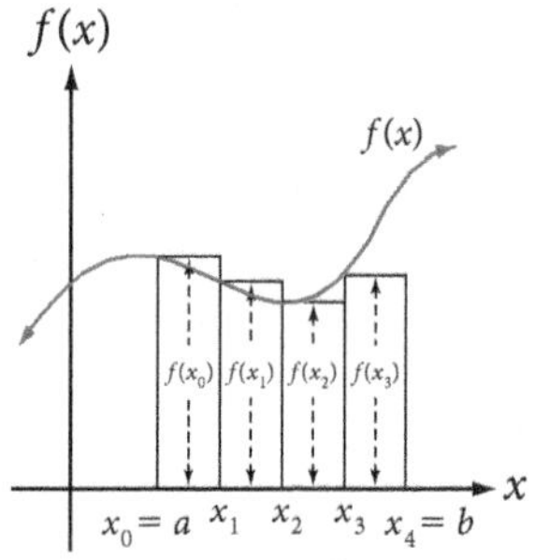

Figure 4

Then, letting S_4 represent the sum of the areas of all four rectangles, we get

$$\begin{aligned} S_4 &= (\text{area of rectangle 1}) + (\text{area of rectangle 2}) \\ &\quad + (\text{area of rectangle 3}) + (\text{area of rectangle 4}) \\ &= f(x_0)\Delta x + f(x_1)\Delta x + f(x_2)\Delta x + f(x_3)\Delta x \end{aligned}$$

and, factoring Δx from each term

$$= [f(x_0) + f(x_1) + f(x_2) + f(x_3)]\Delta x$$

Thus, a four-rectangle approximation to the region is given by

$$S_4 = [f(x_0) + f(x_1) + f(x_2) + f(x_3)]\Delta x$$

We will now look at an actual example of this situation.

VIDEO EXAMPLES

SECTION 5.4

Example 1 Use four rectangles to approximate the area of the planar region under the graph of

$$f(x) = 0.3x^3 - 1.8x^2 + 15$$

over the interval [1, 6]. (See Figure 5.)

Solution We have $a = 1$ and $b = 6$, so that the width of each rectangle is

$$\frac{6-1}{4} = \frac{5}{4} = 1.25 \text{ units}$$

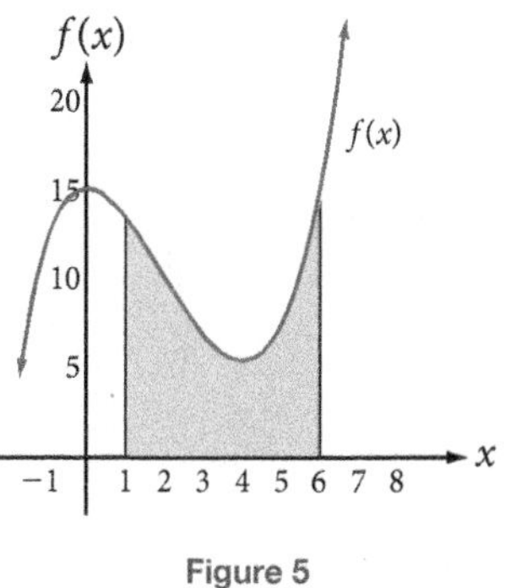

Figure 5

Then,

$$\begin{aligned} x_0 &= 1 \\ x_1 &= 1 + 1.25 = 2.25 \\ x_2 &= 2.25 + 1.25 = 3.50 \\ x_3 &= 3.50 + 1.25 = 4.75 \\ x_4 &= 4.75 + 1.25 = 6 \end{aligned}$$

(x_4 is only used to check that the entire interval is accounted for.) See Figure 6. We will use the formula

$$S_4 = [f(x_0) + f(x_1) + f(x_2) + f(x_3)]\Delta x$$

to approximate the area. The computations for the heights of the rectangles are shown in the following table. Then, $S_4 = (35.15)(1.25) = 43.9375$.

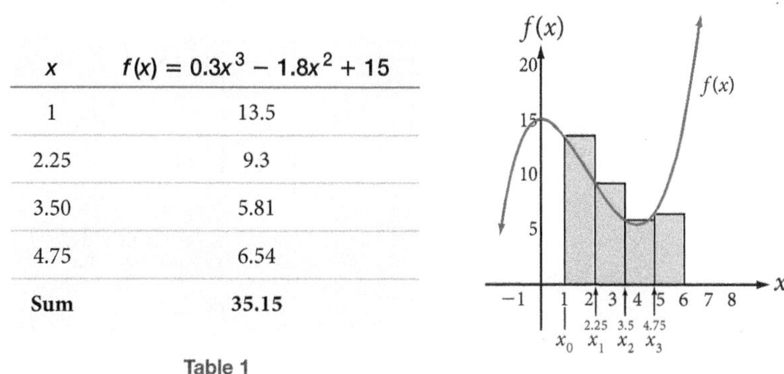

x	$f(x) = 0.3x^3 - 1.8x^2 + 15$
1	13.5
2.25	9.3
3.50	5.81
4.75	6.54
Sum	**35.15**

Table 1

Figure 6

Thus, using four rectangles, we would approximate the area of this planar region as 43.9375 square units.

Using only one rectangle would give

$$\begin{aligned} S_1 &= f(x_0)(b - a) \\ &= f(1)(6 - 1) \\ &= (13.5)(5) \\ &= 67.5 \end{aligned}$$

■

The actual area is 43.125 square units (you will compute this for yourself very soon). Comparing the two results, it is apparent that using four rectangles provides a better approximation to the actual area. What if we were to use 40 or 4,000 rectangles?

Defining the Area of a Planar Region

Suppose we subdivide the interval $[a, b]$ into n subintervals of equal length,

$$\Delta x = \frac{b - a}{n}$$

and construct rectangles using the functional value at c_i in each subinterval, $[x_i, x_{(i+1)}]$, as the height of the rectangle. Then, the approximating area is

$$S_n = [f(c_0) + f(c_1) + \cdots + f(c_{n-1})]\Delta x$$

This sum is called a **Riemann sum**. The larger the value of n, the better the number S_n approximates what we would consider to be the area of the planar region. Figure 7 illustrates this. This leads us to define the area of the planar region below the nonnegative function $f(x)$, above the x-axis, and between the vertical lines $x = a$ and $x = b$, to be

$$\lim_{n \to \infty} S_n = \lim_{n \to \infty} [f(c_0) + f(c_1) + \cdots + f(c_{n-1})]\Delta x$$

provided this limit exists. For the functions we consider in this text, this limit will always exist, but Figure 8 shows a planar region for which it does not. The line $x = b$ is a vertical asymptote to the function $f(x)$. The planar region bounded by the function $f(x)$, the x-axis, and the lines $x = a$ and $x = b$ is said to have undefined area.

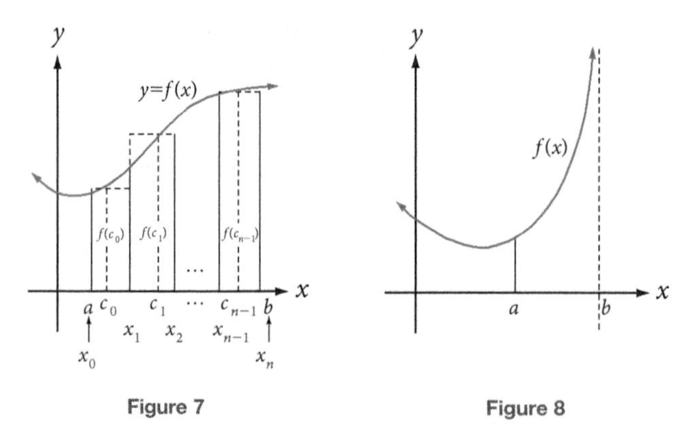

Figure 7 Figure 8

Relating Area and the Definite Integral

Let's analyze what we have done so far. We have defined the area under a curve that lies above the input axis and between two vertical lines using the *accumulated* areas of rectangles. The word *accumulated* makes us think of the definite integral, and, in fact, we draw the following relationship between area and the definite integral.

Area and the Definite Integral

If $f(x)$ is a continuous, nonnegative function on the interval $[a, b]$, the area A (in square units) bounded by the function $f(x)$, the x-axis, and the vertical lines $x = a$ and $x = b$ is precisely equal to the value of the definite integral of $f(x)$, from $x = a$ to $x = b$. That is,

$$A = \int_a^b f(x)\, dx$$

Example 2 Find the area bounded by the function

$$f(x) = 0.3x^3 - 1.8x^2 + 15$$

the x-axis, and the lines $x = 1$ and $x = 6$. (This is our function from Example 1.)

Solution The function is displayed in Figure 5 in Example 1. Notice that over the interval $[1, 6]$, $f(x)$ lies completely above the x-axis so that it is always nonnegative here. Since it is nonnegative on $[1, 6]$, the required area can be found by evaluating the related definite integral.

$$A = \int_1^6 (0.3x^3 - 1.8x^2 + 15)\, dx$$

$$= \left(\frac{0.3x^4}{4} - \frac{1.8x^3}{3} + 15x\right)\Big|_1^6$$

$$= \left(\frac{0.3 \cdot 6^4}{4} - \frac{1.8 \cdot 6^3}{3} + 15 \cdot 6\right) - \left(\frac{0.3 \cdot 1^4}{4} - \frac{1.8 \cdot 1^3}{3} + 15 \cdot 1\right)$$

$$= (97.2 - 129.6 + 90) - (0.075 - 0.6 + 15)$$

$$= 57.6 - 14.475 = 43.125$$

Interpretation

The area bounded by $f(x) = 0.3x^3 - 1.8x^2 + 15$, the x-axis, and the lines $x = 1$ and $x = 6$ is exactly 43.125 square units.

If the function $f(x)$ is negative somewhere in $[a, b]$, the definite integral will not record area. The definite integral acts as an accumulator for area only when the function is nonnegative throughout the interval. Recall that $f(x)$ represents the heights of the rectangles and if these values are negative, a negative value for the area will result, which would not make sense. So, if we want the area of a region where $f(x)$ is negative, we must take the opposite of $f(x)$; that is,

$$A = \int_a^b (-f(x))\, dx$$

The evaluation of a definite integral does not necessarily have to involve the computation of an area and, therefore, can result in values less than or equal to zero. But when the definite integral is being used specifically to find an area, we must pay close attention to the graph of the function and account carefully for any areas below the input axis. The next two examples will illustrate this difference.

Example 3 Find the value of $\int_2^5 (x^2 - 5x + 4)\, dx$.

Solution Notice that we are not asked to find any area. So we will evaluate the integral without looking at its graph.

$$\begin{aligned}\int_2^5 (x^2 - 5x + 4)\, dx &= \left(\frac{x^3}{3} - \frac{5x^2}{2} + 4x\right)\Big|_2^5 \\ &= \left(\frac{5^3}{3} - \frac{5 \cdot 5^2}{2} + 4 \cdot 5\right) - \left(\frac{2^3}{3} - \frac{5 \cdot 2^2}{2} + 4 \cdot 2\right) \\ &= \left(-\frac{5}{6}\right) - \left(\frac{2}{3}\right) \\ &= -\frac{3}{2}\end{aligned}$$

Interpretation

Now surely the area bounded by this curve from $x = 2$ to $x = 5$ is not $-3/2$ square units. Let's look at the graph of $f(x) = x^2 - 5x + 4$ in Figure 9. We see that this definite integral does not accumulate area because the function is not always nonnegative on $[2, 5]$. The function is negative from $x = 2$ to almost $x = 4$. (We say 'almost' because the function is actually zero at $x = 4$.) Recall that the relationship between area and the definite integral exists only when the function is nonnegative over the entire given interval (that is, when the function lies on or above the x-axis throughout the interval). Consequently, the value of the definite integral is $\frac{-3}{2}$, but it does not represent the area of the associated region. We will look at that area in the next example.

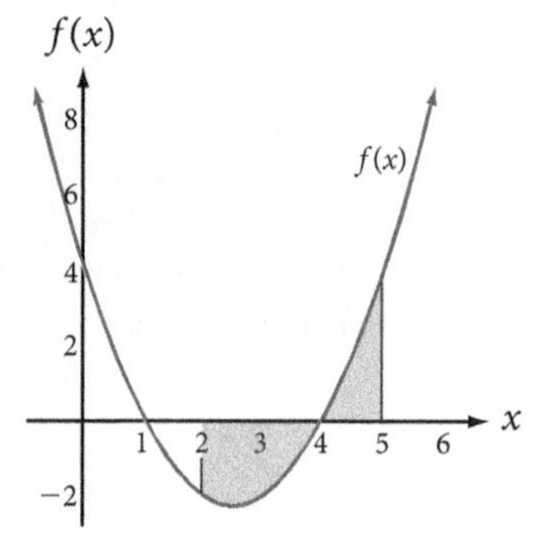

Figure 9

Example 4 Find the area of the region bounded by $f(x) = x^2 - 5x + 4$, the x-axis, and the lines $x = 2$ and $x = 5$.

Solution Figure 9 shows the graph of $f(x)$. To find the area, we will need to write separate integrals for the region below the x-axis, where $f(x) \le 0$, and the region above the x-axis, where $f(x) \ge 0$.

For $f(x) \le 0$, $A = \int_2^4 [-f(x)]\, dx$.

For $f(x) \ge 0$, $A = \int_4^5 f(x)\, dx$.

The accumulated area A is:

$$\begin{aligned}
A &= \int_2^4 [-f(x)]\, dx + \int_4^5 f(x)\, dx \\
&= \int_2^4 [-(x^2 - 5x + 4)]\, dx + \int_4^5 (x^2 - 5x + 4)\, dx \\
&= \left(-\frac{x^3}{3} + \frac{5x^2}{2} - 4x\right)\Big|_2^4 + \left(\frac{x^3}{3} - \frac{5x^2}{2} + 4x\right)\Big|_4^5 \\
&= \left[\left(-\frac{4^3}{3} + \frac{5 \cdot 4^2}{2} - 4 \cdot 4\right) - \left(-\frac{2^3}{3} + \frac{5 \cdot 2^2}{2} - 4 \cdot 2\right)\right] \\
&\quad + \left[\left(\frac{5^3}{3} - \frac{5 \cdot 5^2}{2} + 4 \cdot 5\right) - \left(\frac{4^3}{3} - \frac{5 \cdot 4^2}{2} + 4 \cdot 4\right)\right] \\
&= \left[\left(\frac{8}{3}\right) - \left(-\frac{2}{3}\right)\right] + \left[\left(-\frac{5}{6}\right) - \left(-\frac{8}{3}\right)\right] \\
&= \left(\frac{10}{3}\right) + \left(\frac{11}{6}\right) \\
&= \frac{31}{6}
\end{aligned}$$

Interpretation

The area of the bounded region is $\frac{31}{6}$. Again, notice the difference between this example and Example 3.

Using Technology 5.4

You can use Wolfram|Alpha to visualize and find the area under a curve over a closed interval.

Example 5 Find the area of the region bounded by the curve

$$f(x) = 0.3x^3 - 1.8x^2 + 15$$

the x-axis, and the lines $x = 1$ and $x = 6$.

Solution Let's first get a picture of the region for which we are trying to find the area. Go online to www.wolframalpha.com. Into the entry field enter

Graph 0 < 0.3x^3 – 1.8x^2 + 15 from x = 1 to x = 6

In order to get the shading below the curve and above the x-axis, you must set the function as being positive by describing it as being greater than 0. Do so by starting its description as

"0 < (the function)"

Wolfram|Alpha returns

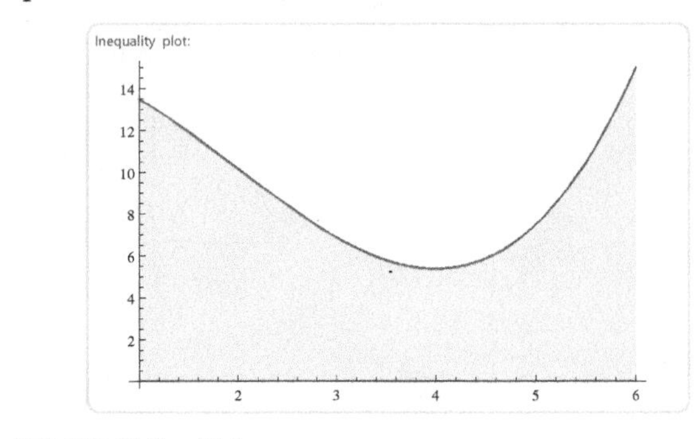

Wolfram Alpha LLC. 2012. Wolfram|Alpha
http://www.wolframalpha.com/
(access November 30, 2012)

Now, to compute the area, enter

Integrate 0.3x^3 – 1.8x^2 + 15 from x = 1 to x = 6

Wolfram|Alpha returns an area of 43.125 square units, which agrees with our answer in Example 2. ■

Getting Ready for Class

After reading through the preceding section, respond in your own words and in complete sentences.

A. Intuitively, what is "area"?

B. What is the name given to the sum

$$S_n = [f(c_0) + f(c_1) + \cdots + f(c_{n-1})]\Delta x$$

C. In the formula for area, $A = \int_a^b f(x)\,dx$, what do the $f(x)$ and the dx represent relative to rectangles?

Problem Set 5.4

Skills Practice

For Problems 1-16, determine the area of the region bounded by the given function, the x-axis, and the given vertical lines. Each region lies above the x-axis.

1. $f(x) = -6x^2 + 26x$, $x = 0$ and $x = 3$
2. $f(x) = \frac{1}{x^3}$, $x = 1$ and $x = 2$
3. $f(x) = 3x^2 - 3x + 5$, $x = 0$ and $x = 2$
4. $f(x) = e^x$, $x = 0$ and $x = 1$
5. $f(x) = \frac{1}{x}$, $x = 1$ and $x = 4$
6. $f(x) = x^2 - 10$, $x = 4$ and $x = 6$
7. $f(x) = 2{,}500e^{0.2x}$, $x = 0$ and $x = 4$
8. $f(x) = 300e^{-0.03x}$, $x = 5$ and $x = 10$
9. $f(x) = 2\sqrt{x}$, $x = 1$ and $x = 4$
10. $f(x) = e^x(e^x + 2)^3$, $x = 0$ and $x = 2$
11. $f(x) = \frac{x}{(x^2 + 8)}$, $x = 3$ and $x = 5$
12. $f(x) = e^{-x+2}$, $x = 1$ and $x = 5$
13. $f(x) = \frac{800}{\sqrt[3]{(x + 3)^2}}$, $x = 10$ and $x = 25$
14. $f(x) = 420e^{0.01x} - 20 + 0.2x$, $x = 0$ and $x = 10$
15. $f(x) = \frac{x^2}{2} - \frac{x}{3}$, $x = 5$ and $x = 20$
16. $f(x) = x^{2/3}$, $x = 0$ and $x = 5$

For Problems 17-20, part of the given function lies below the x-axis and part lies above it. Find the area of the specified region.

17. The function

$$f(x) = x^2 - 7x + 10$$

is positive on $(0, 2)$ and $(5, 10)$ and negative on $(2, 5)$. Find the area of the region bounded by $f(x)$ and the vertical lines $x = 0$ and $x = 10$.

18. The function

$$f(x) = 6x^2 - 42x + 72$$

is positive on $(0, 3)$ and $(4, 5)$ and negative on $(3, 4)$. Find the area of the region bounded by $f(x)$ and the vertical lines $x = 0$ and $x = 5$.

19. The function

$$f(x) = \frac{x - 5}{x^2 + 10x}$$

is negative on (4, 5) and positive on (5, 6). Find the area of the region bounded by $f(x)$ and the vertical lines $x = 4$ and $x = 6$. Round your result to two decimal places.

20. The function

$$f(x) = \frac{1}{x - 3}$$

is negative on (1, 2) and positive on (4, 5). Find the total area of the region bounded by $f(x)$ and the vertical lines $x = 1$ and $x = 2$ as well as the lines $x = 4$ and $x = 5$. Round your result to two decimal places.

Check Your Understanding

21. Suppose $f(x)$ is the function pictured below. Use two definite integrals to express the area from $x = a$ to $x = b$ using a, b, and c.

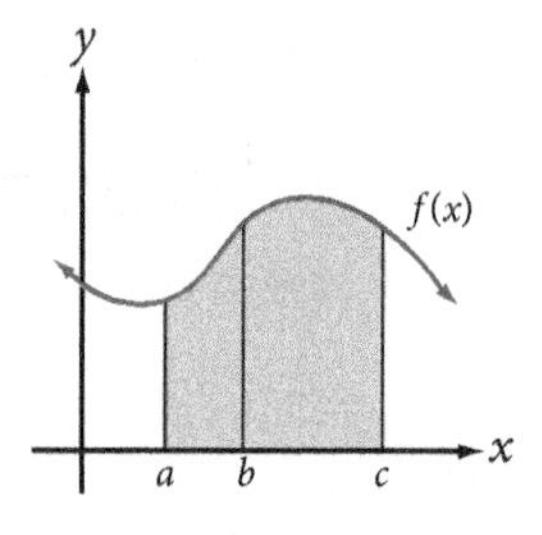

22. Suppose $f(x)$ is the function pictured below. Use two definite integrals to describe the area of the shaded region. (More than one answer is possible.)

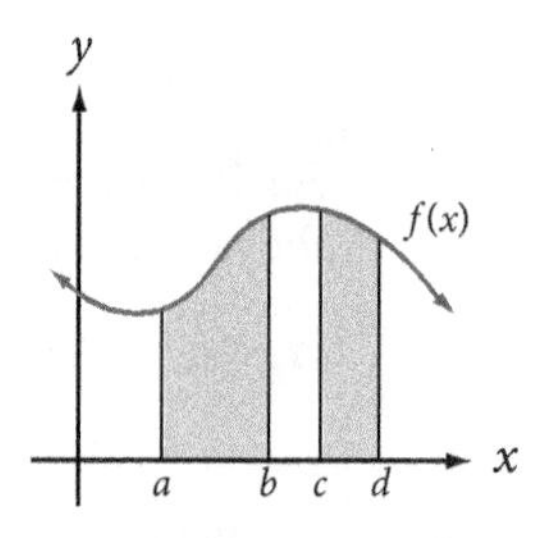

23. Suppose $f(x)$ is the function pictured below. Use two definite integrals to describe the area of the shaded region using a, b, and c as limits of integration.

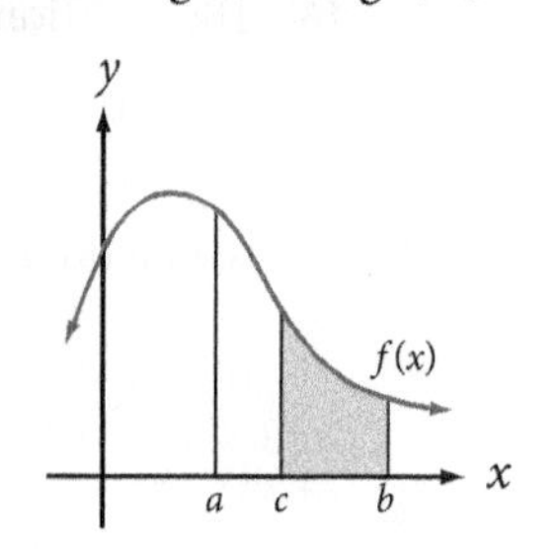

24. The function $f(x)$ pictured below is not always nonnegative over $[a, b]$. It is nonnegative over (a, c), but negative over (c, b). Use two definite integrals to describe the area bounded by the curve $f(x)$ and the vertical lines $x = a$ and $x = b$. (Think about the height of a representative rectangle between c and b. As it is drawn, is it positive or negative?)

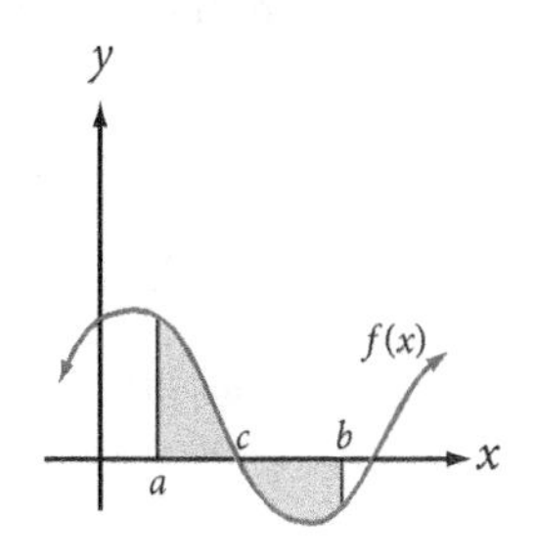

Modeling Practice

© Marcus Clackson/iStockPhoto

25. Economics: Social Utility of Consumption To sell x units of a product each month, a company must price each unit according to the demand function

$$D(x) = \frac{350}{1 + 0.04x}$$

dollars. The figure below shows the graph of this demand function, which we introduced at the beginning of this section. To review, economists call the shaded region bounded by the function, the x-axis, and the vertical lines $x = 0$ and $x = 400$ (or some value a) the **social utility of consumption**, and it is the accumulated value the public places on its consumption of these 400 (or, in general, these a) units per unit of time. Since the social utility of consumption is an accumulated value, it is defined by an integral. In general, the social utility of a units of a product to the public is defined as

$$\text{Social utility of consumption} = \int_0^a D(x)\,dx$$

Determine the social utility of consumption associated with the sale of 400 units of this product.

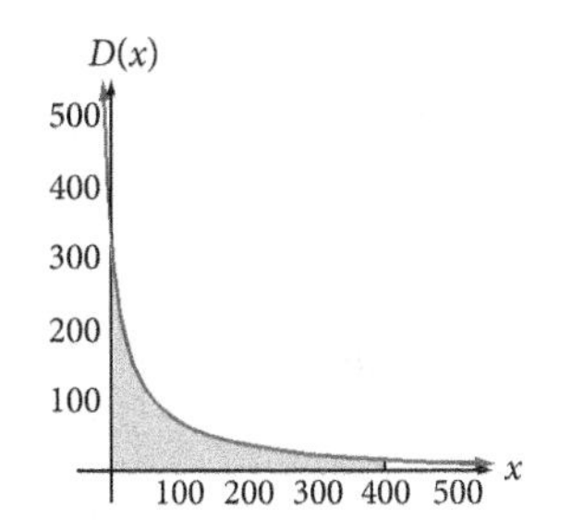

26. Economics: Revenue Stream A person's investment is expected to produce a continuous stream of revenue, R, of

$$R(t) = 2{,}500\sqrt{t + 10}$$

dollars over the next 8 years. The accumulated value of this **revenue stream** from time $t = a$ to $t = b$ is given by the integral

$$\int_a^b R(t)\, dt$$

Find the total revenue generated by this revenue stream over the 8-year period.

27. Business: Revenue Stream A person's investment is expected to produce a continuous stream of revenue R of

$$R(t) = 400e^{0.08t}$$

dollars over the next 8 years. Find the total revenue generated by this revenue stream 5 to 8 years into the investment. (Hint: See Problem 26.)

© Mark Wragg/iStockPhoto

28. Education: Accumulation of Chalk Dust Back in the old days when chalk and blackboards were common in classrooms, students might have noticed that as their instructor erased the chalk from the blackboard, chalk dust would fall from the board and settle into the tray. Knowing that the definite integral represents accumulation, they are interested in knowing the accumulated amount of dust at the end of one erasing. The instructor, after taking a sample and using some curve-fitting techniques, determines that at each point x feet from the left end of the chalk tray,

$$C(x) = \frac{140}{(x + 80)^2}$$

grams of dust fall to the tray. If the tray is 20 feet long, and the board is written on in such a way that the function approximates the way the chalk falls, how many grams of chalk dust will have accumulated in the tray after one board erasure?

29. Business: Capital Accumulation of an Investment The **capital accumulation** of an investment over a time period $t = a$ to $t = b$, is the integral, over that time period, of the rate of investment $\frac{dI}{dt}$, if the cash flow into the investment is continuous. Find the capital accumulation of an investment over a 3-year time period if

$$\frac{dI}{dt} = 2{,}650(4t + 2)^{1/3}$$

30. Social Science: Death Rates In a particular country, the function

$$D(t) = -0.57t^2 + 0.4t + 6$$

approximates the death rate (for each 1,000 people who are t years of age). Find the total number of deaths per 1,000 people of those who are between 0 and 5 years old.

31. **Manufacturing: Worker Production** An average worker can assemble

$$N(t) = -2.7t^2 + 26t$$

units of a product in hour t of the working day. Determine the number of units the average worker will assemble in an 8-hour working day.

32. **Business: Production Savings** A large company has installed new robotic equipment in its assembly areas. This new equipment produces a savings that accumulates at the rate of

$$S(t) = 36e^{0.18t}$$

thousand dollars per month. Determine the savings over the first 1-year period.

Using Technology Exercises

33. The number N of gallons of a toxic chemical being dumped into a river t weeks from the beginning of the year is approximated by the function

$$N(t) = \frac{615}{(0.65t + 3.6)^{3/2}} \qquad 0 \le t \le 20$$

The area under the curve $N(t)$ measures the total amount of the chemical that has entered the river since the beginning of the year. Graph this region and find the total amount of chemical that has entered the river between weeks 5 and 10. Round your result to the nearest gallon.

34. A time-release drug produced by a drug manufacturer keeps A mg of the drug in the bloodstream t hours after it is first ingested. The relationship between A and t is given by the function

$$A(t) = 26.8t^{2.48} - 206.5t^{1.48} + 437.3t^{0.48}$$

where $0 \le t \le 3.5$. The area of the region bounded by the function, the x-axis, the y-axis, and the line $x = 3.5$ gives the total amount of the drug that has been in the bloodstream through the first 3.5 hours. Find the total amount of drug that has been in the bloodstream through the first hour. Round your result to the nearest milligram.

Source: suggested from
Interactive Molecules — Explain it with Molecules
EdInformatics.com

35. The rate at which the mortgage debt outstanding changes for multifamily residences in the United States since 2007 can be modeled by the function

$$M'(t) = 243t^4 + 2{,}739t^3 - 31{,}797t^2 + 81{,}861t + 784{,}628$$

where M is the mortgage debt outstanding (in millions of dollars) and t is the number of years since 2007. The area of the region bounded by this curve, the x-axis, the y-axis, and the line $x = 4$ gives the total amount of mortgage debt outstanding through the years 2007 to 2011. Find that amount.

Source: The Board of Governors of the Federal Reserve System

36. The number of fish in a lake at any given time, t, is approximated by the function

$$N(t) = \frac{15{,}000}{1 + 22e^{-0.2t}}$$

where t is the number of months measured from January 1. Estimate the total number of fish that have populated this lake during the year. Round your estimate to the nearest thousand.

Getting Ready for the Next Section

37. Evaluate both $\int_{1}^{100{,}000} \frac{1}{x}\,dx$ and $\int_{1}^{2} \frac{1}{x}\,dx$.

38. Evaluate $\int_{5}^{b} \frac{1}{x^{3/2}}\,dx$.

39. Evaluate $\int_{b}^{1} \frac{1}{(x-2)^3}\,dx$.

40. Compute $-2{,}700 \cdot \lim_{b\to\infty} \left[\frac{1}{(b+1.5)^2} - \frac{1}{(0+1.5)^2}\right]$.

41. Compute $100 \cdot \lim_{b\to\infty} [(-2e^{-0.5(b)} + 1.25e^{-0.8(b)}) - (-2e^{-0.5(0)} + 1.25e^{-0.8(0)})]$.

Improper Integrals

5.5

© Microgen/iStockPhoto

There are many situations in which it is necessary to accumulate quantities over infinitely long intervals. For example, we may wish to determine how much of a pollutant will seep into the ground near a dump site if it is allowed to discharge into the ground indefinitely, or how many barrels of oil a well will produce if it operates indefinitely, or what total profit a company could realize from the sale of an unlimited number of units of some product. Definite integrals involved in such accumulations are called **improper integrals** and differ from the integrals we have worked with so far in that either the interval of integration is infinite or the integrand (the function being integrated) is unbounded at one or more points in the interval.

Improper Integrals

We can motivate a definition for improper integrals by evaluating $\int_1^b f(x)\,dx$ as b gets bigger and bigger. Table 1 shows the values of $\int_1^b \frac{1}{x^2}\,dx$ for increasingly large values of b, and Figure 1 shows the areas for some of these values.

Note Evaluating the definite integral

$$\int_1^b \frac{1}{x^2}\,dx$$

gives us

$$\frac{-1}{x}\Big|_1^b = \frac{-1}{b} - (-1)$$

For convenience, we have written this last expression as

$$1 - \frac{1}{b}$$

b	$1 - \frac{1}{b}$
2	0.5
3	$0.\overline{6}$
5	0.8
10	0.9
100	0.99
1,000	0.999
10,000	0.9999
100,000	0.99999

Table 1

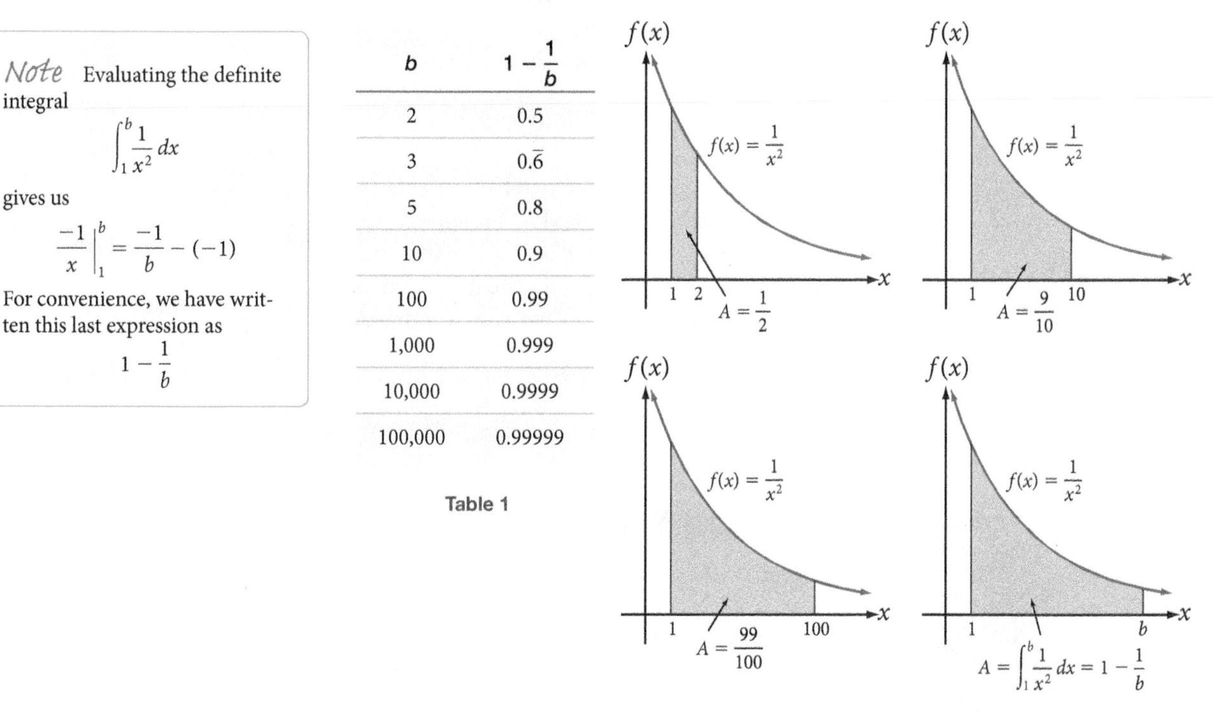

Figure 1

Apparently, as b gets bigger and bigger—that is, as b approaches infinity—$\int_1^b \frac{1}{x^2}\,dx$ approaches 1. We can symbolize this as

$$\lim_{b \to \infty} \int_1^b \frac{1}{x^2}\,dx = 1$$

and make the following definitions.

Improper Integrals

1. $\int_a^\infty f(x)\,dx = \lim_{b\to\infty} \int_a^b f(x)\,dx$

2. $\int_{-\infty}^b f(x)\,dx = \lim_{a\to-\infty} \int_a^b f(x)\,dx$

3. $\int_{-\infty}^\infty f(x)\,dx = \lim_{a\to-\infty} \int_a^c f(x)\,dx + \lim_{b\to\infty} \int_c^b f(x)\,dx$, where $a \le c \le b$

If these limits exist, the improper integral is said to **converge** to the computed value. Otherwise, it **diverges**.

Our first example showed that $\int_1^\infty \frac{1}{x^2}\,dx$ converges to 1; that is, though the region goes on forever, its area is finite and never exceeds 1.

However, the integral $\int_1^\infty \frac{1}{x}\,dx$ does not converge, since

$$\begin{aligned}\int_1^\infty \frac{1}{x}\,dx &= \lim_{b\to\infty} \int_1^b \frac{1}{x}\,dx \\ &= \lim_{b\to\infty} \ln|x| \Big|_1^b \\ &= \lim_{b\to\infty} (\ln b - \ln 1) \\ &= \lim_{b\to\infty} (\ln b - 0) \\ &= \lim_{b\to\infty} \ln b\end{aligned}$$

which does not exist; remember, $\ln x$ is an increasing function, as noted by the approximations in Table 2. It is interesting to note that such an apparently small change in the integrand, in this case from $\frac{1}{x^2}$ to $\frac{1}{x}$, can have such a drastic effect.

b	$\int_1^b \frac{1}{x}\,dx = \ln b$
2	0.6931
3	1.0986
5	1.6094
10	2.3026
100	4.6052
1,000	6.9078
10,000	9.2103
100,000	11.5129

Table 2

$$\int_1^\infty \frac{1}{x^2}\,dx \quad \text{converges, whereas} \quad \int_1^\infty \frac{1}{x}\,dx \text{ diverges.}$$

VIDEO EXAMPLES

SECTION 5.5

Example 1 Evaluate $\int_5^\infty \frac{1}{x^{3/2}}\,dx$.

Solution

$$\begin{aligned}\int_5^\infty \frac{1}{x^{3/2}}\,dx &= \lim_{b\to\infty} \int_5^b \frac{1}{x^{3/2}}\,dx \\ &= \lim_{b\to\infty} \int_5^b x^{-3/2}\,dx \\ &= \lim_{b\to\infty} (-2x^{-1/2}) \Big|_5^b \\ &= \lim_{b\to\infty} -2(b^{-1/2} - 5^{-1/2})\end{aligned}$$

$$= -2\left(0 - \frac{1}{5^{1/2}}\right)$$

$$= \frac{2}{\sqrt{5}}$$

Interpretation

The improper integral $\int_5^\infty \frac{1}{x^{3/2}}\,dx$ converges to $\frac{2}{\sqrt{5}}$. Figure 2 shows the graph of $f(x) = \frac{1}{x^{3/2}}$. Since $f(x)$ lies above the x-axis when $x \geq 5$, the integral can be associated with the area under the curve.

Therefore, the area of the region below $f(x)$, above the x-axis, and to the right of $x = 5$ converges to $\frac{2}{\sqrt{5}} \approx 0.894$ as x increases without bound.

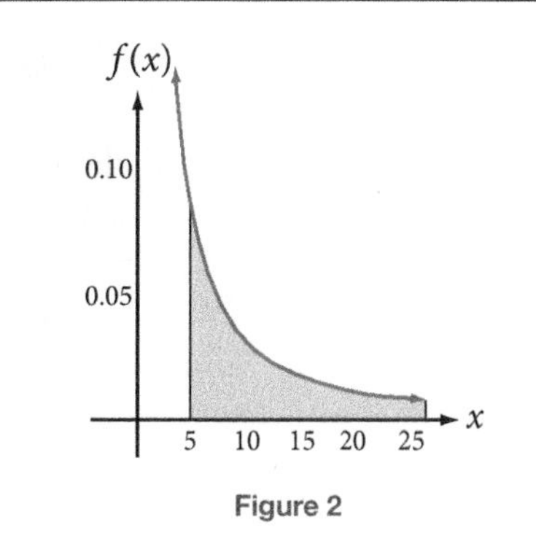

Figure 2

Example 2 Evaluate $\int_{-\infty}^{1} \frac{1}{(x-2)^3}\,dx$.

Solution We let $u = x - 2$ and $du = dx$. The new limits of integration are:

when $x = 1$, then $u = 1 - 2 = -1$ and

when $x = -\infty$, then $u = -\infty$

Therefore,

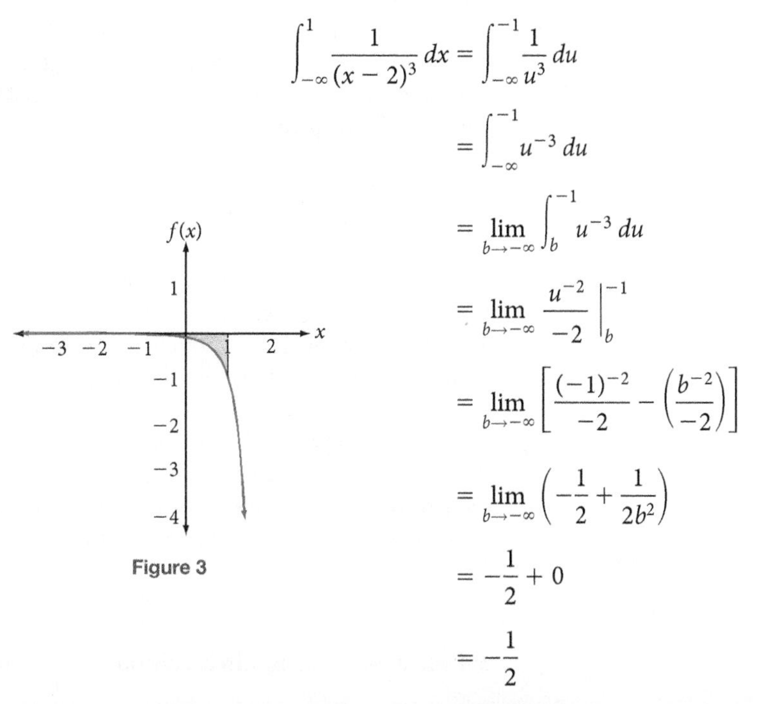

$$\int_{-\infty}^{1} \frac{1}{(x-2)^3}\,dx = \int_{-\infty}^{-1} \frac{1}{u^3}\,du$$

$$= \int_{-\infty}^{-1} u^{-3}\,du$$

$$= \lim_{b\to-\infty} \int_b^{-1} u^{-3}\,du$$

$$= \lim_{b\to-\infty} \left.\frac{u^{-2}}{-2}\right|_b^{-1}$$

$$= \lim_{b\to-\infty} \left[\frac{(-1)^{-2}}{-2} - \left(\frac{b^{-2}}{-2}\right)\right]$$

$$= \lim_{b\to-\infty} \left(-\frac{1}{2} + \frac{1}{2b^2}\right)$$

$$= -\frac{1}{2} + 0$$

$$= -\frac{1}{2}$$

Figure 3

Interpretation

The improper integral $\int_{-\infty}^{1} \frac{1}{(x-2)^3}\,dx$ converges to $-\frac{1}{2}$. Figure 3 shows the graph of $f(x) = \frac{1}{(x-2)^3}$. Since $f(x)$ lies below the x-axis when $x \le 1$, the integral can be associated with the opposite of the area of the region above the curve, below the x-axis, and to the left of $x = 1$. Therefore, the opposite of the area of this region converges to $-\frac{1}{2} = -0.5$ as x decreases without bound.

Example 3 Evaluate $\displaystyle\int_{-\infty}^{\infty} x\,dx.$

Solution

$$\begin{aligned}\int_{-\infty}^{\infty} x\,dx &= \lim_{a\to-\infty}\int_a^0 x\,dx + \lim_{b\to\infty}\int_0^b x\,dx\\ &= \lim_{a\to-\infty}\frac{x^2}{2}\bigg|_a^0 + \lim_{b\to\infty}\frac{x^2}{2}\bigg|_0^b\\ &= \lim_{a\to-\infty}\left(\frac{0^2}{2}-\frac{a^2}{2}\right) + \lim_{b\to\infty}\left(\frac{b^2}{2}-\frac{0^2}{2}\right)\\ &= \lim_{a\to-\infty}\left(-\frac{a^2}{2}\right) + \lim_{b\to\infty}\left(\frac{b^2}{2}\right)\\ &= -\infty + \infty\end{aligned}$$

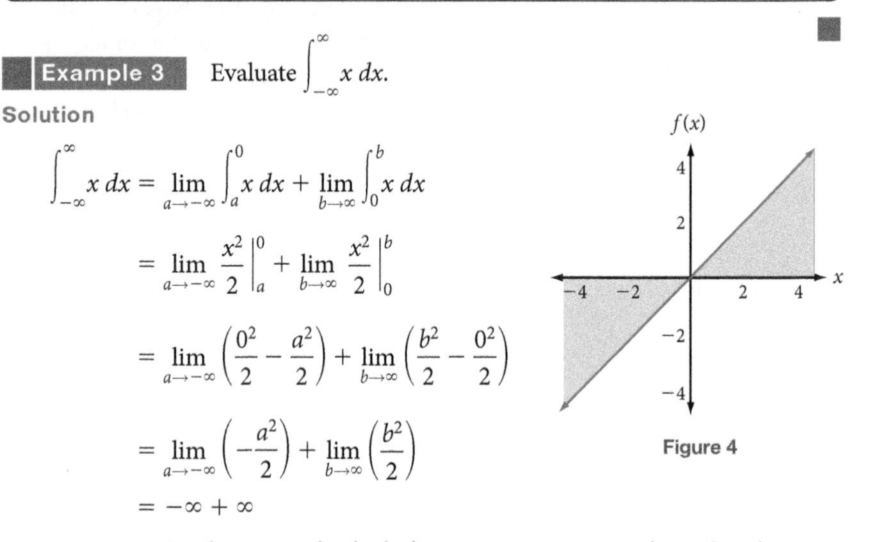

Figure 4

= So, this integral, which does not represent a real number does not exist. That is, diverges.

Interpretation

The improper integral $\int_{-\infty}^{\infty} x\,dx$ diverges since the associated limits do not exist. The graph of $f(x) = x$ is shown in Figure 4. Notice that in the interval from 0 to ∞ the curve is above the x-axis, and therefore, the integral is associated with the area under the curve, which obviously continues to get larger as x increases without bound. In the interval from $-\infty$ to 0, the curve is always below the x-axis, and therefore, the integral is associated with the opposite of the area between the curve and the x-axis that decreases as x decreases without bound. Consequently, the integral does not converge on any particular number; it diverges.

The notation $\infty - \infty$ is an example of an indeterminate form. Although it is clear that when adding a very large number to another very large number, the result is a very large number, it is not clear what happens when subtracting a very large number from a very large number. If the two very large numbers are equal, then the subtraction produces 0. If they are not equal, is the difference a large number or a small number? Since the answer is not clear, the result is said to be indeterminate.

Other indeterminate forms are $\frac{0}{0}$, $\frac{\infty}{\infty}$, $0\cdot\infty$, 0^0, ∞^0, $1^{\pm\infty}$. Indeterminate forms (which we discussed in Section 1.4) can be deceptive and typically arise from limits of discontinuous functions. For example

$$\lim_{x\to1}\frac{2x-2}{x-1} = 2$$

but when evaluated by direct substitution produces the form $\frac{0}{0}$.

Example 4 Determine whether or not the integral $\int_{-1}^{1} \frac{1}{x^2}\,dx$ converges.

Solution Since the function $\frac{1}{x^2}$ is undefined at $x = 0$, we can write the original improper integral as the sum of two separate integrals.

$$\int_{-1}^{1} \frac{1}{x^2}\,dx = \lim_{b \to 0^-} \int_{-1}^{b} \frac{1}{x^2}\,dx + \lim_{a \to 0^+} \int_{a}^{1} \frac{1}{x^2}\,dx$$

Then

$$\int_{-1}^{1} \frac{1}{x^2}\,dx = \lim_{b \to 0^-} \int_{-1}^{b} \frac{1}{x^2}\,dx + \lim_{a \to 0^+} \int_{a}^{1} \frac{1}{x^2}\,dx$$

$$= \lim_{b \to 0^-} \left[\frac{-1}{x} \Big|_{-1}^{b} \right] + \lim_{a \to 0^+} \left[\frac{-1}{x} \Big|_{a}^{1} \right]$$

$$= \lim_{b \to 0^-} \left[\frac{-1}{b} - \frac{-1}{-1} \right] + \lim_{a \to 0^+} \left[\frac{-1}{1} - \frac{-1}{a} \right]$$

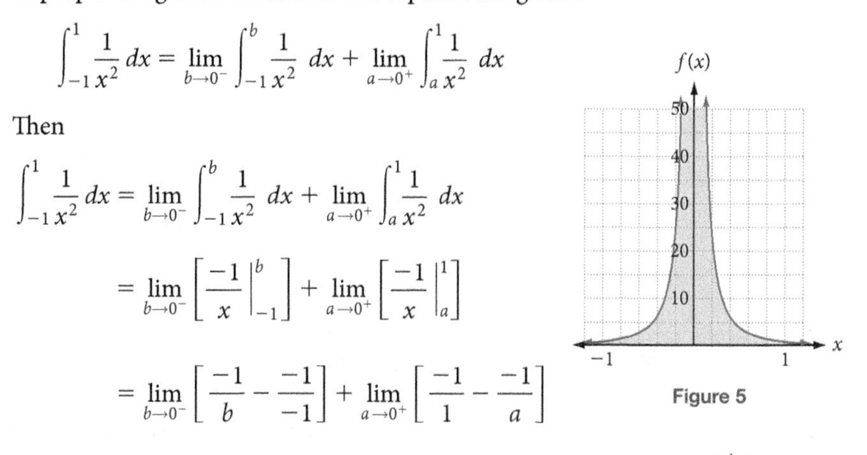

Figure 5

Both these limits are undefined, and we conclude that the integral $\int_{-1}^{1} \frac{1}{x^2}\,dx$ does not converge. ■

Example 5 Suppose a pollutant is seeping into the ground near a dump site at the rate of

$$f(t) = \frac{5{,}400}{(t + 1.5)^3}$$

liters per month, where t denotes the time from now in months. If the seepage of the pollutant continues indefinitely into the future, what is the total amount of pollutant that will seep into the ground?

Solution Since we wish to find the total amount of pollutant, we will use the definite integral as an accumulator. Furthermore, since we wish to accumulate over an interval with an unspecified upper bound, we will need to use an improper integral.

$$\text{Total amount of pollutant} = \int_0^{\infty} \frac{5{,}400}{(t + 1.5)^3}\,dt$$

$$= \lim_{b \to \infty} \int_0^{b} \frac{5{,}400}{(t + 1.5)^3}\,dt$$

$$= \lim_{b \to \infty} \int_0^{b} 5{,}400(t + 1.5)^{-3}\,dt$$

$$= \lim_{b \to \infty} \frac{5{,}400(t + 1.5)^{-2}}{-2} \Big|_0^{b}$$

$$= \lim_{b \to \infty} \frac{-2{,}700}{(t + 1.5)^2} \Big|_0^{b}$$

$$= -2{,}700 \lim_{b \to \infty} \left(\frac{1}{(b + 1.5)^2} - \frac{1}{(0 + 1.5)^2} \right)$$

$$= -2{,}700\left(0 - \frac{1}{2.25}\right)$$

$$= -2{,}700\left(-\frac{1}{2.25}\right)$$

$$= 1{,}200$$

Interpretation
If the seepage of the pollutant goes unchecked and continues indefinitely into the future, 1,200 liters will collect in the ground near the dump site.

Modeling: Bioavailability of a Drug

The bioavailability of a drug is the amount of the drug that is actually absorbed from a given dose. That is, the bioavailability of a drug is the amount of the drug that is available to the target tissue after administration. It is largely determined by the properties of the dosage form. Before reaching its target tissue, orally administered drugs must pass through the intestinal wall and then to the liver. Some amount of the drug is absorbed before it reaches the target tissue. On the other hand, one-hundred percent of drugs administered intravenously reach the target tissue.

The bioavailability of a drug measures the amount of a drug given orally that reaches the target tissue compared to the amount of the drug that would reach the target tissue if it were given intravenously. That is, the bioavailability of a drug measures how effective an oral dose of a drug is relative to the intravenous dose.

If $C_{po}(t)$ represents the concentration of a drug in the bloodstream at time t for an oral dose of the drug and $C_{iv}(t)$ represents the concentration of a drug in the bloodstream at time t for an intravenous dose of the drug, the bioavailability of an oral dose of the drug is defined to be

$$\text{Bioavailability} = F = \frac{\int_0^\infty C_{po}(t)\,dt}{\int_0^\infty C_{iv}(t)\,dt}$$

Source: The Merck Manual and www.boomer.org

Example 6 Suppose the concentration of a particular drug that is administered orally is given by $C_{po}(t) = 100(e^{-0.5t} - e^{-0.8t})$. The concentration of that drug when administered intravenously is given by $C_{iv}(t) = 150e^{-0.5t}$. This function considers both absorption and elimination of the drug over time. Determine the bioavailability, F, of the drug.

Solution Using the formula for bioavailability we get

$$F = \frac{\int_0^\infty 100(e^{-0.5t} - e^{-0.8t})\,dt}{\int_0^\infty 150e^{-0.5t}\,dt}$$

$$= \frac{\lim_{b\to\infty} \int_0^b 100(e^{-0.5t} - e^{-0.8t})\, dt}{\lim_{b\to\infty} \int_0^b 150e^{-0.5t}\, dt}$$

$$= \frac{100 \lim_{b\to\infty} \int_0^b (e^{-0.5t} - e^{-0.8t})\, dt}{150 \lim_{b\to\infty} \int_0^b e^{-0.5t}\, dt}$$

$$= \frac{100 \lim_{b\to\infty} \left(\frac{1}{-0.5} e^{-0.5t} - \frac{1}{-0.8} e^{-0.8t} \right) \Big|_0^b}{150 \lim_{b\to\infty} \frac{1}{-0.5} e^{-0.5t} \Big|_0^b}$$

$$= \frac{100 \lim_{b\to\infty} (-2e^{-0.5t} + 1.25e^{-0.8t}) \Big|_0^b}{150 \lim_{b\to\infty} (-2e^{-0.5t}) \Big|_0^b}$$

$$= \frac{100 \lim_{b\to\infty} [(-2e^{-0.5(b)} + 1.25e^{-0.8(b)}) - (-2e^{-0.5(0)} + 1.25e^{-0.8(0)})]}{150 \lim_{b\to\infty} [-2(e^{-0.5b} - e^{-0.5(0)})]}$$

$$= \frac{100[(0 + 0) - (-2 + 1.25)]}{150[-2(0 - 1)]}$$

$$= \frac{100[0.75]}{150[2]}$$

$$= \frac{75}{300}$$

$$= 0.25$$

That is, bioavailability $= F = 0.25$, which means that if the dose is taken orally it will be only about 25% as effective as a dose taken intravenously. ■

Using Technology 5.5

You can use your calculator or Wolfram|Alpha to evaluate improper integrals. Let's work Example 1 of this section using both tools.

Example 7 Use your calculator to approximate $\int_5^\infty \frac{1}{x^{3/2}}\, dx$.

Solution We can approximate the integral by choosing a large value for the upper limit of integration. For example, we might enter

```
Y1 = 1/x^(3/2)
Y9 = fnInt(Y1,X,A,B)
```

Store 5 for A and 100,000 for B, and then evaluate as before. Your calculator returns approximately 0.8881, which is close to the exact answer of $\frac{2}{\sqrt{5}}$ that we found in Example 1. If we use an upper limit of 1,000,000, we can get even closer.

It is hard to tell what is a large enough value to enter for the upper limit. The key is to try a few, increasingly large values to see if your result approaches some limiting value. ■

Example 8 Use Wolfram|Alpha to evaluate $\int_5^{\infty} \frac{1}{x^{3/2}}\, dx$.

Solution Go online to www.wolframalpha.com. Into the entry field enter

integrate 1/x^(3/2) from x = 5 to infinity

Wolfram|Alpha returns the exact area as $\frac{2}{\sqrt{5}}$ and the approximate area as a decimal number. Wolfram|Alpha is able to compute the exact area because it is a symbolic computer algebra system. Most calculators are not and, therefore, produce only approximate results.

Definite integral:

$$\int_5^{\infty} \frac{1}{x^{3/2}}\, dx = \frac{2}{\sqrt{5}} \approx 0.894427$$

Wolfram Alpha LLC. 2012. Wolfram|Alpha
http://www.wolframalpha.com/
(access October 24, 2012)

■

Getting Ready for Class

After reading through the preceding section, respond in your own words and in complete sentences.

A. What is the name given to definite integrals that are defined on infinitely long intervals?
B. What makes the definite integral $\int_a^{\infty} f(x)\, dx$ an improper integral?
C. Under what condition will the improper integral $\int_a^{\infty} f(x)\, dx$ converge?
D. Explain why the integral $\int_1^5 \frac{x+3}{x-4}\, dx$ is improper.

Problem Set 5.5

Skills Practice

For Problems 1-24, evaluate each improper integral if it is convergent. If it is not convergent, write *divergent.*

1. $\int_1^\infty \frac{1}{x^4}\,dx$

2. $\int_3^\infty \frac{5}{x^3}\,dx$

3. $\int_e^\infty \frac{3}{x}\,dx$

4. $\int_0^\infty \frac{1}{x+3}\,dx$

5. $\int_0^\infty \frac{1}{3x+4}\,dx$

6. $\int_0^\infty e^{-3x}\,dx$

7. $\int_0^\infty e^{-4x}\,dx$

8. $\int_{-\infty}^0 e^{-x}\,dx$

9. $\int_{-\infty}^0 e^{2x}\,dx$

10. $\int_5^\infty e^{2x}\,dx$

11. $\int_0^\infty \frac{1}{(x+1)^3}\,dx$

12. $\int_{10}^\infty \frac{8}{(x-8)^2}\,dx$

13. $\int_1^\infty \frac{6}{x^7}\,dx$

14. $\int_{-\infty}^\infty x^{2/3}\,dx$

15. $\int_0^\infty xe^{-x^2}\,dx$

16. $\int_{100}^\infty \frac{2x+3}{x^2+3x}\,dx$

17. $\int_0^\infty \frac{6x^2}{x^3+6}\,dx$

18. $\int_{200}^\infty \frac{1}{x\ln x}\,dx$

19. $\int_e^\infty \frac{1}{x(\ln x)^2}\,dx$

20. $\int_{\sqrt{e}}^\infty \frac{\ln x}{x}\,dx$

21. $\int_0^\infty ae^{-ax}\,dx \quad a > 0$

22. $\int_{-1}^1 \frac{1}{x^{2/3}}\,dx$

23. $\int_{-1}^1 \frac{1}{x^{4/5}}\,dx$

24. $\int_0^4 \frac{1}{(x-3)^3}\,dx$

Check Your Understanding

25. What does it mean if the bioavailability, F, of a drug taken orally is 85% that of the same drug taken intravenously?

26. What does it mean if the bioavailability, F, of a drug taken orally is 55% that of the same drug taken intravenously?

27. Business: Income Flow The function $f(t)$ relates the rate of flow f (in dollars) of income from a rental property to the time t in years from now. Interpret

$$\int_0^\infty f(t)\,dt$$

28. Ecology: Pollution The function $P(t)$ relates the rate P (in parts per million) of a toxic pollutant that is seeping into the ground near a city's water supply to the time, t, in years from now. Interpret

$$\int_5^\infty P(t)\,dt$$

29. **Ecology: Energy Conservation** The function $E(t)$ relates the rate at which energy, E, (in kilowatt hours) is being conserved in a community to the time, t, in years from now. Interpret

$$\int_{1.5}^{\infty} E(t)\, dt$$

30. **Political Science: Immigration** The function $I(t)$ relates the rate, I, (in thousands of people) at which people from Eastern European countries are immigrating to Western European countries to the time, t, in years from 1970. Interpret

$$\int_{10}^{\infty} I(t)\, dt$$

31. **Business: Quality Control** The function $Q(x)$ relates the rate, Q, (in units) at which a company improves the quality of its product to the number, x, (in units) of the product it produces. Interpret

$$\int_{0}^{\infty} Q(x)\, dx$$

32. **Business: Advertising** The function $P(x)$ relates the rate, P, (in thousands of units) at which consumers purchase a product to the number, x, of people who have been introduced to the product through television advertisements. Interpret

$$\int_{0}^{\infty} P(x)\, dx$$

Modeling Practice

33. **Ecology: Pollution** The rate A (in tons per year) at which a pollutant is being dumped near a city is given by the function

$$A(t) = 1{,}200e^{-0.04t}$$

where t represents the time in years from now. If the dumping continues unchecked, what will be the total amount of pollutant dumped near the city?

34. **Business: Income Flow** The rate I (in dollars per year) at which income from a rental property flows into an account in which interest is compounded continuously at 8% is given by the function

$$I(t) = 200{,}000e^{-0.08t}$$

where t is the time in years from now. If the income is not disrupted and continues indefinitely, how much money will accumulate in this account?

35. Political Science: Waiting Time Telephone calls to a U.S. Congresswoman's local office between 9:00 A.M. and 12:00 noon are received approximately at the rate of 15 per hour. The probability that there is a time period of 5 minutes or more between calls is given by

$$\int_{1/12}^{\infty} 15e^{-15t}\,dt$$

Determine this probability. (The $\frac{1}{12}$ comes from 5 minutes $= \frac{5}{60} = \frac{1}{12}$ hours.)

© Microgen/iStockPhoto

36. Ecology: Pollution In a particular area of the country, manufacturing plants are located alongside a large river that flows into the ocean. As these plants operate, they release chemical pollutants into the river. The river carries the pollutants downstream and eventually into the ocean. If the rate, A, (in tons per year) at which the chemical pollutants are deposited into the ocean is given by

$$A(t) = \frac{1{,}200}{(0.8 + 0.025t)^2}$$

where t is the number of years from now, and the situation continues unabated, how many tons of the chemical pollutant will be deposited into the ocean?

37. Business: Production A company claims it can produce a particular product at the rate of

$$N(t) = 400.4e^{-0.26t}$$

units per year, where t is the time in years from now. Assuming the company's claim is correct, how many units of the product will the company eventually produce?

38. Health Science: Dieting A diet program claims that if the average person stays on its program long into the future, he or she will lose 20% of his or her starting weight. Program analysts use the model $W(t) = Ae^{-5t}$ where A is the starting weight, to determine how much weight (in pounds) is lost t months after starting the diet.

a. If a 250-pound person goes on this diet and stays on it long into the future, the program claims he or she will lose 20% of his or her starting weight. Do you accept or reject the program's claim? Substantiate your decision with a mathematical argument.

b. Use an improper integral to show the claim is true for a person of weight A in pounds.

39. Business: Capital Value The capital value of an asset is the annual rate at which earnings are produced by an asset at time t. It is defined as the improper integral

$$\text{Capital Value} = \int_0^{\infty} f(t)e^{-rt}\,dt$$

where the function $f(t)$ models the annual rate of a continuous flow of income and r is the annual interest rate, compounded continuously. Suppose that the function $f(t) = 12{,}000e^{0.05t}$ models the continuous flow of income produced by a business at time t. If money can be invested at 10% compounded continuously, find the capital value of the business.

40. **Business: Capital Value** Refer to the previous problem. Suppose that the function $f(t) = 12{,}000e^{0.05t}$ models the continuous flow of income produced by a business at time t. If money can be invested at 14% compounded continuously, find the capital value of the business.

41. **Business: Present Value of a Perpetuity** A perpetuity is an annuity that provides an endless stream of cash payments. That is, a perpetuity is an annuity in which periodic payments begin on a fixed date and continue forever. The present value of a perpetuity is the value right now in time of the total future cash flow and is defined as the improper integral

$$\text{Present Value} = P\int_0^\infty e^{-rt}\,dt$$

where P represent the amount of each of the annual payments in dollars, r the annual compounded continuously interest rate, and t the time in years from the start of payments.

a. Demonstrate, by evaluating the improper integral, that

$$\text{Present Value} = P\int_0^\infty e^{-rt}\,dt = \frac{P}{r}$$

b. Find the present value of a perpetuity that pays \$30,000 annually beginning now. Assume an interest rate of 5% compounded continuously.

42. **Biology: Bioavailability of a Drug** Suppose the concentration of a particular drug that is administered intravenously is given by

$$C_{po}(t) = 500(e^{-0.2t} - e^{-1.0t})$$

The concentration of that drug when administered orally is given by

$$C_{iv}(t) = 450e^{-0.2t}$$

Determine the bioavailability, F, of the drug.

43. **Biology: Bioavailability of a Drug** Suppose the concentration of a particular drug that is administered intravenously is given by

$$C_{po}(t) = 300(e^{-0.6t} - e^{-1.5t})$$

The concentration of that drug when administered orally is given by

$$C_{iv}(t) = 300e^{-0.6t}$$

Determine the bioavailability, F, of the drug.

Using Technology Exercises

44. Use your calculator or Wolfram|Alpha to evaluate the improper integral

$$\int_{-\infty}^{1} \frac{1}{(x-2)^3}\,dx$$

45. Use your calculator or Wolfram|Alpha to evaluate the improper integral $\int_{-\infty}^{\infty} \frac{1}{x}\,dx$. Before entering this integral into your technology, notice that the integrand, $\frac{1}{x}$, is undefined at $x = 0$, yet 0 is in the interval on which the integral is to be evaluated. What might we expect as the outcome?

46. An oil well is expected to produce oil at the rate of $6{,}765e^{-0.13t}$ barrels per year for the next t years.

a. Find the total amount of oil the well is expected to produce in its first 10 years of operation.

b. If the well is operated long into the future, find the total number of barrels of oil it is expected to produce over its lifetime.

47. A company expects income to flow continuously into it at a rate of $36e^{-0.22t}$ millions of dollars per year for the next t years. If the company operates long into the future, how much money can it expect to make?

48. When the curve $f(x) = \frac{1}{x}$ from $x = 1$ to $x = \infty$ is revolved about the x-axis, a solid object, called Gabriel's Horn, is produced. The figure below shows the graph of $f(x) = \frac{1}{x}$ from $x = 1$ to the right.

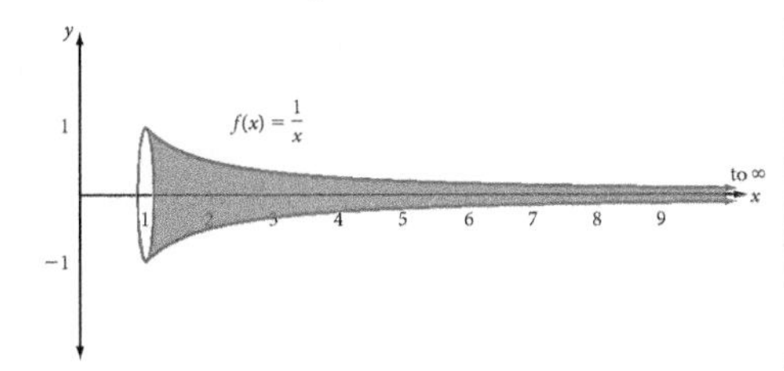

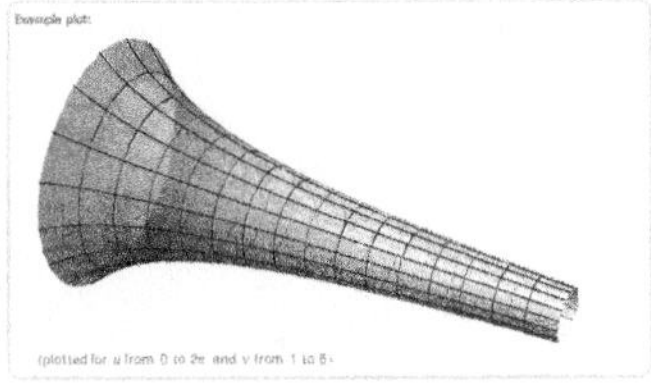

Wolfram Alpha LLC. 2012. Wolfram|Alpha
http://www.wolframalpha.com/
(access October 24, 2012)

a. The improper integral

$$V = \int_1^\infty \pi \left(\frac{1}{x}\right)^2 dx$$

gives the exact volume of this solid. Use Wolfram|Alpha or your calculator to show that the volume of Gabriel's Horn is exactly π cubic units.

b. The integral

$$SA = \int_1^\infty 2\pi \frac{\sqrt{x^4 + 1}}{x^3} dx$$

gives the exact surface area of this solid. Use Wolfram|Alpha to show that this integral does not converge. That is, show that Gabriel's Horn has infinite surface area.

c. Compare parts a and b.

You can see a nice demonstration of the creation of Gabriel's Horn at

www.demonstrations.wolfram.com/GabrielsHorn

49. Use Wolfram|Alpha to evaluate $\int_{-\infty}^{\infty} 2xe^{-x^2}\, dx$. Graph the function to see why the result makes sense.

Getting Ready for the Next Section

For Problems 50 and 51, write an expression for

$$uv - \int v \cdot du$$

Do not evaluate this expression.

50. Given $u = x$, $dv = e^{4x}\,dx$, $du = dx$, and $v = \frac{1}{4}e^{4x}$.

51. Given $u = \ln x$, $dv = x^4\,dx$, $du = \frac{1}{x}\,dx$, and $v = \frac{1}{5}x^5$.

52. Given $u = t$, $dv = e^{-0.08t}\,dt$, $du = dt$, and $v = -12.5e^{-0.08t}$

a. write an expression for $uv - \int v \cdot du$.

b. evaluate the expression you got in part a.

5.6 Integration by Parts

© Kenneth Cheung/iStockPhoto

The following function represents the number, N, of kilowatt hours used by a family each day, where t is the time in hours from 6:00 a.m.

$$N(t) = 8te^{-0.8t}$$

In this section, we will investigate a method for handling integrals of products of functions called integration by parts. This method will help us differentiate parts of functions similar to the one above.

We have seen that sums and differences of functions can be differentiated term by term, but their products and quotients cannot be. That is,

$$\frac{d}{dx}[f(x) \pm g(x)] = \frac{d}{dx}f(x) \pm \frac{d}{dx}g(x), \quad \text{but}$$

$$\frac{d}{dx}[f(x) \cdot g(x)] \neq \frac{d}{dx}f(x) \cdot \frac{d}{dx}g(x), \quad \text{and}$$

$$\frac{d}{dx}\left[\frac{f(x)}{g(x)}\right] \neq \frac{\frac{d}{dx}f(x)}{\frac{d}{dx}g(x)}$$

The situation is similar for integration. That is,

$$\int [f(x) \pm g(x)]\,dx = \int f(x)\,dx \pm \int g(x)\,dx, \quad \text{but}$$

$$\int [f(x) \cdot g(x)]\,dx \neq \int f(x)\,dx \cdot \int g(x)\,dx, \text{ and}$$

$$\int \frac{f(x)}{g(x)}\,dx \neq \frac{\int f(x)\,dx}{\int g(x)\,dx}$$

Integrating Products of Functions Using Integration by Parts

The product rule for derivatives states that for differentiable functions $u = u(x)$ and $v = v(x)$,

$$\frac{d}{dx}[u \cdot v] = u \cdot \frac{dv}{dx} + v \cdot \frac{du}{dx}$$

We can undo the derivative of $u \cdot v$ by antidifferentiating both sides.

Integrating both sides of this equation with respect to x produces

$$\int \frac{d}{dx}[u \cdot v]\,dx = \int \left[u \cdot \frac{dv}{dx} + v \cdot \frac{du}{dx}\right] dx$$

$$u \cdot v = \int \left[u \cdot \frac{dv}{dx}\right] dx + \int \left[v \cdot \frac{du}{dx}\right] dx$$

$$u \cdot v = \int u\,dv + \int v\,du$$

Solving this equation for $\int u\,dv$ produces a formula called **integration by parts.**

Integration by Parts
If $u = u(x)$ and $v = v(x)$ are differentiable functions, then

$$\int u\,dv = uv - \int v\,du$$

This formula can be used for integrating products of two functions, one representing u and the other dv. If the selections of u and dv are made carefully, the integral on the right side, $\int v\,du$, will be easier to integrate than the one on the left side, $\int u\,dv$. We will illustrate this by the following examples.

VIDEO EXAMPLES

SECTION 5.6

Example 1 Evaluate $\int x^4 \ln x\,dx$.

Solution Since this integral involves a product, we will try to solve it using integration by parts. We need to identify one part as u and the other part as dv; u should be something that is easy to differentiate and dv should be something that is easy to integrate. Since $\ln x$ is easy to differentiate but not to integrate, we will try the following substitutions.

Let $u = \ln x$ and $dv = x^4\,dx$, then

$$du = \frac{1}{x}\,dx \quad \text{and} \quad v = \int x^4\,dx = \frac{x^5}{5}$$

Now, substituting into the formula $\int u\,dv = uv - \int v\,du$, we get

$$\int x^4 \ln x\,dx = \int \underbrace{\ln x}_{u} \cdot \underbrace{x^4\,dx}_{dv} = \underbrace{\ln x}_{u} \cdot \underbrace{\frac{x^5}{5}}_{v} - \int \underbrace{\frac{x^5}{5}}_{v} \cdot \underbrace{\frac{1}{x}\,dx}_{du}$$

$$= \frac{1}{5}x^5 \ln x - \frac{1}{5}\int x^4\,dx$$

$$= \frac{1}{5}x^5 \ln x - \frac{1}{5} \cdot \frac{x^5}{5} + C$$

$$= \frac{1}{25}x^5(5\ln x - 1) + C$$

Thus, $\int x^4 \ln x\,dx = \frac{1}{25}x^5(5\ln x - 1) + C$. ■

Example 2 Evaluate $\int xe^{4x}\,dx$.

Solution Since both parts are easy to both integrate and differentiate, we can make our substitution either way. If we let $u = x$ and $dv = e^{4x}\,dx$, then

$$du = dx \quad \text{and} \quad v = \int e^{4x}\,dx = \frac{1}{4}e^{4x}$$

Now, substituting into the formula $\int u\,dv = uv - \int v\,du$, we get

$$\int xe^{4x}\,dx = x \cdot \frac{1}{4}e^{4x} - \int \frac{1}{4}e^{4x}\,dx$$

$$= \frac{1}{4}xe^{4x} - \frac{1}{4}\int e^{4x}\,dx$$

$$= \frac{1}{4}xe^{4x} - \frac{1}{4} \cdot \frac{1}{4}e^{4x} + C$$

$$= \frac{1}{16}e^{4x}(4x - 1) + C$$

Thus, $\int xe^{4x}\,dx = \frac{1}{16}e^{4x}(4x - 1) + C.$ ■

Example 3 Evaluate $\int \ln x\,dx$.

Solution This integral can also be viewed as the product of $\ln x$ and dx and, therefore, evaluated by integration by parts. Since $\ln x$ does not fit any of our integration formulas, we will let it be the part that we differentiate. If we let $u = \ln x$ and $dv = dx$, then

$$du = \frac{1}{x}dx \quad \text{and} \quad v = \int dx = x$$

Now, substituting into the formula $\int u\,dv = uv - \int v\,du$, we get

$$\int \ln x\,dx = \ln x \cdot x - \int x \cdot \frac{1}{x}dx$$

$$= x \ln x - \int 1\,dx$$

$$= x \ln x - x + C$$

Thus, $\int \ln x\,dx = x \ln x - x + C.$ ■

Example 4 A manufacturer estimates that the relationship between the number of months, t, from now and the number, N, (in thousands) of units of a product she can produce per month is given by

$$N(t) = 14te^{-0.08t}$$

Find the function that estimates the total production by the manufacturer if the total production initially is 0.

Solution The total production is the accumulated production, which we can find by integration.

$$\text{Total Production} = \int N(t)\,dt = \int 14te^{-0.08t}\,dt = 14\int te^{-0.08t}\,dt$$

Integrating by parts, if we let $u = t$ and $dv = e^{-0.08t}\,dt$, then

$$du = dt \quad \text{and} \quad v = \frac{e^{-0.08t}}{-0.08} = -12.5e^{-0.08t}$$

Now, substituting into the formula $\int u\,dv = uv - \int v\,du$, we get

$$\int 14te^{-0.08t}\,dt = 14\left[t \cdot (-12.5)e^{-0.08t} - \int (-12.5e^{-0.08t})\,dt\right]$$

$$= 14\left(-12.5te^{-0.08t} + 12.5\int e^{-0.08t}\,dt\right)$$

$$= 14\left(-12.5te^{-0.08t} + 12.5\frac{e^{-0.08t}}{-0.08} + C\right)$$

$$= 14(-12.5te^{-0.08t} - 156.25e^{-0.08t} + C)$$

$$= -175e^{-0.08t}(t + 12.5) + C$$

Initially, at $t = 0$, the total production is 0. Therefore,

$$\text{Total Production} = 0 = -175e^{-0.08 \cdot 0}(0 + 12.5) + C$$

$$0 = -175e^{0}(12.5) + C$$

$$0 = -175 \cdot 1(12.5) + C$$

$$0 = -2{,}187.5 + C$$

$$2{,}187.5 = C$$

Consequently, Total Production $= -175e^{-0.08t}(t + 12.5) + 2{,}187.5$, which is the function we were asked to find. ■

Using Technology 5.6

You can use Wolfram|Alpha to solve problems involving integration by parts.

Example 5 Use integration by parts to evaluate $\int x^2 \ln x \, dx$.

Solution This integration requires parts. Doing this by hand, we would let $u = \ln x$ and $dv = x^2\,dx$. Wolfram|Alpha can do the computations and evaluation for us.

Go online to www.wolframalpha.com. Into the entry field enter

integrate x^2ln(x)

Wolfram|Alpha returns

Indefinite integral:

$$\int x^2 \log(x)\, dx = \frac{1}{9} x^3 (3 \log(x) - 1) + \text{constant}$$

■

Getting Ready for Class

After reading through the preceding section, respond in your own words and in complete sentences.

A. The integral of a sum or difference of two or more functions is the sum or difference of the integrals of the individual functions. That is,

$$\int[f(x) \pm g(x)]\,dx = \int f(x)\,dx \pm \int g(x)\,dx$$

Is the same true of integrals of products and quotients of two or more functions? That is, is

$$\int[f(x) \cdot g(x)]\,dx = \int f(x)\,dx \cdot \int g(x)\,dx \text{ and } \int \frac{f(x)}{g(x)}\,dx = \frac{\int f(x)\,dx}{\int g(x)\,dx}\ ?$$

B. The integrals of products of two or more functions must be integrated using which method?

C. When using the method of integration by parts, you substitute expressions for u and dv, then use the integration by parts formula. Specify that formula.

Problem Set 5.6

Skills Practice

Evaluate each integral. Integration by parts may not be necessary for some problems. (Hint: Sometimes you may have to use integration by parts more than once.)

1. $\int xe^x\,dx$
2. $\int xe^{7x}\,dx$
3. $\int xe^{-8x}\,dx$
4. $\int x^2e^{-x}\,dx$
5. $\int x^3e^{3x}\,dx$
6. $\int \ln(4x)\,dx$
7. $\int \ln(x^3)\,dx$
8. $\int 5x^3 \ln x\,dx$
9. $\int (\ln x)^3\,dx$
10. $\int \frac{\ln x}{x^2}\,dx$
11. $\int \frac{1}{x}\ln x\,dx$
12. $\int x\sqrt{x-1}\,dx$
13. $\int x(\ln x)^3\,dx$
14. $\int xe^{x^2}\,dx$
15. $\int \ln(7x-4)\,dx$
16. $\int \frac{x}{e^x}\,dx$
17. $\int x(1-x)^{3/2}\,dx$
18. $\int x(3x+4)^2\,dx$
19. $\int e^{x+3}(2x+1)\,dx$
20. $\int x\ln(x^2)\,dx$
21. $\int (2x+1)^2\ln(2x+1)\,dx$

Check Your Understanding

In Problem 22-27, an integral is given that can be evaluated using integration by parts. Along with the integral are suggested expressions for u, du, dv, and v. Write an expression for $uv - \int v\,du$. Do not evaluate the integral in the expression.

22. For $\int 5xe^{3x}\,dx$, let $u = 5x$, $dv = e^{3x}\,dx$, $du = 5\,dx$ and $v = \frac{1}{3}e^{3x}$.
23. For $\int 48xe^{0.16x}\,dx$, let $u = 48x$, $dv = e^{0.16x}\,dx$, $du = 48\,dx$ and $v = 6.25e^{0.16x}$.
24. For $\int x^2e^{-2x}\,dx$, let $u = x^2$, $dv = e^{-2x}\,dx$, $du = 2x\,dx$ and $v = -\frac{1}{2}e^{-2x}$.
25. For $\int x^3e^{-6x}\,dx$, let $u = x^3$, $dv = e^{-6x}\,dx$, $du = 3x^{-4}\,dx$ and $v = -\frac{1}{6}e^{-6x}$.
26. For $\int 5x^4\ln x\,dx$, let $u = \ln x$, $dv = 5x^4\,dx$, $du = \frac{1}{x}\,dx$ and $v = x^5$.
27. For $\int 6x^5\ln 6x\,dx$, let $u = \ln 6x$, $dv = 6x^5\,dx$, $du = \frac{1}{x}\,dx$ and $v = x^6$.

In Problems 28-31, an integral is given that can be evaluated using integration by parts. Write expressions for u, du, dv, and v, as well as $uv - \int v\,du$. Do not evaluate the integral in the expression.

28. $\int x^4 e^x$

29. $\int x^{-1} e^{-2x}\,dx$

30. $\int 4x \ln x\,dx$

31. $\int 8x^7 \ln 8x\,dx$

Modeling Practice

32. Business: Rate of Sales A manufacturer has determined that t weeks after the end of an advertising campaign, the rate of change of sales (in 100,000 units per week) is given by the function

$$f(t) = te^{-0.2t}$$

At the end of the advertising campaign, sales were 105,000 units per week. Find the function that describes the total number of sales after t weeks after an advertising campaign.

33. Business: Marginal Cost The marginal cost for a manufacturer for the production of x units of a commodity is given by the function

$$C'(x) = x \ln(x + 1)$$

Find the function that describes the total change in the cost of production in terms of x if the total cost was 0 initially.

34. Manufacturing: Demand for a Product A manufacturer believes that the demand for one of its products over the next 5 years is given by the function

$$N(t) = 420(15 + te^{-0.1t})$$

where t is the number of years from now. Find the function that describes the total number of the product that is demanded after t years when initially the number in demand is 6,300.

35. Business: Revenue The daily revenue R, in dollars, for a company for one of its products is given by the function

$$R(t) = 12{,}500 + 120.5t^2e^{-t/2} \quad 0 \le t \le 365$$

where t is the number of days from now. Find the function that describes the total revenue realized by the company t days from now if the total revenue was \$12,500 initially.

36. Medicine: Drug Assimilation The amount A (in milligrams) of a drug assimilated into an adult male's bloodstream t minutes after ingesting the drug is given by the function

$$A(t) = 4te^{-0.8t}$$

Find the function that describes the total amount of the drug assimilated into the male's bloodstream t minutes after ingestion. Of course, initially, nothing was assimilated.

© Kenneth Cheung/iStockPhoto

37. Economics: Energy Usage A city has determined that the number, N, of kilowatt hours used by a family each day is given by the function

$$N(t) = 8te^{-0.8t}$$

where t is the time in hours from 6:00 a.m. Find the function that describes the total usage by the family in terms of t.

Using Technology Exercises

38. Find $\int x^2e^x\,dx$ using both pencil-and-paper and Wolfram|Alpha. Compare the results.

39. Use Wolfram|Alpha to find $\int x\ln(6x)\,dx$. Before you use the technology, what do you think is the appropriate substitution?

40. Use Wolfram|Alpha to find $\int (\ln x)^2\,dx$. Now, evaluate the same integral but with exponents of 3, 4, 5, and 6. What observations can you make about the integrals? Can you identify any patterns?

41. Use Wolfram|Alpha to evaluate each of the four integrals:

$$\int \frac{\ln x}{x^2}\,dx, \quad \int \frac{\ln x}{x^3}\,dx, \quad \int \frac{\ln x}{x^4}\,dx, \quad \text{and} \quad \int \frac{\ln x}{x^5}\,dx$$

Use the information you get from your observation of these integrals to guess the integral $\int \frac{\ln x}{x^6}\,dx$. Make a conjecture about $\int \frac{\ln x}{x^n}\,dx$. Check your conjecture using Wolfram|Alpha.

42. A well is expected to produce oil at the rate of $14{,}150e^{-0.85t}$ barrels per year t years from now. Write a function that gives the total number of barrels of oil produced by this well at the end of t years. Find the total amount of oil this well is expected to produce over the next 5 years.

Getting Ready for the Next Section

For Problems 43-46, evaluate the integral.

43. $\int (-x^2 + 10x - 16)\,dx$

44. $\int (-2x^2 + 14x - 20)\,dx$

45. $\int [(-x^2 + 4x + 63) - (x^2 + 12x + 39)]\,dx$

46. $\int_{-6}^{2} (-2x^2 - 8x + 24)\,dx$

For Problems 47 and 48, find the values of x for which the two curves intersect.

47. $f(x) = x^3 - 10x + 25$ and $g(x) = 6x + 25$.

48. $f(x) = -x^2 + 4x + 63$ and $g(x) = x^2 + 12x + 39$.

Chapter 5 Summary

EXAMPLES

Antidifferentiation [5.1]

Suppose that $F(x)$ and $f(x)$ are two functions such that $f(x)$ is the derivative of $F(x)$; that is, $F'(x) = f(x)$. Then $F(x)$ is an **antiderivative** or **integral** of $f(x)$.

Indefinite Integral [5.1]

Suppose that $F(x)$ is any one of the infinitely many members of the family of antiderivatives of $f(x)$. Then,

$$\int f(x)\,dx = F(x) + C$$

where C is an arbitrary constant. The function $F(x) + C$ is called the **antiderivative** or **indefinite integral** of $f(x)$. The number C is called the **constant of integration**, the function $f(x)$ is called the **integrand**, and dx is called the **differential**.

Basic Integration Formulas [5.1]

1. $\int 6\,dx = 6x + C$

2. $\int x^5\,dx = \frac{x^{5+1}}{5+1} + C = \frac{x^6}{6} + C$

3. $\int 7x^5\,dx = 7\int x^5\,dx = \frac{7x^6}{6} + C$

4. $\int (x^3 - 4x)\,dx =$ $\int x^3\,dx - \int 4x\,dx =$ $\frac{x^4}{4} - 2x^2 + C$

1. If k is any constant, then $\int k\,dx = kx + C$. To integrate a constant function, affix the independent variable to the constant and add an arbitrary constant.
2. If r is any real number other than -1, then $\int x^r\,dx = \frac{x^{r+1}}{r+1} + C$. To integrate a power function, increase the exponent by 1, divide the resulting power by the new exponent, and add an arbitrary constant.
3. If k is any constant, then $\int kf(x)dx = k\int f(x)\,dx$. To integrate a constant times a function, integrate the function, then multiply that result by the constant, and add an arbitrary constant.
4. For any two functions $f(x)$ and $g(x)$, $\int [f(x) \pm g(x)]\,dx = \int f(x)\,dx \pm \int g(x)\,dx$. To integrate a sum or difference of two functions, integrate each function individually, then add or subtract those results.
5. For $f(x) = e^x$, $\int e^x\,dx = e^x + C$. The antiderivative, or integral, of the exponential function $f(x) = e^x$ is the function itself.
6. For $f(x) = \frac{1}{x}$, $\int \frac{1}{x}\,dx = \ln|x| + C$.

Integration by Substitution [5.2]

The method of integration called **integration by substitution** involves substituting a variable (any variable other than the original can be used; we often use u) for some part of the integrand and then expressing the rest of the integrand in terms of that variable. A beneficial substitution often involves a choice of u for which $\frac{du}{dx}$ is a constant multiple of one of the factors in the integrand. We evaluate that integral, and then resubstitute to get to our destination, a solution in x.

5. To evaluate $\int 6x(3x^2+5)^3\,dx$, let u = (base of the power).

$$u = 3x^2 + 5$$

Then, $du = 6x\,dx$. This transforms the original integral to the new integral

$$\int u^3 \cdot du$$

Then,

$$\int u^3 \cdot du = \frac{u^4}{4} + C$$

Resubstituting gives

$$\int 6x(3x^2+5)^3\,dx = \frac{(3x^2+5)^4}{4} + C$$

Integrating the Exponential Function [5.2]

If $f(x) = e^{ax}$, then

$$\int e^{ax}\,dx = \frac{e^{ax}}{a} + C$$

The Fundamental Theorem of Calculus [5.3]

If $f(x)$ is a continuous function on an interval $[a, b]$, where $a < b$, and if $F(x)$ is an antiderivative of $f(x)$, then the **definite integral** of $f(x)$ from a to b is

$$\int_a^b f(x)\,dx = F(x)\Big|_a^b = F(b) - F(a)$$

and represents the total amount accumulated by $F(x)$ as x increases from a to b. The number a is called the **lower limit of integration** and the number b the **upper limit of integration.**

The Definite Integral [5.3]

The definite integral $\int_a^b f(x)\,dx$ is an *accumulator* and represents a *real number,* a constant that records the total amount accumulated by a particular member of a family of antiderivatives.

6. $\int_0^5 (4x-3)\,dx$

$$\begin{aligned}
&= (2x^2 - 3x)\Big|_0^5 \\
&= (2\cdot 5^2 - 3\cdot 5) - (2\cdot 0^2 - 3\cdot 0) \\
&= (50 - 15) - (0 - 0) \\
&= 35 - 0 \\
&= 35
\end{aligned}$$

Properties of the Definite Integral [5.3]

The definite integral has the following properties.

1. $\int_a^b kf(x)\,dx = k\int_a^b f(x)\,dx$, where k is a constant.

2. $\int_a^a f(x)\,dx = 0$ No accumulation without change.

3. $\int_a^b f(x)\,dx = -\int_b^a f(x)\,dx$

4. $\int_a^b [f(x) \pm g(x)]\,dx = \int_a^b f(x)\,dx \pm \int_a^b g(x)\,dx$

5. $\int_a^b f(x)\,dx = \int_a^c f(x)\,dx + \int_c^b f(x)\,dx$, where $a \leq c \leq b$.

Approximating Area [5.4]

To find the area of planar regions that lie in a rectangular coordinate system, above the input axis, and below some specified nonnegative function, we can divide it into rectangular regions. More regions gives us a better approximation for area. A four-rectangle approximation to the total region is given by

$$S_4 = [f(x_0) + f(x_1) + f(x_2) + f(x_3)]\Delta x$$

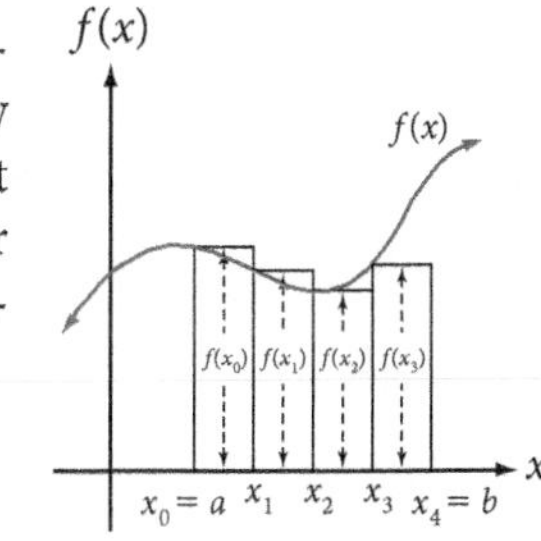

Riemann Sum [5.4]

If we subdivide the interval $[a, b]$ into n subintervals of equal length,

$$\Delta x = \frac{b-a}{n}$$

and construct rectangles using the functional value at c_i in each subinterval, $[x_i, x_{(i+1)}]$, as the height of the rectangle, the approximating area is

$$S_n = [f(c_0) + f(c_1) + \cdots + f(c_{n-1})]\Delta x$$

This sum is called a **Riemann sum**.

The Definite Integral and Area [5.4]

If $f(x)$ is a continuous, nonnegative function on the interval $[a, b]$, the area A (in square units) bounded by the function $f(x)$, the x-axis, and the vertical lines $x = a$ and $x = b$ is precisely equal to the value of the definite integral of $f(x)$, from $x = a$ to $x = b$. That is,

$$A = \int_a^b f(x)\,dx$$

7. To find the area of the region bounded above by the curve $f(x) = x^2 + 2$, below by the x-axis, and the vertical lines $x = 1$ and $x = 4$, we evaluate

$$A = \int_1^4 (x^2 + 2)\,dx$$
$$= \left(\frac{x^3}{3} + 2x\right)\Bigg|_1^4$$
$$= \left(\frac{4^3}{3} + 2 \cdot 4\right) - \left(\frac{1^3}{3} + 2 \cdot 1\right)$$
$$= \left(\frac{64}{3} + 8\right) - \left(\frac{1}{3} + 2\right)$$
$$= 27$$

The area of the region is 27 square units.

Improper Integrals [5.5]

1. $\displaystyle\int_a^{\infty} f(x)\,dx = \lim_{b\to\infty} \int_a^b f(x)\,dx$
2. $\displaystyle\int_{-\infty}^{b} f(x)\,dx = \lim_{a\to-\infty} \int_a^b f(x)\,dx$
3. $\displaystyle\int_{-\infty}^{\infty} f(x)\,dx = \lim_{a\to-\infty} \int_a^c f(x)\,dx + \lim_{b\to\infty} \int_c^b f(x)\,dx$, where $a \le c \le b$

If these limits exist, the improper integral is said to **converge** to the computed value. Otherwise, it **diverges**.

8. Evaluate $\int_0^{\infty} 4e^{-2x}\,dx$

The integral is improper because the upper limit of integration is infinity.

$$\int_0^{\infty} 4e^{-2x}\,dx = \lim_{b\to\infty} \int_0^b 4e^{-2x}\,dx$$
$$= \lim_{b\to\infty} [-2e^{-2x}]\Big|_0^b$$
$$= \lim_{b\to\infty} [(-2e^{-2b}) - (-2e^{-2(0)})]$$
$$= \lim_{b\to\infty} \left[\frac{-2}{e^{2b}} + 2e^0\right]$$
$$= 0 + 2$$
$$= 2$$

Integration by Parts [5.6]

If $u = u(x)$ and $v = v(x)$ are differentiable functions, then

$$\int u\,dv = uv - \int v\,du$$

9. Evaluate $\int xe^{4x}\,dx$. Because this integral involves a product, x times e^{4x}, integration by parts may be a good approach. Let $u = x$ and $dv = e^{4x}\,dx$. Then $du = dx$ and $v = \frac{1}{4}e^{4x}$. Now, substitute into the integration by parts formula $uv - \int v\,du$:

$$\int xe^{4x}\,dx =$$
$$x \cdot \frac{1}{4}e^{4x} - \int \frac{1}{4}e^{4x}\,dx =$$
$$\frac{1}{4}xe^{4x} - \frac{1}{16}e^{4x} + C$$

Chapter 5 Test

1. **Business: Quality Control** A manufacturer's quality control engineers believe the efficiency E (in percent) of its pumps changes as the temperature (in degrees Fahrenheit) of the surrounding air changes. The rate at which the pump's efficiency changes is described by the function $E'(T)$, where T represents the temperature of the air surrounding the pump. Interpret both $E'(10) = -3.2$ and its integral $E(10) = 64$. [5.1]

For Problems 2-5, evaluate the indefinite integral. [5.1]

2. $\int \frac{16}{x}\,dx$

3. $\int (3x^2 - 5x + 2)\,dx$

4. $\int 4\sqrt[5]{x^3}\,dx$

5. $\int 40e^{-0.02x}\,dx$

6. **Business: Advertising and Sales Volume** The management of a chain of retail stores believes that t days after the end of an advertising campaign, the rate at which the volume V (in dollars) of sales is changing is approximated by the function $V'(t) = -12{,}480e^{-0.24t}$. On the day the advertising campaign ends, the sales volume is \$104,000. Find and interpret both $V'(5)$ and $V(5)$. [5.1]

For Problems 7-16, evaluate the indefinite integral. [5.1, 5.2, 5.6]

7. $\int x(2x^2 - 5)^{2/3}\,dx$

8. $\int \frac{x}{x^2 + 1}\,dx$

9. $\int 2(x + 2)\sqrt{x^2 + 4x - 1}\,dx$

10. $\int (5x^4 + 2)e^{x^5 + 2x}\,dx$

11. $\int \frac{e^x}{2e^x + 3}\,dx$

12. $\int \frac{5\ln x}{x}\,dx$

13. $\int \frac{x}{x - 1}\,dx$

14. $\int \frac{3x}{x - 4}\,dx$

15. $\int 52e^{-0.13x}\,dx$

16. $\int x^4 e^{x^5}\,dx$

17. **Psychology: Spread of News** If $H(t)$ represents the rate at which a population hears news between days a and b, interpret

$$\int_a^b H(t)\,dt$$

[5.3]

For Problems 18-25, evaluate the definite integral. [5.2, 5.3, 5.5, 5.6]

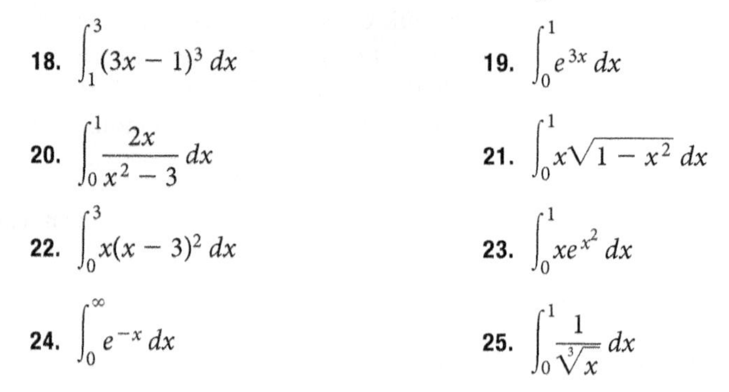

18. $\int_1^3 (3x-1)^3\, dx$

19. $\int_0^1 e^{3x}\, dx$

20. $\int_0^1 \frac{2x}{x^2-3}\, dx$

21. $\int_0^1 x\sqrt{1-x^2}\, dx$

22. $\int_0^3 x(x-3)^2\, dx$

23. $\int_0^1 xe^{x^2}\, dx$

24. $\int_0^\infty e^{-x}\, dx$

25. $\int_0^1 \frac{1}{\sqrt[3]{x}}\, dx$

26. Business: Investment in Technology A company's investment in new technology is expected to save it money at the rate of

$$S(t) = 5{,}000(1 - e^{-0.04t})$$

thousands of dollars each year after the technology is purchased. What is the accumulated savings this company would realize for the first 3 years after purchase? [5.3]

Applications of Integration

6

Chapter Outline

© Jean Gill/iStockPhoto

Make money as you sleep. Some types of money-making vehicles are called income streams because once they are created, they produce an essentially continuous flow of income. Some examples are the creation and selling of eBooks and eCourses, membership programs, iPhone and iPad Apps, and the packaging and selling of podcasts.

It is sometimes possible, by collecting and analyzing data, to construct functions that model the rate at which passive income flows into a business. Once such functions are developed, algebra and calculus can help us to answer questions about the streams of money.

Algebra How much money will be in a business account 1 year after the start of a money stream?

Calculus If $100,000 flows continuously into an account that pays 5% interest compounded continuously, how much money will have accumulated at the end of 5 years?

In order to answer the above question, a company would calculate the present value and future value of the income stream, using the following integrals:

$$\text{Present value} = \int_0^T f(t)e^{-rt}\,dt$$

$$\text{Future value} = e^{rT}\int_0^T f(t)e^{-rt}\,dt$$

where $f(t)$ is the rate of flow of money into an account that pays an annual interest rate r compounded continuously during some time interval $0 \le t \le T$. In this section, we will work with a variety of applications, including those that use these integrals.

Note When you see this icon next to an example or problem in this chapter, you will know that we are using the topics in this chapter to model situations in the world around us.

Success Skills

Never mistake activity for achievement.

— John Wooden, legendary UCLA basketball coach

You may think that this John Wooden quote has to do with being productive and efficient, or using your time wisely, but it is really about being honest with yourself. I have had students come to me after failing a test saying, "I can't understand why I got such a low grade after I put so much time in studying." One student even had help from a tutor and felt she understood everything that we covered. After asking her a few questions, it became clear that she spent all her time studying with a tutor and the tutor was doing most of the work. The tutor can work all the homework problems, but the student cannot. She has mistaken activity for achievement.

Can you think of situations in your life when you are mistaking activity for achievement?

How would you describe someone who is mistaking activity for achievement in the way they study for their math class?

Which of the following best describes the idea behind the John Wooden quote?

- Always be efficient.
- Don't kid yourself.
- Take responsibility for your own success.
- Study with purpose.

Area of Regions in the Plane

6.1

© Susan Wood/ iStockPhoto

In this section we will consider areas of planar regions that are not restricted to areas under a curve as we saw in Section 5.4, but are areas of planar regions between two curves as in Figure 2. As we noted in Section 5.4, many applied problems can be solved by relating their characteristics to area. In this section we will investigate how the area between two curves relates to problems such as revenue flow, advertising, birth rates of insects, and the savings realized from the purchase of new equipment. In subsequent sections, we will relate area to problems involving consumer's and producer's surplus, annuities and money streams, and probability.

The Area Bounded by Two Curves

We know that when $f(x)$ is nonnegative (not below the x-axis) on $[a, b]$, the definite integral $\int_a^b f(x)\,dx$ can be used to determine the area of a planar region that is bounded by the curve $f(x)$, the x-axis, and the two vertical lines $x = a$ and $x = b$. In Section 5.4, we developed this relationship using rectangles. Figure 1 shows such a planar region with a representative rectangle. The height of the rectangle is $f(x)$ and the width is dx.

We can extend this technique to determine the area of a planar region bounded by a curve $f(x)$ and another curve $g(x)$. Figure 2 shows such a region. Look back for a moment at Figure 1 and imagine $f(x)$ as the upper curve and the x-axis as the lower curve. The x-axis is a horizontal line with constant output 0. If we let $g(x)$ represent an output value, the x-axis can be described by $g(x) = 0$. Then the height of the representative rectangle in Figure 1 can be expressed as

Figure 1

$$f(x) \text{ or } f(x) - 0 \text{ or } f(x) - g(x)$$

The area of the planar region is then expressed as

$$A = \int_a^b [f(x) - g(x)]\,dx$$

The height of the representative strip is then the (height of the top of the rectangle, which is $f(x)$) – (height of the bottom of the rectangle, which is $g(x)$).

But now, instead of restricting $g(x)$ to be the x-axis, if we let $g(x)$ represent a curve that lies below the curve $f(x)$, the area of the more general planar region between the two curves (Figure 2) can also be expressed as

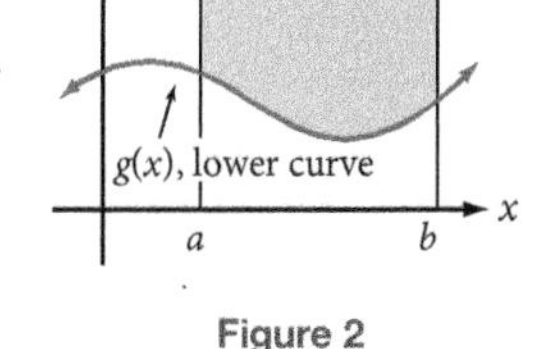

Figure 2

$$A = \int_a^b [f(x) - g(x)]\,dx$$

where $f(x)$ is the upper curve and $g(x)$ is the lower curve. Figure 3 illustrates this idea using a representative rectangle.

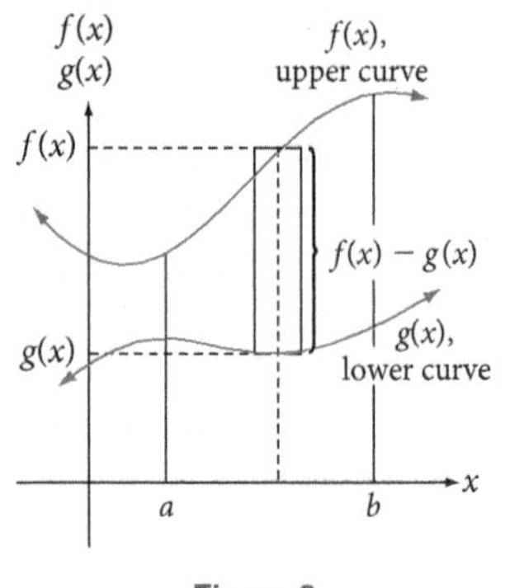

Figure 3

> **Area Bounded by Two Curves**
> If $f(x)$ and $g(x)$ are two continuous functions for which $f(x) \geq g(x)$ on $[a, b]$, then the area bounded by the two functions and the vertical lines $x = a$ and $x = b$ on $[a, b]$ is
>
> $$A = \int_a^b [f(x) - g(x)]\, dx$$

VIDEO EXAMPLES

SECTION 6.1

Example 1 Find the area bounded by the line $f(x) = 2x + 5$, the curve $g(x) = x^2 - 8x + 21$, and the two vertical lines $x = 2$ and $x = 5$.

Solution Figure 4 shows the region bounded by these lines and this curve. The graph shows that on the interval $[2, 5]$, $f(x) \geq g(x)$. That is, the graph of $f(x)$ lies above the graph of $g(x)$, so that $f(x)$ will be the upper curve and $g(x)$ the lower curve.

$$\begin{aligned}
A &= \int_a^b [f(x) - g(x)]\, dx \\
&= \int_2^5 [(2x + 5) - (x^2 - 8x + 21)]\, dx \\
&= \int_2^5 [2x + 5 - x^2 + 8x - 21]\, dx \\
&= \int_2^5 [-x^2 + 10x - 16]\, dx \\
&= \left[\frac{-x^3}{3} + 5x^2 - 16x\right]\Bigg|_2^5 \\
&= \left[\frac{-(5)^3}{3} + 5(5)^2 - 16(5)\right] - \left[\frac{-(2)^3}{3} + 5(2)^2 - 16(2)\right] \\
&= \frac{10}{3} - \left(\frac{-44}{3}\right) = \frac{10}{3} + \frac{44}{3} = \frac{54}{3} = 18
\end{aligned}$$

Figure 4

> **Interpretation**
> The area of the region bounded by the line $f(x) = 2x + 5$, the curve $g(x) = x^2 - 8x + 21$, and the two vertical lines $x = 2$ and $x = 5$, is 18 square units.

Example 2 Find the area of the region bounded by the two curves $f(x) = x^2 - 6x + 10$ and $g(x) = -x^2 + 8x - 10$, and the two vertical lines $x = 3$ and $x = 4$.

Solution Figure 5 shows the region bounded by these curves. The graph shows that on the interval $[3, 4]$, $g(x) \geq f(x)$. That is, the graph of $g(x)$ lies above the graph of $f(x)$, so that $g(x)$ will be the upper curve and $f(x)$ the lower curve.

$$A = \int_a^b [g(x) - f(x)]\,dx$$

$$= \int_3^4 [(-x^2 + 8x - 10) - (x^2 - 6x + 10)]\,dx$$

$$= \int_3^4 [-x^2 + 8x - 10 - x^2 + 6x - 10)]\,dx$$

$$= \int_3^4 [-2x^2 + 14x - 20]\,dx$$

$$= \left[\frac{-2x^3}{3} + 7x^2 - 20x\right]\Big|_3^4$$

$$= \left[\frac{-2(4)^3}{3} + 7(4)^2 - 20(4)\right] - \left[\frac{-2(3)^3}{3} + 7(3)^2 - 20(3)\right]$$

$$= \frac{-32}{3} - (-15) = \frac{13}{3}$$

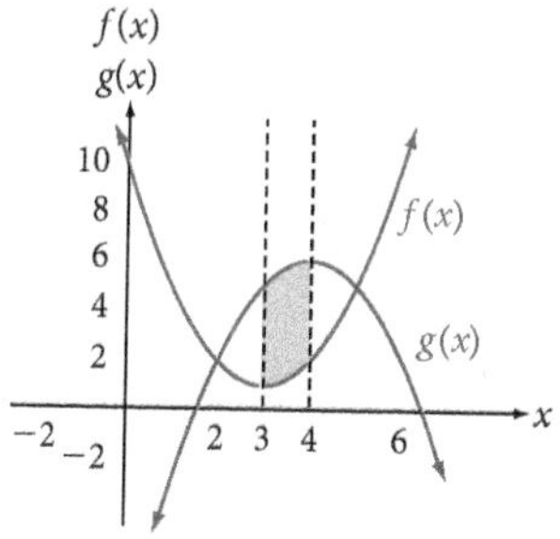

Figure 5

Interpretation

The area of the region bounded by the two curves $f(x) = x^2 - 6x + 10$ and $g(x) = -x^2 + 8x - 10$, and the two vertical lines $x = 3$ and $x = 4$, is $\frac{13}{3}$ square units.

In the next example, we will see a situation where two curves bound a region without any vertical lines. They intersect in such a way that they naturally enclose a region. With this situation, we will need to find the limits of integration by finding the x-values of the points of intersection between the two curves.

Example 3 Find the area of the region that is completely bounded by the two curves $f(x) = -x^2 + 4x + 63$ and $g(x) = x^2 + 12x + 39$.

Solution Figure 6 shows the region bounded by these curves. To find the limits of integration, we must find the points at which these two curves intersect. To do so, we set them equal to each other and solve for x.

$$f(x) = g(x)$$

$$-x^2 + 4x + 63 = x^2 + 12x + 39$$

$$0 = 2x^2 + 8x - 24$$

$$0 = 2(x^2 + 4x - 12)$$

$$0 = 2(x + 6)(x - 2)$$

$$x = -6, 2$$

The two curves intersect at $x = -6$ and at $x = 2$. These x-values are our limits of integration.

The graph shows that on the interval $[-6, 2]$, $f(x) \geq g(x)$. That is, the graph of $f(x)$ lies above the graph of $g(x)$, so that $f(x)$ will be the upper curve and $g(x)$ the lower curve.

$$A = \int_a^b [f(x) - g(x)]\, dx$$

$$= \int_{-6}^{2} [(-x^2 + 4x + 63) - (x^2 + 12x + 39)]\, dx$$

$$= \int_{-6}^{2} [-x^2 + 4x + 63 - x^2 - 12x - 39]\, dx$$

$$= \int_{-6}^{2} [-2x^2 - 8x + 24]\, dx$$

$$= \left[\frac{-2x^3}{3} - 4x^2 + 24x\right]\Bigg|_{-6}^{2}$$

$$= \left[\frac{-2(2)^3}{3} - 4(2)^2 + 24(2)\right] - \left[\frac{-2(-6)^3}{3} - 4(-6)^2 + 24(-6)\right]$$

$$= \frac{80}{3} - (-144) = \frac{512}{3}$$

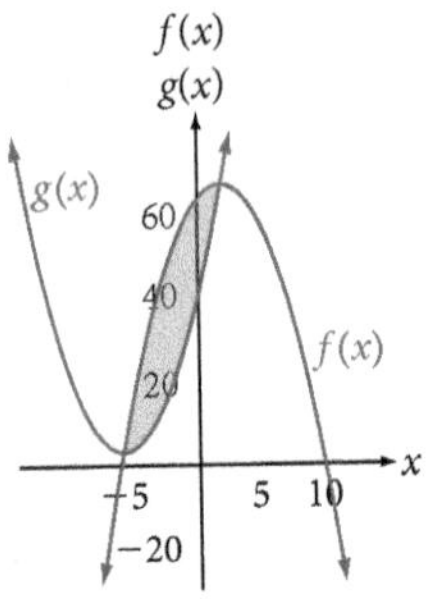

Figure 6

Interpretation

The area of the region bounded by the two curves $f(x) = -x^2 + 4x + 63$ and $g(x) = x^2 + 12x + 39$ is $\frac{512}{3}$ square units.

Example 4 Find the area of the region that is completely bounded by the two curves $f(x) = x^3 - 10x + 25$ and $g(x) = 6x + 25$.

Solution Figure 7 shows the region bounded by these curves. To find the limits of integration, we must find the points at which these two curves intersect. To do so, we set them equal to each other and solve for x.

$$f(x) = g(x)$$

$$x^3 - 10x + 25 = 6x + 25$$

$$x^3 - 16x = 0$$

$$x(x^2 - 16) = 0$$

$$x(x + 4)(x - 4) = 0$$

$$x = 0, -4, 4$$

The two curves intersect at $x = -4$, $x = 0$, and $x = 4$. These x-values are our limits of integration.

The graph shows that on the interval $[-4, 0]$, $f(x) \geq g(x)$. That is, the graph of $f(x)$ lies above the graph of $g(x)$, so that $f(x)$ will be the upper curve and $g(x)$ the lower curve. But then on the interval $[0, 4]$, $g(x) \geq f(x)$. That is, the graph of $g(x)$ lies above the graph of $f(x)$, so that $g(x)$ is the upper curve and $f(x)$ is the lower curve.

To find the area bounded by these curves, we find the area of each region, and add them together.

$$A = \int_{-4}^{0} [f(x) - g(x)]\, dx + \int_{0}^{4} [g(x) - f(x)]\, dx$$

$$= \int_{-4}^{0} [(x^3 - 10x + 25) - (6x + 25)]\, dx$$

$$+ \int_{0}^{4} [(6x + 25) - (x^3 - 10x + 25)]\, dx$$

$$= \int_{-4}^{0} [x^3 - 16x]\, dx + \int_{0}^{4} [16x - x^3]\, dx$$

$$= \left[\frac{x^4}{4} - 8x^2\right]\Bigg|_{-4}^{0} + \left[8x^2 - \frac{x^4}{4}\right]\Bigg|_{0}^{4}$$

$$= \left[\left(\frac{0^4}{4} - 8(0)^2\right) - \left(\frac{(-4)^4}{4} - 8(-4)^2\right)\right] + \left[\left(8(4)^2 - \frac{(4)^4}{4}\right) - \left(8(0)^2 - \frac{(0)^4}{4}\right)\right]$$

$$= [(0) - (-64)] + [(64) - (0)]$$

$$= 64 + 64$$

$$= 128$$

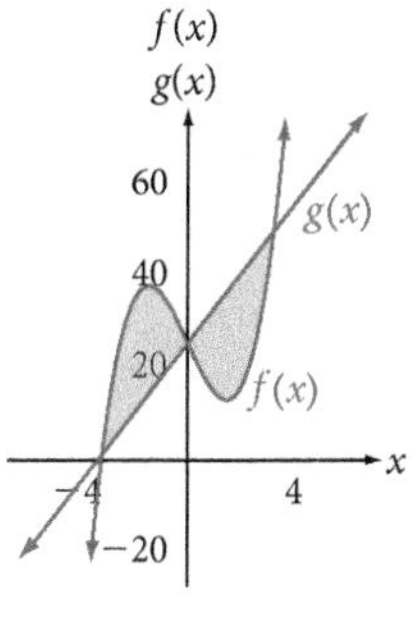

Figure 7

Interpretation

The area of the region that is completely bounded by the two curves $f(x) = x^3 - 10x + 25$ and $g(x) = 6x + 25$ is 128 square units.

Example 5 Over the next 10 years, a company projects its continuous flow of revenue to be $R(x) = 80e^{0.10x}$ thousands of dollars and its costs to be $C(x) = 0.9x^2 + 70$ thousands of dollars, where x represents the number of years from now. Approximate, to the nearest dollar, the profit this company can expect over the next 10 years.

Solution Figure 8 shows the region bounded by these curves and that on the interval [0, 10], $R(x) \geq C(x)$. Profit, we know, is equal to revenue minus cost. That is, $P(x) = R(x) - C(x)$. For us, then,

$$P(x) = 80e^{0.10x} - (0.9x^2 + 70)$$

The profit that can be expected over the next 10 years is then

$$\text{Profit} = \int_{0}^{10} P(x)\, dx = \int_{0}^{10} [R(x) - C(x)]\, dx$$

$$= \int_{0}^{10} [(80e^{0.10x}) - (0.9x^2 + 70)]\, dx$$

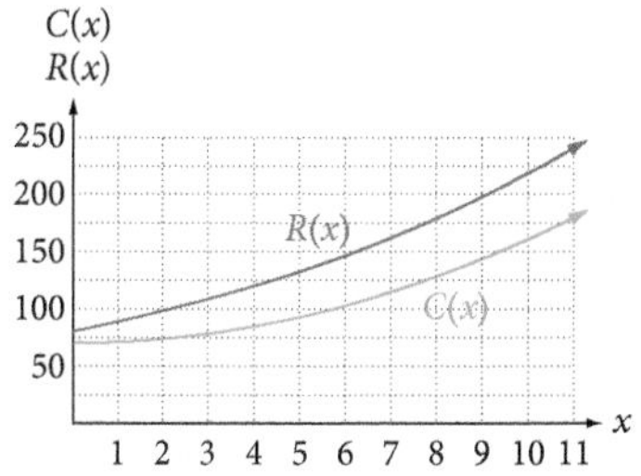

$$= \int_{0}^{10} [80e^{0.10x} - 0.9x^2 - 70)]\, dx$$

Figure 8

$$= [800e^{0.10x} - 0.3x^3 - 70x)] \Big|_0^{10}$$

$$= [800e^{0.10(10)} - 0.3(10)^3 - 70(10)] - [800e^{0.10(0)} - 0.3(0)^3 - 70(0)]$$

$$\approx 1{,}174.625 - 800$$

$$\approx 374.625 \text{ thousands of dollars}$$

Interpretation
If circumstances for this company remain as they are projected to be, the company can expect a profit of about $374,625 dollars over the next 10 years.

■

Using Technology 6.1

You can use Wolfram|Alpha to help you find the area bounded by two curves.

Example 6 In Example 3 we found the area of the region that is completely bounded by the two curves

$$f(x) = -x^2 + 4x + 63 \quad \text{and} \quad g(x) = x^2 + 12x + 39$$

Use Wolfram|Alpha to find this area.

Solution Go to www/wolframalpha.com and in the entry field type

find the area between the -x^2 + 4x + 63 and x^2 + 12x + 39

Wolfram|Alpha displays your entry, the graph, and the area between the two curves.

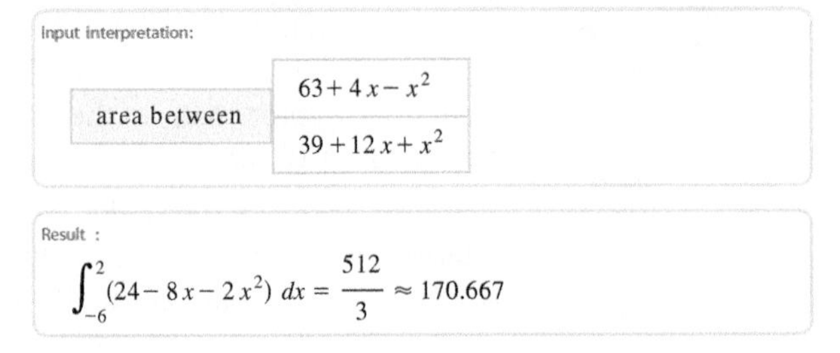

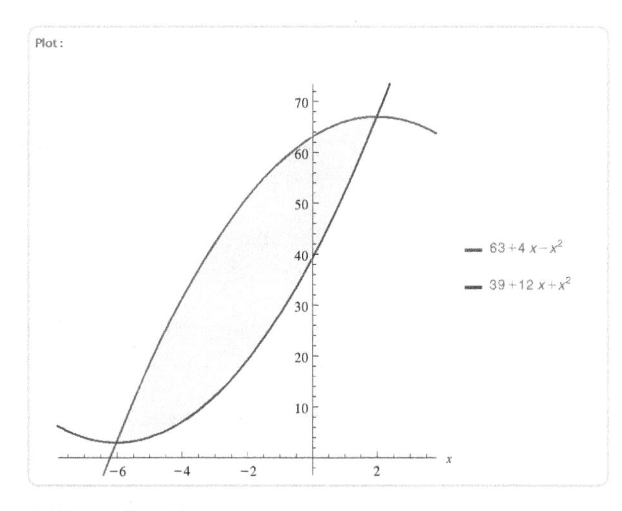

Wolfram Alpha LLC. 2012. Wolfram|Alpha
http://www.wolframalpha.com/
(access December 18, 2012)

Getting Ready for Class

After reading through the preceding section, respond in your own words and in complete sentences.

A. What does it mean for a function $f(x)$ to be nonnegative on some interval $[a, b]$?

B. What does $f(x) \geq g(x)$ mean graphically?

C. Suppose you wish to find the area bounded by two curves $f(x)$ and $g(x)$ over some interval $[a, b]$. At the leftmost part of $[a, b]$, $f(x) \geq g(x)$. But then $f(x)$ and $g(x)$ intersect each other at some point, say $x = c$, in $[a, b]$ and after that point $g(x) \geq f(x)$. Write the two integrals that give the area bounded by these two curves.

D. When finding the complete area bounded by two curves, what is one of the first things, if not the first thing, you should do?

Problem Set 6.1

Skills Practice

For Problems 1-14, find, or approximate to two decimal places, the described area.

1. The area bounded by the functions $f(x) = x + 3$, $g(x) = x^2 + 1$, and the lines $x = 0$ and $x = 2$.

2. The area bounded by the functions $f(x) = x + 5$, $g(x) = x^2 - 2x + 2$, and the lines $x = 0$ and $x = 3$.

3. The area bounded by the functions $f(x) = x + 4$, $g(x) = e^{0.5x}$, and the lines $x = 1$ and $x = 3$.

4. The area bounded by the functions $f(x) = -3x^2 + 6x + 4$, $g(x) = -x + 3$, and the lines $x = 0$ and $x = 2$.

5. The area bounded by the functions $f(x) = -x^2 + 8x - 10$, $g(x) = e^{-0.2x}$, and the lines $x = 3$ and $x = 5$.

6. The area bounded by the functions $f(x) = 9 - x^2$, $g(x) = x^2 + 1$, and the lines $x = 0$ and $x = 2$.

7. The area bounded by the functions $f(x) = x^2 - 5x + 11$, $g(x) = 0.7e^{0.7x}$, and the lines $x = 0$ and $x = 2$.

8. The area bounded by the functions $f(x) = e^{2x}$, $g(x) = e^x$, and the lines $x = 0$ and $x = 2$.

9. The area bounded by the functions $f(x) = x^4$, $g(x) = x^3$, and the lines $x = 0$ and $x = 1$.

10. The area bounded by the functions $f(x) = e^x$, $g(x) = e^{-x}$, and the lines $x = 0$ and $x = 1$.

11. The area completely enclosed by $f(x) = x + 2$ and $g(x) = x^2 - 4$.

12. The area completely enclosed by $f(x) = 2x - 3$ and $g(x) = -x^2 + 4x$.

13. The area completely enclosed by $f(x) = 2 - x^2$ and $g(x) = x$.

14. The area completely enclosed by $f(x) = x$ and $g(x) = \sqrt[3]{x}$.

Check Your Understanding

In Problems 15-20, set up an integral that represents the area of the planar region shaded in each figure.

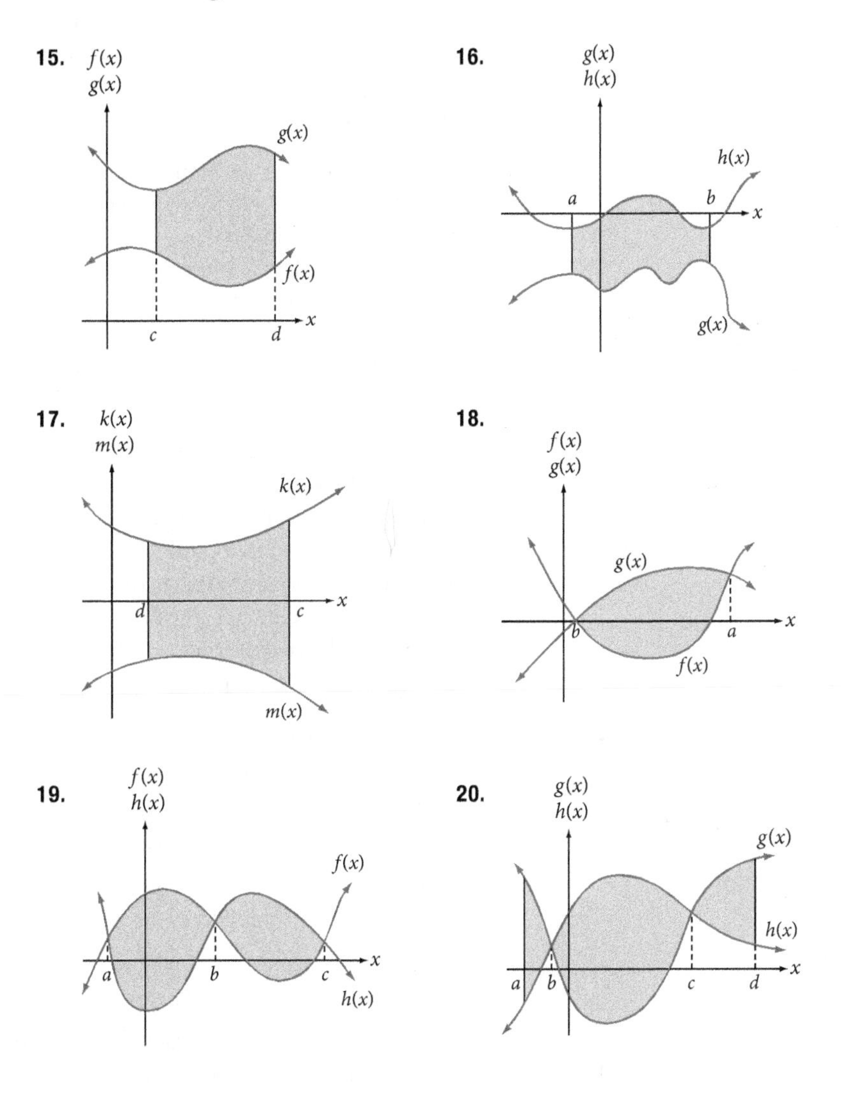

Modeling Practice

21. Business: Accumulated Revenue The current revenue for a company is \$6 million. The chief financial officer (CFO) of the company estimates that over the next 5 years, revenues will increase at a continuous rate of between 4% and 5% each year. The function that models the estimated 4% increase is $R_{4\%}(t) = 6e^{0.04t}$ and the function that models the estimated 5% increase is $R_{5\%}(t) = 6e^{0.05t}$. For each model, t represents the number of years from now. Approximate the total difference in revenue under the two models. (Hint: Which curve lies above the other?)

22. **Business: Accumulated Revenue** The current revenue for a company is $10 million. The chief financial officer (CFO) of the company estimates that over the next 5 years, revenues will decline at a continuous rate of between 2.5% and 5% each year. The function that models the estimated 2.5% decline is $R_{2.5\%}(t) = 10e^{-0.025t}$ and the function that models the estimated 5% decline $R_{5\%}(t) = 10e^{-0.05t}$. For each model, t represents the number of years from now. Approximate the total difference in revenue under the two models. (Hint: Which curve lies above the other?)

23. **Business: Accumulated Profit** The current profit for a company is $9 million. The chief financial officer (CFO) of the company estimates that over the next 5 years, profits will decrease at a continuous rate of between 2% and 3% each year. The function that models the estimated 2% decline is $P_{2\%}(t) = 9e^{-0.02t}$ and the function that models the estimated 3% decline is $P_{3\%}(t) = 9e^{-0.03t}$. For each function, t represents the number of years from now. Approximate the total difference in profit under the two models. (Hint: Which curve lies above the other?)

24. **Business: Accumulated Profit** The current profit for a company is $20 million. The chief financial officer (CFO) of the company estimates that over the next 5 years, profits will decline at a continuous rate of between 2% and 3% each year. The function that models the estimated 2% decline is $P_{2\%}(t) = 20e^{-0.02t}$ and the function that models the estimated 3% decline is $P_{3\%}(t) = 20e^{-0.03t}$. For each function, t represents the number of years from now. Approximate the total difference in profit under the two models. (Hint: Which curve lies above the other?)

25. **Business: Accumulated Profit** At the beginning of the year 2012, a company bought a computer-aided assembly machine that it estimates will help it to generate

$$R(t) = 9{,}200 - 50t^2$$

dollars of revenue t years since purchase. For the time since purchase, the company also estimates that the cost of operating and maintaining the assembly machine to be

$$C(t) = 2{,}000 + 150t^2$$

dollars. The company will keep the machine until the time when the cost to operate and maintain it equals the revenue it produces. Then the company will scrap it.

a. How many years from the time of purchase will the company keep the machine before scrapping it?

b. Knowing that profit = revenue − cost, what is the profit produced by the machine until it is scrapped?

26. Business: Accumulated Profit At the beginning of the year 2012, a company bought a machine that it estimates will help it to generate

$$R(t) = 655{,}000 - 8{,}000t^2$$

dollars of revenue t years since purchase. For the time since purchase, the company also estimates that the cost of operating and maintaining the machine to be

$$C(t) = 15{,}000 + 2{,}000t^2$$

dollars. The company will keep the machine until the time when the cost to operate and maintain it equals the revenue it produces. Then the company will scrap it.

a. How many years from the time of purchase will the company keep the machine before scrapping it?

b. Knowing that profit = revenue − cost, what is the profit produced by the machine until it is scrapped?

27. Business: Difference in Revenue Projections Mr. Azar, a financial analyst, believes the exponential model $R_A(t) = 600{,}000e^{-0.012t}$ accurately predicts a social media's online advertising revenue over the next 5 years. Mr. Hielo, also a financial analyst, believes the exponential model $R_H(t) = 600{,}000e^{-0.018t}$ accurately predicts the social media's online advertising revenue over the next 5 years. The variable t represents the number of years from now. What is the difference in their total projected revenue over the next 5 years?

28. Physical Science: Distance Traveled by a Dragster The rational function

$$V(t) = \frac{340t}{t+3}$$

models the velocity $V(t)$, in miles per hour, of a dragster t seconds since it starts its quarter-mile run. The integral of the velocity function from time $t = a$ to time $t = b$, gives the total distance traveled by the dragster through the time interval a to b. If the driver changes the dragster's velocity function from its current function to

$$V(t) = \frac{341t}{t+3}$$

what is the difference in the distance covered in the first 5 seconds of the dragster's run?

29. Business: Online Advertising Revenue Projections In the year 2010, analysts at a new economics professional's social networking site modeled the revenue (in thousands of dollars) the site could realize from online advertising over the next x years with the function $R(x) = -x^4 + 72x^2$. In the year 2012, the analysts revised their estimate of the revenue the company could realize over the next x years and modeled that new estimate with the function $R(x) = -x^4 + 80x^2$. What is the difference between the original projection and the new, revised projection in the online advertising revenue for the years 2012 to 2015? Both function are still based from the year 2010.

30. **Business: Revenue Flow and Profit** Over the next 10 years, a company projects its continuous flow of revenue to be $R(t) = 54e^{0.06t}$ and its costs $C(t) = 0.6t^2 + 20$, where t is in years from now and both $R(t)$ and $C(t)$ are in millions of dollars. Approximate the profit this company can expect over the next 10 years to the nearest thousand.

31. **Business: Savings Realized from New Equipment** A company can save money by buying new equipment, but at the same time will have to spend money to maintain it. The savings realized by the new equipment are

$$S(t) = 3.6t + 8$$

and the cost of maintaining the equipment is

$$C(t) = 4.8t$$

where t is the number of years since the purchase of the equipment and $S(t)$ and $C(t)$ are in thousands of dollars. Find the area between the curves $S(t)$ and $C(t)$ and interpret the result. (Hint: To find the area you need to establish the limits of integration. Determine where the curves intersect by solving the equation $S(t) = C(t)$.)

32. **Business/Life Science: Birth Rate of Insects** The research staff at a chemical company has determined that, over a period of a few days and in a particular southwestern region, the birth rate of an insect is approximated by

$$B_1(t) = 16e^{0.04x} - 5$$

After an application of its new pesticide, the birth rate is approximated by

$$B_2(t) = 16e^{0.02x} - 5$$

By how many insects will an application of the pesticide have reduced the population over a 3-day period?

Source: *Estimation of Dynamic Rate Parameters in Insect Populations Undergoing Sublethal Exposure to Pesticides*, University of North Carolina

33. **Business: Advertising and Sales** The owner of a hardware store estimates that with extensive radio advertising, store sales could be increasing at the rate of $500e^{0.12t}$ dollars per month t months from now, instead of at the current rate of $(500e^{0.09t})$ dollars per month t months from now. Find the additional amount of money the owner would expect to get in sales over the next 12 months by implementing the radio advertising.

Using Technology Exercises

34. Find the area of the region that is completely bounded by the two curves

$$f(x) = x^2 + 8x + 30 \quad \text{and} \quad g(x) = 2x + 46$$

35. Find the area of the region that is bounded by the two curves

$$f(x) = 200e^{0.12x}, g(x) = 200e^{0.10x}$$

and the vertical lines $x = 0$ and $x = e$. Round your result to the nearest whole number.

36. Find the area of the region bounded by the functions $f(x) = 8$ and $g(x) = 3$ and the vertical lines $x = 4$ and $x = 7$. What shape is this region? Does the area you found using calculus match the area you can find using the area formula for this shape?

37. Researchers found that in a particular part of a casino, t hours after 2:00 PM, income, I, from a group of slot machines flows in continuously at the rate of $I_1(t) = 450e^{0.08t}$ dollars per hour over a 3-hour period. They found that in another part of the casino, income from another group of slot machines flows in at the rate $I_2(t) = 100e^{0.5t}$ dollars per hour over the same 3-hour period. Using these income models, determine the total difference between the income flows of the two groups of machines from 2:00 PM to 5:00 PM.

Source: based on an article in GoErie.com, March 29, 2012

38. An analyst for a company projects annual revenues over the next 5 years to decline according to $R(x) = 200e^{-0.12x}$. When submitting his report he inadvertently omits the negative sign and writes the revenue function as $R(x) = 200e^{0.12x}$. If $R(x)$ is in millions of dollars and x represents the number of years from now, what is the total error in revenue over the next 5 years? Round your result to the nearest million dollar.

Source: KGWN TV

Getting Ready for the Next Section

39. Solve $250 = 1{,}450 - 3x^2$.

40. Evaluate $\int_0^{20} [1{,}200 - 3x^2]\,dx$.

41. To two decimal places, approximate the solution to $0.004x^2 = 25 - 0.005x^2$.

42. Evaluate $\int_0^{52.70} [13.89 - 0.005x^2]\,dx$. Round your result to two decimal places.

43. Evaluate $\int_0^{52.70} [11.11 - 0.004x^2]\,dx$. Round your result to two decimal places.

44. Solve $12x^2 + 300x - 540{,}000 = 0$.

45. Evaluate $\int_0^{200} [114 - (0.0002x^2 + 0.03x + 100)]\,dx$. Round your result to two decimal places.

Consumer's and Producer's Surplus

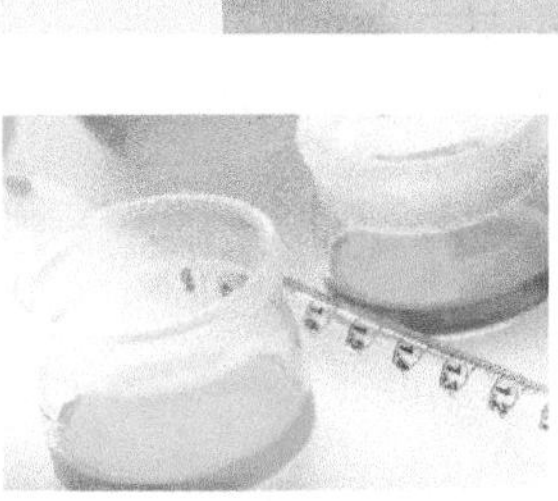

Both the supply and demand for an item depend on the price of the item (we'll see an example of this with chemical glassware later in this section). As the price p increases, the number of units x supplied (the supply) tends to increase, and the number of units x sought after for purchase (the demand) tends to decrease. In this sense, price drives both supply and demand so that p is the input variable and x is the output variable, and x is a function of p. Notice, however, that as the supply x of an item increases, the price p of the item tends to decrease. Also, as the demand x of an item increases, the price p tends to increase. In this sense, supply and demand drive price so that x is the input variable and p is the output variable, and p is a function of x.

Let S represent supply and D represent demand,

$$p = S(x) \quad \text{(price is a function of supply)}$$

$$p = D(x) \quad \text{(price is a function of demand)}$$

The function $p = S(x)$ is called a **supply function** and represents the price per unit at which producers are willing to produce and supply x units of a product. The supply function is of interest to producers.

The function $p = D(x)$ is called a **demand function** and represents the price per unit consumers are willing to pay when x units of the item are made available in the market. The demand function is of interest to consumers.

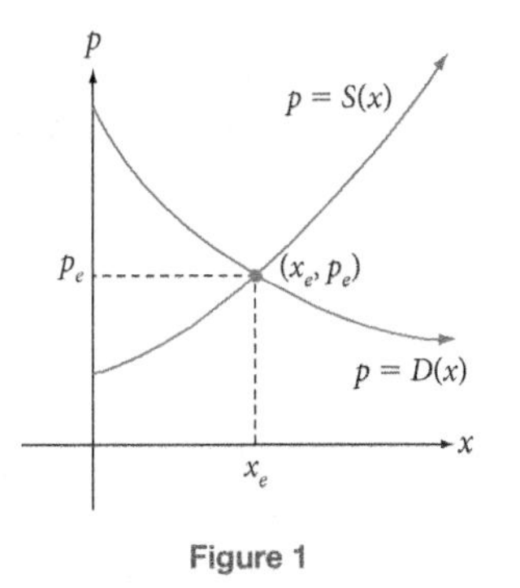

Figure 1

The point (x_e, p_e) at which supply equals demand is called the **equilibrium point** and represents the price, p, that consumers are willing to pay when x_e units of the item are produced. Figure 1 shows the equilibrium point for a pair of supply and demand functions. The equilibrium point (x_e, p_e) is a point of satisfaction for both the consumer and supplier. As the graph in Figure 1 indicates, when the number of items supplied to the market is less than the equilibrium number, both the consumer and supplier benefit. The consumer benefits since, although he was willing to buy at a price higher than the equilibrium price, he was able to buy at a lower price and thus record a savings, or surplus. This surplus is called the **consumer's surplus**, and we will denote it by CS. The supplier benefits since, although he was willing to sell at a price lower than the equilibrium price, he was able to sell at a higher price, and thus record a profit, or surplus. This surplus is called the **producer's surplus**, and we will denote it by PS.

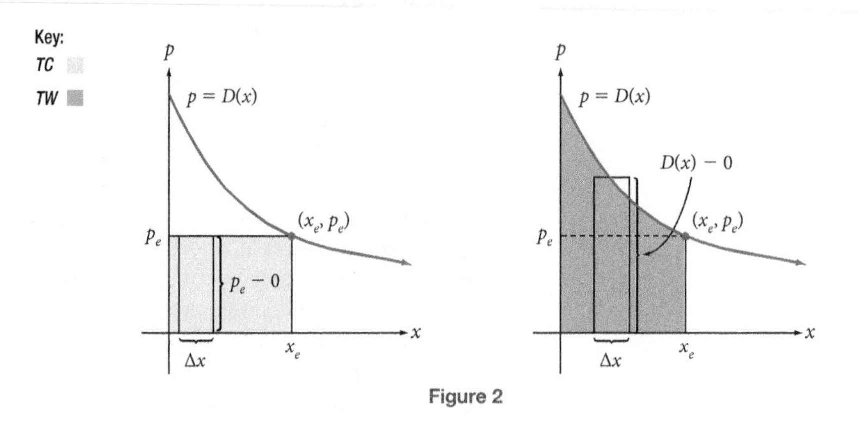

Figure 2

We can measure these surpluses using the definite integral as an accumulator of surplus. First for the consumer, for a production of x units up to and including the equilibrium value, the total cost, TC, to the consumer is the area of the rectangular region shown in the leftmost illustration of Figure 2. This illustration shows a representative rectangle of height p_e, which is actually $p_e - 0$, and width Δx. The area of this rectangle is $[p_e - 0]\Delta x$. The sum of n such rectangles is a Riemann sum and the infinite limit is the definite integral

$$TC = \int_0^{x_e} [p_e - 0]\, dx$$

Similarly, the definite integral

$$TW = \int_0^{x_e} [D(x) - 0]\, dx$$

measures the total amount consumers are willing to spend on up to x_e units. The rightmost illustration of Figure 2 shows the area that corresponds to the consumer's total willingness to buy, and also shows a representative rectangle of height $D(x) - 0$ and width Δx.

Consumer's Surplus

The difference between the total amount a consumer is willing to spend and the total amount actually expended (that is, $TW - TC$) is called the **consumer's surplus** and measures the amount of surplus money that results when a purchase is made at a price that is lower than the maximum the consumer was willing to pay. Figure 3 shows the area that corresponds to the consumer's surplus. The figure also shows a representative rectangle of height $D(x) - p_e$ and width Δx. The definite integral associated with this area, that is, the measure of consumer's surplus, is given as follows.

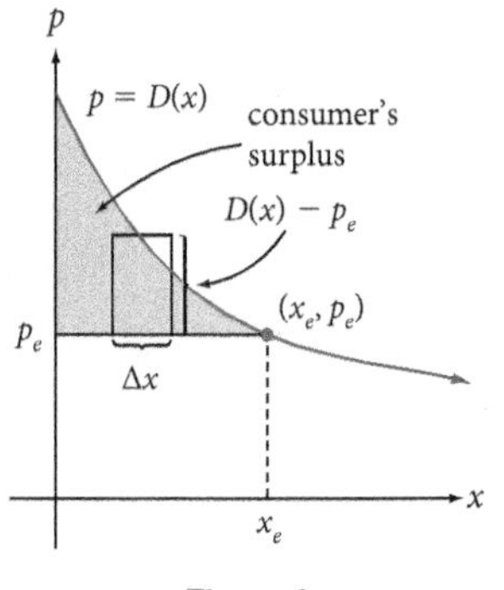

Figure 3

Consumer's Surplus

Consumer's Surplus

The **consumer's surplus** is defined as the definite integral

$$CS = \int_0^{x_e} [D(x) - p_e]dx$$

and measures the amount of surplus money that results when a purchase is made at a price that is lower than the maximum the consumer was willing to pay.

Notice that in computing this integral, the demand function $D(x)$ must be known as well as the equilibrium point (x_e, p_e).

VIDEO EXAMPLES

SECTION 6.2

Example 1 The demand for a particular item is given by the function $D(x) = 1{,}450 - 3x^2$. Find the consumer's surplus if the equilibrium price of a unit is \$250.

Solution We will use the definite integral as an accumulator of surplus. Here $D(x) = 1{,}450 - 3x^2$, and $p_e = 250$. But p_e is only one coordinate of the equilibrium point. We need the other coordinate, x_e. When $p = 250$,

$$250 = 1{,}450 - 3x^2$$
$$3x^2 = 1{,}200$$
$$x^2 = 400$$
$$x = -20, 20$$

Since $x > 0$, we will choose $x = 20$. Thus the equilibrium point is (20, 250), so that when $p_e = 250$ and $x_e = 20$, we have

$$= \int_0^{x_e} [D(x) - p_e]\, dx$$
$$= \int_0^{20} [1{,}450 - 3x^2 - 250]\, dx$$
$$= \int_0^{20} [1{,}200 - 3x^2]\, dx$$
$$= [1{,}200x - x^3] \Big|_0^{20}$$
$$= [1{,}200(20) - 20^3] - [1{,}200(0) - 0^3]$$
$$= [24{,}000 - 8{,}000] - [0]$$
$$= 16{,}000$$

Interpretation

When the cost to the consumer of a unit of a particular product is \$250 and the demand for the product is given by $D(x) = 1{,}450 - 3x^2$, the consumer's surplus is \$16,000. Put another way, when $x_e = 20$ units of a product are demanded and purchased and the equilibrium price is $p_e = \$250$, the total savings to the consumer is \$16,000.

■

Producer's Surplus

Similar to the consumer's surplus, the **producer's surplus** (PS) is a measure of the additional money contributed by consumers on the sale of x_e units of a product. The producer's surplus is the difference between the total amount consumers actually spend and the total amount producers are willing to accept for x_e units. Figure 4 displays the region that corresponds to the producer's surplus. The figure also displays a representative rectangle with height $p_e - S(x)$ and width Δx. The definite integral associated with this area, that is, the measure of producer's surplus, is given as follows.

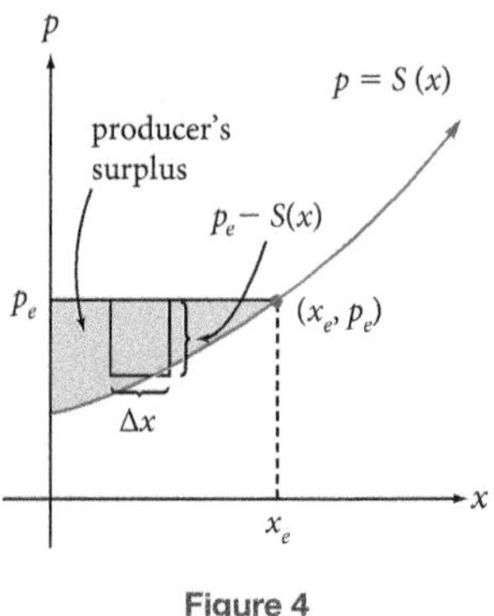

Figure 4

Producer's Surplus

Producer's Surplus

The **producer's surplus** is defined as the definite integral

$$PS = \int_0^{x_e} [p_e - S(x)]dx$$

and measures the additional money contributed by consumers on the sale of x_e units of a product.

Notice that in computing this integral, the supply function $S(x)$ must be known as well as the equilibrium point (x_e, P_e).

Example 2 Find both the consumer's and producer's surplus for a product $D(x) = 25 - 0.005x^2$ and $S(x) = 0.004x^2$.

Solution Since both surpluses require that we know the equilibrium point, we will start by finding its coordinates, x_e and p_e. The equilibrium point is located where supply equals demand; that is, where $S(x) = D(x)$.

$$S(x) = D(x)$$
$$0.004x^2 = 25 - 0.005x^2$$
$$0.009x^2 = 25$$
$$x^2 = \frac{25}{0.009}$$
$$x \approx \pm 52.70$$

Since x must be positive, we will choose $x \approx 52.70$. We may substitute 52.70 for x into either $S(x)$ or $D(x)$ (since $p = S(x)$ and $p = D(x)$). We will use $p = S(x)$ here.

$$p = 0.004x^2$$
$$p \approx 0.004(52.70)^2$$
$$p \approx 11.11$$

Therefore, $x_e \approx 52.70$ and $p_e \approx 11.11$.

We will begin with the consumer's surplus.

$$CS = \int_0^{x_e} [D(x) - p_e]\, dx$$

$$\approx \int_0^{52.70} [25 - 0.005x^2 - 11.11]\, dx$$

$$\approx \int_0^{52.70} [13.89 - 0.005x^2]\, dx$$

$$\approx \left[13.89x - 0.005\frac{x^3}{3}\right]\Bigg|_0^{52.70}$$

$$\approx [488.06] - [0]$$

$$\approx 488.06$$

Interpretation
When the price to the consumer for a unit of a particular product is about \$11.11, and the demand for the product is given by $D(x) = 25 - 0.005x^2$, the consumer's surplus is approximately \$488.06.

We will now find the producer's surplus.

$$PS = \int_0^{x_e} [p_e - S(x)]\, dx$$

$$\approx \int_0^{52.70} [11.11 - 0.004x^2]\, dx$$

$$\approx \left[11.11x - 0.004\frac{x^3}{3}\right]\Bigg|_0^{52.70}$$

$$\approx [390.35] - [0]$$

$$\approx 390.35$$

Interpretation
When the price to the consumer for a unit of a particular product is about \$11.11, and the supply function for the product is given by $S(x) = 0.004x^2$, the producer's surplus is approximately \$390.35.

■

Example 3 The demand and supply for a particular type of chemical glass -ware is modeled by $D(x) = -0.001x^2 + 154$ and $S(x) = 0.0002x^2 + 0.03x + 100$, respectively, where x is the number of glassware units (in thousands) demanded and supplied, and both $D(x)$ and $S(x)$ are the glassware unit prices in dollars. Find the consumer's surplus and the producer's surplus.

Solution Figure 5 shows the region bounded by the demand and supply curves. The graph shows that the two curves intersect at a point near $x = 200$. This point of intersection is the equilibrium point and we can find it by setting the two functions equal to each other and solving for x.

$$S(x) = D(x)$$

$$0.0002x^2 + 0.03x + 100 = -0.001x^2 + 154$$

$$0.0012x^2 + 0.03x - 54 = 0 \quad \text{Multiply by 10,000}$$

$$12x^2 + 300x - 540{,}000 = 0 \quad \text{Divide by the common factor 12}$$

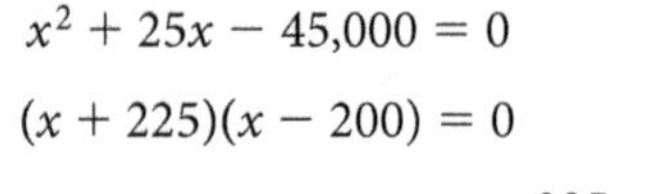

$$x^2 + 25x - 45{,}000 = 0$$
$$(x + 225)(x - 200) = 0$$
$$x = -225 \text{ and } 200$$

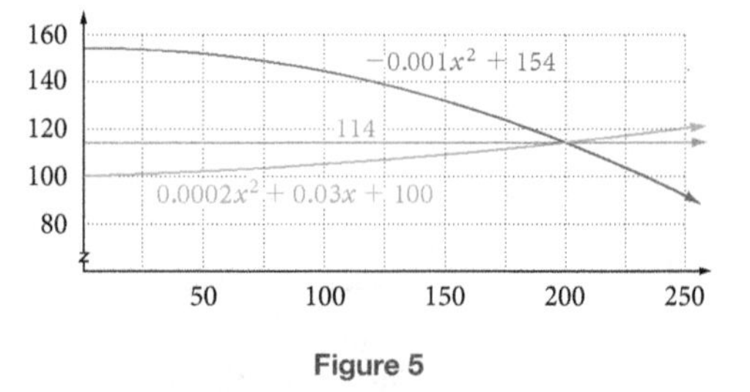

Figure 5

Because x represents the number of thousands of units of the glassware, it cannot be a negative number. We discard the x-value -225 and conclude that the demand and supply curves intersect at $x = 200$.

Substituting 200 for x into either the demand or supply function will give us the corresponding equilibrium price, p_e. Substituting 200 for x in the demand function gives $p_e = D(200) = -0.001(200)^2 + 154 = 114$.

The equilibrium point is (200, 114). We have all we need to compute both the consumer's and producer's surplus.

$$CS = \int_0^{x_e} [D(x) - p_e]\, dx$$
$$= \int_0^{200} [-0.001x^2 + 154 - 114]\, dx$$
$$\approx 5{,}333.33$$

and

$$PS = \int_0^{x_e} [p_e - S(x)]\, dx$$
$$= \int_0^{200} [114 - (0.0002x^2 + 0.03x + 100)]\, dx$$
$$\approx 1{,}666.67$$

When $x = 200$ (that is, when 200,000 units of glassware are demanded and supplied), we conclude that $CS \approx \$5{,}333.33$ and $PS \approx \$1{,}666.67$. ■

Interpretation

The consumer's surplus is the total difference between the price that consumers are willing to pay for a chemical glassware product and the price that they actually do pay for the glassware product, that is, the market price of the glassware product. The $5,333.33 is the total amount of money saved by consumers of the glassware product when they buy each glassware unit for the market price of $114 rather than at a higher price they would be willing to pay.

The producer's surplus is the total difference between the price at which producers are willing to supply the chemical glassware product and the price they actually receive for it, that is, the market price. The $1,666.67 is the total amount of money realized by producers of the glassware product when they sell each glassware unit for the market price of $114 rather than at a lower price they would be willing to accept.

Using Technology 6.2

You can use Wolfram|Alpha to help you find both the consumer's and producer's surplus.

Example 4 In Example 3 we used pencil and paper and our algebra skills to find an equilibrium point, then we used our skill with integration to evaluate the integral that produced the consumer's surplus. Wolfram|Alpha can help us through these computations.

The demand and supply for a particular type of chemical glassware is modeled by $D(x) = -0.001x^2 + 154$ and $S(x) = 0.0002x^2 + 0.03x + 100$, respectively, where x is the number of glassware units (in thousands) demanded and supplied, and both $D(x)$ and $S(x)$ are the glassware unit prices in dollars.

Solution Begin by setting the demand and supply equations equal to each other to find the equilibrium point. Go to www.wolframalpha.com and in the entry field enter

Solve -0.001x^2 + 154 = 0.0002x^2 + 0.03x + 100

Wolfram|Alpha displays your entry, its interpretation of your entry, and the solution to the equation.

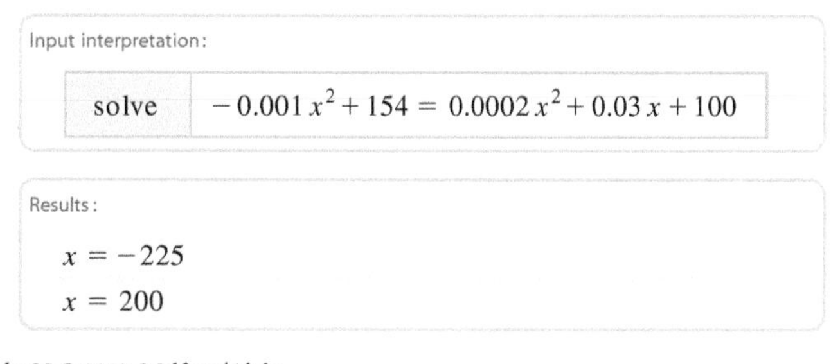

Wolfram Alpha LLC. 2012. Wolfram|Alpha
http://www.wolframalpha.com/
(access December 12, 2012)

Substituting 200 for x into either the demand or supply function will give us the corresponding equilibrium price, p_e. Enter

evaluate -0.001(200)^2 + 154

Click on the = sign or just press the return key. Wolfram|Alpha returns the value 114. The equilibrium point is (200, 114).

Now, evaluate the integral that defines the consumer's surplus. In the entry field, type

Integrate -0.001x^2 + 154 – 114 from x = 0 to 200

Wolfram|Alpha returns the consumer's surplus 5333.33. This agrees with our result in Example 3. ■

Getting Ready for Class

After reading through the preceding section, respond in your own words and in complete sentences.

A. What are the units of the supply function and the demand function?

B. What is the meaning of the equilibrium point?

C. Explain what is meant by consumer's surplus.

D. Explain what is meant by producer's surplus.

Problem Set 6.2

Skills Practice

1. The demand for a particular item is given by the demand function

$$D(x) = 750 - 2x^2$$

Find the consumer's surplus if the equilibrium point (x_e, p_e) is (15, 300).

2. The demand for a particular item is given by the demand function

$$D(x) = 300 - x^2$$

Find the consumer's surplus if the equilibrium point (x_e, p_e) is (10, 200).

3. The supply for a particular item is given by the supply function

$$S(x) = x^2 + 5x + 20$$

Find the producer's surplus if the equilibrium point (x_e, p_e) is (18, 434).

4. The supply for a particular item is given by the supply function

$$S(x) = \sqrt{x + 15}$$

Find the producer's surplus if the equilibrium point (x_e, p_e) is (35, 7.07).

5. Find both the consumer's and producer's surplus if for a product

$$D(x) = 50 - 2x, \quad S(x) = 12 + x$$

and the equilibrium point is $\left(12\frac{2}{3}, 24\frac{2}{3}\right)$.

Check Your Understanding

6. The demand for a particular product is given by the function $D(x)$. The equilibrium point $(x_e, p_e) = (14, 85)$. Interpret

$$\int_0^{14} [D(x) - 85]\, dx = 8{,}500$$

7. The demand for a particular product is given by the function $D(x)$. The equilibrium point $(x_e, p_e) = (46, 240)$. Interpret

$$\int_0^{46} [D(x) - 240]\, dx = 581{,}250$$

8. The supply for a particular product is given by the function $S(x)$. The equilibrium point $(x_e, p_e) = (35, 12)$. Interpret

$$\int_0^{35} [12 - S(x)]\, dx = 1{,}645.15$$

9. The supply for a particular product is given by the function $S(x)$. The equilibrium point $(x_e, p_e) = (925, 1{,}226)$. Interpret

$$\int_0^{925} [1{,}226 - S(x)]\, dx = 45.85$$

10. Complete the figure below by filling in the appropriate English phrase at the top of the drawing, and the appropriate integral at the bottom of the drawing.

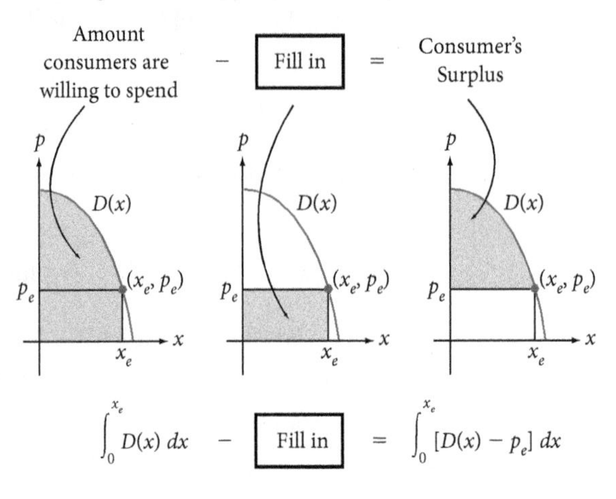

Modeling Practice

11. Business: Consumer's Surplus The demand for a particular product is modeled by the function $D(x) = 10{,}000 - 3x^2$, where the demand is in dollars and x is the number of units demanded each month. Find the consumer's surplus if the equilibrium price per unit of the product is \$2,500.

12. Business: Consumer's Surplus The demand for a particular product is modeled by the function $D(x) = 4{,}000 - 6x^2$, where the demand is in dollars and x is the number of units demanded each week. Find the consumer's surplus if the equilibrium price per unit of the product is \$1,600.

13. Business: Consumer's Surplus The demand for a particular product is modeled by the function $D(x) = -0.01x^2 - 0.2x + 8$, where the demand is in dollars and x is the number of thousands of units demanded each week. Find the consumer's surplus if the equilibrium price per unit of the product is \$5.

14. Business: Consumer's Surplus The demand for a particular product is modeled by the function $D(x) = -0.06x^2 - 0.6x + 140$, where the demand is in dollars and x is the number of thousands of units demanded each week. Find the consumer's surplus if the equilibrium price per unit of the product is \$10.

15. Business: Consumer's Surplus The demand for a particular product is modeled by the function

$$D(x) = \frac{400}{x + 5}$$

where the demand is in dollars and x is the number of thousands of units demanded each month. Find the consumer's surplus if the equilibrium price per unit of the product is \$20.

16. Business: Consumer's Surplus The demand for a particular product is modeled by the function $D(x) = \sqrt{580 - 5x}$, where the demand is in dollars and x is the number of thousands of units demanded each month. Find the consumer's surplus if the equilibrium price per unit of the product is \$20.

17. **Business: Consumer's Surplus** The demand for a particular product is modeled by the function $D(x) = \sqrt{285 - 2.4x}$, where the demand is in dollars and x is the number of thousands of units demanded each month. Find the consumer's surplus if the equilibrium price per unit of the product is \$15.

18. **Business: Consumer's Surplus** The demand for a particular product is modeled by the function $D(x) = 800e^{-0.04x}$, where the demand is in dollars and x is the number of thousands of units demanded each month. Find the consumer's surplus if the equilibrium price per unit of the product is \$241.

19. **Business: Consumer's Surplus** The demand for a particular product is modeled by the function $D(x) = 150 + 200e^{-0.001x}$, where the demand is in dollars and x is the number of thousands of units demanded each month. Find the consumer's surplus if the equilibrium price per unit of the product is \$298.

20. **Business: Producer's Surplus** A manufacturer will make x thousands of units of a product available each year when the price for each unit of the product is set at $S(x) = 12x^2$ dollars. Find the producer's surplus if the equilibrium price per unit of the product is \$75.

21. **Business: Producer's Surplus** A manufacturer will make x thousands of fire pit gas tube units available each year when the price for each gas tube is set at

$$S(x) = 11 + 0.36x^2$$

dollars. Find the producer's surplus if the equilibrium price per unit of the product is \$47.

22. **Business: Producer's Surplus** A manufacturer will make x thousands of guitar stands available each year when the price for each stand is set at

$$S(x) = 0.01 + 0.10\sqrt[3]{x^2}$$

dollars. Find the producer's surplus if the equilibrium price per stand is \$1.61.

23. **Business: Producer's Surplus** A manufacturer produces cardboard tubes onto which paper is rolled and packs them in boxes of 100. The manufacturer will make x thousands of boxes of tubes available each month when the price for each box is set at $S(x) = 0.8e^{0.002x}$ dollars. Find the producer's surplus when he produces and sells 1,500,000 boxes. (Keep in mind that the number of boxes produced is counted in the thousands.)

24. **Business: Consumer's and Producer's Surplus** The monthly demand for a particular item is related to the price of the item by the function $D(x) = -x^2 + 34.8x + 1{,}928$, where the price is in dollars and x is in thousands of units. A manufacturer will supply x thousands of units of the item when the price of the item is $S(x) = 1.4x^2 - 50x + 1{,}480$ dollars. Find both the consumer's and producer's surplus.

25. **Business: Consumer's and Producer's Surplus** The monthly demand for a particular item is related to the price of the item by the function $D(x) = -0.01x^2 + 220$, where the price is in dollars and x is in thousands of units. A manufacturer will supply x thousands of units of the item when the price of the item is $S(x) = 0.03x^2 - 1.6x + 136$ dollars. Find both the consumer's and producer's surplus.

Using Technology Exercises

26. **Business: Consumer's and Producer's Surplus** The monthly demand for a particular item is related to the price of the item by the function $D(x) = 500e^{-0.02x}$, where the price is in dollars and x is in thousands of units. A manufacturer will supply x thousands of units of the item when the price of the item is $S(x) = 100e^{0.02x}$ dollars. Find both the consumer's and producer's surplus.

27. Use Wolfram|Alpha to find the producer's surplus from Example 4.

28. Find the equilibrium point if the demand function is

$$D(x) = 5{,}400 - 4x^2$$

and the equilibrium price is \$500.

29. Find the equilibrium point if the supply function is given by

$$S(x) = 200 + 40e^{0.04x}$$

and the equilibrium price is \$495.56.

Getting Ready for the Next Section

30. Evaluate $\int_0^3 128{,}000\, dt$.

31. Evaluate $\int_0^6 100\, e^{0.005t}\, dt$.

32. Compute the value of Pe^{rt} for $P = 128{,}000$, $r = 0.07$, $t = 3$.

33. Compute the value of Ae^{-rt} for $A = 15{,}000$, $r = 0.08$, $t = 10$.

34. Evaluate $\int_0^5 10{,}000\, dt$.

35. Evaluate $\int_0^5 10{,}000\, e^{-0.07t}\, dt$.

36. Evaluate $e^{0.07(5)}\int_0^5 10{,}000\, e^{-0.07t}\, dt$.

37. Evaluate $\int_0^\infty 70{,}000\, e^{-0.065t}\, dt$.

Annuities and Money Streams

6.3

Typically, a company will reinvest some of its income. The reinvestment may be into an interest-earning account or a non-interest-earning account. The reinvestment may be made in one lump sum, or in equal sums over specified time intervals, or continuously over some time interval. The continuous investment into an interest-earning account is the most interesting investment, and that is our main focus in this section.

In most cases, the company's income does not come in all at once at the end of the year, but rather flows in somewhat continuously throughout the year. For example, money flows into fast-food restaurants, retail stores, airline travel operations, toll roads and bridges, hotels and rental properties, and savings accounts and investments, somewhat continuously and in differing amounts throughout the year. Regardless of how the money flows in, if it is reinvested into a non-interest-earning account, the accumulated amount at the end of some specified time period can be determined by simply adding together the individual amounts.

© Gluca B./iStockPhoto

Accumulated Value of a Non-Interest-Earning Stream

If there are a great many individual amounts, such as the individual transactions at a fast-food restaurant throughout a one-year period, it is convenient to think of the income flow as a **continuous stream of money**. In some cases, it is possible to approximate the rate of flow of the stream by a function $f(t)$. The function $f(t)$ is called the **rate of flow function** and the definite integral can be used as an accumulator to determine the total value of the stream over any specified time period. It is worth noting that the definite integral only approximates the accumulated value since, although the income flow is *actually* discrete (in individual parts), it is being described using a continuous function.

Accumulated Value of a Non-Interest-Earning Stream

If over some interval $0 \leq t \leq T$, money flows continuously into a pool at a rate approximated by the function $f(t)$ dollars per time period, and if the money is reinvested in a non-interest-earning account, then the total amount of money, A, (in dollars) accumulated at the end of T years is

$$A = \int_0^T f(t)\, dt$$

The total accumulated amount of money corresponds to the area bounded by the curve $f(t)$, the t-axis, and the vertical lines $t = 0$ and $t = T$. Figure 1 illustrates this idea.

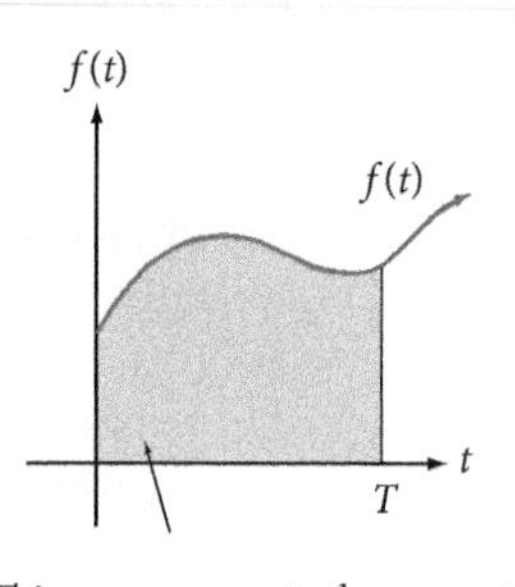

This area represents the amount of money that accumulated over the time interval 0 to T.

Figure 1

Example 1 Revenue flows continuously into a retail operation at the constant rate of $f(t) = 128{,}000$ dollars per year. As the money is received, it is placed into a non-interest-earning account. Determine how much money has accumulated in the operation's account at the end of a 3-year time period.

Solution The total amount of money accumulated is

$$\begin{aligned} A &= \int_0^T f(t)\, dt \\ &= \int_0^3 128{,}000\, dt \\ &= 128{,}000t \Big|_0^3 \\ &= 128{,}000(3) - 128{,}000(0) \\ &= 384{,}000 \end{aligned}$$

Thus, $\int_0^3 128{,}000\, dt = 384{,}000$.

Interpretation
If revenue flows into the account of a retail operation at the constant rate of $f(t) = 128{,}000$ dollars per year, and if it is deposited into a non-interest-earning account, the amount of money that will have accumulated at the end of a 3-year period will be \$384,000. Notice that this is the same computation as 128,000 per year for 3 years: $3 \cdot 128{,}000 = 384{,}000$. In other words, this situation was so basic ($f(t) =$ a constant) that the integration was unnecessary.

■

The Accumulated Value of a One-Time, Lump-Sum, Interest-Earning Investment

We will consider now the case where money is reinvested at some annual interest rate r that is compounded continuously. When P dollars is invested at an annual interest rate r that is compounded continuously, the total amount A of money that will have accumulated at the end of t years is given by

$A = Pe^{rt}$

$$A = Pe^{rt}$$

The amount A is called the **future value of the investment.**

Example 2 If a one-time, lump-sum investment of $P = 128{,}000$ dollars is made into an account paying 7% interest compounded continuously, the amount A of money that will have accumulated at the end of 3 years is

$$\begin{aligned} A &= Pe^{rt} \\ &= 128{,}000e^{0.07(3)} \\ &\approx 157{,}910.79 \end{aligned}$$

Interpretation

If a one-time, lump-sum investment of $P = 128{,}000$ dollars is made into an account paying 7% interest compounded continuously, the amount, A, of money that will have accumulated in the account at the end of 3 years is approximately $157,910.79.

The future value of $128,000 placed into an account paying 7% interest compounded continuously for 3 years is approximately $157,910.79. The $128,000 is the amount of money that would have to be invested now to get $157,910.79 three years from now. Therefore, we say the future value of $128,000 is $157,910.79.

The Accumulated Value of an Interest-Earning Stream: Annuities

We will now consider the case where money is reinvested periodically at some rate of interest r that is compounded continuously. As we have just noted, when P dollars is invested into an interest-earning account at an annual interest rate r that is compounded continuously, the total amount A of money that will have accumulated at the end of t years is given by

$$A = Pe^{rt}$$

where the amount A is called the future value of the investment.

If the same amount of money P (P for *principal*) is regularly deposited into an interest-earning account at the end of equal periods of time, the sequence of deposits is called an **annuity**. The **amount of the annuity**, or the **future value of the annuity**, is the final amount in the account at the end of all the time periods. It is the total amount of the deposits plus the total amount of interest earned from all the deposits. The time between deposits is the **deposit or payment period.**

Example 3 Suppose that at the end of each month, for 6 months, $100 is put into an account paying 6% annual interest compounded continuously. Except for the last $100, which earns no interest at all, each $100 earns interest over a different period of time. Using the continuous compounding formula $A = Pe^{rt}$, we can find the total amount of money in the account at the end of 6 months. The interest rate r is the annual rate and, since we are working in months, the corresponding monthly rate is

$$\frac{0.06}{12} = 0.005$$

1. The first $100 earns interest for 5 months, so at the end of 6 months it is worth

$$A_1 = Pe^{rt} = 100e^{0.005(5)} \approx 100(1.0253) \approx \$102.53$$

2. The second \$100 earns interest for 4 months, so at the end of 6 months it is worth

$$A_2 = Pe^{rt} = 100e^{0.005(4)} \approx 100(1.0202) \approx \$102.02$$

3. The third \$100 earns interest for 3 months, so at the end of 6 months it is worth

$$A_3 = Pe^{rt} = 100e^{0.005(3)} \approx 100(1.0151) \approx \$101.51$$

4. The fourth \$100 earns interest for 2 months, so at the end of 6 months it is worth

$$A_4 = Pe^{rt} = 100e^{0.005(2)} \approx 100(1.0101) \approx \$101.01$$

5. The fifth \$100 earns interest for 1 month, so at the end of 6 months it is worth

$$A_5 = Pe^{rt} = 100e^{0.005(1)} \approx 100(1.0050) \approx \$100.50$$

6. The sixth \$100 earns interest for 0 month, so at the end of 6 months it is worth

$$A_6 = Pe^{rt} = 100e^{0.005(0)} = 100(1) = \$100.00$$

Then the total amount of money that has accumulated at the end of the 6-month period is just the sum of the individual totals, A_1, A_2, A_3, A_4, A_5, and A_6. Thus,

$$\begin{aligned} A &= A_1 + A_2 + A_3 + A_4 + A_5 + A_6 \\ &= 102.53 + 102.02 + 101.51 + 101.01 + 100.50 + 100 \\ &= 607.57 \end{aligned}$$

Interpretation
The future value of \$100 invested at the end of each month, for 6 months, in an account paying 6% annual interest compounded continuously is \$607.57. That is, the \$600 investment is actually now worth \$607.57. (If the \$100 had been deposited into a non-interest-earning account each month, it would be worth \$600 at the end of 6 months.) Figure 2 illustrates the process.

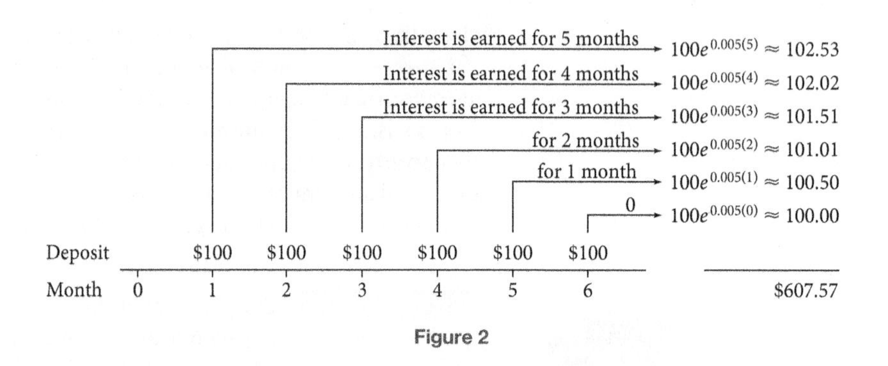

Figure 2

Notice that the amount of the annuity is a sum. In fact, it is a Riemann sum with $\Delta t = 1$ month. This means that the amount of the annuity in this example can be approximated by the definite integral

$$\int_0^6 100e^{0.005t}\,dt$$

■

Future Value of an Annuity

The amount or **future value,** A, of an annuity at the end of n payment periods is approximated by

$$A = \int_0^n Pe^{rt}\,dt$$

where P is the number of dollars invested at the end of each payment period, and r is the interest rate per time period. The integral measures the value of the investment in the future.

Example 4 Approximate the future value of the annuity in Example 3 over

a. a 6-month period, and **b.** a 20-year period.

Solution

a. Over a 6-month period, the future value of the annuity is

$$\begin{aligned} A &= \int_0^6 100e^{0.005t}\,dt \\ &= 20{,}000e^{0.005t}\Big|_0^6 \\ &= 20{,}000(e^{0.005(6)} - e^{0.005(0)}) \\ &\approx 20{,}000(1.03045 - 1) \approx 609 \end{aligned}$$

Interpretation

The future value of \$100 invested at the end of each month, for 6 months, in an account paying 6% annual interest compounded continuously is approximately \$609. That is, the \$600 investment is actually worth approximately \$609.

Comparing the actual sum to the approximated sum we see we are in error by about \$609 − \$607.56 = 1.44. In this case the error is somewhat large compared to the amount of interest that the investment generated, but over longer periods of time, like 5, 10, or 20 years, does the error become less significant as the number of years increase?

b. Over a 20-year period, the future value of the annuity is approximated by the integral

$$\int_0^{240} 100e^{0.005t}\,dt$$

Notice that the upper limit of integration is 240. This comes from the fact that there are $20 \cdot 12 = 240$ payment periods.

$$\begin{aligned} A &= \int_0^{240} 100e^{0.005t}\,dt \\ &= 20{,}000e^{0.005t}\Big|_0^{240} \\ &= 20{,}000(e^{0.005(240)} - e^{0.005(0)}) \\ &\approx 20{,}000(3.320117 - 1) \approx 46{,}402.34 \end{aligned}$$

Interpretation
The future value of \$100 invested at the end of each month, for 20 years (240 months), in an account paying 6% annual interest compounded continuously is approximately \$46,402.34. That is, the \$24,000 investment is actually worth approximately \$46,402.34.

■

To obtain the future value of a one-time, lump-sum, interest-earning investment, we used the formula $A = Pe^{rt}$, where A denoted the future value, P the principal (amount invested), r the annual interest rate, and t the time period. If the formula is solved for P, we get

$$P = Ae^{-rt}$$

(Can you show this?) The value P is called the **present value** of A, and it represents the amount, P, that must be invested now ($t = 0$) so that the amount, A, will be accumulated in the future. This situation would be of interest to us, for example, if we wanted to know how much to invest now so that we could have a particular amount of money after a specific time period to invest in a child's education.

Example 5 How much money must be invested now at 8% interest compounded continuously, so that \$15,000 will be available in 10 years?

Solution We wish to determine the present value of \$15,000.

$$\begin{aligned} P &= Ae^{-rt} \\ &= 15{,}000e^{-0.08\cdot 10} \\ &\approx 6{,}739.93 \end{aligned}$$

Thus, the present value of \$15,000 is \$6,739.93.

Interpretation
If \$6,739.93 is invested now in an account that pays 8% interest compounded continuously, it will be worth \$15,000 in 10 years.

■

The Accumulated Value of an Interest-Earning Continuous Stream

We can extend the concept of the future value of an annuity and the present value of an investment to a continuous income stream.

The future value measures the total accumulated amount of income that will be realized at the end of the time period $[0, T]$. The present value measures the amount of money that would have to be deposited now to produce the future value.

The future and present values of a continuous income stream are useful in the analysis of investment opportunities. In Example 6, we will interpret the present and future value of an income stream in terms of the opportunity for a company to buy a machine that will generate income.

Present and Future Value of an Income Stream
If $f(t)$ is the rate of flow of money into an account that pays an annual interest rate r compounded continuously during some time interval $0 \le t \le T$, then

1. The **future value** of the income stream is

$$\text{Future value} = e^{rT}\int_0^T f(t)e^{-rt}\,dt$$

2. The **present value** of the income stream is

$$\text{Present value} = \int_0^T f(t)e^{-rt}\,dt$$

Example 6 A company can buy a machine that is expected to increase the company's net income by \$10,000 each year for the 5-year life of the machine. The company also estimates that for the next 5 years, the money they invest from that continuous stream will earn 7% compounded continuously. Find and interpret

a. the total amount of money produced only by the machine over its 5-year life,

b. the future value of the income stream, and

c. the present value of the income stream.

d. Use these values to determine the conditions under which the company should and should not buy the machine.

Solution In this situation, the rate of flow of income is given by $f(t) = 10{,}000$.

a. In 5 years, and without any interest, the \$10,000 income stream will produce

$$\int_0^5 10{,}000\,dt = 10{,}000t\Big|_0^5 = \$50{,}000$$

Interpretation
Without being invested into an interest-earning account, the \$10,000 five-year income stream will produce \$50,000.

b. The future value of the income stream will tell us how much money, including direct income and interest, the machine will generate.

$$\begin{aligned}\text{Future value} &= e^{rT}\int_0^T f(t)e^{-rt}\,dt\\ &= e^{0.07(5)}\int_0^5 10{,}000e^{-0.07t}\,dt\\ &= e^{0.35}\cdot\frac{10{,}000}{-0.07}e^{-0.07t}\Big|_0^5\\ &\approx 59{,}866.79\end{aligned}$$

Interpretation
If the \$10,000 continuous stream is invested immediately as it is received into an account paying 7% interest compounded continuously for 5 years, it will be worth \$59,866.79 at the end of 5 years. (This is similar to an annuity, but rather than investing a specific amount at the end of each time period, the investment is made continuously.)

c. The present value of the continuous income stream that the machine will produce is the value of the stream *right now* to the company.

$$\begin{aligned}\text{Present value} &= \int_0^T f(t)e^{-rt}\,dt \\ &= \int_0^5 10{,}000e^{-0.07t}\,dt \\ &= \frac{10{,}000}{-0.07} \cdot e^{-0.07t}\Big|_0^5 \\ &= \frac{10{,}000}{-0.07} \cdot (e^{-0.07(5)} - 1) \\ &\approx 42{,}187.42\end{aligned}$$

Interpretation
The present value of this income stream is \$42,187.42. This means that \$42,187.42 would have to be invested today to produce \$59,866.79 at the end of 5 years. That is, at the end of 5 years, the machine will bring in \$50,000 as direct income, but if the income is invested, the machine will bring in \$59,866.79. The present value of \$42,187.42 is the value of the income stream to the company *right now* in the following sense: If the company were to make a one-time, lump-sum investment of \$42,187.42 now into an account paying 7% interest compounded continuously for 5 years, the total amount of money it would have would be \$59,866.79. (Verify this using the method of Example 2.)

d. Should the company buy the machine? It depends on the price, of course.

1. If the machine sells for exactly \$42,187.42, the present value of the machine, the company may realize a net income, including interest, of \$59,866.79, from the income stream it generates. But, this stream may not be guaranteed. (The machine may not perform as expected, or the company may have overestimated its ability to increase income, for example.) The company can guarantee the \$59,866.79 income by placing the \$42,187.42 into a 7% interest-earning account for 5 years.

2. If the machine sells for more than \$42,187.42, the company would do better to invest the income in a 7% interest-earning account. Otherwise, it has to invest more than \$42,187.42 to generate the same \$59,866.79 income. For example, if the company could buy the machine for \$45,000, it could generate \$59,866.79 from that \$45,000. But it can generate the \$59,866.79 from only \$42,187.42. Consequently, a \$45,000 investment for the machine is not worth it.

3. If the machine sells for less than \$42,187.42, the company should buy it because it can generate the \$59,866.79 income using less than \$42,187.42. For example, if the company could buy the machine for \$25,000, it could generate \$59,866.79 from that \$25,000. Thus, the \$25,000 presently has the same earning capacity to the company and therefore the same value as \$42,187.42 invested in the bank.

The Present Value of a Perpetual Income Stream

In some situations, it is reasonable to consider the income as flowing forever. An income stream that flows essentially forever is called a **perpetual income stream** or a perpetuity. Improper integrals can be used to approximate the present value of a perpetuity.

Example 7 Rental income from a piece of property, upon which there is an indefinite lease, flows at the rate of \$70,000 per year. The income is invested immediately into an account paying 6.5% annual interest compounded continuously. Find and interpret the present value of the flow.

Solution The rate of flow function is $f(t) = 70{,}000$, and since the lease is indefinite, we can, for all practical purposes, consider the time period infinite. Then,

$$\begin{aligned}
\text{Present value} &= \int_0^T f(t)e^{-rt}\,dt \\
&= \int_0^\infty 70{,}000e^{-0.065t}\,dt \\
&= \lim_{b\to\infty}\int_0^b 70{,}000e^{-0.065t}\,dt \\
&= \frac{70{,}000}{-0.065}\cdot \lim_{b\to\infty} e^{-0.065t}\Big|_0^b \\
&= \frac{70{,}000}{-0.065}\cdot \lim_{b\to\infty} (e^{-0.065b} - e^{-0.065(0)}) \\
&= \frac{70{,}000}{-0.065}\cdot \lim_{b\to\infty}\left(\frac{1}{e^{0.065b}} - 1\right) \\
&= \frac{70{,}000}{-0.065}\cdot (0-1) \\
&\approx 1{,}076{,}923.077
\end{aligned}$$

Interpretation

To match the total income realized from the lease of this property, approximately \$1,076,923.08 would have to be invested now into an account that pays annual interest at 6.5% compounded continuously forever.

Using Technology 6.3

You can use your graphing calculator to determine the future value of an annuity.

Example 8 In Example 6, we found that the future value of a $10,000 investment made at 7% for years was $59,866.79. Use your calculator to make this future value computation.

Solution The following entries show how the future value problem of Example 6 is solved.

```
Y1 = 10000e^(-0.07X)
Y2 = fnInt(Y1,X,A,B)
0→A
5→B
e^(0.07*5)*Y2
```

The result, 59,866.79, agrees with the example. ■

Getting Ready for Class

After reading through the preceding section, respond in your own words and in complete sentences.

A. Revenues earned by large corporations such as water or electric companies flow into the corporation essentially all the time. What are such revenues called?

B. What is the name given to an income stream that flows essentially forever?

C. What is an annuity?

D. What is meant by the present value of future income?

Problem Set 6.3

Skills Practice

For Problems 1-14, round to the nearest cent.

1. Find the amount of money that accumulates when \$200,000 is placed each year into a non-interest-earning account for 3 years.

2. Find the amount of money that accumulates when \$140,000 is placed each year into a non-interest-earning account for 2 years.

3. Find the amount of money that accumulates when a one-time lump-sum investment of $P = \$90{,}000$ is made into an account paying 10% interest compounded continuously for 6 years.

4. Find the amount of money that accumulates when a one-time lump-sum investment of $P = \$10{,}000$ is made into an account paying 2% interest compounded continuously for 30 years.

5. At the end of each month, for 24 months, \$300 is put into an account paying 6% annual interest compounded continuously. Find the future value of this account.

6. At the end of each month, for 120 months, \$1,000 is put into an account paying 8% annual interest compounded continuously. Find the future value of this account.

7. How much money must be invested now at 10% interest compounded continuously so that \$120,000 will be available in 5 years?

8. How much money must be invested now at 7% interest compounded continuously so that \$1,400,000 will be available in 10 years?

9. Find the future value of a \$20,000 annual income stream if it is invested immediately as it is received into an account paying 12% interest compounded continuously for 10 years.

10. Find the future value of a \$50,000 annual income stream if it is invested immediately as it is received into an account paying 8% interest compounded continuously for 20 years.

11. Find the present value of a \$20,000 annual income stream if it is invested immediately as it is received into an account paying 12% interest compounded continuously for 10 years.

12. Find the present value of a \$50,000 annual income stream if it is invested immediately as it is received into an account paying 8% interest compounded continuously for 20 years.

13. Find the present value of a \$1,000 annual income stream if it is invested immediately as it is received into an account paying 5% interest compounded continuously forever.

14. Find the present value of a \$12,000 annual income stream if it is invested immediately as it is received into an account paying 6% interest compounded continuously forever.

Check Your Understanding

15. **Business: Future and Present Values for Decision Making** A company considers buying a machine that will increase its annual net income by an estimated $70,000 per year for the next 6 years. Over the next 6 years, company economists believe the money they invest from that continuous stream will earn 5%. The future and present values of this stream over the next 6 years are $489,802.33 and $362,854.49, respectively. If the cost of the machine is $375,000, should the company buy the machine or invest the present value of the machine?

16. **Business: Future and Present Values for Decision Making** A company considers buying a machine that will increase its annual net income by an estimated $106,500 per year for the next 8 years. Over the next 8 years, company economists believe the money they invest from that continuous stream will earn 7.25%. The future and present values of this stream over the next 8 years are $1,154,663.35 and $646,494.12, respectively. If the cost of the machine is $550,750, should the company buy the machine or invest the present value of the machine?

17. **Business: Future and Present Values for Decision Making** A company considers buying a machine that will increase its annual net income by an estimated $35,000 per year for the next 5 years. Over the next 5 years, company economists believe the money they invest from that continuous stream will earn 6%. The future and present values of this stream over the next 5 years are $204,084.30 and $151,189.37, respectively. Should the company buy the machine or invest the present value of the machine?

18. **Business: Future and Present Values for Decision Making** A company considers buying a machine that will increase its annual net income by an estimated $84,000 per year for the next 12 years. Over the next 12 years, company economists believe the money they invest from that continuous stream will earn 5.5%. The future and present values of this stream over the next 12 years are $1,427,682.84 and $737,899.78, respectively. Should the company buy the machine or invest the present value of the machine?

19. **Economics: Future and Present Values of an Investment** For 10 months, a person invests $300 at the end of each month into an account that pays an annual interest rate of 7% compounded continuously. The future and present values are, respectively, $3,089.23 and $2,914.18. Interpret these values.

20. **Economics: Future and Present Values of an Investment** For 24 months, a person invests $500 at the end of each month into an account that pays an annual interest rate of 5.5% compounded continuously. The future and present values are $12,684.88 and $11,363.55, respectively. Interpret these values.

21. **Economics: Future and Present Values of an Investment** For 15 years, a person invests $100 at the end of each month into an account that pays an annual interest rate of 6% compounded continuously. The future and present values are $29,192.06 and $11,868.61, respectively. Interpret these values.

22. **Economics: Future and Present Values of an Investment** For 10 years, a person invests $1,500 at the end of each month into an account that pays an annual interest rate of 8% compounded continuously. The future and present values are $275,746.71 and $123,900.98, respectively. Interpret these values.

Modeling Practice

For each of the following problems, round to the nearest dollar.

23. **Business: Non-Interest-Earning Revenue Flow** Revenue flows continuously into a retail operation at the constant rate of $f(t) = \$200{,}000$ per year. As the money is received, it is placed into a non-interest-earning account. Determine how much money will have accumulated in the operation's account at the end of a 3-year time period.

24. **Economics: One-Time Lump-Sum Investment** What will be the future value if a one-time lump-sum investment of $P = \$94{,}432$ is made into an account paying 5% interest compounded continuously and left for 10 years?

25. **Economics: One-Time Lump-Sum Investment** What will be the future value if a one-time lump-sum investment of $P = \$69{,}880.70$ is made into an account paying 6% interest compounded continuously and left for 20 years?

26. **Economics: Future Value of an Annuity** At the end of each month, for 120 months (10 years), $1,000 is put into an annuity account paying 5% annual interest compounded continuously. Compute the future value of this annuity.

27. **Economics: Future Value of an Annuity** At the end of each month, for 240 months (20 years), $500 is put into an annuity account paying 6% annual interest compounded continuously. Compute the future value of this annuity.

28. **Economics: Present Value of an Annuity** How much money must be invested now at 6% interest compounded continuously so that $100,000 will be available in 5 years?

29. **Economics: Present Value of an Annuity** How much money must be invested now at 5% interest compounded continuously so that $5,000,000 will be available in 3 years?

30. **Business: Future and Present Value of an Income Stream** A continuous flow of money comes into a business at the rate of $80,000 per year and is invested immediately as it is received into an account paying 8% interest compounded continuously for 4 years. Find and interpret
    a. the future value of the income stream.
    b. the present value of the income stream.

31. **Business: Future and Present Value of an Income Stream** A continuous flow of money comes into a business at the rate of $200,000 per year and is invested immediately as it is received into an account paying 5% interest compounded continuously for 3 years. Find and interpret
    a. the future value of the income stream.
    b. the present value of the income stream.

32. **Business: Future and Present Value of an Income Stream** A continuous flow of money comes into a business at the rate of $400,000 per year and is invested immediately as it is received into an account paying 6% interest compounded continuously for 2 years. Find and interpret
    a. the future value of the income stream.
    b. the present value of the income stream.

33. **Business: Future and Present Value of an Income Stream** A continuous flow of money comes into a business at the rate of $1,000,000 per year and is invested immediately as it is received into an account paying 10% interest compounded continuously for 6 years. Find and interpret
   a. the future value of the income stream.
   b. the present value of the income stream.

34. **Business: Future and Present Value of an Income Stream** A company can buy a machine that could increase its net income by $50,000 each year for the 5-year life of the machine. The company estimates that for the next 5 years, the money they invest from that continuous stream will earn 6% compounded continuously. Find and interpret
   a. the total amount of money produced only by the machine over its 5-year life.
   b. the future value of this income stream.
   c. the present value of this income stream.
   d. if the machine costs $200,000, should the company buy it or put the $200,000 into an account that pays 6% interest compounded continuously for 5 years?

35. **Business: Future and Present Value of an Income Stream** A company can buy a machine that could increase its net income by $100,000 each year for the 3-year life of the machine. The company estimates that for the next 3 years, the money they invest from that continuous stream will earn 4% compounded continuously. Find and interpret
   a. the total amount of money produced only by the machine over its 3-year life.
   b. the future value of this income stream.
   c. the present value of this income stream.
   d. Iif the machine costs $250,000, should the company buy it or put the $250,000 into an account that pays 4% interest compounded continuously for 3 years?

36. **Business: Annuities and Comparing Investments** Revenue flows continuously into a retail operation at the constant rate of $f(t) = \$10,000$ per month.
   a. If, at the end of each month, it is placed into a non-interest-earning account, determine how much money will have accumulated in the operation's account at the end of a 5-year time period. (Five years is 60 months.)
   b. If, at the end of each month, it is placed into an annuity account paying 8% annual interest compounded continuously, determine how much money will have accumulated in the operation's account at the end of a 5-year time period. (Five years is 60 months.)
   c. How much more money accumulated in the annuity account than in the non-interest-earning account?

37. **Business: Annuities and Comparing Investments** Revenue flows continuously into a retail operation at the constant rate of $f(t) = \$10,000$ per month.
   a. If, at the end of each month, it is placed into a non-interest-earning account, determine how much money will have accumulated in the operation's account at the end of a 5-year time period. (Five years is 60 months.)

b. If, at the end of each month, it is placed into an annuity account paying 4% annual interest compounded continuously, determine how much money will have accumulated in the operation's account at the end of a 5-year time period. (Five years is 60 months.)

c. How much more money accumulated in the annuity account than in the non-interest-earning account?

© Katherine Heistand Shields

38. Business: Present Value of a Perpetual Income Stream Rental income from a beachfront home in Pismo Beach, CA, comes in to the owner of the property at the rate of $80,000 per year. There is a long-term indefinite lease on the property and the property owner immediately invests the income into an account paying 4% interest compounded continuously. Find and interpret the present value of the income stream.

39. Business: Present Value of a Perpetual Income Stream Rental income from a beachfront home in Kapa'a on the island of Kaua'i in Hawai'i comes in to the owner of the property at the rate of $70,000 per year. There is a long-term indefinite lease on the property and the property owner immediately invests the income into an account paying 7% interest compounded continuously. Find and interpret the present value of the income stream.

Using Technology Exercises

For each of the following problems, round to the nearest dollar.

40. Income flows continuously into a manufacturing company at the rate of $124,000 per year. If the income is invested at 5.5% compounded continuously, find

a. the future value of the money stream, and

b. the present value of the money stream.

41. Income flows continuously into a company at the rate of $36{,}450e^{-0.052t}$ dollars per year, where t is the number of years from now. If the income is invested at 8.5% compounded continuously, find

a. the future value of the money stream, and

b. the present value of the money stream.

42. Money flows continuously into a mutual fund at the rate of $1,000 per month. If the annual interest rate is 11%, find the future value of the annuity over

a. the first 12-month period, and

b. the first 20-year period.

43. Rental income from a piece of property, upon which there is an indefinite lease, flows at the rate of $62,575 per year. The income is invested immediately into an account paying 8.5% annual interest compounded continuously. Find and interpret the present value of the flow.

44. A company expects its profits over the next 6 years to be $f(t) = 22 + 8.6e^{0.03t}$ thousand dollars per year. If they immediately reinvest their profits at 7.4% interest, compounded continuously, what is the present value of the company?

Getting Ready for the Next Section

45. Evaluate $\int \frac{x}{1+x^2}\,dx$.

46. Find $\frac{dy}{dx}$ for $y = \ln(1 + x^2)^{1/2} + C$.

47. Solve $\frac{-1}{y} = x^3 + C$ for C when $x = 2$, $y = -0.1$.

48. Solve $\frac{-1}{y} = x^3 + 2$ for y.

49. For the equation $y^4 = 10x^2 + C$, first, find the value of C when $x = -4$ and $y = 3$, and rewrite the equation using the value of C. Second, solve the equation for y.

50. For the equation $\ln y = \frac{x^3}{3} + C$, first, find the value of C when $x = 0$ and $y = e^5$, and rewrite the equation using the value of C. Second, solve the equation for y.

51. Knowing that $p = 100$ when $t = 0$, solve the equation $200p^{1/2} = t + C$ for p.

Differential Equations

6.4

Many phenomena in business, life science, and the social sciences are governed by particular principles. So that we can predict the behavior of these phenomena, we wish to discover and then mathematically model these underlying principles. Discovery is often through observation and data collection. We can then construct mathematical models (functions or equations) by applying statistical or mathematical methods to the collected data, or by assuming the quantities involved, and/or their rates of change, are related by some proportion. (A quantity y is said to be proportional to a quantity x if $y = kx$, where k is a constant called the constant of proportionality. The quantity y is jointly proportional to x and z if $y = kxz$, and it is inversely proportional to x if $y = \frac{k}{x}$.)

Terminology

We often find that observations or assumptions of proportionality involve changes of functions so that the direct modeling function is unknown and must be found. Functions or equations that involve one dependent variable and one or more of its derivatives (or differentials) with respect to an independent variable are called **differential equations**. A function is a solution to the differential equation if it

1. relates the variables involved in the phenomena, and
2. satisfies the differential equation.

The following equations are examples of differential equations.

$$\frac{dy}{dx} = -\frac{2y}{3x}$$

$$(1 + x^2)y' = x$$

$$ty^3\,dt + e^{t^2}\,dy = 0$$

In each of these equations, the dependent variable is y, and the independent variable is either x or t. Also, each equation is an **ordinary differential equation** because there is only one independent variable. We will restrict our attention to ordinary differential equations.

Separation of Variables

Since solutions to differential equations can be challenging to find, we will examine only a special class of differential equations: the **separable variables** class. The solutions to these equations can be found relatively easily by separating the dependent and independent variables to opposite sides of the equal sign and integrating.

VIDEO EXAMPLES

SECTION 6.4

Example 1

a. Show that the differential equation $\frac{dy}{dx} = 15x^2y$ is separable.

b. Verify, by differentiation, that the function $y = Ce^{5x^3}$ is a solution to the separable differential equation $\frac{dy}{dx} = 15x^2y$.

Solution

a. The variables x and y in the equation $\frac{dy}{dx} = 15x^2y$ can be separated by multiplying each side by dx and then dividing each side by y.

$$\frac{dy}{dx} \cdot dx = 15x^2y \cdot dx$$

$$\frac{dy}{dx} \cdot dx = 15x^2y \cdot dx \qquad \text{Multiply each side by } dx$$

$$dy = 15x^2y\,dx \qquad \text{Divide each side by } y$$

$$\frac{dy}{y} = 15x^2\,dx$$

b. Differentiate $y = Ce^{5x^3}$.

$$\frac{dy}{dx} = Ce^{5x^3} \cdot 15x^2$$

$$= 15x^2 \cdot Ce^{5x^3}$$

Now, since we know that $y = Ce^{5x^3}$, we can substitute y for Ce^{5x^3} in the equation, getting $\frac{dy}{dx} = 15x^2y$, the original differential equation. Thus, Ce^{5x^3} is a solution to the differential equation. ■

A few examples of differential equations that are not separable are:

$$\frac{dy}{dx} = \frac{3x^2 - y^2}{2xy} \quad \text{and} \quad \frac{dy}{dx} = 2^{xy}$$

It is not possible to separate the variables in these equations. Our work in this course is concerned only with the separable variable class of differential equations.

Example 2 Solve $(1 + x^2)\frac{dy}{dx} = x$, and verify the solution.

Solution Dividing each side of the equation by $1 + x^2$ and multiplying each side by dx will separate the variables so that the dependent variable y appears on the left side of the equal sign and the independent variable x appears on the right side.

$$dy = \frac{x}{1 + x^2}\,dx$$

Now, we integrate each side.

$$\int dy = \int \frac{x}{1 + x^2}\,dx$$

$$y + C_1 = \int \frac{x}{1 + x^2}\,dx$$

To evaluate the integral $\int \frac{x}{1 + x^2}\,dx$, we let $u = 1 + x^2$, so that $du = 2x\,dx$ and $x\,dx = \frac{1}{2}\,du$. Then

$$y + C_1 = \int \frac{\frac{1}{2}\,du}{u}$$

$$y + C_1 = \frac{1}{2}\ln|u| + C_2$$

$$y + C_1 = \ln |u^{1/2}| + C_2$$

$$y + C_1 = \ln(1 + x^2)^{1/2} + C_2$$

$$y = \ln(1 + x^2)^{1/2} + \underbrace{C_2 - C_1}_{C}$$

$$y = \ln(1 + x^2)^{1/2} + C$$

Note We can drop the absolute value sign because $1 + x^2$ is always positive.

To verify that this is indeed a solution, we will differentiate it.

$$\frac{dy}{dx} = \frac{1}{(1 + x^2)^{1/2}} \cdot \frac{1}{2}(1 + x^2)^{-1/2} \cdot 2x$$

$$\frac{dy}{dx} = \frac{x}{1 + x^2}$$

Multiplying each side by $1 + x^2$ produces the original equation $(1 + x^2)\frac{dy}{dx} = x$, and we conclude that $y = \ln(1 + x^2)^{1/2} + C$ is a solution to the differential equation. ■

General Solutions

In Example 2, the solution $y = \ln(1 + x^2)^{1/2} + C$ involves an arbitrary constant C, which, we know from our examination of the indefinite integral, results in a *family of curves*. Solutions of differential equations that involve one or more arbitrary constants are called **general solutions** and represent a family of solutions. Figure 1 shows some of the members of the family $y = \ln(1 + x^2)^{1/2} + C$; we have shown the graphs for which $C = 0, 5, -5, 10, -10, 20, -20, 30$, and -30.

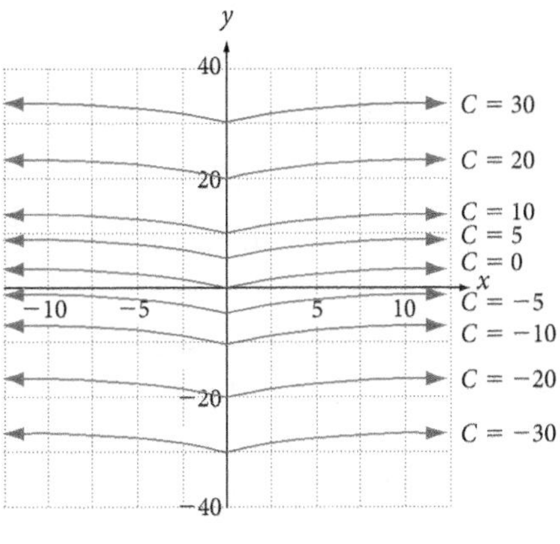

Figure 1

Initial Values and Particular Solutions

Often in applied situations, we will know some output value that corresponds to some input value; that is, we will know some initial condition. When this is the case, these values can be substituted into the general solution to obtain a value for the arbitrary constant C. Then, from all of the members of the family of solutions, we can specify the one that suits our particular situation. Solutions that arise from substituting initial conditions into general solutions are called **particular solutions** and the problems that produce them are called **initial-value problems**.

Example 3 Solve the differential equation $y' = 3x^2y^2$ subject to the condition that $y = -0.1$ when $x = 2$.

Solution Since we know that $y = -0.1$ when $x = 2$, this problem is an initial-value problem. Begin by writing y' as $\frac{dy}{dx}$, then separate the variables and integrate.

$$y' = 3x^2y^2$$

$$\frac{dy}{dx} = 3x^2y^2$$

$$\frac{dy}{y^2} = 3x^2\,dx$$

$$\int y^{-2}\,dy = \int 3x^2\,dx$$

$$\frac{y^{-1}}{-1} = x^3 + C$$

$$\frac{-1}{y} = x^3 + C$$

Now we substitute 2 for x and -0.1 for y to find the value of C, and hence, the particular solution.

$$\frac{-1}{-0.1} = 2^3 + C$$

$$10 = 8 + C$$

$$2 = C$$

Thus,

$$\frac{-1}{y} = x^3 + 2 \quad \text{and} \quad y = \frac{-1}{x^3 + 2}$$

Therefore, $y = \dfrac{-1}{x^3 + 2}$ is the particular solution. ■

Example 4 Solve the differential equation $y' = \frac{5x}{y^3}$ subject to the condition that $y = 3$ when $x = -4$.

Solution Since we know that $y = 3$ when $x = -4$, this problem is an initial-value problem. Begin by writing y' as $\frac{dy}{dx}$, then separate the variables and integrate.

$$y' = \frac{5x}{y^3}$$

$$\frac{dy}{dx} = \frac{5x}{y^3}$$

$$y^3\,dy = 5x\,dx$$

$$\int y^3\,dy = \int 5x\,dx$$

$$\frac{y^4}{4} = \frac{5x^2}{2} + C$$

$$y^4 = 10x^2 + C$$

Now we substitute -4 for x and 3 for y to find C.

$$3^4 = 10(-4)^2 + C$$

$$81 = 10 \cdot 16 + C$$

$$81 = 160 + C$$

$$-79 = C$$

Thus,

$$y^4 = 10x^2 - 79$$

$$y = \pm \sqrt[4]{10x^2 - 79}$$

Therefore, the particular solution is $y = \pm \sqrt[4]{10x^2 - 79}$.

Example 5 Solve the differential equation $y' = x^2 y$ subject to the condition that $y = e^5$ when $x = 0$.

Solution Since we know that $y = e^5$ when $x = 0$, this problem is an initial-value problem. Begin by writing y' as $\frac{dy}{dx}$, then separate the variables and integrate.

$$y' = x^2 y$$

$$\frac{dy}{dx} = x^2 y$$

$$\frac{dy}{y} = x^2\, dx$$

$$\int \frac{dy}{y} = \int x^2\, dx$$

$$\ln y = \frac{x^3}{3} + C$$

Now we substitute 0 for x and e^5 for y to find C.

$$\ln e^5 = \frac{0^3}{3} + C$$

$$5 = 0 + C$$

$$5 = C$$

Thus,

$$\ln y = \frac{x^3}{3} + 5$$

$$y = e^{\frac{x^3}{3}+5}$$

Therefore, the particular solution is $y = e^{\frac{x^3}{3}+5}$ or $y = e^5 e^{\frac{x^3}{3}}$.

Example 6 The growth of a new restaurant depends on the number of its regular customers as

$$p'(t) = 0.01\sqrt{p(t)}$$

where $p(t)$ represents the number of regular customers at any time t and $p'(t)$ represents the rate at which the number of regular customers is changing at any time t. Solve this differential equation subject to the condition that $p = 100$ when $t = 0$.

Solution Since we know that $p = 100$ when $t = 0$, this problem is an initial-value problem. Begin by writing $p'(t)$ as $\frac{dp}{dt}$ and $p(t)$ as p, then separate the variables and integrate.

$$p'(t) = 0.01\sqrt{p(t)}$$

$$\frac{dp}{dt} = 0.01p^{1/2}$$

$$100\frac{dp}{dt} = p^{1/2}$$

$$100\frac{dp}{p^{1/2}} = dt$$

$$100\int p^{-1/2}\,dp = \int dt$$

$$100 \cdot \frac{p^{1/2}}{1/2} = t + C$$

$$200p^{1/2} = t + C$$

Now we substitute 0 for t and 100 for p to find C.

$$200 \cdot (100)^{1/2} = 0 + C$$

$$C = 2{,}000$$

Thus,

$$200p^{1/2} = t + 2{,}000$$

$$p = \left[\frac{t + 2{,}000}{200}\right]^2$$

Therefore, the particular solution is $p(t) = \left[\frac{t + 2{,}000}{200}\right]^2$. ■

Using Technology 6.4

You can use Wolfram|Alpha to help you solve a differential equation.

Example 7 In Example 2 we solved the differential equation

$$(1 + x^2)\frac{dy}{dx} = x$$

Use Wolfram|Alpha to solve this equation.

Solution Go to www.wolframalpha.com and in the entry field enter

solve (1+x^2)y'(x)=x

Wolfram|Alpha displays your entry, its interpretation of your entry, and the solution to the equation.

Differential equation solution:

$$y(x) = c_1 + \frac{1}{2}\log(x^2 + 1)$$

Wolfram Alpha LLC. 2012. Wolfram|Alpha
http://www.wolframalpha.com/
(access December 12, 2012)

Notice that Wolfram|Alpha writes the 1/2 as a coefficient rather than an exponent and c_1 to represent the constant of integration C. Although Wolfram|Alpha writes the solution to the differential equation differently than we did in Example 2, you can see that the two forms are equivalent. ■

Example 8 In Example 3, we solved the differential equation $y' = 3x^2y^2$ subject to the initial condition that $y = -0.1$ when $x = 2$. Use Wolfram|Alpha to solve this equation.

Solution Go to www.wolframalpha.com and in the entry field enter

solve y'(x) = 3x^2 y^2 , y(2) = -0.1

Wolfram|Alpha uses the notation $y'(x)$ to indicate that the variable y has been differentiated with respect to the variable x and the notation $y(2) = -0.1$ to indicate that $y = -0.1$ when $x = 2$.

Wolfram|Alpha displays your entry, its interpretation of your entry, and the solution to the equation.

Notice that the Wolfram|Alpha form of the solution is different, but equivalent to the form of the solution to Example 3. That is,

$$y = \frac{1}{-x^3 - 2} = \frac{1}{-(x^3 + 2)} = \frac{-1}{x^3 + 2}$$

■

Example 9 You may find that Wolfram|Alpha solves a differential equation and presents the solution in a form very different from the form you develop. In Example 6, we solved the differential equation $p'(t) = 0.01\sqrt{p(t)}$ subject to the initial condition that $p = 100$ when $t = 0$.

Solution The solution to the differential equation as presented in Example 6 was

$$p = \left[\frac{t + 2{,}000}{200}\right]^2$$

Wolfram|Alpha presented this solution as $p(t) = 0.000025t^2 + 0.1t + 100$. These two forms appear to be quite different. But, as a little algebra shows, they are the same.

$$p = \left[\frac{t + 2{,}000}{200}\right]^2 = \left[\frac{t}{200} + \frac{2{,}000}{200}\right]^2 = [0.005t + 10]^2$$
$$= 0.000025t^2 + 0.1t + 100$$

Getting Ready for Class

After reading through the preceding section, respond in your own words and in complete sentences.

A. What is an ordinary differential equation?
B. Describe the method of separation of variables.
C. In the context of a differential equation, what is a general solution and how is one represented graphically?
D. Solutions of differential equations that arise from substituting initial conditions into the general solution are given a name. What is that name? What is the name of the problems that produce such solutions?

Problem Set 6.4

Skills Practice

For Problems 1-12, find the general solution to each differential equation.

1. $y' = x^4$
2. $y' = 7$
3. $\frac{dy}{dx} = 0$
4. $\frac{dy}{dx} = 25 - 4x$
5. $y\frac{dy}{dx} = 2x$
6. $\frac{4}{3} \cdot \frac{dy}{dx} = e^{3x-3}$
7. $y' = xe^{x^2}$
8. $xy' = 1$
9. $(1 + x)\frac{dy}{dx} = x$
10. $(1 + x)\frac{dy}{dx} + (1 + y) = 0$
11. $x\frac{dy}{dx} = y \ln y$
12. $y' = \frac{xy}{x - 1}$

For Problems 13-22, find the particular solution to each differential equation.

13. $y' = \frac{x^2}{y^2}$, and $y = 4$ when $x = 0$.
14. $y' = \frac{x^2}{y^3}$, and $y = 1$ when $x = 1$.
15. $y' = \frac{1}{xy}$, and $y = 4$ when $x = 1$.
16. $y' = \frac{x}{x^2 + 1}$, and $y = 2$ when $x = 0$.
17. $y' = xe^x - x$, and $y = -1$ when $x = 0$.
18. $(x^2 - 3)y' = 4x$, and $y = 5$ when $x = 2$.
19. $2xyy' = 1 + y^2$, and $y = 3$ when $x = 5$.
20. $\frac{dR}{dS} = \frac{k}{S}$, and $R = 0$ when $S = S_0$. (This equation is called the **Weber-Fechner law** and describes the relationship between a stimulus S and a response R.)
21. $\frac{dP}{dt} = kP$, and $P = P_0$ when $t = 0$. (This equation models unlimited growth.)
22. $\frac{dP}{dt} = k(L - P)$, where L is a constant, and $P = P_0$ when $t = 0$. (This equation models limited growth.)

Check Your Understanding

For Problems 23-30, verify that each function is a solution to the corresponding differential equation.

23. $y = 2x^2 + C$, $\frac{dy}{dx} = 4x$
24. $y = 2x^4 + x^2 + C$, $\frac{dy}{dx} = 8x^3 + 2x$
25. $y = Ce^{x^2}$, $\frac{dy}{dx} = 2xy$
26. $y^2 = x^2 + C$, $\frac{dy}{dx} = \frac{x}{y}$

27. $y = Ce^x, \ \frac{dy}{dx} = y$

28. $y = Ce^{kx}, \ \frac{dy}{dx} = ky$

29. $\ln x + \frac{1}{y} = C, \ \frac{dy}{dx} = \frac{y^2}{x}$

30. $y = e^{\frac{-4}{5} \cdot x} + 10, \ 5y' + 4y = 0$

For Problems 31-38, state whether the differential equation is in the separable variable class or not.

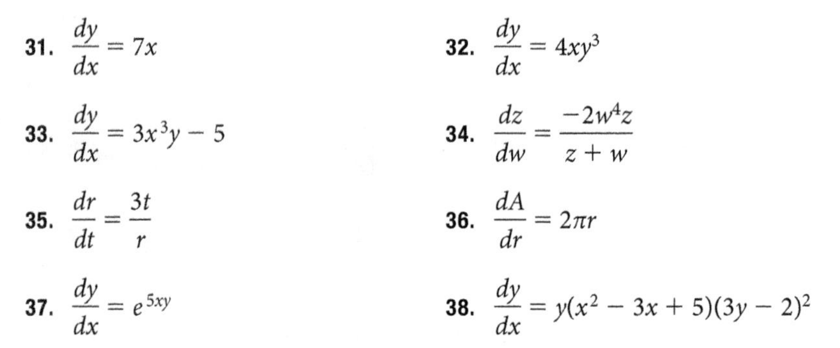

31. $\frac{dy}{dx} = 7x$

32. $\frac{dy}{dx} = 4xy^3$

33. $\frac{dy}{dx} = 3x^3y - 5$

34. $\frac{dz}{dw} = \frac{-2w^4z}{z + w}$

35. $\frac{dr}{dt} = \frac{3t}{r}$

36. $\frac{dA}{dr} = 2\pi r$

37. $\frac{dy}{dx} = e^{5xy}$

38. $\frac{dy}{dx} = y(x^2 - 3x + 5)(3y - 2)^2$

Modeling Practice

39. Education: Rate of Learning In a community college's court reporting program, students are believed to progress at the rate of

$$\frac{dQ}{dt} = k(160 - Q)$$

where Q represents the number of words per minute a student can type t weeks after the start of the program. If the average student cannot type any words initially and can type 35 words per minute 4 weeks into the program, how many words per minute can the average student be expected to type after 18 weeks into the program?

40. Economics: Investment Two years ago, \$6,000.00 was placed into an investment in which the amount of money present, P, grows according to

$$\frac{dP}{dt} = kP$$

Today, the investment is worth \$7,328.00. How much will the investment be worth 5 years from now? (Round to the nearest cent.)

41. Social Science: White-Collar Crime A county's statistics indicate that the number of white-collar crimes is growing according to the formula

$$\frac{dN}{dt} = 0.04N$$

If 135 white-collar crimes were reported in 2009, how many can be expected to be reported in 2014 if this trend continues? (Hint: Use $y = Ce^{kt}$.)

42. Manufacturing: Newton's Law of Cooling Newton's Law of Cooling states that

$$\frac{dT}{dt} = k(T - S)$$

where T represents the temperature of an object t minutes after it is removed from a heater (or refrigerator), and S is the temperature of the surrounding medium. If an object is 225 °F when it is removed from an oven and placed into a room with a constant temperature of 70 °F, and has cooled to 200 °F 20 minutes later, what will be its temperature 60 minutes from the time it is removed from the oven?

43. Ergonomics: Aircraft Cabin Pressure As one leaves the surface of the earth and rises into the atmosphere, the pressure exerted by the atmosphere diminishes according to

$$\frac{dP}{dh} = kP$$

where P represents pressure (in pounds per square inch), and h represents the height (in feet) above sea level. At 18,000 feet above sea level, the pressure is half of what it is at sea level; that is, at 18,000 feet above sea level,

$$P = \frac{1}{2}P_0$$

where P_0 is the pressure at sea level. Find the pressure, as a percentage of P_0, at 10,000 feet.

44. Medicine: Healing Time for Wounds Medical researchers have found that the rate at which the number of square centimeters A of unhealed skin changes t days after receiving a wound is given by

$$\frac{dA}{dt} = kA$$

If a wound is initially 2 square centimeters, and only 1.3 square centimeters after 3 days, what will be the size of the wound 7 days after it is first received? (Round to two decimal places.)

Source: National Center for Biotechnology Information

© Microgen/iStockPhoto

45. Ecology: Pollution The rate of change of the concentration C (in parts per million, ppm) of pollutants that enter a particular river each day is related to the number x of homes near the river by the differential equation

$$\frac{dC}{dx} = kC$$

When there were no homes near the river the number of pollutants entering the river each day is believed to have been 4 ppm, and when there were 50 homes near the river, the number of pollutants entering the river each day was 10 ppm. A contractor is contemplating a new housing site with 20 homes. With these additional homes, how many pollutants will be entering the river each day?

46. Business: Marginal Cost A company's marginal cost for producing x units of a product is given by

$$\frac{dC}{dx} = \frac{k}{C^{1/3}}$$

The company's fixed costs are \$2,300, and it costs \$8,500 to make 25 units.

a. Find a function that gives the cost of producing x units of the product.

b. Find how much it costs to produce 100 units.

47. Political Science: Voting In a particular county, the percentage, P, of registered voters who do not vote in presidential elections is changing at the rate of

$$\frac{dP}{dt} = k\sqrt[3]{P}$$

where t is the time in years, with $t = 0$ being 2000. If in 2000, sixty percent of all registered voters voted, but in 2004 only 53% voted, what percentage can be expected to vote in 2016?

Source: United States Census Bureau

48. Biology: Cell Nutrients A cell receives nutrients through its surface. The function $w'(t) = k[w(t)]^{2/3}$ expresses the relationship between $w(t)$ (the weight of the cell at time t) and $w'(t)$ (the rate at which the weight is changing), where k is a positive constant that depends on the type of cell. Solve this differential equation subject to the condition that the initial weight is 2 units.

Source: Cell Metabolism
www.cell.com/cell-metabolism/home

49. Economics: Reinvestment A farmer reinvests his profits in buying neighboring land. The growth rate of the land he owns is given by $a'(t) = 0.2[a(t)]^{3/4}$, where a is the area, in acres, of the land he owns. Solve this differential equation for a, subject to the condition that $a = 20$ when $t = 0$.

50. Business: Aerospace Employment Employment in aerospace-related industries in a state varies as

$$\frac{dy}{dt} = 125 - 0.03y$$

where y is the number of employees t years after the year 2000.

a. Solve this differential equation for $y(t)$ if $y = 85{,}000$ when $t = 0$.

b. Estimate the number of employees this state had in the year 2010.

c. Estimate the number of employees this state will have in the year 2015.

d. At what rate will employment in aerospace-related industries in this state be changing in the year 2015?

Source: govcentral.monster.com

Using Technology Exercises

Use Wolfram|Alpha to help you solve each differential equation.

51. $y' = -\frac{y}{x^2}$

52. $(x \ln y)y' = y \ln y$

53. $y' = x^2e^x - x$ subject to the initial condition that $y = 2$ when $x = 0$

54. $y' = \frac{1}{xy}$ subject to the initial condition that $y = 4$ when $x = 1$

Getting Ready for the Next Section

55. Solve for P: $\frac{dP}{dt} = kP$.

56. Solve $242{,}000 = 240{,}000e^{kt}$ for k when $t = 4$. Round your result to four decimal places.

57. Evaluate $\int \frac{dP}{L - P}$ where L is a constant.

58. Solve $P = L - Ce^{-kt}$ for C when $t = 0$ and $P = P_0$.

59. Solve $160 = 30 + 190e^{-k(20)}$ for k. Round your result to three decimal places.

60. Approximate $P = 4{,}000 - 3{,}920e^{-0.592t}$ when $t = 6$.

61. To nine decimal places, find the value of k in the equation

$$P = \frac{114{,}000{,}000}{3{,}800 + 26{,}200e^{-30{,}000kt}}$$

when $P = 4{,}500$ and $t = 24$.

Spotlight on **Success**

Student Instructor Stefanie

Never confuse a single defeat with a final defeat.
—F. Scott Fitzgerald

The idea that has worked best for my success in college, and more specifically in my math courses, is to stay positive and be resilient. I have learned that a 'bad' grade doesn't make me a failure; if anything it makes me strive to do better. That is why I never let a bad grade on a test or even in a class get in the way of my overall success.

By sticking with this positive attitude, I have been able to achieve my goals. My grades have never represented how well I know the material. This is because I have struggled with test anxiety and it has consistently lowered my test scores in a number of courses. However, I have not let it defeat me. When I applied to graduate school, I did not meet the grade requirements for my top two schools, but that did not stop me from applying.

One school asked that I convince them that my knowledge of mathematics was more than my grades indicated. If I had let my grades stand in the way of my goals, I wouldn't have been accepted to both of my top two schools, and will be attending one of them in the Fall, on my way to becoming a mathematics teacher.

Applications of Differential Equations

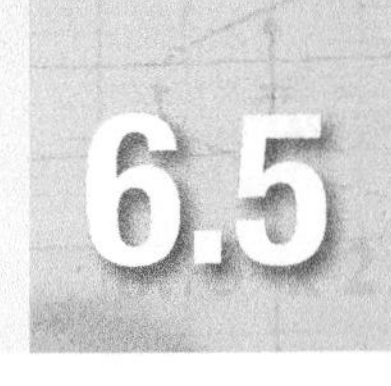

In many phenomena one quantity changes proportionally to another. In the last section, Section 6.4, we saw some examples of such changes. The examples of this section illustrate how mathematical models that can be used to predict the behavior of these phenomena can be constructed using the idea of proportionality. We develop three models: the unlimited growth model, the limited growth model, and the logistic growth model.

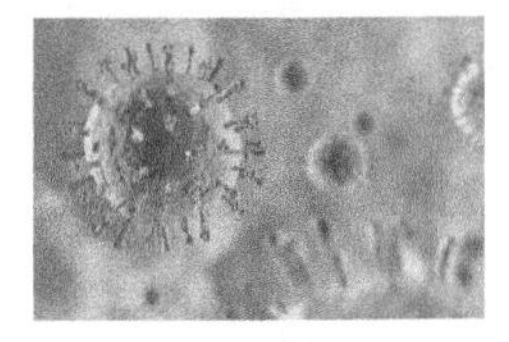

Models of Unlimited Growth

Many populations (not just biological ones) grow in such a way that the larger the population is, the faster it grows. The **law of natural growth** states that *the rate at which a population changes over time is proportional to the current size of the population.* We can symbolize this law and construct a mathematical model that represents it by letting P represent the size of the population at time t. Since the law assumes no limit to the size of the population, we refer to the model as the **unlimited growth model.**

> **Unlimited Growth Model**
> If P represents the size of a population, t represents time, and k is a constant, then
>
> $$\frac{dP}{dt} = kP$$

We can solve for the unknown function, P, by separating the variables and integrating.

$$\frac{dP}{dt} = kP$$

$$\frac{dP}{P} = k\,dt$$

$$\int \frac{dP}{P} = \int k\,dt$$

$$\ln P = kt + C_1$$

To solve for P, we need to eliminate the ln. So we take exponentials of each side.

$$e^{\ln P} = e^{kt+C_1}$$

Then, using the properties of exponents,

$$P = e^{kt} \cdot e^{C_1}$$

Since e^{C_1} represents a constant raised to a constant, it is itself a constant that we will simply call C. This gives us

$P = Ce^{kt}$

$$P = Ce^{kt}$$

Thus, the solution to the differential equation $\frac{dP}{dt} = kP$ is the exponential function

$$P = Ce^{kt}$$

The constant k is called the **growth constant** and it represents a positive number. (Keep in mind that if k is positive, then $-k$ represents its opposite and is a negative number, and that might remind us of decay.) Figure 1 shows the graphs of each model. (These should look familiar; you studied them in Chapter 4.)

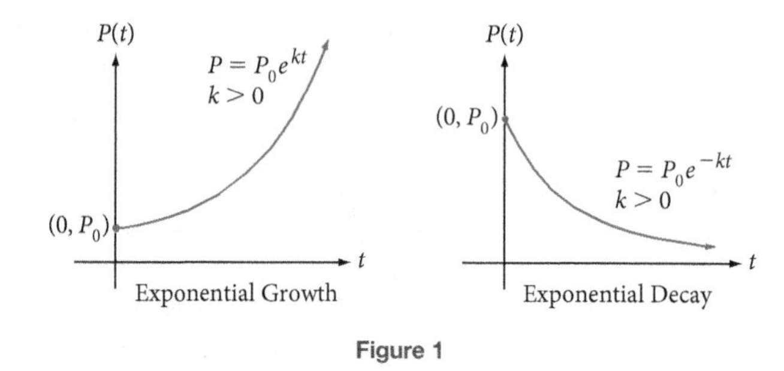

Figure 1

If, at some initial observation time t, the population size P is P_0, that is, if $P = P_0$ when $t = 0$, then $P = P_0e^{kt}$. This fact is illustrated in Example 1 and you will be asked to prove it in general in the exercises. Now the size of the population at $t = 0$ can be immediately substituted for the arbitrary constant.

Example 1 The value of real estate in a city is increasing, without limitations, at a rate proportional to its current value. A piece of property in the city appraised at \$240,000 4 months ago appraises at \$242,000 today. At what value will the property be appraised 12 months from now?

VIDEO EXAMPLES

SECTION 6.5

Solution The unlimited growth model applies to this situation. Let P represent the value of the property at time t, where $t = 0$ is the time at which the property was appraised at \$240,000. Then the solution to the unlimited growth differential equation

$$\frac{dP}{dt} = kP$$

is $P = 240{,}000e^{kt}$.

We can solve for k, the growth constant, using the fact that when $t = 4$ (which corresponds to today—4 months from the initial appraisal), $P = 242{,}000$. Then the solution to this particular situation (the particular solution) can be specified.

$$242{,}000 = 240{,}000e^{k(4)}$$

$$\frac{242{,}000}{240{,}000} = e^{4k}$$

$$\ln\frac{242{,}000}{240{,}000} = \ln e^{4k}$$

$$\ln\frac{242{,}000}{240{,}000} = 4k$$

$$k = \frac{\ln\frac{242{,}000}{240{,}000}}{4}$$

$$k \approx 0.0021$$

Thus, the particular solution is $P = 240{,}000e^{0.0021t}$. Figure 2 shows $P(t)$.

Now, using this mathematical model we can predict the value of this piece of real estate at any time t. Specifically, 12 months from now is $t = 4 + 12 = 16$ months from when $t = 0$, so that $t = 16$. Then

$$P = 240{,}000e^{0.0021(16)}$$

$$P \approx 248{,}201$$

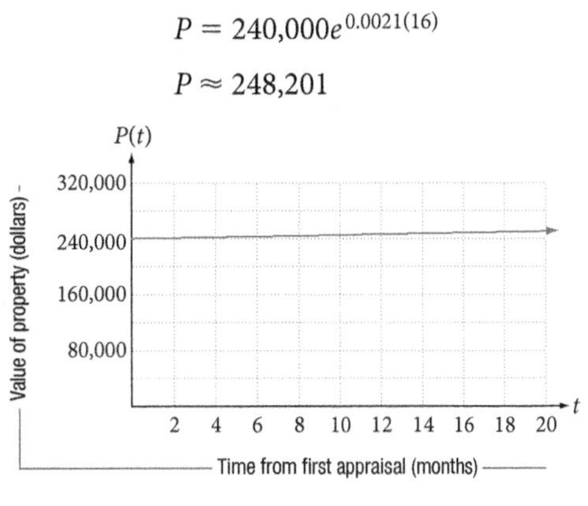

Figure 2

Interpretation
Twelve months from now, this piece of property will appraise at approximately \$248,201. The growth constant k in this case is 0.0021. This means that the value of this property is increasing (growing) at the rate of 0.21% per month.

Models of Limited Growth

Many populations grow in such a way that the rate of increase is large when the population is small and then decreases (because of limitations on growth) as the population increases. Let P represent the size of the population at time t and L be a fixed upper (or lower) limit to the size of the population. When the number L represents an upper limit, it is called the **carrying capacity** of the population and represents the maximum size of the population.

Limited Growth Model
When the rate at which a population changes with respect to time is proportional to the difference between the carrying capacity and the current size of the population, the mathematical model for growth (or decay) is

$$\frac{dP}{dt} = k(L - P)$$

If P_0 is the size of the population at some initial observation time $t = 0$, then the particular solution is

$$P = L + (P_0 - L)e^{-kt}$$

We can derive this particular solution by separating the variables and integrating as before.

$$\frac{dP}{dt} = k(L - P)$$

$$\frac{dP}{L-P} = k\,dt$$

$$\int \frac{dP}{L-P} = \int k\,dt$$

$$-\ln(L-P) + C_1 = kt + C_2$$

$$\ln(L-P) + C_1 = -kt - C_2$$

$$\ln(L-P) = -kt \underbrace{- C_2 - C_1}_{C_3}$$

$$\ln(L-P) = -kt + C_3$$

$$e^{\ln(L-P)} = e^{-kt+C_3}$$

$$L - P = e^{-kt} \cdot \underbrace{e^{C_3}}_{C}$$

$$L - P = Ce^{-kt}$$

$$-P = -L + Ce^{-kt}$$

$$P = L - Ce^{-kt}$$

Now, from the initial condition that $P = P_0$ when $t = 0$,

$$P_0 = L - Ce^{-k(0)}$$

$$P_0 = L - Ce^0$$

$$P_0 = L - C$$

$$C = L - P_0$$

and thus,

$$P = L - (L - P_0)e^{-kt}$$

Factor -1 from $(L - P_0)$ to get

$$P = L + (P_0 - L)e^{-kt}$$

Figure 3 shows the graph of this particular solution for the cases when L is a fixed upper limit and when L is a fixed lower limit, along with phenomena these cases might model.

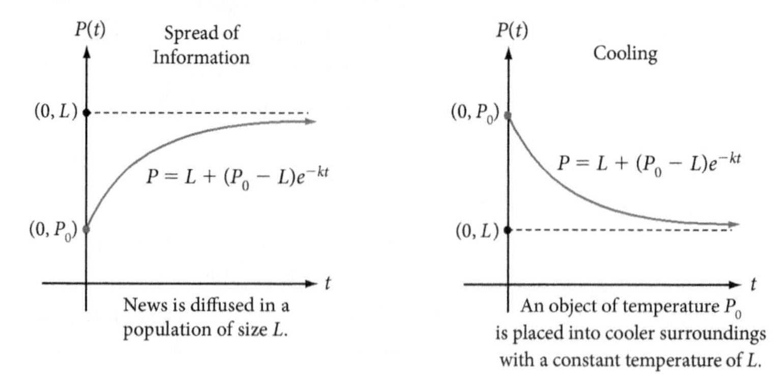

Figure 3

Example 2 The temperature of a machine when it is first shut down after operating is 220 °C. The surrounding air temperature is 30 °C. After 20 minutes, the temperature of the machine is 160 °C. Find a function that gives the temperature of the machine at any time, t, and then find the temperature of the machine 30 minutes after it is shut down.

Solution The limited decay model applies in this situation because there is a lower limit to the temperature of the machine. Let P represent the temperature of the machine at time t and let $t = 0$ be the time when the machine is shut down after operating. At $t = 0$, the initial temperature of the machine is $P_0 = 220$ °C, and the limiting temperature is 30 °C, the temperature of the surrounding air. Then

$$\begin{aligned} P &= 30 + (220 - 30)e^{-kt} \\ &= 30 + 190e^{-kt} \end{aligned}$$

We can use the initial condition that $P = 160$ when $t = 20$ to find the value of the growth constant (decay constant, in this case) k.

$$\begin{aligned} 160 &= 30 + 190e^{-k(20)} \\ 130 &= 190e^{-20k} \\ \frac{130}{190} &= e^{-20k} \\ \ln\frac{13}{19} &= \ln e^{-20k} \\ \ln\frac{13}{19} &= -20k \\ \frac{\ln\frac{13}{19}}{-20} &= k \\ 0.019 &\approx k \end{aligned}$$

Then $P = 30 + 190e^{-0.019t}$ models this particular situation and gives the temperature of the machine at any time t. Figure 4 shows $P(t)$. Now, using this mathematical model we can predict the temperature of the machine at any time, t. Specifically, 30 minutes from the time the machine is shut down means that $t = 30$. Then

$$\begin{aligned} P &= 30 + 190e^{-0.019(30)} \\ P &\approx 137.45 \end{aligned}$$

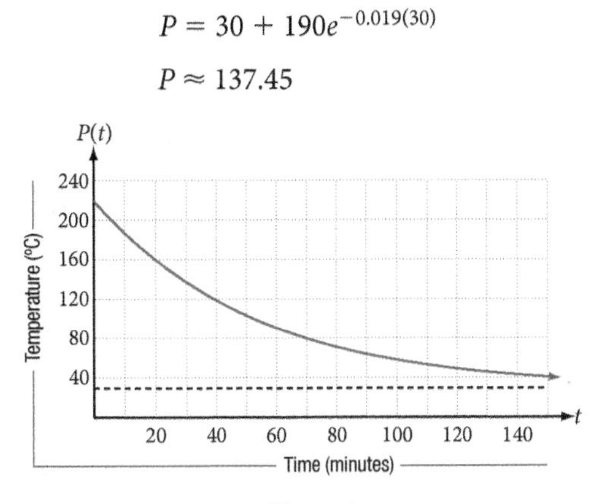

Figure 4

> **Interpretation**
> Thirty minutes after the machine is shut down, its temperature will be approximately 137.45 °C. The growth constant k in this case is 0.019. This means that the temperature of the machine is decreasing at the rate of 1.9% each minute.

■

The differential equation $\frac{dP}{dt} = k(L - P)$ also models phenomena such as the diffusion of information into a population and the learning of new tasks. Both these phenomena grow quickly at first, then more slowly. Information spreads quickly at first and then more slowly as there are fewer people to hear it. In a similar way, people learn or become better at a new task quickly at first and then more slowly as time goes by.

Example 3 A government agency with 4,000 employees has introduced a new software application on its personal computers. Initially, 80 employees knew how to use the application very well, but after two weeks, 2,800 employees knew how to use it very well. The number of people who learn how to use the application increases at a rate that is proportional to the difference between the total number of employees at the agency and the number of employees who have learned to use it at any time t (in weeks). Find a function that gives the number of employees who have learned to use the software at any time t, then find the number of employees who have learned to use it 6 weeks after its introduction into the agency.

Solution The differential equation $\frac{dP}{dt} = k(L - P)$ with solution

$$P = L + (P_0 - L)e^{-kt}$$

is the appropriate model for this application. In this case, $L = 4{,}000$, and $P_0 = 80$. Thus,

$$\begin{aligned} P &= 4{,}000 + (80 - 4{,}000)e^{-kt} \\ &= 4{,}000 - 3{,}920e^{-kt} \end{aligned}$$

We can solve for k using the initial condition that $P = 2{,}800$ when $t = 2$. (You should be able to solve for k using Examples 1 and 2 as guides.)

$$\begin{aligned} 2{,}800 &= 4{,}000 - 3{,}920e^{-k(2)} \\ k &\approx 0.592 \end{aligned}$$

Then, $P = 4{,}000 - 3{,}920e^{-0.592t}$ models this particular situation and gives the number of employees who have learned the software application at any time t. Figure 5 shows the graph of $P(t)$.

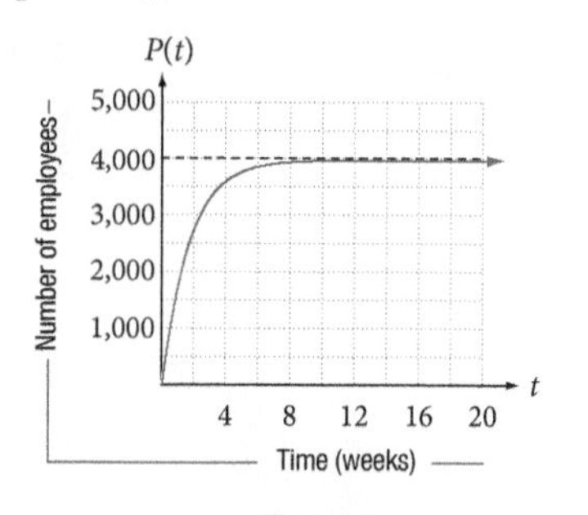

Figure 5

Specifically, in 6 weeks, $t = 6$, so that

$$P = 4{,}000 - 3{,}920e^{-0.592(6)}$$
$$P \approx 4{,}000 - 3{,}920(0.0287)$$
$$P \approx 4{,}000 - 112.504$$
$$P \approx 3{,}888$$

Interpretation
Six weeks from the time the software application is introduced into the agency, approximately 3,888 of the 4,000 employees have learned to use it.

Models of Logistic Growth

When bounds are placed on population growth, the **logistic growth** function may best model the phenomena. Phenomena that exhibit logistic growth grow exponentially at first and then more slowly, as in the limited growth model. Figure 6 shows the graph of the logistic growth curve. We saw the logistic growth function in Example 7 in Section 4.1.

As in limited growth, the number L is called the *carrying capacity* of the population.

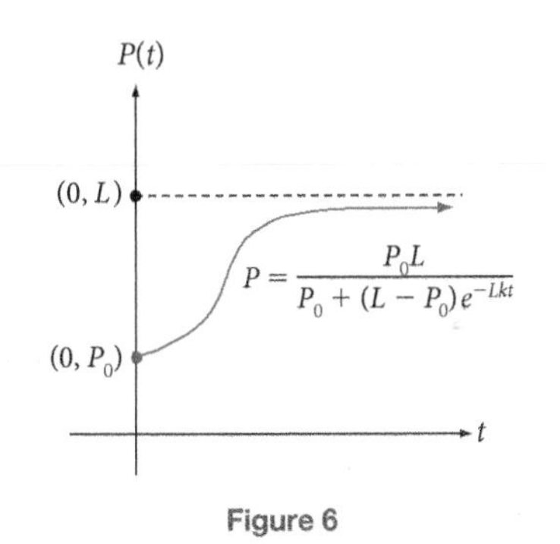

Figure 6

Logistic Growth Model
When the rate at which a population changes is proportional to the current size of the population and the difference between the carrying capacity and the current size of the population, the mathematical model is

$$\frac{dP}{dt} = kP(L - P)$$

If P_0 is the size of the population at time $t = 0$, then the particular solution is

$$P = \frac{P_0 L}{P_0 + (L - P_0)e^{-Lkt}}$$

The derivation of this model is beyond the scope of this course, but we will illustrate its use with an example.

Example 4 Twenty-four months ago, the membership of a business marketer's professional organization was 3,800. Today it is 4,500. The number of members increases at a rate that is proportional to the current size of the membership and the difference between the total number of potential members and the current membership. If there are 30,000 potential members, find a function that gives the number of members, and find the number of members 4 months from now.

Solution The differential equation $\frac{dP}{dt} = kP(L - P)$ with the particular solution

$$P = \frac{P_0 L}{P_0 + (L - P_0)e^{-Lkt}}$$

is the appropriate model for this situation. In this case, $L = 30{,}000$, $P_0 = 3{,}800$, and, since t is measured from 24 months ago, $t = 24$ is the present. Thus,

$$\begin{aligned} P &= \frac{3{,}800(30{,}000)}{3{,}800 + (30{,}000 - 3{,}800)e^{-30{,}000kt}} \\ &= \frac{114{,}000{,}000}{3{,}800 + 26{,}200e^{-30{,}000kt}} \end{aligned}$$

We can solve for k using the initial condition that $P = 4{,}500$ when $t = 24$.

$$\begin{aligned} 4{,}500 &= \frac{114{,}000{,}000}{3{,}800 + 26{,}200e^{-30{,}000k(24)}} \\ 4{,}500 &= \frac{114{,}000{,}000}{3{,}800 + 26{,}200e^{-720{,}000k}} \\ 4{,}500(3{,}800 + 26{,}200e^{-720{,}000k}) &= 114{,}000{,}000 \\ 17{,}100{,}000 + 117{,}900{,}000e^{-720{,}000k} &= 114{,}000{,}000 \\ e^{-720{,}000k} &= \frac{114{,}000{,}000 - 17{,}100{,}000}{117{,}900{,}000} \\ e^{-720{,}000k} &\approx 0.8219 \\ -720{,}000k &\approx \ln(0.8219) \\ k &\approx \frac{\ln(0.8219)}{-720{,}000} \\ k &\approx 0.000000272 \end{aligned}$$

Then $P \approx \dfrac{114{,}000{,}000}{3{,}800 + 26{,}200e^{-30{,}000(0.000000272)t}}$, so that,

$$\begin{aligned} P &\approx \frac{114{,}000{,}000}{3{,}800 + 26{,}200e^{-0.00816t}} \\ &\approx \frac{1{,}140{,}000}{38 + 262e^{-0.00816t}} \end{aligned}$$

models this particular situation and gives the number of members at any time, t.

Four months from now, $t = 24 + 4 = 28$, so that

$$P \approx \frac{1{,}140{,}000}{38 + 262e^{-0.00816(28)}}$$

$$\approx 4{,}625$$

Interpretation

Four months from now, the professional organization can expect to have approximately 4,625 members.

■

Using Technology 6.5

Wolfram|Alpha can be helpful in finding the general solution to a differential equation, then evaluating the result at particular values.

Example 5 Use Wolfram|Alpha to help you find the general solution to the differential equation

$$\frac{dP}{dt} = kP$$

Solution Go to www.wolframalpha.com and in the entry field enter

solve p' (t) = k*p(t)

Rather than entering this equation using the capital letter "P," use the lowercase "p." Wolfram|Alpha reserves the capital P for other quantities in its knowledge base. Enclosing the variable t in parentheses, tells Wolfram|Alpha that P is a function of t. You must tell Wolfram|Alpha that you want to differentiate with respect to t. Otherwise, Wolfram|Alpha may think you wish to solve $\frac{dP}{dt} = k$ rather than $\frac{dP}{dt} = kP$. It is a good idea, when two letters are involved, to indicate a multiplication by using the asterisk symbol "*." k*p(t) makes it clear to Wolfram|Alpha that k and $p(t)$ are to be multiplied. Wolfram|Alpha may see $kp(t)$ as the name of one variable.

Wolfram|Alpha displays your entry, its interpretation of your entry, and the solution to the equation.

Differential equation solution:

$$p(t) = c_1 e^{kt}$$

Wolfram Alpha LLC. 2012. Wolfram|Alpha
http://www.wolframalpha.com/
(access December 12, 2012)

Notice that Wolfram|Alpha uses c_1 to represent the constant of integration C.

■

Example 6 In Example 1, we solved the differential equation $\frac{dP}{dt} = kP$ subject to the conditions that $P = 240{,}000$ when $t = 0$ and $P = 242{,}000$ when $t = 4$. Use Wolfram|Alpha to solve this equation.

Solution Go to www.wolframalpha.com and in the entry field enter

solve p'(t) = k*p(t) , p(0) = 240,000, real

Wolfram|Alpha uses the notation $p'(t)$ to indicate that the variable p has been differentiated with respect to the variable t and the notation $p(0) = 240{,}000$ to indicate that $p = 240{,}000$ when $t = 0$.

Wolfram|Alpha displays your entry, its interpretation of your entry, and the solution to the equation.

Differential equation solution:

$$p(t) = 240\,000\, e^{k\,t}$$

Wolfram Alpha LLC. 2012. Wolfram|Alpha
http://www.wolframalpha.com/
(access December 12, 2012)

Use the condition that $p = 242{,}000$ when $t = 4$ to solve for the growth constant k.

In the Wolfram|Alpha entry field, enter

solve 242,000 = 240,000e^(k*4), real

Wolfram|Alpha returns the value of k.

Real solution:

$$k = \frac{1}{4}\log\left(\frac{121}{120}\right) \approx 0.0020747$$

Wolfram Alpha LLC. 2012. Wolfram|Alpha
http://www.wolframalpha.com/
(access December 12, 2012)

You now have the solution to the differential equation.

$$p(t) = 240{,}000e^{0.0020747t}$$

You can evaluate this equation when $t = 12$ by entering

evaluate 240,000e^(0.0020747*12)

Wolfram|Alpha returns 246,050. Since we entered a value of k accurate to more decimal places, Wolfram|Alpha's answer differs slightly from our answer in Example 1. ■

Getting Ready for Class

After reading through the preceding section, respond in your own words and in complete sentences.

A. What is the differential equation that represents the unlimited growth model and what is its solution?

B. What is the differential equation that represents the limited growth model and what is its solution?

C. What is the differential equation that represents the logistic growth model and what is its solution?

D. What is the differential equation that is used when the rate at which a population changes with respect of time is proportional to the difference between the carrying capacity and the current size of the population?

Problem Set 6.5

Skills Practice

1. In the derivation of the solution of the unlimited growth model, we stated that $P = Ce^{kt}$ became $P = P_0e^{kt}$ if $P = P_0$ when $t = 0$. Show that this is true.

2. In the derivation of the solution to the limited growth model, the integral

$$\int \frac{dP}{L - P}$$

occurred. Show that this integral is equivalent to

$$-\ln (L - P) + C$$

3. In Example 3, the differential equation $\frac{dP}{dt} = k(L - P)$ with solution

$$P = L + (P_0 - L)e^{-kt}$$

occurred. The conditions $L = 4{,}000$ and $P_0 = 80$ resulted in the equation $P = 4{,}000 - 3{,}920e^{-kt}$. Then using the initial conditions that $P = 2{,}800$ when $t = 2$, the value of k was determined to be approximately 0.592. Show that this is true.

Check Your Understanding

For Problems 4-11, choose the model (unlimited growth, limited growth, or logistic growth) that is appropriate for the situation and explain your answer.

4. **Physics: Temperature Control** The control that manages the internal temperature of a building is set at 68 °F. A person enters the building with a cup of coffee that has a temperature of 180 °F. We are interested in knowing the temperature of the coffee at any time t.

5. **Economics: Population Growth of a City** The population of a particular city is increasing at a rate proportional to itself. We are interested in knowing the size of the population at any time t.

6. **Manufacturing: Newton's Law of Cooling** An object is heated to 140 °C and then placed into a water bath that is 6 °C. We are interested in knowing the temperature of the object at any time t.

7. **Biology: Growth of a Mold** A mold is growing at a rate that is proportional to itself. We are interested in knowing the size of the mold at any time t.

8. **Physical Science: Radioactive Decay** A radioactive substance is decaying at a rate proportional to the current amount present. We are interested in knowing how much of the substance remains at any time t.

9. **Health Science: Spread of a Virus** Five hundred fifty students, all of whom are susceptible to an infectious virus, live in a college dormitory. The virus grows at a rate that is jointly proportional to the number of infected students and the number of uninfected students. (Recall that a quantity y is jointly proportional to the quantities x and z if $y = kxz$.) We are interested in knowing how many students have been infected at any time t.

10. **Chemistry: Chemical Conversion** A substance A is being converted, through a chemical reaction, to another substance B. The rate of the conversion of substance A is proportional to the amount of substance A. We are interested in knowing how much of substance A has been converted at any time t.

11. **Business: Learning a Technical Procedure** A company believes that its employees learn a new, complicated technical procedure at a rate that is jointly proportional to the total number of employees who need to learn the procedure and the number of employees who already know it. We are interested in knowing how many employees have learned the new procedure at any time t.

Modeling Practice

12. **Manufacturing: Newton's Law of Cooling** An object that is heated by a manufacturing process to 90 °C is placed into the surrounding air which is 25 °C. Twenty minutes after being removed from the heater, the temperature of the object is 75 °C.
    - **a.** Find a function that gives the temperature of the object at any time, t, in minutes after removal from the heater.
    - **b.** Find the temperature of the object 60 minutes after it is removed from the heater.

13. **Manufacturing: Newton's Law of Cooling** A chemical company is experimenting with a chemical that it hopes will keep liquids that have been refrigerated, then placed into a warmer environment, from warming too rapidly. A liquid that has been treated with the company's chemical is refrigerated to 40 °F and then placed into the surrounding air, which is 68 °F. Fifteen minutes later, the liquid has warmed to 45 °F.
    - **a.** Find a function that gives the temperature of the liquid at any time, t, in minutes after removal from the refrigerator.
    - **b.** Find the temperature of the liquid 30 minutes after it is removed from the refrigerator and placed into the surrounding air.

14. **Health Science: Spread of Disease** A college dormitory houses 700 students, all of whom are susceptible to a particular infectious virus. On a particular day, the infectious virus had been contracted by 25 students. Three days later, 125 students had contracted the virus. The number of infected students grows at a rate that is jointly proportional to the current number of students who have contracted the virus and the difference between the total number of students in the dorm and the number who have contracted the virus.
    - **a.** Find a function that gives the number of students who have contracted the virus at any time t.
    - **b.** Find the number of students who have contracted the virus 5 days after its initial outbreak.

15. **Business: Worker Production** A worker's production increases at a rate that is jointly proportional to the difference between 650 units per month and the number of units the worker currently produces per month. A worker with no experience can produce 45 units per month and after 2 weeks' experience, can produce 160 units per month.
    - **a.** Find a function that gives the number of units an experienced worker can produce at any time t.
    - **b.** Find the number of units a worker with 1-month experience can produce.

16. Physical Science: Radioactive Decay A radioactive substance decays at a rate that is proportional to the current amount present. There are initially 1,500 grams of the substance present, but 1,350 grams are present after 1 hour.

a. Find a function that gives the number of grams present at any time t.

b. Find the number of grams present after 10 hours.

17. Social Science: Spread of a Rumor In a group of 2,500 people, one person starts a rumor. Two hours later, 40 people have heard the rumor. The rumor spreads at a rate that is jointly proportional to the number of people who have heard it and the number of people who have not yet heard it.

a. Find a function that gives the number of people who have heard the rumor at any time t.

b. Find the number of people who have heard the rumor 6 hours after it was started.

18. Political Science: Political Polls A polling group has determined that the percentage of voters registered as members of a new political party is increasing at a rate that is proportional to the percentage of registered voters who are not members of the new party. In 2010, 0.2% of all registered voters were members of the new party and 0.6% were members in 2011. In this situation, L represents the total percentage of registered voters. As a decimal, $L = 1$.

a. Find a function that gives the percentage of all registered voters who are members of the new party at any time t.

b. Find the percentage of all registered voters that the new party can expect to have registered in the year 2016.

Source: California Secretary of State, Political Party Qualification.

19. Physical Science: Carbon-14 Dating The half-life of carbon-14, a radioactive isotope of carbon, is 5,770 years. This means that if P_0 is the current amount of carbon-14, then the amount present in 5,770 years will be $\frac{1}{2}P_0$. Suppose an archaeologist finds a human bone that has only one-fifth the amount of carbon-14 it originally contained. If it is known that carbon-14 decays at a rate that is proportional to the current amount present, how old is the bone? (Hint: You need to first use the information about the half-life of carbon-14 to establish the value of k, then the fact that one-fifth the original amount P_0 can be expressed as $\frac{1}{5}P_0$.)

Source: The Collaboration for NDT Education
Iowa State University, www.ndt-ed.org

20. Business: Investment of Capital In their investment of capital, a company uses the strategy that whenever the capital C decreases below a baseline value of ten million dollars, it is invested at a rate proportional to the difference between ten million and C. Suppose that the initial capital is two million dollars and that 6 months later it is five million dollars. What is the expected capital 12 months later if this investment strategy is adhered to?

21. Social Science: Spread of News The news media of a city announced an important story at noon. At this time, 6% of the population had heard it. 4 hours later 60% of the city's people had heard it. How long will it take for 99% of the city's people to hear the story?

Using Technology Exercises

Use Wolfram|Alpha to help you solve each differential equation.

22. The value of real estate in a city is decreasing, without limitations, at a rate proportional to its current value. A piece of property in the city on January 1, 2008 appraised at $510,000 and, four years later on January 1, 2012 at $290,000. According to this model, at what value will this piece of property be appraised in January 2015?

23. Solve the differential equation $\frac{dP}{dt} = k(L - P)$ subject to the condition that $P = P_0$ when $t = 0$. That is, $P(0) = P_0$. This differential equation represents the limited growth model. Rather than trying to enter P_0 into Wolfram|Alpha, use the letter A in its place. Use $P(0) = A$ rather than $P(0) = P_0$. Then in your solution, replace A with P_0.

In the text, the solution is presented as $P = L + (P_0 - L)e^{-kt}$. This form is different from that produced by Wolfram|Alpha. Use your algebra skills to show they are equivalent by converting the Wolfram|Alpha form to the form of the text.

24. Coffee at 190 °F is poured from a coffee maker into a cup that is sitting in a room at 68 °F. After sitting for 5 minutes, the coffee in the cup has cooled to 150 °F.

a. Use the limited growth model to find the value of the growth (decay) constant.

b. Write an equation that gives the temperature of the coffee at any time, t.

c. If the coffee is safe to drink when its temperature is 143 °F, how many minutes after the coffee is poured into the cup will it take this cup of coffee to be safe to drink?

Source: Physics Forums

25. Use Wolfram|Alpha to solve the differential equation associated with the logistic growth model. That is, solve $\frac{dP}{dt} = kP(L - P)$ subject to the condition that $P = P_0$ when $t = 0$. That is, $P(0) = P_0$. As you did in Using Technology Exercise 23, use $P(0) = A$ rather than trying to enter P_0 into Wolfram|Alpha. Then in your solution, replace A with P_0.

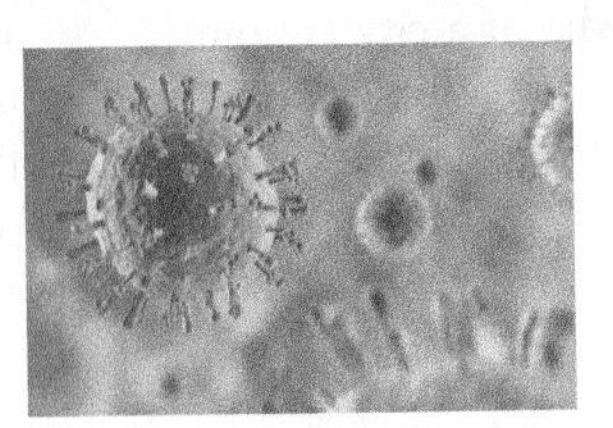
© Antti-Pekka Lehtinen/iStockPhoto

26. Health Science: Outbreak of a Virus H1N1, sometimes called "swine flu," is a flu virus that was first detected in people in the United States in April, 2009. The virus grows according to the logistic growth model. Suppose that 20 days ago in a city of 800,000 people, 200 people woke up one morning with H1N1. Nothing was done at that time to curtail the outbreak of the virus. Today, 1,850 people in the city have the virus. If nothing is done to prevent the virus from spreading, eventually everyone in the city will have contracted it.

a. Use the logistic growth model

$$P = \frac{P_0 L}{P_0 + (L - P_0)e^{-Lkt}}$$

and the initial conditions to find the value of the growth constant k. (Use nine decimal places for k.)

b. If nothing is done to slow or stop the spread of the virus, how many days from now will it take for 500,000 people in the city to contract the virus? Keep in mind that $t = 0$ represents 20 days in the past.

Source: squareCircleZ, inmath.com

Getting Ready for the Next Section

27. Calculate the sum of the numbers 0.6957, 0.2618, 0.0394, 0.0030, 0.0001, and 0.000002

For Problems 28-31, evaluate. Round your result to 4 decimal places.

28. $\frac{3}{124}\int_1^5 x^2\,dx$

29. $\frac{3}{124}\int_1^3 x^2\,dx$

30. $\int_{100}^{120} \frac{1}{120}\,dx$

31. $\int_{120}^{150} 0.01e^{-0.01x}\,dx$

32. Use the formula $z = \frac{x - \mu}{\sigma}$ to compute z when $x = 5.5$, $\mu = 6.0$, and $\sigma = 0.5$.

Probability

6.6

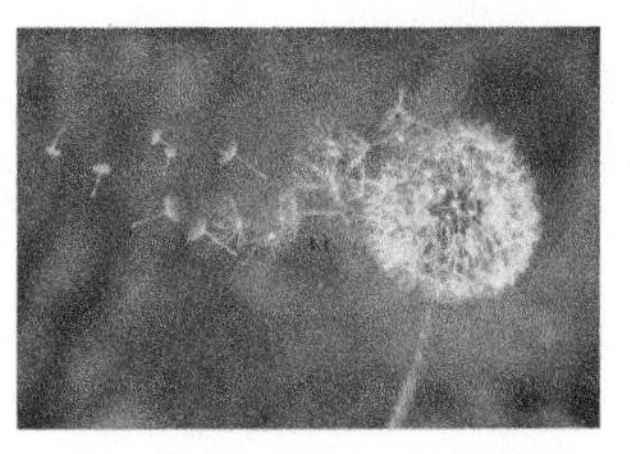

© Diane Diederich/iStockphoto

We have seen that differentiation measures the amount of change in a physical system and that integration measures the amount of accumulation. We will now see that integration can be used to measure the *chance* that a physical system will assume a particular state when the system is allowed to operate in some prescribed manner that allows for some degree of variability (that is, due to chance). The measure of chance is called **probability** and its value reflects the percentage of a large number of trials that can be expected to result in a particular state of a system.

Random Variables

In probability and statistics we work with a type of variable called a random variable.

VIDEO EXAMPLES

SECTION 6.6

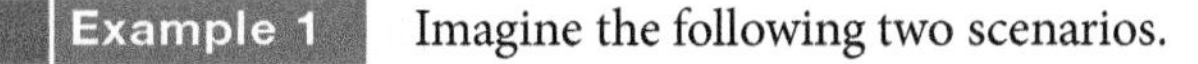

Imagine the following two scenarios.

1. In her book *Are You Normal?*, Bernice Kanner notes that about 7% of Americans say that they have flossed their teeth with their hair. Imagine an experiment in which 5 randomly selected Americans are each asked if he or she has ever flossed his or her teeth with his or her hair. The outcomes of the experiment would be a collection of "yes" or "no" answers.
2. A 2011 post from *The Information Diet: Come Hungry, Leave Happy*, notes that from a survey of 70,000 songs lasting from just above 0 second to about 425 seconds, the average length of a song is 242 seconds, almost exactly 4 minutes. Imagine an experiment in which one of the 70,000 songs in the library is selected and its playing time recorded. The outcome of the experiment would be a number in the interval $(0, 425)$. ■

In both these experiments, we can represent the outcomes as numbers. In the first experiment, we might ask a question such as, "In a randomly selected group of 5 Americans, what is the probability that 4 or 5 of them have flossed their teeth with their hair?" That is, what is the probability that in the collection of 5 yes and no outcomes, 4 or 5 of them are "yesses"? We are counting the number of "yes" outcomes and that count can be any of the numbers in the set $\{0, 1, 2, 3, 4, 5\}$.

In the second experiment, the outcomes are already numbers, music play times. We might ask the question, What is the probability that a randomly selected song is between 150 and 200 seconds? We are representing outcomes of an experiment with numbers.

Some experiments, such as these two, have outcomes that are due to chance and, therefore, outcomes that can vary. One group of 5 people may produce a different number of yes and no responses than another group of 5 people to the question about flossing their teeth with their hair. The length of one randomly selected song may differ from the length of another. The outcomes of both experiments are subject to chance because the items sampled are selected at random.

To answer a question that involves outcomes that are subject to chance, we introduce a **random variable.**

Random Variable
A random variable is a numerical description of the outcome of an experiment. The outcomes of the experiment are subject to variations that are due to chance. A random variable is commonly represented by a capital letter such as X (as opposed to the lowercase x).

Example 2

1. In the first experiment, the possible values of X are the numbers in the set $\{0, 1, 2, 3, 4, 5\}$, the numbers of possible "yes" answers to the question, "Have you ever flossed your teeth with your hair."
   X is a random variable because its value results from an experiment with outcomes that can vary from experiment to experiment. The number of "yes" answers can vary from one experiment to another due to chance.
2. In the second experiment, the possible values of X are the infinitely many numbers in the interval $(0, 425)$. The numbers represent the time length of a song.
   X is a random variable because its value, the length of a randomly selected song, results from an experiment with outcomes that can vary from experiment to experiment. The lengths of selected songs can vary from one experiment to another due to chance. ■

Discrete and Continuous Random Variables

There are two types of random variables, discrete and continuous.

1. A **discrete** random variable is one that assumes only a countable number of distinct values. Discrete random variables are most often counts.
   The random variable of the first experiment is discrete as it can assume only one or more of the 6 numbers in the set $\{0, 1, 2, 3, 4, 5\}$. The number of possible values for the random variable X is countable.
2. A **continuous** random variable is one that can assume an infinite number of possible values. Continuous random variables are most often measurements (time, weight, height, area, volume, etc.).
   The random variable of the second experiment is continuous as it can assume an infinite number of values in the continuous interval $(0, 425)$. The measured length, in seconds, of a randomly chosen song might be somewhere in the interval $(60, 120)$. There are infinitely many numbers in this interval.

Probability Distributions and Probability Density Functions

In the first experiment, there is a 0% chance of randomly selecting 5 people and getting 6 "yes" responses and a 100% chance of getting 0 or 1 or 2 or 3 or 4 or 5 "yes" responses.

In the second experiment, there is a 0% chance of randomly selecting one of the 70,000 songs and getting a song play time of 0 second, and a 100% chance of getting a play time of somewhere between 0 and 425 seconds.

In each experiment, there is a 0% chance that nothing happens and a 100% chance that something does happen. Probability is a number between and including 0 and 1.

Imagine 1 square unit of probability as shown in Figure 1.

Figure 1

A primary goal in probability theory is to determine how to divide up the 1-unit of probability and assign the pieces to the outcomes of an experiment. There is one way of doing this for discrete distributions and another way for continuous distributions.

Associated with discrete random variables are **probability distributions.** A probability distribution is a listing of the probabilities associated with each value of the random variable. For example, for the first experiment we could write the table that displays how the 1-unit of probability is divided and then distributed to each value of the random variable. The table appears below. The probabilities were determined by the binomial formula, a formula that we will not cover in this text.

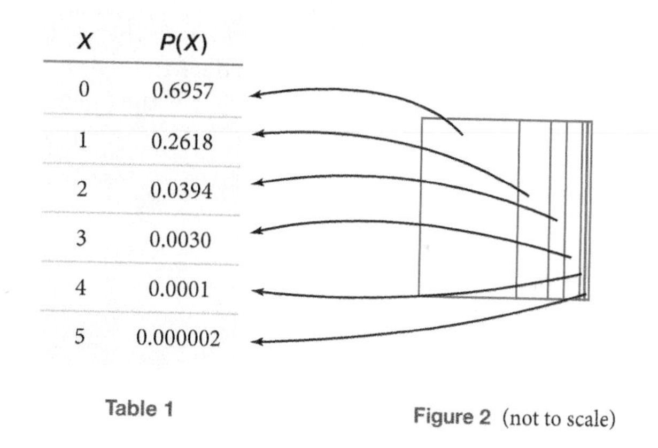

X	$P(X)$
0	0.6957
1	0.2618
2	0.0394
3	0.0030
4	0.0001
5	0.000002

Table 1

Figure 2 (not to scale)

This listing of probabilities is called a probability distribution because probabilities are being distributed to all the individual values of the random variable.

Associated with continuous random variables are **probability density functions.** Now imagine the 1 square unit of probability not being cut up into individual pieces as was done for discrete random variables, but rather stretched and pulled and reshaped into a new form and placed on and above the number line. The new shape is determined by a particular function that is associated with an experiment. Figure 3 shows such a reshaping.

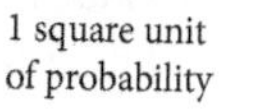

1 square unit of probability

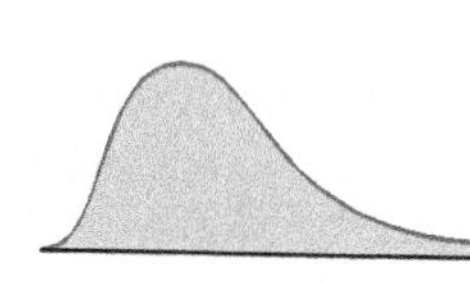

1 square unit of probability

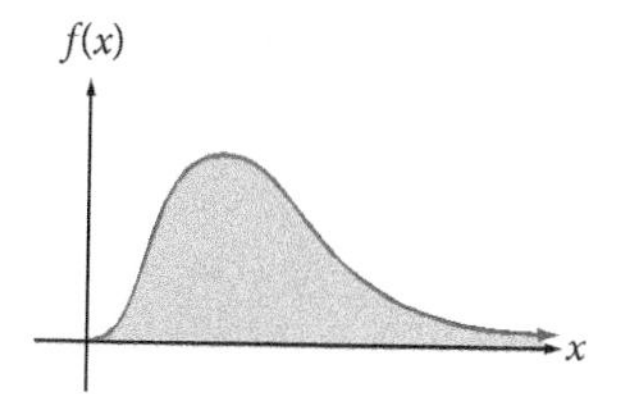

Figure 3

For continuous distributions, we do not ask about the probability that the random variable assumes any one specific number. Since there are infinitely many numbers on the number line, the probability that any particular one occurs is essentially 0. We ask, rather, what is the probability that the random variable assumes a number within a particular interval. That probability is given by the area bounded by the curve $f(x)$, the interval, and the vertical lines at the endpoints of the interval. Figure 4 shows such an area. Mathematically, area and probability are equivalent. The measure of an area is the measure of the probability of an outcome. Measuring areas, you know, can be made using calculus.

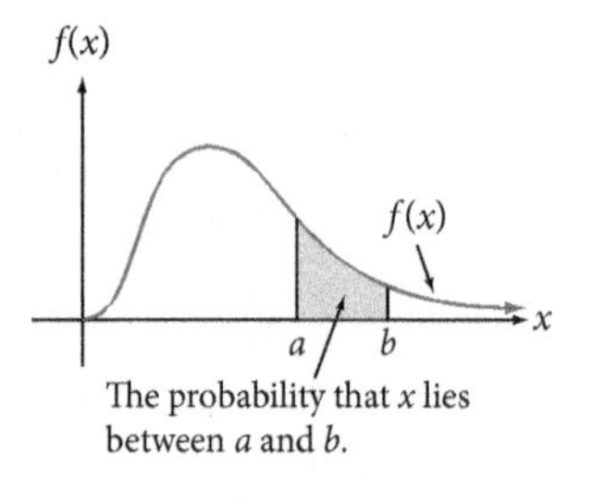

Figure 4

The probability density function pictured in Figure 5 may help you to understand why the term "density" is used. Each of the displayed intervals in the graph is 1 unit in length. The leftmost interval has more area associated with it than does the rightmost interval. That is, the leftmost interval has more probability associated with it than does the rightmost interval. The leftmost interval is more dense with probability than is the rightmost interval.

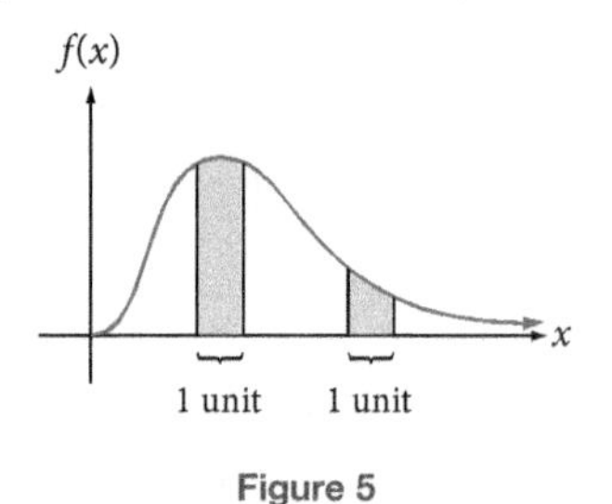

Figure 5

Imagine two bars of music, one with many more notes in it than the other as illustrated in Figure 6. You might say that one bar of music is more *dense* than the other.

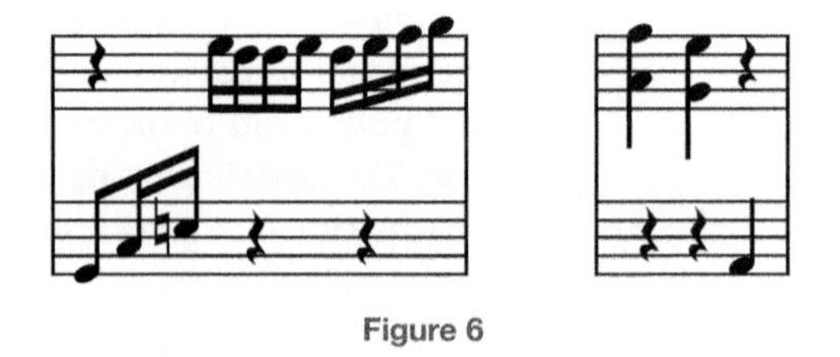

Figure 6

The probability that a randomly chosen value of X lies in some interval $[a, b]$ is the area of the region that lies below the curve $f(x)$ and above the interval $[a, b]$. Figure 7 shows this probability. The notation $P(a \le X \le b)$ represents the probability that the random variable X takes on a value between a and b, inclusive. But this area, we know, is the value of the definite integral over this interval. Thus,

$$P(a \le X \le b) = \int_a^b f(x)\,dx$$

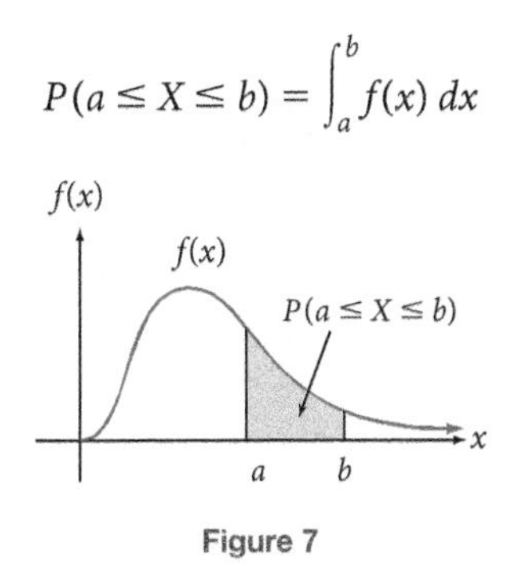

Figure 7

Probability Density Function

A probability density function is any function $f(x)$ that satisfies the two conditions

1. $f(x) \ge 0$ for each value of x, and
2. $\int_{-\infty}^{\infty} f(x)\,dx = 1$

The first condition states that a probability function is a function that produces only nonnegative numbers as output values. The second condition states that when an experiment is performed, there is a 100% chance that something will happen.

Example 3 Show that $f(x) = \frac{3x^2}{124}$ is a valid probability density function for each x in $[1, 5]$ given that $f(x) = 0$ for every x outside of $[1, 5]$.

Solution To demonstrate that $f(x)$ is a valid probability density function, we must show that $f(x)$ satisfies both conditions of the definition.

1. Because x^2 is positive for every x in $[1, 5]$, $\frac{3x^2}{124}$ is positive for every x in $[1, 5]$, and condition 1 is satisfied.
2. We now demonstrate that $\int_{-\infty}^{\infty} f(x)\,dx = 1$. Because $f(x) = 0$ for every x outside of $[1, 5]$, we need show only that $\int_1^5 f(x)\,dx = 1$.

$$\begin{aligned}\int_1^5 \frac{3x^2}{124}\,dx &= \frac{3}{124}\int_1^5 x^2\,dx \\ &= \frac{3}{124}\cdot\frac{x^3}{3}\bigg|_1^5 \\ &= \frac{3}{124}\left[\frac{5^3}{3} - \frac{1^3}{3}\right] \\ &= \frac{3}{124}\left[\frac{125}{3} - \frac{1}{3}\right] \\ &= \frac{3}{124}\cdot\frac{124}{3} \\ &= 1\end{aligned}$$

Since both conditions for a probability density function are satisfied, $f(x) = \frac{3x^2}{124}$ is a valid probability density function. ■

Example 4 Find the probability that a randomly selected value of X lies in the interval $[1, 3]$ if $f(x) = \frac{3}{124}x^2$. Round the probability to 4 decimal places. (In Example 3 we showed that this function is a valid probability density function.)

Solution We are asked to find $P(1 \le X \le 3)$. We use the fact that

$$P(a \le X \le b) = \int_a^b f(x)\, dx$$

$$\begin{aligned} P(1 \le X \le 3) &= \frac{3}{124}\int_1^3 x^2\, dx \\ &= \frac{3}{124} \cdot \frac{x^3}{3}\bigg|_1^3 \\ &= \frac{3}{124} \cdot \left[\frac{3^3}{3} - \frac{1^3}{3}\right] \\ &= \frac{3}{124} \cdot \left[\frac{27}{3} - \frac{1}{3}\right] \\ &= \frac{3}{124} \cdot \frac{26}{3} \\ &\approx 0.2097 \end{aligned}$$

Interpretation
If the experiment described by the probability function $f(x) = \frac{3}{124}x^2$ were performed a great many times, we would expect about 21% of the outcomes to have values between 1 and 3, inclusive.

Probability Models

Some probability density functions have been found to model particular physical and theoretical phenomena. In fact, many seemingly different phenomena are described by the same type of probability model. Figure 8 shows some of the commonly used probability models.

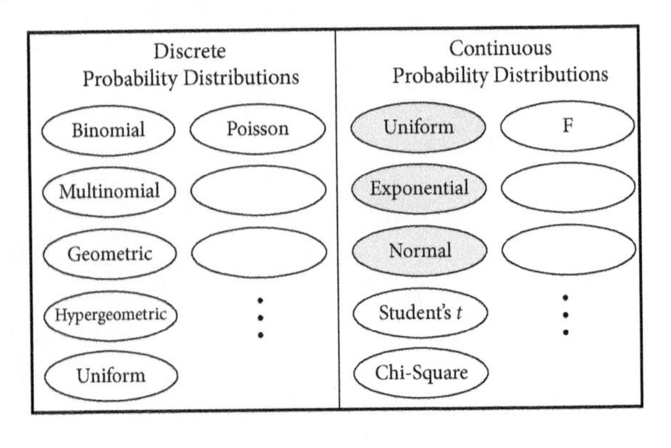

Figure 8

The probability models on the left side of the diagram are functions of discrete random variables and are studied in some detail in introductory statistics courses.

The probability models on the right side of the diagram are functions of continuous random variables. In this section, we will examine the three models that are highlighted. The others are studied in statistics courses.

The Uniform Probability Model

A random variable, X, is uniformly distributed in the interval $[a, b]$, if the probability that it assumes a value in any one subinterval of the interval $[a, b]$, is the same as the probability that it assumes a value in any other subinterval of $[a, b]$ the same length. Figure 9 illustrates this idea graphically. In the graph, both intervals are the same length and both rectangular regions have the same area.

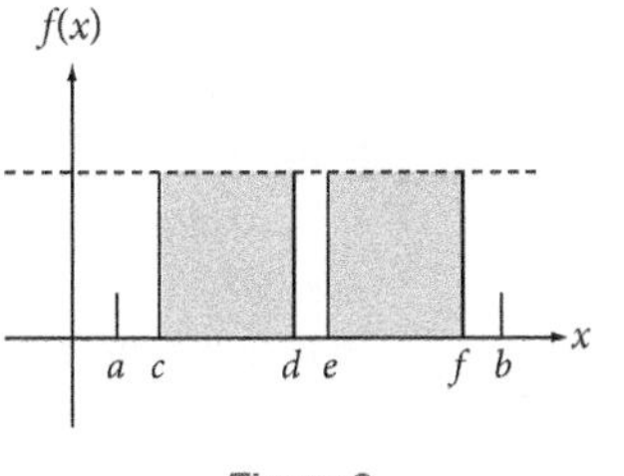

Figure 9

The Uniform Probability Model
A random variable X is uniformly distributed on the interval $[a, b]$ if its probability density function $f(x)$ is defined by

$$f(x) = \begin{cases} \dfrac{1}{b-a} & \text{for } x \text{ in } [a, b] \\ 0 & \text{for } x \text{ outside } [a, b] \end{cases}$$

Example 5 A state's weather service provides a continuously running 120-second recorded telephone message about weather conditions throughout the state. If a person calls and gets connected to the message he or she hears the message at any point in the 120-second run. If a person calls and gets connected, what is the probability that he or she will hear at most 20 seconds of the message before it repeats?

Solution Let X represent the amount of time the message is heard by the caller. Since a phone call can be connected any time within the 120-second message, all connect times are equally likely so that x is uniformly distributed on $[0, 120]$, with the probability density function

$$f(x) = \frac{1}{120 - 0} = \frac{1}{120}$$

A caller will hear at most 20 seconds of the message if he or she is connected any time after the message has played 100 seconds. Thus, we want to compute the probability that X lies in the interval $[100, 120]$.

$$P(100 \le X \le 120) = \int_{100}^{120} \frac{1}{120}\,dx$$

$$= \frac{1}{120}x \Big|_{100}^{120}$$

$$= \frac{1}{120}[120 - 100]$$

$$= \frac{1}{120}[20]$$

$$= \frac{20}{120}$$

$$\approx 0.1667$$

Interpretation
If a person calls and is connected to the weather service message, there is about a 0.1667 chance that he or she will hear at most 20 seconds of the message before it repeats. That is, out of every 10,000 who call and get connected, we would expect about 1,667 of them to hear the last 20 seconds of the message.

The Exponential Probability Model

Figure 10 shows the graph of an exponential probability distribution. The exponential probability model is often an appropriate model for situations in which values of the random variable are more likely to be small than large.

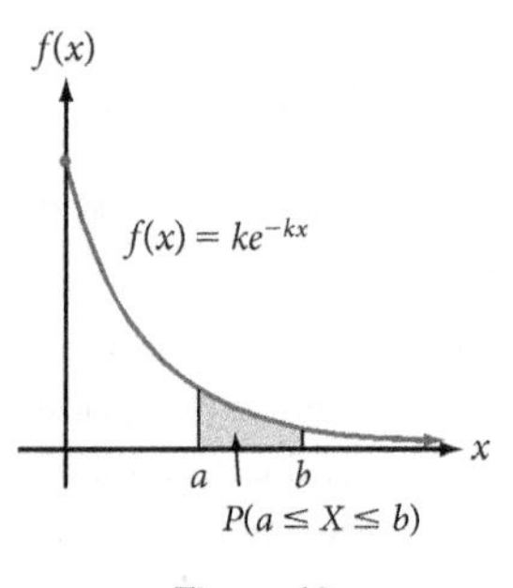

Figure 10

Notice from the graph, the exponential probability model is defined for only non-negative values of X and that the curve exhibits the same decreasing behavior as the exponential decay function. In fact, the exponential probability distribution is an exponential decay function and is defined in the following way.

The Exponential Probability Model
A random variable X is exponentially distributed on the interval $[0, \infty]$ if its probability density function $f(x)$ is defined by

$$f(x) = \begin{cases} ke^{-kx} \text{ for } x \text{ in } [0, \infty) \\ 0 \text{ for } x \text{ outside } [0, \infty) \end{cases}$$

where k is a constant determined by the situation and is called the *rate parameter.*

Exponentially distributed random variables are characterized by a waiting time model and include such models as the reliability or lifespan of a product (waiting time for failure), the duration of a signal (the waiting time for a signal to start or stop), the intervals of time between successive parts appearing on an assembly line, and the amount of time it takes to learn a task (the waiting time until a task is successfully completed).

Example 6 The research and development department of an audio equipment manufacturer has determined that its new wave-shaper circuit has a lifespan (in months) that is exponentially distributed with probability density function

$$f(x) = 0.01e^{-0.01x}$$

Find the probability that a new and randomly selected wave-shaper made by this manufacturer will last between 120 and 150 months under continuous use.

Source: Sonic Lion

Solution Let X represent the lifespan of the wave-shaper. Since X is exponentially distributed,

$$\begin{aligned} P(120 \le X \le 150) &= \int_{120}^{150} 0.01e^{-0.01x}\,dx \\ &= -e^{-0.01x}\Big|_{120}^{150} \\ &= -e^{-0.01(150)} - (-e^{-0.01(120)}) \\ &\approx -0.2231 + 0.3012 \\ &\approx 0.0781 \end{aligned}$$

Interpretation
The manufacturer can expect that approximately 7.81% of its new wave-shapers will last between 120 and 150 months.

The Normal Probability Model

The normal probability model is a common model in statistical analysis. Its graph is the familiar bell-shaped curve pictured in Figure 11. The normal curve is characterized by two numbers, the mean, represented by the Greek letter mu (μ) and the standard deviation, represented by the Greek letter sigma (σ). The mean is the center (an average) of the distribution and the high point of the curve, and the standard deviation is a measure of how the data values are dispersed around the mean.

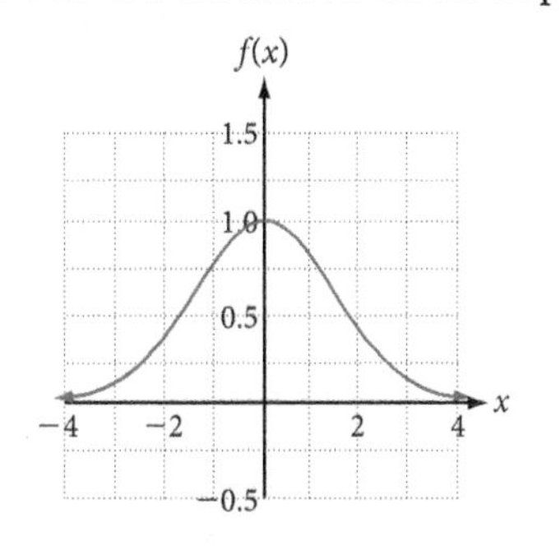

Figure 11

The Normal Probability Model

A random variable X with mean, μ, and standard deviation, σ, is normally distributed on the interval $(-\infty, \infty)$ if its probability density function $f(x)$ is defined by

$$f(x) = \frac{1}{\sigma\sqrt{2\pi}} \cdot e^{-\frac{(x-\mu)^2}{2\sigma^2}}$$

The probability that a randomly selected value X from a normal distribution lies in an interval $[a, b]$ is given by the integral

$$P(a \leq X \leq b) = \frac{1}{\sigma\sqrt{2\pi}} \cdot \int_a^b e^{-\frac{(x-\mu)^2}{2\sigma^2}}\, dx$$

The integral is a great challenge to evaluate using pencil-and-paper so it is commonly evaluated using a computer-generated table or using some computer technology directly. Table 2, located on the following page, is one such computer-generated table. Infinitely many normal curves exist, one for each pair of values μ and σ. Statisticians developed a formula that transforms these infinitely many curves into one single curve called the **standard normal curve**. It has mean $\mu = 0$ and $\sigma = 1$. When a normal distribution is normalized, the random variable associated with it must also be normalized. The formula

$$Z = \frac{X - \mu}{\sigma}$$

converts values of a random variable X to values of a new random variable Z, the random variable used in the standard normal curve. Figure 12 shows a data-generated normal curve, its corresponding standardized curve, and the relationship between the corresponding random variables and probabilities.

Note Notice that the formula for converting X values to Z values is composed of capital letters rather than the lowercase letters x and z, that mathematical formulas typically use. Lowercase letters in formulas refer to constants and a formula is a set of instructions that converts one or more constants into another constant. By using capital letters in the normal probability conversion formula, we are indicating to the user that we are converting from one random variable to another.

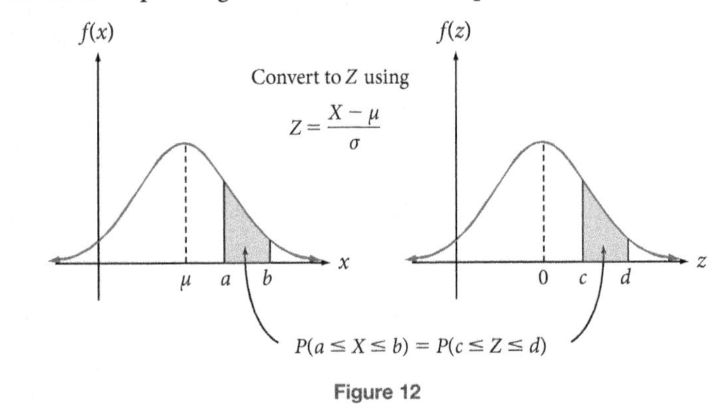

Figure 12

z^*	.00	.01	.02	.03	.04	.05	.06	.07	.08	.09
.0	.0000	.0040	.0080	.0120	.0160	.0199	.0239	.0279	.0319	.0359
.1	.0398	.0438	.0478	.0517	.0557	.0596	.0636	.0675	.0714	.0753
.2	.0793	.0832	.0871	.0910	.0948	.0987	.1026	.1064	.1103	.1141
.3	.1179	.1217	.1255	.1293	.1331	.1368	.1406	.1443	.1480	.1517
.4	.1554	.1591	.1628	.1664	.1700	.1736	.1772	.1808	.1844	.1879
.5	.1915	.1950	.1985	.2019	.2054	.2088	.2123	.2157	.2190	.2224
.6	.2257	.2291	.2324	.2357	.2389	.2422	.2454	.2486	.2517	.2549
.7	.2580	.2611	.2642	.2673	.2704	.2734	.2764	.2794	.2823	.2852
.8	.2881	.2910	.2939	.2967	.2995	.3023	.3051	.3078	.3106	.3133
.9	.3159	.3186	.3212	.3238	.3264	.3289	.3315	.3340	.3365	.3389
1.0	.3413	.3438	.3461	.3485	.3508	.3531	.3554	.3577	.3599	.3621
1.1	.3643	.3665	.3686	.3708	.3729	.3749	.3770	.3790	.3810	.3830

Table 2 z-scores

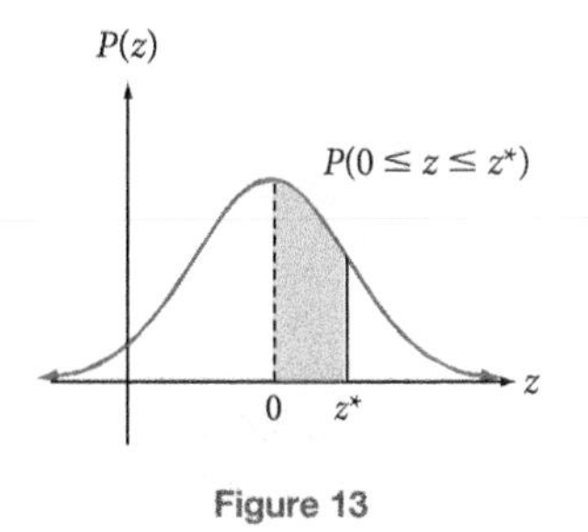

Figure 13

The table gives a probability that a statistic is between 0 and z^*. Because the curve is symmetric about 0, the probability that a statistic is between 0 and $-z^*$ is the same as the probability that it is between 0 and z^*.

Note Table 2 is an excerpt from the full table of z-scores that you will find in the appendix.

Example 7 A machine is adjusted to dispense 6 ounces of coffee into a 7-ounce cup. The amount of coffee dispensed is normally distributed with mean $\mu = 6$ ounces and standard deviation $\sigma = 0.5$ ounces. Find the probability that a randomly selected cup is filled with between 5.5 and 5.8 ounces of coffee.

Solution Let X represent the number of ounces of coffee that is dispensed by the machine. We wish to find $P(5.5 \le X \le 5.8)$. We can do this using Table 2. To use the table, we need to convert the X-scores to Z-scores. We do so using the conversion formula $Z = \frac{X - \mu}{\sigma}$.

$$P(5.5 \le X \le 5.8) = P\left(\frac{5.5 - 6.0}{0.5} \le Z \le \frac{5.8 - 6.0}{0.5}\right)$$
$$= P(-1.00 \le Z \le -0.40)$$

Using Table 2, we find this area using the fact that the curve is symmetric about $Z = 0$. The area to the right of 0 is 0.5 and to the left of 0 is also 0.5 (so they combine to produce 1). We subtract the area for the interval 0 to -0.40 from the area for the interval 0 to -1.00. See Figure 14.

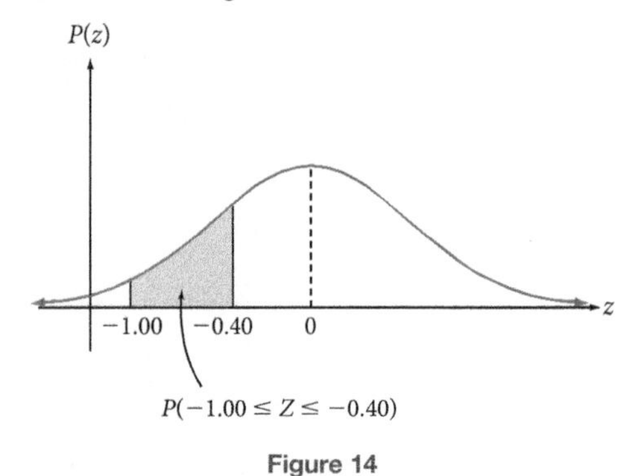

Figure 14

$$\begin{aligned}(\text{Area from 0 to } -1.00) - (\text{Area from 0 to } -0.40) &= 0.3413 - 0.1554 \\ &= 0.1859\end{aligned}$$

Or, as a fraction, $\dfrac{1,859}{10,000}$.

We conclude that $P(5.5 \le X \le 5.8) = 0.1859$.

Interpretation

With its present adjustment, the machine dispenses between 5.5 and 5.8 ounces of coffee into 7-ounce cups about 18.59% of the time. That is, out of every 10,000 cups of coffee the machine dispenses, it dispenses between 5.5 and 5.8 ounces of coffee into about 1,859 of them.

■

Using Technology 6.6

You can use both your calculator and Wolfram|Alpha to help you solve problems involving probability.

© Alexandru Romanciuc/iStockPhoto

Example 8 A machine shop produced steam turbine parts on one of its lathes that had a mean length of 40.05 mm with a standard deviation of 0.03 mm. If the lengths of the parts are normally distributed, use your calculator to find the probability that a randomly selected part has length

a. between 40.06 mm and 40.08 mm.

b. greater than 40.07 mm.

Solution

a. Access the normal probability distribution with [2nd] then [VARS]. This brings up the collection of probability distributions. Select the second one, normalcdf(.

This is the normal cumulative density function. It computes probabilities over intervals. The syntax for this command is

normalcdf(lower limit, upper limit, mean, standard deviation)

Enter

normalcdf(40.06, 40.08, 40.05, 0.03)

The calculator returns .2107861441. Rounding this number to 4 decimal places gives 0.2108. Thus, $P(40.06 \leq X \leq 40.08) = 0.2108$.

b. Enter

normalcdf(40.07, E99, 40.05, 0.03)

E99 represents 10^{99} and is the calculator's way of representing infinity. $P(X \geq 40.07) = 0.2525$ to four decimal places. ■

Example 9 The research and development department of a lightbulb manufacturer produces lightbulbs that have lifespans (in weeks) that are exponentially distributed according to $f(x) = 0.02e^{-0.02x}$. Using Wolfram|Alpha, find what percent of this manufacturer's lightbulbs will last at most 100 weeks under continuous use.

Solution Go to www.wolframalpha.com and in the entry field enter

integrate 0.02e^(−0.02x) from x = 0 to 100

Wolfram|Alpha displays your entry, its interpretation of your entry, and the value of the integral.

$$P(X \leq 100) = 0.8647$$

We conclude that about 86.47% of this manufacturer's lightbulbs will last at most 100 weeks under continuous use. ■

Getting Ready for Class

After reading through the preceding section, respond in your own words and in complete sentences.

A. Give an example of a discrete random variable and a continuous random variable.

B. Specify the two conditions that are required for a function to be a probability density function.

C. Specify the names of the three probability distributions we studied in this section.

Problem Set 6.6

Skills Practice

For Problems 1-6, the given function is not a probability density function. Show that at least one probability density function condition, 1 or 2, fails.

1. $f(x) = \frac{x-2}{16}$, on $[1, 5]$

2. $f(x) = \frac{5}{96}(x-4)$, on $[0, 10]$

3. $f(x) = 4x^2$, on $[-2, 2]$

4. $f(x) = \frac{1}{21}x^2$, on $[-1, 1]$

5. $f(x) = \frac{5}{12}x$, on $[-12, 12]$

6. $f(x) = \frac{1}{5}x$, on $[-2, 2]$

For Problems 7-12, find a value for k that will make the given function a probability density function.

7. $f(x) = kx^{1/3}$, on $[1, 8]$

8. $f(x) = kx^{2/3}$, on $[0, 8]$

9. $f(x) = kx$, on $[1, 5]$

10. $f(x) = kx$, on $[-3, 0]$

11. $f(x) = \frac{k}{\sqrt{x}}$, on $[4, 9]$

12. $f(x) = \frac{k}{\sqrt[4]{x}}$, on $[1, 16]$

Each function in Problems 13-20 is a probability density function. Compute each indicated probability. (Round to 4 decimal places.)

13. $f(x) = \frac{xe^{-x^2/2}}{2}$, find $P(0 \le X \le 1)$.

14. $f(x) = e^{-x}$, find $P(0 \le X \le 1)$.

15. $f(x) = \frac{1}{9x}$, find $P(1 \le X \le 10)$.

16. $f(x) = \frac{1}{x^6}$, find $P(1 \le X \le 32)$.

17. $f(x) = 0.04e^{-0.04x}$, find $P(5 \le X \le 10)$.

18. $f(x) = 0.02e^{-0.02x}$, find $P(0 \le X \le 20)$.

19. $f(x) = \frac{3}{40}(x^2 - 2x)$, find $P(2 \le X \le 4)$.

20. $f(x) = \frac{1}{10}(1 - e^{-3x})$, find $P(0 \le X \le 2)$.

Check Your Understanding

For Problems 21-30, interpret the probability statement in terms of the given situation.

21. Health Science: Exposure to Radiation The probability that a randomly selected air traveler is exposed to between 5.00 and 5.50 mrem of radiation while flying across the continental United States is 0.0952.

22. Health Science: Effectiveness of Insecticide The probability that a newly developed insecticide will kill at least 85 of 100 insects is 0.866.

23. Meteorology: Amount of Rain The probability that a particular county will receive between 2.4 and 4.4 inches of rain during the month of February is 0.571.

24. Health Science: Clotting of Blood The probability that, after receiving a wound in which blood flows, a person's blood clots within 35 seconds is 0.722.

25. Social Science: Length of a Phone Call The probability that, in a certain city, a telephone call will last longer than 16 minutes is 0.151.

26. Social Science: Dinner at Home The probability that, of a group of working adults who had eaten dinner at home the previous night, the dinner had been prepared at home is 0.734.

27. Psychology: Reaction Time to a Stimulus The probability that a person's reaction time to a particular stimulus is more than 0.2 seconds is 0.006.

28. Biology: Germination of Seedlings The probability of germination for a particular type of seedling is 0.899.

29. Mathematics: Probability Functions Explain why the function $p(X) = \frac{3}{100}x^2$ over the interval $[2, 5]$ is *not* a probability function.

30. Mathematics: Probability Functions Explain why the function $p(X) = \frac{3}{e}e^{2x-1}$ over the interval $[0, 1]$ is *not* a probability function.

Modeling Practice

31. Government: Probability and Telephone Messages A state's weather service provides a continuously running 150-second recorded telephone message about weather conditions throughout the state. If a person calls and gets connected to the message he or she hears the message at any point in the 150-second run. The distribution of connections to the service is uniform over the time interval $[0, 150]$. Let the random variable X represent the amount of time the message is heard. If a person calls and gets connected, what is the probability that he or she will hear

a. at most 20 seconds of the message before it repeats?

b. at least 20 seconds of the message before it repeats?

c. at most half the message before it repeats?

d. at least half the message before it repeats?

32. Government: Probability and Telephone Messages A state's highway service provides a continuously running 90-second recorded telephone message about highway conditions throughout the state. If a person calls and gets connected to the message he or she hears the message at any point in the 90-second run. The distribution of connections to the service is uniform over the time interval $[0, 90]$. Let the random variable X represent the amount of time the message is heard. If a person calls and gets connected, what is the probability that he or she will hear

a. at most 10 seconds of the message before it repeats?

b. at least 10 seconds of the message before it repeats?

c. at most half the message before it repeats?

d. at least half the message before it repeats?

33. Business: Probability and Arrival Times of Web Server Requests The arrival time (in seconds) of requests per hour to a particular web server can be modeled with a uniform distribution in the interval [0, 3,600] where time is measured in seconds. If 1,000 requests are made to this server, about how many would be expected to arrive in the first 12 minutes of the hour?

34. Biology: Probability and the Urge to Visit the Restroom A typical person's urinary bladder can hold anywhere from 0 to 600 mL of urine. The urge to urinate is usually experienced when the bladder contains about 150 mL. Suppose that in any large group of people, the amount of urine in their bladders is uniformly distributed over the interval [0, 600]. Let the random variable X represent the number of milliliters of urine in a person's bladder. In a group of 2,000 people at a concert, about how many of them can be expected to have the urge to visit the restroom?

Source: *Human Anatomy and Physiology*. 2nd ed., Hole

35. Biology: Probability and the Time to Digest Food The length of time to digest food and eliminate it from the body depends on the individual person. For most healthy adults, that time is usually between 24 and 72 hours. Suppose that for a large group of healthy adults, the time for complete digestion of food and elimination from the body is uniformly distributed on the interval [24, 72]. Let the random variable X represent the time it takes for complete digestion. What percent of people can be expected to completely digest and eliminate a meal in the final 6 hours of the typical time?

Source: www.mayoclinic.com

36. Business: Probability and Stir-Fry Choices Sampling suggests that customers at a Mongolian barbecue in the food court at a shopping mall load their plates with between 4 and 20 ounces of items such as chicken, lamb, scallops, and carrots and noodles. If the number of ounces of food people take is uniformly distributed over the interval [4, 20], what percent of customers can be expected to take between 15 and 20 ounces of food items?

For Problems 37-41, make use of the exponential probability distribution.

37. Business: Probability and the Lifespan of a Mechanical Device The probability density function for the 20,000-hour operating lifespan of a mechanical device in miniature electrical rotary switches is

$$f(x) = 0.0002e^{-0.0002x} \quad x > 0$$

Let the random variable X represent the lifespan of a mechanical device. Find and interpret the probability that a randomly selected device lasts less than 100 hours.

38. Business: Probability and the Effective Life of an Electrical Component The probability density function for the effective life (in years from the time of purchase), of an electrical component is modeled by

$$f(x) = 0.5e^{-0.5x} \quad x > 0$$

Let the random variable X represent the effective life of a component. Find and interpret the probability that a randomly selected device has an effective life of

a. less than 5 years.

b. between 5 and 10 years.

Source: ElectroSwitch

39. **Business: Probability and the Failure of Equipment** Many equipment breakdowns follow an exponential distribution. Suppose a particular system contains a component for which the time, in years, to failure is modeled by the exponential density function

$$f(x) = \frac{1}{5}e^{-\frac{1}{5}x}$$

If the component fails, the system breaks down. Let the random variable X represent the number of years to failure. What is the probability that the system breaks down

a. sometime in the first year of operation?

b. sometime after 1 year?

40. **Transportation: Distance Between Consecutive Cars** A city's transportation department believes that the distance, in feet, between two consecutive cars on the city's main freeway during the morning commute can be modeled by the exponential probability density function

$$f(x) = 0.03e^{-0.03x}$$

Let the random variable X represent the distance between two consecutive cars. Find the probability that for two consecutive cars making the morning commute, the distance between them

a. is less than 40 feet.

b. is between 50 and 100 feet.

© Tim McCaig/iStockPhoto

For Problems 41-46, make use of the normal probability distribution. For each problem, convert to z-scores then use the normal probability table (Table 1-A in the appendix) to find the specified probability.

41. **Marketing: Probability and Battery Life** A company advertises that the batteries it sells have a mean life of 20 hours with a standard deviation of 4 hours. Let the random variable X represent battery life in hours. What is the probability that a randomly selected battery sold by this company lasts

a. less than 10 hours?

b. between 20 and 35 hours?

42. **Marketing: Probability and Operational Life** A company advertises that the molecular mechanical devices it manufactures have a mean operational life of 5,000 hours with a standard deviation of 600 hours. Let the random variable X represent operational life in hours. What is the probability that a randomly selected device manufactured by this company lasts

a. less than 4,000 hours?

b. between 6,000 and 6,500 hours?

43. **Psychology: Probability and Eye Contact Time** In a study of facial behavior, people in a control group were timed for eye contact in a 5-minute period. Their eye contact times are normally distributed with a mean of 180 seconds and a standard deviation of 50 seconds. Let the random variable X represent eye contact time in seconds. What is the probability that an individual person selected at random from the control group has an eye contact time between 190 and 200 seconds?

44. Business: Probability and Weights of Quarters U. S. quarters have a mean weight of 5.670 grams and a standard deviation of 0.062 grams. A vending machine is configured to accept only those quarters with weights between 5.550 g and 5.790 g.

a. What percent of quarters will the vending machine accept?

b. If 10,000 quarters are eventually put into the vending machine, about how many could be expected to be accepted?

Source: www.usmint.gov

© Sarun Laowong/iStockPhoto

45. Manufacturing: Probability and Seat Design Anthropometry is the science that defines physical measures of a person's size, form, and functional capacities. Designers of seats for theaters and airplanes use data obtained from anthropometric studies to design seats that fit a range of people of some prescribed size. Anthropometric studies indicate that for women in United States, the length from the uncompressed buttocks to the popliteal angle at the back of the knee is normally distributed with a mean of 19.29 inches and standard deviation of 1 inch. What percent of U.S. women will fit comfortably in seats that are between 17.65 in and 20.93 in?

Source: www.cdc.gov

46. Psychology: Probability and Interpersonal Violence Resulting in Psychological Stress Interpersonal violence often leads to psychological stress for the victim. The distribution of the time elapsed between the violent incident and the initial sign of stress has a mean of 5.1 years and a standard deviation of 6.1 years. What is the probability that a randomly selected person who has experienced a violent incident shows the first sign of stress

a. less than 1 year after the incident?

b. more than 6 years after the incident?

Source: www.deepdyve.com

Using Technology Exercises

47. The probability that a person has to wait t minutes to be seated at a restaurant is given by the function

$$f(t) = \frac{16.535}{16(t+1)^2} \qquad 0 \le t \le 60$$

Find the probability that any person entering the restaurant will have to wait

a. between 10 and 15 minutes to be seated.

b. at most 5 minutes to be seated.

c. between 30 and 60 minutes to be seated.

48. As part of a research effort, a medical researcher introduces cholesterol into blood samples in such a way that the cholesterol levels are uniformly distributed over the interval [50, 500]. What is the probability that a randomly selected sample has a cholesterol level between 175 and 250?

49. A manufacturer claims that the probability that one of its machine components will last t months after it is put into use is given by the exponential probability function

$$f(t) = 0.024e^{-0.024t}$$

What is the probability that a randomly selected component lasts for

a. at most 3 years?

b. at least 10 years?

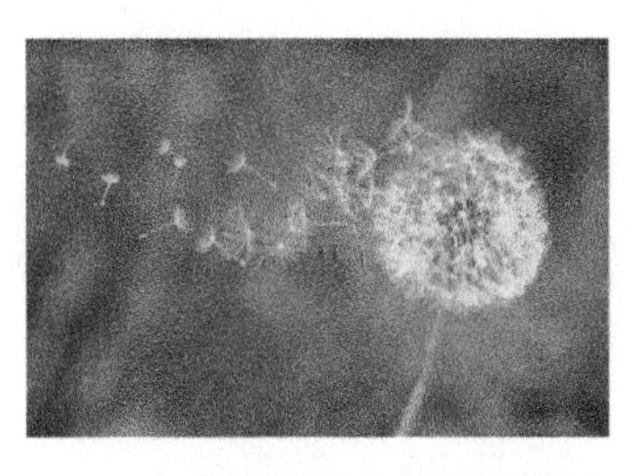

50. The seeds of plants in a certain area are dispersed by the wind in such a way that the probability that a seed travels x feet from the plant is given by the function

$$f(x) = 0.013e^{-0.013x}$$

a. Find the probability that a seed is not blown more than 10 feet from the plant.

b. Find the probability that a seed is blown from 25 to 50 feet from the plant.

c. Find the probability that a seed is blown more than 1,000 feet from the plant.

51. A company sells a product that it claims has an expected lifespan, X, that is normally distributed with a mean of 12 years and a standard deviation of 3.3 years. Find the probability that a randomly selected unit lasts

a. less than 5 years.

b. between 8 and 12 years.

c. more than 20 years.

52. The 68, 95, and 99.7% Rule The standard normal probability distribution has mean 0 and standard deviation 1. About what percentage of all an experiment's data will lie

a. within 1 standard deviation of the mean?

b. within 2 standard deviations of the mean?

c. within 3 standard deviations of the mean?

Getting Ready for the Next Section

53. Evaluate the expression $2{,}300x + 1{,}400y + 900z - 3x^2 - y^2 - z^2$ for $x = 80$, $y = 16$, and $z = 20$.

54. Evaluate the expression $2x^3 - 5y^2 + 3z$ for $x = 2$, $y = 1$, and $z = -5$.

55. Evaluate the expression $4x^{3/5}y^{2/5}$ for $x = 85$ and $y = 20$. Round your result to two decimal places.

56. Evaluate the expression $4x^{3/5}y^{2/5}$ for $x = 84$ and $y = 25$. Round your result to two decimal places.

57. Show that if x is replaced with kx and y is replaced with ky, the expression $4x^{3/5}y^{2/5}$ becomes $k \cdot 4x^{3/5}y^{2/5}$. (Hint: You will use the exponent property $(a \cdot b)^n = a^n \cdot b^n$.)

Chapter 6 Summary

EXAMPLES

Areas of Regions in the Plane [6.1]

If $f(x)$ and $g(x)$ are two continuous functions for which $f(x) \geq g(x)$ on $[a, b]$, then the area bounded by the two functions and the vertical lines $x = a$ and $x = b$ on $[a, b]$ is

$$A = \int_a^b [f(x) - g(x)]$$

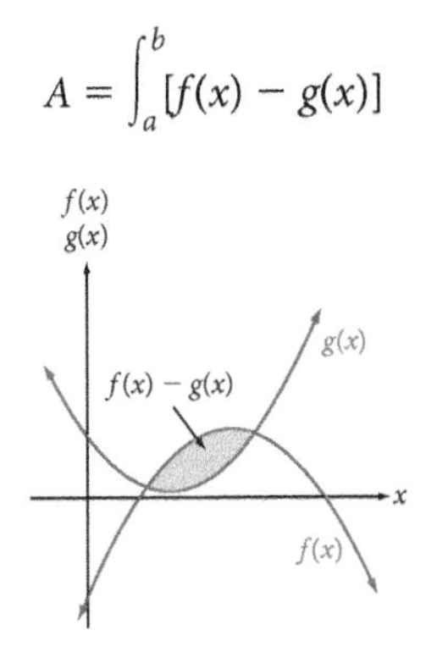

When preparing to make this computation, we choose $f(x)$ to be the higher curve and $g(x)$ to be the lower curve.

1. The area of the region bounded by the graph's upper curve

$$f(x) = -x^2 + 14x - 29$$

and the lower curve

$$g(x) = x^2 - 8x + 18$$

and the vertical lines $x = 4$ and $x = 7$ is

$$\begin{aligned} A &= \int_a^b [f(x) - g(x)]\,dx \\ &= \int_4^7 [(-x^2 + 14x - 29) - (x^2 - 8x + 18)]\,dx \\ &= \int_4^7 [-2x^2 + 22x - 47]\,dx \\ &= 36 \end{aligned}$$

The area of the region is 36 square units.

Supply and Demand Functions [6.2]

The **supply function** represents the price per unit at which producers are willing to produce and supply x units of a product. The supply function is of interest to producers.

$$p = S(x)$$

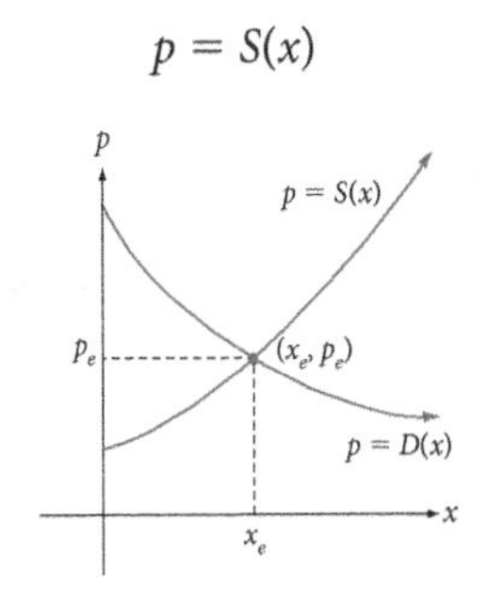

The **demand function** represents the price per unit consumers are willing to pay when x units of the item are made available in the market. The demand function is of interest to consumers.

$$p = D(x)$$

The **equilibrium point,** (x_e, p_e), is the point at which supply equals demand, and represents the price p_e consumers are willing to pay when x_e units of the item are produced.

2. The demand and supply for a particular item is modeled by the functions

$$D(x) = 220 - 0.6x^2$$

and

$$S(x) = 0.3x^2$$

respectively. Find the equilibrium point.

$$\begin{aligned} D(x) &= S(x) \\ 220 - 0.6x^2 &= 0.3x^2 \\ 220 &= 0.9x^2 \\ x^2 &\approx 244.4444 \\ x &\approx \pm 15.6347 \end{aligned}$$

Use only the positive value for $x_e = 15.63$. Substitute the unrounded value into one of the equations to get $p_e = 73.33$. The equilibrium point is (15.63, 73.33).

Consumer's Surplus [6.2]

The **consumer's surplus** is the definite integral which measures the amount of surplus money that results when a purchase is made at a price that is lower than the maximum the consumer was willing to pay.

$$CS = \int_0^{x_e} [D(x) - p_e]\, dx$$

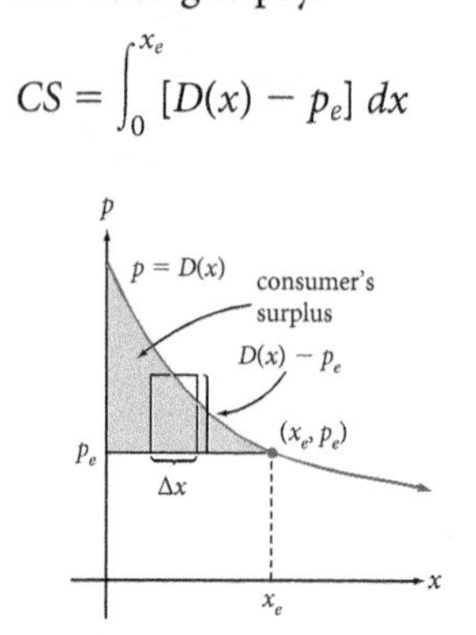

3. The demand and supply for a particular item is modeled by the functions

$$D(x) = 220 - 0.6x^2$$

and

$$S(x) = 0.3x^2$$

respectively. Find the consumer's surplus.

From Example 2, we have $x_e = 15.63$ and $p_e = 73.33$. Then,

$$CS = \int_0^{x_e} [D(x) - p_e]\, dx$$

$$= \int_0^{15.63} [220 - 0.6x^2 - 73.33]\, dx$$

$$\approx 1{,}528.78$$

The $1,528.78 is the total amount of money saved by consumers of the product when they buy each unit for the market price of $73.33 rather than at a higher price they would be willing to pay.

Producer's Surplus [6.2]

The **producer's surplus** is the definite integral which measures the additional money contributed by consumers on the sale of x_e units of a product.

$$PS = \int_0^{x_e} [p_e - S(x)]\, dx$$

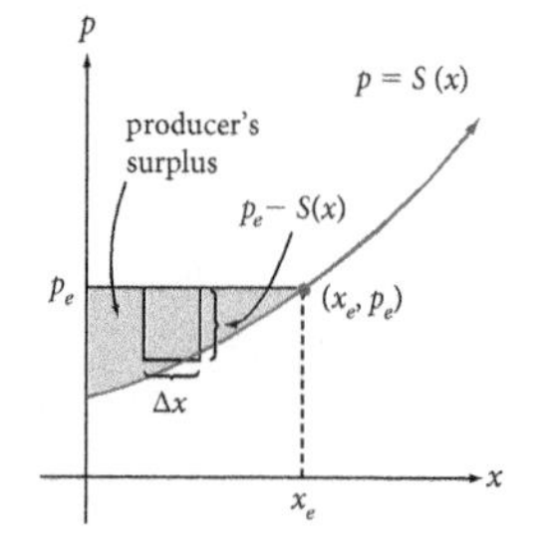

4. The demand and supply for a particular item is modeled by the functions

$$D(x) = 220 - 0.6x^2$$

and

$$S(x) = 0.3x^2$$

respectively. Find the producer's surplus.

From Example 2, we have $x_e = 15.63$ and $p_e = 73.33$. Then,

$$PS = \int_0^{x_e} [p_e - S(x)]\, dx$$

$$= \int_0^{15.63} [73.33 - 0.3x^2]\, dx$$

$$\approx 764.31$$

The $764.31 is the total amount of money realized by producers of the product when they sell each unit for the market price of $73.33 rather than at a lower price they would be willing to accept.

Accumulated Value of a Non-Interest-Earning Stream [6.3]

If over some interval $0 \leq t \leq T$, money flows continuously into a pool at a rate approximated by the function $f(t)$ dollars per time period, and if the money is reinvested in a non-interest-earning account, then the total amount of money A accumulated at the end of T years is

$$A = \int_0^T f(t)\, dt$$

This is known as **accumulated value of a non-interest-earning stream.**

5. Revenue flows continuously into a business at the rate of \$300,000 per year. If the money is not invested into an interest bearing account, how much accumulates in 5 years?

$$A = \int_0^T f(t)\, dt$$

$$= \int_0^5 300{,}000\, dt$$

$$= 1{,}500{,}000$$

In 5 years, \$1,500,000 accumulates.

Future Value of the Investment [6.3]

Future value of the investment is defined as

$$A = Pe^{rt}$$

where P dollars is invested at an annual interest rate r that is compounded continuously into an interest-earning account, and A is the total amount of money that will have accumulated at the end of t years.

6. A one-time lump-sum investment of

$$P = \$40{,}000$$

is made into an account that pays 5% interest compounded continuously. How much money will have accumulated in this account at the end of 4 years?

$$A = Pe^{rt}$$

$$= 40{,}000e^{0.05(4)}$$

$$\approx 48{,}856.11$$

In four years, \$48,856.11 will have accumulated in this account.

Future Value of the Annuity [6.3]

An **annuity** is a sequence of equal payments in which a particular amount of money is regularly deposited into an interest-earning account at the end of equal time periods. The amount of the annuity is called the **future value** of the annuity and it is the final amount of money in the account at the end of all the time periods. The future value of the annuity is the total amount of the payments plus the total amount of interest earned from all the payments. The time period between payments is called the **payment period.** The future value of an annuity at the end of n payment periods is approximated by

$$A = \int_0^n Pe^{rt}\, dt$$

P is the number of dollars invested at the end of each time period and r is the interest rate per time period. The value of r is computed by dividing the annual interest rate by the number of payment periods per year.

7. At the end of each month, \$500 is put into an account paying 4% interest compounded continuously. How much money has accumulated in the account at the end of 30 years?

$$A = \int_0^n Pe^{rt}\, dt$$

$$= \int_0^{(12)(30)} 500e^{\frac{0.04}{12}t}\, dt$$

$$\approx 348{,}018$$

In 30 years, \$348,018 will have accumulated in this account.

Present Value of the Annuity [6.3]

The **present value**, P, of an annuity is the amount of money that must be invested now at the annual rate r, compounded continuously, so that an amount, A, will be accumulated in the future. It is defined as the function

$$P = Ae^{-rt}$$

8. How much money must be invested now at 5% interest compounded continuously so that \$25,000 will be available in 15 years?

$$\begin{aligned} P &= Ae^{-rt} \\ &= 25{,}000e^{-0.05(15)} \\ &\approx 11{,}809 \end{aligned}$$

Right now, \$11,809 must be invested to have \$25,000 in 15 years.

Future Value of an Interest-Earning Income Stream [6.3]

Some flows of income can be considered to be continuous income streams. If $f(t)$ is the rate of flow of money into an account that pays an annual rate of interest, r, compounded continuously during some time interval $0 \le t \le T$, then the future value, FV, of the income stream is

$$FV = e^{rT}\int_0^T f(t)e^{-rt}\,dt$$

9. At the end of each year for five years, \$200,000 is invested in an account that pays 3% interest compounded continuously. The future value of the income stream is

$$\begin{aligned} FV &= e^{rT}\int_0^T f(t)e^{-rt}\,dt \\ &= e^{0.03(5)}\int_0^5 200{,}000e^{-0.03t}\,dt \\ &\approx 1{,}078{,}895 \end{aligned}$$

The \$200,000 annual money stream will be worth \$1,078,895 in five years.

Present Value of an Interest-Earning Income Stream [6.3]

The present value of an income stream is the amount of money that would have to be invested right now as a one-time, lump-sum, to produce an amount equal to the future value of the income stream. If $f(t)$ is the rate of flow of money into an account that pays an annual rate of interest r, compounded continuously during some time interval $0 \le t \le T$, then the present value, PV, of the income stream is

$$PV = \int_0^T f(t)e^{-rt}\,dt$$

10. At the end of each year for five years, \$200,000 is invested in an account that pays 3% interest compounded continuously. The future value of the income stream is \$1,078,895. What is the present value of the income stream?

$$\begin{aligned} PV &= \int_0^T f(t)e^{-rt}\,dt \\ &= \int_0^5 200{,}000e^{-0.03t}\,dt \\ &\approx 928{,}613 \end{aligned}$$

To produce the future value amount \$1,078,895, the company would have to invest \$928,613 now into an account paying 3% interest compounded continuously.

The Present Value of a Perpetual Income Stream [6.3]

The present value of a perpetual income stream is the amount of money that would have to be invested right now as a one-time, lump sum to produce an amount equal to the future value of an income stream that flows into an account forever. It can be approximated using an improper integral.

$$\text{Present value} = \int_0^\infty f(t)e^{-rt}\,dt$$

11. Find the present value of a $200,000 annual income if it is invested immediately as it is received into an account that pays 5% annual interest compounded continuously forever.

$$PV = \int_0^\infty f(t)e^{-rt}\,dt$$

$$= \int_0^\infty 200{,}000e^{-0.05t}\,dt$$

$$= 4{,}000{,}000$$

To match the total income realized from an annual stream of $200,000 that continues indefinitely, approximately $4,000,000 would have to be invested now into an account that pays annual interest at 5% compounded continuously forever.

Differential Equations and Their Solutions [6.4]

An ordinary differential equation is an equation that involves one dependent variable and one or more of its derivatives with respect to one independent variable.

12. The equations $(1 + x^2)\frac{dy}{dx} = x$ and $y' = 3x^2y^2$ are differential equations. A function is a solution to a differential equation if it

1. relates the variables involved, and
2. satisfies the differential equation.

13. The functions $y = \ln(1 + x^2)^{1/2} + C$ and $y = \frac{-1}{x^3 + C}$ are general solutions to the differential equations given above.

General Solutions to Differential Equations [6.4]

Solutions of differential equations that involve one or more arbitrary constants are called general solutions and they represent families of solutions. The graph shows a family of solutions to the differential equation

$$(1 + x^2)\frac{dy}{dx} = x$$

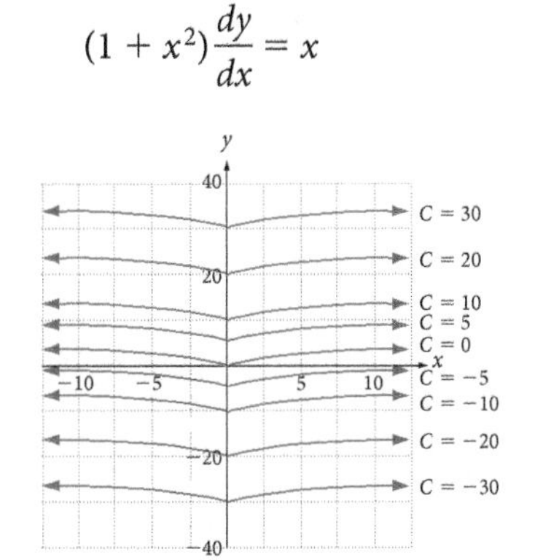

14. The function $y = \ln(1 + x^2)^{1/2} + C$ is a general solution to the differential equation $(1 + x^2)\frac{dy}{dx} = x$.

Solving Differential Equations Using the Method of Separation of Variables [6.4]

15. The equation $(1 + x^2)\frac{dy}{dx} = x$ is separable to $dy = \frac{x}{1+x^2}\,dx$. Integrating produces the general solution $y = \ln(1 + x^2)^{1/2} + C$.

A special class of differential equations is called the separable variable class. Solutions to these equations can be found by separating the dependent and independent variables to opposite sides of the equal sign then integrating. The integration produces an arbitrary constant, c, and therefore, produces a general solution and a family of curves.

Initial Values and Particular Solutions [6.4]

16. The differential equation $y' = x^2y$ has the general solution $\ln y = \frac{x^3}{3} + C$. If we know that $y = e^5$ when $x = 0$, we can substitute these values into the general solution to get

$$\ln e^5 = \frac{0^3}{3} + C$$

$$C = 5$$

and the particular solution $y = e^{\frac{x^3}{3}+5}$.

General solutions to differential equations involve an arbitrary constant, typically denoted by the letter C. They form a family of solutions, one member for each value of C. If some initial condition is known, that is, if some output value that corresponds to a particular input value is known, those values can be substituted into the general solution and the particular value of the arbitrary constant C can be determined. Then from all the members of the family of solutions, the one particular member that suits a particular situation can be specified.

Models of Unlimited Growth [6.5]

17. A population of bacteria left uncontrolled, grows according to the unlimited growth model. Initially there are 200,000 bacteria. Twenty hours later, that are 250,000 bacteria. The system becomes negatively affected when there are 10,000,000,000 (ten billion) bacteria present. How long will it take for this population to grow to be 10,000,000,000 (ten billion) in size? Use the initial conditions to solve $P = Ce^{kt}$ for the growth constant k.

$$250{,}000 = 200{,}000e^{k(20)}$$

$$1.25 = e^{20k}$$

$$\ln 1.25 = 20k$$

$$k = \frac{\ln 1.25}{20} = 0.0112$$

Substitute k into the solution formula and solve $10{,}000{,}000{,}000 = 200{,}000e^{0.0112t}$ for t.

$$10{,}000{,}000{,}000 = 200{,}000e^{0.0112t}$$

$$50{,}000 = e^{0.0112t}$$

$$\ln 50{,}000 = 0.0112t$$

$$t = \frac{\ln 50{,}000}{0.0112} \approx 966$$

The initial population of 250,000 bacteria will grow to ten billion bacteria in about 966 hours.

The law of natural growth states the rate at which a population changes over time is proportional to the current size of the population. If P represents the current size of a population at time t and k is a constant, then the differential equation below models the unlimited growth of the population.

$$\frac{dP}{dt} = kP$$

The solution to this differential equation is $P = Ce^{kt}$.

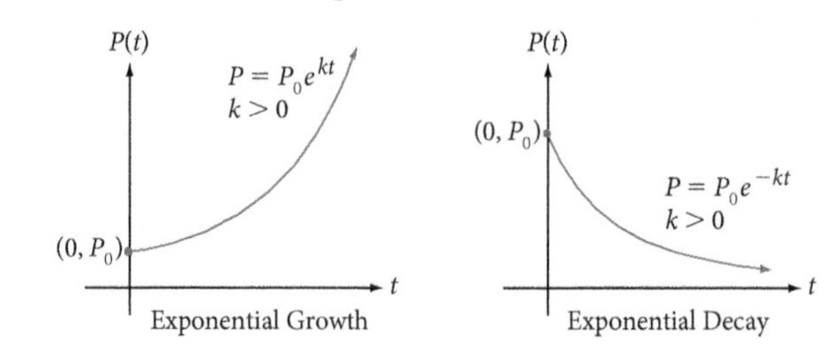

Models of Limited Growth [6.5]

Many populations grow in such a way that the rate of increase is large when the population is small and then decreases (because of limitations on growth) as the population increases. Let P represent the size of the population at time t and L be a fixed upper (or lower) limit to the size of the population. When the number L represents an upper limit, it is called the **carrying capacity** of the population and represents the maximum size of the population.

$$\frac{dP}{dt} = k(L - P)$$

The solution to this equation is

$$P = L + (P_0 - L)e^{-kt}$$

where P_0 is the initial size of the population. That is, P_0 is the size of the population when $t = 0$.

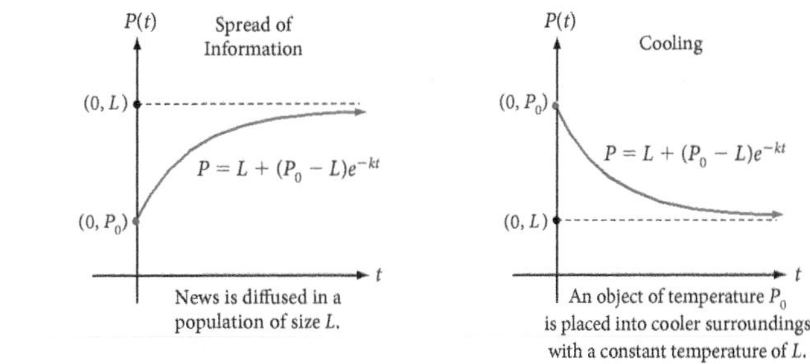

18. A component of an electronic system is programmed to switch off if its temperature reaches 100 °C. Tests show that in a certain environment, a particular component heats from 30 °C to 40 °C in 500 hours. Use the limited growth model to determine how long it would take for this component to heat from 30 °C to 90 °C in this environment.

First, use the initial conditions to solve for the growth constant, k.

$$40 = 100 + (30 - 100)e^{-k(500)}$$
$$-60 = -70e^{-500k}$$
$$k = \frac{\ln\frac{60}{70}}{-500} \approx 0.00031$$

Substitute k into the solution formula and solve $90 = 100 + (30 - 100)e^{-0.00031t}$ for t.

$$90 = 100 + (30 - 100)e^{-0.00031t}$$
$$\frac{10}{70} = e^{-0.00031t}$$
$$t = \frac{\ln\frac{10}{70}}{-0.00031} \approx 6{,}277$$

This component will reach 90 °C in about 6,277 hours (262 days).

Models of Logistic Growth [6.5]

When bounds are placed on population growth, the logistic growth function may best model the phenomena. Phenomena that exhibit logistic growth grow exponentially at first and then more slowly, as in the limited growth model. As in limited growth, the number L is called the carrying capacity of the population. The differential equation for logistic growth is

$$\frac{dP}{dt} = kP(L - P)$$

which has as its solution

$$P = \frac{P_0 L}{P_0 + (L - P_0)e^{-Lkt}}$$

P(t)
(0, L)
$P = \frac{P_0 L}{P_0 + (L - P_0)e^{-Lkt}}$
(0, P₀)
t

19. Twenty-four months ago, the membership of a business marketer's professional organization was 500. Today it is 1,200. The number of members increases at a rate that is proportional to the current size of the membership and the difference between the total number of potential members and the current membership. We want to find a function that gives the number of members 4 years from now if there are currently 6,000 potential members.

First, use the initial conditions to solve for the growth constant k.

$$1{,}200 = \frac{500(6{,}000)}{500 + (6{,}000 - 500)e^{-6{,}000k(2)}}$$

$$k \approx 0.000084$$

Substitute k into the logistic growth formula and solve for t when $P = 4{,}000$.

$$P = \frac{500(6{,}000)}{500 + (6{,}000 - 500)e^{-6{,}000k(0.000084)t}}$$

$$t \approx 6.1$$

The membership of the organization will reach 4,000 in about 6 years from twenty-four months ago, that is, 4 years from now.

Probability [6.6]

We can use integration to measure the chance that a physical system will assume a particular state when the system is allowed to operate in some prescribed manner that allows for some degree of variability. A measure of chance is called probability and its value reflects the percentage of a large number of trials that can be expected to result in a particular state of a system.

Probability and Random Variables [6.6]

A random variable is a numerical description of the outcome of an experiment. The outcomes of the experiment are subject to variations that are due to chance. They are commonly represented with capital letters (such as with X as opposed to the lowercase x). Random variables can be discrete or continuous.

Discrete Random Variables [6.6]

20. Let X represent the number of times a person attends a Santana concert. The values of X could be 0, 1, 2, 3, 4, ...

A discrete random variable is one that assumes only a countable number of distinct values. They are most often counts.

Continuous Random Variables [6.6]

21. Let X represent the length of a randomly selected song from an iTunes catalog. It could be that

0 minutes $< X \leq$ 5 minutes

A continuous random variable is one that can assume an infinite, non-countable number of values. They are most often measurements such as height, weight, time, area, or volume.

Probability Density Functions [6.6]

22. The function $P(x) = \frac{3}{875}x^2$ for each X in [5, 10] is a valid probability density function since

1. $P(x) \geq 0$ for each x, and

2. $\int_5^{10} \frac{3}{875}x^2\,dx = 1$

A probability density function is a function $f(x)$ that satisfies the two conditions

1. $f(x) \geq 0$ for each value of x, and

2. $\int_{-\infty}^{\infty} f(x) = 1$

The first condition states that a probability function is a function that produces only nonnegative numbers as output values. The second condition states that when an experiment is performed, something must happen.

Probability density functions can be integrated to find the probability that a random variable X lies in some interval $[a, b]$. The notation $P(a \leq X \leq b)$ represents the probability that the random variable X takes on a value between a and b, inclusive.

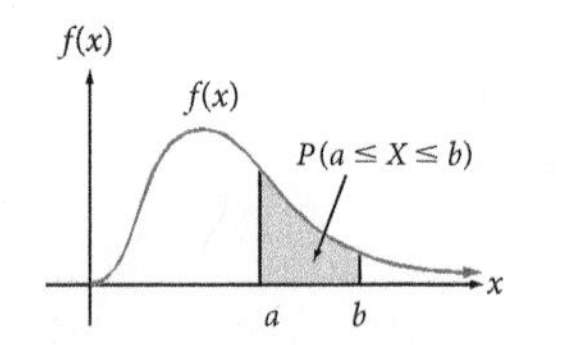

Probability Models [6.6]

Some probability density functions have been found to model particular physical and theoretical phenomena. In fact, many seemingly different phenomena are described by the same type of probability model. Some commonly used models are the uniform distribution, the exponential distribution, and the normal distribution.

The Uniform Probability Distribution [6.6]

A random variable, X, is uniformly distributed in the interval $[a, b]$ if the probability that it assumes a value in any one subinterval of the interval $[a, b]$, is the same as the probability that it assumes a value in any other subinterval of $[a, b]$ of the same length.

A random variable X is uniformly distributed on the interval $[a, b]$ if its probability density function $f(x)$ is defined by

$$f(x) = \begin{cases} \dfrac{1}{b-a} \text{ for } x \text{ in } [a, b] \\ 0 \text{ for } x \text{ outside } [a, b] \end{cases}$$

23. A traffic light stays red for 45 seconds and does so uniformly throughout the day. What percent of people arriving at the light and finding it red, will have to wait at most 10 seconds for the light to turn green?

$$P(x) = \int_{35}^{45} \frac{1}{45-0}\,dx$$
$$= \frac{2}{9} \approx 0.22$$

About 22% of people arriving at the light and finding it red will have to wait at most 10 seconds for the light to turn green.

The Exponential Probability Distribution [6.6]

The exponential probability model is often an appropriate model for situations in which values of the random variable are more likely to be small than large. A random variable X is exponentially distributed on the interval $[0, \infty]$ if its probability density function $f(x)$ is defined by

$$f(x) = \begin{cases} ke^{-kx} \text{ for } x \text{ in } [0, \infty) \\ 0 \text{ for } x \text{ outside } [0, \infty) \end{cases}$$

The rate parameter k represents a constant that is determined by the situation.

24. A manufacturer of halogen light bulbs determines that the lifespan of the bulbs is exponentially distributed with probability density function $f(x) = 0.005e^{-0.005x}$. What is the probability that a randomly selected bulb from this manufacturer lasts between 100 and 200 hours?

$$P(100 \le X \le 200) = \int_{100}^{200} 0.005e^{-0.005x}\,dx$$
$$= \frac{0.005}{-0.005}e^{-0.005x}\Big|_{100}^{200}$$
$$\approx 0.2387$$

The probability that a randomly selected bulb will last between 100 and 200 hours is about 24%.

The Normal Probability Distribution [6.6]

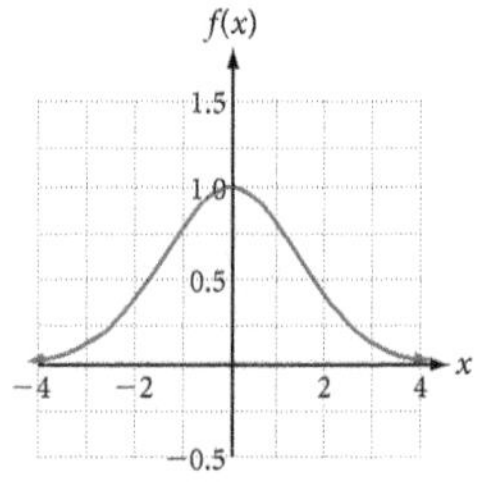

The normal probability model is a common model in statistical analysis. Its graph is the familiar bell-shaped curve. The normal curve is characterized by two numbers, the mean, μ, and the standard deviation, σ. The mean is the center (an average) of the distribution and the high point of the curve, and the standard deviation is a measure of how the data values are dispersed around the mean.

A random variable X with μ and σ is normally distributed on the interval $(-\infty, \infty)$ if its probability density function $f(x)$ is defined by

$$f(x) = \frac{1}{\sigma\sqrt{2\pi}} \cdot e^{-\frac{(x-\mu)^2}{2\sigma^2}}$$

The probability that a randomly selected value X from a normal distribution lies in an interval $[a, b]$ is given by the integral

$$P(a \le X \le b) = \frac{1}{\sigma\sqrt{2\pi}} \cdot \int_a^b e^{-\frac{(x-\mu)^2}{2\sigma^2}}\, dx$$

The Normal Probability Distribution and Z-Scores [6.6]

The integral for the normal probability distribution is commonly evaluated using a computer-generated table or using some computer technology directly. A formula that transforms the infinitely many normal curves into one single curve called the standard normal curve has mean $\mu = 0$ and $\sigma = 1$. When a normal distribution is normalized, the random variable associated with it must also be normalized. The formula

$$Z = \frac{X - \mu}{\sigma}$$

converts values of a random variable X to values of a new random variable Z, the random variable used in the standard normal curve. Once the z-scores have been found, the probability of the event can be determined using a table of probabilities.

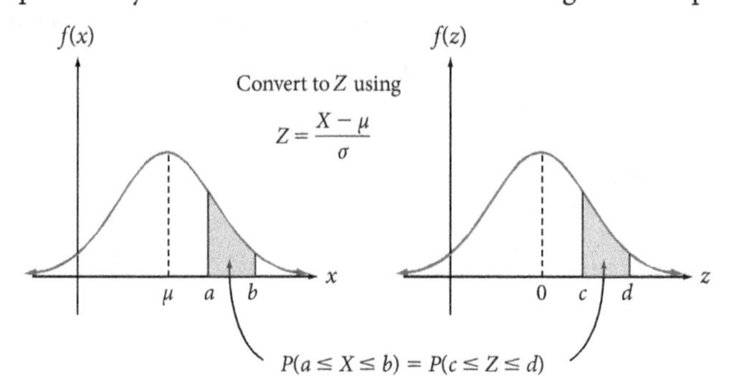

25. If the random variable X is normally distributed with mean 65 and standard deviation 2.5, find $P(60 \le X \le 72)$.

Convert the X-scores to Z-scores.

$$P(60 \le X \le 72) = P\left(\frac{60-65}{2.5} \le Z \le \frac{72-65}{2.5}\right)$$

$$= P(-2 \le Z \le 2.8)$$

Using Table 1-A from the appendix, we get

$$P(60 \le X \le 72) = P(-2 \le Z \le 2.8)$$

$$= 0.4974 + 0.4772$$

$$= 0.9746$$

Chapter 6 Test

1. Find the area of the region bounded by the curves $f(x) = -x^2 + 4$ and $g(x) = x^2$ from $x = 0$ to $x = 1$. [6.1]

2. Find the area of the region bounded by the curves $f(x) = 4x^2 + x - 3$ and $g(x) = 3x^2 + 3x + 5$ from $x = 0$ to $x = 1$. [6.1]

3. The supply and demand for a particular product is modeled by the functions $D(x) = 300 - 0.5x^2$ and $S(x) = 0.25x^2$ where in each case, x is the number of units that are produced. Find the equilibrium point. [6.2]

4. The demand and supply for a particular item is modeled by the functions $D(x) = 300 - 0.5x^2$ and $S(x) = 0.25x^2$, respectively. Find and interpret the consumer's surplus. [6.2]

5. The demand and supply for a particular item is modeled by the functions $D(x) = 300 - 0.5x^2$ and $S(x) = 0.25x^2$, respectively. Find and interpret the producer's surplus. [6.2]

6. Revenue flows continuously into a business at the rate of \$100,000 per year. If the money is not invested into an interest-bearing account, how much accumulates in 3 years? [6.3]

7. A one-time lump-sum investment of $P = \$5,000$ is made into an account that pays 6% interest compounded continuously. How much money will have accumulated in this account at the end of 10 years? [6.3]

8. At the end of each month, \$1,000 is put into an account paying 5% interest compounded continuously. How much money has accumulated in the account at the end of 30 years? [6.3]

9. How much money must be invested now at 5% interest compounded continuously so that \$100,000 will be available in 18 years? [6.3]

10. At the end of each year for 10 years, \$50,000 is invested in an account that pays 4% interest compounded continuously. Find the future value of this income stream. [6.3]

11. At the end of each year for 10 years, \$100,000 is invested in an account that pays 5% interest compounded continuously. What is the present value of the income stream? [6.3]

12. Find the present value of a \$20,000 annual income if it is invested immediately as it is received into an account that pays 5% annual interest compounded continuously forever. [6.3]

13. By differentiating the function $x^2 - y^2 = 5$ determine if it is or is not a solution to the differential equation $\frac{dy}{dx} = \frac{y}{x}$. [6.4]

14. Show that the differential equation

$$\frac{dy}{dx} = \frac{y}{x}\sqrt{1 - x^2}$$

is a separable differential equation. [6.4]

15. Find the particular solution to the differential equation $\frac{dy}{dx} = \frac{2x}{y^2}$ if $y = 4$ when $x = 1$. [6.4]

16. Find the particular solution to the differential equation $y' = y^2(1 + e^x)$ if $y = 1$ when $x = 0$. [6.4]

17. **Biology: Bacteria Growth** A population of bacteria left uncontrolled, grows according to the unlimited growth model. Initially there are 500,000 bacteria. Ten hours later, there are 1,000,000 bacteria. To the nearest hour, how long will it take for this population to grow to be 2,000,000 in size? [6.5]

18. **Engineering: Electronic Systems** A component of an electronic system is programmed to switch off if its temperature reaches 120 °C. Tests show that in a certain environment, a particular component heats from 50 °C to 60 °C in 200 hours. Use the limited growth model to determine how long in this environment would it take for this component to heat to 70 °C? [6.5]

19. **Biology: Bacteria Growth** A population of bacteria growing beneath a rock near a pool of stagnant water increases according to the logistic growth model with a carrying capacity of 5,000,000. Initially, 10,000 bacteria are present and two days later, there are 80,800 bacteria present. Use the logistic growth model to determine how long it will take this population of bacteria to reach 4,000,000 in number. [6.5]

20. The probability that a randomly selected male in the age group 15-20 will buy a particular style of advertised clothes is 0.1855. Interpret this value. [6.6]

21. Find a value of k that will make

$$f(x) = \frac{k\sqrt{2}}{x^2}$$

a probability function over $[5, 10]$. [6.6]

22. **Manufacturing: Batteries** The manufacturer of low amp recreational boat batteries guarantees they will provide at least 5 hours of use and will replace, at no charge to a buyer, any battery that does not. The probability density function for the life, x, of the batteries is

$$f(x) = \frac{4}{789}(15 - \sqrt[3]{x})$$

over $[0, 15]$. To the nearest one percent, what percentage of batteries can this manufacturer expect to replace? [6.6]

23. A traffic light stays red for 35 seconds and does so uniformly throughout the day. If a person arrives at the light and finds it red, what is the probability that he or she will have to wait less than 15 seconds for it to turn green? [6.6]

24. The probability density function for the number of minutes a customer must wait to be connected to a service representative of a big box retailer is $f(x) = 0.36e^{-0.36x}$. What is the probability a randomly selected customer must wait more than 5 minutes? [6.6]

25. Manufacturing: Lane Dividers A company that makes reflective yellow paint that a city uses on its streets for lane dividers claims that the deterioration time of the paint is normally distributed with a mean of life of 5,000,000 vehicle crossings and a standard deviation of 1,000,000 crossings. Find the probability that a randomly selected line will withstand up to 6,000,000 vehicle crossings before it deteriorates and has to be restored. [6.6]

Calculus of Functions of Several Variables

7

Chapter Outline

Note When you see this icon next to an example or problem in this chapter, you will know that we are using the topics in this chapter to model situations in the world around us.

Manufacturers constantly make decisions about how to allocate their resources. How much money should be allocated to capital and how much to labor? An airline company might weigh the benefit of hiring more booking agents or investing in sophisticated booking software. Both cost money. Agents require salaries and benefits and specialized software can cost millions of dollars. The cost of salaries and benefits typically rise and the cost of software typically falls.

There is often an interaction between capital and labor; capital cannot do everything labor can do. Some types of labor complement capital and as capital gets cheaper, that labor gets more in demand.

Economists use mathematics to describe the interaction between capital and labor and are able to ask and answer questions about them. Some questions can be answered using only algebra, but some questions require calculus for their answers.

Algebra If, this year, we allocate x dollars for labor and y dollars for capital, what would be the level of our productivity?

Calculus This year we have a finite amount of money available that we can allocate between labor and capital. How much should we assign to labor and how much to capital in order to maximize our productivity?

We could answer the above question using a Cobb-Douglas production function, such as

$$f(x, y) = 4x^{3/5}5y^{2/5}$$

where x is the units of labor to produce the product, y is the units of capital used, and $f(x, y)$ is the units of product produced. This function gives a graph on a 3-dimensional coordinate system, which is a different form than we have seen previously. In this chapter, we will work with functions of several variables and their graphs, such as the example above.

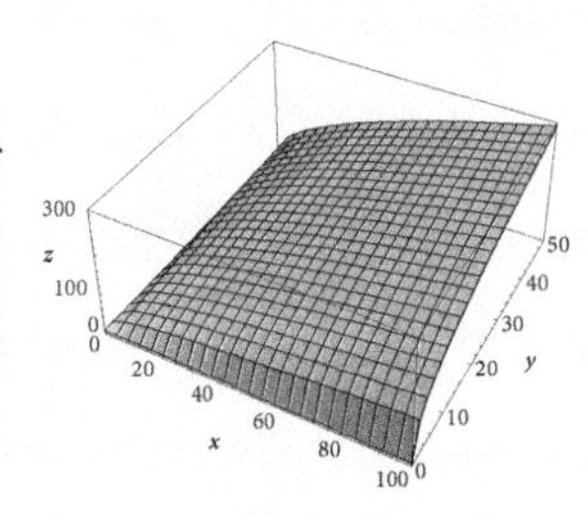

Success Skills

Dear Student,

Now that you are close to finishing this course, I want to pass on a couple of things that have helped me a great deal with my career. I'll introduce each one with a quote:

Do something for the person you will be 5 years from now.

I have always made sure that I arranged my life so that I was doing something for the person I would be 5 years later. For example, when I was 20 years old, I was in college. I imagined that the person I would be as a 25-year-old, would want to have a college degree, so I made sure I stayed in school. That's all there is to this. It is not a hard, rigid philosophy. It is a soft, behind the scenes, foundation. It does not include ideas such as "Five years from now I'm going to graduate at the top of my class from the best college in the country." Instead, you think, "five years from now I will have a college degree, or I will still be in school working towards it."

This philosophy led to a community college teaching job, writing textbooks, doing videos with the textbooks, then to MathTV and the book you are reading right now. Along the way there were many other options and directions that I didn't take, but all the choices I made were due to keeping the person I would be in 5 years in mind.

It's easier to ride a horse in the direction it is going.

I started my college career thinking that I would become a dentist. I enrolled in all the courses that were required for dental school. When I completed the courses, I applied to a number of dental schools, but wasn't accepted. I kept going to school, and applied again the next year, again, without success. My life was not going in the direction of dental school, even though I had worked hard to put it in that direction. So I did a little inventory of the classes I had taken and the grades I earned, and realized that I was doing well in mathematics. My life was actually going in that direction so I decided to see where mathematics would take me. It was a good decision.

It is a good idea to work hard toward your goals, but it is also a good idea to take inventory every now and then to be sure you are headed in the direction that is best for you.

I wish you good luck with the rest of your college years, and with whatever you decide to do for a career.

Pat McKeague
Owner of XYZ Textbooks

Functions of Several Variables

7.1

© Yang Yu/iStockPhoto

The functions we have worked with up to this point have all been functions of one variable. For example, the sales function

$$N(x) = 2{,}400x - 3x^2$$

is a function of the one variable x. If x represents the amount of money a company spends on television advertisements, then $N(x)$ indicates that the number, N, of sales depends *only* on the amount of money spent on television advertisements. In familiar terminology, $N(x)$ indicates that the dependent variable N depends only on the one independent variable x.

Suppose, however, that the company not only advertises on television, but also on radio and in newspapers. If x represents the amount of money spent on television advertisements, y the amount of money spent on radio advertisements, and z the amount of money spent on newspaper advertisements, then $N(x, y, z)$ indicates that N depends on all three variables x, y, and z. In familiar terminology, $N(x, y, z)$ indicates that the dependent variable N depends on the three independent variables x, y, and z.

Notation

When working with functions of one variable, we saw (in Chapter 1) that a function such as $y = 3x + 4$ was better described using the f notation. Rather than writing $y = 3x + 4$, we replaced y with $f(x)$ and wrote $f(x) = 3x + 4$ and kept in mind that y and $f(x)$ both represented the same rule. Sometimes it is convenient to use y, other times, $f(x)$. Similarly, a function of two variables x and y might be represented by $f(x, y)$ or, maybe, just z. If both notations can be used, then we keep in mind that z and $f(x, y)$ both represent the same rule; that is, that $z = f(x, y)$. Likewise, if w and $f(x, y, z)$ represent the same rule, then $w = f(x, y, z)$. These alternate notations are often useful in labeling axes when graphing.

A Function of Two Variables
A function of two variables is a rule that assigns to each ordered pair (x, y) in a set D a unique number $f(x, y)$. The set of ordered pairs is called the domain of the function. The rule lays out the procedure for producing the new number $f(x, y)$ from the two numbers x and y. Functions of three, four, or even more variables are defined in a similar way.

Evaluation of Functions of Several Variables

Suppose that the sales function $N(x, y, z)$ is defined by the expression

$$2{,}300x + 1{,}400y + 900z - 3x^2 - y^2 - z^2$$

Since a function is a rule that prescribes how an output value is obtained from an input value—or from several input values—we can evaluate a function of several variables by substituting the input values for their corresponding variables and computing.

VIDEO EXAMPLES

SECTION 7.1

Example 1 For the television, radio, and newspaper sales function

$$N(x, y, z) = 2{,}300x + 1{,}400y + 900z - 3x^2 - y^2 - z^2$$

find the number of sales, N, if $x = 80$, $y = 16$, and $z = 20$, where x, y, and z are in thousands of dollars.

Solution We substitute 80 for x, 16 for y, and 20 for z and compute.

$$N(80, 16, 20) = 2{,}300(80) + 1{,}400(16) + 900(20) - 3(80)^2 - (16)^2 - (20)^2$$
$$= 204{,}544$$

Interpretation
We conclude that if \$80,000 is spent on television ads, \$16,000 on radio ads, and \$20,000 on newspaper ads, the company will make 204,544 sales of the advertised item.

Graphs of Functions of Several Variables

We know that a function of one variable requires two coordinate axes, a horizontal one from which the input values are obtained, and a vertical one from which the output values are read. The two axes are constructed perpendicular to each other and form a flat surface (a plane), and the graph is some type of curve. For example, the graph of $f(x) = x^2 - 4x + 1$ appears in Figure 1 in its $f(x)$ and y forms, respectively.

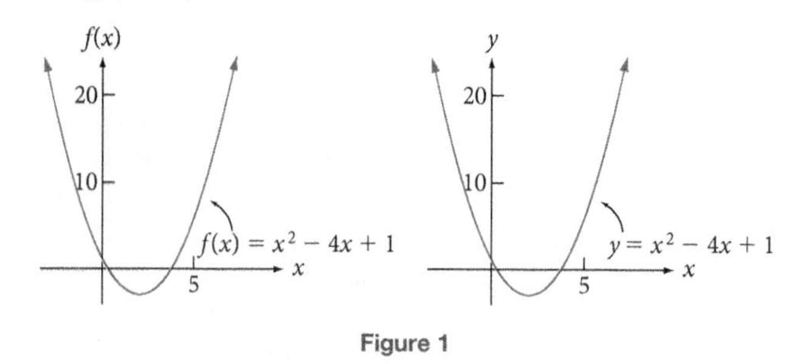

Figure 1

To graph a function of two variables requires three coordinate axes, two from which the values of the independent variables are obtained and one from which the values of the dependent variable are read. The three axes are constructed mutually perpendicular to each other and form what we call **space**. Figure 2 shows this 3-dimensional coordinate system in both its $f(x, y)$ and z forms, respectively. Try to imagine these 2-dimensional drawings as 3-dimensional.

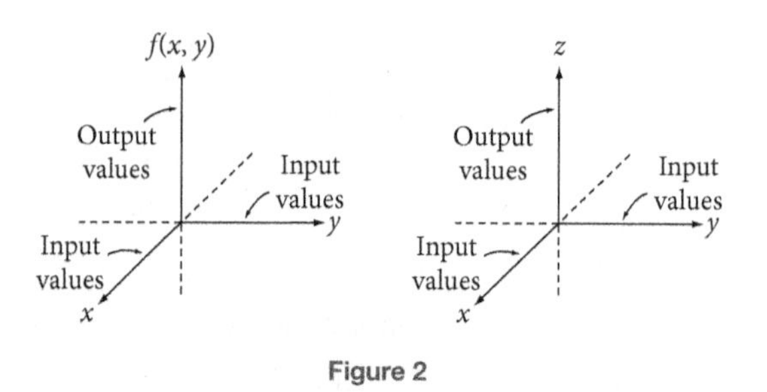

Figure 2

Graphs in space are no longer just curves, but various types of surfaces. Since they can be hard to draw with pencil and paper, we produce them using computers. Graphs of functions of two variables can be very interesting to look at but at the same time difficult to interpret. Figure 3 illustrates two graphs of functions of two variables.

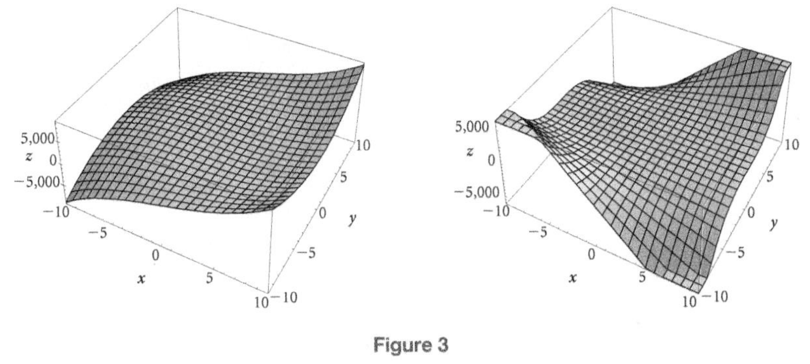

Figure 3

Interpretations

We have a lot of experience interpreting graphs of functions of one variable. For example, we can conclude from the function illustrated in Figure 4 that if the input value 6 is increased by 1 unit to 7, the output value will decrease. You just have to place your pencil on the curve at $x = 6$ and notice that to get to $x = 7$, you must move your pencil downward along the curve. Also, if the input value 11 is increased by 1 unit to 12, the output value will increase. In fact, it is the derivative $f'(x)$ that approximates this decrease or increase in $f(x)$.

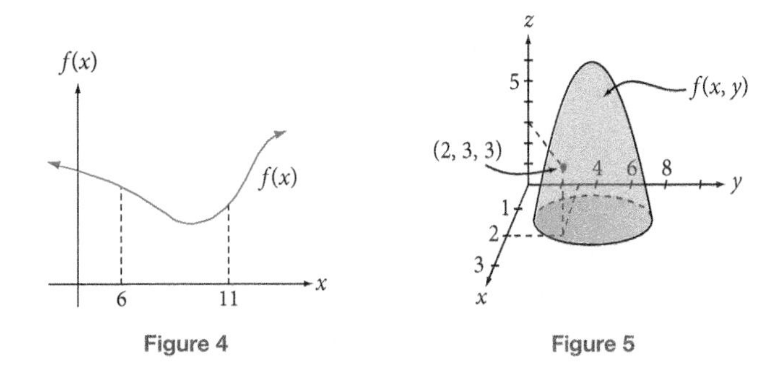

Figure 4

Figure 5

To interpret graphs of functions of two variables we will have to look and think harder. For example, consider the graph of the function $f(x, y)$ illustrated in Figure 5. The dot on the surface represents the output value $z = 3$ corresponding to the input values $x = 2$ and $y = 3$. To see this, place your pencil at the origin and then move it 2 units along the x-axis. Now move your pencil 3 units in the y direction (parallel to the y-axis). This point in the xy-plane represents the input values. Now move your pencil vertically 3 units, stopping when you get to the surface (at the dot). This vertical distance is the output value associated with the two input values. The **ordered triple** associated with this point is (2, 3, 3).

Now, located at this point on the surface, we can change our position in one of three ways. We can move in the x direction only, thus changing only the x coordinate and keeping the y coordinate constant. Or, we can move in the y direction only, thus changing only the y coordinate and keeping the x coordinate constant.

Or, we can move diagonally, thus changing both the x and y coordinates. Figure 6a shows a change of 5 units in the x direction only, and Figure 6b shows a change of 3 units in the y direction only. Figure 7 illustrates changes in both the x and y directions, a 5-unit change in the x direction and a 3-unit change in the y direction.

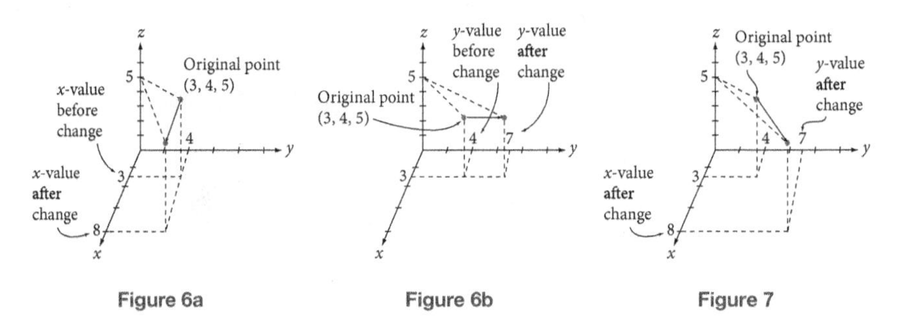

Figure 6a Figure 6b Figure 7

You can interpret the effect of one of these input changes on the output by physically making the change with your pencil. As you try to physically work your way through the next few examples, you may find it difficult to actually see where you are located on the surface; fortunately, we seldom actually have to do this. Just try your best.

Changing x and keeping y constant Refer to Figure 8 and place your pencil on the surface at the dot. To change only the x-value by 1 unit, move your pencil parallel to the x-axis. Move 1 unit in the positive x direction by moving parallel to the x-axis and the same distance as the distance between two consecutive tick marks on the x-axis. To get back onto the surface, you must move your pencil downward $1\frac{1}{4}$ units.

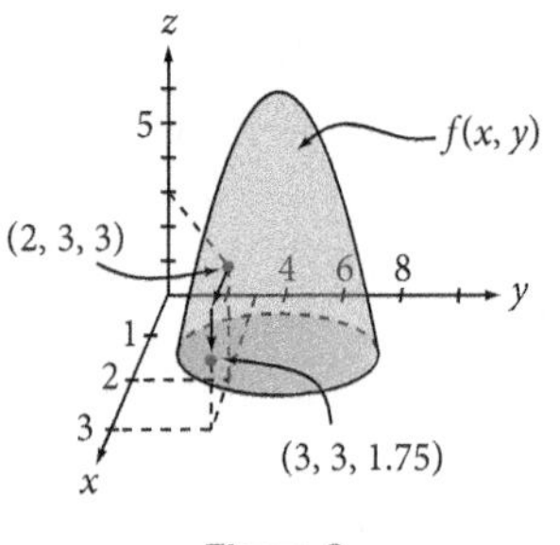

Figure 8

Interpretation
If, when $x = 2$ and $y = 3$, so that $z = 3$, x is increased by 1 unit and y is kept constant, then the function decreases by $1\frac{1}{4}$ units.

Changing y and keeping x constant Refer to Figure 9 and place your pencil on the surface at the dot. To change only the y-value by 1 unit, move your pencil parallel to the y-axis. Move 1 unit in the positive y direction by moving parallel to the y-axis and the same distance as the distance between two consecutive tick marks on the y-axis. To get back onto the surface, you must move your pencil downward $1\frac{3}{4}$ units.

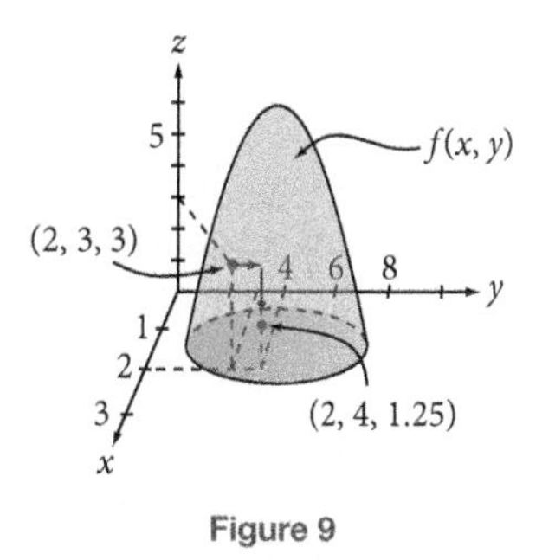

Figure 9

Interpretation
If, when $x = 2$ and $y = 3$, so that $z = 3$, y is increased by 1 unit and x is kept constant, then the function decreases by $1\frac{3}{4}$ units.

Changing both the x and y coordinates Refer to Figure 10 and place your pencil on the surface at the dot. To change both the input values, change one of them first, and then before moving to the surface, change the other. Then after both changes are made, move vertically to the surface. Place your pencil on the surface at the dot. To change the x-value by 1 unit and the y-value by 1 unit, first change x according to Figure 8 and then change y according to Figure 9. To get back onto the surface you must move your pencil downward 2 units.

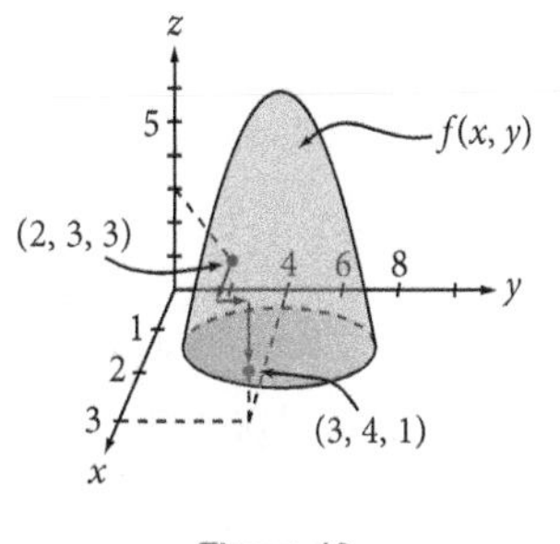

Figure 10

Interpretation
If, when $x = 2$ and $y = 3$, so that $z = 3$, x is increased by 1 unit and y is increased by 1 unit, then the function decreases by 2 units.

Partial Derivatives and the Total Differential

The derivative of a function of several variables approximates the change in the output value for a change in one or all of the input values. If only one of the input values is changed (as in Figures 8 and 9 previously), the derivative is called a **partial derivative** of the function. We will see how to compute partial derivatives in Section 7.2. If all the input values are changed (as in Figure 10 above), we get the **total differential**. (There is no total derivative.) We will see how to compute the total differential in Section 7.5.

Cobb-Douglas Production Functions

In the 1920s, Charles W. Cobb and Paul H. Douglas constructed a power function useful for production studies. The function relates the number of units of a product a company can produce to the number of units of labor and capital it uses to produce the product. The function is of the form

$$f(x, y) = Cx^a y^b$$

where x represents the number of units of labor used to produce the product and y the number of units of capital used. $f(x, y)$ represents the number of units of the product produced. The numbers C, a, and b are constants.

The impact of the work done by Cobb and Douglas was so great that these power functions are now referred to as **Cobb-Douglas production functions.** These power functions answer questions of **returns to scale**; that is, questions that ask how a proportionate increase or decrease in all the input values will affect the output value, the total production. If the proportional increase in all the input values is equal to the proportional increase in the output, then the *returns to scale are constant.* For example, if returns to scale are constant, a doubling of the number of units of both labor and capital results in a doubling of production. If the proportional increase in output is greater than the proportional increase in all the inputs, then the *returns to scale are increasing.* If the proportional increase in output is less than the proportional increase in all the inputs, then the *returns to scale are decreasing.* Cobb-Douglas production functions having the form

$$f(x, y) = Cx^a y^{1-a}$$

where the exponents on the input variables add to 1, always exhibit a constant returns to scale.

Example 2 The manufacturing process of a company is described by the Cobb-Douglas production function $f(x, y) = 4x^{3/5}y^{2/5}$.

a. Approximate, to two decimal places, the number of units produced when 85 units of labor and 20 units of capital are used.

b. Find the change in the level of production if the number of units of labor is decreased by 1 from 85 units to 84 units, and the number of units of capital is increased by 5 from 20 to 25.

c. Notice that the exponents on the variables x and y add to one. Show that this function exhibits constant returns to scale.

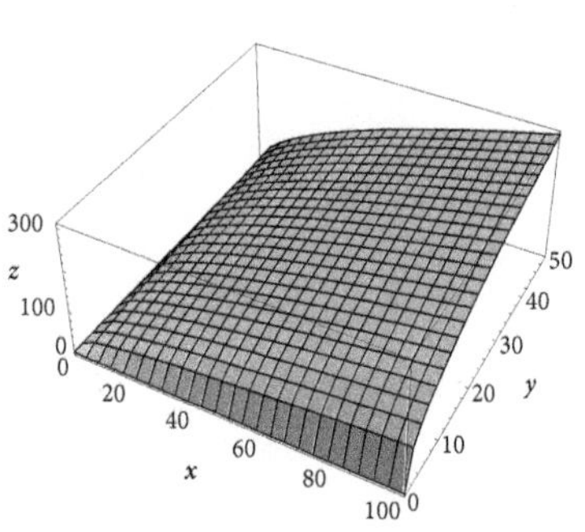

Figure 11

Solution

a. Substituting 85 for x and 20 for y gives

$$f(85, 20) = 4(85)^{3/5}(20)^{2/5}$$
$$\approx 190.60$$

Figure 11 shows the graph of this Cobb-Douglas production function.

Interpretation
When 85 units of labor and 20 units of capital are used in a manufacturing process, approximately 190.6 units will be produced.

b. Substituting 84 for x and 25 for y gives

$$\begin{aligned} f(84, 25) &= 4(84)^{3/5}(25)^{2/5} \\ &\approx 206.92 \end{aligned}$$

Then, the change in the production level is

$$\begin{aligned} f(84, 25) - f(85, 20) &\approx 206.92 - 190.60 \\ &\approx 16.32 \end{aligned}$$

Interpretation
If the number of units of labor is decreased by 1 from 85 to 84, and the number of units of capital is increased by 5 from 20 to 25, the level of production will increase by approximately 16.32 units.

c. To show that the returns to scale is constant, we need to show that if both the input values x and y are multiplied by some constant, say k, then the output value $f(x, y)$ is multiplied by the same constant. We begin by noting that

$$f(x, y) = 4x^{3/5}y^{2/5}$$

We now multiply each input value by k.

$$\begin{aligned} f(kx, ky) &= 4(kx)^{3/5}(ky)^{2/5} \\ &= 4k^{3/5}x^{3/5}k^{2/5}y^{2/5} \\ &= 4k^{3/5}k^{2/5}x^{3/5}y^{2/5} \\ &= 4k^{3/5+2/5}x^{3/5}y^{2/5} \\ &= 4k^{5/5}x^{3/5}y^{2/5} \\ &= 4kx^{3/5}y^{2/5} \\ &= k \cdot 4x^{3/5}y^{2/5} \\ &= k \cdot f(x, y) \end{aligned}$$

Thus, if both the input values x and y are multiplied by some constant, say, k, then the output value $f(x, y)$ is multiplied by the same constant, and we conclude that this Cobb-Douglas function exhibits a constant returns to scale. ■

Note Just before Example 2, we noted that Cobb-Douglas functions in which the exponents add to 1 always exhibit a constant returns to scale. Does our work in part c agree with this statement?

Finally, we make a general comment about functions of more than two variables. These functions are very difficult to interpret graphically (try to imagine a fourth axis perpendicular to the other three, etc.). Fortunately for our work and your work, most functions that you will need to deal with are of only one or two variables.

Using Technology 7.1

Wolfram|Alpha can be helpful in evaluating functions of several variables as well as in graphing functions of two variables. We'll use Wolfram|Alpha in the following two examples.

Note Notice that Wolfram|Alpha produces 3-D graphs on coordinate systems that are oriented differently than the ones we used earlier in this section.

Example 3 Evaluate functions of several variables.

Solution Go to www.wolframalpha.com and in the entry field enter

Evaluate 2x^3 – 5y^2 + 3z for x = 2, y = 1, z = –5

Wolfram|Alpha responds with -4. ■

Example 4 Construct the graph of the given function. (The surface described by this function is called a hyperboloid.)

$$f(x, y, z) = x^2 + y^2 - z^2 = 1 \text{ on } -4 \le x \le 4, -4 \le y \le 4, \text{ and } -4 \le z \le 4$$

Solution Go to www.wolframalpha.com and in the entry field enter

Graph x^2 + y^2 - z^2 = 1 for -4<=x<=4, -4<=y<=4, -4<=z<=4

Wolfram|Alpha returns the graph of $x^2 + y^2 - z^2 = 1$.

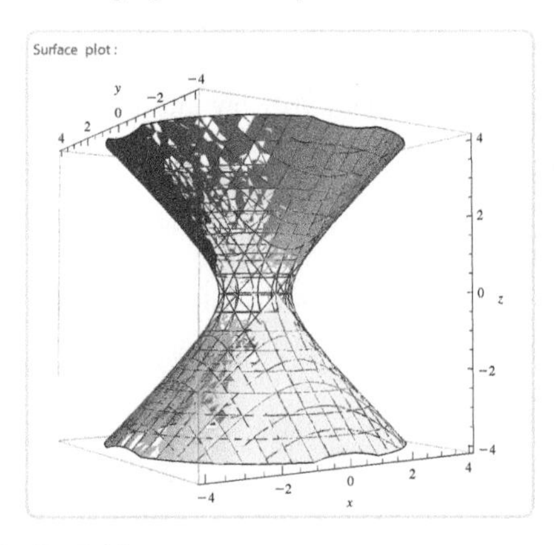

Wolfram Alpha LLC. 2013. Wolfram|Alpha
http://www.wolframalpha.com/
(access January 14, 2013) ■

Getting Ready for Class

After reading through the preceding section, respond in your own words and in complete sentences.

A. Explain the rule for functions of two variables.
B. Explain why the graph of a function of two variables requires three coordinate axes.
C. Explain why we interpret a graph of a function of two variables differently than we interpret a graph of a function of one variable.
D. What is a Cobb-Douglas production function?

Problem Set 7.1

Skills Practice

For Problems 1-12, find the exact or approximate (to two decimal places) value of each function.

1. Find $f(1, 4)$ if $f(x, y) = 2x^3 - y^2$.
2. Find $f(3, 2)$ if $f(x, y) = 50x + 25y - 2x^2 - 3y^2$.
3. Find $f(200, 60)$ if $f(x, y) = 4x^{1/3}y^{2/3}$.
4. Find $f(465, 106)$ if $f(x, y) = 25x^{2/5}y^{3/5}$.
5. Find $P(20, 27)$ if $P(s, t) = \sqrt{3s^3 - 2t^2}$.
6. Find $N(15, 7)$ if $N(x, y) = \ln(2x + y) + \frac{y}{x}$.
7. Find $f(5, 8)$ if $f(x, y) = e^{0.01(3x-5y)} \ln(5y - 3x)$.
8. Find $f(2, 1, 3)$ if $f(x, y, z) = e^x e^y \ln(z)$.
9. Find $P(12, 2, 20)$ if $P(x, y, z) = e^{\sqrt{x+3z}} - e^{\sqrt{y+\ln(y)}}$.
10. Find $f(10, 4, 35, 0)$ if $f(x, y, z, w) = \sqrt[3]{\ln(x) + 200e^{-y} + 8z - w}$.
11. Find $f(0, 1)$ if $f(x, y) = y^{e^x+\ln(y)}$.
12. Find $P(2{,}000, 10)$ if $P(A, t) = Ae^{0.08t}$.

For Problems 13-14, your answer will be in terms of h.

13. Find $\frac{f(x + h, y) - f(x, y)}{h}$ if $f(x, y) = x^2 - y^2$.
14. Find $\frac{f(x, y + h) - f(x, y)}{h}$ if $f(x, y) = 3x + 5y - 8x^2 - y^2$.

For Problems 15-16, find the requested limit.

15. Find $\lim_{h \to 0} \frac{f(x, y + h) - f(x, y)}{h}$ if $f(x, y) = y^2 + 4xy$.
16. Find $\lim_{h \to 0} \frac{f(x + h, y) - f(x, y)}{h}$ if $f(x, y) = 5x^2 + 2y^2 - 6x + 20$.

Check Your Understanding

For Problems 17-20, answers may vary.

17. **Manufacturing: Shipping Costs** To reduce shipping distances between its manufacturing facilities and one of its major consumers, a manufacturer is looking into producing cell phone components in one of its U.S. facilities. The manufacturer's shipping cost, C, depends on the price, x, of fuel. Specify another variable, label it y and then use several variable function notation to indicate that the cost of shipping depends on the identified variables.

18. **Economics: Education** The number, N, of students attending a community college is a function of the number, x, of unemployed people in the community. But it is also a function of other variables. Specify another one, name it y, and then use several variable function notation to indicate that the number of students attending a community college depends on the identified two variables.

19. **Animal Science: Anatomy** The surface area, A, of a mammal is a function of the weight, w, of the mammal. But it is also a function of other variables. Specify another one, name it y, and then use several variable function notation to indicate that the surface area of a mammal depends on the identified variables.

20. **Manufacturing: Cost** The cost, C, of producing a precision timing instrument is a function of the cost, u, of unskilled labor and the cost, i, of an integrated circuit chip. But it is also a function of other variables. Specify another one, name it with a letter, and then use several variable function notation to indicate that the cost of producing a precision timing instrument depends on the identified variables.

For each of Problems 21-26, construct a 3-dimensional coordinate system and plot the given point.

21. $(2, 3, 4)$
22. $(4, 2, 2)$
23. $(5, 5, 1)$
24. $(-2, 3, 4)$
25. $(-1, -3, -5)$
26. $(0, 0, 4)$

Modeling Practice

27. **Economics: Cobb-Douglas Production** For a particular product, the Cobb-Douglas production function is

$$f(x, y) = 600x^{0.4}y^{0.6}$$

   a. Find the number of items produced if 120 units of labor, x, and 50 units of capital, y, are used.
   b. Find the change in the level of production if the number of units of labor is increased by 1 unit from 120 to 121 units.
   c. Find the change in the level of production if the number of units of capital is decreased by 5 from 50 to 45 units, while units of labor is still 120.
   d. This function exhibits a constant returns to scale. Show that, if both the costs of labor and capital are tripled, the level of production will also be tripled.

28. **Medicine: Drug Effect** The effect E of a drug on a human patient is a function of both the amount x (in milligrams) of the drug administered and the amount of time t (in hours) that has passed since the drug was administered. For a particular drug, the function that relates these quantities is

$$E(x, t) = 18.5x^{1.20}e^{-0.04t}$$

   a. Find the effect that 200 mg of the drug has on a patient 2 hours after it has been administered.
   b. Find the change in the effect of the drug on the patient if the dosage had been 190 mg rather than 200 mg.
   c. Find the change in the effect of the drug 3 hours after 200 mg has been administered.

29. Psychology: Depression A psychology researcher believes that the function

$$D(x, y, z) = 10x^{2/3} + 8y^{4/3} - 7z^{3/2} - \ln(x^2 + y^2 - z^2)$$

relates the number, D, of units (on the researcher's scale) of depression a college student feels on a particular day to the number, x, of hours of sleep over 6 hours the student had the previous night, the number, y, of semester units beyond 15 the student is enrolled in, and the number z of good quality conversations with friends or family during the previous 2 days.

a. Find the number of depression units for a student who the previous night had 13 hours of sleep, who is carrying 21 units, and who had no good quality conversations for the past 2 days with friends or family.

b. What would be the change in the number of depression units for this student if, in the next 2 days, he has 5 good quality conversations with friends or family, but his amount of sleep and number of units he is enrolled in remain constant?

30. Business: Waiting Time The function

$$W(a, s) = \frac{1}{2(s - a)} \qquad a < s$$

relates the average waiting time W, in hours, in a line to the average service rate, s, expressed in the number of customers each hour, and the average customer arrival rate, a, expressed in the number of customers per hour.

a. What is the average waiting time in a line if the average service rate is 20 customers per hour and the average arrival rate is 14 customers per hour?

b. What would be the change in the average waiting time if the average service rate were to increase by 3 customers per hour, from 20 to 23, and the average arrival rate were to increase by 5 customers per hour, from 14 to 19 customers per hour?

Using Technology Exercises

Use Wolfram|Alpha to help you evaluate or graph each function.

31. A stockbroker charges commission, C, based on the number, x, of shares purchased or sold and the price y of each share according to the function

$$C(x, y) = 30 + 0.02x + 0.001xy$$

a. What is the stockbroker's commission on the purchase of 1,500 shares of a stock that sells for $48.50 per share?

b. Graph this function on the intervals $0 \le x \le 2{,}000$ and $0 \le y \le 100$.

32. A manufacturer subjects its products to three inspections. The cost, in dollars, of repairing a defective unit depends on the number of defective parts found and is given by the function

$$C(x, y, z) = 0.12x^{1.6} + 1.1x^{0.45}y^{0.68}z^{0.33} + 1.45x + 0.25y^{0.25} + 0.34z^{1.6}$$

where x represents the number of defective parts found in the first inspection, y represents the number of defective parts found in the second inspection, and z represents the number of defective parts found in the third inspection. Find the cost of repairing a unit if 15 defective parts are found in the first inspection, 7 in the second, and 0 in the third.

33. A retail department store bases its projections of customer motivation M to purchase on the index function

$$M(x, y) = -2.5x^2 - 3.0y^2 + 205x + 217y + 1.6xy - 10{,}000$$

where x represents the Fahrenheit temperature of the store and y the decibel level of music in the store. An index of 1,000 represents total motivation. Find the customer motivation index if the store temperature is 70 °F and the music level is 45 decibels.

Source: Helium

34. Minor changes in the description of a function can have interesting effects. To see this, follow the instructions in parts a and b.

a. Construct the graph of the function $f(x, y) = xy \cdot \dfrac{x^2 - y^2}{x^2 + y^2}$.

b. Then change the function by making the numerator the addition and the denominator the subtraction. That is, construct the graph of the function

$$f(x, y) = xy \cdot \frac{x^2 + y^2}{x^2 - y^2}.$$

35. Natural log functions often occur in applied problems. You know what the graph of $f(x) = \ln(x)$ looks like. You might wonder if the graph of $f(x, y) = \ln(x \cdot y)$ exhibits the same increasing behavior as does $f(x) = \ln(x)$. Graph both functions to see. Make a statement about their behavior.

Getting Ready for the Next Section

For Problems 36-44, find each derivative. In each case, j and k represent constants.

36. Find $\frac{dy}{dx}$ for $y = 8x^3 + 5k^2 + 6x - 2k$.

37. Find $f'(y)$ for $f(y) = 8k^3 + 5y^2 + 6k - 2y$.

38. Find $\frac{df}{dx}$ for $f(x) = x^4 + 8x^2k^3 + 5k^4$.

39. Find $\frac{df}{dy}$ for $f(y) = k^4 + 8k^2y^3 + 5y^4$.

40. Find $\frac{dy}{dx}$ for $y = (5x^3 - 8k^2)^4$.

41. Find $f'(y)$ for $f(y) = (5k^3 - 8y^2)^4$.

42. Find $\frac{df}{dx}$ for $f(x) = e^{x+3j} + 5\ln(jkx)$.

43. Find $\frac{df}{dy}$ for $f(y) = e^{j+3k} + 5\ln(jky)$.

44. Find $\frac{df}{dz}$ for $f(z) = e^{j+3z} + 5\ln(jkz)$.

45. If at $x = 2$, $f'(x) > 0$ and $f''(x) < 0$, what can you say about the behavior of $f(x)$ at $x = 2$?

46. If at $x = 170$, $f'(x) < 0$ and $f''(x) < 0$, what can you say about the behavior of $f(x)$ at $x = 170$?

Partial Derivatives

7.2

As we know, differentiation is the language of change. A natural question is, "What effect on the output of a function of more than one variable does an increase (or decrease) in *one* of the input variables have?" The answer is given by the *partial derivative* of the function. For example, suppose the inequalities below give information about the rate of destruction of a country's rain forest, using the function $N(x, y)$ where x is the amount of money (in millions of dollars) the World Bank loans to the country and y is the population (in millions of people) of the country.

$$\frac{\partial}{\partial y} N(40, 70) > 0 \quad \text{and} \quad \frac{\partial^2}{\partial y^2} N(40, 70) > 0$$

These inequalities use notation that represents the partial derivation for functions with more than one variable. In this section, we examine the notation of a partial derivative and the process of finding one.

The Partial Derivative

We will investigate partial differentiation by finding the partial derivatives of the function $f(x, y) = 7x^2 + 3y^2 + 5xy + 12x - 8y$.

Suppose we wish to know how the function changes as x changes and y is held constant at some particular value, say, k. Substituting k for y in the function produces the function of the one variable x

$$f(x, k) = 7x^2 + 3k^2 + 5xk + 12x - 8k$$

Now, keeping in mind that the derivative of a constant is zero,

$$\begin{aligned} \frac{d}{dx} f(x, k) &= 14x + 0 + 5k + 12 - 0 \\ &= 14x + 5k + 12 \end{aligned}$$

Now replace k with y and obtain

$$\frac{d}{dx} f(x, y) = 14x + 5y + 12$$

Of course, we could have saved ourselves some energy by not replacing y with k at all, and just *visualizing* y as a constant. We will do this in all the following examples. Since it may be hard to visualize y as a constant (since all through algebra it was a variable), you may want to replace it with k until you feel more comfortable with the process. After all, k looks more like a constant than does y.

Notation for Partial Derivatives

The derivative notations $\frac{d}{dx}$ and $\frac{d}{dy}$ are usually reserved for functions of one variable. For functions of more than one variable, we use

$$\frac{\partial f}{\partial x} \quad \text{or} \quad f_x \quad \text{or} \quad f_x(x, y)$$

to indicate the partial derivative of f with respect to x, and

$$\frac{\partial f}{\partial y} \quad \text{or} \quad f_y \quad \text{or} \quad f_y(x, y)$$

to indicate the partial derivative of f with respect to y.

$\frac{\partial f}{\partial x}$ represents the change in the value of the function with respect to x when the value of x is changed and the value of y is held constant; that is, it represents the rate of change of the function *in the x direction.* $\frac{\partial f}{\partial y}$ represents the change in the value of the function with respect to y when the value of y is changed and the value of x is held constant; that is, it represents the rate of change of the function *in the y direction.*

VIDEO EXAMPLES

SECTION 7.2

Example 1 The following examples will illustrate the meaning of

$$\frac{\partial}{\partial x} f(x, y) = 14x + 5y + 12$$

for several different values of x and y.

a. $\frac{\partial}{\partial x} f(2, 3) = 14(2) + 5(3) + 12 = 55$

> **Interpretation**
> If, when $x = 2$ and $y = 3$, x is increased by 1 unit, from 2 to 3, then the value of the function will increase by approximately 55 units.

b. $\frac{\partial}{\partial x} f(8, 3) = 14(8) + 5(3) + 12 = 139$

> **Interpretation**
> If, when $x = 8$ and $y = 3$, x is increased by 1 unit, from 8 to 9, then the value of the function will increase by approximately 139 units.

For the above derivative, we held y constant and differentiated with respect to x. As noted, we can also hold x constant. Using the function from the beginning of this section, $f(x, y) = 7x^2 + 3y^2 + 5xy + 12x - 8y$, and differentiating with respect to y, we have

$$\begin{aligned} \frac{\partial}{\partial y} f(x, y) &= 0 + 6y + 5x + 0 - 8 \\ &= 6y + 5x - 8 \end{aligned}$$

(You may find it helpful to verify this by replacing x with k, differentiating with respect to the variable y, and then replacing k with x.)

The Geometry of Partial Derivatives

Figure 1 illustrates the geometric meaning of $\frac{\partial f}{\partial x}$ and $\frac{\partial f}{\partial y}$.

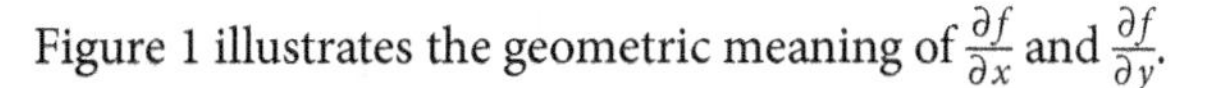

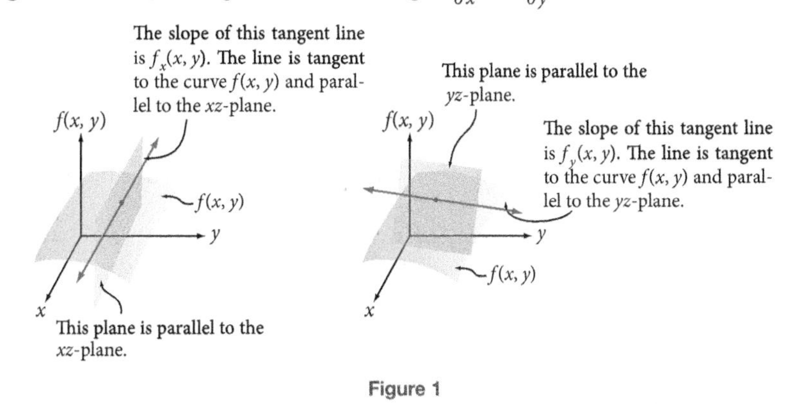

Figure 1

In the leftmost illustration of Figure 1, the slope of the tangent line to the curve $f(x, y)$—that lies in a plane that is parallel to the x-$f(x, y)$ plane—is given by $f_x(x, y)$ or $\frac{\partial f}{\partial x}$. Similarly, in the rightmost illustration, the slope of the tangent line to the curve $f(x, y)$—that lies in a plane that is parallel to the y-$f(x, y)$ plane—is given by $f_y(x, y)$ or $\frac{\partial f}{\partial y}$.

As you study each of the following examples, you may wish to follow along with pencil and paper. Until you feel confident that you can treat a letter that is usually a variable as a constant, you may wish to include the step of substituting k for the value to be held constant.

Example 2 Find $\frac{\partial f}{\partial x}$ and $\frac{\partial f}{\partial y}$ for $f(x, y) = 8x^3 + 5y^2 + 6x - 2y$ and evaluate each at $(2, 7)$.

Solution

a. To find $\frac{\partial f}{\partial x}$, we keep y constant and differentiate with respect to x.

$$\begin{aligned}\frac{\partial f}{\partial x} &= \frac{\partial}{\partial x}[8x^3] + \frac{\partial}{\partial x}[5y^2] + \frac{\partial}{\partial x}[6x] - \frac{\partial}{\partial x}[2y] \\ &= 24x^2 + 0 + 6 - 0 \\ &= 24x^2 + 6\end{aligned}$$

Thus, $\frac{\partial f}{\partial x} = 24x^2 + 6$, and this rate of change depends only on x. Then

$$\begin{aligned}\left.\frac{\partial f}{\partial x}\right|_{(2,7)} &= 24(2)^2 + 6 \\ &= 102\end{aligned}$$

Interpretation

If, when $x = 2$ and $y = 7$, the value of x is increased by 1 unit, from 2 to 3, the value of the function will increase by approximately 102 units. (Remember, the change in the function value is only approximate since the derivative involves a variable.)

b. To find $\frac{\partial f}{\partial y}$, we keep x constant and differentiate with respect to y.

$$\begin{aligned}\frac{\partial f}{\partial y} &= \frac{\partial}{\partial y}[8x^3] + \frac{\partial}{\partial y}[5y^2] + \frac{\partial}{\partial y}[6x] - \frac{\partial}{\partial y}[2y] \\ &= 0 + 10y + 0 - 2 \\ &= 10y - 2\end{aligned}$$

Thus, $\frac{\partial f}{\partial y} = 10y - 2$, and this rate of change depends only on y. Then

$$\begin{aligned}\left.\frac{\partial f}{\partial y}\right|_{(2,\,7)} &= 10(7) - 2 \\ &= 68\end{aligned}$$

Interpretation
If, when $x = 2$ and $y = 7$, the value of y is increased by 1 unit, from 7 to 8, the value of the function will increase by approximately 68 units.

Example 3 Find and interpret $f_x(2, 3)$ and $f_y(2, 3)$ for

$$f(x, y) = x^4 + 8x^2y^3 + 5y^4.$$

Solution

a. To find f_x, we keep y constant and differentiate with respect to x.

$$\begin{aligned}f_x &= \frac{\partial}{\partial x}[x^4] + \frac{\partial}{\partial x}[8x^2y^3] + \frac{\partial}{\partial x}[5y^4] \\ &= 4x^3 + 8y^3 \cdot 2x + 0 \\ &= 4x^3 + 16xy^3\end{aligned}$$

Thus, $f_x(x, y) = 4x^3 + 16xy^3$, and this rate of change depends on the values of both x and y.

$$\begin{aligned}f_x(2, 3) &= 4(2)^3 + 16(2)(3)^3 \\ &= 896\end{aligned}$$

Interpretation
When the input values are $x = 2$ and $y = 3$, if x is increased by 1 unit from 2 to 3, the output value will increase by approximately 896 units.

b. To find f_y, we keep x constant and differentiate with respect to y.

$$\begin{aligned}f_y &= \frac{\partial}{\partial y}[x^4] + \frac{\partial}{\partial y}[8x^2y^3] + \frac{\partial}{\partial y}[5y^4] \\ &= 0 + 8x^2 \cdot 3y^2 + 20y^3 \\ &= 24x^2y^2 + 20y^3\end{aligned}$$

Thus, $f_y(x, y) = 24x^2y^2 + 20y^3$, and this rate of change depends on the values of both x and y.

$$f_y(2, 3) = 24(2)^2(3)^2 + 20(3)^3$$
$$= 1{,}404$$

Interpretation
When the input values are $x = 2$ and $y = 3$, if y is increased by 1 unit from 3 to 4, the output value will increase by approximately 1,404 units.

Example 4 Find f_x and f_y for $f(x, y) = (5x^3 - 8y^2)^4$.

Solution

a. To find f_x, we will keep y constant and differentiate with respect to x. Viewing this function globally, we see a power, so we will begin by using the general power rule.

$$f_x = \underbrace{4 \cdot (5x^3 - 8y^2)^3}_{(\text{outside})'} \cdot \underbrace{(15x^2 - 0)}_{(\text{inside})'}$$
$$= 4(5x^3 - 8y^2)^3(15x^2)$$
$$= 60x^2(5x^3 - 8y^2)^3$$

Thus, $f_x = 60x^2(5x^3 - 8y^2)^3$, and this derivative depends on both x and y.

b. To find f_y, we will keep x constant and differentiate with respect to y. Viewing this function globally, we see a power, so we will begin by using the general power rule.

$$f_y = \underbrace{4 \cdot (5x^3 - 8y^2)^3}_{(\text{outside})'} \cdot \underbrace{(0 - 16y)}_{(\text{inside})'}$$
$$= 4(5x^3 - 8y^2)^3(-16y)$$
$$= -64y(5x^3 - 8y^2)^3$$

Thus, $f_y = -64y(5x^3 - 8y^2)^3$, and this derivative depends on both x and y.

Example 5 Find f_x, f_y, and f_z for $f(x, y, z) = e^{x+3z} + 5\ln(xyz)$.

Solution

a. To find f_x, we will treat both y and z as constants and differentiate with respect to x.

$$f_x = \frac{\partial}{\partial x}[e^{x+3z}] + \frac{\partial}{\partial x}[5\ln(xyz)]$$
$$= e^{x+3z} \cdot \frac{\partial}{\partial x}[x + 3z] + 5 \cdot \frac{1}{xyz} \cdot \frac{\partial}{\partial x}[xyz]$$
$$= e^{x+3z} \cdot (1) + 5 \cdot \frac{1}{xyz} \cdot (yz)$$
$$= e^{x+3z} + \frac{5}{x}$$

Thus, $f_x = e^{x+3z} + \frac{5}{x}$, and its value depends on the values of both x and z.

b. To find f_y, we will treat both x and z as constants and differentiate with respect to x.

$$\begin{aligned} f_y &= \frac{\partial}{\partial y}[e^{x+3z}] + \frac{\partial}{\partial y}[5\ln(xyz)] \\ &= e^{x+3z} \cdot \frac{\partial}{\partial y}[x+3z] + 5 \cdot \frac{1}{xyz} \cdot \frac{\partial}{\partial y}[xyz] \\ &= e^{x+3z} \cdot (0) + 5 \cdot \frac{1}{xyz} \cdot (xz) \\ &= \frac{5}{y} \end{aligned}$$

Thus, $f_y = \frac{5}{y}$, and its value depends only on the value of y.

c. To find f_z, we will treat both x and y as constants and differentiate with respect to z.

$$\begin{aligned} f_z &= \frac{\partial}{\partial z}[e^{x+3z}] + \frac{\partial}{\partial z}[5\ln(xyz)] \\ &= e^{x+3z} \cdot \frac{\partial}{\partial z}[x+3z] + 5 \cdot \frac{1}{xyz} \cdot \frac{\partial}{\partial z}[xyz] \\ &= e^{x+3z} \cdot (3) + 5 \cdot \frac{1}{xyz} \cdot (xy) \\ &= 3e^{x+3z} + \frac{5}{z} \end{aligned}$$

Thus, $f_z = 3e^{x+3z} + \frac{5}{z}$, and its value depends on the values of both x and z. ■

Example 6 Find all the points (x, y) for which both f_x and f_y equal zero, where $f(x, y) = x^2 + y^2 - xy + y - 8$.

Solution We begin by finding each partial derivative.

$$f_x = 2x - y \quad \text{and} \quad f_y = 2y - x + 1$$

To find the points where $f_x = 0$ and $f_y = 0$, we need to solve the system

$$\begin{cases} f_x = 0 & \ldots(1) \\ f_y = 0 & \ldots(2) \end{cases} \quad \text{that is,} \quad \begin{cases} 2x - y = 0 & \ldots(1) \\ 2y - x + 1 = 0 & \ldots(2) \end{cases}$$

We will solve this system using the method of elimination by substitution. Solving equation (1) for y, we get $y = 2x$. Then, substituting $2x$ for y in equation (2), we get

$$\begin{aligned} 2(2x) - x + 1 &= 0 \\ x &= -\frac{1}{3} \end{aligned}$$

Then, since $y = 2x$ and $x = -\frac{1}{3}$, we get $y = 2\left(-\frac{1}{3}\right) = -\frac{2}{3}$. Thus, for the function $f(x, y) = x^2 + y^2 - xy + y - 8$, both partial derivatives f_x and f_y equal zero at the point $\left(-\frac{1}{3}, -\frac{2}{3}\right)$. ■

Higher-Order Partial Derivatives

Just as it was useful to differentiate the derivative $f'(x)$ of a function of one variable $f(x)$, it is useful to differentiate the partial derivatives of a function of two variables $f(x, y)$. Since a function of two variables has *two* partial derivatives, and each partial derivative may, in turn, be a function of two variables, each partial derivative can be differentiated again. Thus, a function $f(x, y)$ of two variables has four partial derivatives.

Notation:

$$\frac{\partial}{\partial x}\left(\frac{\partial f}{\partial x}\right) = \frac{\partial^2 f}{\partial x^2} = f_{xx}$$

$$\frac{\partial}{\partial x}\left(\frac{\partial f}{\partial y}\right) = \frac{\partial^2 f}{\partial x \partial y} = f_{yx}$$

$$\frac{\partial}{\partial y}\left(\frac{\partial f}{\partial y}\right) = \frac{\partial^2 f}{\partial y^2} = f_{yy}$$

$$\frac{\partial}{\partial y}\left(\frac{\partial f}{\partial x}\right) = \frac{\partial^2 f}{\partial y \partial x} = f_{xy}$$

The partial derivatives f_{xx} and f_{yy} are called **second-order partial derivatives** of f with respect to x and y, respectively. The partial derivatives f_{yx} and f_{xy} are called **mixed partial derivatives.**

It is important to note the order of differentiation implied by the notations $\frac{\partial^2 f}{\partial y \partial x}$, f_{xy} and $\frac{\partial^2 f}{\partial x \partial y}$, f_{yx}.

1. $\underbrace{f_{xy}}_{\text{left-to-right}} = \underbrace{\frac{\partial^2 f}{\partial y \partial x}}_{\text{right-to-left}}$ and indicates that f is to be differentiated first with respect to x, then with respect to y, and

2. $\underbrace{f_{yx}}_{\text{left-to-right}} = \underbrace{\frac{\partial^2 f}{\partial x \partial y}}_{\text{right-to-left}}$ and indicates that f is to be differentiated first with respect to y, then with respect to x.

Example 7 Find the second partial derivatives of

$$f(x, y) = x^3 + 2x^2y - 5xy^2 + 3y^3$$

Solution We begin by finding the first partial derivatives.

$$f_x = \frac{\partial}{\partial x}(x^3 + 2x^2y - 5xy^2 + 3y^3) = 3x^2 + 4xy - 5y^2, \quad \text{and}$$

$$f_y = \frac{\partial}{\partial y}(x^3 + 2x^2y - 5xy^2 + 3y^3) = 2x^2 - 10xy + 9y^2, \quad \text{so that}$$

$$f_x = 3x^2 + 4xy - 5y^2 \quad \text{and} \quad f_y = 2x^2 - 10xy + 9y^2$$

Then, taking the derivatives of these functions with respect to x and y,

$$f_{xx} = \frac{\partial}{\partial x}(f_x) = \frac{\partial}{\partial x}(3x^2 + 4xy - 5y^2) = 6x + 4y, \quad \text{and}$$

$$f_{yy} = \frac{\partial}{\partial y}(f_y) = \frac{\partial}{\partial x}(2x^2 - 10xy + 9y^2) = -10x + 18y, \quad \text{and}$$

$$f_{xy} = \frac{\partial}{\partial y}(f_x) = \frac{\partial}{\partial y}(3x^2 + 4xy - 5y^2) = 4x - 10y, \quad \text{and}$$

$$f_{yx} = \frac{\partial}{\partial x}(f_y) = \frac{\partial}{\partial x}(2x^2 - 10xy + 9y^2) = 4x - 10y$$

Thus, $f_{xx} = 6x + 4y$, $f_{yy} = -10x + 18y$, $f_{xy} = 4x - 10y$, $f_{yx} = 4x - 10y$.

Notice that f_{xy} and f_{yx} are equal. This is no coincidence; it turns out that all functions of two variables for which the mixed partial derivatives are continuous on an open set (such as the interior of a circle) have equal mixed partial derivatives at any point in the open set. All the functions we will examine in this text meet such conditions and will have equal mixed partial derivatives. ■

Second-Order Partial Derivatives as Rates of Change

As the second derivative $f''(x)$ measures the concavity of the curve $f(x)$, f_{xx} and f_{yy} measure the concavity in the x direction and the y direction, respectively, of the surface $f(x, y)$. In fact, the signs of f_x and f_{xx}, f_y and f_{yy}, give us information about the behavior of the function. (This is analogous to our discussion in Section 2.3. You may wish to examine that discussion again.)

Two partial derivatives of the same kind, f_x, f_{xx}, and f_y, f_{yy}, with the *same sign* indicate that the function f changes at an *increasing rate*.

$f_x > 0$ and $f_{xx} > 0 \rightarrow f$ is increasing at an increasing rate in the x-direction

$f_y > 0$ and $f_{yy} > 0 \rightarrow f$ is increasing at an increasing rate in the y-direction

$f_x < 0$ and $f_{xx} < 0 \rightarrow f$ is decreasing at an increasing rate in the x-direction

$f_y < 0$ and $f_{yy} < 0 \rightarrow f$ is decreasing at an increasing rate in the y-direction

Two partial derivatives of the same kind, f_x, f_{xx}, and f_y, f_{yy}, with the *opposite sign* indicate that the function f changes at a *decreasing rate*.

$f_x > 0$ and $f_{xx} < 0 \rightarrow f$ is increasing at a decreasing rate in the x-direction

$f_y > 0$ and $f_{yy} < 0 \rightarrow f$ is increasing at a decreasing rate in the y-direction

$f_x < 0$ and $f_{xx} > 0 \rightarrow f$ is decreasing at a decreasing rate in the x-direction

$f_y < 0$ and $f_{yy} > 0 \rightarrow f$ is decreasing at a decreasing rate in the y-direction

Example 8 To produce batteries, a manufacturer uses x units of chemical A and y units of chemical B. The amount P of pollution washed into the cleaning water is given by the pollution function

$$P(x, y) = 0.05x^2 + 0.009xy + 0.03y^2$$

If the company is currently using 16 units of chemical A and 10 units of chemical B, find the rate at which the amount of pollution washed into the cleaning water is changing as the number of units of chemical B changes. Determine if this rate is changing at an increasing or a decreasing rate.

Solution

a. To find the rate at which the amount of pollution is changing as the amount of chemical B changes, we need to find

$$\frac{\partial}{\partial y}P(x, y) = \frac{\partial}{\partial y}(0.05x^2 + 0.009xy + 0.03y^2).$$

$$\begin{aligned}\frac{\partial}{\partial y}P(x, y) &= \frac{\partial}{\partial y}(0.05x^2 + 0.009xy + 0.03y^2)\\ &= 0.009x + 0.06y \quad \text{and}\end{aligned}$$

$$\frac{\partial}{\partial y}P(16, 10) = 0.009(16) + 0.06(10) = 0.744$$

Thus, $\frac{\partial}{\partial y}P(16, 10) = 0.744.$

b. To find if this rate is changing at an increasing or a decreasing rate, we need to find and interpret $\frac{\partial^2}{\partial y^2}P(16, 10)$.

$$\begin{aligned}\frac{\partial^2}{\partial y^2}P(x, y) &= \frac{\partial}{\partial y}(0.009x + 0.06y)\\ &= 0.06\end{aligned}$$

Thus, $\frac{\partial^2}{\partial y^2}P(16, 10) = 0.06$. Since 0.06 is always positive, the rate is changing at an increasing rate.

Interpretation

If, when 16 units of chemical A and 10 units of chemical B are being used to produce batteries, the number of units of chemical B is increased by 1, from 10 units to 11, then the amount of pollutant washed into the cleaning water will increase by approximately 0.744 units. Furthermore, at this point, the amount of pollution is increasing at an increasing rate.

Using Technology 7.2

Wolfram|Alpha can be helpful in differentiating functions of two variables.

Example 9 Use Wolfram|Alpha to find both $\frac{\partial f}{\partial x}$, $\frac{\partial f}{\partial y}$, and $\frac{\partial f}{\partial y \partial x}$ for the function $f(x, y) = x^2 - 2y^3 + 5xy$.

Solution Go to www.wolframalpha.com and in the entry field enter

d/dx (x^2 – 2y^3 + 5xy)

Wolfram|Alpha responds with $2x + 5y$. Now enter

d/dy (x^2 – 2y^3 + 5xy)

Wolfram|Alpha responds with $5x - 6y^2$. To compute the second order partial $\frac{\partial f}{\partial y \partial x}$, recall that it means $\frac{\partial f}{\partial y}\left[\frac{\partial f}{\partial x}\right]$. Enter

d/dy d/dx (x^2 - 2y^3 + 5xy)

Wolfram|Alpha returns 5. ■

Example 10 Use Wolfram|Alpha to find $f_x(4, 7)$ and $f_{xx}(4, 7)$ for the function $f(x, y) = x^4 + 5x^2y$.

Solution Go to www.wolframalpha.com and in the entry field enter

d/dx (x^4 + 5x^2y) for x = 4, y = 7

Wolfram|Alpha responds with 536.

Now enter

d/dx d/dx (x^4 + 5x^2y) for x = 4, y = 7

Wolfram|Alpha returns 262. ■

Getting Ready for Class

After reading through the preceding section, respond in your own words and in complete sentences.

A. Describe how a partial derivative differs from an ordinary derivative.

B. Explain what is meant about a function $f(x, y)$ if $f_x > 0$ and $f_{xx} > 0$.

C. Describe how $f_x(x, y)$ differs geometrically from $f_y(x, y)$.

Problem Set 7.2

Skills Practice

1. For $f(x, y) = 5x^2 + 6xy + 8y^3$, find

   a. $\frac{\partial f}{\partial x}$

   b. $\frac{\partial f}{\partial y}$

   c. $\frac{\partial}{\partial x} f(2, 1)$

   d. $\frac{\partial}{\partial y} f(4, 2)$

2. For $f(x, y) = 6x^2 + 2y^2 - 2xy + 25$, find

   a. $\frac{\partial f}{\partial x}$

   b. $\frac{\partial f}{\partial y}$

   c. $\frac{\partial}{\partial x} f(-1, 2)$

   d. $\frac{\partial}{\partial y} f(2, -2)$

3. For $f(x, y) = x^3 - 4y^2 + 3x^3y^2$, find

   a. f_x

   b. f_y

   c. $f_x(0, -1)$

   d. $f_y(-2, 1)$

4. For $f(x, y) = 10x + 2y - x^2 - y^2 + 4x^2y^4$, find

   a. f_x

   b. f_y

   c. $f_x(4, 1)$

   d. $f_y(0, 2)$

5. For $f(x, y) = e^{x+y}$, find

   a. f_x

   b. f_y

   c. $f_x(1, 1)$

   d. $f_y(2, 1)$

6. For $f(x, y) = e^{2x-y}$, find

   a. f_x

   b. f_y

   c. $f_x(2, 0)$

   d. $f_y(2, 0)$

7. For $f(x, y) = \ln(2 + 4x^2y^2)$, find

   a. f_x

   b. f_y

   c. $f_x(0, 0)$

   d. $f_y(0, 0)$

8. For $f(x, y) = \ln(3x^4 - 4y^3)$, find

   a. f_x

   b. f_y

   c. $f_x(1, 1)$

   d. $f_y(-1, -1)$

9. For $f(x, y) = \frac{x^2 + 3y}{5x - 3y^2}$, find

   a. f_x

   b. f_y

   c. $f_x(0, 2)$

   d. $f_y(0, 1)$

10. For $f(x, y) = \dfrac{6x^2 - 6y^2}{x^2 + y^2}$, find

a. f_x **b.** f_y

c. $f_x(1, 1)$ **d.** $f_y(1, 1)$

11. For $f(x, y) = x^2 e^{3y}$, find

a. f_x **b.** f_y

c. $f_x(0, 0)$ **d.** $f_y(0, 0)$

12. For $f(x, y) = (y^2 - 3)\ln(x + e^y)$, find

a. f_x **b.** f_y

c. $f_x(0, 0)$ **d.** $f_y(0, 0)$

13. For $f(x, y) = 2x^2 + 5y^2 - 3x - 4y + 10$, find

a. f_{xx} **b.** f_{yy}

c. f_{xy} **d.** f_{yx}

14. For $f(x, y) = 8x^3 + 5y^3 - 8x^2 - y^2 + 100$, find

a. f_{xx} **b.** f_{yy}

c. f_{xy} **d.** f_{yx}

15. For $f(x, y) = 10x^2 + 5y^2 + 4xy - 8$, find

a. f_{xx} **b.** f_{yy}

c. f_{xy} **d.** f_{yx}

16. For $f(x, y) = x^2 - 25xy + 2y^2 + 180$, find

a. f_{xx} **b.** f_{yy}

c. f_{xy} **d.** f_{yx}

17. For $f(x, y) = 9x^2 - 6xy + y^2 + 1{,}400$, find

a. f_{xx} **b.** f_{yy}

c. f_{xy} **d.** f_{yx}

18. For $f(x, y) = -5xe^y$, find

a. f_{xx} **b.** f_{yy}

c. f_{xy} **d.** f_{yx}

19. For $f(x, y) = 8ye^{2x}$, find

a. f_{xx} **b.** f_{yy}

c. f_{xy} **d.** f_{yx}

20. For $f(x, y, z) = 3x^2 + 5xz + 2z^3$, find

a. f_{xx} b. f_{yy}

c. f_z d. f_{zy}

21. For $f(x, y, z) = 6y^3 - 6xy - z^2 + 15$, find

a. f_{xx} b. f_{yy}

c. f_z d. f_{zy}

22. For the function $f(x, y) = 4x^2 + 3y^2 + 2xy - 1$, find the values of x and y so that both $f_x(x, y) = 0$ and $f_y(x, y) = 0$.

23. For the function $f(x, y) = x^2 + 7y^2 + 3x + 4y + 10$, find the values of x and y so that both $f_x(x, y) = 0$ and $f_y(x, y) = 0$.

24. For the function $f(x, y) = 5x^2 + 3y^2 - x^3 - y^3$, find the values of x and y so that both $f_x(x, y) = 0$ and $f_y(x, y) = 0$.

25. For the function $f(x, y) = 500 + 3x + 2y - x^3 - y^3$, find the values of x and y so that both $f_x(x, y) = 0$ and $f_y(x, y) = 0$.

Check Your Understanding

26. Suppose that $P(x, y)$ is a function of the two variables x and y. Interpret

$$\frac{\partial}{\partial x}P(3, 8) = 12 \quad \text{and} \quad \frac{\partial}{\partial y}P(5, 6) = -9.2$$

27. Suppose that $N(x, y)$ is a function of the two variables x and y. Interpret

$$\frac{\partial}{\partial x}N(100, 600) = 30 \quad \text{and} \quad \frac{\partial}{\partial y}N(80, 420) = 16$$

28. Suppose that $T(x, y, z)$ is a function of the three variables x, y, and z. Interpret

$$\frac{\partial}{\partial x}T(2, 3, 11) = 0.06, \quad \frac{\partial}{\partial y}T(3, 4, 10) = 0.008, \quad \text{and} \ \frac{\partial}{\partial z}T(5, 5, 8) = 0.4$$

29. Suppose that $S(x, y, z, w, m)$ is a function of the five variables x, y, z, w, and m. Interpret

$$\frac{\partial}{\partial y}S(2, 1, 3, 3, 4) = -65$$

30. **Business: Profit** Suppose $P(x, y)$ represents the monthly profit for a company when x thousands of dollars are spent on advertising and y number of salespeople are working. What information about the profit of this company is contained in the inequalities

$$\frac{\partial}{\partial x}P(15, 6) > 0 \quad \text{and} \quad \frac{\partial^2}{\partial x^2}P(15, 6) < 0$$

31. Economics: World Bank Loans A scientific organization believes that the function $N(x, y)$ relates the number of acres of destroyed rain forest to x, the amount of money (in millions of dollars) the World Bank loans to the country in which the forest is located, and y, the population (in millions of people) of the country. What information about the rate of destruction of the country's rain forest is contained in the inequalities

$$\frac{\partial}{\partial y}N(40, 70) > 0 \quad \text{and} \quad \frac{\partial^2}{\partial y^2}N(40, 70) > 0$$

32. Biology: Blood Flow The function $R(l, r)$ relates the resistance to blood flow in a vessel to the length l (in centimeters) and the radius r (in millimeters) of the vessel. What information about blood flow in the vessel is contained in the inequalities

$$\frac{\partial}{\partial r}R(10, 2) < 0 \quad \text{and} \quad \frac{\partial^2}{\partial r^2}R(10, 2) > 0$$

33. Psychology: IQ A person's IQ (intelligence quotient) is a function of the person's mental age M (in years) and chronological age C (in years); that is, $IQ(M, C)$. What information about a person's IQ is contained in the inequalities

$$\frac{\partial}{\partial c} IQ(M, C) < 0 \quad \text{and} \quad \frac{\partial^2}{\partial c^2} IQ(M, C) > 0$$

34. Business: Quality Control Costs For a high-tech electronics company, the cost C of quality control is related to the number of inspections x, y, and z made at three inspection points R, S, and T. What information about the cost C of quality control is contained in the inequalities $C_y(x, y, z) > 0$ and $C_{yy}(x, y, z) > 0$?

35. Health Science: Illness Recovery Time The time T (in days) it takes a person to recover from pneumonia is related to the age t (in years) of the person and the number n of units of penicillin the person receives each day; that is, $T(t, n)$. What information about the time it takes a person to recover from pneumonia is contained in the inequalities $T_t(t, n) < 0$ and $T_{tt}(t, n) < 0$?

36. Social Science: Welfare The proportion P of the population applying for welfare each month in a particular state is related to the monthly unemployment rate x and the average number of years of education y of the adults in the state; that is, $P(x, y)$. What information about the proportion of the population applying for welfare is contained in the inequalities $P_x(x, y) > 0$ and $P_{xx}(x, y) > 0$?

37. Social Science: City Arson Suppose a study shows that the number N of arsons in a city is related to the concentration x of residents of public housing in the city and the amount y (in thousands of dollars) spent by the city on maintenance of the housing. What information about the number of arsons is contained in the inequalities

$$\frac{\partial}{\partial x} N(x, y) > 0 \quad \text{and} \quad \frac{\partial^2}{\partial x^2} N(x, y) > 0$$

38. Health Science: Radiation Treatment In the treatment of cancer, the dose D of radiation is related to the width w of the radiating ray and the depth d of penetration of the ray; that is, $D(w, d)$. What information about the dose of radiation in the treatment of cancer is contained in the inequalities

$$\frac{\partial}{\partial w} D(w, d) > 0 \quad \text{and} \quad \frac{\partial^2}{\partial w^2} D(w, d) > 0$$

For Problems 39 and 40, recall that the Cobb-Douglas production function

$$f(x, y) = Cx^a y^{1-a}$$

relates the number of units of labor x and the number of units of capital y used in a particular production process to the number of units f that are produced from the process.

39. Business: Cobb-Douglas Production Suppose that for a company that produces kitchen sink faucets, the production function is $f(x, y) = x^{0.3}y^{0.7}$. For this function,

$$\frac{\partial}{\partial x} f = 0.3x^{-0.7}y^{0.7} \quad \text{and} \quad \frac{\partial^2}{\partial x^2} f = -0.21x^{-1.7}y^{0.7}$$

What information about production is contained in the partial derivatives

$$\frac{\partial}{\partial x} f(20, 40) \quad \text{and} \quad \frac{\partial^2}{\partial x^2} f(20, 40)?$$

Round your answer to two decimal places.

40. Business: Cobb-Douglas Production Suppose that for a company that produces tool sets, $f(x, y) = x^{0.25}y^{0.75}$. For this function,

$$\frac{\partial}{\partial y} f = 0.75x^{0.25}y^{-0.25} \quad \text{and} \quad \frac{\partial^2}{\partial y^2} f = -0.1875x^{0.25}y^{-1.25}$$

What information about production is contained in the partial derivatives

$$\frac{\partial}{\partial y} f(30, 100) \quad \text{and} \quad \frac{\partial^2}{\partial y^2} f(30, 100)?$$

Round your answer to three decimal places.

41. Economics: Demand/Price The demand for computer printers is given by $D(p_1, p_2)$, where p_1 is the price of the printers and p_2 is the price of ink cartridges. Interpret

$$\frac{\partial D}{\partial p_1} < 0 \quad \text{and} \quad \frac{\partial D}{\partial p_2} < 0$$

42. Economics: Demand/Price The demand for a particular type of graphing calculator depends on both the price p_1 of the calculator and the price p_2 of its only competitor. The demand is given by $D(p_1, p_2)$. Interpret

$$\frac{\partial D}{\partial p_1} < 0 \quad \text{and} \quad \frac{\partial D}{\partial p_2} > 0$$

Modeling Practice

43. Business: Revenue The total revenue in dollars realized by a company for the sale of x units of product A and y units of product B is

$$R(x, y) = 350x + 600y - 4x^2 - 3y^2$$

a. Specify the total revenue if 20 units of product A and 15 units of product B are sold.

b. How will the company's revenue change if x increases one unit, from 20 to 21, while y remains constant at 15?

c. How will the company's revenue change if y decreases one unit, from 15 to 14, while x remains constant at 20?

44. Business: Number of Sales The number of units of a particular product that a company sells each month depends on the number of thousands of dollars spent on advertisements in newspapers x, on television y, and on radio z, and is given by the function

$$N(x, y, z) = 400x + 550y + 120z - 20x^2 - 20y^2 - 10z^2$$

a. Specify the total number of units sold if the company spends \$2,000 on newspaper ads, \$8,000 on TV ads, and \$3,000 on radio ads.

b. How will the number of units sold change if the amount spent on newspaper ads is increased by \$1,000, from \$2,000 to \$3,000?

c. How will the number of units sold change if the amount spent on radio ads is decreased by \$1,000, from \$3,000 to \$2,000?

45. Animal Science: Oxygen Consumption The oxygen consumption C of a well insulated, non-sweating animal is approximately related to the animal's internal body temperature, T (in °C), the temperature F (in °C) of the animal's fur, and the animal's weight W (in kilograms (kg)) by the function

$$C(T, F, W) = 2.5(T - F)W^{-0.67}$$

a. Find the oxygen consumption of an animal weighing 36 kg, with an internal temperature of 40 °C, and a fur temperature of 22 °C.

b. How will the animal's oxygen consumption change if its fur temperature is increased by 1 °C, from 22 °C to 23 °C, and its internal temperature and weight remain constant at 40 °C and 36 kg, respectively?

c. How will the animal's oxygen consumption change if its fur and internal temperature remain constant at 22 °C and 40 °C, respectively, but its weight decreases by 1 kg, from 36 kg to 35 kg?

46. Psychology: IQ A person's IQ (intelligence quotient) is a function of the person's mental age m (in years) and chronological age c (in years) and is defined by the function

$$IQ(m, c) = \frac{100m}{c}$$

a. Find the IQ of a 22-year-old person who has a mental age of 30 years.

b. How will a person's IQ change if his chronological age increases by one year, from 22 years to 23 years, but his mental age remains constant at 30 years?

47. Business: Cobb-Douglas Production Suppose that x units of labor and y units of capital are needed to produce

$$f(x, y) = 75x^{2/3}y^{1/3}$$

units of a particular commodity.

a. Find the number of units produced if 27 units of labor and 64 units of capital are used.

b. How will the number of units produced change if the number of units of labor used increases by 1 unit, from 27 units to 28?

c. How will the number of units produced change if the number of units of capital used decreases by 1 unit, from 64 units to 63?

Using Technology Exercises

48. Use Wolfram|Alpha to help you find $f(4, 2)$, $f_x(x, y)$, $f_y(x, y)$, $f_x(4, 2)$, and $f_y(4, 2)$ for the function $f(x, y) = 2x^4 + 6y^2 + 3xy$.

49. Use Wolfram|Alpha to help you find $f(1, 0)$, $f_x(x, y)$, $f_y(x, y)$, $f_x(1, 0)$, and $f_y(1, 0)$ for the function $f(x, y) = e^{-xy} + e^{y/x}$.

50. Use Wolfram|Alpha to help you find $f(0, 0)$, $f_x(x, y)$, $f_y(x, y)$, $f_x(0, 0)$, and $f_y(0, 0)$ for the function $f(x, y) = 2x^2 - e^{y^2}$.

51. For a manufacturer of oriented strand board, the Cobb-Douglas production function is $f(x, y) = 35x^{0.43}y^{0.57}$. Both cost and revenue are in dollars. Find and interpret both $\frac{\partial f}{\partial x}$ and $\frac{\partial f}{\partial y}$ when 15 units of labor and 21 units of capital are used.

Source: Georgia-Pacific

52. A manufacturer produces x units of product X and y units of product Y. The cost function for these products is

$$C(x, y) = \sqrt{0.6x^{1.2}y^{0.8} + 14.2}$$

the revenue function is

$$R(x, y) = 2.6x^{1.1} + 1.3y^{1.2}$$

and the profit function is

$$P(x, y) = R(x, y) - C(x, y)$$

Find and interpret

$$P(600, 440), P_x(600, 440), \text{ and } P_{xx}(600, 440)$$

Getting Ready for the Next Section

53. Find both $f_x(x, y)$ and $f_y(x, y)$ for $f(x, y) = \frac{3}{2}x^2 + y^2 + 6x - 8y + 9$.

54. Find both $f_x(x, y)$ and $f_y(x, y)$ for $f(x, y) = \frac{x^2}{2} + \frac{y^2}{2} - 3x + 9y + 5xy + 6$.

55. Find both $f_x(x, y)$ and $f_y(x, y)$ for $f(x, y) = \frac{3x^2}{2} + \frac{y^4}{2} - y^2 - 3$.

56. Solve the system of equations

$$\begin{cases} x + 5y - 3 = 0 \\ 5x + y + 9 = 0 \end{cases}$$

57. Given

$$f(x, y) = 2x^2 + 3y^2 + 8x - 12y + 3$$

evaluate

$$f_{xx}(x, y) \cdot f_{yy}(x, y) - f_{xy}^2(x, y)$$

58. Given

$$f(x, y) = \frac{2x^3}{3} + \frac{4y^3}{3} - 8y^2 - 50x + 1$$

evaluate

$$f_{xx}(x, y) \cdot f_{yy}(x, y) - f_{xy}^2(x, y)$$

59. Find the values of x and y for which both $f_x(x, y) = 0$ and $f_y(x, y) = 0$ for

$$f(x, y) = \frac{2x^3}{3} + \frac{4y^3}{3} - 8y^2 - 50x + 1$$

60. Find the values of x and y for which both $f_x(x, y) = 0$ and $f_y(x, y) = 0$ for

$$f(x, y) = 82x - \frac{x^2}{18} + 118y - \frac{y^2}{28} - 45{,}000$$

Optimization of Functions of Two Variables

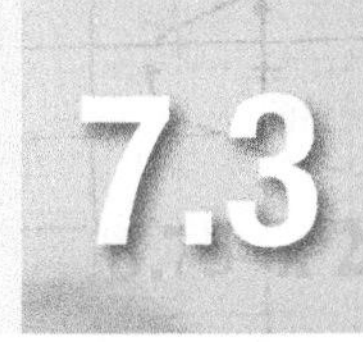

© Andrey Tsidvintsev/iStockPhoto

Just as derivatives of one-variable functions are helpful in locating relative extreme points, partial derivatives are helpful in locating relative extreme points of functions of several variables. We will restrict our attention to two-variable functions for which all second-order partial derivatives exist and are continuous, as it is functions of two variables that occur most often in business, life sciences, and social sciences. For example, suppose a marketing department is tasked with maximizing revenue derived from radio and newspaper advertisements. The marketers could use the following function to find relative extrema:

$$R(x, y) = -0.07x^2 - 100y^2 + 4x + 5y + 2xy$$

where x is the thousands of dollars spent on radio advertisements and y is the thousands of dollars spent on newspaper advertisements. We will work with similar functions in this section.

Relative Extrema

We begin with the definition of relative extrema for a function of two variables.

Relative Extrema
The point (a, b) produces a **relative maximum** of the function $f(x, y)$ if for every point (x, y) *near* (a, b), $f(a, b) \geq f(x, y)$.
The point (a, b) produces a **relative minimum** of the function $f(x, y)$ if for every point (x, y) *near* (a, b), $f(a, b) \leq f(x, y)$.

Figure 1a illustrates a function with a relative maximum at the point $(a, b, f(a, b))$. Notice that all the points (x, y) near (in the small disk) the point (a, b) produce function values smaller than that produced by the point (a, b). Figure 1b illustrates a function with a relative minimum at $(a, b, f(a, b))$. Notice that all the points (x, y) near (in the small disk) the point (a, b) produce function values greater than that produced by the point (a, b).

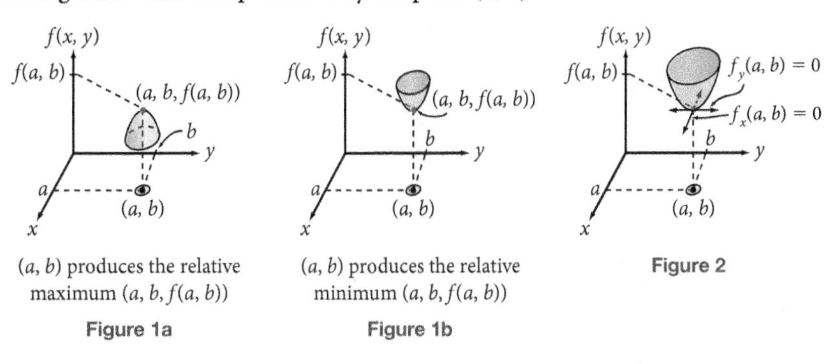

(a, b) produces the relative maximum $(a, b, f(a, b))$

Figure 1a

(a, b) produces the relative minimum $(a, b, f(a, b))$

Figure 1b

Figure 2

Figure 2 displays a function of two variables with a relative minimum at $(a, b, f(a, b))$ and in which the tangent lines in both the x and y directions have been sketched in. Notice that each of the tangent lines is horizontal relative to the xy plane. This means, of course, that each has zero slope; that is, that $f_x(a, b) = 0$ and $f_y(a, b) = 0$.

This is an important fact that we will use when we attempt to locate relative extreme points and critical points of functions of two variables.

Location of Relative Extrema/Critical Points

The fact that at a relative extreme point, both partial derivatives must simultaneously be zero helps us to locate such points.

Location of Relative Extrema/Critical Points
If $f(x, y)$ is a function with a relative extreme point at $(a, b, f(a, b))$ and both $f_x(a, b)$ and $f_y(a, b)$ exist, then

$$f_x(a, b) = 0 \quad \text{and} \quad f_y(a, b) = 0$$

The points (a, b) for which $f_x(a, b) = 0$ and $f_y(a, b) = 0$ are called **critical points** of the function $f(x, y)$.

The fact that the partial derivatives are both zero at a point (a, b) is not sufficient to guarantee a relative extremum at that point. Recall that for a function of one variable, it was possible that at some point $x = c, f'(c) = 0$, but $x = c$ did not produce a relative extremum. Instead, it produced a point of inflection. Figure 3 shows just such a case.

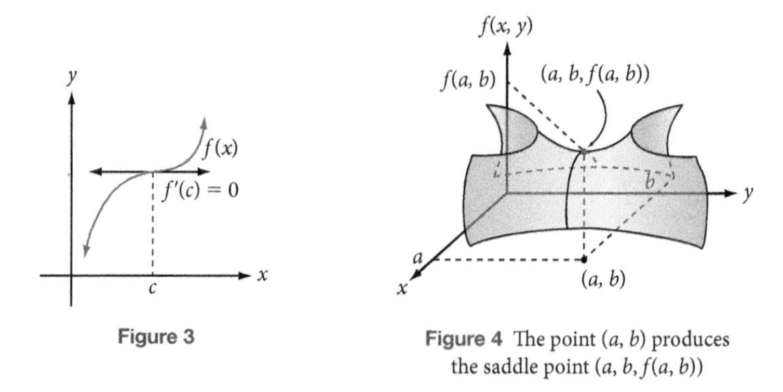

Figure 3

Figure 4 The point (a, b) produces the saddle point $(a, b, f(a, b))$

An analogous situation is possible for functions of two variables. Figure 4 illustrates a function for which both $f_x(a, b) = 0$ and $f_y(a, b) = 0$, but the point (a, b) does not produce a relative extreme point. The point $(a, b, f(a, b))$ is called a *saddle point.* A **saddle point** is a point that is a minimum in one direction (the y direction, in this case) and a maximum in the other (the x direction, in this case).

The following Examples 1-4 illustrate how to locate critical points. A graph of each function is presented to better convince you that the method we use actually does produce all the critical points of a function.

VIDEO EXAMPLES

SECTION 7.3

Example 1 Find the critical points of the function

$$f(x, y) = \frac{3}{2}x^2 + y^2 + 6x - 8y + 9$$

Solution By the previous discussion, critical points are points (a, b) for which $f_x(a, b) = 0$ and $f_y(a, b) = 0$. We begin by finding f_x and f_y.

$$f_x(x, y) = 3x + 6 \quad \text{and} \quad f_y(x, y) = 2y - 8$$

Since these partial derivatives need to be zero simultaneously, we need to solve the system

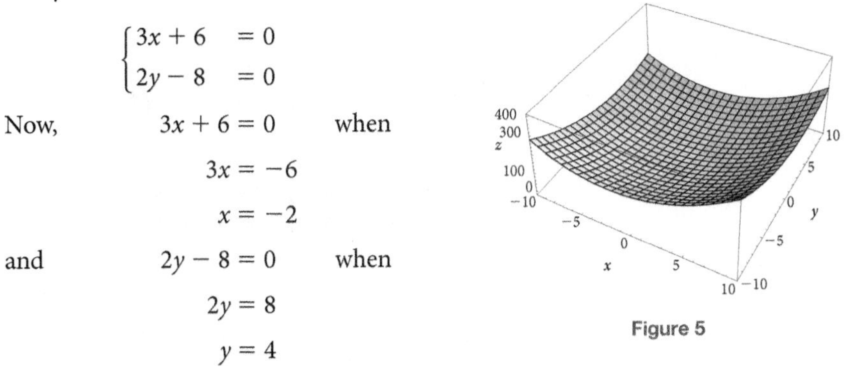

$$\begin{cases} 3x + 6 &= 0 \\ 2y - 8 &= 0 \end{cases}$$

Now, $\quad 3x + 6 = 0 \quad$ when

$$3x = -6$$
$$x = -2$$

and $\quad 2y - 8 = 0 \quad$ when

$$2y = 8$$
$$y = 4$$

Figure 5

This function has only one critical point at $(-2, 4)$.

Interpretation
The critical point $(-2, 4)$ may produce a relative maximum, a relative minimum, or a saddle point. The graph of $f(x, y)$ is displayed in Figure 5. It seems to indicate that the critical point $(-2, 4)$ produces a relative minimum.

Example 2 Find the critical points of the function

$$f(x, y) = \frac{x^2}{2} + \frac{y^2}{2} - 3x + 9y + 5xy + 6$$

Solution We begin by finding f_x and f_y.

$$f_x(x, y) = x - 3 + 5y \quad \text{and} \quad f_y(x, y) = y + 9 + 5x$$

Since critical points are located where these partial derivatives are simultaneously zero, we need to solve the system

$$\begin{cases} x - 3 + 5y &= 0 \\ y + 9 + 5x &= 0 \end{cases} \rightarrow \begin{cases} x + 5y &= 3 \quad \ldots(1) \\ y + 5x &= -9 \quad \ldots(2) \end{cases}$$

Solving equation 1 for x produces $x = 3 - 5y$. Substituting $3 - 5y$ for x in equation 2 will give us the value of y.

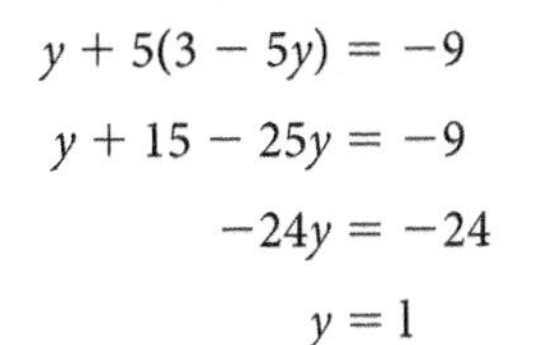

$$y + 5(3 - 5y) = -9$$
$$y + 15 - 25y = -9$$
$$-24y = -24$$
$$y = 1$$

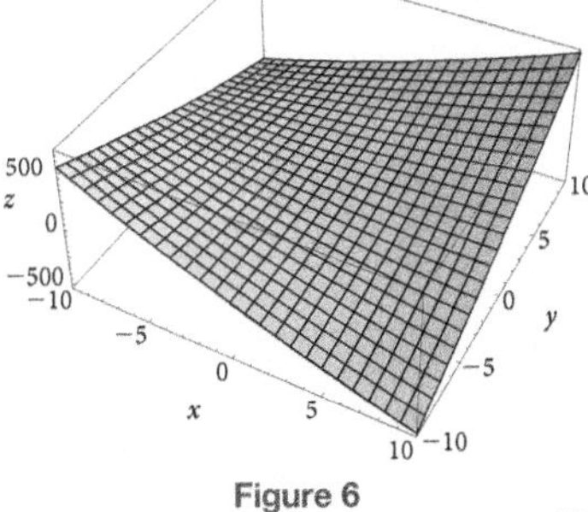

Figure 6

Now, $x = 3 - 5y$ and with $y = 1, x = 3 - 5(1)$ so that $x = -2$. This function has only one critical point at $(-2, 1)$.

Interpretation
The critical point $(-2, 1)$ may produce a relative maximum, a relative minimum, or a saddle point. The graph of $f(x, y)$ is displayed in Figure 6. It seems to indicate that the critical point $(-2, 1)$ produces a saddle point.

Example 3 Find the critical points of the function

$$f(x, y) = \frac{3x^2}{2} + \frac{y^4}{2} - y^2 - 3$$

Solution We begin by finding f_x and f_y.

$$f_x(x, y) = 3x \quad \text{and} \quad f_y(x, y) = 2y^3 - 2y$$

Since critical points are located where these partial derivatives are simultaneously zero, we need to solve the system

$$\begin{cases} 3x = 0 & \ldots(1) \\ 2y^3 - 2y = 0 & \ldots(2) \end{cases}$$

Solving equation 1 for x produces $x = 0$. Solve equation 2 for y produces

$$2y^3 - 2y = 0$$
$$2y(y^2 - 1) = 0$$
$$2y(y + 1)(y - 1) = 0$$
$$y = 0, -1, 1$$

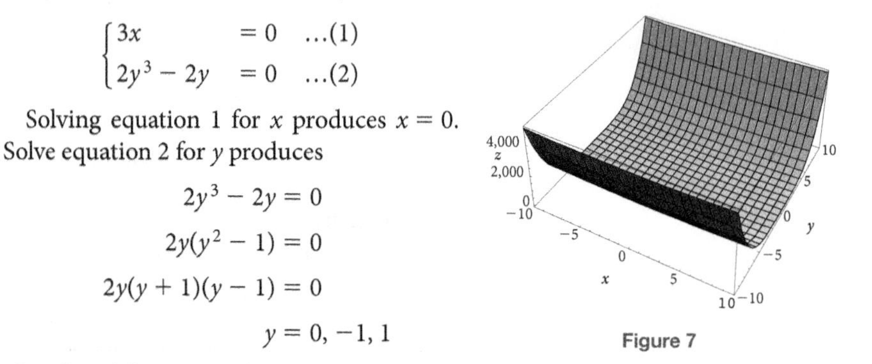

Figure 7

Thus f_x and f_y are zero when $x = 0$ and $y = 0$, -1, 1, and consequently, this function has three critical points, (0, 0), (0, −1) and (0, 1).

> **Interpretation**
> The critical points could produce relative maxima, relative minima, saddle points, or some combination of these. The graph of $f(x, y)$ is displayed in Figure 7. It seems to indicate that the critical point (0, 0) produces a saddle point and that (0, −1), and (0, 1) produce relative minima.

■

The next example illustrates the meaning of a system of equations that has no solution.

Example 4 Find the critical points of the function $f(x, y) = 4x - 9y + 2$.

Solution We begin by finding f_x and f_y.

$$f_x(x, y) = 4 \quad \text{and} \quad f_y(x, y) = -9$$

Since critical points are located where these partial derivatives are simultaneously zero, we need to solve the system

$$\begin{cases} 4 = 0 & \ldots(1) \\ -9 = 0 & \ldots(2) \end{cases}$$

But this system has no solutions; that is, f_x and f_y are never zero.

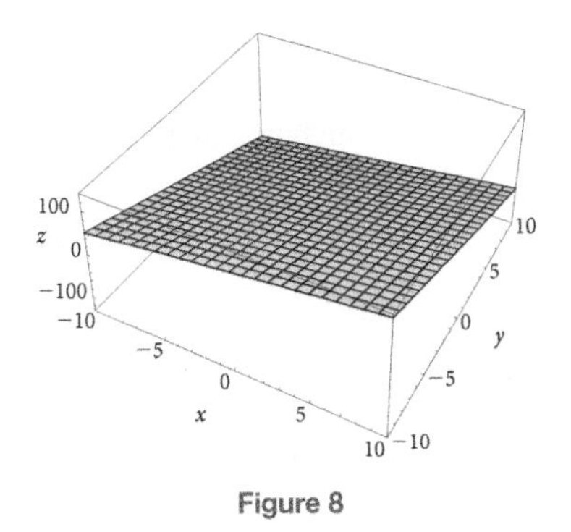

Figure 8

Interpretation
This function has no critical points. The graph of $f(x, y)$ is displayed in Figure 8. It seems to indicate that the function is a plane (a flat surface) which would lead us to believe that no relative extreme points could exist.

The Second Derivative Test

The Second Derivative Test provides a way of classifying critical points as relative maxima, relative minima, or saddle points.

Second Derivative Test
Suppose that (a, b) is a critical point of the function $f(x, y)$ and that $D(x, y)$ represents $f_{xx}(a, b) \cdot f_{yy}(a, b) - f_{xy}^2(a, b)$. Then,

1. if $D(a, b) > 0$ and $f_{xx}(a, b) > 0$, then (a, b) produces a relative minimum.
2. if $D(a, b) > 0$ and $f_{xx}(a, b) < 0$, then (a, b) produces a relative maximum.
3. if $D(a, b) < 0$, then (a, b) is a saddle point.
4. if $D(a, b) = 0$, then the test provides no information about what (a, b) might produce.

Note that if D, when evaluated at a critical point (a, b), is positive, the critical point will necessarily be a relative extreme point. The critical point will be a saddle point only when D is negative.

The following examples illustrate the use of the Second Derivative Test. The critical points are located using the techniques demonstrated in Examples 1-4.

Example 5 Locate the relative extrema, if any exist, of the function

$$f(x, y) = 2x^2 + 3y^2 + 8x - 12y + 3$$

Solution The Second Derivative Test indicates that we need to determine the partial derivatives f_x, f_y, f_{xx}, f_{yy}, and f_{xy}.

$$f_x(x, y) = 4x + 8, \quad f_y(x, y) = 6y - 12, \quad f_{xx} = 4, \quad f_{yy} = 6, \quad f_{xy} = 0$$

We will use the first-order partials, f_x and f_y, to generate any critical points. Setting $f_x = 0$ and $f_y = 0$ produces the system

$$\begin{cases} 4x + 8 &= 0 \\ 6y - 12 &= 0 \end{cases}$$

which produces the single critical point $(-2, 2)$.

Now, let

$$D(x, y) = f_{xx}(a, b) \cdot f_{yy}(a, b) - f_{xy}^2(a, b)$$

$D(x, y) = 4 \cdot 6 - 0^2$, so that $D(x, y) = 24$. Since $D(x, y) = 24$, it is *always* positive. In particular, $D(-2, 2) = 24 > 0$, so that the critical point $(-2, 2)$ is necessarily a relative extreme point. To determine which type (maximum or minimum), we compute $f_{xx}(-2, 2)$.

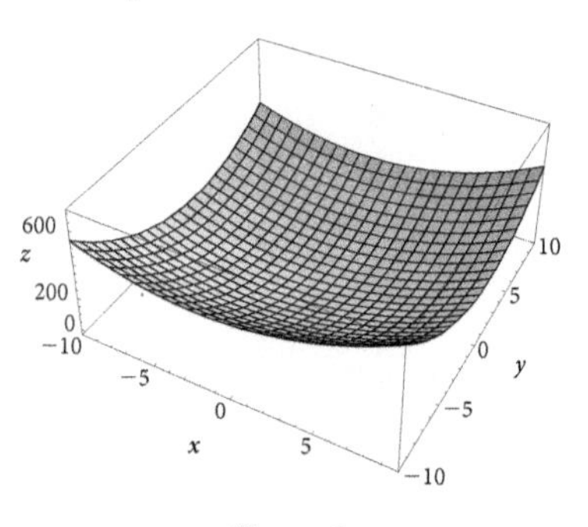

Figure 9

$f_{xx} = 4$ for all points in this example; therefore, $f_{xx} > 0$, leading us to conclude that $(-2, 2)$ produces a relative minimum. To locate the point, we must compute the output value for the input values $x = -2$ and $y = 2$. Substituting -2 for x and 2 for y into the original function, we get

$$\begin{aligned} f(-2, 2) &= 2(-2)^2 + 3(2^2) + 8(-2) - 12(2) + 3 \\ &= -17 \end{aligned}$$

Interpretation

The critical point $(-2, 2)$ produces the relative minimum point $(-2, 2, -17)$. The graph of $f(x, y)$ is displayed in Figure 9.

Notice that the partial derivatives f_{xx}, f_{yy}, and f_{xy} are each constant. This makes their evaluation at the critical point very easy (since no substitution or computation is necessary).

Example 6 Locate the relative extrema, if any exist, of

$$f(x, y) = \frac{2x^3}{3} + \frac{4y^3}{3} - 8y^2 - 50x + 1$$

Solution The Second Derivative Test indicates that we need to determine the partial derivatives f_x, f_y, f_{xx}, f_{yy}, and f_{xy}.

$f_x(x, y) = 2x^2 - 50, \quad f_y(x, y) = 4y^2 - 16y, \quad f_{xx} = 4x, \quad f_{yy} = 8y - 16, \quad f_{xy} = 0$

We will use the first-order partials, f_x and f_y, to generate any critical points. Setting $f_x = 0$ and $f_y = 0$ produces the system

$$\begin{cases} 2x^2 - 50 &= 0 \\ 4y^2 - 16y &= 0 \end{cases}$$

which produces the critical points $(-5, 0)$, $(-5, 4)$, $(5, 0)$, and $(5, 4)$.

Now, using the partial derivatives of $f(x, y)$, compute $D(x, y)$.

$$\begin{aligned} D(x, y) &= f_{xx}(x, y) \cdot f_{yy}(x, y) - f_{xy}^2(x, y) \\ &= 4x \cdot (8y - 16) - 0^2 \\ &= 32xy - 64x \end{aligned}$$

To classify these four critical points as relative maxima, minima, or saddle points, we will employ the Second Derivative Test. To do so, we need to compute $D(x, y)$ and $f_{xx}(x, y)$ at each of these points. We summarize our computations in the following tables.

CP	*D*	$f_{xx}(x)$	Conclusion
$(-5, 0)$	$32(-5)(0) - 64(-5) > 0$	$4(-5) < 0$	Produces rel max
$(-5, 4)$	$32(-5)(4) - 64(-5) < 0$		Produces saddle pt
$(5, 0)$	$32(5)(0) - 64(5) < 0$		Produces saddle pt
$(5, 4)$	$32(5)(4) - 64(5) > 0$	$4(5) > 0$	Produces rel min

Table 1

Placing these input values into the function produces the output values and the relative extrema and saddle points.

$f(a, b)$	Conclusion
$f(-5, 0) \approx 167.67$	Rel max at approx $(-5, 0, 167.67)$
$f(-5, 4) = 125$	Saddle pt at $(-5, 4, 125)$
$f(5, 0) \approx -165.67$	Saddle pt at approx $(5, 0, -165.67)$
$f(5, 4) \approx -208.33$	Rel min at approx $(5, 4, -208.33)$

Table 2

Figure 10

The graph of $f(x, y)$ is displayed in Figure 10. ■

Example 7 A manufacturer markets a product in two states, A and B, and, because of the different economies of the states, must price the product differently in each. (Such pricing is called *price discrimination* or *differential pricing.*) The manufacturer wishes to sell x units of the product in state A and y units of the product in state B. To do so, the manufacturer must set the price in state A at $86 - \frac{x}{18}$ dollars and in state B at $122 - \frac{y}{28}$ dollars. The cost of producing all $x + y$ items is $45{,}000 + 4(x + y)$ dollars. How many items should be produced for states A and B, respectively, to maximize the manufacturer's profit, and what is that maximum profit?

Solution Since we wish to maximize profit, we need to construct a profit function.

$$\text{Profit} = (\text{revenue}) - (\text{cost})$$

$$= [(\text{revenue from state A}) + (\text{revenue from state B})] - (\text{cost})$$

Since (revenue) = (price) · (quantity), we have

$$P(x, y) = \left[\left(86 - \frac{x}{18}\right)x + \left(122 - \frac{y}{28}\right)y\right] - [45{,}000 + 4(x + y)]$$

$$P(x, y) = 86x - \frac{x^2}{18} + 122y - \frac{y^2}{28} - 45{,}000 - 4x - 4y$$

$$P(x, y) = 82x - \frac{x^2}{18} + 118y - \frac{y^2}{28} - 45{,}000$$

To use the Second Derivative Test to locate the extreme points of this function we need P_x, P_y, P_{xx}, P_{yy}, and P_{xy}.

$$P_x(x, y) = 82 - \frac{x}{9}, \quad P_y(x, y) = 118 - \frac{y}{14}, \quad P_{xx}(x, y) = -\frac{1}{9},$$

$$P_{yy}(x, y) = -\frac{1}{14}, \quad P_{xy}(x, y) = 0$$

We use the first partial derivatives P_x and P_y to find any critical points. Setting $P_x = 0$ and $P_y = 0$ produces the following equivalent systems

$$\begin{cases} 82 - \frac{x}{9} & = 0 \\ 118 - \frac{y}{14} & = 0 \end{cases} \rightarrow \begin{cases} \frac{x}{9} & = 82 \\ \frac{y}{14} & = 118 \end{cases} \rightarrow \begin{cases} x = 738 \\ y = 1{,}652 \end{cases}$$

which produce the single critical point (738, 1,652).

We now compute $D(x, y) = P_{xx}(x, y) \cdot P_{yy}(x, y) - P_{xy}^2(x, y)$ to determine its sign (+ or −) at the critical point (738, 1,652).

$$D(738, 1{,}652) = \left(-\frac{1}{9}\right) \cdot \left(-\frac{1}{14}\right) - 0^2 > 0$$

so that the critical point (738, 1,652) necessarily produces an extreme point.

To classify which type (maximum or minimum), we compute $P_{xx}(738, 1{,}652)$.

$$P_{xx}(738, 1{,}652) = -\frac{1}{9} < 0$$

leading us to conclude that (738, 1,652) produces a relative maximum. Then, substituting 738 for x and 1,652 for y into the original function, we get

$$P(738, 1{,}652) = 82(738) - \frac{738^2}{18} + 118(1{,}652) - \frac{1{,}652^2}{28} - 45{,}000$$

$$= 82{,}726$$

Interpretation

The manufacturer should produce and sell 738 units of the product in state A and 1,652 units of the product in state B to obtain the maximum profit of $82,726.

Using Technology 7.3

Wolfram|Alpha can be helpful in finding the relative extreme points of functions of two variables.

Example 8 Use Wolfram|Alpha to find the relative maxima and relative minima of the function

$$f(x, y) = \frac{3}{2}x^2 + y^2 + 6x - 8y + 9$$

Solution Go to www.wolframalpha.com and in the entry field enter

max and min of (3/2)x^2 + y^2 + 6x - 8y + 9

Wolfram|Alpha responds with

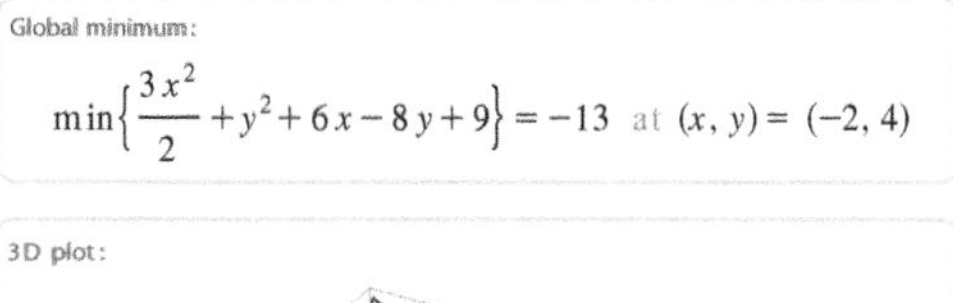

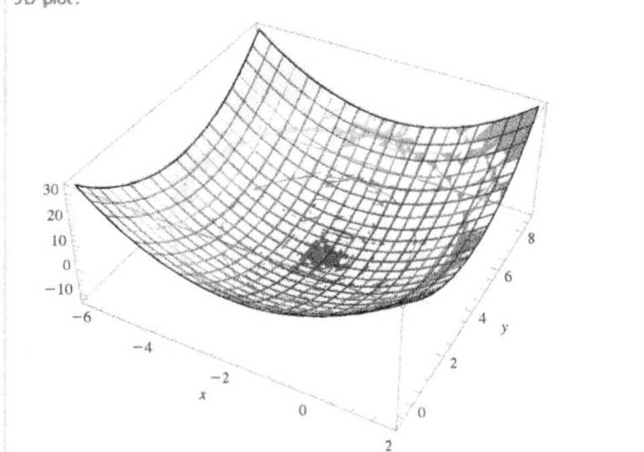

Wolfram Alpha LLC. 2012. Wolfram|Alpha
http://www.wolframalpha.com/
(access December 18, 2012)

We conclude that there is a relative minimum at $(-2, 4, -13)$. Wolfram|Alpha draws the graph of the function. The graph indicates (look closely) a relative minimum at $(-2, 4)$. ■

Getting Ready for Class

After reading through the preceding section, respond in your own words and in complete sentences.

A. The definition of a relative maximum of the function $f(x, y)$ states that if the point (a, b) produces a relative maximum of $f(x, y)$, then for every point (x, y) near the point (a, b), $f(a, b) > f(x, y)$. Explain what this means geometrically.

B. Explain how it is possible for both $f_x(x, y)$ and $f_y(x, y)$ to equal zero at a point (a, b), but for the point (a, b) to not produce a relative maximum or a relative minimum.

C. The Second Derivative Test for a function of two variables states that for the critical point (a, b), if $D(a, b) > 0$ and $f_{xx}(a, b) > 0$, the point (a, b) produces a relative minimum. Explain why $f_{xx}(a, b) > 0$ makes this so. (Hint: Think about what $f''(x) > 0$ tells you about the concavity of the function $f(x)$.)

Problem Set 7.3

Skills Practice

For Problems 1-13, find and classify, if possible, all the relative extreme points and saddle points.

1. $f(x, y) = \frac{3}{2}x^2 + y^2 + 15x - 8y + 6$
2. $f(x, y) = 3x^2 - y^2 - 12x + 16y + 21$
3. $f(x, y) = \frac{5}{2}y^2 - 2x^2 - 12x - 20y + 7$
4. $f(x, y) = -x^2 - \frac{3}{2}y^2 + 6x + 21y + 8$
5. $f(x, y) = -x^2 - y^2 - xy + x + 6y + 12$
6. $f(x, y) = x^2 - \frac{3}{2}y^2 - 5xy + 11x + 3y - 8$
7. $f(x, y) = x^3 - y^2 - 3x + 4y + 5$
8. $f(x, y) = x^2 - 2y^3 + 6y + 8$
9. $f(x, y) = x^2 - 2y^3 + 6y - 10$
10. $f(x, y) = x^3 - 2xy + 4y + 6$
11. $f(x, y) = x^3 - 3xy + y^3$
12. $f(x, y) = e^{xy}$
13. $f(x, y) = x^{2/3} + y^{2/3}$

Check Your Understanding

In each of Problems 14-20, the first-order, second-order, and mixed partial derivatives of a function are given along with the critical point(s) of the function. Use this information to classify, if possible, the critical point(s) as a point(s) that produces a relative maximum, relative minimum, or a saddle point.

14. $f_x(x, y) = 3x + 12$
    $f_y(x, y) = 2y - 8$
    $f_{xx} = 3, \quad f_{yy} = 2, \quad f_{xy} = 0$
    $(-4, 4)$

15. $f_x(x, y) = 5x - 15$
    $f_y(x, y) = 3y + 6$
    $f_{xx} = 5, \quad f_{yy} = 3, \quad f_{xy} = 0$
    $(3, -2)$

16. $f_x(x, y) = 9x - 18$
    $f_y(x, y) = -4y + 20$
    $f_{xx} = 9, \quad f_{yy} = -4, \quad f_{xy} = 0$
    $(2, 5)$

17. $f_x(x, y) = -8x + 16$
    $f_y(x, y) = y + 6$
    $f_{xx} = -8, \quad f_{yy} = 1, \quad f_{xy} = 0$
    $(2, -6)$

18. $f_x(x, y) = 3x + 2y - 19$

$f_y(x, y) = 4x - 2y - 2$

$f_{xx} = 3, \quad f_{yy} = -2, \quad f_{xy} = 2$

$(3, 5)$

19. $f_x(x, y) = -2x + 8y - 2$

$f_y(x, y) = 4x + 2y + 4$

$f_{xx} = -2, \quad f_{yy} = 2, \quad f_{xy} = 8$

$(-1, 0)$

20. $f_x(x, y) = -2x - 4$

$f_y(x, y) = -2y$

$f_{xx} = -2, \quad f_{yy} = -2, \quad f_{xy} = 0$

$(-2, 0)$

Modeling Practice

21. Marketing: Revenue from Advertising The marketing department of a company has determined that if it spends x thousands of dollars on radio advertisements and y thousands of dollars on newspaper advertisements, the company's revenue (in thousands of dollars) will be

$$R(x, y) = -0.07x^2 - 100y^2 + 4x + 5y + 2xy$$

How much money should this company spend on radio advertisements and newspaper advertisements to maximize its revenue?

22. Manufacturing: Maximizing Profit A manufacturer markets a product in two states, A and B, and prices it differently in each state. The manufacturer wishes to sell x units of the product in state A and y units in state B. To do so, it must set the price in state A at $678 - \frac{x}{22}$ dollars per unit and in state B at $151 - \frac{y}{16}$ dollars per unit. The cost of producing all $x + y$ units of the product is

$$\$85{,}000 + 6(x + y)$$

How many units of the product should be produced for states A and B, respectively, to maximize the manufacturer's profit?

23. Manufacturing: Minimizing Surface Area A manufacturer of boxes wishes to construct a rectangular box having a volume of 64 cubic inches. What dimensions of the box will have the minimal surface area?

24. Business: Construction Costs A rectangular building with a flat roof is to be constructed so as to enclose 31,250 cubic feet. The cost of the roof is \$8.00 per square foot, the cost of the sides is \$8.00 per square foot, and the cost of the floor is \$16.00 per square foot. What dimensions of the building will minimize construction costs? (Assume there are no doors or windows.)

25. Business: Postal Regulations Postal service regulations require that the (length) + (girth) of a rectangular package be no more than 84 inches. (The girth is the distance around the middle of the package.) Find the dimensions of the rectangular package of largest volume that meet these postal requirements.

Using Technology Exercises

26. Use Wolfram|Alpha to help you find the relative minimum of the function

$$f(x, y) = x^2 + 3y^3 + 4x - 9y + 11$$

27. Use Wolfram|Alpha to help you find the relative maximum of the function

a. $f(x, y) = \dfrac{300}{1 + x^2 + y^2}$

b. $f(x, y) = \dfrac{300}{100 + x^2 + y^2}$

c. $f(x, y) = \dfrac{300}{300 + x^2 + y^2}$

d. $f(x, y) = \dfrac{300}{n + x^2 + y^2}$

28. Use Wolfram|Alpha to help you find the saddle point of the function

$$f(x, y) = (x - 4)^2 - (y - 6)^2 + 50$$

To find the saddle point, use the command

saddle point of (x - 4)^2 - (y - 6)^2 + 50

29. The profit $P(x)$, in dollars, from the sale of x units of Product A and y units of Product B is modeled by the function

$$P(x, y) = 22x + 73y - 1.2x^2 - 1.5y^2$$

How many units of each product should be produced and sold to maximize the profit? What is the maximum profit?

30. A manufacturer can produce

$$N(x, y) = 15x + 12y - 1.4x^4 - 1.3y^4$$

thousands of units of a product using x units of labor and y units of capital. What is the number of units of labor and capital that determine the maximum number of units the manufacturer can produce?

Getting Ready for the Next Section

31. Solve the equation $x - 2y = 4$ for x and substitute that resulting expression into $f(x, y) = x^2 - 3y^2 + 2x + 4y$ so as to make $f(x, y)$ a function of y only.

32. Find the value of y that produces the relative minimum of

$$f(y) = y^2 + 24y + 24$$

33. Find $F_x(x, y, \lambda)$, $F_y(x, y, \lambda)$, and $F_\lambda(x, y, \lambda)$ for

$$F(x, y, \lambda) = x^2 + 10y^2 + \lambda x - \lambda y - 18\lambda$$

where λ represents a constant.

34. Find $F_x(x, y, \lambda)$, $F_y(x, y, \lambda)$, and $F_\lambda(x, y, \lambda)$ for

$$F(x, y, \lambda) = 30x^{1/5}y^{4/5} + 30{,}000\lambda x + \lambda y - 4{,}500{,}000\lambda$$

where λ represents a constant.

35. By first eliminating λ, solve the system of equations

$$\begin{cases} 2x + \lambda = 0 \\ 20y - \lambda = 0 \\ x - y = 18 \end{cases}$$

36. By first eliminating λ, solve the system of equations

$$\begin{cases} 6x^{-4/5}y^{4/5} + 30{,}000\lambda = 0 \\ 24x^{1/5}y^{-1/5} + \lambda = 0 \\ 30{,}000x + y - 4{,}500{,}000 = 0 \end{cases}$$

Constrained Maxima and Minima

7.4

An important application of differentiation is optimization. Using differentiation, we are able to determine if any relative extrema of a function exist, and if they do, to locate and classify them as relative maxima, relative minima, or saddle points. In Section 7.3, we used the Second Derivative Test to find all the relative extreme points of a function of two variables. This method involved no restriction on the values of the input variables.

In this section, we study the method of Lagrange multipliers, a method for optimizing a function when there *are* restrictions on the values of the input variables. Optimization is used in the real world in areas such as manufacturing. Suppose a company wants to know how to allocate money to labor and capital in order to maximize productivity in the current year. The company can use a Cobb-Douglas production function, which we introduced in Section 7.1, along with its fixed labor cost, length of a work day, number of annual work days, and its annual budget for labor and capital to determine how to allocate funds. We will revisit this application later in the section where you'll practice applying Lagrange multipliers. First, let's take a closer look at unconstrained and constrained relative extrema.

Unconstrained and Constrained Relative Extrema

Figure 1 shows the graph of a function $f(x, y)$ of two variables with a relative maximum at the point (a, b, c_1). This relative maximum is called an **unconstrained relative maximum** because there are no constraints (restrictions) on the values of the input variables x and y.

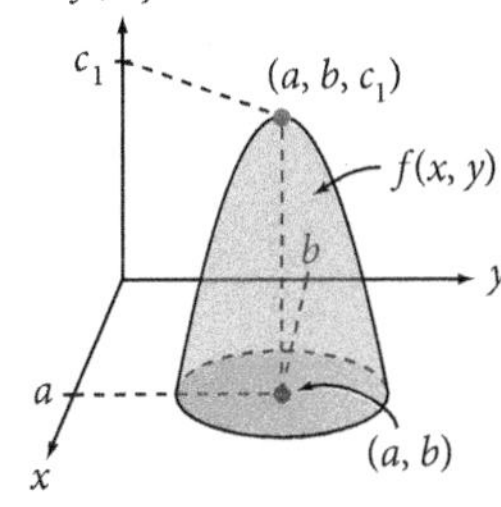

Figure 1

However, very often in applied situations, there is a constraint on the input variables that must be considered when optimizing a function $f(x, y)$. A constraint is a restriction that limits the values of the input variables and, therefore, the value of the output variable. Constraints occur when the input variables are related.

There are many different types of constraints, but an example in business might involve constraints due to a certain amount of money budgeted for labor costs and investments. Constraints can, therefore, be described as functions, and Figure 2 shows the graph of a function $f(x, y)$ with a constrained relative maximum at the point (a, b, c_2), this point lies on the surface of $f(x, y)$. The relative maximum

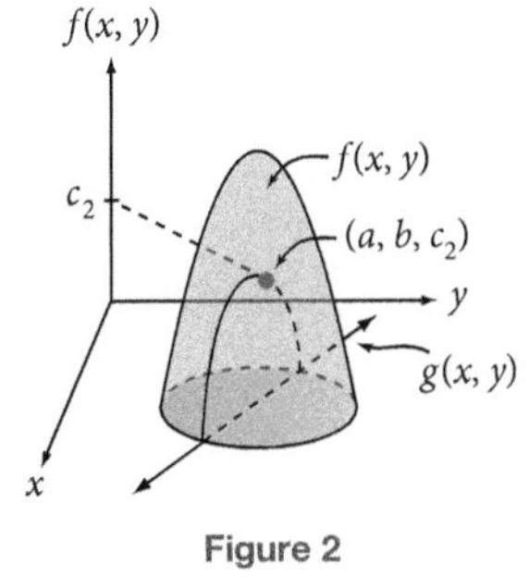

Figure 2

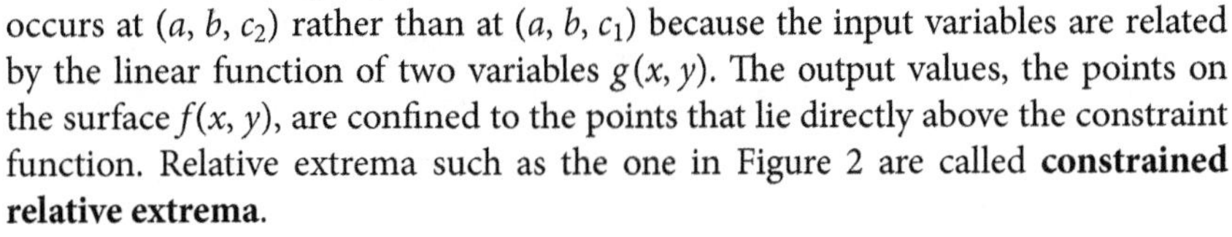

occurs at (a, b, c_2) rather than at (a, b, c_1) because the input variables are related by the linear function of two variables $g(x, y)$. The output values, the points on the surface $f(x, y)$, are confined to the points that lie directly above the constraint function. Relative extrema such as the one in Figure 2 are called **constrained relative extrema**.

The Method of Algebraic Substitution

If the constraint is easily solved for one of the variables, any relative extreme points can be located and classified using substitution. The optimization technique discussed in Section 3.3 can then be used.

VIDEO EXAMPLES

SECTION 7.4

Example 1 Use the substitution method to find and classify the relative extreme points of the function $f(x, y) = x^2 - 3y^2 + 2x + 4y$ subject to the condition that $x - 2y = 4$.

Solution Notice first that the input variables are related by the function $x - 2y = 4$. That is, the value of x minus twice the value of y is always equal to 4. We can easily solve for x in terms of y. Adding $2y$ to each side produces $x = 2y + 4$. Substitute the expression $2y + 4$ for x into the original function. Since all the x's will be eliminated and replaced with y's, the original function of two variables will be reduced to a function of only one variable.

$$f(x, y) = x^2 - 3y^2 + 2x + 4y$$

$$f(y) = (2y + 4)^2 - 3y^2 + 2(2y + 4) + 4y$$

$$= y^2 + 24y + 24$$

We can now use the first derivative test method introduced in Section 3.3 to identify the critical values.

$$f'(y) = 2y + 24$$

a. Where is $f'(y) = 0$? (smooth maximum and minimum points)

$$2y + 24 = 0 \rightarrow 2(y + 12) = 0 \rightarrow y = -12$$

b. Where is $f'(y)$ undefined? (cusp/corner maximum and minimum points)

Since $f(y)$ is a polynomial function, it is never undefined. The only critical value is $y = -12$.

Use the first derivative test by choosing a point to the left of -12 and substituting it into $f'(y) = 2(y + 12)$ to determine the sign of the first derivative. Then, choose a point to the right of -12 and substitute it into $f'(y) = 2(y + 12)$ to again determine the sign of the first derivative.

$$f'(y) = 2(y + 12)$$

f' $(+)(-) = (-)$ $(+)(+) = (+)$

decreasing -12 increasing y

relative min

Figure 3

Since $f'(y)$ is negative to the left of -12 and positive to the right, we conclude that -12 produces a relative minimum.

The value of x that corresponds to $y = -12$ is found by substituting -12 for y in the constraint equation $x = 2y + 4$.

$$x = 2(-12) + 4 = -24 + 4 = -20$$

The critical point of the original function is $(-20, -12)$. Substituting these values into the original function produces the constrained relative minimum $(-20, -12, -120)$.

The Method of Lagrange Multipliers

Sometimes it may be difficult (or even impossible) to solve the constraint for one of the variables. Also, even when the constraint can be solved for one of the variables, a complicated function may result from the substitution. In the eighteenth century, the French mathematician Joseph Lagrange (1736-1813) constructed a method for optimizing functions of several variables that are subject to a constraint, $g(x, y) = 0$, that eliminates the need for the substitution. In Lagrange's honor, the method is called the **method of Lagrange multipliers**. The method involves creating a new function, called the **Lagrangian function**, which, for a function of the two variables x and y, is defined as

$$\boldsymbol{F(x, y, \lambda) = f(x, y) + \lambda \cdot g(x, y)}$$

The Greek letter lambda, λ, is called the **Lagrange multiplier** and serves to eliminate the dependency of one variable on another. The Lagrange multiplier always multiplies the constraint function $g(x, y)$. The Lagrangian function, $F(x, y, \lambda)$, is an unconstrained function of the three independent variables x, y, and λ. Because of the independence of the variables, a form of the Second Derivative Test may be used to locate and classify any critical points as relative extrema.

The Method of Lagrange Multipliers

To locate potential relative extreme points of a function $f(x, y)$ subject to the constraint $g(x, y) = 0$,

1. Set the constraint equal to zero so that it is in the form $g(x, y) = 0$.
2. Construct the Lagrangian function
$$F(x, y, \lambda) = f(x, y) + \lambda g(x, y)$$
3. Find the first-order partial derivatives F_x, F_y, and F_λ.
4. Locate critical points by solving the system of equations
$$\begin{cases} F_x = 0 & \ldots(1) \\ F_y = 0 & \ldots(2) \\ F_\lambda = 0 & \ldots(3) \end{cases}$$
for x, y, and λ. Each ordered pair solution, (x, y), is a critical point. The relative extreme points are among these critical points.
5. Evaluate the original function $f(x, y)$ at each critical point obtained in Step 4. The point that produces the greatest value produces a relative maximum. The point that produces the least value produces a relative minimum.

Note The system in step 4 can often be solved using equations 1 and 2 and eliminating λ. This will produce a new equation that involves only x and y. Then, using equation 3 and this new equation, we obtain the values of x and y. These values can then be substituted into either of equations 1 or 2 to obtain the corresponding value of λ. Although it is not necessary to obtain the value of λ to locate and classify relative extreme points, λ has an important and interesting interpretation (which we will discuss after we present an example).

Example 2 Use the method of Lagrange multipliers to find the minimum of the function $f(x, y) = x^2 + 10y^2$ subject to the constraint $x - y = 18$.

Solution Although this constraint makes the function a good candidate for the method of algebraic substitution, we will use the method of Lagrange multipliers as a point of illustration.

a. We begin by writing the constraint in the form $g(x, y) = 0$.

$$x - y = 18 \quad \rightarrow \quad \underbrace{x - y - 18}_{g(x,y)} = 0, \text{ so that } g(x, y) = x - y - 18$$

b. Next, we construct the Lagrangian function $F(x, y, \lambda) = f(x, y) + \lambda g(x, y)$.

$$F(x, y, \lambda) = x^2 + 10y^2 + \lambda(x - y - 18)$$

or

$$F(x, y, \lambda) = x^2 + 10y^2 + \lambda x - \lambda y - \lambda 18$$

c. Now we find the first-order partial derivatives F_x, F_y, and F_λ.

$$F_x = 2x + \lambda, \quad F_y = 20y - \lambda, \quad F_\lambda = x - y - 18$$

Notice that F_λ equals $g(x, y)$. The Lagrangian function is constructed so that this will always be true. (Try differentiating $F(x, y, \lambda) = f(x, y) + \lambda g(x, y)$ with respect to λ.)

d. We solve the system

$$\begin{cases} F_x = 0 & \ldots(1) \\ F_y = 0 & \ldots(2) \\ F_\lambda = 0 & \ldots(3) \end{cases} \quad \rightarrow \quad \begin{cases} 2x + \lambda & = 0 \quad \ldots(1) \\ 20y - \lambda & = 0 \quad \ldots(2) \\ x - y - 18 & = 0 \quad \ldots(3) \end{cases}$$

We will eliminate λ by adding equations (1) and (2). Symbolically, our method will be (1) + (2), and we will label the resulting equation (4).

$$\begin{cases} 2x + \lambda = 0 & \ldots(1) \\ 20y - \lambda = 0 & \ldots(2) \end{cases}$$
$$\overline{2x + 20y = 0 \quad \ldots(4)}$$

Now we form a system that involves only x and y using equations (3) and (4).

$$\begin{cases} x - y - 18 = 0 & \ldots(3) \\ 2x + 20y = 0 & \ldots(4) \end{cases} \qquad \begin{cases} x - y = 18 & \ldots(3) \\ 2x + 20y = 0 & \ldots(4) \end{cases}$$

Next we eliminate x by adding -2 times equation (3) to equation (4). This will produce the value of y.

$$\begin{cases} -2x + 2y = -36 & (-2 \text{ times equation } (3)) \\ 2x + 20y = 0 & \ldots(4) \end{cases}$$
$$\overline{22y = -36}$$

$$y = \frac{-36}{22}$$

$$y = \frac{-18}{11}$$

Then, if $y = \frac{-18}{11}$, substitution into equation (3) produces

$$x - \frac{-18}{11} - 18 = 0$$

$$x = \frac{180}{11}$$

Then, if $x = \frac{180}{11}$ and $y = \frac{-18}{11}$, substitution into equation (1) produces

$$2 \cdot \frac{180}{11} + \lambda = 0$$

$$\lambda = \frac{-360}{11}$$

Thus, $x = \frac{180}{11}, y = \frac{-18}{11}, \lambda = \frac{-360}{11}$; and $\left(\frac{180}{11}, \frac{-18}{11}\right)$ is the only critical point. (We will discuss the meaning of λ in the next subsection.)

e. Evaluate $f(x, y)$ at $\left(\frac{180}{11}, -\frac{18}{11}\right)$ to obtain the minimum.

$$f\left(\frac{180}{11}, -\frac{18}{11}\right) = \left(\frac{180}{11}\right)^2 + 10\left(\frac{-18}{11}\right)^2$$

$$= \frac{3{,}240}{11}$$

Interpretation
The function $f(x, y) = x^2 + 10y^2$, when subjected to the constraint $x - y - 18 = 0$, has a relative minimum at $\left(\frac{180}{11}, \frac{-18}{11}, \frac{3{,}240}{11}\right)$.

The Significance of the Lagrange Multiplier

In Example 2, we found the value of the Lagrange multiplier, λ, but did not discuss its significance. The Lagrange multiplier provides useful information.

The Significance of the Lagrange Multiplier (λ)
Suppose that M is the constrained maximum or minimum value of the function $f(x, y)$ when it is subjected to the constraint $g(x, y) = c$, where c is some constant. Then,

$$\frac{dM}{dc} = \lambda$$

That is, λ is the rate at which the constrained maximum or minimum value of the function $f(x, y)$ changes with respect to c.

Interpretation If c is increased by 1 unit, from c to $c + 1$, then M changes by approximately λ units.

In Example 2, the function to be optimized was

$$f(x, y) = x^2 + 10y^2$$

subject to the constraint $x - y = 18$. The maximum value of $f(x)$ was $\frac{3{,}240}{11}$ and λ was $\frac{-360}{11}$. In this case, $c = 18$, and we conclude that if the constant in the constraint is increased by 1 unit, from 18 to 19, then the maximum value of $f(x, y)$ will decrease by approximately $\frac{360}{11}$ units, from $\frac{3{,}240}{11}$ to $\frac{2{,}880}{11}$.

Example 3 The production of a manufacturer is given by the Cobb-Douglas production function

$$f(x, y) = 30x^{1/5}y^{4/5}$$

© Patrick Heagney/iStockPhoto

where x represents the number of units of labor (in hours) and y represents the number of units of capital (in dollars) invested. Labor costs \$15 per hour and there are 8 hours in a working day, and 250 working days in a year. The manufacturer has allocated \$4,500,000 this year for labor and capital. How should the money be allocated to labor and capital to maximize productivity this year?

Solution The function to be maximized is $f(x, y) = 30x^{1/5}y^{4/5}$. We need to develop the appropriate constraint function. The number of units of labor to be used is x. The annual cost of labor is then

$$(\$15 \text{ dollars per hour})(8 \text{ hours per day})(250 \text{ days})x = \$30{,}000x$$

Thus, labor's contribution to the total cost is \$30,000x. Capital's contribution to the total cost is simply y (since y represents the number of dollars invested). Therefore, the constraint function is

$$30{,}000x + y = 4{,}500{,}000$$

Hence, we need to maximize $f(x, y) = 30x^{1/5}y^{4/5}$ subject to the constraint

$$30{,}000x + y = 4{,}500{,}000$$

a. We express the constraint function in the form $g(x, y) = 0$.

$$30{,}000x + y = 4{,}500{,}000 \quad \rightarrow \quad 30{,}000x + y - 4{,}500{,}000 = 0$$

so that $g(x, y) = 30{,}000x + y - 4{,}500{,}000$

b. We construct the Lagrangian function $F(x, y, \lambda) = f(x, y) + \lambda g(x, y)$.

$$F(x, y, \lambda) = 30x^{1/5}y^{4/5} + \lambda(30{,}000x + y - 4{,}500{,}000)$$

or

$$F(x, y, \lambda) = 30x^{1/5}y^{4/5} + 30{,}000\lambda x + \lambda y - 4{,}500{,}000\lambda$$

c. We find the first-order partial derivatives F_x, F_y, and F_λ.

$$F_x = 6x^{-4/5}y^{4/5} + 30{,}000\lambda$$

$$F_y = 24x^{1/5}y^{-1/5} + \lambda$$

$$F_\lambda = 30{,}000x + y - 4{,}500{,}000$$

d. We solve the system

$$\begin{cases} F_x = 0 & \ldots(1) \\ F_y = 0 & \ldots(2) \\ F_\lambda = 0 & \ldots(3) \end{cases} \rightarrow \begin{cases} 6x^{-4/5}y^{4/5} + 30{,}000\lambda & = 0 \quad \ldots(1) \\ 24x^{1/5}y^{-1/5} + \lambda & = 0 \quad \ldots(2) \\ 30{,}000x + y - 4{,}500{,}000 & = 0 \quad \ldots(3) \end{cases}$$

Since this system involves an xy term, we will need to solve it differently than we did the system in Example 2. We will solve equations (1) and (2) for λ, set those resulting expressions equal, and establish a relation between x and y.

From (1),

$$30{,}000\lambda = \frac{-6y^{4/5}}{x^{4/5}}$$

$$\lambda = \frac{-y^{4/5}}{5{,}000x^{4/5}}$$

From (2),

$$\lambda = \frac{-24x^{1/5}}{y^{1/5}}$$

Therefore, $\dfrac{-y^{4/5}}{5{,}000x^{4/5}} = \dfrac{-24x^{1/5}}{y^{1/5}}$.

We eliminate the denominators by multiplying both sides by $5{,}000x^{4/5}y^{1/5}$.

$$5{,}000x^{4/5}y^{1/5} \cdot \frac{-y^{4/5}}{5{,}000x^{4/5}} = 5{,}000x^{4/5}y^{1/5} \cdot \frac{-24x^{1/5}}{y^{1/5}}$$

$$-y = -120{,}000x$$

$$y = 120{,}000x$$

Substitute $120{,}000x$ for y in equation (3) to obtain a single equation in x.

$$30{,}000x + 120{,}000x - 4{,}500{,}000 = 0$$

$$150{,}000x - 4{,}500{,}000 = 0$$

$$150{,}000x = 4{,}500{,}000$$

$$x = 30$$

Since each unit of labor is valued at \$30,000, thirty units of labor amounts to \$900,000.

Now determine y. Since $y = 120{,}000x$ and $x = 30$,

$$y = 120{,}000(30) = 3{,}600{,}000$$

Determine the value of λ by substituting 30 for x and 3,600,000 for y in

$$\lambda = \frac{-24x^{1/5}}{y^{1/5}} = \frac{-24(30)^{1/5}}{(3{,}600{,}000)^{1/5}} = -2.31$$

We now have $x = 30$, $y = 3{,}600{,}000$, and $\lambda = -2.31$. The only critical point is (30, 3,600,000).

Evaluate $f(x, y) = 30x^{1/5}y^{4/5}$ at (30, 3,600,000) to get a maximum value.

$$f(30, 3{,}600{,}000) = 30(30)^{1/5}(3{,}600{,}000)^{4/5}$$

$$\approx 10{,}413{,}279.04$$

Interpretation
To maximize production at 10,413,279.04 units, the manufacturer should allocate 30 units to labor (that is, \$900,000 to labor) and \$3,600,000 to capital. Also, if the number of dollars allocated to labor and capital is increased by 1 from \$4,500,000 to \$4,500,001, the number of units produced will decrease by approximately 2.31, from about 10,413,279.04 units to about 10,413,276.73 units.

Using Technology 7.4

Wolfram|Alpha can help you find the constrained relative extreme points of functions of two variables.

Example 4 Use Wolfram|Alpha to find the relative maxima and relative minima of the function

$$f(x, y) = -3x^2 - 2y^2 + 4xy + 13x + 8y + 80$$

subject to the constraint

$$x + 2y = 20$$

Solution Go to www.wolframalpha.com and in the entry field enter

max and min of -3x^2 - 2y^2 + 4xy + 13x + 8y + 80 subject to x + 2y = 20

Wolfram|Alpha responds with

Global maximum:

$$\max\{-3x^2 - 2y^2 + 4xy + 13x + 8y + 80 \mid x + 2y = 20\} = \frac{3881}{22} \text{ at } (x, y) = \left(\frac{69}{11}, \frac{151}{22}\right)$$

Global minima:

(no global minima found)

Wolfram Alpha LLC. 2012. Wolfram|Alpha
http://www.wolframalpha.com/
(access December 27, 2012)

This function has a relative maximum value of $\frac{3,881}{22} \approx 176.41$ when $x = \frac{69}{11} \approx 6.27$ and $y = \frac{151}{22} \approx 6.86$.

Now that we know it's a maximum, take out "and min" from the entry field and hit enter again. Wolfram|Alpha draws the graph of the function with the constraint. The graph indicates (look closely) a relative maximum at (6.27, 6.86, 176.41).

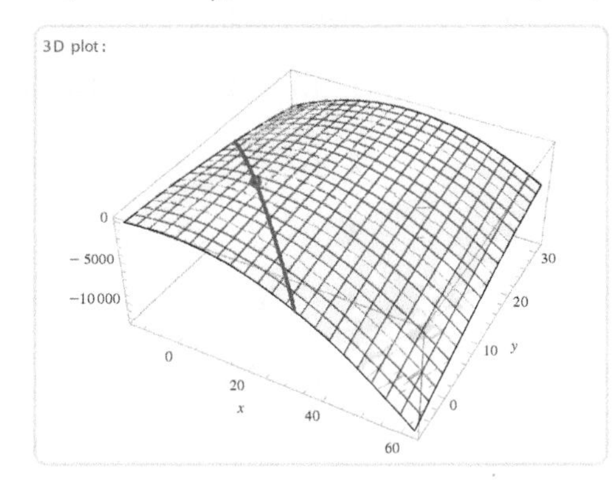

Wolfram Alpha LLC. 2012. Wolfram|Alpha
http://www.wolframalpha.com/
(access December 27, 2012)

Getting Ready for Class

After reading through the preceding section, respond in your own words and in complete sentences.

A. Describe the difference between unconstrained and constrained relative extrema.

B. Describe the strategy you would use to solve a problem such as Problem 1 or 2 in the Problem Set.

C. Explain, as if you were writing a letter to a friend who is also studying calculus, the meaning of the Lagrange multiplier λ.

Problem Set 7.4

Skills Practice

For Problems 1-4, use the substitution method to find and classify the relative extreme points of each function subject to the given constraint.

1. $f(x, y) = x^2 - 8y^2 + 5x + 9y + 8$ subject to $x - y = 7$.
2. $f(x, y) = x^2 - 2y^2 - 5x - 26y - 24$ subject to $x - 2y = 8$.
3. $f(x, y) = x^2 + y^2 + 3xy + 7x + 63y - 60$ subject to $x + 4y = 5$.
4. $f(x, y) = x^2 - 11y^2 + 2xy + 4x - 44y + 4$ subject to $x - 3y = -2$.
5. Maximize $f(x, y) = 25 - x^2 - y^2$ subject to $x + y = -1$.
6. Maximize $f(x, y) = 100 - x^2 - y^2$ subject to $2x + y = -5$.
7. Minimize $f(x, y) = y^2 + 6x$ subject to $y - 2x = 0$.
8. Minimize $f(x, y) = 2x^2 - y^2$ subject to $x - y = 2$.
9. Maximize $f(x, y) = 6x^{1/3}y^{2/3}$ subject to $4x + 3y = 36$.
10. Maximize $f(x, y) = e^{2xy}$ subject to $x^3 + y^3 = 2$.

Check Your Understanding

In each of Problems 11-15, a value of λ is given along with a critical point that produces a constrained relative maximum or minimum, and the corresponding maximum or minimum value. Interpret the meaning of λ as it applies to the given situation.

11. **Education: Achievement** In a large, four-unit general anthropology course given at a college, the function $f(x, y)$ measures student achievement. The number of hours devoted to lecture is represented by x and the number of hours devoted to discussion by y. The course requires 72 hours of instruction. The constraint function is $x + y = 72$, achievement is maximized at 85 at $(54, 18)$, and the corresponding value of λ is 1.3.

12. **Economics: Pesticide Application** Scientists believe that the function $N(x, y)$ approximates the number (in thousands) of insects killed when sprayed with x pounds of pesticide A and y pounds of pesticide B. Because too much pesticide is harmful to the plants that live in the vicinity of the insects, no more than 250 pounds of the pesticide can be used in one spraying. The constraint function is then $x + y = 250$. $N(x, y)$ is maximized at $(180, 70)$ and, furthermore, $N(180, 70) = 250$. The corresponding value of λ is 3.2.

13. **Business: Container Size** The typical beverage can has a volume of about 21.66 cubic inches. The function $S(h, r)$ describes the surface area of the can in terms of the can's height and radius. A radius of 1.51 inches and a height of 3.02 inches produces the minimum surface area of approximately 42.98 square inches. The constraint function is $\pi r^2 h = 21.66$, and the corresponding value of λ is 1.32.

14. **Business: Cobb-Douglas Production** For a certain company, the Cobb-Douglas production function $f(x, y)$ is maximum at (1,500, 3,500), when the constraint is $3x + 11y = 43{,}000$. The corresponding value of λ is 5.2.

15. **Business: Cost of Quality Control** The cost, $C(x, y)$ (in dollars), of quality control at a manufacturing company is a function of the number x of items tested at point A, and the number y of items tested at point B. Company policy specifies that 35 items be tested each day. The constraint function is $x + y = 35$, and the cost function is minimized at \$355 when $x = 12$ and $y = 23$. The corresponding value of λ is 4.09.

16. **Business: Cobb-Douglas Production** It is true, although we will not prove it here, that a constrained relative extreme point of a Cobb-Douglas production function is always a maximum. Show that the constrained relative maximum of the Cobb-Douglas production function $f(x, y) = Cx^a y^{1-a}$, when it is subjected to the linear constraint $px + qy = k$, is

$$\left(\frac{ak}{p}, \frac{(1-a)k}{q}\right)$$

Modeling Practice

17. **Manufacturing: Cobb-Douglas Production** The production of a manufacturer is given by the Cobb-Douglas production function

$$f(x, y) = 12x^{2/3}y^{1/3}$$

where x represents the number of units of labor (in hours) and y represents the number of units of capital (in dollars) invested. Labor costs \$8 per hour, there are 8 hours in a working day and 260 working days in a year. The manufacturer has allocated \$1,622,400 this year for labor and capital. How should the money be allocated to labor and capital to maximize productivity this year? Find and interpret the marginal productivity of money.

© Patrick Heagney/iStockPhoto

18. **Manufacturing: Cobb-Douglas Production** The production of a manufacturer is given by the Cobb-Douglas production function

$$f(x, y) = 4x^{5/8}y^{3/8}$$

where x represents the number of units of labor (in hours) and y represents the number of units of capital (in dollars) invested. Labor costs \$9.50 per hour, there are 8 hours in a working day and 255 working days in a year. The manufacturer has allocated \$2,232,576 this year for labor and capital. How should the money be allocated to labor and capital to maximize productivity this year? Find and interpret the marginal productivity of money.

19. **Manufacturing: Cobb-Douglas Production** The production of a manufacturer is given by the Cobb-Douglas production function

$$f(x, y) = x^{3/5}y^{2/5}$$

where x represents the number of units of labor (in hours) and y represents the number of units of capital (in dollars) invested. Labor costs \$12 per hour, there are 8 hours in a working day and 22 working days in a month. The manufacturer has allocated \$48,032 this month for labor and capital. How should the money be allocated to labor and capital to maximize productivity this month? Find and interpret the marginal productivity of money.

20. **Manufacturing: Production Costs** A manufacturer has two sites, A and B, at which it can produce a product, and because of certain conditions, site A must produce three times as many units as site B. The total cost of producing the units is given by the cost function

$$C(x, y) = 0.3x^2 - 150x - 610y + 150{,}000$$

where x represents the number of units produced at site A and y represents the number of units produced at site B. How many units should be produced at each site to minimize the cost? What is the minimal cost? Find and interpret the value of the Lagrange multiplier.

21. **Manufacturing: Quality Assurance Costs** A manufacturer finds that the cost of quality assurance is related to the number of inspections, x and y, it makes each day at two points, A and B, respectively, in its assembly process. The manufacturer must make 5 times as many inspections at point A than at point B. The cost of making these inspections is approximated by the function

$$C(x, y) = 6x^2 + 3y^2 - 612x$$

How many inspections at each location should the manufacturer make to minimize costs?

22. **Manufacturing: Container Size** The volume of a right circular cylinder with radius r and height h is given by $\pi r^2 h$, and the circumference of a circle with radius r is $2\pi r$. Use these facts to find the dimensions of a 10-ounce (approximately 5.74π cubic inches) can with a top and bottom that can be made using the least amount of material.

23. **Business: Advertising** A franchise operation in a city has determined that if it buys x radio and y newspaper advertisements each day, the number of people it can reach and possibly influence is approximated by the function

$$N(x, y) = 30x(4y + 40)^{2/3}$$

If the franchise operation has \$42,000 budgeted each day for advertisements and each radio advertisement costs \$100 per day and each newspaper advertisement costs \$50 per day, how many radio and newspaper ads should it buy each day to reach the maximum number of people? What is that maximum number of people? Find and interpret the value of the Lagrange multiplier.

24. Business: Publishing Revenue A publisher believes it can sell 26,600 copies of its new book. The publisher will first make the book available in hardback and then, some time later, in paperback. If it sells each paperback book for x dollars, and each hardback book for y dollars, the publisher believes it can sell $5{,}150 - 350x + 700y$ paperback copies and $9{,}200 + 1{,}050x - 350y$ hardback copies. Find the price of each type of book that will maximize the publisher's revenue. Find and interpret the value of the Lagrange multiplier. (Hint: Recall that revenue equals price times quantity.)

25. Business: Agricultural Pesticide To keep pests from destroying a particular crop, agriculturists spray a mix of x pounds of pesticide A and y pounds of pesticide B onto the crop. They have determined that the percentage of pests killed by this application is approximated by the function

$$f(x, y) = 1 - \frac{1}{4}e^{-x/5} - \frac{1}{4}e^{-y/70}$$

If 2,000 pounds of pesticide are to be applied, how many pounds of each type of pesticide should be used to maximize the percentage of pests killed by an application of the pesticide? (Hint: You will need to use the natural logarithm to solve for x or y.)

Using Technology Exercises

26. Use Wolfram|Alpha to help you *approximate* the maximum of the function

$$f(x, y) = -3x^4 - 2y^4 + 4x^2y^2 - 13x - 8y + 8{,}000$$

subject to the constraint $x + 2y = 20$.

27. Use Wolfram|Alpha to help you with each of the following computations.

a. Construct the graph of $f(x) = x$ from $x = 0$ to $x = 5$.

b. Construct the graph of $f(x) = xe^{-0.3x}$ from $x = 0$ to 5, $y = 0$ to 5. (Enter "graph x*e^(-0.3x) for x = 0 to 5, y = 0 to 5" into the command line.)

c. Compare the graphs you constructed in parts a and b. What effect does multiplying x by $e^{-0.3x}$ have on the graph of $y = x$?

d. Construct the graph of $f(x, y) = xe^{-0.3x} + ye^{-0.3y}$ for $x = 0$ to 5, $y = 0$ to 5. Does your conclusion for the question of part c hold for the effect multiplying y by $e^{-0.3y}$ has in the direction of the y-axis?

e. Find the unconstrained max of the function $f(x) = xe^{-0.3x} + ye^{-0.3y}$.

f. Find the maximum of the function under the constraint that $y = x$.

28. Find the maximum value of the function $f(x, y) = 900x^{0.85}y^{0.15}$ subject to the constraint $20x + 10y = 1{,}200$. Wolfram|Alpha produces not only the maximum value and where it occurs, but also the graph of the function with the constraint.

29. The profit $P(x)$, in thousands of dollars, from the sale of x units of product A and y units of product B is modeled by the function

$$P(x, y) = 22x + 73y - 1.2x^2 - 1.5y^2$$

How many units of each product should be produced and sold to maximize the profit if the company can produce only 30 units total? What is the maximum profit?

30. Consider the function $f(x, y) = 1{,}000x^{0.25}y^{0.75}$ with constraint $x + y = 50$.

a. Find the values of x and y that maximize the function.

b. Form the Lagrangian function $F(x, y, \lambda) = f(x, y) + \lambda g(x, y)$. To form the Lagrangian in Wolfram|Alpha, use the lowercase letter "a" in place of the Greek letter λ.

c. Find $F_x(x, y, \lambda)$, $F_y(x, y, \lambda)$, and $F_\lambda(x, y, \lambda)$.

d. Set $F_x(x, y, \lambda)$, $F_y(x, y, \lambda)$, and $F_\lambda(x, y, \lambda)$ equal to zero to solve for λ, the Lagrange multiplier.

e. Interpret the value of the Lagrange multiplier.

Getting Ready for the Next Section

31. Given $f(x, y) = 42x^{3/7}y^{4/7}$, find $f_x(x, y)$.

32. Given $f(x, y) = 42x^{3/7}y^{4/7}$, find $f_y(x, y)$.

33. Use the result of Problem 31 to evaluate $f_x(x, y)\, dx$ when $(x, y) = (2{,}187, 128)$ and $dx = -4$.

34. Use the result of Problem 32 to evaluate $f_y(x, y)\, dy$ when $(x, y) = (2{,}187, 128)$ and $dy = 2$.

The Total Differential

7.5

For a function $f(x, y)$, the partial derivative f_x, can be interpreted as the approximate change in the function value as the value of x changes and the value of y remains constant. The partial derivative for f_y can be similarly interpreted. In this section, we investigate how to approximate the change in the function value as both x and y change.

Suppose a city is analyzing its public transportation system, and is using the following function

$$N(x, y) = \frac{0.1x}{0.5 + \ln y}.$$

© Pavlina Perry/iStockPhoto

where N is the number of riders each day (in thousands), x is the number of people living in the city (in thousands), and y is the price of a ticket (in dollars). If the population changes at the same time as the ticket price changes, we can approximate the change in the number of riders by using what we call the *total differential.* We will answer this question later in the problem set. For now, let's discuss the details of this method.

The Total Differential

Just as there is a tangent line at a particular x-value associated with a function $f(x)$ of one variable, there is a tangent plane at a particular point associated with a function $f(x, y)$ of two variables. Imagine being located at a point (x, y) in the xy plane and moving small distances dx and dy in the x and y directions, respectively, to the new point $(x + dx, y + dy)$. To get to the new point, we could first move dx units parallel to the x-axis, and from there, dy units parallel to the y-axis. The product

$$f_x(x, y) \cdot dx$$

approximates the change in the function as x changes value, and the product

$$f_y(x, y) \cdot dy$$

approximates the change in the function as y changes value.

The total change in the function is then approximated by

$$f_x(x, y)dx + f_y(x, y)dy$$

We denote this sum of individual differentials by df (f being the name of the function) and call it the **total differential** of the function $f(x, y)$.

The Total Differential

The total differential of a function of two variables $f(x, y)$ is

$$df = f_x(x, y)dx + f_y(x, y)dy$$

and it approximates the change in the function as x changes a small amount by dx units and y changes a small amount by dy units.

The exact change in the function is $f(x + dx, y + dy) - f(x, y)$, and thus,

$$df \approx f(x + dx, y + dy) - f(x, y)$$

The differential formula can be extended to functions of more than two variables. For example, for a function of three variables, x, y, and z, the differential formula is

$$df = f_x\,dx + f_y\,dy + f_z\,dz$$

VIDEO EXAMPLES

SECTION 7.5

Example 1 The productivity of a company (the number of units of a product a company is able to produce) is described by the Cobb-Douglas production function

$$f(x, y) = 42x^{3/7}y^{4/7}$$

where x represents the number of units of labor and y the number of units of capital utilized. Approximate the change in output if the number of units of labor is decreased from 2,187 to 2,183, and the number of units of capital is increased from 128 to 130.

Solution We will begin by finding the partial derivatives, f_x and f_y, and the changes in x and y.

$$f_x(x, y) = \frac{18y^{4/7}}{x^{4/7}} \quad \text{and} \quad f_y(x, y) = \frac{24x^{3/7}}{y^{3/7}}$$

$$dx = 2{,}183 - 2{,}187 = -4 \quad \text{and} \quad dy = 130 - 128 = 2$$

Substituting the partial derivatives into the differential formula $df = f_x\,dx + f_y\,dy$ produces the differential for the function.

$$df(x, y) = \frac{18y^{4/7}}{x^{4/7}}dx + \frac{24x^{3/7}}{y^{3/7}}dy$$

We then evaluate the differential at $(x, y) = (2{,}187, 128)$, $dx = -4$, and $dy = 2$.

$$\begin{aligned} df(2{,}187, 128) &= \frac{18(128)^{4/7}}{(2{,}187)^{4/7}} \cdot (-4) + \frac{24(2{,}187)^{3/7}}{(128)^{3/7}} \cdot (2) \\ &\approx -14.2 + 162 \\ &\approx 147.8 \end{aligned}$$

Interpretation

If the number of units of labor is decreased by 4 units, from 2,187 to 2,183, and the number of units of capital is increased by 2 units, from 128 to 130, the number of units of output will increase by approximately 147.8. (Remember, the change is approximate because there is a variable in the derivative expression.)

■

The exact value of the change in the function is given by the functional value at the new point minus the functional value at the original point.

$$\begin{aligned} f(2{,}183, 130) - f(2{,}187, 128) &= 42(2{,}183)^{3/7}(130)^{4/7} - 42(2{,}187)^{3/7}(128)^{4/7} \\ &\approx 147.1 \end{aligned}$$

Thus, the difference between the approximate value and the actual value is

$$147.8 - 147.1 = 0.7$$

a relatively small error.

Using Technology 7.5

You can use your graphing calculator to find the total differential of a function of two variables.

Example 2 Find the total differential for the function $f(x, y) = 42x^{3/7}y^{4/7}$.

Solution The following entries illustrate the procedure:

Y1 = 42X^(3/7) Y^(4/7)
Y5 = nDeriv(Y1, X, X, .001)
Y7 = nDeriv(Y1,Y, Y, .001)
Y9 = Y5*C + Y7*D

The functions Y5 and Y7 are, respectively, f_x and f_y. The entries C and D represent, respectively, the differentials dx and dy. In the computation window, store 2187 to X, 128 to Y, −4 to C, and 2 to D.

2187 [STO▸] X
128 [STO▸] Y
−4 [STO▸] C
2 [STO▸] D

Then compute Y9. Your result should be about 147.77....

Getting Ready for Class

After reading through the preceding section, respond in your own words and in complete sentences.

A. Describe how the total differential of a function $f(x, y)$ differs from a partial derivative of $f(x, y)$.

B. Describe the strategy you would use to solve a problem such as Problem 5 or 6 in the Problem Set.

Problem Set 7.5

Skills Practice

For Problems 1-12, find the total differential of the given function.

1. $f(x, y) = 8x^2 - 3y^2 + 4x + 5y + 6$, if $dx = 1$ and $dy = 2$

2. $f(x, y) = 16x^2 + 4y^2 - 10xy$, if $dx = 3$ and $dy = 1$

3. $f(x, y) = x^2 - 15y^2 + 4e^{xy}$, if x increases from 3 to 4, and y increases from 1 to 2

4. $f(x, y) = 3x + 2y + 6\ln(x^2y)$, if x decreases from 5 to 4, and y increases from 10 to 11

5. $f(x, y) = \sqrt{xy}$, if x increases from 9 to 10, and y decreases from 4 to 3

6. $f(x, y, z) = x^2 + y^2 - z^2$, if x increases from 10 to 11, y increases from 25 to 27, and z decreases from 1 to 0

7. $f(x, y, z) = 5x^2e^{2y-z^2}$, if x decreases from 3 to 1, y increases from 1 to 2, and z decreases from 1 to 0

8. $f(x, y) = 300x^{2/3}y^{1/3}$, if x increases from 27 to 28, and y decreases from 64 to 61

9. $f(x, y) = \frac{y}{x}$, if x increases from 150 to 155, and y increases from 20 to 21

10. $f(x, y) = \ln(y - x)$, if x decreases from 5 to 4, and y increases from 60 to 61

11. $f(x, y, z) = y\ln x + z^2$, if x increases from 1 to 2, y increases from 0 to 2, and z increases from 0 to 1

12. $f(x, y, z) = \frac{x + z}{y - z}$, if x decreases from 100 to 96, y decreases from 25 to 24, and z increases from 10 to 11

Check Your Understanding

13. Suppose that for a function $f(x, y)$, $dx = 1$ and $dy = 2$ at the point (a, b). What conditions do $f_y(a, b)$ and dy have to satisfy so that $df(a, b) < 0$?

14. Suppose that for a function $f(x, y)$, $f_x(a, b) > 0$ and $dx > 0$ at the point (a, b). What conditions do $f_y(a, b)$ and dy have to satisfy so that $df(a, b) < 0$?

15. Suppose that for a function $f(x, y) = e^{x+y}$, $dx < 0$ and $dy < 0$ at the point (a, b). Is it possible that $df(a, b) > 0$? If so, how?

16. If you think the differential of a function f of four variables x, y, z, and w could exist, specify its form.

Modeling Practice

17. **Business: Cobb-Douglas Production** The output of a company is given by the Cobb-Douglas production function $f(x, y) = 90x^{2/3}y^{1/3}$, where x represents the number of thousands of dollars invested in labor and y the number of thousands of dollars invested in capital.
    a. What is the approximate change in the output if the amount of money invested in labor is increased from \$27,000 to \$28,000 and the amount invested in capital is decreased from \$27,000 to \$26,000?
    b. Compare this approximate change to the actual change.

© Pavlina Perry/iStockPhoto

18. **Economics: City Transportation** The number N (in thousands) of riders of a city's public transportation system each day is related to both the number x (in thousands) of people living in the city and the price y (in dollars) of a ticket by the function

$$N(x, y) = \frac{0.1x}{0.5 + \ln y}$$

    a. What is the approximate change in the number of riders if the number of people living in the city increases from 163,000 to 164,000 and the price of a ticket increases from \$3.00 to \$3.10? (Remember, x is in thousands, so that x increases from 163 to 164.)
    b. Compare this approximate change to the actual change.

19. **Business: Quality Assurance Costs** The cost C (in dollars) of quality assurance at a company is related to the number of inspections x made at point A in a manufacturing process, and the number y made at point B by the function

$$C(x, y) = 6x^2 + 4y^2 - x - 3y$$

    a. What is the approximate change in the cost of quality assurance if the number of inspections at point A is increased from 14 to 16, and the number at point B is decreased from 12 to 10?
    b. Compare this approximate change to the actual change.

20. **Business: Cost of Production** A company produces two products, A and B. The cost C in dollars of producing x units of A and y units of B is

$$C(x, y) = \frac{x^3}{600} + \frac{y^3}{600} + 25x + 10y + 10xy$$

    a. What is the approximate change in the cost of producing these two products if the company increases production of product A from 45 units to 50 units, and decreases the production of product B from 30 units to 28 units?
    b. Compare this approximate change to the actual change.

21. **Business: Production Output** The output (in thousands of units each week) of a company is given by the production function $f(x, y) = 0.8xe^{0.03y}$, where x represents the number of hours allocated each week to labor and y the number of thousands of dollars allocated each week to capital. What is the approximate change in the output of the company if the number of weekly hours allocated to labor is decreased from 200 to 195, and the amount of money allocated for capital is increased from \$40,000 to \$40,500?

Using Technology Exercises

22. Physical Science: Wind Chill Wind chill is often used to measure the effect that wind has on a person's perception of temperature. Suppose a wind chill function is

$$WC(v, T) = 35.74 + 0.6215T - 35.75v^{0.16} + 0.4275Tv^{0.16}$$

The variables WC, v, and T measure the wind chill, the wind velocity in miles per hour, and the temperature in degrees Fahrenheit, respectively. Approximate the change in the wind chill if the wind velocity increases from 22 miles per hour to 24.5 miles per hour and the temperature decreases from 20.3 °F to 18.5 °F.

Source: www.srh.noaa.gov

23. Psychology: Reaction Time An industrial psychologist has determined that reaction time R, in reaction units, to x units of a particular drug t hours after it is introduced into the bloodstream is approximated by the function

$$R(x, t) = 3.2x(8.6 - 1.3x)t^{1.2}e^{-1.8t}$$

Find the approximate change in a person's reaction if the number of units of the drug is decreased from 150 units to 145 units and the time since introduction increases from 0.75 hours to 0.88 hours.

24. Manufacturing: Cobb-Douglas Production For a manufacturer, the Cobb-Douglas production function is

$$f(x, y) = 245x^{0.36}y^{0.64}$$

Approximate the change in production if the amount of money invested in labor is decreased from \$650 to \$645 and if the amount of money invested in capital is decreased from \$825 to \$820.

25. Business: Advertising The number N, in thousands of people reached by an advertiser each week through the use of x radio advertisements and y television advertisements, is approximated by the function

$$N(x, y) = 16.5x(9.3y + 18.2)^{0.45}$$

Approximate the change in the number of people reached by the advertiser if the number of radio ads is decreased by 3 from 20 and the number of TV ads is increased by 4 from 40.

Getting Ready for the Next Section

For Problems 26-33, evaluate the integral. Assume k is a constant and approximate the answer to two decimal places where appropriate.

26. $\int 12kx\, dx$

27. $\int_3^{10} (k + 4y)\, dy$

28. $\int_2^5 (7x + 182)\, dx$

29. $\int_{x^3}^{4x} (3k^2 + 6ky^2)\, dy$

30. $\int_0^2 (12x^3 + 128x^4 - 3x^5 - 2x^{10})\, dx$

31. $\int_{9.999}^{10.001} 10{,}000\, dy$

32. $\int_{0.999}^{1.001} 20\, dx$

33. $-40(e^{-9} - 1)\int_0^{12} (e^{-0.5x})\, dx$

Double Integrals as Volume

Partial differentiation, you know, is the process of differentiating a function of several variables with respect to one of its variables while treating the other variables as if they were constants. For example, if $f(x, y) = 6x^2y^3$, then $f_x(x, y) = 12xy^3$. You can integrate a function of several variables in a similar way, integrating with respect to one of the variables while treating the others as constants. The process of integrating a function of several variables with respect to only one of the variables is called **partial integration**.

Example 1 If $f_x(x, y) = 12xy^3$, find $f(x, y)$ by evaluating $\int 12xy^3\,dx$.

Solution Notice the dx in the integral. The dx indicates that we are to integrate with respect to the variable x and that we are treat the variable y as if it were a constant.

$$f_x(x, y) = 12xy^3$$

so that

$$\begin{aligned} f(x, y) &= \int 12xy^3\,dx \\ &= 12y^3 \int x\,dx \\ &= 12y^3 \cdot \frac{1}{2}x^2 + g(y) \\ &= 6x^2y^3 + g(y) \end{aligned}$$

Because y is treated as a constant, any expression involving y is considered a constant. Since the "constant" of integration could involve y rather than only a real number, we represent it using the function notation $g(y)$ rather than just the letter C. ■

From Two Dimensions to Three—The Double Integral

Your first geometric experience with the definite integral was with area under a curve. You saw that by adding the areas of many thin rectangular strips it was possible to approximate the area under a curve and then, by letting the number of rectangles tend to infinity, produce the exact area. See Figure 1.

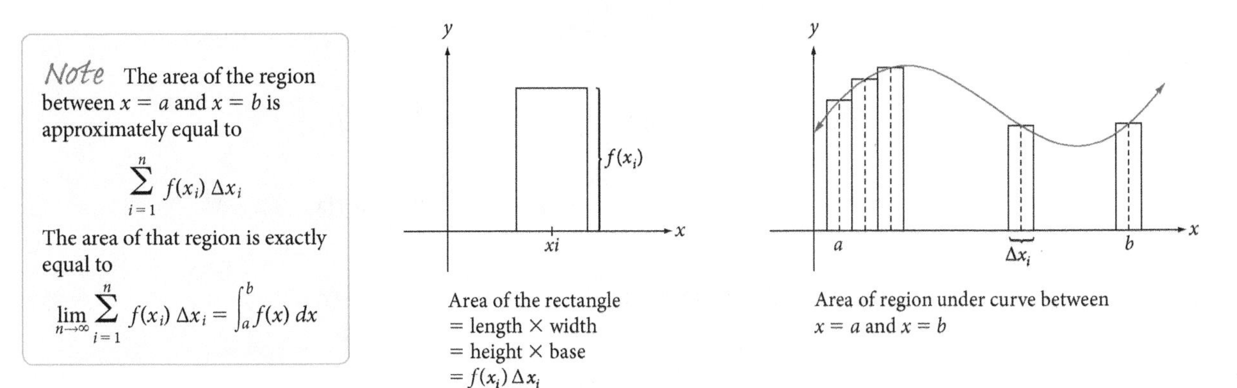

Area of the rectangle
= length × width
= height × base
= $f(x_i)\,\Delta x_i$

Area of region under curve between $x = a$ and $x = b$

Figure 1

Except for an increase in dimension, from 2-D to 3-D, the experience is nearly identical for computing the volume under a surface. Approximate the volume by adding together the volumes of many thin rectangular boxes, then produce the exact volume by letting the number of boxes tend to infinity. For a function of a single variable, we use an interval $[a, b]$. For a function of two variables, we use a bounded region R. See Figure 2.

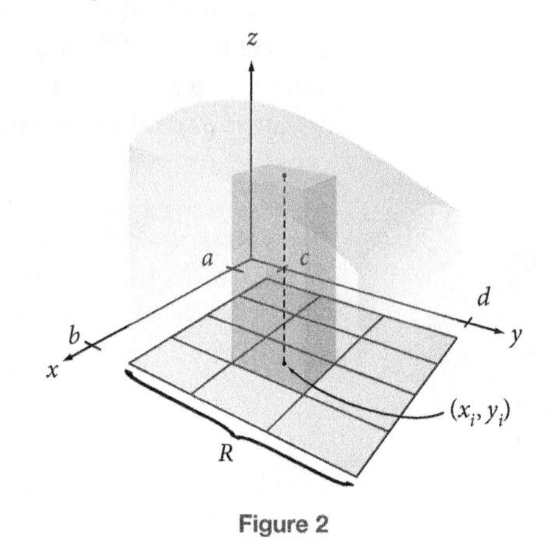

Figure 2

Figure 2 illustrates that the total volume of the space under the surface and over the region R in the xy-plane is approximately equal to the sum of the volumes of the individual boxes. This sum is

$$\sum_{i=1}^{n} f(x_i, y_i)\,\Delta y_i\,\Delta x_i$$

or equivalently,

$$\sum_{i=1}^{n} f(x_i, y_i)\,\Delta A_i$$

where ΔA_i represents the area of the rectangular bases of the boxes and the function $f(x, y)$ is positive over the region. That is, the graph of $f(x, y)$ lies above the xy-plane.

The volume of the space under the surface and over the region R in the xy-plane is exactly equal to

$$\lim_{n\to\infty}\sum_{i=1}^{n} f(x_i, y_i)\,\Delta A_i = \lim_{n\to\infty}\sum_{i=1}^{n} f(x_i, y_i)\,\Delta y_i\,\Delta x_i$$

Just as $\lim_{n\to\infty}\sum_{i=1}^{n} f(x_i)\,\Delta x_i$ is represented by $\int_a^b f(x)\,dx$,

$$\lim_{n\to\infty}\sum_{i=1}^{n} f(x_i, y_i)\,\Delta A_i = \lim_{n\to\infty}\sum_{i=1}^{n} f(x_i, y_i)\,\Delta y_i\,\Delta x_i$$

is represented by the double integral $\iint_R f(x, y)\,dydx$.

How to Calculate the Double Integral to Get Volume

Just as it is not practical to find the area under a curve by computing and summing the areas of rectangles, it is not practical to compute and sum the volumes of rectangular boxes. We use rules and techniques of integration to calculate areas. We use the same rules and techniques to evaluate double integrals; we just evaluate each integral, one after the other. Here is one way to evaluate a double integral to find volume.

Step 1 Moving parallel to the yz-plane from the backmost part of the region R to the frontmost, slice the region below the surface into many thin pieces. Figure 3 shows one such slice made at the x-value x_i.

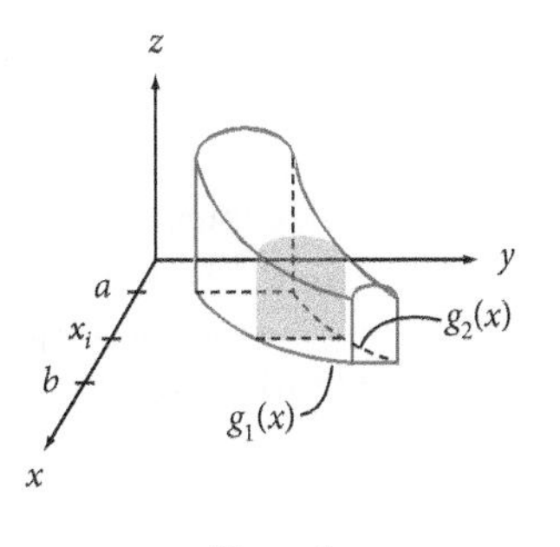

Figure 3

Step 2 Represent the area of this slice with $S(x_i)$. (The S reminds us of slice.) We know how to compute this area. It is just a single integral we have used before, but instead of the curve $f(x)$, we have the surface $f(x, y)$.

$$S(x_i) = \int_{g_1(x)}^{g_2(x)} f(x_i, y)\, dy$$

This single integral represents the area at the single x-value, x_i. The variable in the integral is y, and we are computing the area between the smaller y-value, $g_1(x)$, and the larger y-value, $g_2(x)$. Figure 4 shows a top-view of the region and the slice in the xy-plane.

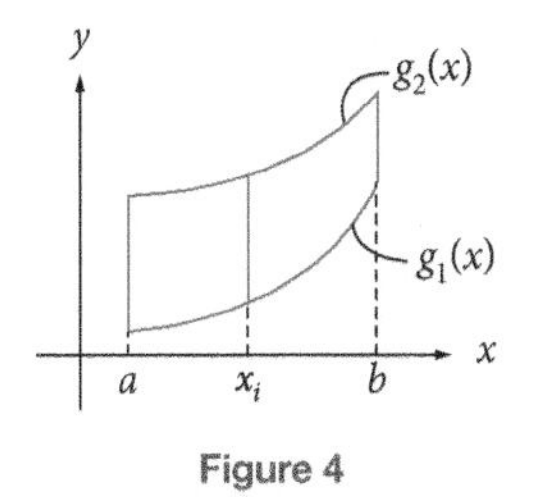

Figure 4

Step 3 The distance between two very close together consecutive slices is Δx_i. Two consecutive slices produce a slab with volume approximated by

(area of the face of the slab) $\times$ (width of the slab)

That is,

$$\begin{aligned} \text{Volume of slab} &= (\text{area of slab}) \times (\text{length of base}) \\ &= \left[\int_{g_1(x)}^{g_2(x)} f(x_i, y)\, dy\right] \Delta x_i \end{aligned}$$

Then, the approximate total volume is the sum of the volumes of the many slabs.

$$V \approx \sum_{i=1}^{n} \left[\int_{g_1(x)}^{g_2(x)} f(x_i, y)\, dy \right] \Delta x_i$$

and the exact volume is

$$V = \lim_{n \to \infty} \sum_{i=1}^{n} \left[\int_{g_1(x)}^{g_2(x)} f(x_i, y)\, dy \right] \Delta x_i$$

$$= \int_a^b \left[\int_{g_1(x)}^{g_2(x)} f(x, y)\, dy \right] dx$$

$$= \int_a^b \int_{g_1(x)}^{g_2(x)} f(x, y)\, dy\, dx$$

These integrals are called **iterated** integrals, and they are evaluated one after the other. The inner integral is evaluated first, then the outer. If we had taken slices parallel to the xz-plane rather the yz-plane, we would generate the same volume but with the iterated integrals

$$V = \int_c^d \int_{h_1(y)}^{h_2(y)} f(x, y)\, dx\, dy$$

Computing Volume Using Double Integrals

1. If $g_1(x)$ and $g_2(x)$ are continuous functions on the interval $[a, b]$ running along the x-axis and R is a region in the xy-plane bounded by the lines $x = a$ and $x = b$, and the curves $g_1(x)$ and $g_2(x)$, then

$$V = \int_a^b \int_{g_1(x)}^{g_2(x)} f(x, y)\, dy\, dx$$

(see Figure 5a below).

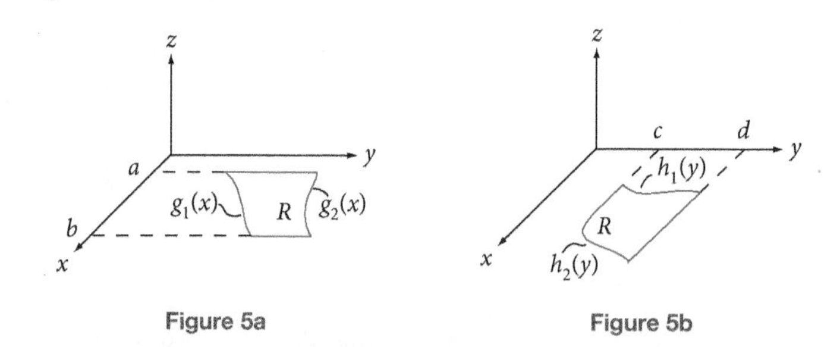

Figure 5a

Figure 5b

2. If $h_1(y)$ and $h_2(y)$ are continuous functions on the interval $[c, d]$ running along the y-axis and R is a region in the xy-plane bounded by the lines $y = c$ and $y = d$, and the curves $h_1(y)$ and $h_2(y)$, then

$$V = \int_c^d \int_{h_1(y)}^{h_2(y)} f(x, y)\, dx\, dy$$

(see Figure 5b above).

Example 2 Find the volume of the space under the surface

$$f(x, y) = x + 4y$$

and over the rectangular region bounded by $2 \leq x \leq 5$ and $3 \leq y \leq 10$. The rectangular region is pictured in Figure 6. The view is from above looking down onto the xy-plane from the z-axis.

Solution The rectangular region lies in the xy-plane and the function surface resides over it. The region in the xy-plane is illustrated in Figure 6a.

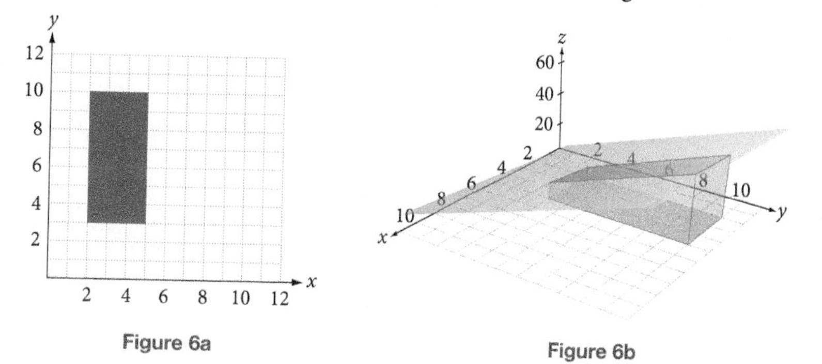

Figure 6a

Figure 6b

To compute the volume, we use

$$V = \int_a^b \int_{g_1(x)}^{g_2(x)} f(x, y)\, dy\, dx$$

where $g_1(x)$ and $g_2(x)$ are simply the lines $y = 3$ and $y = 10$.

$$V = \int_a^b \int_{g_1(x)}^{g_2(x)} f(x, y)\, dy\, dx$$

$$= \int_2^5 \int_3^{10} (x + 4y)\, dy\, dx$$

$$= \int_2^5 \left[\int_3^{10} (x + 4y) dy \right] dx \qquad \text{Integrate treating } x \text{ as a constant}$$

$$= \int_2^5 \left[xy + 2y^2 \Big|_3^{10} \right] dx$$

$$= \int_2^5 [(x \cdot 10 + 2(10)^2) - (x \cdot 3 + 2(3)^2)]\, dx$$

$$= \int_2^5 [(10x + 200) - (3x + 18)]\, dx$$

$$= \int_2^5 [7x + 182]\, dx$$

$$= \frac{7}{2}x^2 + 182x \Big|_2^5$$

$$= \left(\frac{7}{2}(5)^2 + 182(5)\right) - \left(\frac{7}{2}(2)^2 + 182(2)\right)$$

$$= \frac{1{,}239}{2} = 619.5$$

The volume of the space below the surface $f(x, y) = x + 4y$ is 619.5 cubic units. ■

Example 3 Find the volume of the space under the surface

$$f(x, y) = 3x^2 + 6xy^2$$

and over the region bounded by the curve $g_1(x) = x^3$, the line $g_2(x) = 4x$, and between $x = 0$ and $x = 2$. The region is pictured in Figure 7. The view is from above looking down onto the xy-plane from the z-axis.

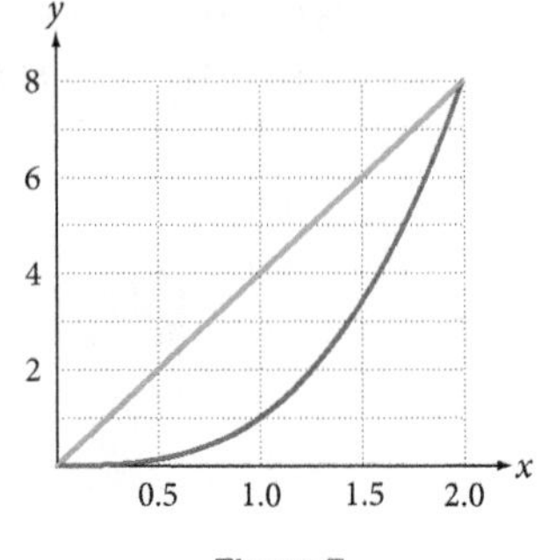

Figure 7

Solution

$$V = \int_a^b \int_{g_1(x)}^{g_2(x)} f(x, y)\, dydx$$

$$= \int_0^2 \int_{x^3}^{4x} (3x^2 + 6xy^2)\, dydx$$

$$= \int_0^2 \left[\int_{x^3}^{4x} (3x^2 + 6xy^2)\, dy \right] dx$$ Integrate treating x as a constant

$$= \int_0^2 \left[3x^2y + 2xy^3 \Big|_{x^3}^{4x} \right] dx$$

$$= \int_0^2 [(3x^2(4x) + 2x(4x)^3) - (3x^2(x^3) + 2x(x^3)^3)]\, dx$$

$$= \int_0^2 [(12x^3 + 2x(64x^3)) - (3x^5 + 2x(x^9))]\, dx$$

$$= \int_0^2 [(12x^3 + 128x^4) - (3x^5 + 2x^{10})]\, dx$$

$$= \int_0^2 [(12x^3 + 128x^4 - 3x^5 - 2x^{10})\, dx$$

$$= \left[3x^4 + \frac{128}{5}x^5 - \frac{3}{6}x^6 - \frac{2}{11}x^{11} \right] \Big|_0^2 dx$$

$$= \left(3(2)^4 + \frac{128}{5}(2)^5 - \frac{3}{6}(2)^6 - \frac{2}{11}(2)^{11} \right) - \left(3(0)^4 + \frac{128}{5}(0)^5 - \frac{3}{6}(0)^6 - \frac{2}{11}(0)^{11} \right)$$

$$= \frac{25{,}456}{55} \approx 462.8$$

The volume of the space below the surface $f(x, y) = 3x^2 + 6xy^2$ is about 462.8 cubic units.

The next example shows that the *order of integration* does not matter when finding the volume of the space below a surface and over a region R. That is, Example 4 illustrates how

$$V = \int_a^b \int_{g_1(x)}^{g_2(x)} f(x, y)\, dy\, dx = \int_c^d \int_{h_1(y)}^{h_2(y)} f(x, y)\, dx\, dy$$

Example 4 In Example 3 we used the double integral

$$\int_0^2 \int_{x^3}^{4x} (3x^2 + 6xy^2)\, dydx$$

to determine the volume of the space below the surface

$$f(x, y) = 3x^2 + 6xy^2$$

Show that the same volume can be found by changing the order of integration to *dxdy*.

Note If you are wondering how we got the new limits when we changed the order of integration, look back at Figure 7. Can you see that going in the x-direction first, we go from the curve $\frac{y}{4}$ to $y^{1/3}$ and then going in the y-direction, we go from 0 to 8?

Solution Note first that $y = x^3$ means that $x = y^{1/3}$ and that $y = 4x$ means that $x = \frac{y}{4}$. Then

$$V = \int_c^d \int_{h_1(y)}^{h_2(y)} f(x, y)\, dxdy$$

$$= \int_0^8 \int_{\frac{y}{4}}^{y^{1/3}} (3x^2 + 6xy^2)\, dxdy$$

$$= \int_0^8 \left[\int_{\frac{y}{4}}^{y^{1/3}} (3x^2 + 6xy^2)\, dx \right] dy$$

$$= \int_0^8 \left[y + 3y^{8/3} - \frac{y^3}{64} - \frac{3y^4}{16} \right] dy$$

$$= \left[\frac{y^2}{2} + \frac{9y^{11/3}}{11} - \frac{y^4}{256} - \frac{3y^5}{80} \right] \Bigg|_0^8$$

$$= \frac{25{,}456}{55} \approx 462.8$$

■

Example 5 A manufacturer produces laser-cut glass rods that are to have a radius of 1 inch and are to be 10 inches long. Because of some variability in the production process, the glass rods have radii that are uniformly distributed between 0.995 and 1.005 inches and lengths that are uniformly distributed between 9.995 and 10.005 inches. The cuts for the radius and length are independent of each other. What is the probability that a randomly selected glass rod will have a radius between 0.999 inches and 1.001 inches and a length between 9.999 inches and 10.001 inches?

Solution Since the measures of the radius and length of a rod are uniformly distributed we can use the uniform probability density model for each one. That is, if X represents the measure of a radius, and Y represents the measure of a length, then

$$P(X) = \frac{1}{b - a} = \frac{1}{1.005 - 0.995} = \frac{1}{0.01} = 100 \quad \text{and}$$

$$P(Y) = \frac{1}{b-a} = \frac{1}{10.005 - 9.995} = \frac{1}{0.01} = 100$$

We are asked to find $P(0.999 \leq X \leq 1.001 \text{ and } 9.999 \leq Y \leq 10.001)$. Because X and Y are independent, the probability function we are looking for, the joint probability density function, is the product of the two individual density functions.

$$P(X \text{ and } Y) = P(X) \cdot P(Y) = 100 \cdot 100 = 10{,}000$$

Then,

$$P(0.999 \leq X \leq 1.001 \text{ and } 9.999 \leq Y \leq 10.001) =$$

$$= \int_{0.999}^{1.001} \int_{9.999}^{10.001} 10{,}000 \, dydx$$

$$= \int_{0.999}^{1.001} \left[\int_{9.999}^{10.001} 10{,}000 \, dy \right] dx$$

$$= \int_{0.999}^{1.001} \left[[10{,}000y] \Big|_{9.999}^{10.001} \right] dx$$

$$= \int_{0.999}^{1.001} [10{,}000(10.001) - 10{,}000(9.999)] \, dx$$

$$= \int_{0.999}^{1.001} 20 \, dx$$

$$= 20x \Big|_{0.999}^{1.001}$$

$$= 0.04$$

Interpretation

The probability that a randomly selected glass rod will have a radius between 0.999 inches and 1.001 inches and a length between 9.999 inches and 10.001 inches is 0.04. That is, the manufacturer can expect about 4% of the glass rods that are produced will have both a radius and length that differ by less than 0.001 inches from their expected values.

■

Population Density and Population Size

If the function $P(x, y)$ represents the population density, that is, the number of individuals in a population per square unit, then

$$P(x, y)dA = P(x, y) \, dydx$$

represents the number of individuals in the population. For example, if there are

$$20 \text{ million molds per square inch} = \frac{20{,}000{,}000 \text{ molds}}{1 \text{ square inch}}$$

then

$$\frac{20{,}000{,}000 \text{ molds}}{1 \text{ square inch}} \times (1 \text{ square inch}) = 20{,}000{,}000 \text{ molds}$$

in that particular part of the region.

This suggests that given a population density we can find the total population by multiplying by the total area. If the population density is variable over some region, we can find the total population by integration.

> To find the number of individuals in a population scattered over a region, R, compute the double integral
>
> $$\iint_R P(x, y)\, dA = \int_a^b \int_{g_1(x)}^{g_2(x)} P(x)\, dydx$$

Example 6 Molds are microscopic organisms found both outside the home and inside the home. Outside, they help to break down plant and animal matter. Inside, however, they can cause health problems such as skin rash, eye irritation, or chronic cough, and they can put people with respiratory disease at risk for lung infection. Imagine a wall covered with a population of mold. Take the lower left corner of the wall as the origin of a Cartesian coordinate system with the x-axis running along the bottom of the wall and the y-axis running along the left side of the wall. See Figures 8 and 9.

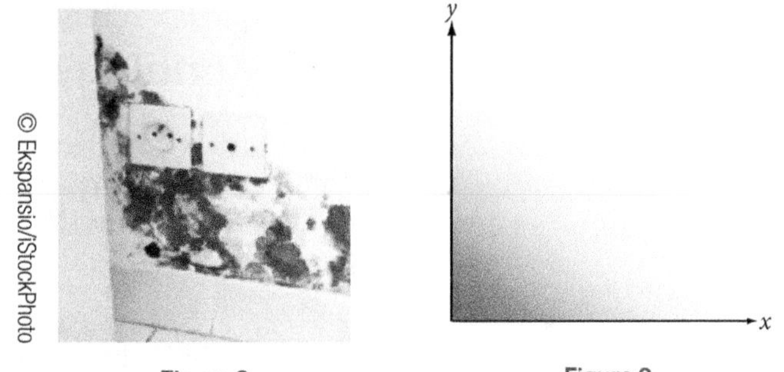

Figure 8 **Figure 9**

Calculate the mold population on a 12-foot-long by 9-foot-high wall if the population density function for the mold is

$$P(x, y) = 40e^{-0.5x-y},$$

where x and y are in feet and P is in millions.

Source: U. S. Environmental Protection Agency, www.epa.gov

Solution Evaluate the double integral to get the size of the population.

$$\iint_R P(x, y)\, dA = \int_0^{12}\int_0^9 40\, e^{-0.5x-y}\, dydx$$

From the law of exponents, $e^{a+b} = e^a e^b$

$$= 40\int_0^{12}\int_0^9 e^{-0.5x}e^{-y}\, dydx$$

$$= 40\int_0^{12}\left[\int_0^9 e^{-0.5x}e^{-y}dy\right] dx$$

Remember, $\int_a^b e^{ax}\, dx = \left.\frac{e^{ax}}{a}\right|_a^b$

$$= 40\int_0^{12}\left[\left.\frac{e^{-0.5x}e^{-y}}{-1}\right|_0^9\right] dx$$

$$= -40\int_0^{12} [(e^{-0.5x}e^{-9}) - (e^{-0.5x}e^{-0})]\, dx$$

$$= -40\int_0^{12} [e^{-0.5x}e^{-9} - e^{-0.5x}]\, dx$$

$$= -40\int_0^{12} [e^{-0.5x}(e^{-9} - 1)]\, dx$$

$$= -40(e^{-9} - 1)\int_0^{12} [e^{-0.5x}]\, dx$$

$$= -40(e^{-9} - 1)\left[\frac{e^{-0.5x}}{-0.5}\Bigg|_0^{12}\right]$$

$$= 80(e^{-9} - 1)[e^{-0.5(12)} - e^{-0.5(0)}]$$

$$= 80(e^{-9} - 1)[e^{-6} - e^{0}]$$

$$\approx 79.79$$

Interpretation
The mold population on the wall is about 80 million.

Using Technology 7.6

Wolfram|Alpha can help you evaluate integrals that involve more than one variable as well as find the volume of the space below a surface that exists over some region in the xy-plane.

Example 7 Use Wolfram|Alpha to evaluate the integral $\int 12xy^3\, dy$.

Solution In the Wolfram|Alpha entry field enter

integrate 12xy^3 dy

Wolfram|Alpha responds with

Indefinite integral:

$$\int 12\, x\, y^3\, dy = 3\, x\, y^4 + \text{constant}$$

Wolfram Alpha LLC. 2012. Wolfram|Alpha
http://www.wolframalpha.com/
(access December 27, 2012)

Recall that since we are integrating with respect to y, the constant may involve an expression in x. To indicate this possibility, we denote the constant with $g(x)$. We conclude that

$$\int 12xy^3\, dy = 3xy^4 + g(x)$$

Example 8 Use Wolfram|Alpha to find the volume of the space that exists below the surface $f(x, y) = 3x^2 + 6xy^2$ and over the region bounded by the curve $g_1(x) = x^3$, the line $g_2(x) = 4x$, and between $x = 0$ and $x = 2$. This is the function and region we used in Example 3.

Solution In the Wolfram|Alpha entry field enter

integrate 3x^2 + 6xy^2 dydx from y = x^3 to 4x and x = 0 to 2

Wolfram|Alpha responds with

Definite integral:

$$\int_0^2 \int_{x^3}^{4x} (3x^2 + 6xy^2)\, dy\, dx = \frac{25456}{55} \approx 462.836$$

Wolfram Alpha LLC. 2012. Wolfram|Alpha
http://www.wolframalpha.com/
(access December 27, 2012)

The volume of the space is about 462.8 cubic units.

Example 9 A manufacturer produces laser-cut glass rods and advertises that they have a mean radius of 4 cm and a mean length of 10 cm. Because of some variability in the production process, the radii are normally distributed with a mean of 4 cm and standard deviation of 0.01 cm. The lengths of the rods are also normally distributed with a mean of 10 cm and standard deviation of 0.02 cm. What is the probability that a randomly selected glass rod will have a radius or length that varies from the mean by less than 0.03 cm?

Solution Since the measures of the radius and length of a rod are normally distributed, we can use the normal probability density model for each one. That is, if X represents the measure of a radius, and Y represents the measure of a length, then

$$P(X) = \frac{1}{\sigma\sqrt{2\pi}} \cdot e^{-\frac{(x-\mu)^2}{2\sigma^2}} = \frac{1}{0.01\sqrt{2\pi}} \cdot e^{-\frac{(x-4)^2}{2(0.01)^2}} \quad \text{and}$$

$$P(Y) = \frac{1}{0.02\sqrt{2\pi}} \cdot e^{-\frac{(y-10)^2}{2(0.02)^2}}$$

We are asked to find

$$P(3.97 \le X \le 4.03 \text{ and } 9.97 \le Y \le 10.03)$$

Because X and Y are independent, the probability function we are looking for, the joint probability density function, is the product of the two individual density functions.

$$P(X \text{ and } Y) =$$

$$P(X) \cdot P(Y) = \frac{1}{0.01\sqrt{2\pi}} e^{-\frac{(x-4)^2}{2(0.01)^2}} \cdot \frac{1}{0.02\sqrt{2\pi}} e^{-\frac{(y-10)^2}{2(0.02)^2}}$$

Then,

$$P(3.97 \le X \le 4.03 \text{ and } 9.97 \le Y \le 10.03) =$$

$$\int_{3.97}^{4.03} \int_{9.97}^{10.03} \left[\frac{1}{0.01\sqrt{2\pi}} e^{-\frac{(x-4)^2}{2(0.01)^2}} \cdot \frac{1}{0.02\sqrt{2\pi}} e^{-\frac{(y-10)^2}{2(0.02)^2}} \right] dydx$$

In the Wolfram|Alpha entry field enter

integrate 1/(0.01*sqrt(2*pi))e^(-(x - 4)^2/(2(0.01)^2))*1/(0.02*sqrt(2*pi)) e^(-(y - 10)^2/(2(0.02)^2)) dydx from y = 9.97 to 10.03 and x = 3.97 to 4.03

Wolfram|Alpha responds with

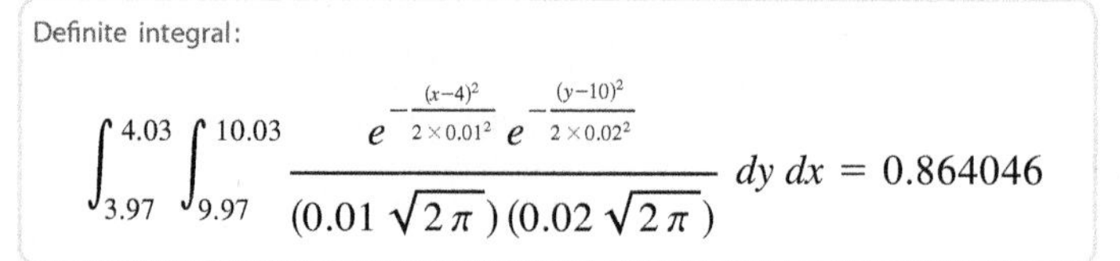

Wolfram Alpha LLC. 2012. Wolfram|Alpha
http://www.wolframalpha.com/
(access December 27, 2012)

Interpretation

The probability that a randomly chosen rod will have a radius or length that differs from the advertised mean by less than 0.03 cm is 0.8640. Put another way, about 86.40% of the glass rods produced by this manufacturer will differ from the advertised mean by less than 0.03 cm.

Getting Ready for Class

After reading through the preceding section, respond in your own words and in complete sentences.

A. Describe in geometric terms, the meaning of the expression $f(x, y)\, dxdy$.

B. Describe, in geometric terms, why
$$\int_a^b \int_c^d f(x, y)\, dydx = \int_c^d \int_a^b f(x, y)\, dxdy$$

C. What is the name given to integrals such as
$$\int_a^b \int_c^d f(x, y)\, dydx$$
and how are they evaluated?

D. When evaluating an integral in which the integrand (the expression to be integrated) involves two variables x and y, a term such as $g(y)$ or $h(x)$ is produced. Describe the meaning of this type of term.

Problem Set 7.6

Skills Practice

For Problems 1-8, find the volume of the space between the surface $f(x, y)$ and the rectangular region R in the xy-plane.

1. $f(x, y) = 7xy$, where $0 \le y \le 1$ and $0 \le x \le 2$.

2. $f(x, y) = x + y$, where $0 \le y \le 2$ and $0 \le x \le 3$.

3. $f(x, y) = x - y$, where $0 \le y \le 2$ and $0 \le x \le 3$.

4. $f(x, y) = y - x$, where $0 \le x \le 1$ and $0 \le y \le 2$.

5. $f(x, y) = x + 3y$, where $0 \le x \le 2$ and $0 \le y \le 1$.

6. $f(x, y) = x + y$, where $0 \le y \le 2$ and $1 \le x \le 3$.

7. $f(x, y) = 6x^2 + y^2$, where $0 \le x \le 4$ and $0 \le y \le 3$.

8. $f(x, y) = xe^{xy}$, where $0 \le y \le 1$ and $0 \le x \le 2$.

For Problems 9-17, find the volume of the space between the surface $f(x, y)$ and the region R in the xy-plane. Note that the region R may not always be rectangular.

9. $f(x, y) = xy$, where $0 \le y \le 4$ and $0 \le x \le 1$.

10. $f(x, y) = x^2$, where $0 \le y \le 4$ and $0 \le x \le 2$.

11. $f(x, y) = 4ye^y$, where $x \le y \le 3$ and $0 \le x \le 2$.

12. $f(x, y) = 4ye^x$, where $0 \le x \le y^2$ and $0 \le y \le 1$.

13. $f(x, y) = \dfrac{4x}{y}$, where $0 \le x \le \sqrt{y}$ and $0 \le y \le 5$.

14. $f(x, y) = \dfrac{12x^3}{y^2}$, where $0 \le x \le \sqrt{y}$ and $6 \le y \le 7$.

15. $f(x, y) = e^{x+y}$, where $0 \le x \le 1$ and $0 \le y \le 1$.

16. $f(x, y) = 100{,}000$, where $0.195 \le x \le 0.205$ and $5.05 \le y \le 5.15$.

17. $f(x, y) = \dfrac{20{,}000e^y}{1 + x}$, where $0 \le y \le 4$ and $0 \le x \le 20$.

Check Your Understanding

For Problems 18-21, evaluate the integral.

18. $\int 24x^2y^3\,dx$

19. $\int (12xy^3 + 8xy)\,dx$

20. $\int (x^2y^4 - xy^5)\,dy$

21. $\int 4xe^x\,dy$

For Problems 22-30, the limits on each integral are constant. Evaluate each double integral.

22. $\int_0^1 \int_1^2 (1 + x^2)\, dydx$

23. $\int_1^2 \int_0^1 xy\, dydx$

24. $\int_1^3 \int_0^2 dydx$

25. $\int_0^2 \int_1^3 dxdy$

26. $\int_0^2 \int_0^1 xe^{xy}\, dydx$

27. $\int_1^2 \int_0^1 (y - x)\, dxdy$

28. $\int_0^1 \int_0^1 2xye^{-x^2}\, dxdy$

29. $\int_1^2 \int_2^3 \frac{2y}{x}\, dydx$

30. $\int_3^4 \int_1^e \frac{x}{3y}\, dydx$

For Problems 31-40, the limits on the inner integral are variable and the limits on the outer integral are constant. Evaluate each double integral.

31. $\int_1^2 \int_0^y (x + 3y)\, dxdy$

32. $\int_0^2 \int_0^{3x} xy^2\, dydx$

33. $\int_0^1 \int_0^x e^{x^2}\, dydx$

34. $\int_0^2 \int_y^{y^2} (6x^2 + y^2)\, dxdy$

35. $\int_0^1 \int_0^{x/2} e^{2y-x}\, dydx$

36. $\int_0^2 \int_x^{3x} xy\, dydx$

37. $\int_0^3 \int_0^{x+2} \frac{y}{x+2}\, dydx$

38. $\int_0^1 \int_0^x \sqrt{1 - x^2}\, dydx$

39. $\int_0^1 \int_0^{\sqrt{1-x^2}} (2x + y)\, dydx$

40. $\int_0^2 \int_{-y}^0 e^{x-y}\, dxdy$

Modeling Practice

41. Manufacturing: Probability and Glass Rod Dimensions A manufacturer produces laser-cut glass rods that are to have a radius of 2 inches and are to be 1 inch long. Because of some variability in the production process, the glass rods have radii that are uniformly distributed between 1.9 and 2.1 inches and lengths that are uniformly distributed between 0.9 and 1.1 inches. The cuts for the radius and length are independent of each other. What is the probability that a randomly selected glass rod will have a radius between 1.95 inches and 2.05 inches and a length between 0.95 inches and 1.05 inches?

42. Health Science: Mold Population Size Calculate the mold population on a 10" high and 10" long surface if the probability density function for the mold is given by $P(x, y) = 600e^{-0.1x-0.3y}$, where x and y are in inches and P is in millions.

© Jesus Ayala/iStockPhoto

43. **Manufacturing: Probability and Quartz Crystal Diameters** A manufacturer of quartz crystals produces crystals that are to be 10 mm in diameter and 0.200 mm in thickness. Because of some variability in the production process, the crystals have diameters that are uniformly distributed between 9 mm and 11 mm and thicknesses that are uniformly distributed between 0.190 and 0.210. The cuts for the diameter and thickness are independent of each other. What is the probability that a randomly selected quartz crystal produced by this company will have a diameter between 9.5 and 10.5 mm and a thickness between 0.195 and 0.205 mm?

Source: Structure Probe, Inc

44. **Health Science: Diabetes Test** Diabetes glucose test strips are rectangular strips of plastic coated with an enzyme reagent such as glucose oxidase or glucose dehydrogenase. When a drop of blood is placed on the strip in contact with the reagent, the enzyme chemically reacts with the glucose in the blood allowing one to determine the level of glucose in the blood. Suppose a manufacturer of glucose test strips produces them in such a way that the layer of enzyme coating the strip is applied to a 20-mm-long and 10-mm-wide end section of the strip according to the function $f(x, y) = 0.001e^{-x}$, where $f(x, y)$ too is measured in millimeters. See the figure below. Determine the volume of enzyme applied to each test strip. Round your result to six decimal places.

Source: livestrong.com

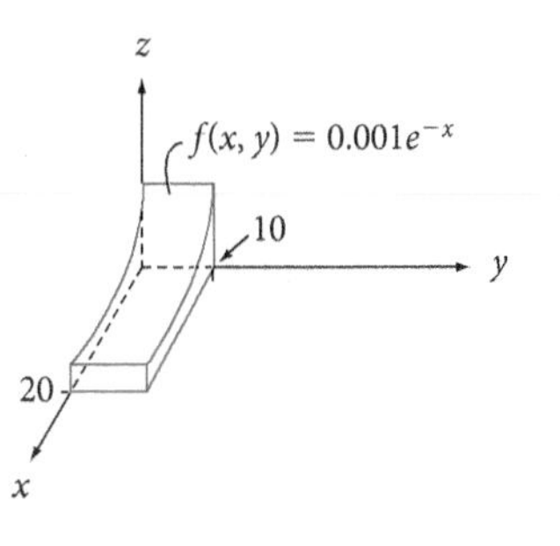

Using Technology Exercises

45. Use Wolfram|Alpha to help you evaluate $\int (3x^2 + 6xy^2)\, dy$.

46. Evaluate both $\int_0^3 \int_0^2 (3x^2 + 6xy^2)\, dxdy$ and $\int_0^2 \int_0^3 (3x^2 + 6xy^2)\, dydx$ and compare the results. Are the results the same or different? What would you expect them to be?

47. Find the volume of the space bounded above by the surface

$$f(x, y) = 900x^{0.85}y^{0.15}$$

and below by the region defined by $0 \le y \le 3$ and $0 \le x \le 2$.

48. Manufacturing: Scalpel Blade Length A manufacturer produces surgical scalpel blades and advertises that they have mean length of 1.27 mm and a mean feather (blade cut angle) of 15°. Because of some variability in the production process, the lengths are normally distributed with a mean of 1.27 mm and standard deviation of 0.04 mm. The feathers are also normally distributed with a mean of 15° and standard deviation of 1°. What is the probability that a randomly selected scalpel blade will have a length that differs by less than 0.07 mm from the mean and a feather that differs by less than 2° from the mean?

Source: Electron Microscopy Sciences and Dimensions Guide

49. Health Science: Mold Population Size Calculate the mold population on a 5-foot-long by 3-foot-wide region of a wooden floor if the population density function for the mold is

$$P(x, y) = 300e^{-0.2x - 0.9y}$$

where x and y are in feet and P is in millions.

50. Health Science: Mold Population Size Calculate the mold population on a circular dish of radius 10 cm if the population density function for the mold is

$$P(x, y) = 30e^{-0.2x - 0.2y}$$

where x and y are in cm and measured from the center of the circular plate, and P is in thousands. The distance from the center of the circular plate to the edge of the plate is $y = \sqrt{10^2 - x^2}$ cm.

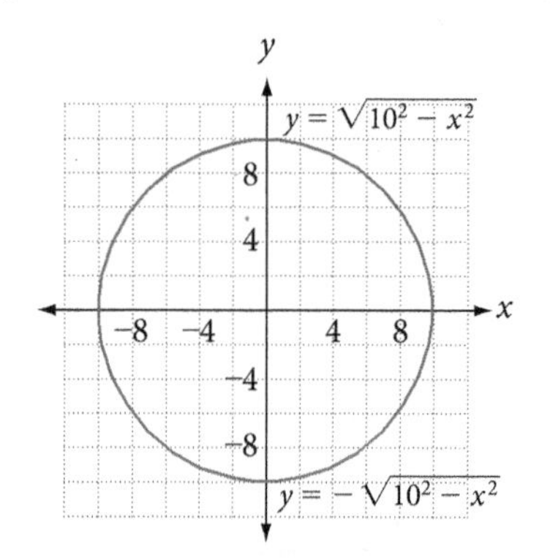

Getting Ready for the Next Section

51. Find the average of the five numbers 2, 4, 6, 8, 10

52. Find the average of the six numbers 12, 15, 16, 20, 22, 30

53. Suppose there are five rectangles each having a base 1 unit in length and heights given by the function $f(x) = 7x$, where $x = 2, 4, 6, 8$, and 10. Find the average heights of the five rectangles.

54. Suppose there are five rectangles each having a base 1 unit in length and heights given by the function $f(x) = x^2$, where $x = 2, 4, 6, 8$, and 10. Find the average area of the five rectangles.

For Problems 55-58, evaluate.

55. $\frac{1}{6-3}\int_3^6 3x^2\,dx$

56. $\frac{1}{2-0}\int_0^2 5x^4\,dx$

57. $\frac{1}{7-3}\int_3^7 1{,}000e^{-0.50x}\,dx$

58. $\frac{1}{5-2}\int_2^5 900e^{-0.50x}\,dx$

59. Find the value of $\frac{f(0)+f(1)+f(2)+f(3)}{4}$ for $f(x) = x + 5$

60. Find the value of $\frac{f(0)+f(1)+f(2)+f(3)}{4}$ for $f(x) = x^2$

Spotlight on **Success**

Instructor Edwin

You never fail until you stop trying.
—Albert Einstein

Coming to the United States at the age of 10 and not knowing how to speak English was a very difficult hurdle to overcome. However, with hard work and dedication I was able to rise above those obstacles. When I came to the U.S. our school did not have a strong English development program as it was known at that time, English as a Second Language (ESL). The approach back then was "sink or swim." When my self-esteem was low, my mom and my three older sisters were always there for me and they would always encourage me to do well. My mom was a single parent, and her number one priority was that we would receive a good education. My mother's perseverance is what has made me the person I am today. At a young age I was able to see that she had overcome more than what my situation was, and I would always tell myself, "if Mom can do it, I could also do it." Not only did she not have an education, but she also saved us from a civil war that was happening in my home country of El Salvador.

When things in school got hard, I would always reflect on all the hard work, sacrifice and effort of mother. I would just tell myself that I should not have any excuses and that I needed to keep going. If my mother, who worked as a housekeeper, could send all four of her kids to college doesn't motivate you, I don't know what does. It definitely motivated me. The day everything began to change for me was when I was in eighth grade. I was sitting in my biology class not paying attention to the teacher because I was really focusing on a piece of paper on the wall. It said, "You never fail until you stop trying." I read it over and over, trying to digest what the quote meant. With my limited English I was doing my best to translate what it meant in my native language. It finally clicked! I was able to figure out what those seven words meant. I memorized the quote and began to apply it to my academics and to real-life situations. I began to really focus in my studies. I wanted to do well in school, and most important I wanted to improve my English. To this day I always reflect to that quote when I feel I can't do something.

I was able to finish junior high successfully. Going to high school was a lot easier and I ended up with very good grades and eventually I was accepted to an excellent college. I was never the smartest student on campus, but I always did well because I never quit. I earned my college degree and now I teach at a dual immersion elementary school. I have that same quote in my classroom and I constantly remind my students to never stop trying.

The Average Value of a Function

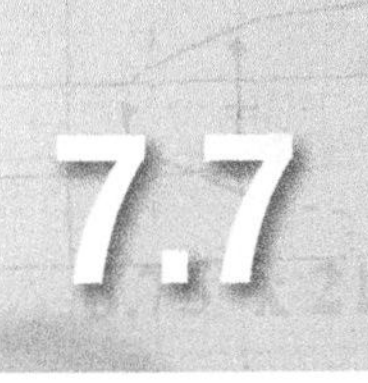

The average, as we normally use it, is a measure that describes the middle of a set of data values. In this section we extend the concept of average to functions of one and two variables. We develop the average value geometrically and then examine a problem in which the average value of the function offers useful information.

Average Value of a Function of One Variable

Suppose that the function displayed in Figure 1 represents a person's wealth, $f(t)$, at time t (in years from 2008). We could use the definite integral

$$\int_a^b f(t)\, dt$$

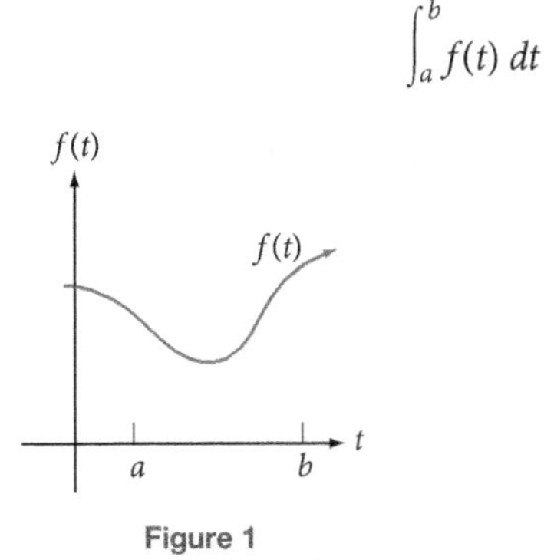

Figure 1

© Alexander Novikov/iStockPhoto

as an accumulator to determine the wealth this person accumulated between years a and b.

We could also ask, what is this person's *average* wealth over these $(b - a)$ years?

You probably know that the average of a discrete collection of numbers is the sum (accumulated value) of those numbers divided by the number of numbers. That is, to determine the average value of the numbers $y_1, y_2, y_3, \ldots, y_n$, you would compute

$$\text{Average} = \frac{y_1 + y_2 + y_3 + \cdots + y_n}{n}$$

To determine the average value of a continuous function on some interval $[a, b]$, we make a similar computation. Choose n evenly spaced values of x in $[a, b]$, say, $x_1, x_2, x_3, \ldots, x_n$, and compute their corresponding output values $f(x)$, say, $f(x_1)$, $f(x_2), f(x_3), \ldots, f(x_n)$. The average value of these output values is then the sum (accumulated value) of the output values divided by the number of output values. That is,

$$\text{Average} = \frac{f(x_1) + f(x_2) + f(x_3) + \cdots + f(x_n)}{n}$$

This will serve as an approximation to our expectation of the actual average value of the function. Also, as n, the number of selected x-values, gets larger and larger, the approximation gets better and better.

In both cases, the average is computed by considering an accumulated value. This makes us think of the definite integral. In fact, we can use the definite integral of a function to produce the definition of the average value of a function in the following way.

Subdivide the interval $[a, b]$ into n subintervals of length

$$\Delta x = \frac{b-a}{n}$$

Choose an x in each subinterval, say, x_1 in the first, x_2 in the second, ... , and x_n, in the nth. The x's may be the left endpoints of the subintervals or any point in the interval. The average associated with these x-values is

$$\text{Average value} = \frac{f(x_1) + f(x_2) + f(x_3) + \cdots + f(x_n)}{n}$$

$$= f(x_1) \cdot \frac{1}{n} + f(x_2) \cdot \frac{1}{n} + f(x_3) \cdot \frac{1}{n} + \cdots + f(x_n) \cdot \frac{1}{n}$$

where $f(x_1), f(x_2), f(x_3), \ldots$, and $f(x_n)$ represent the heights of rectangles. We can get the widths, Δx, of the rectangles into this expression using the fact that Δx and n are related by

$$\Delta x = \frac{b-a}{n}$$

$$\Delta x = (b-a) \cdot \frac{1}{n}$$

so that

$$\frac{1}{n} = \frac{\Delta x}{b-a}$$

Then, the average value of the function $f(x)$

$$= f(x_1) \cdot \frac{1}{n} + f(x_2) \cdot \frac{1}{n} + f(x_3) \cdot \frac{1}{n} + \cdots + f(x_n) \cdot \frac{1}{n}$$

$$= f(x_1) \cdot \frac{\Delta x}{b-a} + f(x_2) \cdot \frac{\Delta x}{b-a} + f(x_3) \cdot \frac{\Delta x}{b-a} + \cdots + f(x_n) \cdot \frac{\Delta x}{b-a}$$

$$= f(x_1)\Delta x \cdot \frac{1}{b-a} + f(x_2)\Delta x \cdot \frac{1}{b-a} + f(x_3)\Delta x \cdot \frac{1}{b-a} + \cdots + f(x_n)\Delta x \cdot \frac{1}{b-a}$$

$$= \frac{1}{b-a}[f(x_1)\Delta x + f(x_2)\Delta x + f(x_3)\Delta x + \cdots + f(x_n)\Delta x]$$

The expression $[f(x_1)\Delta x + f(x_2)\Delta x + f(x_3)\Delta x + \cdots + f(x_n)\Delta x]$ is a Riemann sum and approaches

$$\int_a^b f(x)\, dx \text{ as } n \to \infty$$

Considering the infinitely many points in the interval $[a, b]$ helps us make the following definition of the average value of a function.

Average Value of a Function of One Variable

If $f(x)$ is a continuous function on $[a, b]$, then the **average value** of $f(x)$ on $[a, b]$ is

$$\text{Average value} = \frac{1}{b-a}\int_a^b f(x)\, dx$$

Geometrically, $f(x)$ represents the height of a rectangle from a point x in the interval $[a, b]$ to the curve. The average value of $f(x)$ is then the average of all such heights and represents the average height of the graph over $[a, b]$.

VIDEO EXAMPLES

SECTION 7.7

Example 1 The function $S(x) = 20{,}380e^{-0.25x}$ represents the number of people per week surging to shopping malls to buy a new product x weeks after it is made available to the market.

a. Approximate the total number of people surging to buy the new product between weeks 3 and 7.

b. Approximate the average number of people surging to buy the new product each week for weeks 3 through 7.

Solution

a. To find the total, we can use the definite integral as an accumulator.

$$\begin{aligned}
\text{Total} &= \int_3^7 S(x)\,dx \\
&= \int_3^7 20{,}380e^{-0.25x}\,dx \\
&= \frac{20{,}380}{-0.25}e^{-0.25x}\Big|_3^7 \\
&= -81{,}520e^{-0.25x}\Big|_3^7 \\
&= [-81{,}520e^{-0.25(7)}] - [-81{,}520e^{-0.25(3)}] \\
&\approx 24{,}341
\end{aligned}$$

Interpretation
Between weeks 3 and 7, 24,341 people can be expected to surge into the shopping malls to buy the new product.

b. The approximate average number of people surging into the shopping malls each week between weeks 3 and 7 is

$$\begin{aligned}
\text{Average value} &= \frac{1}{7-3}\int_3^7 20{,}380e^{-0.25x}\,dx \\
&= \frac{1}{4}(24{,}341) \\
&\approx 6{,}085
\end{aligned}$$

Interpretation
On the average, between weeks 3 and 7 after the new product is introduced on the market, approximately 6,085 people per week can be expected to surge to the malls to buy the new product.

Average Value of a Function of Two Variables

Thus far, we have defined the average value of a function of one variable over an interval $[a, b]$ to be the definite integral

$$\frac{1}{b-a}\int_a^b f(x)\,dx$$

and have interpreted it as the average height of the graph over $[a, b]$. The formula represents the sum of all the heights over $[a, b]$ divided by the length of the interval $[a, b]$.

It is natural to want to extend this idea to functions of two variables. We wish to know the average height of the surface $f(x, y)$ over the rectangular region R. Geometrically, $f(x, y)$ represents the height of a rectangular box from a point (x, y) in a rectangular region R to a surface over R. Extending the definition of average value of functions of one variable to functions of two variables, we present a formula that sums the heights of all such rectangular boxes over R and divides it by the *area* of the region R.

Average Value of a Function of Two Variables

The average value of the function $f(x, y)$ over a rectangular region R is given by

$$\text{Average value} = \frac{1}{(b-a)(d-c)}\int_c^d\int_a^b f(x, y)\,dx\,dy$$

where $a \le x \le b$ and $c \le y \le d$, and $(b-a)(d-c)$ represents the area of the rectangular region R.

Example 2 The weekly revenue realized by a company is approximated by the revenue function $R(x, y) = 350x + 600y - 4x^2 - 3y^2$ for the sale of x units per week of product A and y units per week of product B. Over the year the company produces between 20 and 30 units of product A each week and between 80 and 110 units of product B each week. Approximate this company's average weekly revenue from products A and B over the year.

Solution To determine the average weekly revenue, we need to find the average value of a function of two variables.

$$\text{Average value} = \frac{1}{(b-a)(d-c)}\int_c^d\int_a^b f(x, y)\,dx\,dy$$

$$= \frac{1}{(30-20)(110-80)}\int_{80}^{110}\int_{20}^{30}[350x + 600y - 4x^2 - 3y^2]dx\,dy$$

$$= \frac{1}{(10)(30)}\int_{80}^{110}\left[175x^2 + 600xy - \frac{4x^3}{3} - 3xy^2\right]\Big|_{20}^{30} dy$$

$$= \frac{1}{300}\int_{80}^{110}\left[6{,}000y + \frac{186{,}500}{3} - 30y^2\right]dy$$

$$= \frac{1}{300}\left[3{,}000y^2 + \frac{186{,}500}{3}y - 10y^3\right]\Big|_{80}^{110} dy$$

$$= \frac{1}{300} \cdot 10{,}775{,}000$$

$$= \frac{107{,}750}{3}$$

$$\approx 35{,}916.67$$

Interpretation

On the average, this company realizes a revenue of approximately \$35,916.67 each week on the sale of x units of product A and y units of product B. Geometrically, 35,916.67 is the average height of the surface

$$R(x, y) = 350x + 600y - 4x^2 - 3y^2$$

above the rectangular region $20 \le x \le 30$ and $80 \le y \le 110$. Figure 2 displays this surface. (Does 35,916.67 seem like a reasonable average height to you?)

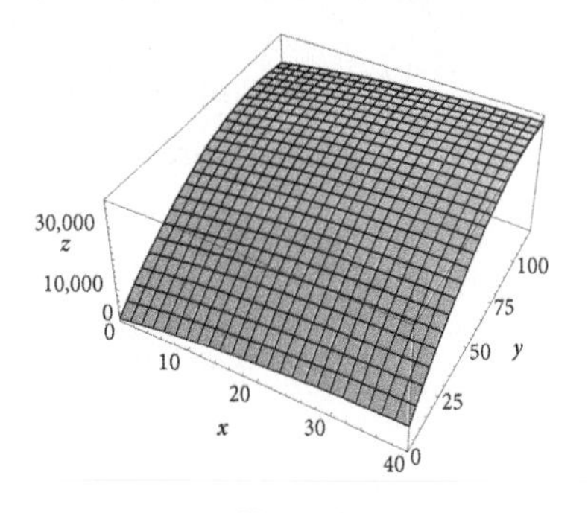

Figure 2

Using Technology 7.7

Wolfram|Alpha can help you compute the average value of functions of one or two variables.

Example 3 Use Wolfram|Alpha to find the average value of the function $f(x) = 10xe^{-0.3x}$ on the interval $[0, 5]$.

Solution In the Wolfram|Alpha entry field enter

integrate 1/(5 - 0) *10xe^(-0.3x) from x = 0 to 5

Wolfram|Alpha responds with

Definite integral:

$$\int_0^5 \frac{10\,xe^{-0.3\,x}}{5-0}\,dx = 9.8261$$

Wolfram Alpha LLC. 2013. Wolfram|Alpha
http://www.wolframalpha.com/
(access January 23, 2013)

Wolfram|Alpha indicates that for values of x between 0 and 5, the average value of the function is about 9.8. We can interpret this as meaning that the average length of the infinitely many vertical lines drawn from the x-axis to the curve is about 9.8 units. If we had to pick a vertical line in the interval $[0, 5]$ that extends from the x-axis to the curve, our best guess for its length would be 9.8 units.

Look at the graph of $f(x) = 10xe^{-0.3x}$ on $[0, 5]$. (Hint: Use the same entry field, but delete the $\frac{1}{(5-0)}$ part in order to get the graph of the function itself.) Does the 9.8 value produced by Wolfram|Alpha look like a reasonable average height? ■

Example 4 In Example 2, we used pencil-and-paper and a calculator to determine the company's average weekly revenue from the sales of its products A and B over the year. We can use Wolfram|Alpha to help us with this computation.

Solution In the Wolfram|Alpha entry field enter

integrate 1/((30 - 20)(110 - 80))*(350x + 600y - 4x^2 - 3y^2) dxdy
for x = 20 to 30 and y = 80 to 110

Wolfram|Alpha responds with $\frac{107{,}750}{3}$.

On the average, this company realizes a weekly revenue of about

$$\$\frac{107{,}750}{3} \approx \$35{,}916.67$$

on the sale of x units of product A and y units of product B over the year. ■

Getting Ready for Class

After reading through the preceding section, respond in your own words and in complete sentences.

A. Describe, in geometric terms, the meaning of the formula

$$\text{Average value} = \frac{1}{(b-a)(d-c)} \int_c^d \int_a^b f(x, y)\, dx\, dy$$

B. Describe the strategy you would use to solve Problems 5 and 6 in the Problem Set.

Problem Set 7.7

Skills Practice

For Problems 1-8, find the average value of each function on the indicated interval.

1. $f(x) = x^2 + 1$ on $[0, 4]$
2. $f(x) = 8 - x^3$ on $[0, 2]$
3. $f(x) = \sqrt{x}$ on $[0, 16]$
4. $f(x) = \sqrt[3]{x}$ on $[1, 8]$
5. $f(x) = 440e^{-0.04x}$ on $[20, 50]$
6. $f(x) = \dfrac{10}{x}$ on $[1, 5]$
7. $f(x) = x\sqrt{x^2 - 1}$ on $[1, 5]$
8. $f(x) = \dfrac{x^2}{(1 - x^3)^{4/3}}$ on $[100, 150]$

Check Your Understanding

9. **Ecology: Pollution from Manufacturing** The function $P(x)$ describes the level of air pollution emitted from a manufacturing facility. The variable x represents the distance in miles from the facility and P represents the number of pollution units. Interpret

$$\frac{1}{10 - 3} \int_3^{10} P(x)\, dx$$

10. **Management: Stress in Sales** The human resources department of a manufacturing company believes that the function $A(t)$ relates the amount of stress (in stress units) experienced by pharmaceutical salespeople after visiting a medical agency for a sale of pharmaceutical products and the time t (in days) until they hear that they have or have not made the sale. Interpret

$$\frac{1}{16 - 9} \int_9^{16} A(t)\, dt$$

11. **Economics: Depreciation of Equipment** The function $D(t)$ represents the depreciation of a piece of equipment t months after it was purchased. Interpret

$$\frac{1}{24 - 12} \int_{12}^{24} D(t)\, dt$$

12. **Economics: Population of a City** City supervisors believe that the function $P(t)$ approximately describes the population (in thousands of people) of their city t years from now. Interpret

$$\frac{1}{10 - 5} \int_5^{10} P(t)\, dt$$

13. **Economics: Value of Art** A piece of art is thought to increase in value according to the function $V(t)$, where V is in dollars and t is the time in years since the piece was introduced into the market. Interpret

$$\frac{1}{20 - 0} \int_0^{20} V(t)\, dt$$

© Alexander Novikov/iStockPhoto

14. Business: Employee's Earnings The management of a company believes that, for its sales division, the function $E(x)$ approximates its employees' earnings, where E is in dollars and x is the number of sales made each week. Interpret

$$\frac{1}{35-25}\int_{25}^{35} E(x)\,dx$$

15. Business: Worker Production For a particular company, the average worker's production level L (in units completed per day) depends on the number of days t the worker has been on the job, and is approximated by the function $L(t)$. Interpret

$$\frac{1}{120-60}\int_{60}^{120} L(t)\,dt$$

16. Biology: Pulse Rate During heavy physical activity, a person's pulse rate R (in beats per minute) is determined by the amount of time t (in minutes) since the beginning of the activity. Interpret

$$\frac{1}{40-30}\int_{30}^{40} R(t)\,dt$$

Modeling Practice

17. Economics: Property Value In recessionary times, the value of a house in a particular county is approximated by the function $V(t) = 230{,}000e^{-0.02t}$ where t is the number of years since the beginning of the recession. Find and interpret the average value of a house in this county over the first 4 years of a recession.

18. Business: Revenue A small business estimates that its annual revenue over the next 5 years will be $R(t) = 120{,}000e^{0.08t}$ dollars. Find and interpret the average revenue for this business over the next 5 years.

19. Business: Marginal Cost The marginal cost associated with a particular product is $MC(x) = -0.06x + 35$ dollars. Find and interpret the average marginal cost for units 400 to 500 in a production run of 500 units.

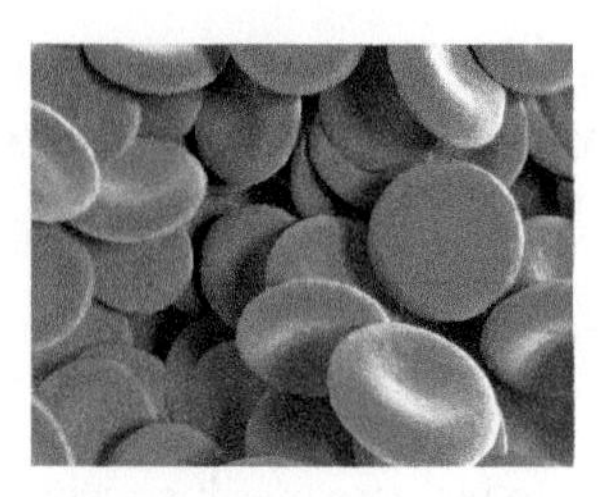
© Andrew Dorey/iStockPhoto

20. Health Science: Drug Concentration The amount A (in milligrams) of a drug in a patient's bloodstream t hours after it is injected into the patient's body is $A(t) = 200e^{-0.4t}$. Find and interpret the average amount of drug in the bloodstream

a. during the first three hours after injection.

b. over hours eight, nine, and ten after injection.

21. Business: Cobb-Douglas Production Using x thousand worker-hours of labor and y million dollars of capital, a manufacturer can produce

$$f(x, y) = 800x^{2/3}y^{1/3}$$

units of a product each month. What is the average number of units that can be produced each month if the number of thousands of worker-hours ranges from 8 to 12 each month, and the number of millions of dollars of capital ranges from 1 to 3 each month?

22. **Business: Cobb-Douglas Production** Using x thousand worker-hours of labor and y million dollars of capital, a manufacturer can produce

$$f(x, y) = 3{,}500x^{0.6}y^{0.4}$$

units of a product each month. What is the average number of units that can be produced each month if the number of thousands of worker-hours ranges from 4 to 10 each month, and the number of millions of dollars of capital ranges from 50 to 63 each month?

23. **Business: Average Revenue** A company sells two products, A and B, having demand functions $x_A = 430 - 4p_A$ and $x_B = 560 - 3p_B$, respectively, where p_A is the price per unit of product A, and p_B is the price per unit of product B. The total revenue realized each month on the sale of these two products is then $R(x_A, x_B) = x_A p_A + x_B p_B$. Approximate the average revenue if the price of product A ranges between \$80 and \$110, and the price of product B ranges between \$40 and \$50.

24. **Economics: Average Property Value** From a particular intersection of two streets A and B just outside a city, the property value at any point (x, y) in a rectangular region is

$$V(x, y) = 1{,}000xe^{2x+2y}$$

dollars per square mile. The number of miles due east of the intersection is represented by x and the number of miles due north by the letter y. Approximate the average property value if the rectangular region is bounded by a street 2 miles due east of the intersection of streets A and B, and by a street 1 mile due north of the intersection of streets A and B.

25. **Ecology: Average Amount of Pollution from Production** To produce safety glass, a manufacturer uses x units of chemical A and y units of chemical B. The amount P of pollution washed into the cleaning water each day during the manufacturing process is given by the function

$$P(x, y) = 0.04x^2 + 0.08xy + 0.006y^2$$

What is the average amount of pollution washed into the cleaning water each day if over a 1-year period the amount of chemical A used varies between 4 and 5 units each day, and the amount of chemical B varies between 16 and 20 units each day?

26. **Health Science: Recovery Time from Illness** A medical journal article stated that for a particular illness, the recovery time, R, (in days) for a person t years old who is given d milligrams each day of medication is approximated by the function

$$R(t, d) = \frac{500t^{1.2}}{d^{1.4}}$$

Approximate the average recovery time for people between 20 and 30 years old who are given between 250 and 400 milligrams per day of the medication.

27. **Business: Average Revenue** The annual revenue, in hundreds of dollars, realized by a company for the sale of x units of product A and y units of product B is

$$R(x, y) = 120x + 270y - x^2 - y^2$$

Over the year the company produces between 50 and 70 units of product A each week and between 200 and 350 units of product B each week. Approximate this company's average weekly revenue from products A and B over the year.

28. **Business: Advertising and Average Revenue** The marketing department of a company has determined that if it spends x thousands of dollars per month on radio advertisements and y thousands of dollars per month on newspaper advertisements, it will realize a revenue of

$$R(x, y) = -0.8x^2 - y^2 + 4x + 5y + 2xy$$

thousands of dollars. If over a 1-year period the company spends between $10,000 and $15,000 each month on radio advertisements and between $4,000 and $9,000 each month on newspaper advertisements, what will be its average monthly revenue over the 1-year period?

29. **Business: Advertising and Influence** A franchise operation in a city believes that if it buys x thousands of dollars of radio advertisements each week and y thousands of dollars of newspaper advertisements each week, it can reach and possibly influence

$$N(x, y) = 20x(4y + 20)$$

people each week. If over a 1-year period, the franchise operation buys between 12 and 20 thousand dollars of radio advertisements each week and between 20 and 24 thousand dollars of newspaper advertisements each week, what is the average number of people this operation can expect to reach and influence each week?

30. **Economics: Welfare** Because of the state of the economy in the country, a state has determined that approximately

$$P(x, y) = 1.6e^{0.04x} - 1.4e^{0.002y}$$

percent of its population will apply for welfare benefits each month. In the function, x represents the unemployment rate during the month, and y is the average number of years of education of people of working age in the state. Approximate the average percentage of the state's population that will apply for welfare benefits each month if unemployment varies between 4% and 7% each month, and the average number of years of education varies between 11 and 14.

31. Business: Average Weekly Production The weekly output, in single units, of a company is given by the production function

$$f(x, y) = 0.6xe^{0.04y}$$

where x represents the number of hundreds of hours allocated each week to labor, and y is the number of thousands of dollars allocated each week to capital. Find the average weekly output if the number of hours allocated each week to labor ranges from 150 to 200 and the number of dollars allocated each week to capital ranges from 30,000 to 38,000. (Watch the units!)

Using Technology Exercises

32. Use Wolfram|Alpha to help you approximate, to two decimal places, the average value of the function $f(x) = e^{-0.05x} \ln x$ on the interval $[1, 5]$.

33. Health Science: Average Amount of a Drug in the Body The amount A, in micrograms per milliliter (μg/mL), of a drug in the human body t hours after the drug is administered is modeled by the function

$$A(x) = 2.5te^{-0.07t}$$

What is the average amount of the drug in the body over

a. the first 12 hours after it is administered?

b. the second 12 hours after it is administered?

c. the first 24 hours after it is administered?

Source: British Journal of Clinical Pharmocology

34. Manufacturing: Average Level of Output By using x thousands of worker-hours of labor and y millions of dollars of capital, a manufacturer is able to produce

$$N(x, y) = 2{,}650x^{0.35}y^{0.65}$$

units of a product. If over the next year, the manufacturer uses between 5 and 10 thousand worker-hours of labor and between 3 and 5 million dollars of capital, determine the manufacturer's average level of output.

Spotlight on **Success**

Howard College

The motto at Howard College in Texas is "Education, for learning, for earning, for life."

The school's website, www.earnmydegree.com poses the question: "Does a college degree pay off?"

You can make much more money by earning a college degree.

The data shows that a college degree correlates directly to your salary range—and the relationship between compensation and education level is becoming even more prominent.

Employers have increasingly use diplomas and degrees as a way to screen applicants. And once you've landed the job you want, your salary will reflect your credentials. On average, a person with a master's degree earns $31,900 more per year than a high school graduate—a difference of as much as 105%!

Average Annual Earnings by Education Level

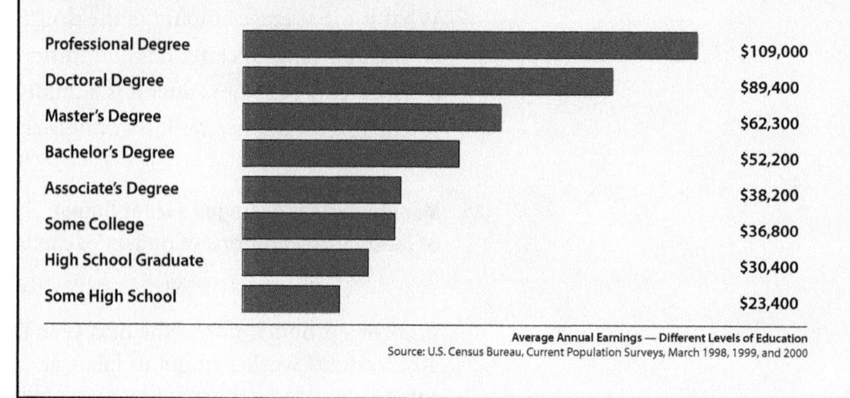

Average Annual Earnings — Different Levels of Education
Source: U.S. Census Bureau, Current Population Surveys, March 1998, 1999, and 2000

Chapter 7 Summary

EXAMPLES

Functions of Several Variables [7.1]

A function of two variables is a rule f that assigns to each ordered pair (x, y) in a set D a unique number $f(x, y)$. The set of ordered pairs is called the domain of the function. The rule lays out the procedure for producing the new number $f(x, y)$ from the two numbers x and y. A function of three variables is a rule f that assigns to each ordered triple (x, y, z) in a set D a unique number $f(x, y, z)$. Functions of four or even more variables are defined in a similar way.

1. $f(x, y) = x^2 + 3y$ is a function of the two variables x and y.

   $f(x, y, z) = 4x^2 - 2xy + 5z$ is a function of the three variables x, y, and z.

Evaluating Functions of Two Variables [7.1]

Evaluate a function of several variables by substituting the input values for their corresponding variables and computing.

2. Evaluate $f(x, y) = 50x^{3/4}y^{1/4}$ for $x = 625$ and $y = 81$.

   $$f(625, 81) = 50(625)^{3/4}(81)^{1/4} = 18{,}750$$

Graphs of Functions of Several Variables [7.1]

Graphs of functions of two variables are surfaces and require three coordinate axes.

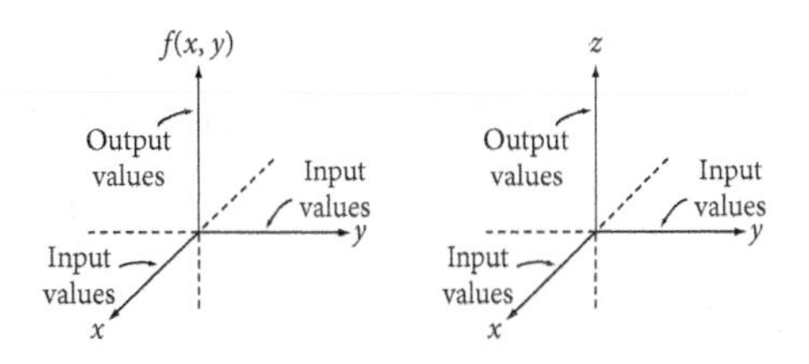

3. $f(x, y) = x^2 - y^2$

Cobb-Douglas Production Functions [7.1]

The **Cobb-Douglas production function** relates the number of units of a product a company can produce to the number of units of labor and capital it uses to produce this product. The function is of the form

$$f(x, y) = Cx^a y^b$$

where x represents the number of units of labor used to produce the product and y the number of units of capital used. $f(x, y)$ represents the number of units of the product produced. The numbers C, a, and b are constants.

4. When 20 units of labor and 10 units of capital are used each week, a company that models its production using the function

   $$f(x, y) = 150x^{3/5}y^{2/5}$$

   can produce

   $$f(20, 10) = 150(20)^{3/5}(10)^{2/5} \approx 2{,}274$$

   units of its product each week.

Returns to Scale [7.1]

5. If both the input values in the Cobb-Douglas production function

$f(x, y) = 50x^{3/4}y^{1/4}$

are tripled, the output is tripled and the function exhibits constant returns to scale. Triple both x and y, then

$$\begin{aligned} f(x, y) &= 50(3x)^{3/4}(3y)^{1/4} \\ &= 50 \cdot 3^{3/4}x^{3/4} \cdot 3^{1/4}y^{1/4} \\ &= 50 \cdot 3^{3/4+1/4}x^{3/4}y^{1/4} \\ &= 50 \cdot 3x^{3/4}y^{1/4} \\ &= 3 \cdot 50x^{3/4}y^{1/4} \\ &= 3 \cdot f(x, y) \end{aligned}$$

Cobb-Douglas production functions can be used to answer questions of returns to scale. That is, they can be used to answer questions about how a proportionate increase or decrease in all the input values will affect the total production.

Constant returns to scale If the proportional increase in all the input values is equal to the proportional increase in the output, then the returns to scale are constant. For example, if returns to scale are constant, a doubling of the number of units of both labor and capital results in a doubling of production.

Increasing returns to scale If the proportional increase in output is greater than the proportional increase in all the inputs, then the returns to scale are increasing. For example, if returns to scale are increasing, a doubling of the number of units of both labor and capital results in more than a doubling of production.

Decreasing returns to scale If the proportional increase in output is less than the proportional increase in all the inputs, then the returns to scale are decreasing. For example, if returns to scale are decreasing, a doubling of the number of units of both labor and capital results in less than a doubling of production.

Partial Derivatives [7.2]

6. For the function

$f(x, y) = 5x^2 + 4y^2 + 8xy^3$

$\frac{\partial f}{\partial x} = 10x + 8y^3$ and

$\frac{\partial f}{\partial y} = 24xy^2 + 8y$

A partial derivative of a function of more than one variable provides the effect on the output of the function when exactly one of the variables is increased or decreased in value and the other variables are held constant. Common notations for partial derivatives of a function of two variables $f(x, y)$ with respect to x are

$$\frac{\partial f}{\partial x} \quad \text{or} \quad \frac{\partial}{\partial x}f \quad \text{or} \quad f_x \quad \text{or} \quad f_x(x, y)$$

Interpreting Partial Derivatives [7.2]

7. For the function

$f(x, y) = 5x^2 + 4y^2 + 8xy^3$

if $x = 3$ and $y = 1$, then

$$\begin{aligned} \frac{\partial}{\partial x}f(3, 1) &= \\ &= 10(3) + 8(1)^3 \\ &= 38 \end{aligned}$$

If, when $x = 3$ and $y = 1$, x is increased by 1 unit, from 3 to 4, then the value of the function will increase by approximately 38 units.

$\frac{\partial f}{\partial x}$ represents the change in the value of the function with respect to x when the value of x is changed and the value of y is held constant; that is, it represents the rate of change of the function in the x direction. $\frac{\partial f}{\partial y}$ represents the change in the value of the function with respect to y when the value of y is changed and the value of x is held constant; that is, it represents the rate of change of the function in the y direction.

Higher-Order Partial Derivatives [7.2]

8. For the function
$f(x, y) = 5x^2 + 4y^2 + 8xy^3$
$f_x = 10x + 8y^3$ and
$f_y = 24xy^2 + 8y$.

Then, $f_{xx} = 10$,
$f_{yy} = 48xy + 8, f_{xy} = 24y^2$,
and $f_{yx} = 24y^2$.

Where the derivative of $f(x)$ is $f'(x)$, a function with two variables has two partial derivatives, which can each be differentiated. Thus, a function $f(x, y)$ has four partial derivatives.

$\frac{\partial^2 f}{\partial x^2}$, or more simply, f_{xx}. f is differentiated first with respect to x, then with respect to x again.

$\frac{\partial^2 f}{\partial y^2}$, or more simply, f_{yy}. f is differentiated first with respect to y, then with respect to y again.

$\frac{\partial^2 f}{\partial x \partial y}$, or more simply, f_{yx}. f is differentiated first with respect to y, then with respect to x.

$\frac{\partial^2 f}{\partial y \partial x}$, or more simply, f_{xy}. f is differentiated first with respect to x, then with respect to y.

Second-Order partial Derivatives as Rates of Change [7.2]

9. For the function $f(x, y)$, if it is known that

a. $f_x(30, 25) > 0$ and $f_{xx}(30, 25) > 0$, we can conclude that the function is increasing at an increasing rate in the x-direction through (30, 25).

b. $f_y(30, 25) > 0$ and $f_{yy}(30, 25) > 0$, we can conclude that the function is increasing at an increasing rate in the y-direction through (30, 25).

c. $f_x(30, 25) < 0$ and $f_{xx}(30, 25) > 0$, we can conclude that the function is decreasing at a decreasing rate in the x-direction through (30, 25).

d. $f_y(30, 25) > 0$ and $f_{yy}(30, 25) < 0$, we can conclude that the function is increasing at a decreasing rate in the y-direction through (30, 25).

As the second derivative $f''(x)$ measures the concavity of the curve $f(x)$, f_{xx} and f_{yy} measure the concavity in the x direction and the y direction, respectively, of the surface $f(x, y)$. Below we list the behavior partial derivatives of the same kind indicate.

$f_x > 0$ and $f_{xx} > 0 \to f$ is increasing at an increasing rate in the x-direction
$f_y > 0$ and $f_{yy} > 0 \to f$ is increasing at an increasing rate in the y-direction
$f_x < 0$ and $f_{xx} < 0 \to f$ is decreasing at an increasing rate in the x-direction
$f_y < 0$ and $f_{yy} < 0 \to f$ is decreasing at an increasing rate in the y-direction

$f_x > 0$ and $f_{xx} < 0 \to f$ is increasing at a decreasing rate in the x-direction
$f_y > 0$ and $f_{yy} < 0 \to f$ is increasing at a decreasing rate in the y-direction
$f_x < 0$ and $f_{xx} > 0 \to f$ is decreasing at a decreasing rate in the x-direction
$f_y < 0$ and $f_{yy} > 0 \to f$ is decreasing at a decreasing rate in the y-direction

Optimization of Functions of Two Variables [7.3]

A point (a, b) produces a **relative maximum** of the function $f(x, y)$ if for every point (x, y) *near* (a, b), $f(a, b) \geq f(x, y)$. A point (a, b) produces a **relative minimum** of the function $f(x, y)$ if for every point (x, y) *near* (a, b), $f(a, b) \leq f(x, y)$.

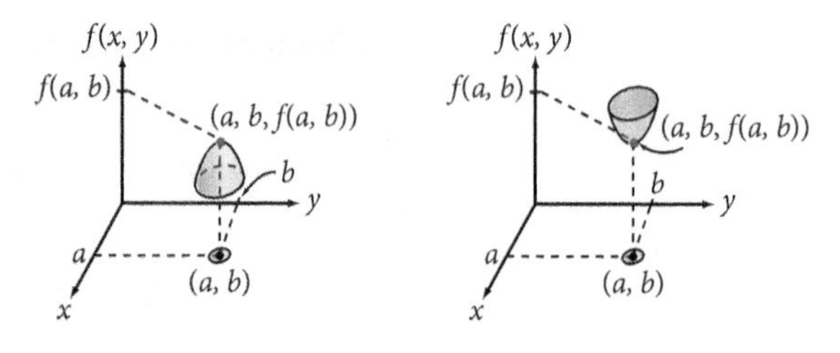

Locating Relative Extrema and Critical Points [7.3]

10. For the function $f(x, y) = 5x^2 - 4y^2 + 8x - 4y$
$f_x(x, y) = 10x + 8$ and $f_y(x, y) = -8y - 4$. Then

$$\begin{cases} f_x(x, y) = 0 \\ f_y(x, y) = 0 \end{cases} \rightarrow \begin{cases} 10x + 8 = 0 \\ -8y - 4 = 0 \end{cases}$$

$$\rightarrow x = \frac{-4}{5}, \quad y = -\frac{1}{2}$$

The only critical point is $\left(\frac{-4}{5}, -\frac{1}{2}\right)$. It may produce a relative maximum, a relative minimum, or a saddle point.

If $f(x, y)$ is a function with a relative extreme point at $(a, b, f(a, b))$ and both $f_x(a, b)$ and $f_y(a, b)$ exist, then

$$f_x(a, b) = 0 \quad \text{and} \quad f_y(a, b) = 0$$

The points (a, b) for which $f_x(a, b) = 0$ and $f_y(a, b) = 0$ are called **critical points** of the function $f(x, y)$. This means you can find critical points (a, b) by setting both $f_x(x, y)$ and $f_y(x, y)$ equal to zero and solving the system

$$\begin{cases} f_x(x, y) = 0 \\ f_y(x, y) = 0 \end{cases}$$

The Second Derivative Test [7.3]

11. For the function in Example 10 above, we found that the only critical point is $\left(\frac{-4}{5}, -\frac{1}{2}\right)$. To determine if this critical point produces a relative maximum, minimum or saddle point, find the second partial derivatives and the mixed partial derivatives:

$f_{xx}(x, y) = 10, f_{yy}(x, y) = -8$, and $f_{xy}(x, y) = 0$.

$D(x, y) =$

$= f_{xx}(a, b) \cdot f_{yy}(x, y) - f_{xy}^2(x, y)$

and evaluate it at $x = \frac{-4}{5}, y = -\frac{1}{2}$.

$D(x, y) = (10)(-8) - 0^2 = -80$.
In particular,

$D\left(\frac{-4}{5}, -\frac{1}{2}\right) = (10)(-8) - 0^2 < 0$.

Since $D(x, y) < 0$, the critical point $\left(\frac{-4}{5}, -\frac{1}{2}\right)$ is necessarily a saddle point.

Suppose that (a, b) is a critical point of the function $f(x, y)$ and that the new function $D(x, y)$ represents $f_{xx}(a, b) \cdot f_{yy}(a, b) - f_{xy}^2(a, b)$. Then,

1. if $D(a, b) > 0$ and $f_{xx}(a, b) > 0$, (a, b) produces a relative minimum.
2. if $D(a, b) > 0$ and $f_{xx}(a, b) < 0$, (a, b) produces a relative maximum.
3. if $D(a, b) < 0$, (a, b) is a saddle point.
4. if $D(a, b) = 0$, the test provides no information about what (a, b) might produce.

   Note that if D, when evaluated at a critical point (a, b) is positive, the critical point will necessarily be a relative extreme point. The critical point will produce a saddle point when D is negative.

Constrained Maxima and Minima [7.4]

Sometimes there are restrictions on the values of the variables in functions. The figure shows the maximum value of the function $f(x, y)$ when the values of x and y are restricted. For example, it may be that for the function $f(x, y)$, x and y are restricted to values that satisfy $x + y = 10$.

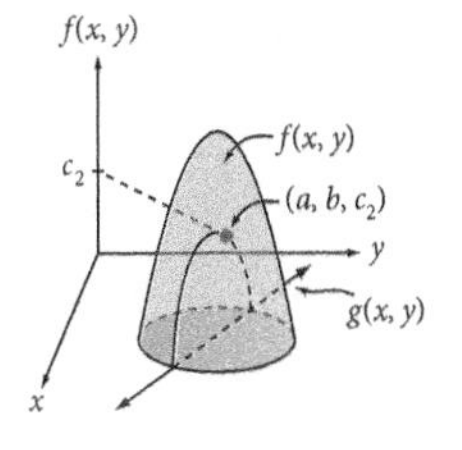

When there are restrictions on the values of x and y, the Second Derivative Test of Section 7.3 cannot be used. There are, instead, two methods that can be used. One is the method of algebraic substitution and the other is the method of Lagrange Multipliers.

Finding Constrained Extrema Using Substitution [7.4]

12. To find the minimum of $f(x, y) = 30x^2 + 20y^2$ subject to the constraint that $x + y = 10$, substitute $y = -x + 10$ for y in $f(x, y) = 30x^2 + 20y^2$

This gives us
$f(x, y) = 30x^2 + 20(-x + 10)^2$
$f(x) = 50x^2 - 400x + 2{,}000$

$f'(x) = 100x - 400$ and $f'(x) = 0$ when $x = 4$. When $x < 4, f'(x) < 0$. When $x > 4, f'(x) > 0$. So, $x = 4$ produces a relative minimum. When $x = 4$, $y = -4 + 10 = 6$. The point $(4, 6)$ produces the relative minimum at $(4, 6, 1{,}200)$.

If the constraint is easily solved for one of the variables, any relative extreme points can be located and classified using substitution and the first derivative test. For example, the constraints on a function $f(x, y) = 30x^2 + 20y^2$ might be that the values of x and y satisfy $x + y = 10$. In this case, either x or y is easily solved for. Solving for y gives $y = -x + 10$. Then the expression $-x + 10$ is substituted for y in the original function $f(x, y) = 30x^2 + 20y^2$ making it a function of only one variable, x, and the first derivative test is applied to any critical points.

Finding Constrained Extrema Using Lagrange Multipliers [7.4]

13. If you are to find the minimum of $f(x, y) = x^2 + 4y^2$ subject to $x - 2y = 40$, then $g(x, y) = x - 2y - 40$. The new Lagrangian function is
$F(x, y, \lambda) = x^2 + 4y^2 + \lambda(x - 2y - 40)$

The method of Lagrange multipliers can be used to locate the extreme points of a function $f(x, y)$. The method involves creating a new function, called the Lagrangian function, which, for a function of the two variables x and y, is defined as

$$F(x, y, \lambda) = f(x, y) + \lambda g(x, y)$$

where the function $g(x, y)$ is the constraint function and λ is a constant, the Lagrange multiplier.

The Method of Lagrange Multipliers [7.4]

14. Find the minimum of the function $f(x, y) = x^2 + 4y^2$ subject to the constraint $x - 2y = 40$.

Step 1: Rewrite constraint as $g(x, y) = x - 2y - 40$

Step 2: Construct Lagrangian function

$$\begin{aligned} F(x, y, \lambda) &= f(x, y) + \lambda g(x, y) \\ &= x^2 + 4y^2 + \lambda(x - 2y - 40) \\ &= x^2 + 4y^2 + \lambda x - 2\lambda y - 40\lambda \end{aligned}$$

Step 3: The first-order partials are

$$F_x = 2x + \lambda,\ F_y = 8y - 2\lambda,\ F_\lambda = x - 2y - 40$$

Step 4: Set the partials equal to 0 and solve the system

$$\begin{cases} 2x + \lambda = 0 \\ 8y - 2\lambda = 0 \\ x - 2y - 40 = 0 \end{cases} \rightarrow \begin{cases} x = 20 \\ y = -10 \\ \lambda = -40 \end{cases}$$

The only critical point is $(20, -10)$.

Step 5: Evaluate the function at $x = 20, y = -10$

$$f(x, y) = (20)^2 + 4(-10)^2 = 800$$

The minimum value of 800 occurs when $x = 20$ and $y = -10$.

To locate potential relative extreme points of a function $f(x, y)$ subject to the constraint $g(x, y) = 0$,

1. Set the constraint equal to zero so that it is in the form $g(x, y) = 0$.
2. Construct the Lagrangian function
$$F(x, y, \lambda) = f(x, y) + \lambda g(x, y)$$
3. Find the first-order partial derivatives F_x, F_y, and F_λ.
4. Locate critical points by solving the system of equations
$$\begin{cases} F_x = 0 & \ldots(1) \\ F_y = 0 & \ldots(2) \\ F_\lambda = 0 & \ldots(3) \end{cases}$$
for x, y, and λ. Each ordered pair solution, (x, y), is a critical point. The relative extreme points are among these critical points.
5. Evaluate the original function $f(x, y)$ at each critical point obtained in Step 4. The point that produces the greatest value produces a relative maximum. The point that produces the least value produces a relative minimum.

The Meaning of the Lagrange Multiplier [7.4]

15. When constrained by the function $x - 2y = 40$, the function $f(x, y) = x^2 + 4y^2$ is minimized at 800 when $x = 20$ and $y = -10$. The constant in the constraint function is 40 and the value of λ is -40. If it were to be increased by 1-unit, from 40 to 41, the minimum of the function would decrease by 40 units, from 800 to 760.

The Lagrange multiplier, λ, provides information about how a function $f(x, y)$ changes as changes are made to the constant c in the constraint equation $g(x, y) = c$. Suppose that M is the constrained maximum or minimum value of the function $f(x, y)$ when it is subjected to the constraint $g(x, y) = c$, where c is some constant. Then,

$$\frac{dM}{dc} = \lambda$$

That is, λ is the rate at which the constrained maximum or minimum value of the function $f(x, y)$ changes with respect to c.

The Total Differential [7.5]

16. Approximate the change in the function $f(x, y) = 84x^{3/7}y^{4/7}$ as x increases by 2 units from 30 to 32 and y decreases by 1 unit from 10 to 9.

$df = f_x(x, y)\,dx + f_y(x, y)\,dy$

$= \dfrac{36y^{4/7}}{x^{4/7}}\,dx + \dfrac{48x^{3/7}}{y^{3/7}}\,dy$

Then for $x = 30$, $y = 10$, $dx = 2$, and $dy = -1$

$df =$

$\dfrac{36(10)^{4/7}}{(30)^{4/7}}(2) + \dfrac{48(30)^{3/7}}{(10)^{3/7}}(-1)$

≈ -38.43

With these changes, the function will decrease in value by about 38.43 units.

The total differential of a function of two variables $f(x, y)$ is

$$df = f_x(x, y)dx + f_y(x, y)dy$$

and it approximates the change in the function as x changes a small amount by dx units and y changes a small amount by dy units.

Partial Integration [7.6]

17. If $f_x(x, y) = 25x^4y^3$, find $f(x, y)$ by evaluating $\int 25x^4y^3\,dx$.

$f(x, y) = \int 25x^4y^3\,dx$

$= 25y^3\int x^4\,dx$

$= 25y^3 \cdot \frac{1}{5}x^5 + g(y)$

$= 5x^5y^3 + g(y)$

Because the variable y is being treated as a constant, the constant of integration could involve y rather than only a real number. The constant of integration is represented using the function notation $g(y)$ rather than just the letter C.

Partial differentiation is the process of differentiating a function of two variables with respect to one of its variables while treating the other variable as if it were a constant. You can integrate a function of two variables in a similar way integrate with respect to one of the variables while treating the other as if were a constant. The process of integrating a function of two variables with respect to only one of the variables is called partial integration. Because one of the variables is considered to be a constant, the constant of integration could involve that variable.

Double Integrals as Volume [7.6]

The integral of a function of one variable $f(x)$ can, under the right conditions, represent the area of some region in the plane. The right conditions are that $f(x)$ be continuous and positive over some interval $[a, b]$. That is,

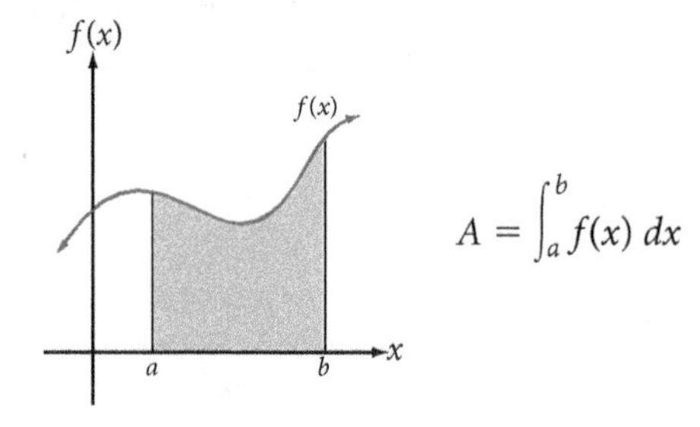

$$A = \int_a^b f(x)\, dx$$

Similarly, the double integral of a function of two variables $f(x, y)$ can, under the right conditions, represent the volume of some space under a surface and above a region in the plane. The right conditions are that the surface $f(x, y)$ be continuous and positive over the region in the plane. The region in the plane can be rectangular, $a \leq x \leq b$, $c \leq y \leq d$, or it can have as its bounds two lines and two curves. The figure below shows the space under a surface and over a rectangular region.

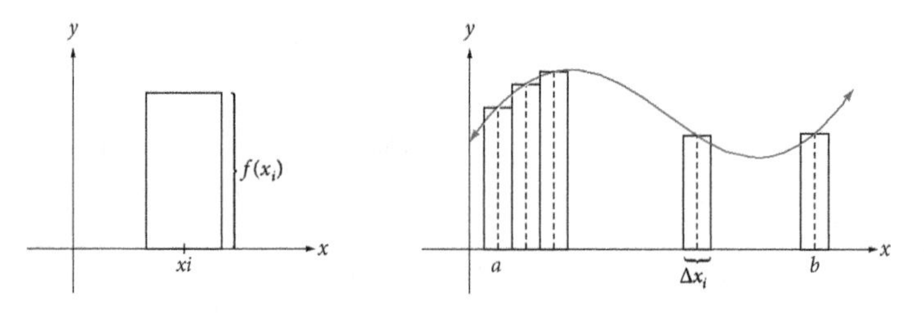

In this case, the volume of the space under the surface and over the region R is

$$\iint_R f(x, y)\, dydx$$

The figure on the left below shows the space under a surface and above a region R. The figure on the right shows a region that is bounded by the two lines $x = a$ and $x = b$ and the two curves $y = g_1(x)$ and $y = g_2(x)$.

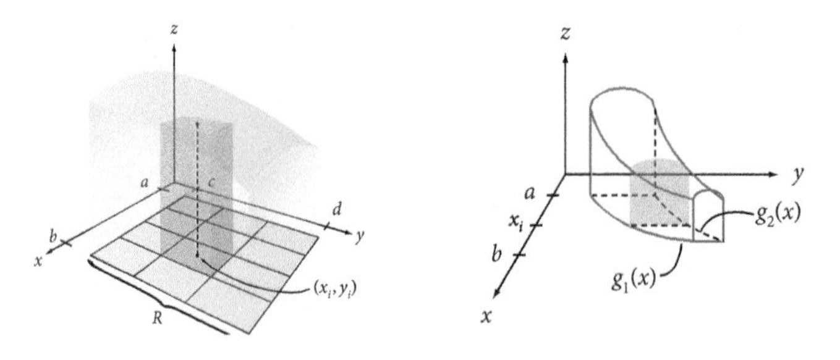

In this case, the volume of the space under the surface and over the region is

$$V = \int_a^b \int_{g_1(x)}^{g_2(x)} f(x, y)\, dy\, dx$$

Computing Volume Using Double Integrals [7.6]

If $g_1(x)$ and $g_2(x)$ are continuous functions on the interval $[a, b]$ running along the x-axis and R is a region in the xy-plane bounded by the lines $x = a$ and $x = b$, and the curves $g_1(x)$ and $g_2(x)$, then

$$V = \int_a^b \int_{g_1(x)}^{g_2(x)} f(x, y)\, dy\, dx$$

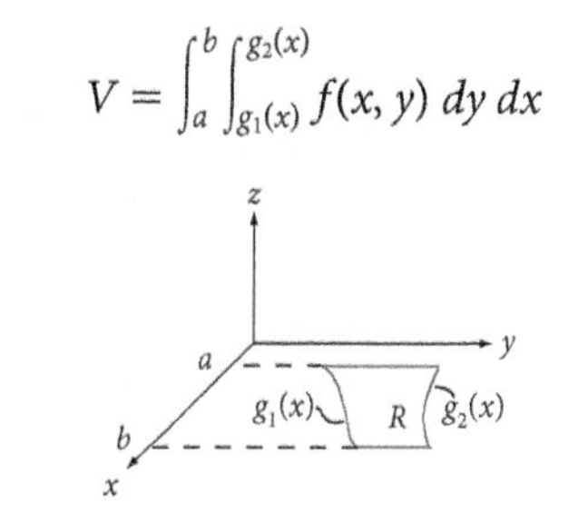

If $h_1(y)$ and $h_2(y)$ are continuous functions on the interval $[c, d]$ running along the y-axis and R is a region in the xy-plane bounded by the lines $y = c$ and $y = d$, and the curves $h_1(y)$ and $h_2(y)$, then

$$V = \int_c^d \int_{h_1(y)}^{h_2(y)} f(x, y)\, dx\, dy$$

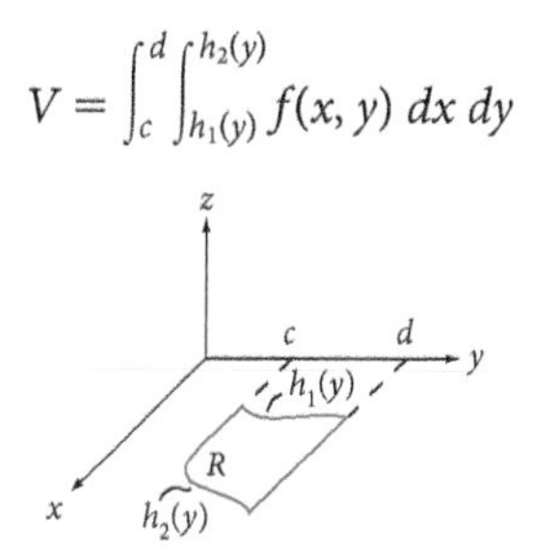

18. Show that
$$\int_0^4 \int_0^1 (4xy)\, dydx = \int_0^1 \int_0^4 (4xy)\, dxdy$$
by computing both volumes.

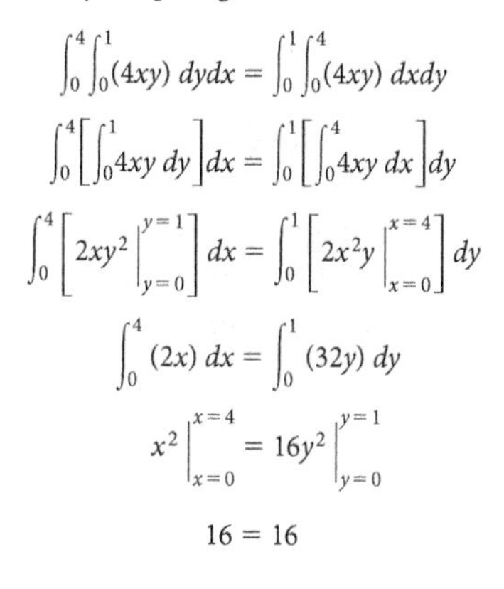

$$\int_0^4 \int_0^1 (4xy)\, dydx = \int_0^1 \int_0^4 (4xy)\, dxdy$$
$$\int_0^4 \left[\int_0^1 4xy\, dy\right] dx = \int_0^1 \left[\int_0^4 4xy\, dx\right] dy$$
$$\int_0^4 \left[2xy^2 \Big|_{y=0}^{y=1}\right] dx = \int_0^1 \left[2x^2y \Big|_{x=0}^{x=4}\right] dy$$
$$\int_0^4 (2x)\, dx = \int_0^1 (32y)\, dy$$
$$x^2 \Big|_{x=0}^{x=4} = 16y^2 \Big|_{y=0}^{y=1}$$
$$16 = 16$$

Computing the Volume of Space Over a Rectangular Region [7.6]

The double integral

$$\int_a^b \int_c^d f(x, y)\, dy\, dx$$

can be used to compute the volume of the space that lies beneath a surface $f(x, y)$ and above the rectangular region bounded by the lines $x = a$ and $x = b$ and $y = c$ and $y = d$. For example, the graph shows the surface

$$f(x, y) = x + 4y$$

over the rectangular region bounded by the lines $x = 2$ and $x = 5$ and $y = 3$ and $y = 10$. This double integral produces the volume of the space under the surface and above the rectangular region.

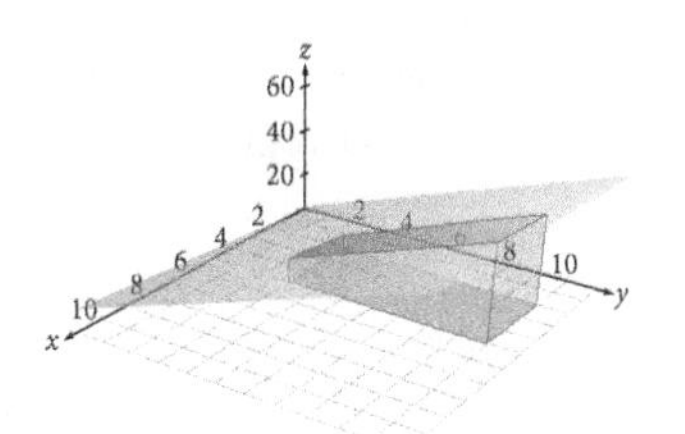

19. Show that the volume of the space under the surface $f(x, y) = 4xy$ and above the rectangular region bounded by the lines $x = 0$ and $x = 4$ and $y = 0$ and $y = 1$ is 16 cubic units.

$$V = \int_0^1 \int_0^4 (4xy)\, dxdy$$
$$= \int_0^1 \left[\int_0^4 4xy\, dx\right] dy$$
$$= \int_0^1 \left[2x^2y \Big|_{x=0}^{x=4}\right] dy$$
$$= \int_0^1 32y\, dy$$
$$= 16y^2 \Big|_{y=0}^{y=1}$$
$$= 16$$

Computing the Volume of Space Over a Non-Rectangular Region [7.6]

The double integral

$$V = \int_a^b \int_{g_1(x)}^{g_2(x)} f(x, y)\, dy\, dx$$

can be used to compute the volume of the space that lies beneath a surface $f(x, y)$ and above the region bounded by the lines $x = a$ and $x = b$ and the curves $y = g_1(x)$ and $y = g_2(x)$. For example, the graph below shows the non-rectangular region in the xy-plane over which the surface $f(x, y) = 4xy$ exists. The region is bounded by the lines $x = 0$ and $x = 1$ and the curves $y = x^2$ and $y = x$. The double integral produces the volume of the space under the surface and above the region.

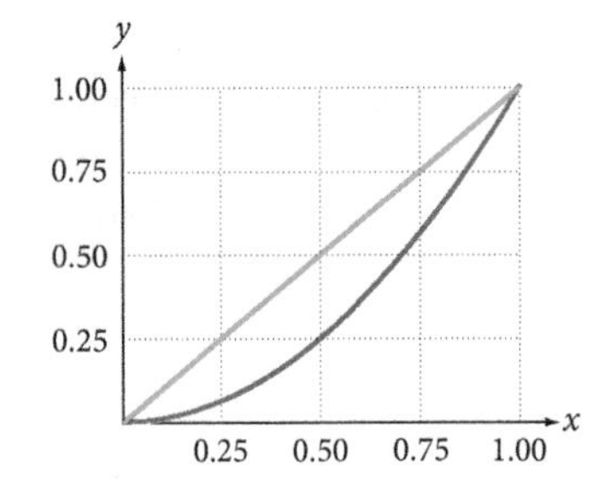

20. Show that the volume of the space under the surface $f(x, y) = 4xy$ and above the region bounded by the lines $x = 0$ and $x = 1$ and the curves $y = x^2$ and $y = x$ is $\frac{1}{6}$ cubic units.

$$V = \int_0^1 \int_{x^2}^{x} (4xy)\, dydx$$

$$= \int_0^1 \left[\int_{x^2}^{x} 4xy\, dy \right] dx$$

$$= \int_0^1 \left[4x \cdot \frac{1}{2} y^2 \Big|_{y=x^2}^{y=x} \right] dx$$

$$= 2 \int_0^1 x[x^2 - x^4]\, dx$$

$$= 2 \int_0^1 [x^3 - x^5]\, dx$$

$$= 2\left[\frac{x^4}{4} - \frac{x^6}{6} \right] \Big|_{x=0}^{x=1}$$

$$= 2\left[\left(\frac{1^4}{4} - \frac{1^6}{6} \right) - \left(\frac{0^4}{4} - \frac{0^6}{6} \right) \right]$$

$$= \frac{1}{6}$$

Average Value of a Function of One Variable [7.7]

If $f(x)$ is a continuous function on $[a, b]$, then the **average value** of $f(x)$ on $[a, b]$ is

$$\text{Average value} = \frac{1}{b - a} \int_a^b f(x)\, dx$$

21. Find the average value of $f(x) = 3x^2$ over the interval $[5, 8]$.

Average Value =

$$= \frac{1}{b - a} \int_a^b f(x)\, dx$$

$$= \frac{1}{8 - 5} \int_5^8 3x^2\, dx$$

$$= 129$$

Average Value of a Function of Two Variables [7.7]

The average value of the function $f(x, y)$ over a rectangular region R is given by

$$\text{Average value} = \frac{1}{(b - a)(d - c)} \int_c^d \int_a^b f(x, y)\, dx\, dy$$

where $a \le x \le b$ and $c \le y \le d$, and $(b - a)(d - c)$ represents the area of the rectangular region R.

22. Find the average value of $f(x, y) = 12x^2y^2$ over the region defined by $1 \le x \le 4$ and $3 \le y \le 5$.

Average Value =

$$= \frac{1}{(b - a)(d - c)} \int_c^d \int_a^b f(x, y)\, dx\, dy$$

$$= \frac{1}{(4 - 1)(5 - 3)} \int_3^5 \int_1^4 12x^2y^2\, dx\, dy$$

$$= 1{,}372$$

The average value is 1,372 units.

Chapter 7 Test

1. **Business: Advertising Revenue** Suppose x represents the number of dollars spent by a company on newspaper advertisements each month and y the number of dollars on radio advertisements. If R represents the revenue realized by the company from advertising, describe the difference between $R(x)$ and $R(x, y)$. [7.1]

2. Find $f(0, 2)$ for the function $f(x, y) = 4xe^{x+y}$. [7.2]

3. Suppose $N(x, y)$ is a function of the two variables x and y. Interpret

$$\frac{\partial}{\partial x}N(4, 10) = 15 \quad \text{and} \quad \frac{\partial}{\partial y}N(4, 10) = -4$$ [7.2]

4. **Manufacturing: Product Demand** The demand, D, realized by a manufacturer for a particular product depends on both the price x the manufacturer charges for the product and the price y the company's competitor charges. The demand is modeled by the function $D(x, y)$. What signs (< 0 or > 0) are most appropriate for $D_x(x, y)$ and $D_y(x, y)$? [7.2]

5. **Health Science: Body Mass Index** The body mass index (BMI) is a number that health providers use as an indicator of body fatness. The BMI was developed in the 1840s by the Belgian mathematician Adolphe Quetelet and is used as a predictor for obesity or starvation levels. It is a function of two variables, weight and height. With w representing a person's weight, in pounds, and h a person's height, in inches

$$\text{BMI}(w, h) = \frac{703w}{h^2}$$

Source: Center for Disease Control and Prevention

Given both w and h are positive numbers, answer the following. [7.2]

   a. Compute $\frac{\partial}{\partial w}\text{BMI}(w, h)$ and note its sign (positive or negative). What does the sign tell you about the behavior of this function?

   b. Compute $\frac{\partial}{\partial h}\text{BMI}(w, h)$ and note its sign (positive or negative). What does the sign tell you about the behavior of this function?

6. For $f(x, y) = 6x^2 - 6y^2 + 5xy$, find both $f_x(2, 1)$ and $f_y(2, 1)$. [7.2]

7. For the function $f(x, y) = x^2 + 7y^2 + 3x + 4y + 60$, find the values of x and y so that both $f_x(x, y) = 0$ and $f_y(x, y) = 0$. [7.3]

8. **Manufacturing: Cobb-Douglas Production** Suppose x units of labor and y units of capital are needed to produce $f(x, y) = 60x^{1/3}y^{2/3}$ units of a product. [7.2]

   a. How will the number of units produced change if the number of units of labor is increased by 1, from 64 to 65, and the number of units of capital are kept constant at 27?

   b. How will the number of units produced change if the number of units of capital is decreased by 1, from 27 to 26, and the number of units of labor are kept constant at 64?

9. Use the information about the first-order, second-order, mixed partial derivatives and the critical point of the function $f(x, y)$ to classify the critical point as a relative maximum, relative minimum, or saddle point. [7.3]

$$f_x(x, y) = 7x + 4, f_y(x, y) = -6y + 5,$$
$$f_{xx}(x, y) = 7, f_{yy}(x, y) = -6, f_{xy}(x, y) = 0$$
$$\left(\frac{-4}{7}, \frac{5}{6}\right)$$

10. Find, if they exist, all the relative extrema and saddle points of the function

$$f(x, y) = 3x^2 - y^2 - 12x + 16y + 5 \qquad [7.3]$$

11. Find, if they exist, all the relative extrema and saddle points of the function

$$f(x, y) = x^4 - 2x^2 + y^2 \qquad [7.3]$$

12. Manufacturing: The Lagrange Multiplier Interpret the meaning of the Lagrange multiplier, λ, as it applies to the following situation. Economists at a manufacturing company developed the function $f(x, y)$ to describe the relationship between the number of units of a product that are produced when x units of labor and y units of capital are used. For a particular product, the number of dollars allocated to labor and capital that will be used is 531,000. So, the constraint function is $x + y = 531{,}000$. The function $f(x, y)$ is maximized by (420, 121) and the corresponding value of λ is 4.3. [7.4]

13. Maximize the function

$$f(x, y) = 300x^{1/3}y^{2/3}$$

subject to constraints

$$x + 2y = 300 \qquad [7.4]$$

14. Economics: Cobb-Douglas Production The Cobb-Douglas production function $f(x, y) = 40x^{1/4}y^{3/4}$ describes the production of a company for which each unit of labor costs \$75 and each unit of capital costs \$125. For a new project, the company has allocated \$60,000 for labor and capital. [7.4]

a. Find the amount of money that the company should allocate to labor and capital to maximize production.

b. What is the maximum level of production?

15. The Total Differential Find the differential of the function

$$f(x, y) = 5x^2 - 4y^2 + 11x - 6y + 4$$

as x changes from 3 to 3.2 and y changes from 8 to 7.9. [7.5]

16. Manufacturing: Cobb-Douglas Production The output of a company is given by the Cobb-Douglas production function $f(x, y) = 400x^{2/5}y^{3/5}$, where x represents the number of thousands of dollars invested in labor and y the number of thousands of dollars invested in capital. [7.5]

a. What is the approximate change in the output if the amount of money invested in labor is decreased from \$11,000 to \$10,000 and the amount in capital is increased from \$40,000 to 42,000?

b. Compare this approximate change to the actual change.

17. Evaluate $\int 16x^3y^2\,dx$. [7.6]

18. Specify another representation for the sum $\lim_{n\to\infty}\sum_{i=1}^{n} f(x_i, y_i)\,\Delta y_i\,\Delta x_i$. [7.6]

For Problems 19 and 20, evaluate the double integral. [7.6]

19. $\int_1^5\int_1^2 x^2y\,dxdy$

20. $\int_1^e\int_0^{\ln x}\frac{y}{x}\,dydx$

21. Find the volume under the surface $f(x, y) = x^2 + xy^3$ and over the rectangular region $1 \le x \le 2$ and $0 \le y \le 1$. [7.6]

22. Find the volume under the surface $f(x, y) = 4xy - y^3$ where $x^3 \le y \le \sqrt{x}$ and $0 \le x \le 1$. [7.6]

23. Health Science: Average Amount of a Drug in the Bloodstream For a typical male weighing between 135 and 175 pounds, the amount, A, (in milligrams) of a drug in his bloodstream t hours after the drug is administered is approximated by the function

$$A(t) = 28e^{-0.4t}$$

Approximate the average amount of the drug in a typical male's bloodstream

a. over the first 6 hours after the drug is administered, to two decimal places.

b. through the second day after the drug is administered, to six decimal places. [7.7]

24. Manufacturing: Average Level of Production Using x thousands of worker-hours and y millions of dollars of capital each month, a manufacturer can produce $f(x, y) = 660x^{2/5}y^{3/5}$ units of a product. What is the average number of units the manufacturer can produce each month if the number of thousands of worker-hours ranges from 6 to 10 each month and the number of millions of dollars of capital ranges from 4 to 5 each month? [7.7]

Appendix

A.1 Rational Expressions

Once a calculus operation has been applied to a function, the new resulting function may need to be algebraically simplified. A common technique for simplifying functions-defining expressions is adding or subtracting rational expressions.

Rational expressions are ratios between polynomial expressions, such as

$$\frac{3x + 2}{4x - 1} \quad \text{and} \quad \frac{3}{2x^2 - 5x + 1}$$

Since they are algebraic fractions, they are operated on using the same rules that are used to operate on arithmetic fractions.

We reduce rational expressions to lowest terms by removing any common factors, other than 1 and -1, between the numerator and denominator. Removing common factors is accomplished by factoring the numerator and denominator and dividing out (canceling) factors that are common. If a common factor contains a variable and could take on the value of zero, then a restriction on all values that would cause this factor to become zero must be stated.

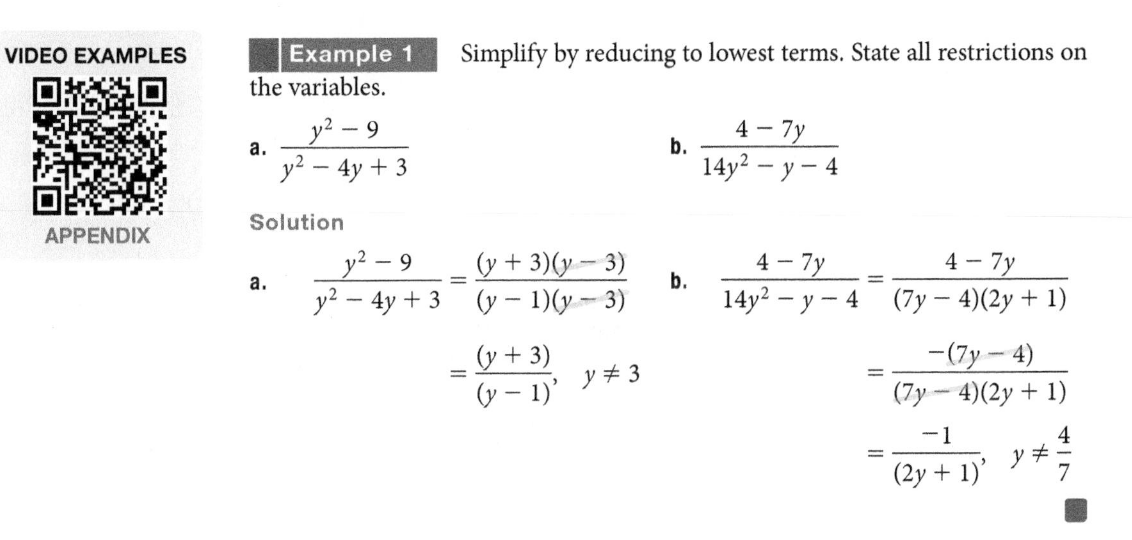

Example 1 Simplify by reducing to lowest terms. State all restrictions on the variables.

a. $\dfrac{y^2 - 9}{y^2 - 4y + 3}$

b. $\dfrac{4 - 7y}{14y^2 - y - 4}$

Solution

a. $$\frac{y^2 - 9}{y^2 - 4y + 3} = \frac{(y + 3)(y - 3)}{(y - 1)(y - 3)} = \frac{(y + 3)}{(y - 1)}, \quad y \neq 3$$

b. $$\frac{4 - 7y}{14y^2 - y - 4} = \frac{4 - 7y}{(7y - 4)(2y + 1)} = \frac{-(7y - 4)}{(7y - 4)(2y + 1)} = \frac{-1}{(2y + 1)}, \quad y \neq \frac{4}{7}$$

We multiply rational expressions by multiplying their numerators and multiplying their denominators and then simplifying the resulting expression as we did above. Division of rational expressions is converted into multiplication by using the reciprocal of the second expression.

Example 2 Perform the multiplication $\frac{x^2-12x+35}{x^2-5x-24}\cdot\frac{x^2-4x-32}{x^2-5x-14}$.

Solution

$$\begin{aligned}\frac{x^2-12x+35}{x^2-5x-24}\cdot\frac{x^2-4x-32}{x^2-5x-14} &= \frac{(x^2-12x+35)(x^2-4x-32)}{(x^2-5x-24)(x^2-5x-14)}\\ &= \frac{(x-5)(x-7)(x-8)(x+4)}{(x+3)(x-8)(x-7)(x+2)},\\ &= \frac{(x-5)(x+4)}{(x+3)(x+2)} \quad x\neq 7, 8.\end{aligned}$$

Example 3 Perform the division $\frac{y-2}{y^2-y-12}\div\frac{y^2-5y+6}{y^2-9}$.

Solution

$$\begin{aligned}\frac{y-2}{y^2-y-12}\div\frac{y^2-5y+6}{y^2-9} &= \frac{y-2}{y^2-y-12}\div\frac{y^2-9}{y^2-5y+6}\\ &= \frac{(y-2)(y^2-9)}{(y^2-y-12)(y^2-5y+6)}\\ &= \frac{(y-2)(y+3)(y-3)}{(y+3)(y-4)(y-2)(y-3)}\\ &= \frac{1}{y-4} \quad y\neq \pm 3, 2.\end{aligned}$$

We add or subtract rational expressions by finding a common denominator, just as with arithmetic fractions, and then we add the numerators together, leaving the denominator as the common denominator. Finally, we simplify the resulting expression by reducing it to lowest terms.

Example 4 Perform the subtraction $\frac{3x-7}{4x+1}-\frac{2x-15}{4x+1}$.

Solution

$$\begin{aligned}\frac{3x-7}{4x+1}-\frac{2x-15}{4x+1} &= \frac{(3x-7)-(2x-15)}{4x+1}\\ &= \frac{3x-7-2x+15}{4x+1}\\ &= \frac{x+8}{4x+1}\end{aligned}$$

Example 5 Perform the addition $\frac{5x}{x-7}+\frac{3x}{x-3}$.

Solution

$$\begin{aligned}\frac{5x}{x-7}+\frac{3x}{x-3} &= \frac{5x(x-3)}{(x-7)(x-3)}+\frac{3x(x-7)}{(x-7)(x-3)}\\ &= \frac{5x^2-15x}{(x-7)(x-3)}+\frac{3x^2-21x}{(x-7)(x-3)}\end{aligned}$$

$$= \frac{5x^2 - 15x + 3x^2 - 21x}{(x-7)(x-3)}$$

$$= \frac{8x^2 - 36x}{(x-7)(x-3)}$$

Which can also be expressed as

$$= \frac{4x(2x-9)}{(x-7)(x-3)}$$

Example 6 Perform the addition $\frac{y+2}{y-1} + \frac{y+8}{y^2 - 5y + 4}$.

Solution

$$\frac{y+2}{y-1} + \frac{y+8}{y^2-5y+4} = \frac{y+2}{y-1} + \frac{y+8}{(y-4)(y-1)}$$

$$= \frac{(y+2)(y-4)}{(y-1)(y-4)} + \frac{y+8}{(y-4)(y-1)}$$

$$= \frac{(y+2)(y-4) + (y+8)}{(y-1)(y-4)}$$

$$= \frac{y^2 + 2y - 8 + y + 8}{(y-1)(y-4)}$$

$$= \frac{y^2 - y}{(y-1)(y-4)}$$

$$= \frac{y(y-1)}{(y-1)(y-4)}$$

$$= \frac{y}{(y-4)}$$

A.2 Logarithms

Many theoretical and applied phenomena are modeled by exponential and logarithmic functions. This section helps you recall logarithms and reminds you of some of their properties. A more detailed examination of natural logarithms is made in Sections 4.2 and 4.3. Logarithms are used for solving exponential equations such as $e^{2x+5} = 12$.

We will begin our recollection of logarithms with the definition of a logarithm.

Logarithm

A **logarithm** is an exponent. Symbolically,

$$y = \log_b x \quad \text{if and only if} \quad b^y = x$$

where $b > 0$, $b \neq 1$, and $x > 0$.

The definition indicates that *a logarithm is the exponent* on a number that produces the number x.

Example 7

1. $\log_3 81 = 4$ since $3^4 = 81$. The exponent 4 on the number 3 produces the number 81. Therefore, 4 is the logarithm of 81, base 3.
2. $\log_{10} 1 = 0$ since $10^0 = 1$. The exponent 0 on the number 10 produces the number 1. Therefore, 0 is the logarithm of 1, base 10.
3. $\log_2 \frac{1}{8} = 3$ since $2^3 = \frac{1}{8}$.

There are some basic properties of logarithms similar to those for exponents (remember, logarithms are exponents). The following properties are true for $b > 0$ and $b \neq 1$, $m > 0$ and $n > 0$, and any real number r.

Logarithm Properties

1. $\log_b 1 = 0 \quad (\text{since } b^0 = 1)$
2. $\log_b b = 1 \quad (\text{since } b^1 = b)$
3. $\log_b b^r = r \quad (\text{since } b^r = b^r)$

   Note the implied order of performing operations is such that exponentiations are applied before function evaluations; thus, $\log_b b^r = \log_b(b^r)$.
4. $\log_b (mn) = \log_b m + \log_b n$
5. $\log_b \left(\frac{m}{n}\right) = \log_b m - \log_b n$
6. $\log_b m^r = r \log_b m$

The proofs of these properties can be found in any algebra book. The following examples illustrate these properties.

Example 8

1. $\log_5 1 = 0$ Property 1
2. $\log_5 5 = 1$ Property 2
3. $\log_5 5^7 = 7$ Property 3
4. $\log_4 (3x) = \log_4 3 + \log_4 x$ Property 4
5. $\log_7 \left(\frac{3}{5}\right) = \log_7 3 - \log_7 5$ Property 5
6. $\log_{10} 5^7 = 7 \log_{10} 5$ Property 6
7. $\log_{14} \sqrt{3} = \log_{14} 3^{1/2} = \frac{1}{2} \log_{14} 3$ Property 6

Example 9 Expand $\log_6 \left(\frac{10mn}{7abc}\right)$ as much as possible.

Solution

$$\begin{aligned}\log_6 \left(\frac{10mn}{7abc}\right) &= \log_6 (10mn) - \log_6 (7abc) \\ &= \log_6 10 + \log_6 m + \log_6 n - (\log_6 7 + \log_6 a + \log_6 b + \log_6 c) \\ &= \log_6 10 + \log_6 m + \log_6 n - \log_6 7 - \log_6 a - \log_6 b - \log_6 c\end{aligned}$$

Example 10 Expand $\log_2\left(\frac{x^2\sqrt{y}}{\sqrt[5]{w^4}}\right)$ as much as possible.

Solution

$$\log_2\left(\frac{x^2\sqrt{y}}{\sqrt[5]{w^4}}\right) = \log_2(x^2\sqrt{y}) - \log_2(\sqrt[5]{w^4})$$

$$= \log_2 x^2 + \log_2 y^{1/2} - \log_2 w^{4/5}$$

$$= 2\log_2 x + \frac{1}{2}\log_2 y - \frac{4}{5}\log_2 w$$

Example 11 Write $2\log_2 x + 5\log_2 y$ as a single logarithmic expression.

Solution $2\log_2 x + 5\log_2 y = \log_2 x^2 + \log_2 y^5 = \log_2(x^2y^5)$

Example 12 Write $\frac{3}{4}\log_5 x - \frac{2}{7}\log_5 y - \frac{1}{4}\log_5 z$ as a single logarithmic expression.

Solution $\frac{3}{4}\log_5 x - \frac{2}{7}\log_5 y - \frac{1}{4}\log_5 z = \log_5 x^{3/4} - \log_5 y^{2/7} - \log_5 z^{1/4}$

$$= \log_5\left(\frac{x^{3/4}}{y^{2/7}z^{1/4}}\right)$$

$$= \log_5\left(\frac{\sqrt[4]{x^3}}{\sqrt[7]{y^2}\sqrt[4]{z}}\right)$$

Two logarithms that are most commonly used are base 10 logarithms, referred to as **common logarithms** ($\log_{10} x$ is expressed as $\log x$) and base e logarithms, referred to as **natural logarithms** ($\log_e x$ is expressed as $\ln x$). Of course, these logarithms have the same properties as mentioned above.

Both common logarithms and natural logarithms are used to solve exponential equations. Some are demonstrated in the next examples.

Example 13 Solve $8^x = 7$ and state its meaning in exponential terms.

Solution

$$8^x = 7$$

$$\log 8^x = \log 7$$

$$x\log 8 = \log 7$$

$$x = \frac{\log 7}{\log 8}$$ Approximate the solution with a calculator

$$x \approx \frac{0.8451}{0.9031}$$

$$x \approx 0.9358$$

This means that $8^{0.9358} \approx 7$. Try this computation on your calculator to see how close the approximation is.

Example 14 Solve $3^{4x-1} = 4$ and state its meaning in exponential terms.

Solution

$$3^{4x-1} = 4$$
$$\ln 3^{4x-1} = \ln 4$$
$$(4x - 1)\ln 3 = \ln 4$$
$$4x - 1 = \frac{\ln 4}{\ln 3}$$
$$4x = \frac{\ln 4}{\ln 3} + 1$$
$$x = \frac{1}{4} \cdot \left(\frac{\ln 4}{\ln 3} + 1\right)$$
$$x \approx \frac{1}{4} \cdot \left(\frac{1.3863}{1.0986} + 1\right)$$
$$x \approx 0.5655$$

This means that $3^{4(0.5655)-1} \approx 4$.

Example 15 Solve $e^{7x-5} = 2$ and state its meaning in exponential terms.

Solution

$$e^{7x-5} = 2$$
$$\ln e^{7x-5} = \ln 2$$
$$(7x - 5)\ln e = \ln 2$$
$$(7x - 5) \cdot 1 = \ln 2$$
$$7x - 5 = \ln 2$$
$$7x = \ln 2 + 5$$
$$x = \frac{\ln 2 + 5}{7}$$
$$x \approx 0.8133$$

This means that $e^{7(0.8133)-5} \approx 2$.

A.3 *z*-scores Table

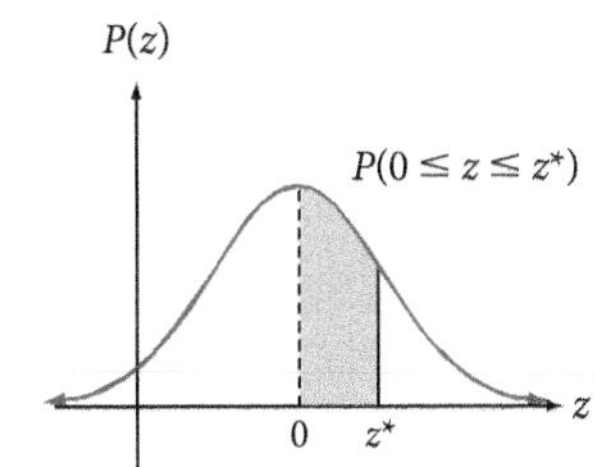

z*	.00	.01	.02	.03	.04	.05	.06	.07	.08	.09
.0	.0000	.0040	.0080	.0120	.0160	.0199	.0239	.0279	.0319	.0359
.1	.0398	.0438	.0478	.0517	.0557	.0596	.0636	.0675	.0714	.0753
.2	.0793	.0832	.0871	.0910	.0948	.0987	.1026	.1064	.1103	.1141
.3	.1179	.1217	.1255	.1293	.1331	.1368	.1406	.1443	.1480	.1517
.4	.1554	.1591	.1628	.1664	.1700	.1736	.1772	.1808	.1844	.1879
.5	.1915	.1950	.1985	.2019	.2054	.2088	.2123	.2157	.2190	.2224
.6	.2257	.2291	.2324	.2357	.2389	.2422	.2454	.2486	.2517	.2549
.7	.2580	.2611	.2642	.2673	.2704	.2734	.2764	.2794	.2823	.2852
.8	.2881	.2910	.2939	.2967	.2995	.3023	.3051	.3078	.3106	.3133
.9	.3159	.3186	.3212	.3238	.3264	.3289	.3315	.3340	.3365	.3389
1.0	.3413	.3438	.3461	.3485	.3508	.3531	.3554	.3577	.3599	.3621
1.1	.3643	.3665	.3686	.3708	.3729	.3749	.3770	.3790	.3810	.3830
1.2	.3849	.3869	.3888	.3907	.3925	.3944	.3962	.3980	.3997	.4015
1.3	.4032	.4049	.4066	.4082	.4099	.4115	.4131	.4147	.4162	.4177
1.4	.4192	.4207	.4222	.4236	.4251	.4265	.4279	.4292	.4306	.4319
1.5	.4332	.4345	.4357	.4370	.4382	.4394	.4406	.4418	.4429	.4441
1.6	.4452	.4463	.4474	.4484	.4495	.4505	.4515	.4525	.4535	.4545
1.7	.4554	.4564	.4573	.4582	.4591	.4599	.4608	.4616	.4625	.4633
1.8	.4641	.4649	.4656	.4664	.4671	.4678	.4686	.4693	.4699	.4706
1.9	.4713	.4719	.4726	.4732	.4738	.4744	.4750	.4756	.4761	.4767
2.0	.4772	.4778	.4783	.4788	.4793	.4798	.4803	.4808	.4812	.4817
2.1	.4821	.4826	.4830	.4834	.4838	.4842	.4846	.4850	.4854	.4857
2.2	.4861	.4864	.4868	.4871	.4875	.4878	.4881	.4884	.4887	.4890
2.3	.4893	.4896	.4898	.4901	.4904	.4906	.4909	.4911	.4913	.4916
2.4	.4918	.4920	.4922	.4925	.4927	.4929	.4931	.4932	.4934	.4936
2.5	.4938	.4940	.4941	.4943	.4945	.4946	.4948	.4949	.4951	.4952
2.6	.4953	.4955	.4956	.4957	.4959	.4960	.4961	.4962	.4963	.4964
2.7	.4965	.4966	.4967	.4968	.4969	.4970	.4971	.4972	.4973	.4974
2.8	.4974	.4975	.4976	.4977	.4977	.4978	.4979	.4979	.4980	.4981
2.9	.4981	.4982	.4982	.4983	.4984	.4984	.4985	.4985	.4986	.4986
3.0	.4987	.4987	.4987	.4988	.4988	.4989	.4989	.4989	.4990	.4990

Table 1-A *z*-scores

For $z > 3.10$, approximate the area with 0.4999.

Selected Answers

Chapter 1

Problem Set 1.1

1. Domain: $\{1, 3, 5, 7\}$; Range: $\{2, 4, 6, 8\}$; a function

3. Domain: $\{0, 1, 2, 3\}$; Range: $\{4, 5, 6\}$; a function

5. Domain: $\{a, b, c, d\}$; Range: $\{3, 4, 5\}$; a function

7. Domain: $\{a\}$; Range: $\{1, 2, 3, 4\}$; not a function

9. A function **11.** A function

13. Domain: $\{x \mid -5 \le x \le 5\}$; Range: $\{y \mid 0 \le y \le 5\}$

15. Domain: $\{x \mid -5 \le x \le 3\}$; Range: $\{y \mid y = 3\}$

17. -1 **19.** -11 **21.** 2 **23.** 4 **25.** $a^2 + 3a + 4$

27. $2a + 7$ **29.** $\frac{3}{10}$ **31.** $\frac{2}{5}$ **33.** undefined

35. **a.** $a^2 - 7$ **b.** $a^2 - 6a + 5$ **c.** $x^2 - 2$ **d.** $x^2 + 4x$
e. $a^2 + 2ab + b^2 - 4$ **f.** $x^2 + 2xh + h^2 - 4$

37. Domain: All real numbers; Range: $\{y \mid y \ge -1\}$; a function

39. Domain: $\{x \mid x \ge -1\}$; Range: All real numbers; not a function

41. Domain: $\{x \mid x \ge 0\}$; Range: All real numbers; not a function

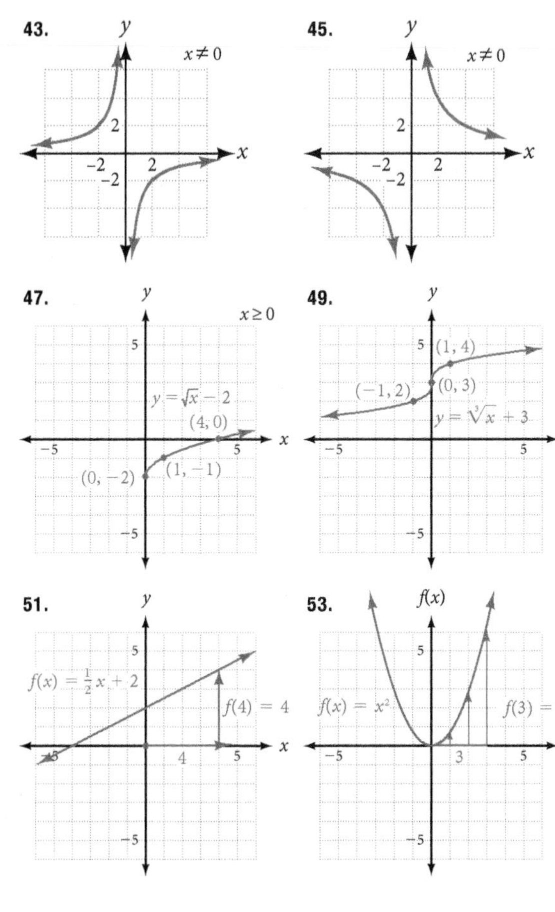

55. **a.** $y = 9.50x \quad \text{for} \quad 10 \le x \le 40$

b.

Hours Worked	Function Rule	Gross Pay ($)
x		y
10	$y = 9.50(10)$	95.00
20	$y = 9.50(20)$	190.00
30	$y = 9.50(30)$	285.00
40	$y = 9.50(40)$	380.00

c.

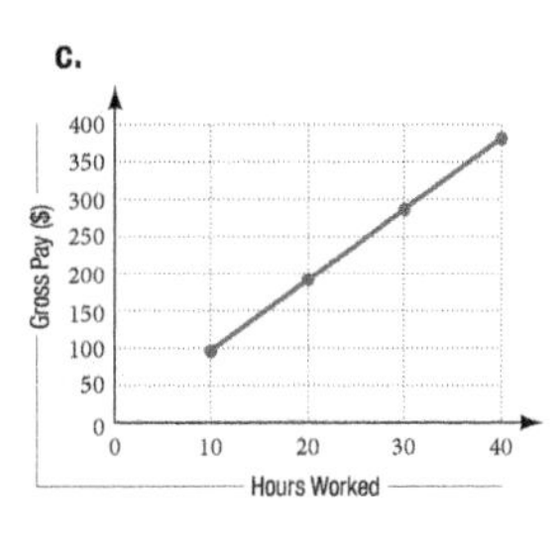

d. Domain: $\{x \mid 10 \le x \le 40\}$; Range: $\{y \mid 95 \le y \le 380\}$
e. Minimum = \$95; Maximum = \$380

57. Domain: $\{x \mid 450 \le x \le 2{,}600\}$; Range: $\{y \mid 10 \le y \le 40\}$

59. **a.**

Year	Revenue
2007	0
2008	\$16,000,000
2009	\$6,000,000
2010	0
2011	\$4,000,000
2012	0

b. $R(6)$ and $R(16)$ are negative. Therefore, revenues will not rise after 2012

61. **a.** True **b.** False **c.** True **d.** False **e.** True

63.

Weeks	Weight
0	200
1	194.3
4	184
12	173.3
24	168

65. **a.** $\{x \mid x \ge 0\}$ **b.** \$11.26

67. $0, \pm 2$ **69.** $-1, 2$

71.

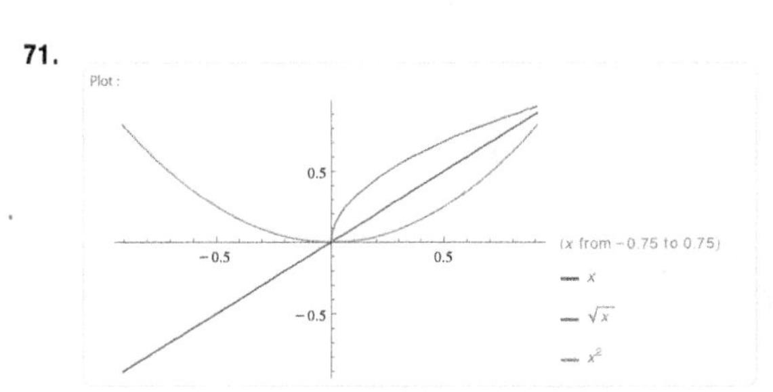

Wolfram Alpha LLC. 2012. Wolfram|Alpha
http://www.wolframalpha.com/
(access December 05, 2012)

73. a. $\{x \mid x \geq 0\}$ **b.** $\{x \mid x \geq 2\}$
c. $\{x \mid x \geq -2\}$

75.

77.

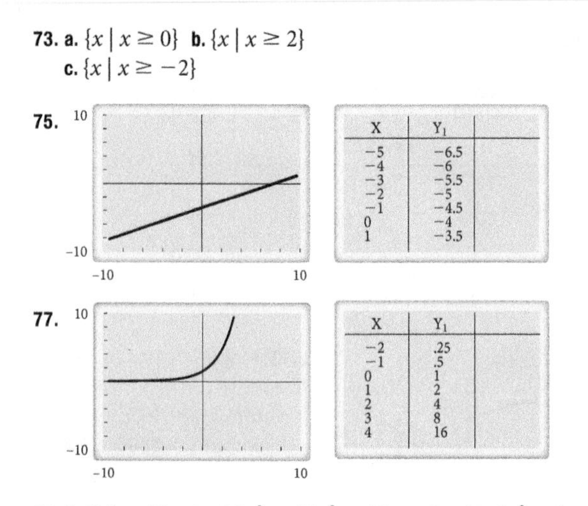

X	Y₁
−5	−6.5
−4	−6
−3	−5.5
−2	−5
−1	−4.5
0	−4
1	−3.5

X	Y₁
−2	.25
−1	.5
0	1
1	2
2	4
3	8
4	16

79. $0.6M - 42$ **81.** $16x^3 - 40x^2 + 33x - 9$ **83.** $4x^2 - 3x$
85. $6x^2 - 2x - 4$ **87.** 11

Problem Set 1.2

1. $6x + 2$ **3.** $-2x + 8$ **5.** $8x^2 + 14x - 15$ **7.** $\frac{2x+5}{4x-3}$
9. $4x - 7$ **11.** $3x^2 - 10x + 8$ **13.** $-2x + 3$
15. $3x^2 - 11x + 10 = h(x)$ **17.** $9x^3 - 48x^2 + 85x - 50$
19. $x - 2 = g(x)$ **21.** $\frac{1}{x-2}$ **23.** $3x^2 - 7x + 3$
25. $6x^2 - 22x + 20$ **27.** 15 **29.** 98 **31.** $\frac{3}{2}$ **33.** 1 **35.** 40
37. 147 **39. a.** 81 **b.** 29 **c.** $(x+4)^2$ **d.** $x^2 + 4$
41. a. -2 **b.** -1 **c.** $16x^2 + 4x - 2$ **d.** $4x^2 + 12x - 1$
43. $(f \circ g)(x) = 5\left[\frac{x+4}{5}\right] - 4 = x + 4 - 4 = x$ $\quad (g \circ f)(x) = \frac{(5x-4)+4}{5} = \frac{5x}{5} = x$
45. a. $M(x) = 220 - x$ **b.** $M(24) = 196$ **c.** 142
d. 135 **e.** 128
47. 100 or 130 DVDs **49.** $f(g(16)) = 76, g(f(16)) = 35.6$
51. $f(g(2)) = 0, g(f(2)) = \frac{4}{9} \approx 0.4$
53. a. $P(x) = -0.02x^3 - 0.6x^2 + 31.5x - 50$
b. $P(19) = \$194.72$ **c.** $P(20) = \$180.00$
d. $P(21) = \$161.68$ **e.** Profit is decreasing
55. a. $F(x) = x - 50$ **b.** $D(x) = 0.8x$
c. $D(F(x)) = 0.8(x - 50)$ **d.** $D(F(500)) = \$360$
57. 5.53 ppm; The daily level of carbon monoxide 15 years from now is 5.53 ppm.
59. $-\frac{6}{100}$ **61.** 1 **63.** $-\frac{4}{3}$ **65.** undefined
67. $y = mx + b$ **69.** $y = -2x - 5$ **71.** 5

Problem Set 1.3

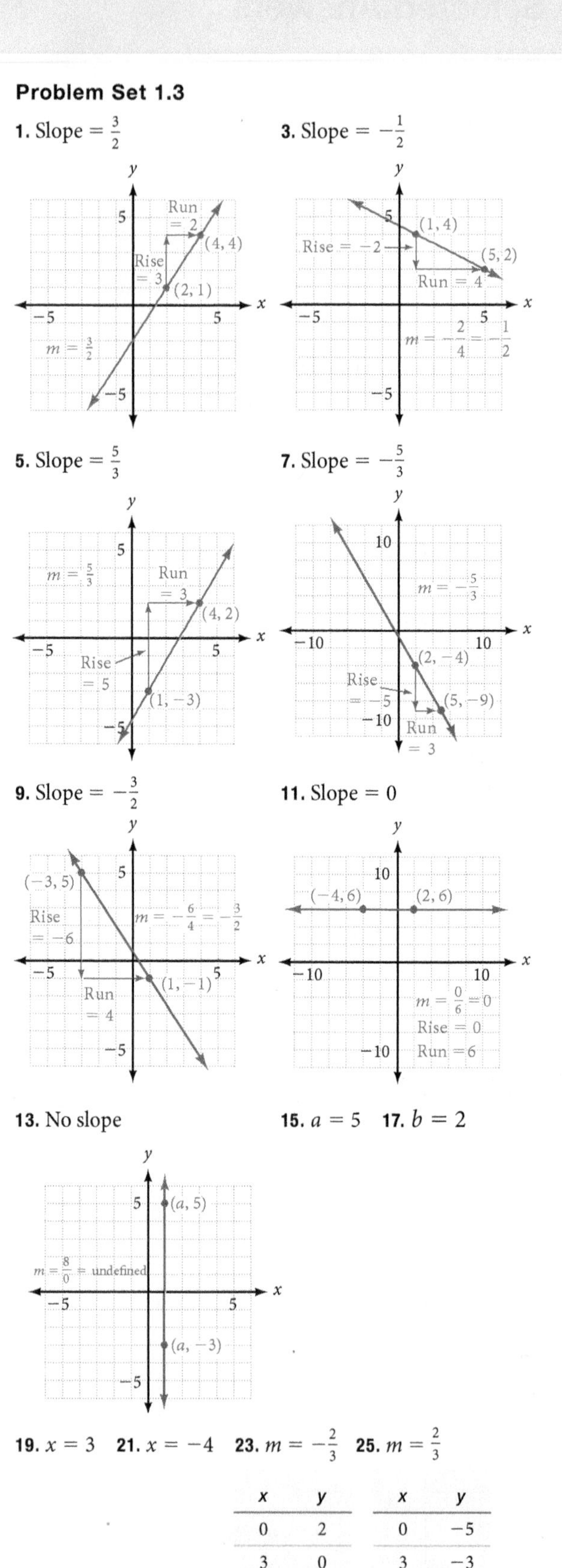

1. Slope $= \frac{3}{2}$ **3.** Slope $= -\frac{1}{2}$

5. Slope $= \frac{5}{3}$ **7.** Slope $= -\frac{5}{3}$

9. Slope $= -\frac{3}{2}$ **11.** Slope $= 0$

13. No slope **15.** $a = 5$ **17.** $b = 2$

19. $x = 3$ **21.** $x = -4$ **23.** $m = -\frac{2}{3}$ **25.** $m = \frac{2}{3}$

x	y
0	2
3	0

x	y
0	−5
3	−3

27. 4 **29.** 5 **31.** $\frac{b^2 - a^2}{b - a} = b + a$ **33.** $\frac{b^2 - a^2}{b - a} = b + a$

35. $\frac{b^2 - a^2 + 3(a - b)}{b - a} = b - a - 3$ **37.** $y = -4x - 3$

39. $y = -\frac{2}{3}x$ **41.** $y = -\frac{2}{3}x + \frac{1}{4}$

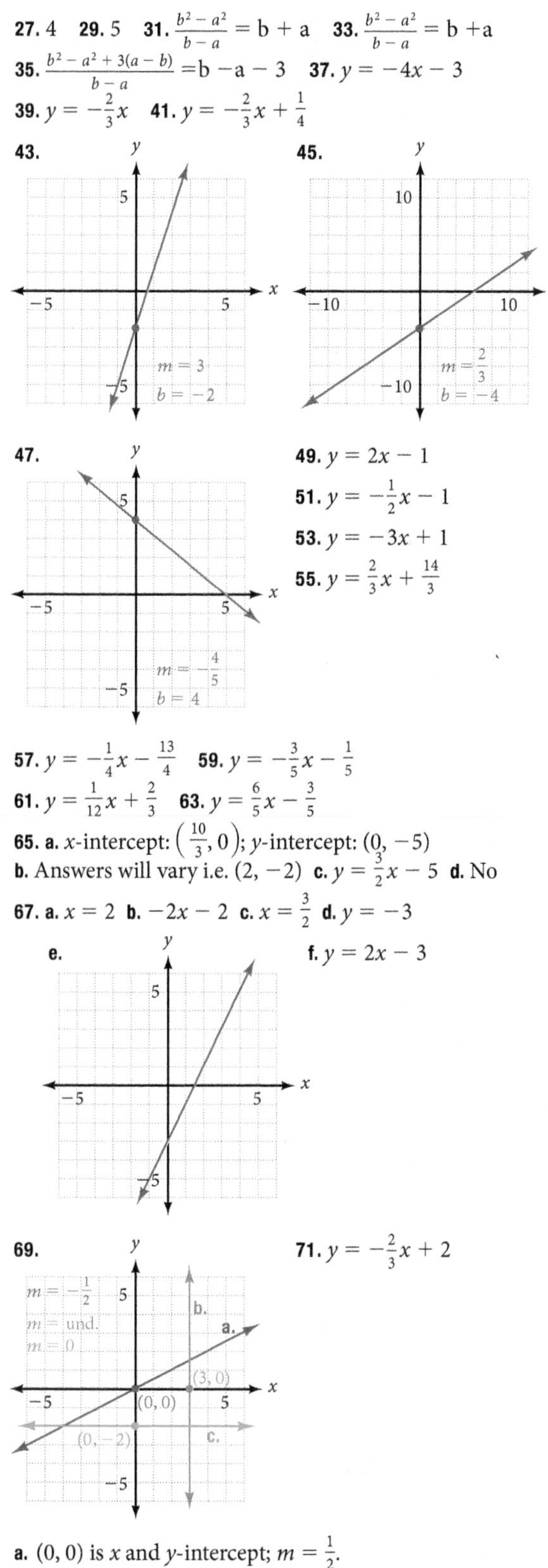

49. $y = 2x - 1$

51. $y = -\frac{1}{2}x - 1$

53. $y = -3x + 1$

55. $y = \frac{2}{3}x + \frac{14}{3}$

57. $y = -\frac{1}{4}x - \frac{13}{4}$ **59.** $y = -\frac{3}{5}x - \frac{1}{5}$

61. $y = \frac{1}{12}x + \frac{2}{3}$ **63.** $y = \frac{6}{5}x - \frac{3}{5}$

65. a. x-intercept: $\left(\frac{10}{3}, 0\right)$; y-intercept: $(0, -5)$
b. Answers will vary i.e. $(2, -2)$ **c.** $y = \frac{3}{2}x - 5$ **d.** No

67. a. $x = 2$ **b.** $-2x - 2$ **c.** $x = \frac{3}{2}$ **d.** $y = -3$

e.

f. $y = 2x - 3$

69.

71. $y = -\frac{2}{3}x + 2$

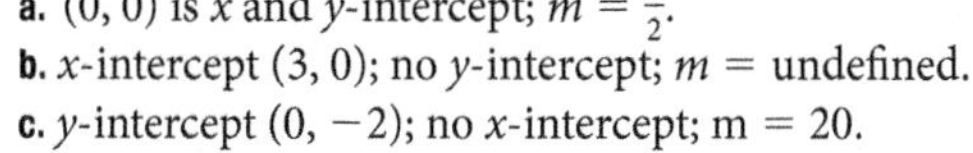

a. $(0, 0)$ is x and y-intercept; $m = \frac{1}{2}$.
b. x-intercept $(3, 0)$; no y-intercept; $m =$ undefined.
c. y-intercept $(0, -2)$; no x-intercept; m = 20.

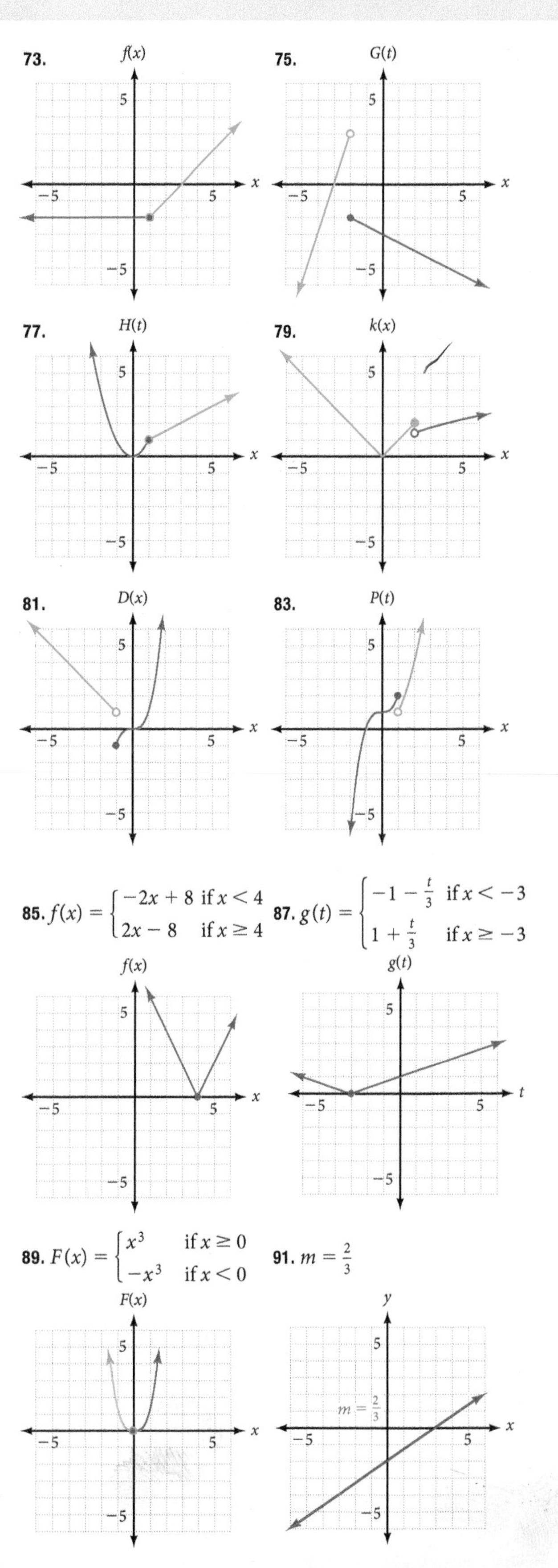

85. $f(x) = \begin{cases} -2x + 8 & \text{if } x < 4 \\ 2x - 8 & \text{if } x \geq 4 \end{cases}$ **87.** $g(t) = \begin{cases} -1 - \frac{t}{3} & \text{if } x < -3 \\ 1 + \frac{t}{3} & \text{if } x \geq -3 \end{cases}$

89. $F(x) = \begin{cases} x^3 & \text{if } x \geq 0 \\ -x^3 & \text{if } x < 0 \end{cases}$ **91.** $m = \frac{2}{3}$

93. $\frac{3}{2}$ **95.** No slope **97.** $\frac{2}{3}$ **99.** $m = \frac{1}{2}; b = -4; y = \frac{1}{2}x - 4$

101. $m = -\frac{2}{3}; b = 3; y = -\frac{2}{3}x + 3$

103. $(0, -4); (2, 0); y = 2x - 4$

105. $(0, 4); (-2, 0); y = 2x + 4$ **107.** 17.5 mph

109. 120 ft/sec

111. a. 10 minutes **b.** 20 minutes

c.

$$y(x) = \begin{cases} 20x - 20 & \text{if } 0 \le x < 1 \\ 0 & \text{if } 1 \le x < 10 \\ 10x - 100 & \text{if } 10 \le x < 20 \\ 100 & \text{if } x \ge 20 \end{cases}$$

d. $m = 20$, °C/min **e.** $m = 10$, °C/min **f.** 1st minute

113. a. \$16 million/yr **b.** \$3 million/yr

115. a. $\frac{11}{215} \approx 0.051$ watts/lumen **b.** $\frac{3}{215} \approx 0.014$ watts/lumen
c. Energy efficient light bulbs are better because less energy is used to produce the same light output.

117. 23.18 mph/sec

119. a. $S(0) = 0$; $S(1) = 5{,}000$; $S(2) = 10{,}000$. No books were sold at the beginning. After 1 year, 5,000 books were sold. After 2 years, 10,000 books were sold.
b. 5,000 books/year. This is half of $S(2)$. **c.** 7,500 books/year.

121. a. $E(A) = 132 - 0.6A$ **b.** 119 beats per minute.

c.

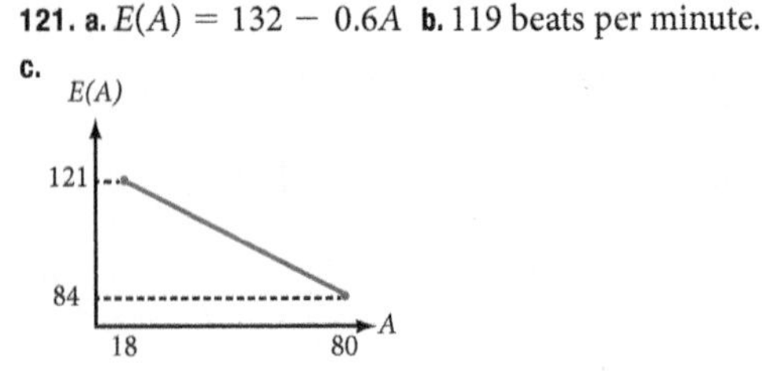

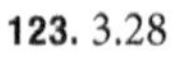

123. 3.28

125.

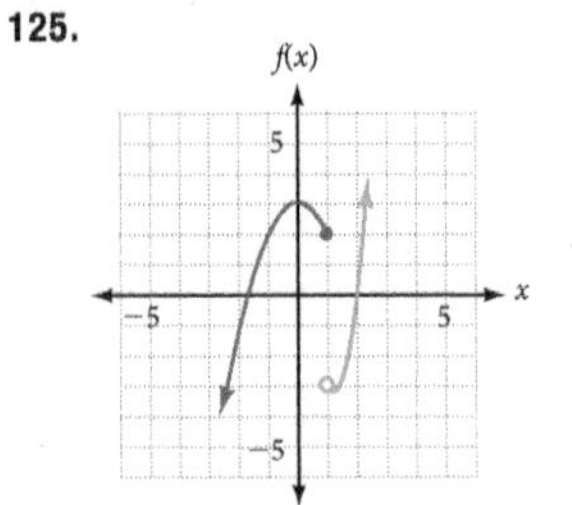

127.

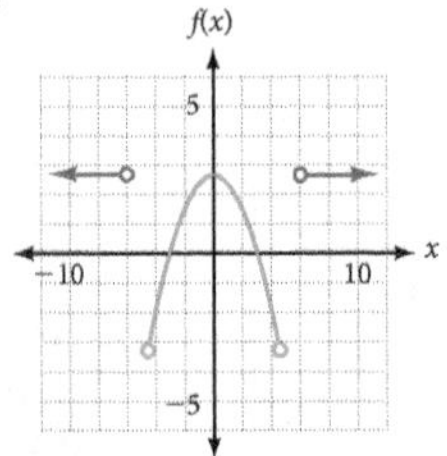

129. a. $\{x \mid x < 2 \text{ or } x > 2\}$
b. $f(1) = 7; f(1.5) = 8.5; f(1.99) = 9.97$. **c.** 4. **d.** $-\frac{4}{3}$.

131. a. $\{x \mid x < 3 \text{ or } x > 3\}$ **b.** $f(2) = 1, f(5) = \frac{1}{4}$ **c.** None
d.

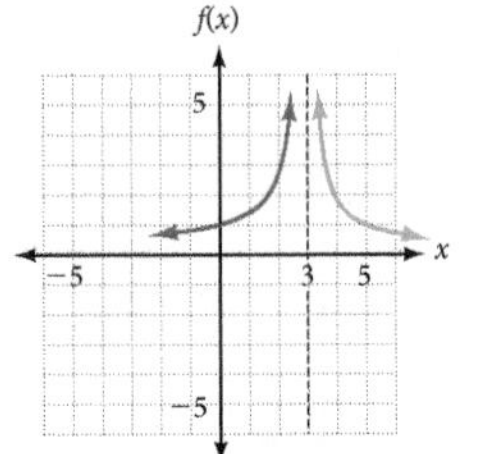

133. a. $\frac{\sqrt{2}}{2}$ **b.** $\sqrt{2} - 1$ **c.** $3 + 2\sqrt{2}$

Problem Set 1.4

1. $\lim_{x \to a} y = P$

3. a. 6 **b.** 4

5. $\lim_{x \to 8^-} f(x) = 13$; $\lim_{x \to 8^+} f(x) = 13$; $\lim_{x \to 8} f(x) = 13$

7. $\lim_{x \to 6^-} f(x) = 2$; $\lim_{x \to 6^+} f(x) = 4$; $\lim_{x \to 6} f(x)$ does not exist

9. $\lim_{x \to 1^-} f(x) = -3$; $\lim_{x \to 1^+} f(x) = -3$; $\lim_{x \to 1} f(x) = -3$

11. 5 **13.** 2 **15.** -3 **17.** -11 **19.** -4 **21.** $-\frac{1}{3}$

23. Does not exist

25. $\frac{2}{3}$ **27.** 0 **29.** ∞ **31.** 5 **33.** 850 **35.** 1

37. a. -1 **b.** -1 **c.** Does not exist

39. a. ∞ **b.** ∞ **c.** 4

41. a. 2 **b.** 0 **c.** False **43. a.** 2 **b.** -1 **c.** True

45. a. $\lim_{x \to 210} C(x)$ does not exist because $\lim_{x \to 210^-} C(x) \ne \lim_{x \to 210^+} C(x)$

b. $\lim_{x \to 340^-} C(x) = \lim_{x \to 340^+} C(x) = 27.20$

c. $\lim_{x \to 450^-} C(x) = \lim_{x \to 450^+} C(x) = 38.20$

47. $\lim_{x \to \infty} \overline{C}(x) = 0.89$; As the company produces more and more tubes of toothpaste, the average cost per tube approaches \$0.89.

49. $\lim_{p \to \infty} N(p) = 0$, All of the fish would die.

51. a. \$17.20 **b.** \$3.00

53. The concentration approaches 0 mg/L.

55. Horizontal asymptote is $V(t) = 340$, vertical asymptote is $t = -3$.

57. a. 3 **b.** 3 **c.** 3 **d.** 3

Problem Set 1.5

1. Continuous at $x = -1$ **3.** Continuous at $x = 3$

5. Discontinuous at $x = 2$ because $f(2)$ does not exist

7. $f(x)$ is discontinuous at $x = 1$ because $f(1)$ does not exist

9. $f(x)$ is continuous at $x = 0$

11. $f(x)$ is discontinuous at $x = -4$ because $f(-4) \neq \lim_{x \to -4} f(x)$.

13. Continuous at $x = 3$

15. Discontinuous; $\lim_{x \to a} f(x)$ does not exist

17. Discontinuous; $f(a)$ does not exist

19. Continuous at $x = a$ **21.** Continuous at $x = a$

23. Discontinuous; $\lim_{x \to a} f(x) \neq f(a)$

25. Discontinuous; $f(a)$ does not exist

27. $A(a)$ is discontinuous. The jump at $t = a$ might be the time when it all makes sense.

29. $P(t)$ is discontinuous at $t = a$. The sudden drop between $t = a$ and $t = b$ might be caused by the firing of a chief executive officer.

31. $P(t)$ is discontinuous at $t = a$. The sudden jump at a might be caused by a leaking oil tank.

33. The actual graph might be a set of disconnected points above or below the curve.

35. a. Domain is $\{x \mid x \geq 0\}$. **b.** The discontinuity is at $x = -150$ which is not in the domain of the function.

37. **39.** **41.**

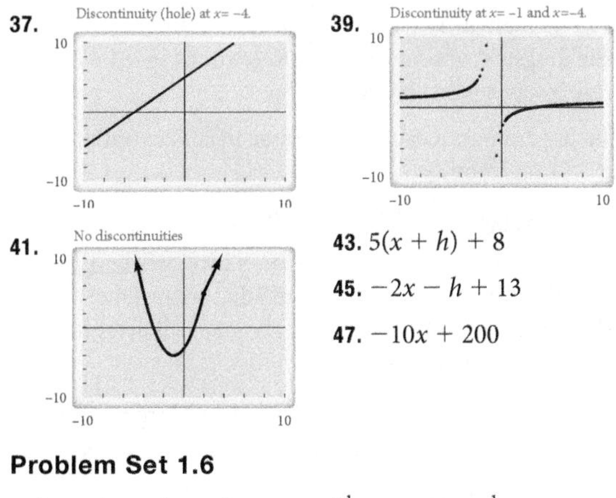

43. $5(x + h) + 8$

45. $-2x - h + 13$

47. $-10x + 200$

Problem Set 1.6

1. 5 **3.** 4 **5.** 1 **7.** 3 **9.** $\frac{-1}{\sqrt{x} + \sqrt{x+h}}$ **11.** $\frac{1}{x(x+h)}$

13. 6

15. 4

17. 4

19. -5

21. 16,000 **23.** -1.5

25. The average rate of change is $-\frac{7}{80}$. This means that the concentration of this drug in the bloodstream is decreasing at an average rate of 7 mg/cc every 80 hours.

27. For a selling price between 75 cents and 80 cents, the average profit increases 60 cents for every 1 cent increase in price.

29. Instantaneous rate of change of y when $x = 4$ is 31. When $x = 4$, y increases by 31 for every unit increase in x.

31. The average rate-of-change is 1680. This means that, on the average, we can expect about 1680 new jobs in the physician assistant occupation each year during the years from 2010 and 2020.

33. a. \$17 **b.** $-$\$3.

35. a. .08 years **b.** .08 years **c.** .08 years

37. 40.8 mph per sec **39. a.** 0.641 **b.** 0.658 **41.** 53

43. $2x + 6$ **45.** $y = 4x + 5$ **47.** 3

Chapter 1 Test

1. $\{x \mid x \neq 3 \text{ or } x \neq -3\}$ **2. a.** 13 **b.** 13 **c.** 11

3. $(3, 0)$; $(0, 6)$; $m = -2$ **4.** $y = 2x + 5$ **5.** $3x + 7y = 5$

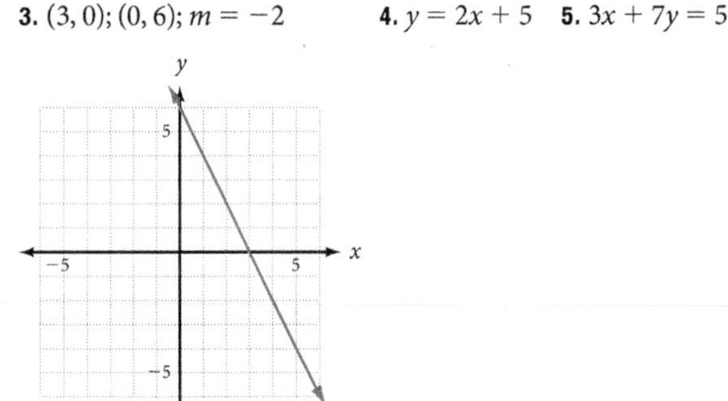

6.

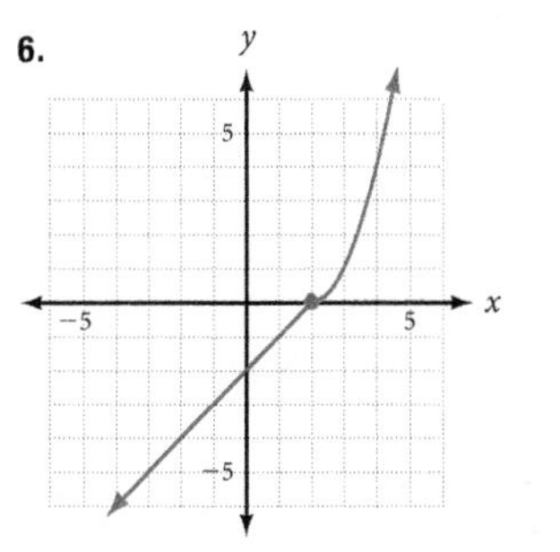

7. $\lim_{x \to 13} f(x) = 2$

8. a. 4 **b.** 6 **9.** 8 **10.** 0 **11.** 4 **12.** -4 **13.** $\frac{7}{8}$ **14.** 0 **15.** ∞

16. Discontinuous at $x = a$ because $\lim_{x \to a} f(x)$ does not exist

17. Discontinuous at $x = a$ because $f(a)$ does not exist

18. Continuous at $x = 6$

19. Discontinuous at $x = 1$ because $\lim_{x \to 1} f(x)$ does not exist

20. 1 **21.** $-6x - 3h + 1$ **22.** 22 **23.** -8

24. a. 16,875 **b.** 600 **c.** 2,400

25. a. $\{x \mid x \geq 0\}$ **b.** \$11.26 **c.** $-$\$0.0183 **d.** \$3; in the long run, the average cost will be \$3 **e.** $\lim_{x \to -150} \overline{C}(x)$ does not exist. As x approaches -150 from the right the graph approaches ∞.

Chapter 2

Problem Set 2.1

1. $f'(x) = 2x + 2$ **3.** $y' = 8$ **5.** $f'(x) = -\frac{3}{x^2}$

7. $f'(x) = 30x^2 - 12$ **9.** $f'(x) = 20x^{3/2}$

11. $y' = -9x^{-5/2} + 2x^{-1/2} + 12x^{1/2}$ **13.** $y' = \frac{-1}{x^2}$

15. $f'(x) = -\frac{6}{x^4} - \frac{6}{x^3} - \frac{3}{x^2}$ **17.** $f'(x) = \frac{4}{5\sqrt[5]{x^7}} + \frac{1}{\sqrt[3]{x^4}}$

19. $f(3) = 23; f'(3) = 12; f(4) - f(3) = 13$

21. $\left.\frac{dy}{dx}\right|_{x=0} = 1$ **23.** $x = -3$

25. a. 5 **b.** -8 **c.** $x = 2$ **d.** $y = 8x - 19$

27. $P(60) = 40$ means that when 60 thousand dollars is spent on advertising, the resulting profit is 40 thousand dollars. $\left.\frac{dP}{dx}\right|_{x=60} = 2$ means that if the amount spent on advertising is increased from 60 to 61 thousand, profit will increase by approximately $2,000.

29. $N(150) = 1{,}200$ means that it requires 150 hours to build 1,200 objects. $N'(150) = 21.25$ means that if the number of hours increases from 150 to 151, the number of objects built increases by approximately 21.25.

31. $S(44) = 87$ means that if there are 44 teachers with degrees in their teaching area, the average exam score will be 87. $S'(44) = 1.5$ means that if the number of teachers with degrees increases from 44 to 45, the average score on the exam will increase by about 1.5 points.

33. $f(2) = 7$ means that the cloud has traveled 7 miles in 2 hours since its release. $f'(2) = 3$ means that between $t = 2$ hours and $t = 3$ hours the cloud travels approximately 3 miles.

35. a. $R(4) = 4$ means that the revenue was $4 million. **b.** $R'(4) = 5$ means that the revenue was changing by about $5 million per year in 2011.

37. a. $A(54) = 1.57153 \text{ m}^2$ **b.** $A'(54) = 0.0194 \text{ m}^2/\text{kg}$ (For a 54-kg man, a 1-kg increase in weight means an increase in surface area of about 0.0194 m^2.)

39. a. 125 beats/min **b.** 62.5 beats/min **c.** $h'(81) \approx -0.26$ beats/min²; For an 81-lb child, an increase of one pound means a decrease in heart rate of approximately 0.26 beats/min.

41. a. Approximate change = $46.02 **b.** $637,287.43

43. Decreasing by 0.848528 mg/hr **45.** 0.10081

47. The calculator responds with ERR:NONREAL ANS and Wolfram|Alpha with (indeterminate). The reason for these responses is that for the function $f(x)$, $f'(x) = \frac{9}{4\sqrt[4]{x}} + \frac{2}{3\sqrt[3]{x}}$ which is not defined at $x = 0$.

49. $5x^5 + 4x^3 + 55x^2 + 44$ **51.** $3x^4 - x^2 - 2$

53. $4t^3 - 6t$ **55.** $t = \pm 2$ **57.** $\frac{-2}{9}$ **59.** -1 **61.** $y = -x + 7$

63. $y' = 10x$ **65.** $y' = 2$ **67.** $y' = 2t$

69. $y' = 25x^4 + 12x^2 + 110x$

Problem Set 2.2

1. $16x - 6$ **3.** $3x^2 - 4x + 4$ **5.** $4x^3 + 2x$ **7.** $\frac{-7}{(5x+2)^2}$

9. $\frac{14}{(3x+2)^2}$ **11.** $\frac{3x^4 - 29x^2 - 2x + 28}{(x^2-4)^2}$ **13.** $\frac{x^6 + 4x^3 + 18x^2}{(x^3+1)^2}$

15. $\frac{28x^3 + 5x^2 - 4x + 55}{(7x+2)^2}$ **17.** $\frac{-16x - 24}{(x^2+3x+4)^2}$ **19.** $\frac{-\frac{\sqrt{x}}{2} - \frac{2}{\sqrt{x}}}{(x-4)^2}$

21. At $x = 4$, the function is increasing 20 units for each unit increase in x.

23. At $t = 10$, the function is decreasing 0.00027 units for each unit increase in t.

25. $12x^3(x^3 + 1) + 3x^4(3x^2)$

27. $(3x + 7)\left(\frac{5}{2\sqrt{5x+2}}\right) + \sqrt{5x+2} \cdot 3$

29. $\frac{(8x+1) \cdot 5 - (5x-7) \cdot 8}{(8x+1)^2}$

31. $\frac{1}{4}(x^2 + 3)^{-3/4}(2x)(2x + 7) + 2(x^2 + 3)^{1/4}$

33. $N(1) = 3.2$; $N(3) \approx 1.02$ One hour after memorizing them, a person remembers about 3 facts, and 3 hours after memorizing, only 1 fact is remembered. $N'(1) \approx -10.24$; $N'(3) \approx -0.116$ One hour after memorizing, a person is forgetting about 10 facts per hour, and 3 hours after memorizing, the person is forgetting about 0.1 facts per hour.

35. a. 194.3 mph **b.** $V'(t) = \frac{1{,}020}{(t+3)^2}$ **c.** 20.8 mph per sec **d.** If the dragster runs for one more second, its speed will increase by approximately 20.8 mph.

37. The physician can expect a decrease in concentration of about 0.41 mg/mm³.

39. a. It is $\frac{3}{5}$ as big **b.** $\frac{-12x^2}{(-x^2+9x)^2}$ or $\frac{-12}{(x-9)^2}$ **c.** $-\frac{12}{25}$ **d.** $-\frac{12}{25} = -0.48$. As the number of years increases by 1 from 2011 to 2012, the size of the financial advisors revenue compared to the size of the accountants revenue will decrease by approximately 0.48 million dollars.

41. a. Marginal revenue: $R'(x) = 95 - 0.16x$; Marginal cost: $E'(x) = 35 - 0.08x$; Marginal profit: $P'(x) = 60 - 0.08x$ **b.** $R'(601) = -1.16$, $R'(900) = -49$; As production increases, marginal revenue is decreasing. **c.** $C'(601) = -13.08$, $C'(900) = -37$; As production increases, marginal cost is decreasing. **d.** $P'(601) = 11.92$, $R'(900) = -12$; at $x = 601$, as production increases, marginal profit is increasing, and at $x = 900$, as production increases, marginal profit is decreasing.

43. a. The marginal profit is given by the derivative of the profit function. Applying the quotient rule,

$$P'(t) = \frac{-100(5t^2 + 6t - 50)}{(t^2 + 10)^2}$$

b. At $t = 5$, $P'(5) \approx -8.571$. The publisher can expect the profit realized on the sale of the book to decrease by about \$8,571 from month 5 to month 6. **c.** At $t = 10$, $P'(10) \approx -4.215$. The publisher can expect the profit realized on the sale of the book to decrease by about \$4,215 from month 10 to month 11. **d.** For all values of $t > 3$, $P'(t) < 0$, demonstrating that the profit function $P(t)$ is always decreasing.

45. a. This quotient indicates that 2 years from now, the revenue predicted by analyst Star will be about 1.04 times as big as that predicted by analyst Moon. **b.** This quotient indicated that 2 years from now, the revenue predicted by analyst Star will be increasing by about 3.12 times as fast as that predicted by analyst Moon. **c.** This means that in year 2, the ratio of Star's predicted value to Moon's predicted value will increase by a factor of about 0.02. In year 2 then, we expect, Star's predicted value to be approximately $1.04 + 0.02 = 1.06$ times Moon's predicted value.

47. $f'(x) = \frac{11}{(x+6)^2}$

49. $f'(64) = \frac{1}{216}$

51. $P(3) = 0.1875 = 18.75\%$ and $P'(3) \approx -0.02 = -2\%$ when 3 liters of distilled water is added to 5 liters of a 30% acid solution, the 30% acid solution is diluted to an 18.75% acid solution. If the number of liters of distilled water that is to be added to 5 liters of a 30% acid solution increases by 1, from 3 to 4, the 30% acid solution will be diluted to about 2%.

53. $f(2) = -8$

55. $x = 0, \pm 3$ **57.** $g'(x) = 6x^2 - 8$ **59.** $f'(x) = 3x^2 - 9$

61. $y' = -\frac{1}{x^2}$

Problem Set 2.3

1. $f'(x) = 6x^5 - 10x^4 + 6; f''(x) = 30x^4 - 40x^3;$ $f'''(x) = 120x^3 - 120x^2; f^{(4)} = 360x^2 - 240x$

3. $f'(x) = 4x^3 + 6x^2 - 11x^{-2}; f''(x) = 12x^2 + 12x + 22x^{-3};$ $f'''(x) = 24x + 12 - 66x^{-4}; f^{(4)} = 24 + 264x^{-5}$

5. $f'(x) = 3x^2 + 12x - 2x^{-1/2}; f''(x) = 6x + 12 + x^{-3/2};$ $f'''(x) = 6 - \frac{3}{2}x^{-5/2}; f^{(4)} = \frac{15}{4}x^{-7/2}$

7. $f'(x) = \frac{4}{x^2}; f''(x) = \frac{-8}{x^3}$

9. $f'(x) = \frac{-4x-21}{x^4}; f''(x) = \frac{12x+84}{x^5}$

11. $f'(x) = 2x + 5; f'(1) = 7$ means that at $x = 1$ the function is increasing 7 units for every unit increase in x; $f''(x) = 2; f''(1) = 2$ means that at $x = 1$ the function is increasing at an increasing rate

13. a. $f(1) = -3$ **b.** $f'(1) = 2$ **c.** $f''(1) = 16$ **d.** $x = \frac{1}{4}, \frac{3}{2}$ **e.** $x = \frac{7}{8}$

15. Since the first derivative is positive and the second derivative is positive when 4,500 pillows are produced, the cost is increasing at an increasing rate.

17. Since the first derivative and the second derivative are both negative 230 days after the beginning of the year, the flow rate is decreasing at an increasing rate.

19. Since the first derivative is positive and the second derivative is negative at a complexity index of 7, the useful life of the toy is increasing at a decreasing rate.

21. Since the first and second derivatives are both negative at month 15, the number of alcohol-related accidents is decreasing at an increasing rate.

23. Since both first and second derivatives are positive 32 years after 1980, the global manufacturing value added is increasing at an increasing rate.

25. Since the first derivative is positive and the second derivative is negative when 40 mg of the drug is administered, the strength of the reaction is increasing at a decreasing rate.

27. At 4 hours after introduction of the toxin, the population is decreasing at a decreasing rate.

29. $N'(7)$ is negative and $N''(7)$ is negative, so 7 days after the end of the advertising campaign, the number of items sold is decreasing at an increasing rate.

31. $R'(5)$ is negative and $R''(5)$ is negative, so in 2012 the revenue is decreasing at an increasing rate.

33. $f''(x) = -\frac{22}{(x+6)^3}$

35. $f'(64) = 0.00148148$ and $f''(64) = -0.00004064$; since the first derivative is positive and the second derivative is negative we conclude that $f(x)$ is increasing at a decreasing rate.

37. $8x - 30$ **39.** $(7x - 8)^{3/5}$ **41.** 2.57 **43.** 441.62

45. $y' = 2$ **47.** $y' = 7$ **49.** $y' = 12x^2 + 5$ **51.** $y' = 6x^5$

53. $y' = \frac{3}{5x^{2/5}}$

Problem Set 2.4

1. 36 **3.** $-20x$ **5.** $300x^3 + 240x$ **7.** $750x^5 - 600x^3 + 60x$

9. $6(5x+1)(5x^2 + 2x - 6)^2$ **11.** $\frac{6x}{5\sqrt[5]{(x^2+7)^2}}$

13. $2(x^2+3)^2(2x-1)^4(11x^2 - 3x + 15)$

15. $\frac{x^3(21x+8)}{(5x+2)^{4/5}}$ **17.** $\frac{2x^6(31x+7)}{(4x+1)^{1/4}}$ **19.** $\frac{-1}{(x+3)^2}$ **21.** $\frac{-24}{(x-4)^4}$

23. $\frac{-24}{(2x+7)^5}$ **25.** $\frac{(x^2+6)^6(13x^2 - 28x - 6)}{(x-2)^2}$ **27.** $\frac{3(-x^2+4x-1)}{(x+2)^4}$

29. $\frac{90(9x-4)^2}{(3x+2)^4}$ **31.** $5x^5 + 4x^3 + 55x^2 + 44$

33. $x^4 - 17x^2 - 8x - 6$ **35.** $4x^2 - 30x + 54$ **37.** yes **39.** no

41. At $x = 2$, y changes 4 units for each unit change in x.

43. At $E = 204$, the population increases 4,800 for each billion dollar increase in the state's economy.

45. A decrease in government spending of approximately $453,000,000.

47. a. $\left(\frac{dc}{dd}\right)\left(\frac{dd}{dp}\right) = \frac{dc}{dp}$; $\frac{dc}{dp} = \frac{2{,}800}{(420p^{-2})^{4/3} \cdot p^3}$ **b.** When the price increases from $50 to $51 per unit, the cost of production increases by less than one cent. (0.002076)

49. a. $\frac{dN}{dc} = -4.5C + 4.5$ **b.** $\left.\frac{dN}{dc}\right|_{c=1.70} = -3.15$; A decrease of approximately 3,150 pads.

51. $D'(40) = -2.25$; Demand will decrease by approximately 2,250 tablets.

53. $30x(3x^2 + 4)^4$.

55. At $x = 7$, a denominator is 0, therefore the slope is undefined.

57. $y = \pm\sqrt{4 - x^2}$

59. a. $360{,}000\pi$ ft² $\approx 1{,}130{,}973$ ft² **b.** 7,536

Problem Set 2.5

1. $y' = \frac{-x^2}{2y^3}$ **3.** $y' = \frac{1 - 20x^3}{9y^2 - 4y}$ **5.** $y' = \frac{8x + 1}{8(y + 6)^7}$

7. $y' = \frac{2x - 5x(2y^3 + 5x^2)^2}{3y^2(2y^3 + 5x^2)^2}$ **9.** $y' = \frac{7 - 12x - 8xy^4}{16x^2y^3}$

11. $y' = \frac{6xy^2 - 10x + 2}{-6x^2y - 3y^2}$ **13.** $y' = -10(y + 1)^{3/5}$

15. $y' = 0$ **17.** $y'(-2, 1) = \frac{96}{13}$ **19.** $y'(-2, -3) = -\frac{81}{16}$

21. a. ± 3 **b.** $\pm\frac{4}{3}$ **c.** $(0, 5), (0, -5)$ **d.** 24π cm²/sec

23. Step 1: $12y^2y' - 16x + y' = 0$

25. Step 1: $4(2y^3 - 5x^2)^3(6y^2y' - 10x) + 6y' = 0$

27. Sales will decrease by approximately 1,150 collars.

29. $84,292 approximate increase per month.

31. Monthly revenue is increasing by approximately $467.71.

33. $y' = -\frac{x}{y^2}$ **35.** $y'(2, 3) = -\frac{6}{41}$

37. $\frac{dp}{dt} = 0.84$ At the point in time when the pressure on 500 in³ of air is 30 lb/in² and the volume of air is decreasing at the rate of 10 in³/sec, the pressure is increasing at the rate of 0.84 lb/in² per second.

39. Cost is increasing by approximately $381 per month.

41. $\left(0, -\frac{1}{2}\right), \left(-\frac{2}{5}, 0\right)$ **43.** $\frac{5}{3}$

45. $(1, \infty)$ or $x > 1$ **47.** $f'(x) = \frac{-26}{(3x - 4)^2}$; $f''(x) = \frac{156}{(3x - 4)^3}$

Chapter 2 Test

1. $f'(x) = 3x^2 + 5$ **2.** $f'(x) = 0$ **3.** $f'(x) = -4(5 - 3x)^{1/3}$

4. $f'(x) = 3x^2 - 8x + 3$ **5.** $f'(x) = 12x - 13$

6. $f'(x) = 3x(5x + 4)^2\,(25x + 8)$

7. $f'(x) = 5(x - 4)(x + 1)^2\,(x - 2)$ **8.** $f'(x) = \frac{-2}{(3x + 4)^{4/3}}$

9. $f'(x) = \frac{2}{(x + 3)^2}$ **10.** $f'(x) = \frac{(x - 16)(x + 2)^2}{(x - 4)^3}$

11. Any 3rd degree trinomial will work. **12.** $\frac{dy}{dx} = 3$

13. $\frac{dy}{dx} = 8x + 12$ **14.** $y' = \frac{-2x}{9y^2}$ **15.** $y' = \frac{-(2xy^2 + 5)}{2x^2y - 3y^2}$

16. a. $f'(x) = 12x^2 + 6x - 1$ **b.** $f''(x) = 24x + 6$ **c.** $f'''(x) = 24$

17. a. $f'(x) = 3(x - 3)^2$ **b.** $f''(x) = 6(x - 3)$

18. a. $f(2) = 10$ **b.** $f'(2) = 11$ **19.** $m = 0$ **20.** $x = \frac{-3}{2}$

21. $y = 2x - 5$

22. $f(18) = 10$ means that a cup 18 mm thick has a breaking strength of 10 lb/in². $f'(18) = 0.8$ means that, for a cup 18 mm thick, increasing the thickness by 1 mm will increase the breaking strength by about 0.8 lb/in².

23. When the ferrite grain size is 8 $(d/\mu m)^{1/2}$, the ductile/brittle transition temperature is 50 °C and is increasing at a decreasing rate.

24. a. $f'(x) = -\frac{7}{320}x + \frac{7}{4}$ **b.** $f'(80) = 0$. When the human cannonball is 80 horizontal feet from the end of the cannon, he is at his maximum height.

25. $16\pi \approx 50.3$ cm²/min

26. Decrease of $0.02

27. $P'(10) \approx 23.61$. At year 10, the marginal profit is $23,610/yr; $P'(11) \approx 18.81$. At year 11, the marginal profit is about $18,810/yr.

Chapter 3

Problem Set 3.1

1.

f' (+) (−) (+) (−) (−) (+)

x: 1 3 5 6 7

3.

f' (−) (+) (+) (−) (−)

x: 4 6 7 9

5.

Int/Val	f′(x)	f(x)	Behavior of f(x)
$(-\infty, 5)$	(−)		$f(x)$ is decreasing
5	0	4	(5, 4) is a relative and absolute minimum
$(5, \infty)$	(+)		$f(x)$ is increasing

$f(x)$ has no absolute maximum

7.

Int/Val	$f'(x)$	$f(x)$	Behavior of $f(x)$
0		16	(0, 16) is an absolute maximum
(0, 11)	(−)		$f(x)$ is decreasing
11	0	3	(11, 3) is a relative and absolute minimum
(11, ∞)	(+)		$f(x)$ is increasing
			$y = 10$ is a horizontal asymptote.

9. None **11.** $x = \frac{3}{4}$ **13.** $x = -4, 0, 1$ **15.** $x = -3, 0$

17. None **19.** None **21.** $x = -4, 0$ **23.** $x = 0$

25. Increasing on $(-\infty, \infty)$

27. Increasing on $(-\infty, 0)$ and $\left(\frac{8}{3}, \infty\right)$; decreasing on $\left(0, \frac{8}{3}\right)$

29. Decreasing on $(-\infty, -3)$; increasing on $(-3, 0)$ and $(0, \infty)$

31. Increasing on $(-\infty, 0)$; and on $(0, \infty)$

33. Decreasing on $(-\infty, 6)$; increasing on $(6, \infty)$

35. Increasing on $(5, \infty)$

37. Relative minimum at $(-6, -36)$

39. Relative maximum $(4, 64)$

41. Relative maximum $(0, 10)$; relative minimum at $(6, -1934)$

43. Relative maximum $(-2, -4)$; relative minimum at $(2, 4)$

45. No relative extrema

47. Absolute maximum at $(5, 11)$; absolute minimum at $(3, -17)$

49. Absolute maximum at $(3, 1)$; no absolute minimum

51. About 2 inches **53.** 150 feet

55. About 7 inches **57.** About 3.9 weeks.

59. a. The claim is false: $V'(x) = \frac{1020}{(x+3)^2}$ and $V(x)$ is never 0 or undefined in the interval [0, 4] and therefore there is no increasing then decreasing behavior in the first 4 seconds. **b.** The absolute max occurs at 4 seconds and about 194.29 mph.

61. a. $C'(x) = \frac{x^2 - 144}{180x^2} = 0$ when $x = 12$. Cost is minimum when 1200 systems are manufactured.

b. After 1,200 units have been manufactured

$$C'(x) = \frac{x^2 - 144}{180x^2} > 0$$

for all values of x. Therefore, cost will only increase.

63. a. $S'(x) = \frac{-3(x - 200)}{4\sqrt{300 - x}} = 0$ when $x = 200$. So, the reaction is max at 200 milligrams of the drug. **b.** Above 200 mg, $S'(x) = \frac{-3(x - 200)}{4\sqrt{300 - x}} < 0$ for all values of $x < 300$ so that $S(x)$ is always decreasing.

65. $R'(p) = \frac{-(d + c)}{p^2}$ indicating that $R(p)$ is an always decreasing function.

67. $M'(d) = \frac{-10}{d^3}$ indicating that $M(d)$ is an always decreasing function. Thus, the modulus of rupture decreases as the beam depth increases.

69. There is a relative maximum at the point $\left(-\frac{2}{\sqrt{3}}, 5 + \frac{16}{3\sqrt{3}}\right)$ and a relative minimum at the point $\left(\frac{2}{\sqrt{3}}, 5 - \frac{16}{3\sqrt{3}}\right)$.

71. The maximum concentration is $\frac{1}{600}$ mg/cm³ and it occurs 3 hours after the drug is administered.

73. $f'(x) = -3x^2 + 18x$ **75.** $f'(x) = \frac{-6}{25x^{7/5}}$

77. $f'(x) = \frac{-26}{(3x - 4)^2}$ **79.** $x = \frac{4}{3}$ **81.** $f''(x) = 6x + 9$

Problem Set 3.2

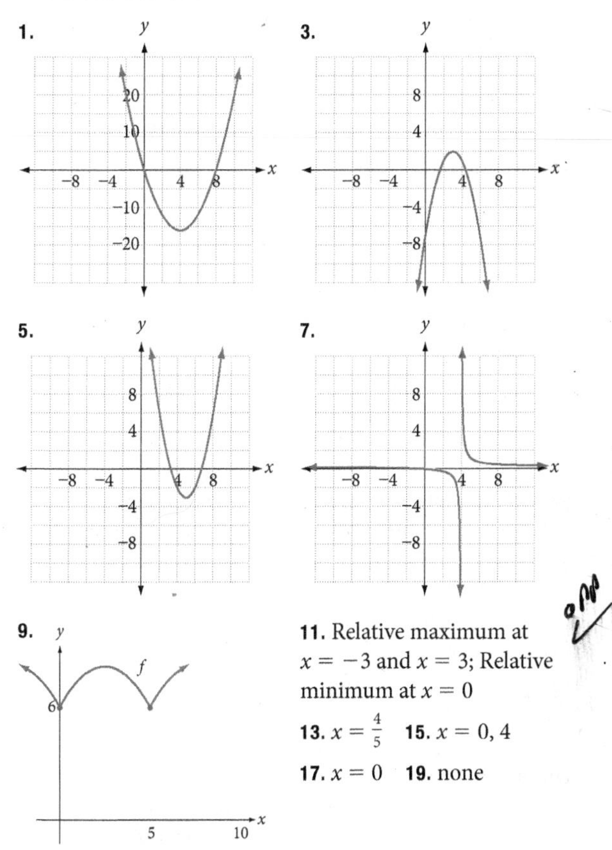

11. Relative maximum at $x = -3$ and $x = 3$; Relative minimum at $x = 0$

13. $x = \frac{4}{5}$ **15.** $x = 0, 4$

17. $x = 0$ **19.** none

21. Concave downward on $(-\infty, 0)$ and concave upward on $(0, \infty)$.

23. Concave upward on $(-\infty, -1)$ and $(1, \infty)$ and concave downward on $(-1, 1)$.

25. Concave upward on $(-\infty, -1)$ and concave downward on $(-1, \infty)$.

27. Concave upward on $(-\infty, \infty)$ and never concave downward.

29. Concave downward on $(-\infty, 1)$ and concave upward on $(1, \infty)$.

31. Concave upward on $(-\infty, 1)$ and concave downward on $(1, \infty)$.

33. Concave upward on $(-\infty, -1)$ and concave downward on $(-1, \infty)$.

35. Concave upward on $(-\infty, -1)$ and $(1, \infty)$ and concave downward on $(-1, 1)$.

37. \$5,000,000 **39.** \$300

41. Hypercritical value at $x = 6$. Concave downward on $(-\infty, 3)$ and $(3, 6)$ and concave upward on $(6, \infty)$.

43. Hypercritical values at $x = 6$ and $x = 9$. Concave upward on $[0, 6)$ and $(9, \infty)$ and concave downward on $(6, 9)$.

45. No hypercritical values. Concave upward on $(-\infty, \infty)$.

47. Concave upward **49.** Concave downward

51. Concave downward

53. Yes, it is possible. **55.** Yes, it is possible.

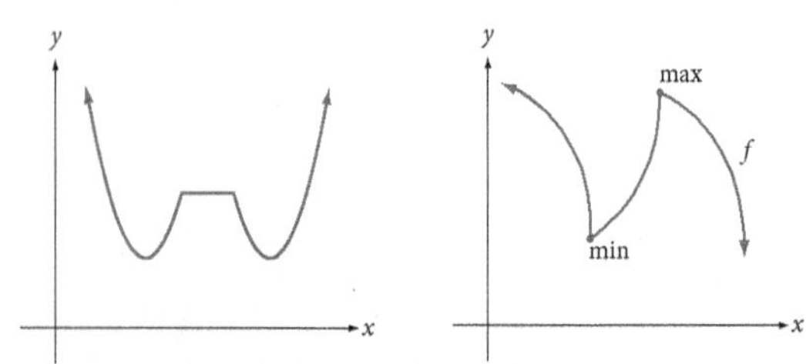

57. As the number of units produced increases toward 1,500, the average cost per unit decreases at a decreasing rate. At 1,500 units produced, the average price reaches a minimum of \$25. Then as the number of units produced increases beyond 1,500, the average price per unit increases at an increasing rate.

59. As the increase in tuition increases from 0 to 70%, the average enrollment increases at an increasing rate toward 24,000 students. At a 70% tuition increase, the average enrollment hits 24,000. As the increase intuition increases beyond 70%, the average enrollment increases at a decreasing rate toward 37,000 students. This number of students represents the maximum number that can enroll at the university.

61. $V'(x) = \frac{1{,}020}{(x+3)^2} > 0$ and $V''(x) = \frac{-2{,}040}{(x+3)^3} < 0$; for values of x in the interval from 0 to 4. Therefore, in the first 4 seconds of its run, the dragster's speed is increasing at a decreasing rate. Reject the claim.

63. a. As x increases from 0 to 200,

$$S'(x) = \frac{-3(x-200)}{4(300-x)^{1/2}} > 0 \text{ and } S''(x) = \frac{3(x-400)}{8(300-x)^{3/2}} < 0.$$

Therefore, the reaction to the drug is increasing at a decreasing rate. Accept the claim.

b. $S''(200) < 0$ when $x = 200$, so $S(x)$ is concave downward. The reaction is max at 200 milligrams of the drug. Accept the claim.

c. As x increases from 200 to 300,

$$S'(x) < 0 \text{ and } S''(x) < 0$$

so that the reaction to the drug is decreasing at an increasing rate. Reject the claim.

65. After 1,200 units ($x > 12$) have been manufactured,

$$C'(x) = \frac{x^2 - 144}{180x^2} > 0 \text{ and } C''(x) = \frac{8}{5x^3} > 0$$

$C'(x) > 0$ for all values of x so the cost increases at an increasing rate.

67. **69.**

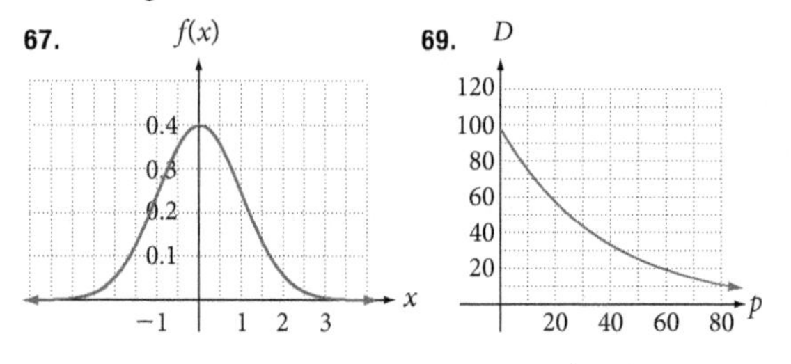

71. There is one hypercritical value at $x = \frac{4}{3}$.

73. Because there is a point of inflection at day 4 of training, after day 4, the trainee's skill level increases at an increasing rate.

75. $P(w) = 4w + 40$ **77.** $w = 20$

79. $4{,}000 + 200r - 20r^2$ **81.** $V'(h) = 160 - 104h + 12h^2$

83. $t = 7$

Problem Set 3.3

1. 4 minutes **3.** 6 inches **5.** 50,000 units **7.** about 1980

9. Lengths of the other two legs are 10 units each.

11. 2,500 magazines

13. 150 ft by 200 ft

15. 10 increases; rate is \$60

17. 35 trees per acre

19. Length $= 33\frac{1}{3}$ inches and width and height $= 16\frac{2}{3}$ inches

21. Rental price = \$40.

23. Absolute minimum at $x = 2, f(2) = -1.76$.

25. The population of insects is minimum at about 239,000 approximately 9 days after the insecticide is applied.

27. The person's level of interest is at its maximum 9 minutes after learning begins.

29. a. $C'(x) = -0.070x + 40$ **b.** -2

31. $R'(x) = -0.06x^2 - 1.2x + 31.5$

33. $16x$ **35.** $4500p^{-2/3}$ **37.** \$13,500

Problem Set 3.4

1. a. $\varepsilon = 2.5$ **b.** elastic **3. a.** $\varepsilon = 0.4$ **b.** inelastic

5. a. $\varepsilon = \frac{2}{3}$ **b.** inelastic **7. a.** $\varepsilon = 1.5$ **b.** elastic

9. A 1% increase per unit increases revenue by \$1,584.

11. A 1% increase in price per unit decreases revenue by \$18,480.

13. $C(250) = 12{,}000$ tells us that it costs the manufacturer \$12,000 to produce 250 units. $C'(250) = -85$ tells us that if the amount of units is increased from 250 to 251, the cost will drop by about \$85.

15. $P(55) = 16$ tells us that if the company sells units at \$55, the profit will be \$16,000. $P'(55) = -4$ tells us that if the price per unit increased from 55 to 56, the profit will drop by about \$4,000.

17. $C(200) = 1{,}500$ tells us that the total inventory cost for a lot size of 200 is \$1,500. $C'(200) = 300$ tells us that if the lot size is increased from 200 to 201, the total inventory cost will increase by about \$300.

19. The demand is elastic. Revenue will decrease.

21. The demand is unit elastic. The revenue will remain the same.

23. The most units the manufacturer can produce and still make a profit is 55. The total cost at this level of production is \$7,732.50.

25. $MP(100) = 30$; $MP(260) = -50$; In order to have the maximum profit, the company should spend \$160,000 on advertising.

27. Maximum profit happens when 4,332 units are sold. At this many units the company has a profit of \$234,227.80.

29. The retailer should reorder 10 times a year and each lot size is 30 to minimize the inventory costs. The total inventory cost is \$300.

31. Nick's conclusion is accurate. Reorder20 times with lot size of 2000 units. Minimum inventory cost is \$8,000.

33. $\sqrt{\frac{2 \cdot 15 \cdot 300}{10}} = 30$ and $\frac{300}{30} = 10$. We verify that the inventory cost is minimized with 10 orders of 30 units each.

35. $\varepsilon = 2.75$; the demand is elastic, which means that if the price is raised by 1% the demand will decrease by 2.75% and the revenue will decrease also.

37. Revenue will decrease by \$441,728

39. $\varepsilon = \frac{1}{3}$; inelastic **41.** $\varepsilon = 1.1$; elastic

43. The revenue increases by \$262.70 **45.** $2^{-3} = \frac{1}{8}$; $2^3 = 8$

47. 20,468 **49.** 6,562.06

Chapter 3 Test

1. $(\infty, 1), f(x)$ is decreasing, 1 is a critical value; $(1, 8), f(x)$ is increasing, 8 is a critical value; $(8, 14), f(x)$ is decreasing, 14 is a critical value; $(14, \infty), f(x)$ is increasing.

2. Concave up on $(-\infty, 8)$ and $(8, 21)$; concave down on $(21, \infty)$; hypercritical values at $x = 8$ and $x = 21$.

3. Absolute and relative maximum at $x = 5$;
Relative minimum at $x = 2$

4. concave down **5.** concave down

6.

$f'(x)$	$(-)$		$(+)$
x		6	
$f''(x)$	$(+)$		$(-)$

7. The cut should be 4 to maximize box volume; max volume $\approx$ 1,150 units^3

8.

Interval/Value	Behavior of $f(x)$
$(-\infty, -5)$	$f(x)$ is increasing, concave up
-5	-5 is a hypercritical value (point of inf)
$(-5, 9)$	$f(x)$ is increasing, concave down
9	9 is a critical value (absolute max)
$(9, 25)$	$f(x)$ is decreasing, concave down
25	25 is a hypercritical value (point of inf)
$(25, \infty)$	$f(x)$ is decreasing, concave up

9. As time increases, the money in the account increases at an increasing rate.

10. As time increases from the injection of the medication, the rate of production of antibodies increases at a decreasing rate for about the first 5 hours. At about 5 hours, the rate reaches a maximum value. From about 5 hours to 6 hours after injection, the rate of antibody production decreases at an increasing rate. From about 6 hours on, the rate decreases at a decreasing rate.

11. $x = \frac{17}{24}$ **12.** $x = 0, 7$

13. Relative min at $x = -2$ and $x = 2$. Relative max at $x = 0$.

14. $\left(-\infty, \frac{5}{3} - \frac{\sqrt{133}}{3}\right)$ increasing; $\left(\frac{5}{3} - \frac{\sqrt{133}}{3}, \frac{5}{3} + \frac{\sqrt{133}}{3}\right)$ decreasing; $\left(\frac{5}{3} + \frac{\sqrt{133}}{3}, \infty\right)$ increasing

15. Relative max: (0, 20); Relative min: (−2, −28) and (8, −2028)

16. Absolute max at (5, 5); No absolute min.

17. $(-\infty, -1)$ concave down; $(-1, \infty)$ concave up

18. The best price to maximize profit is to leave the price at 100 and sell 40 ceramic tiles. Use the following formula: $(40 + 2x)(100 - 5x)$

19. The inventory costs $5,000 for 370 items. If the lot size increases from 370 to 371, the inventory cost will increase by about $200.

20. The company's revenue will increase because point elasticity is $0 < \varepsilon < 1$. Revenue will increase by $101.04.

21. a. Actual cost of 401st unit ≈ $2.60; Actual cost of 2001st unit ≈ −$7.00 **b.** MC(400) ≈ $2.60; MC(2,000) ≈ −$7.00

22. $U'(x)$ is positive; total utility is increasing. Each item adds to the utility of the previous one. $U''(x)$ is negative; the marginal utility of each additional item is decreasing.

23. $h = 6$ in.; $w = 9$ in. **24.** 1,000

Chapter 4

Problem Set 4.1

1. $f(3) = 1{,}000$, growth **3.** $f(6) \approx 0.090$, decay

5. $f(5) \approx 299.462$, growth **7.** $f(3) = 0.063$, decay

9. $f(2) \approx 0.135$, decay **11.** $f(105) \approx 1{,}938{,}676$, growth

13. $f(0.10) \approx 0.030$ **15.** $f(3) \approx 98.846$

17. a. D **b.** B **c.** C **d.** A

19. Algebraic **21.** Exponential **23.** Algebraic

25. Exponential **27.** Exponential **29.** Algebraic

31. $b > 1$, exponential growth; $0 < b < 1$, exponential decay

33. In 5 years, the painting will be worth about $9,000, not $10,000. The owner has overestimated the painting's value. In 10 years, the painting will be worth about $18,000, not $25,000. The owner has overestimated the painting's value.

35. a. $A(60) \approx 8{,}192$ **b.** $A(120) \approx 33{,}556{,}703$

37. a. $N(0) = 15{,}000$ **b.** $N(1) \approx 14{,}198$ **c.** $N(5) \approx 11{,}398$ **d.** $N(24) \approx 4{,}014$

39. a. $842.53 **b.** $860.21 **c.** $864.46 **d.** $866.55

41. a. $2,496.81 **b.** $2,505.09 **c.** $2,509.16 **d.** $2,509.30 **e.** Yes, they are close.

43. a. 120 people **b.** 240 people

45. a. 0.1 telephones **b.** 0.9 telephones

47. a. $A(15) \approx 1.3$ mg **b.** $A(60) \approx 10.0$ mg

49. a. $N(0) \approx 1.154$(1,154 people) **b.** $N(3) \approx 1.521$(1,521 people) **c.** $N(10) \approx 2.867$(2,867 people) **d.** $N(26) \approx 10.678$(10,678 people) **e.** 45,000 people

51. As $x \to \infty$, $f(x)$ approaches e. The function rises sharply at first, then begins to level off.

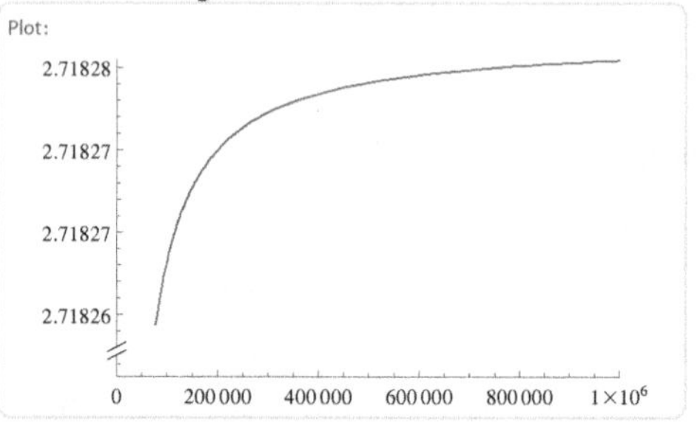

Wolfram Alpha LLC. 2012. Wolfram|Alpha http://www.wolframalpha.com/ (access December 20, 2012)

53. The larger the coefficient, the steeper the graph.

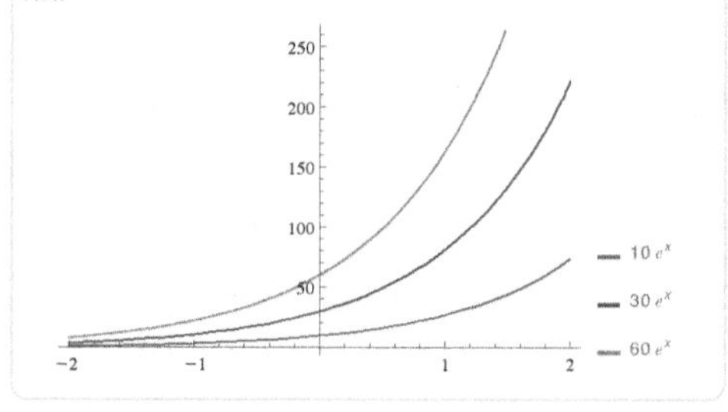

Wolfram Alpha LLC. 2012. Wolfram|Alpha http://www.wolframalpha.com/ (access December 20, 2012)

55. In the first 12 hours, the quickest spreading will happen, maximizing at 600, which also happens to be the numerator.

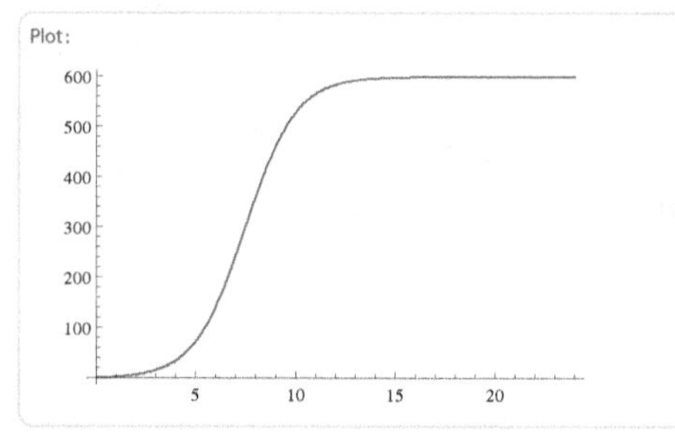

Wolfram Alpha LLC. 2013. Wolfram|Alpha http://www.wolframalpha.com/ (access January 24, 2013)

57. a. $P(200) \approx 19\%$ **b.** $P(1{,}000) \approx 1.5\%$ **59.** exponent

61. $\ln y = 5x$

63. graph of $y = \ln x$

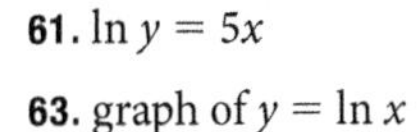

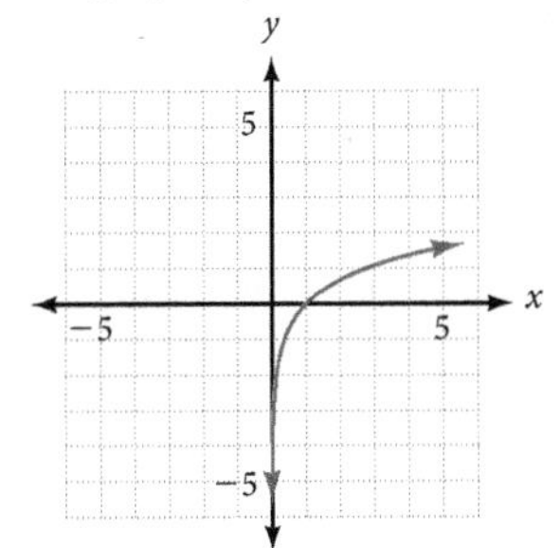

65. $3 \ln x + 2 \ln y - 4 \ln z$

Problem Set 4.2

1. $\ln m = 3n$ **3.** $\ln y = 3x - 5$ **5.** $\ln 4 = 3k + 1$

7. $4x = e^6$ **9.** $y + 1 = e^{x+3}$ **11.** $3x + 4 = e^{-7.18}$

13. $\ln 5 + \ln x$ **15.** $\ln 5 - 2 \ln y$

17. $\ln 10 + \ln x - \ln (x - 6)$ **19.** $\ln (8x)$ **21.** $\ln \frac{x}{y^3 z^5}$

23. $\ln \frac{x+2}{x+7}$ **25.** -1.5050 **27.** 0.9685 **29.** 24.9976

31. About 1.09 cm^2 **33.** About 7.0 lbs/in^2

35. 27.4 million **37.** $N(21) \approx 102$ tasks

39. $N(15) \approx 2{,}352$ people

41. A drug with a 30-hour half-life will decay faster than one with a 40-hour half-life. Comparing the two models, we see that the decay rate $k = -0.0231$ is greater than the decay rate $k = -0.0173$ and therefore the first must be the 30-hour half-life model.

43. $433,319.06 **45.** 80 °F

47. About 3.9 miles above sea level **49.** $t \approx 79.8$ years

51. $t \approx 30$ days **53.** $t \approx 7.07$ minutes **55.** $t \approx 27.95$ years

57. a. 8.7 mg **b.** 20 hrs

59. a. $t \approx 11.55$ years **b.** $t \approx 8.66$ years **c.** $t \approx 6.93$ years

61. $A_0 \approx \$13{,}479.87$ **63.** $t \approx 7.5$ years **65.** 2.04

67. 9.4 years

69. 22 years; No

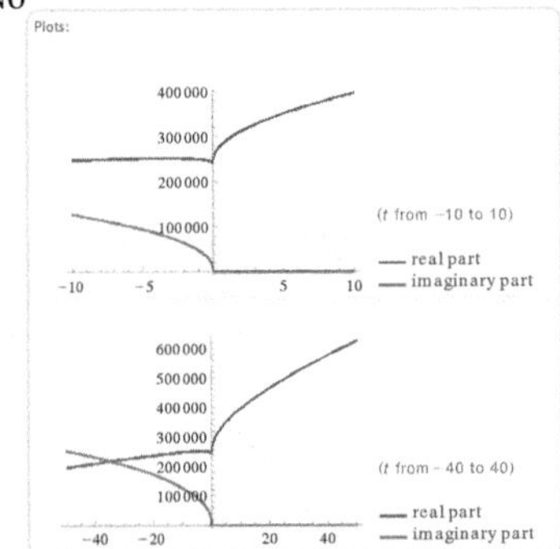

71. e **73.** 4 **75.** $y' = 16x$

77. $y' = 10x + 2$ **79.** 51.723 **81.** 0.020 **83.** $\ln x + 2 \ln y$

Problem Set 4.3

1. $f'(x) = \frac{2}{2x - 7}$ **3.** $f'(x) = \frac{16}{x}$

5. $f'(x) = 2x \ln (3x) + x$

7. $f'(x) = \frac{5}{x + 4}$ **9.** $f'(w) = \frac{3\,[\ln (w + 4)]^2}{w + 4}$

11. $f'(t) = \frac{3}{2(3t + 5)}$ **13.** $f'(u) = \frac{1}{u \ln u}$

15. $f'(t) = 2t \ln (t^2)\,[\ln (t^2) + 2]$ **17.** $f'(x) = \frac{1}{x}$

19. a. $f(e) \approx 80$ **b.** $f'(e) \approx 118$ **c.** $x = e^{-1/3}$

d. $y = 118x - 241$ **21.** $y' = -\frac{9x^3y + y}{2x}$ **23.** $y' = -\frac{2xy^3 + y}{4x^2y^2 + x}$

25. $f'(3) = \frac{8}{15}$ means if x increases from 3 to 4, f will increase by approximately $\frac{8}{15}$.

27. $f'(1) = -2$ means if x increases from 1 to 2, f will decrease by approximately 2.

29. $f'(1) \approx -0.238$ means if x increases from 1 to 2, f will decrease by approximately 0.238.

31. Incorrect **33.** Incorrect **35.** Correct

37. Incorrect **39.** Incorrect

41. Use power property to get $5[\ln (6x^2)]^4$ times the derivative of $\ln (6x^2)$.

43. $C'(60) \approx 11.88$; the cost will increase by approximately $11.88 if production increases from 60 to 61.

45. $\frac{dR}{dr} = \frac{a}{r} - b$

47. $T'(V) = \frac{1.44}{V} > 0$ and $T''(V) = \frac{-1.44}{V^2} < 0$ for all $V > 0$: Because the first derivative is positive and the second derivative is negative, the function increases at a decreasing rate. Accept the claim.

49. $f'(x) = 10x[1 \ln(x^2)]$ **51.** $f''(x) = \frac{1}{x}$

53. $y' = \frac{y(3 - 2x^2y^3)}{3x(x^2y^3 - 1)}$ **55.** e **57.** 0.7634 **59.** 0.7490

61. $y' = 4x$ **63.** $y' = 6x + 5$ **65.** $y = 7x$

Problem Set 4.4

1. $3e^{3x+6}$ **3.** $6e^{6x}$ **5.** $10xe^{5x^2+4}$ **7.** $27e^{3x+1}$

9. $3(5x^2 + 2 + 2e^{-3x})$ **11.** $5xe^{x^2+x}(2x^2 + x + 2)$

13. $\frac{3(2x - 3)e^{\sqrt[4]{x^2 - 3x}}}{4(x^2 - 3x)^{3/4}}$ **15.** $-2e^{-x}\left[\ln (x^2 + 2x) - \frac{2(x + 1)}{x^2 + 2x}\right]$

17. $-2.5e^{-0.05x}$ **19.** $\frac{e^{2x}(e^x + 2)}{(e^x + 1)^2}$ or $\frac{e^{5x} + 2e^{4x}}{(e^x + e^{2x})^2}$ **21.** $y' = -\frac{2x^2 + ye^{3xy}}{xe^{3xy}}$

23. $y' = -\frac{y}{x}$ **25.** $f''(x) = 2e^{x^2+4}(1 + 2x^2)$

27. $f'(1) \approx 21.75$ means if x increases from 1 to 2, f will increase by approximately 21.75.

29. $f'(40) \approx -120.69$ means if x increases from 40 to 41, f will decrease by approximately 120.69.

31. a. $f(0) = 0$ **b.** $f'(0) = 4$ **c.** $x = -1$ **d.** $y = 4x$

33. Incorrect **35.** Incorrect **37.** Correct

39. Correct **41.** Incorrect

43. a. $V'(1) \approx -18{,}875$ sales/day
b. $V'(6) \approx -1{,}892$ sales/day

45. $N'(4) \approx 382$ more people between days 4 and 5.

47. a. 4.7 mg **b.** 96.3 hours **c.** -0.07 mg/hr

49. $A'(10) \approx -0.0144$ grams between years 10 and 11.

51. a. 72.2% **b.** $p'(f) = -16.3e^{-0.163f}$
c. $P'(2) = -16.3e^{(-0.163)(2)} \approx -11.77$; If the number of feet of clay increases by 1, from 2 feet to 3 feet, the percentage of surfactants that remain in the fuel decreases by about 11.77.

53. a. \$600,000 **b.** Mr. Azar: 1.2% and Mr. Hielo: 1.8%
c. $\frac{d}{dt}[R_A(5) - R_H(5)] \approx 3{,}089.75$ means that 5 years from now, the rate at which the difference between the two estimates will increase by about \$3089.75. That is, if the number of years from now, increases by 1 from 5 to 6, the Mr. Azar's estimate of the company's revenue will be \$3089.75 higher than that of Mr. Hielo's estimate. **d.** $\frac{R'_H(s)}{R'_A(s)} \approx 1.46$ means that 5 years from now, the rate at which Mr. Hielo believes the revenue will be decreasing will be about 1.46 times the rate at which Mr. Azar believes the revenue will be decreasing.

55. a. 5 months from now, the profit prediction of Analyst A will be about 1.02 times that of Analyst B. **b.** This quotient indicates that 5 months from now, the profit predicted by analyst A will be increasing at a rate about 1.11 times as fast as that predicted by analyst B. **c.** This means that in month 5, the ratio of A's predicted value to B's predicted value will increase by about 0.004. The ratio in month 5 is 1.02. In month 6, then, the ratio will be approximately $1.02 + 0.004 \approx 1.024$. That is, in month 6, we expect A's prediction to be about 1.024 times that of B's prediction.

57. a. $\frac{R_A(8)}{R_B(8)} = \frac{11.7351}{11.4568} \approx 1.02$. So, 8 weeks from now, the revenue prediction of Analyst A will be about 1.02 times that of Analyst B. The two predictions are almost the same.
b. $\frac{R'_A(8)}{R'_B(8)} = \frac{0.2347}{0.1948} \approx 1.20$. This quotient indicates that 8 weeks from now, the revenue predicted by analyst A will be increasing about 1.20 times as fast as that predicted by analyst B. **c.** This means that in the week 8, the ratio of A's predicted value to B's predicted value will increase by about 0.003 units. The ratio in week 8 is 1.02. In week 9, then the ratio will be approximately $1.02 + 0.003 \approx 1.023$. That is, in week 9, we expect A's revenue prediction to be about 1.023 times that of B's revenue prediction.

59. $\frac{d}{dt} 40.5e^{-0.06931t}\Big|_{t=12} \approx -1.22$ mg/hr If the number of hours after it is administered increases by 1, from 12 to 13, the number of mg of AndroGel 1.62% in the bloodstream will decrease by about 1.22 mg.

61. $f(0) \approx 66.31; f'(0) \approx -8.15; f''(0) \approx 4.08$

63. a. $f'(-30.1) \approx 0.0225; f'(-30) \approx 0;$
$f'(-29.9) \approx -0.0223$; The tangent line is horizontal at $x = -30$ **b.** The function appears to reach a maximum at $x = -30$

65. x^{-4} **67.** $-\frac{1}{3x^3}$ **69.** 3

Chapter 4 Test

1. a. Algebraic **b.** exponential

2. a. $3 \ln x + 4 \ln y - 2 \ln z$ **b.** $\ln\left(\frac{x^5}{y^3 z^4}\right)$

3. $4x = \ln 5$ **4.** $3y - 1 = \ln k$ **5.** $3x = e^5$

6. $x + 6 = e^{-2}$ **7.** 0.9722 **8.** -4.9929 **9.** $y = 3x - 8$

10. $f'(x) = -3e^{-3x}$ **11.** $f'(x) = 5e^{5x+4}$

12. $f'(x) = 2x^3 e^{6x}(3x + 2)$ **13.** $f'(x) = \frac{1}{x}$

14. $f'(x) = \frac{9}{3x-2}$ **15.** $f'(x) = 20e^{20x-24}$

16. $f'(x) = \frac{4}{x+1} + 70e^{7x+7}$ **17.** $f'(x) = \frac{3xe^{3x} - e^{3x} - 1}{x^2}$

18. $y' = -\frac{12x^2}{e^y}$ **19.** $f''(x) = 9e^{3x}$ **20.** $f''(x) = \frac{1}{x}$

21. $f'(1) = -\frac{3}{e} \approx -1.1$ means y is decreasing at a rate of approximately 1.1 from $x = 1$ to $x = 2$.

22. $f'(11) = \frac{2}{1} = 2$ means y is increasing at a rate of 2 from $x = 11$ to $x = 12$.

23. 14 days **24.** 4 days

25. Increasing at a rate of 24,237 people per year

26. Decreasing at a rate of approximately 0.0003 grams per year

Chapter 5

Problem Set 5.1

1. $\frac{x^6}{6} + C$ **3.** $-\frac{1}{3x^3} + C$ **5.** $\frac{40x^{3/5}}{3} + C$ **7.** $6e^x + C$

9. $\frac{x^4}{2} - \frac{3}{x} + x + C$ **11.** $-\frac{1}{2}x^{-2} - x^{-1} - \ln|x| + C$

13. $0.03x + 0.24x^{1/2} + C$ **15.** $x^5 - x^4 + 4x^2 + 5$

17. $3e^x + 4 \ln x + 2e$

19. Between days 50 and 51, the number of pants the company produces and sells increases by approximately 30. The number of pants sold on the 50th day of the year is 550.

21. If the speed at which the units are produced increases by 1, from 170 units per minutes to 171 units per minutes, the number of units that fail will increase by approximately 0.008. When the rate of production is 170 units per minute, the total number of units that can be expected to fail is approximately 3.

23. If the number of negative advertisements in a week sponsored by the candidate's opponent increases by 1, from 10 to 11, the candidates favorable rating will decrease by about 0.5%. When the number of negative advertisements sponsored each week by the candidate's opponent is 10, the candidate's favorable rating will be at 45%.

25. If the number of minutes from 5:00 AM increases by 1, from 120 to 121, the number of cars entering the boulevard will increase by about 3. When the number of minutes from 5:00 AM is 120, the number of cars entering the boulevard is 12.

27. $f(x) = F'(x) = 3x^4 - 5x^3 + 4x$

29. $f(x) = F'(x) = 2x(x^2 + 4)^2(x^2 - 5)^3(7x^2 + 1)$

31. Correctly evaluated

33. Not correctly evaluated;

$$\int (16x^3 + 9x^2)\, dx = 4x^4 + 3x^3 + C$$

35. Not correctly evaluated;

$$e^{4x} + \ln|x| + C$$

37. Not correctly evaluated;

$$\int \left(e^{2x} + \frac{3}{x} + 3\right) dx = \frac{e^{2x}}{2} + \ln x^3 + 3x + C$$

39. Correctly evaluated

41. a. $P(t) = 200\sqrt[5]{t^6} + 170t + 8{,}400$ **b.** $P(1) \approx 8{,}770$ **c.** $P'(6) \approx 513$; $P(6) \approx 11{,}137$; This means that 6 years after 2013, the population is about 11,137 people and is changing at a rate of about 513 people/yr.

43. a. 1.7 grams/day **b.** 3.9 grams/day **c.** 194.5 grams **d.** 180.88 grams

45. a. 1.752 m^2 **b.** 0.01825 m^2/kg **c.** $A'(64)$ means that at 64 kg, the surface area of a person is changing at 0.01825 m^2/kg.

47. $C'(30) = 0$, $C(30) = \$1{,}200$. When the chain orders lots of size 30 fifteen times a year, its inventory cost is $1,200. If the chain increases the lot size by 1, from 30 to 31, its inventory cost will neither increase nor decrease.

49. $C'(2) = -1.02$, $C(2) = 1.875$. Two years after the study, the national consumption of the chemical was down to 1.875 million gallons and was decreasing by about 1.02 million gallons per year.

51. 160 **53.** d/dx (int $5x^3 + 2x$) **55.** $12x + 4$

57. $(12x + 4)\, dx$ **59.** $\frac{x}{3x^2 + 5}$

61. $\frac{du}{dx} = 10x$, $du = 10x\, dx$, and $dx = \frac{1}{10x}\, du$

63. $\frac{1}{6}\ln|u| + C$ **65.** $x\sqrt{x + 3}$

Problem Set 5.2

1. $\frac{1}{5}(x + 2)^5 + C$ **3.** $\frac{1}{25}(5x + 1)^5 + C$ **5.** $\frac{1}{6}(x^3 + 2)^6 + C$

7. $\frac{2}{3}(x^2 + 4x)^{3/2} + C$ **9.** $\frac{8}{3}(5x^3 - 8x)^{3/2} + C$

11. $\ln(2x^2 + 1) + C$ **13.** $\frac{1}{4}\ln(e^{4x} + 2) + C$

15. $\ln(4x^3 + 5x - 6) + C$ **17.** $\frac{1}{2}(\ln|x| + 7)^2 + C$

19. $x - \frac{4}{5} + \frac{4}{5}\ln|5x - 4| + C$ **21.** $-50e^{-0.02x} + C$

23. $300e^{0.15x} + C$ **25.** $\frac{1}{3}e^{3x} - 4e^{-3x} + C$

27. Let $u = ax$ so that $du = a dx$ and $dx = \frac{1}{a}du$

Then, $\int e^{ax}\, dx = \int e^u \frac{1}{a}\, du$

$$= \frac{1}{a}\int e^u\, du$$

$$= \frac{1}{a}(e^u + C_1) =$$

$$= \frac{1}{a}e^u + \frac{1}{a}C_1 =$$

$$= \frac{1}{a}e^{ax} + C, \text{ where } \frac{1}{a}C_1 = C$$

29. $A = 10x$ **31.** $A = 6x^2 + x$ **33.** $A = 10x + 10$ **35.** $A = \frac{2}{x}$

37. a. $N(t) = 300\ln(2t + 3) - 329.58$ **b.** About 440,000 antibodies

39. a. $R(t) = 395{,}000e^{-0.004t} - 9{,}511.83$ **b.** $R(24) \approx 349{,}331.46$ **c.** Eventually, the device will have no resale value

41. a. $21{,}484.375e^{-0.32t}$ **b.** $N(10) \approx 34{,}391.38$ **c.** $\approx 33{,}515.625$

43. a. $P(t) = 7\ln|1 + e^{18t}| + 27{,}995.148$ **b.** $29,255 people **c.** About 126 people

45. a. $R(x) = \frac{2}{3}x\sqrt{100 - x}$ $0 \leq x \leq 100$ **b.** $238,000

47. a. $-\frac{1{,}020}{t + 3} + 340$ **b.** 136 mph

49. $\frac{(x^3 - 7)^6}{6}$; Wolfram|Alpha expands this.

51. $\frac{2}{15}(4 + 5\ln x)^{3/2} + C$; the appropriate substitution is $u = 4 + 5\ln x$

53. $\frac{e^{ax+b}}{a} + C$ **55.** 176,418.7 **57.** $\ln|x - 3| + C$ **59.** 176,418.7

Problem Set 5.3

1. 14 **3.** −16 **5.** 201 **7.** $\frac{8{,}975}{6}$ **9.** 0 **11.** $2(e^8 - 1)$

13. $2 - \frac{2}{e^{10}}$ or $\frac{2e^{10} - 2}{e^{10}}$ **15.** $\frac{1}{5}(e^5 - 1)$ **17.** $\frac{4}{e^2} - 4$ **19.** 3

21. $\frac{3}{2}$ **23.** 35 **25.** ln 4 **27.** ln 6 **29.** $\ln\frac{7}{2}$ **31.** $3\ln\frac{31}{6}$
33. $e^4 - 1$ **35.** 9 **37.** $\frac{2}{\ln(2)}$

39. The error occurs in step 2. The expression is mistakenly evaluate first at the lower limit then at the upper limit rather than first at the upper limit then at the lower limit.

Step 2: $\left(\frac{3(6)^2}{2} - 6(6)\right) - \left(\frac{3(2)^2}{2} - 4(2)\right)$

41. The error is in step 1. It looks like this function was differentiated rather than integrated.

Step 1: $\int_2^3 3e^{3x}\,dx = \frac{3e^{3x}}{3}\Big|_2^3$

43. The integral represents the total amount of money the company spends on advertising between months a and b.

45. The integral represents the total amount of gasoline pumped from the local gas station between days a and b.

47. The amount of chemical leaked into the lake from time a to b.

49. The amount of wealth gained from age a to b.

51. The amount the income stream increased from year a to b.

53. The amount of depreciation from month a to b.

55. The total increase of probability from time a to b.

57. \$680

59. a. 5,520,000 gallons **b.** 155,520,000 gallons

61. \$31.25

63. 495.8 feet

65. a. $\frac{d}{dx}\frac{400x + 1{,}000}{x + 1} = \frac{-600}{(x+1)^2} < 0$ for $0 \le x \le 100$

b. $\lim_{x\to\infty} C(x) = \lim_{x\to\infty}\frac{400x + 1{,}000}{x + 1} = \$400$

c. $\int_0^{100}\frac{400x + 1{,}000}{x + 1}\,dx = \$42{,}769$

67. a. Since $R'(x) = 0.96e^{-0.002x} > 0$ and $R''(x) = -0.00192e^{-0.002x} < 0$ for $0 \le x \le 100$, accept the analyst's claim.

b. Since $\int_0^{1{,}000}(800 - 480e^{-0.002x})\,dx \approx 592{,}480 > 500{,}000$, accept the analyst's claim.

69. a. $\int_0^1 (1.5t + 4)\sqrt{1.5t^2 + 8t}\,dx = \$78{,}000$
b. $\int_9^{10}(1.5t + 4)\sqrt{1.5t^2 + 8t}\,dx = \$2{,}134{,}000 > \$250{,}000$
c. 27.2 times as expensive

71. ≈ 569.61

73. $N(5) \approx 230.20$ means that if the number of weeks of experience of an average worker increases by 1, from 5 to 6, the worker can be expected to assemble about 230 more units each month; $\int_0^5(355 - 145e^{-0.03t}) = 1{,}101.76$ means that the total number of units an average worker can assemble each month after 5 weeks of experience is about 1,101.

75. A: $G(x) \approx 0.4505$; B: $G(x) \approx 0.4012$; Since the Gini index for Country B is closer to 0, the money in that country is more evenly distributed than it is in Country A.

77. 43.125 **79.** $\frac{31}{6}$

Problem Set 5.4

1. 63 square units **3.** 12 square units
5. ln (4) square units **7.** 15,319.26 square units
9. $\frac{28}{3}$ square units **11.** 0.332 or $\frac{1}{2}\ln\frac{33}{17}$ square units
13. 1,644.61 square units **15.** 1,250 square units
17. $\frac{277}{3}$ square units
19. 0.01 square units **21.** $\int_a^b f(x)\,dx + \int_b^c f(x)\,dx$
23. $\int_a^b f(x)\,dx - \int_a^c f(x)\,dx$ **25.** \$24,790.60 **27.** \$2,023.28
29. \$15,513.51 **31.** 371.2 units **33.** $\int_5^{10} N(t) \approx 128$ Gallons

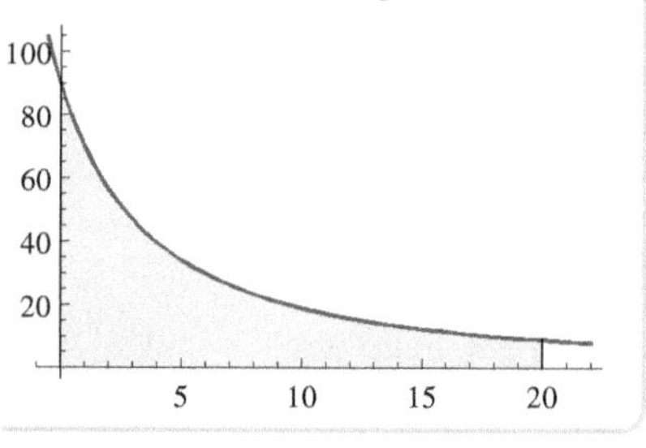

Wolfram Alpha LLC. 2013. Wolfram|Alpha
http://www.wolframalpha.com/
(access January 18, 2013)

35. \$3,340,126.4 million **37.** 11.5129 and 0.6931
39. $-\frac{1}{2} + \frac{1}{2(b-2)^2}$ **41.** 75

Problem Set 5.5

1. $\frac{1}{3}$ **3.** Divergent **5.** Divergent **7.** $\frac{1}{4}$ **9.** $\frac{1}{2}$ **11.** $\frac{1}{2}$ **13.** 1
15. $\frac{1}{2}$ **17.** Divergent **19.** 1 **21.** 1 **23.** 10

25. If the dose is taken orally it will be only about 85% as effective as a dose taken intravenously.

27. The total amount of money received from a rental property until long into the future.

29. The total energy being conserved from now until long into the future.

31. Maximum amount of improvement the company can achieve (at ∞ units produced)

33. 30,000 tons **35.** $\frac{1}{e^{5/4}} \approx 0.2865$ **37.** 1,540

39. Capital Value $= \int_0^{\infty}(12{,}000e^{0.05t} \cdot e^{-0.10t})\,dt = \$240{,}000$

41. a. $P\int_0^{\infty} e^{-rt}\,dt = P \lim_{b\to\infty}\left[\frac{-1}{r}e^{-rt}\right]\Big|_0^b$

$= P \cdot \frac{-1}{r}\lim_{b\to\infty}[e^{-b} - e^0] = \frac{-P}{r}[0-1] = \frac{P}{r}$

b. $\frac{P}{r} = \frac{\$30{,}000}{0.05} = \$600{,}000$

43. Bioavailability $= F = 0.60$ which means that if the dose is taken orally it will be only about 60% as effective as a dose taken intravenously.

45. Undefined

47. 163,636,363

49. 0 **51.** $\frac{x^5 \ln x}{5} - \int \frac{x^4}{5}\,dx$

Problem Set 5.6

1. $e^x(x-1)+C$ **3.** $\frac{-1}{64}e^{-8x}(8x+1)+C$

5. $\frac{e^{3x}}{27}(9x^3 - 9x^2 + 6x - 2) + C$ **7.** $x(\ln(x^3) - 3) + C$

9. $x((\ln x)^3 - 3(\ln x)^2 + 6\ln x - 6) + C$

11. $\frac{(\ln x)^2}{2} + C$ **13.** $\frac{x^2}{8}(4\ln^3 x - 6\ln^2 x + 6\ln x - 3) + C$

15. $\frac{1}{7}(7x-4)(\ln(7x-4) - 1) + C$

17. $-\frac{2}{35}(1-x)^{5/2}(5x+2) + C$

19. $e^{x+3}(2x-1) + C$

21. $\frac{1}{18}(2x+1)^3(3\ln(2x+1) - 1) + C$

23. $48x \cdot 6.25e^{0.16x} - \int 6.25e^{0.16x} \cdot 48\,dx$

25. $x^3 \cdot \frac{-1}{6}e^{-6x} - \int \frac{-1}{6}e^{-6x} \cdot 3x^{-4}\,dx$

27. $x^6 \ln 6x - \int x^5\,dx$

29. $u = x^{-1}$; $du = -x^{-2}\,dx$; $dv = e^{-2x}\,dx$; $v = -\frac{1}{2}e^{-2x}$

31. $u = \ln 8x$; $du = \frac{1}{x}\,dx$; $dv = 8x^7\,dx$; $v = x^8$

33. $C(t) = \int C'(x) = \int x\ln(x+1)\,dx$, $C(0) = 0$, $C(x) = \frac{1}{4}[2(x^2-1)\ln(x+1) - t^2 + 2t]$

35. $12{,}500t - 241e^{-0.5t}(t^2 + 4t + 8) + 14{,}428$

37. $K(t) = -e^{-0.8t}(10t + 12.5) + 12.5$

39. $\frac{1}{4}x^2(2\ln(6x) - 1) + C$

41. $\int \frac{\ln x}{x^n} = \frac{-(n-1)\ln x + 1}{(n-1)^2 x^{n-1}}$ **43.** $\frac{-x^3}{3} + 5x^2 - 16x + C$

45. $\frac{-2x^3}{3} - 4x^2 + 24x + C$ **47.** $x = 0, -4, 4$

Chapter 5 Test

1. If the temperature of the air surrounding the pump increases by 1 degree, from 10 °F to 11 °F, the efficiency of the pump will decrease by about 3.2%. When the temperature of the air surrounding the pump is 10 °F, the efficiency of the pump is 64%.

2. $16\ln x + C$ **3.** $x^3 - \frac{5}{2}x^2 + 2x + C$ **4.** $\frac{5}{2}x^{8/5} + C$

5. $-2{,}000e^{-0.02x} + C$

6. $V'(5) \approx -3{,}758.90$. The rate of sale volume is decreasing at $-\$3{,}758.90$/day by day 5 after campaign; $V(5) \approx 67{,}662.10$. On the fifth day, the store brings in \$67,662.10.

7. $\frac{3}{20}(2x^2 - 5)^{5/3} + C$ **8.** $\frac{1}{2}\ln(x^2+1) + C$

9. $\frac{2}{3}(x^2 + 4x - 1)^{3/2} + C$ **10.** $e^{x^5 + 2x} + C$

11. $\frac{1}{2}\ln(2e^x + 3) + C$ **12.** $\frac{5(\ln x)^2}{2} + C$

13. $x + \ln(x-1) + C$ **14.** $3x + 12\ln(x-4) + C$

15. $-400e^{-0.13x} + C$ **16.** $\frac{e^{x^5}}{5} + C$

17. Total number of people who have heard the news between days a to b.

18. 340 **19.** $\frac{e^3 - 1}{3}$ **20.** $\ln\left(\frac{2}{3}\right)$ **21.** $\frac{1}{3}$ **22.** $\frac{27}{4}$

23. $\frac{1}{2}(e-1) \approx 0.86$ **24.** 1 **25.** $\frac{3}{2}$ **26.** 865,055

Chapter 6

Problem Set 6.1

1. 3.33 **3.** 6.33 **5.** 10.43 **7.** 11.61 **9.** 0.05 **11.** 20.83

13. 4.5 **15.** $\int_c^d [g(x) - f(x)]\,dx$ **17.** $\int_d^c [k(x) - m(x)]\,dx$

19. $\int_b^c [h(x) - f(x)]\,dx + \int_a^b [f(x) - h(x)]\,dx$

21. \$872,636

23. \$1,035,555

25. a. Six years after purchase **b.** \$28,800

27. \$42,813 **29.** \$312,000

31. Area is 26.67 square units. Actual savings realized from purchase of new equipment is \$26,667.

33. \$2,615.75 **35.** 18 square units

37. \$829.44

39. $\pm$ 20

41. $\pm$52.70 **43.** 390.35 **45.** 1,666.67

Problem Set 6.2

1. \$4,500 **3.** \$4,698 **5.** C.S. = \$160.44; P.S. = \$80.22

7. Consumer's surplus is \$581,250

9. Producer's surplus is \$45.85 **11.** \$250,000 **13.** \$16.67

15. \$254.52 **17.** \$23.99 **19.** \$7,436.45 **21.** \$240

23. \$16,468.43 **25.** C.S. = 2,286.67; P.S. = 2,940.00

27. \$1,666.67 **29.** (50, 495.56) **31.** 609.09

33. 6,739.93 **35.** 42,187.42 **37.** 1,076,923.077

39. \$1,000,000; \$1,000,000 would have to be invested now into an account paying 7% interest compounded continuously indefinitely into the future in order to match the total income realized from the lease of the property.

41. Present value = \$131,941; Future value = \$201,816

43. Present value = \$736,176

45. $\frac{1}{2}\ln(x^2+1)+C$ **47.** $C=2$

49. $C=-79; y=\pm\sqrt[4]{10x^2-79}$ **51.** $p=\left[\frac{t+2{,}000}{200}\right]^2$

Problem Set 6.3

1. \$600,000 **3.** \$163,990.69 **5.** \$7,649.81 **7.** \$72,783.68

9. \$386,686.15 **11.** \$116,467.63 **13.** \$20,000

15. Invest \$362,854.49 (present value) because it is a bigger investment to buy the machine.

17. Purchase the machine if it is valued at less than the present value of \$151,189.37.

19. The present value indicates the amount of money the person should have at the moment and avoid paying for up to 10 months. If the person invests \$300 for 10 months with 7% interest, the value will amount to \$3,089.23.

21. If a person deposits \$100 each month into an account that pays an annual interest rate of 6% compounded continuously, they'll have \$29,192.06 after 15 years. If, instead, that person deposits \$11,868.61 and didn't deposit money again, after 15 years they would match the amount of \$29,192.06.

23. \$600,000 **25.** \$232,012 **27.** \$232,012

29. \$4,303,540

31. a. \$647,337; If the \$200,000 income stream is invested immediately as it is received into an account paying 5% interest compounded continuously for 3 years, it will be worth \$647,337 at the end of 3 years. **b.** \$557,168; \$557,168 would have to be invested today into an account paying 5% interest compounded continuously for 3 years to produce \$647,337.

33. a. \$8,221,188; If the \$1,000,000 income stream is invested immediately as it is received into an account paying 10% interest compounded continuously for 6 years, it will be worth \$8,221,188 at the end of 6 years. **b.** \$4,511,884; \$4,511,884 would have to be invested today into an account paying 10% interest compounded continuously for 6 years to produce \$8,221,188.

35. a. \$300,000; Without being invested into an interest-earning account, the \$100,000 three-year income stream will produce \$300,000. **b.** \$318,742 if the \$100,000 continuous income stream is invested immediately as it is received into an account paying 4% interest compounded continuously for 3 years, it will be worth \$318,742 at the end of 3 years. **c.** \$282,699; \$282,699 would have to be invested today at 4% interest compounded continuously to produce \$318,742 at the end of 3 years. **d.** The company should buy the machine since the cost of the machine is less than its present value.

37. a. \$600,000 **b.** \$664,208 **c.** \$64,208

Problem Set 6.4

1. $y=\frac{x^5}{5}+C$ **3.** $y=C$ **5.** $y=\pm\sqrt{2x^2+C}$

7. $y=\frac{e^{x^2}}{2}+C$ **9.** $y=x-\ln|x+1|+C$ **11.** $y=e^{cx}$

13. $y=\sqrt[3]{x^3+64}$ **15.** $y=\pm\sqrt{2\ln|x|+16}$

17. $y=e^x(x-1)-\frac{x^2}{2}$ **19.** $y=\pm\sqrt{2|x|-1}$

21. $P=P_0e^{k\cdot t}$

23. Derivative of function is equal to corresponding differential equation.

25. Derivative of function is equal to corresponding differential equation

27. Derivative of function is equal to corresponding differential equation

29. Derivative of function is equal to corresponding differential equation

31. Separable **33.** Not separable

35. Separable **37.** Not separable

39. About 107 words per minute **41.** About 164

43. $P=68\%$ pressure at sea level **45.** 14 ppm **47.** 34%

49. $a=(0.05t+2.11)^4$

51. $y=Ce^{1/x}$

53. $y(x)=-\frac{x^2}{2}+e^x(x^2-2x+2)-2e^2+2$

55. $P=Ce^{kt}$ **57.** $-\ln|L-P|+C$ **59.** $K=0.019$

61. $K=0.000000272$

Problem Set 6.5

1. When $P(0)=P_0$,

$$P_0=Ce^{k(0)} \Rightarrow P_0=C \Rightarrow P=P_0e^{kt}$$

3. $P=L+(P_0-L)e^{-kt}$

$$P=4{,}000+(80-4{,}000)\,e^{-kt}$$

$$P=4{,}000-(3{,}920)e^{-kt}$$

if $P=2{,}800$ when $t=2$

$$2{,}800 = 4{,}000 - 3{,}920e^{-2k}$$
$$-1{,}200 = -3{,}920e^{-2k}$$
$$0.306 = e^{-2k}$$
$$\ln(0.306) = -2k$$
$$\frac{\ln(0.306)}{-2} = k$$
$$0.592 = k$$

5. Unlimited growth, because it is increasing at a rate proportional to itself, with no limit.

7. Unlimited growth because it is increasing at a rate proportional to itself, with no limit.

9. Logistic growth. The rate at which the population changes is proportional to the current size of the student population and the difference between the total population and current size of the infected population.

11. Logistic growth. The rate at which the number of employees who have learned the procedure is proportional to the current number who know the procedure and the difference between the total number of employees and the current number of employees who know the procedure.

13. a. $T = 68 - 28e^{-0.0131t}$ **b.** $T = 49$ °F

15. a. $p = \frac{29{,}250}{45 + 605e^{-0.74t}}$ **b.** 383 units

17. a. $p = \frac{2{,}500}{1 + 2{,}499e^{-1.8523t}}$ **b.** 2,410 people

19. 13,398 years **21.** About 9 hrs

23.
$$P = e^{-kt}(P_0 + L(e^{kt} - 1))$$
$$= e^{-kt}(P_0 + Le^{kt} - L)$$
$$= P_0e^{-kt} + Le^{kt} \cdot e^{-kt} - Le^{-kt}$$
$$= P_0e^{-kt} + L - Le^{-kt}$$
$$= L + P_0e^{-kt} - Le^{-kt}$$
$$= L + (P_0 - L)e^{-kt}$$

25.
$$P = \frac{P_0Le^{kLt}}{P_0(e^{kLt} - 1) + L}$$
$$= \frac{P_0L}{(P_0e^{kLt} - P_0 + L)e^{-kLt}}$$
$$= \frac{P_0L}{(P_0e^{kLt}e^{-kLt} - P_0e^{-kLt} + Le^{-kLt})}$$
$$= \frac{P_0L}{P_0 - P_0e^{-kLt} + Le^{-kLt}}$$
$$= \frac{P_0L}{P_0 + Le^{-kLt} - P_0e^{-kLt}}$$
$$= \frac{P_0L}{P_0 + (L - P_0)e^{-kLt}}$$

27. ≈ 1 **29.** $\frac{13}{62} \approx 0.2097$ **31.** 0.0781

Problem Set 6.6

1. No, since condition 1 fails. $f(1) = \frac{1-2}{16} = \frac{-1}{16}$ so that not every x in $[1, 5]$ is such that $0 \le f(x) \le 1$

3. No, since condition 1 fails. $f(-2) = 4(-2)^2 = 16 > 1$ so that not every x in $[-2, 2]$ is such that $0 \le f(x) \le 1$

5. No, since condition 2 fails. $\int_{-12}^{12} \frac{5}{12}x\,dx = 0 \ne 1$ condition 1 also fails.

7. $k = \frac{4}{45}$ **9.** $k = \frac{1}{12}$ **11.** $k = \frac{1}{2}$ **13.** $\frac{1}{2} - \frac{1}{2\sqrt{e}} \approx 0.1967$

15. $\frac{\ln 10}{9} \approx 0.2558$ **17.** 0.1484 **19.** $\frac{1}{2}$

21. Of every 10,000 air travelers, we expect about 952 to be exposed to between 5.00 and 5.50 mrem of radiation while flying across the continental United States.

23. We expect rain to occur on about 16 days in February.

25. Out of 1000 phone calls, about 151 are expected to last longer than 16 minutes.

27. Out of 1000 people given a particular stimulus, we expect that about 6 of them will have reaction times of more than 0.2 seconds.

29. Although for each x in $[2, 5]$, $0 \le f(x) \le 1$,
$$\int_2^5 \frac{3}{100}x^2\,dx = 1.17 \ne 1$$

31. a. 0.1333 **b.** 0.8667 **c.** 0.5000 **d.** 0.5000

33. $\frac{1}{5} = 0.2000 = 20\%$ so about $0.20(1{,}000) = 200$ calls.

35. About $P(66 \le X \le 72) = \int_{66}^{72} \frac{1}{72-24}\,dx = \frac{6}{48} = 0.125$ $= 12.5\%$ of the people.

37. 0.0198; About 20 of 1,000 of these devices will have operating life spans that are less than 100 hours.

39. a. $P(0 \le X \le 1) = \int_0^1 \frac{1}{5}e^{-\frac{1}{5}x}\,dx = 0.1812$

b. $P(x \ge 1) = \int_1^\infty \frac{1}{5}e^{-\frac{1}{5}x}\,dx = 0.8187$

41. a. 0.0062 **b.** 0.500 **43.** 0.0761 **45.** 90%

47. a. $P \approx 0.029$ **b.** $P \approx 0.861$ **c.** $P \approx 0.016$

49. a. $P \approx 0.0695$ **b.** $P \approx 0.7866$

51. a. $P \approx 0.0168$ **b.** $P \approx 0.3873$ **c.** 0.0077

53. 204,544 **55.** 190.6

Chapter 6 Test

1. 3.333 square units **2.** 36 square units

3. (20,100)

4. \$2,666.67; The total savings to the consumer is about \$2,667.67.

5. \$1,333.33; The total savings to the producer is about \$1,333.33

6. \$300,000 **7.** \$9,110.59

8. \$835,605 **9.** \$40,656.97 **10.** \$614,780.87

11. \$786,939 **12.** \$400,000 **13.** It is not a solution.

14. $\frac{1}{y}\,dy = \frac{\sqrt{1-x^2}}{x}\,dx$ **15.** $y = (3x^2 + 61)^{1/3}$

16. $y = \frac{1}{2 - x - e^x}$ **17.** 20 hrs **18.** 437 hrs

19. About 7.6 days

20. Of every 10,000 males between the ages of 15 and 20, 1,855 will buy a particular style of advertised clothes.

21. $K = 5\sqrt{2}$ **22.** 35% **23.** 0.4286 **24.** 0.1653

25. 0.8412

Chapter 7

Problem Set 7.1

1. −14 **3.** 358.51 **5.** 150.14 **7.** 2.51 **9.** 4,837.80

11. 1 **13.** $2x + h$ **15.** $4x + 2y$

17. One possibility is $y =$ the distance from the manufacturing facility to the consumer. Another possibility is $y =$ the number of components that need to be shipped to the consumer.

19. $y =$ length of the mammal; $A(w, y) = 3y + w^{1/3}$

21. **23.** **25.**

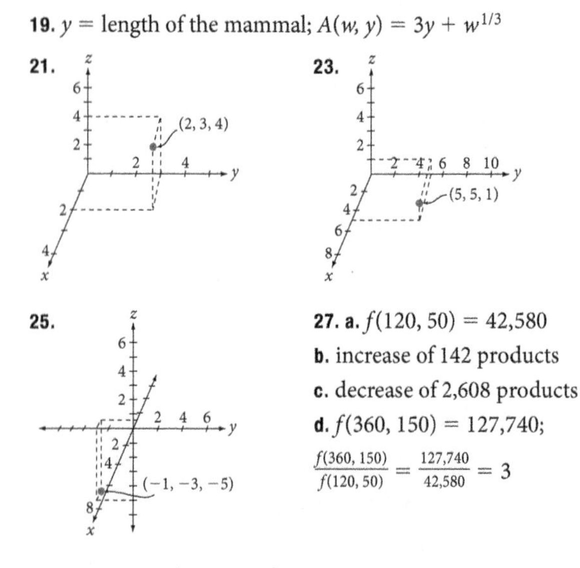

27. a. $f(120, 50) = 42{,}580$
b. increase of 142 products
c. decrease of 2,608 products
d. $f(360, 150) = 127{,}740$;
$\frac{f(360, 150)}{f(120, 50)} = \frac{127{,}740}{42{,}580} = 3$

29. a. 119 **b.** A decrease of 78 units

31. a. $f(1{,}500, 48.50) = 132.75$ The stockbroker's commission is \$132.75 **b.**

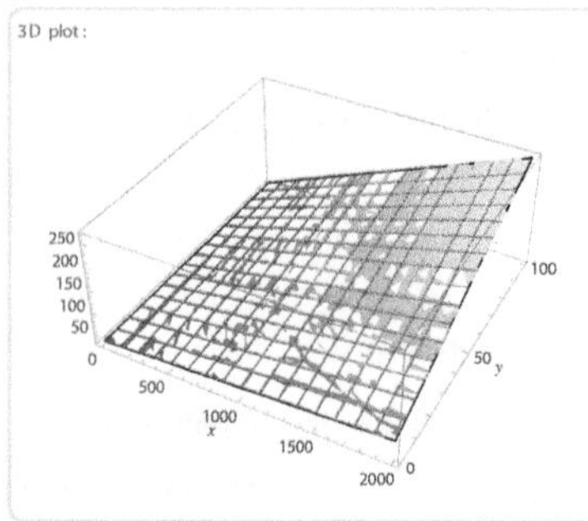

Wolfram Alpha LLC. 2013. Wolfram|Alpha
http://www.wolframalpha.com/
(access January 17, 2013)

33. $M(70, 45) = 830$ **35.** $f(x) = \ln(x)$ increases at a decreasing rate. $f(x, y)$ increasing at a decreasing rate in both the x and y directions.

37. $f'(y) = 10y - 2$ **39.** $\frac{df}{dy} = 24k^2y^2 + 20y^3$

41. $f'(y) = -64y(5k^3 - 8y^2)^3$ **43.** $\frac{df}{dy} = \frac{5}{y}$

45. $f(x)$ is increasing at a decreasing rate.

Problem Set 7.2

1. a. $10x + 6y$ **b.** $6x + 24y^2$ **c.** 26 **d.** 120

3. a. $3x^2 + 9x^2y^2$ **b.** $-8y + 6x^3y$ **c.** 0 **d.** −56

5. a. e^{x+y} **b.** e^{x+y} **c.** $e^2 = 7.389$ **d.** $e^3 = 20.086$

7. a. $\frac{4xy^2}{1 + 2x^2y^2}$ **b.** $\frac{4x^2y}{1 + 2x^2y^2}$ **c.** 0 **d.** 0

9. a. $\frac{5x^2 - 6xy^2 - 15y}{(5x - 3y^2)^2}$ **b.** $\frac{9y^2 + 15x + 6x^2y}{(5x - 3y^2)^2}$ **c.** $-\frac{5}{24}$ **d.** 1

11. a. $2x \cdot e^{3y}$ **b.** $3x^2 \cdot e^{3y}$ **c.** 0 **d.** 0

13. a. 4 **b.** 10 **c.** 0 **d.** 0 **15. a.** 20 **b.** 10 **c.** 4 **d.** 4

17. a. 18 **b.** 2 **c.** −6 **d.** −6

19. a. $32ye^x$ **b.** 0 **c.** $16e^{2x}$ **d.** $16e^{2x}$

21. a. $-\frac{2}{x^2} + y^2e^{xy}$ **b.** $-\frac{2}{y^2} + x^2e^{xy} + z^2e^{yz}$
c. $\frac{2}{z} + ye^{yz}$ **d.** $e^{yz}(yz + 1)$

23. $\left(-\frac{3}{2}, -\frac{2}{7}\right)$ **25.** $\left(\pm 1, \pm\sqrt{\frac{2}{3}}\right)$

27. At the point (100, 600), if x changes from 100 to 101, $N(x, y)$ will increase by about 30; At the point (80, 420), if y changes from 420 to 421, $N(x, y)$ will increase by about 16.

29. When $x = 2, y = 1, z = 3, w = 3$, and $m = 4$, if we keep all the variables the same (except changing y from 1 to 2) $S(x, y, z, w, m)$ will decrease by about 65.

31. If the population of a country increases from 70 million to 71 million, and that country is being lent 40 million dollars, the amount of destroyed rainforest is increasing at an increasing rate.

33. As a person's chronological age is increasing, their IQ is decreasing at a decreasing rate.

35. As a person's age increases, the time in days it takes a person to recover from pneumonia using a particular daily dose of penicillin, will increase at an increasing rate.

37. As the concentration of residents of public housing increases, the number of arsons in that city increases at an increasing rate.

39. As the unit of labor increases from 20 to 21, the production will increase by about 0.49 units, and this rate will decrease by about 0.02.

41. If the price of printers increases and price of ink stays the same, the demand will decrease. If the price of ink increases and the price of printers stays the same, the demand will also increase.

43. a. \$13,725 **b.** increase of \$190 **c.** decrease of \$510

45. a. 4.08 **b.** decreases by 0.23 **c.** increases by 0.08

47. a. 2,700 **b.** increases by 66.67 **c.** decreases by 14.1

49. $f(1, 0) = 2, f_x(x, y) = -ye^{-xy} - \frac{ye^{y/x}}{x^2}$,
$f_y(x, y) = \frac{e^{y/x}}{x} - xe^{-xy}, f_x(1, 0) = 0$, and $f_y(1, 0) = 0$

51. $\frac{\partial f}{\partial x}(15, 21) = 18.23$ and $\frac{\partial f}{\partial x}(15, 21) = 17.26$ If, when 15 units of labor and 21 units of capital are being used, the number of units of labor is increased by 1, from 15 to 16, then the number of units produced by the manufacturer will increase by 18.23. If on the other hand, the number of units of capital is increased by 1, from 21 to 22, then the number of units produced will increase by about 17.26.

53. $f_x(x, y) = 3x + 6; f_y(x, y) = 2y - 8$

55. $f_x(x, y) = 3x; f_y(x, y) = 2y^3 - 2y$

57. 24 **59.** $x = \pm 5; y = 0, 4$

Problem Set 7.3

1. $(-5, 4)$, relative minimum **3.** $(-3, 4)$, saddle point

5. $\left(-\frac{4}{3}, \frac{11}{3}\right)$, relative maximum

7. (1, 2), saddle point; $(-1, 2)$, relative maximum

9. (0, 1), saddle point; $(0, -1)$, relative minimum

11. (0, 0), saddle point; (1, 1), relative minimum

13. no critical points **15.** $(3, -2)$, relative minimum

17. $(2, -6)$, saddle point **19.** $(-1, 0)$, saddle point

21. \$33,750 on radio advertising; \$362.50 on newspaper advertising

23. $l = w = h = 4$ in. **25.** $l = w = h = 28$ in.

27. a. The relative maximum is at (0, 0, 300) **b.** The relative maximum is at (0, 0, 3) **c.** The relative maximum is at (0, 0, 1) **d.** The relative maximum is at $\left(0, 0, \frac{300}{n}\right)$

29. The maximum is at (9.167, 24.333, 989). The profit will be maximized at \$989 when about 9 units of product A and about 24 units of product B are produced and sold.

31. $f(y) = y^2 + 24y + 24$

33. $F_x(x, y, \lambda) = 2x + \lambda, F_y(x, y, \lambda) = 20y - \lambda$,
$F_\lambda(x, y, \lambda) = x - y - 18$

35. $x = \frac{180}{11}, y = \frac{-18}{11}, \lambda = \frac{-360}{11}$

Problem Set 7.4

1. Relative maximum (9, 2, 120).

3. Relative minimum $(9, -1, -5)$.

5. $f(-\frac{1}{2}, -\frac{1}{2}) = \frac{49}{2}$ **7.** $f(-\frac{3}{4}, -\frac{3}{2}) = -\frac{9}{4}$

9. $f(3, 8) = 24\sqrt[3]{3} \approx 34.6$

11. If the course requires 1 more hour of instruction, student achievement will increase by about 1.3 (or up to about 86.3).

13. If the volume of the can is increased by 1, from 21.66 to 22.66, then the surface area will increase by about 1.32, (or to about 44.3 square inches).

15. If the company increases the number of items to be tested each day, by 1, from 35 to 36, the cost will increase by about \$4.09 (or to 359.09).

17. To maximize production at 15,797.49 units, the manufacturer should allocate 65 units to labor, or \$1,081,600 to labor and \$540,800 to capital. Also if the number of dollars allocated to labor and capital is increased from \$1,622,400 to \$1,622,401, the number of units will decrease by approximately 0.0097 (from 15,797.49 to 15,797.48).

19. To maximize production at 247.987 units, the manufacturer should allocate 13.6 units to labor and \$19,212.80 to capital. Also, if the number of dollars allocated to labor and capital is increased by one, the number of units will decrease by approximately 0.005.

21. 50 inspections at A and 10 at B.

23. The operation should buy 255 radio and 330 newspaper ads that will reach about 939,047 people. If the number of ads is increased by 1, the number of people reached will decrease by about 37. ($\lambda = -36.83$)

25. About 1,470.2 pounds of A and 529.8 pounds of B.

27. a. **b.**

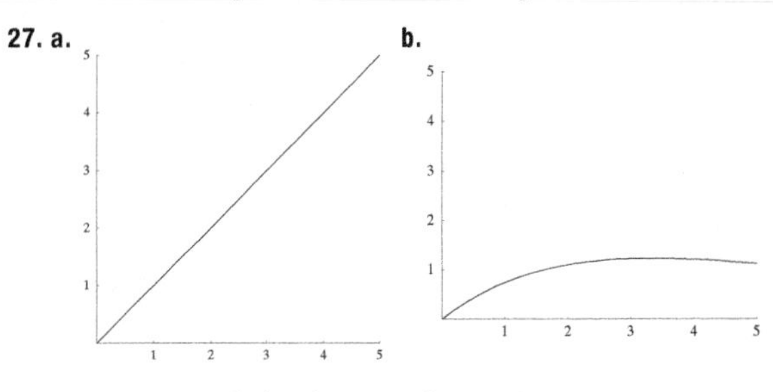

c. It creates a graph that becomes less steep as x increases.

d.

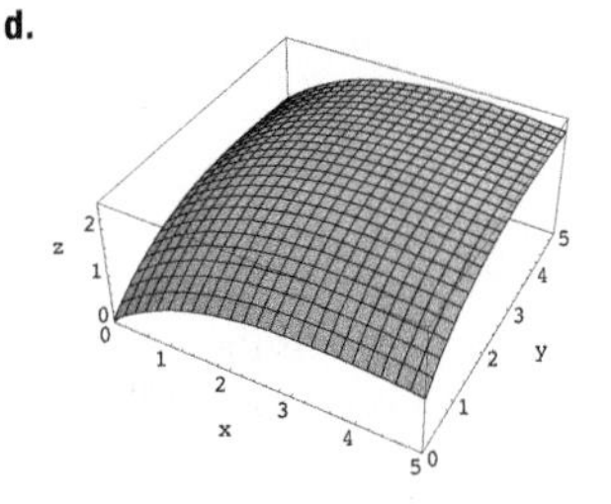

e. 2.45 **f.** 2.45

29. 7 units of product A and 23 units of product B for a maximum profit of \$981 thousand.

31. $f_x(x, y) = \frac{18y^{4/7}}{x^{4/7}}$

33. $-\frac{128}{9} \approx -14.2$

Problem Set 7.5

1. $df = 16x - 12y + 14$ **3.** $df = 16e^3 - 24 \approx 297.4$

5. $df = -\frac{5}{12} \approx -0.4167$ **7.** $df = 120e \approx 362.19$

9. $df = \frac{1}{450} \approx 0.0022$

11. $df = 0$

13. Both $f_x(a, b)$ and $f_y(a, b)$ are negative, $2fy(a, b) < -fx(a, b)$

15. $df = e^{x+y}(dx + dy)$. If $dx < 0$ and $dy < 0$, then $dx + dy < 0$. e^{x+y} is always greater than 0. Therefore, df is always negative.

17. a. Decreases by approximately 30. **b.** Actual decrease of 28.51, with a difference between actual and estimate of 1.49.

19. a. Cost increases by about \$148 **b.** Actual increase of \$188, with a difference of \$40.

21. The output would decrease by about \$5,312.19

23. Increase of about 1,547.16 units

25. An approximate decrease of 518,058 people would be reached.

27. $7k + 182$ **29.** $kx(12k + 128x^2 - 3kx^2 - 2x^8)$ **31.** 20

33. 79.79

Problem Set 7.6

1. 7 units^3 **3.** 3 units^3 **5.** 5 units^3 **7.** 420 units^3

9. 4 units^3 **11.** $16e^3 - 8 \approx 313.369$ units^3 **13.** 10 units^3

15. 2.95 units^3 **17.** 3,263,615.41 units^3

19. $6x^2y^3 + 4x^2y + g(y)$ **21.** $4xe^xy + g(x)$ **23.** $\frac{3}{4}$ **25.** 4

27. 1 **29.** 3.466 **31.** $\frac{49}{6} \approx 8.17$ units^3 **33.** $\frac{e-1}{.2} \approx 0.859$

35. 0.184 **37.** $\frac{21}{4}$ **39.** 1 **41.** $\frac{1}{4}$ **43.** $\frac{1}{4}$

45. $3x^2y + 2xy^3 + g(x)$ **47.** 5,394.7 **49.** 982.73 million

51. 6 **53.** 42 **55.** 63 **57.** 96.47 **59.** 6.5

Problem Set 7.7

1. $\frac{19}{3}$ **3.** $\frac{8}{3}$ **5.** 115.13 **7.** 9.80

9. Solving for the expression given will show us the average number of pollution units per mile between 3 to 10 miles away from a manufacturing facility.

11. This expression tells us by how much per month, on average, the piece of equipment depreciates from month 12 to month 24.

13. This expression tells us the average price the piece is worth for its first 20 years.

15. This expression tells us the average production level of a worker between 60 and 120 days the worker has been on the job.

17. The average value of a house in this country over the first 4 years of recession is \$221,040.

19. The average marginal cost for units 400 to 500 is \$8.

21. 4,625.5 units **23.** \$23,550 **25.** 9.25 units **27.** 31,667

29. 34,560 people **31.** 4.108 units

33. a. 8.74 µg/mL **b.** 12.54 µg/mL **c.** 10.64 µg/mL

Chapter 7 Test

1. $R(x)$ is the revenue realized from only newspaper advertisements. $R(x, y)$ is the revenue realized from both newspaper and radio advertising.

2. $f(0, 2) = 0$

3. $\frac{\partial}{\partial x}N(4, 10) = 15$ tells us that if x increases from 4 to 5 and y stays the same, $N(x, y)$ will increase by about 15; $\frac{\partial}{\partial y}N(4, 10) = -4$ tells us that if y increases from 10 to 11 and x stays the same, $N(x, y)$ will decrease by about 4.

4. $D_x(x, y) < 0$ and $D_y(x, y) > 0$

5. a. $\frac{\partial}{\partial w}\text{BMI}(w, h) = \frac{703}{h^2}$; positive (as weight increases, BMI also increases) **b.** $\frac{\partial}{\partial h}\text{BMI}(w, h) = -\frac{1{,}406}{h^3}$; negative (as height increases, BMI decreases)

6. $f_x(2, 1) = 29; f_y(2, 1) = -2$ **7.** $\left(-\frac{3}{2}, -\frac{2}{7}\right)$

8. a. An increase of about 11.25 units of a product **b.** A decrease of about 53.33 units

9. $\left(-\frac{4}{7}, \frac{5}{6}\right)$ is a saddle point **10.** (2, 8) is a saddle point

11. (−1, 0) local minimum; (0, 0) saddle point; (1, 0) local minimum

12. $\lambda = 4.3$ tells us that if the constant in the constraint function is increased by 1 (to 531,001), then the number of units produced would increase by about 4.3.

13. 30,000 **14. a.** \$15,000 to labor, \$45,000 to capital **b.** 2,432 units

15. $df = 15.2$

16. a. The output will decrease by approximately 60.75 units. **b.** The actual decrease in the output is 84.12 units.

17. $4x^4y^2 + g(y)$ **18.** $\int f(x, y)\, dydx$ **19.** 28 **20.** $\frac{1}{6}$ **21.** $\frac{65}{24}$

22. $\frac{55}{156}$ **23. a.** 10.61 mg **b.** 0.00198 mg

24. 3,727 units of product

Index

R

S

T

U

V